Mathematical Programming
The State of the Art

Bonn 1982

Edited by
A. Bachem M. Grötschel B. Korte

With 30 Figures

Springer-Verlag
Berlin Heidelberg New York Tokyo
1983

Achim Bachem
Bernhard Korte

Institut für Ökonometrie und Operations Research
Universität Bonn
Nassestraße 2
D-5300 Bonn 1

Martin Grötschel

Lehrstuhl für Angewandte Mathematik II
Universität Augsburg
Memminger Straße 6
D-8900 Augsburg

ISBN-13:978-3-642-68876-8   e-ISBN-13:978-3-642-68874-4
DOI: 10.1007/978-3-642-68874-4

Library of Congress Cataloging in Publication Data. Main entry under title: Mathematical programming, Bonn 1982. Based on the 11th International Symposium on Mathematical Programming held Aug. 23–27, 1982 at the University of Bonn. 1. Programming (Mathematics) — Addresses, essays, lectures. I. Bachem, A. II. Grötschel, Martin. III. Korte, B. H. (Bernhard H.), 1938—. IV. International Mathematical Programming Symposium (11th: 1982: University of Bonn)
QA402.5.M3529, 1983. 519.7 83-6809. ISBN-13:978-3-642-68876-8 (U.S.)

This work is subject to copyright. All rights are reserved, whether the whole or part of the material is concerned, specifically those of translation, reprinting, reuse of illustrations, broadcasting, reproduction by photocopying machine or similar means, and storage in data banks. Under § 54 of the German Copyright Law where copies are made for other than private use a fee is payable to "Verwertungsgesellschaft Wort", Munich.

© Springer-Verlag Berlin Heidelberg 1983
Softcover reprint of the hardcover 1st edition 1983

Preface

In the late forties, Mathematical Programming became a scientific discipline in its own right. Since then it has experienced a tremendous growth. Beginning with economic and military applications, it is now among the most important fields of applied mathematics with extensive use in engineering, natural sciences, economics, and biological sciences. The lively activity in this area is demonstrated by the fact that as early as 1949 the first "Symposium on Mathematical Programming" took place in Chicago.

Since then mathematical programmers from all over the world have gathered at the international symposia of the Mathematical Programming Society roughly every three years to present their recent research, to exchange ideas with their colleagues and to learn about the latest developments in their own and related fields. In 1982, the XI. International Symposium on Mathematical Programming was held at the University of Bonn, W. Germany, from August 23 to 27. It was organized by the Institut für Ökonometrie und Operations Research of the University of Bonn in collaboration with the Sonderforschungsbereich 21 of the Deutsche Forschungsgemeinschaft.

This volume constitutes part of the outgrowth of this symposium and documents its scientific activities.

Part I of the book contains information about the symposium, welcoming addresses, lists of committees and sponsors and a brief review about the Fulkerson Prize and the Dantzig Prize which were awarded during the opening ceremony.

The scientific program consisted of state-of-the-art tutorials, invited and contributed talks. The state-of-the-art tutorials constituted the main frame of the symposium. 23 leading experts gave one-hour lectures which were a combined didactic introduction and survey of the most interesting and recent results in rapidly developing areas of mathematical programming. Emphasis was placed on latest research results not covered in monographs and textbooks in order to acquaint scientists with new ideas and an effort was made to bridge the gap between different fields of mathematical programming.

Twentyone of these state-of-the-art tutorials have materialized into comprehensive survey papers which form the core of this volume. They can be found in Part II of the book. These papers document the state of the art of our field in the year 1982.

In addition to these tutorials 500 invited and contributed talks of 25 minutes length each were given. These lectures were grouped into four groups of sessions a day. There were up to twelve sessions in parallel, each session consisting of three talks. A complete listing of all sessions and talks given during the symposium can be found in Part III as well as a list of authors of papers presented in Bonn. There will be no proceedings volume of the invited and contributed papers. Many of these will appear in the scientific literature in the near future.

Finally we would like to thank all those people and organizations without whose enthusiastic and unselfish support this symposium would not have been possible. In particular, we would like to thank Mrs. H. Higgins who was the mainstay of the organisation. We are furthermore very much indebted to the sponsors of this meeting. Special acknowledgments are contained in Part I of this volume. Last but not least our thanks go to Springer Verlag for their excellent cooperation in publishing this volume.

Bonn, May 1983   Achim Bachem
                                                                Martin Grötschel
                                                                Bernhard Korte

Table of Contents

I. About the XIth International Symposium on Mathematical Programming 1

Program and Organizing Committee 1
Welcoming Addresses 3
List of Sponsors 8
The Fulkerson Prize and the Dantzig Prize 1982 9

II. Mathematical Programming: The State of the Art – Bonn 1982 13

Predictor-Corrector and Simplicial Methods for Approximating Fixed Points and Zero Points of Nonlinear Mappings
  E. L. Allgower and *K. Georg* 15
Polyhedral Theory and Commutative Algebra
  L. J. Billera 57
Reminiscences About the Origins of Linear Programming
  G. B. Dantzig 78
Penalty Functions
  R. Fletcher 87
Applications of the FKG Inequality and its Relatives
  R. L. Graham 115
Semi-Infinite Programming and Applications
  S.-Å. Gustafson and *K. O. Kortanek* 132
Applications of Matroid Theory
  M. Iri 158
Recent Results in the Theory of Machine Scheduling
  E. L. Lawler 202
Submodular Functions and Convexity
  L. Lovász 235
Recent developments in Algorithms and Software for Trust Region Methods
  J. J. Moré 258
Variable Metric Methods for Constrained Optimization
  M. J. D. Powell 288
Polyhedral Combinatorics
  W. R. Pulleyblank 312
Generalized Equations
  Stephen M. Robinson 346

Generalized Subgradients in Mathematical Programming
  R. T. Rockafellar 368
Nondegeneracy Problems in Cooperative Game Theory
  J. Rosenmüller 391
Conic Methods for Unconstrained Minimization and Tensor Methods for Nonlinear Equations
  R. B. Schnabel 417
Min-Max Results in Combinatorial Optimization
  A. Schrijver 439
Generalized Gradient Methods of Non-Differentiable Optimization Employing Space Dilatation Operations
  N. Z. Shor 501
The Problem of the Average Speed of the Simplex Method
  S. Smale 530
Solution of Large Linear Systems of Equations by Conjugate Gradient Type Methods
  J. Stoer 540
Stochastic Programming: Solution Techniques and Approximation Schemes
  R. J.-B. Wets 566

III. Scientific Program 605

IV. List of Authors 649

I. About the XIth International Symposium on Mathematical Programming

Program and Organizing Committee

The XIth International Symposium on Mathematical Programming was held under the auspices of the President of the Federal Republic of Germany Professor Dr. Karl Carstens.

International Program Committee

B. Korte (W. Germany) Chairman
J. Abadie (France)
A. Auslender (France)
M. Avriel (Israel)
E. Balas (USA)
M. L. Balinski (France)
E. M. L. Beale (England)
J. F. Benders (Netherlands)
B. Bereanu (Romania)
C. Berge (France)
R. Burkard (Austria)
A. Charnes (USA)
V. Chvátal (Canada)
R. W. Cottle (USA)
G. B. Dantzig (USA)
M. A. H. Dempster (England)
L. C. W. Dixon (England)
K.-H. Elster (DDR)
S. Erlander (Sweden)
J. Evtushenko (USSR)
R. Fletcher (Scotland)
A. Geoffrion (USA)
F. Giannessi (Italy)
J.-L. Goffin (Canada)
E. Golstein (USSR)
R. L. Graham (USA)
P. L. Hammer (Canada)
M. Held (USA)
P. Huard (France)
T. Ibaraki (Japan)
M. Iri (Japan)
R. G. Jeroslow (USA)
E. L. Johnson (USA)
P. Kall (Switzerland)
L. Kantorovich (USSR)
R. M. Karp (USA)
V. Klee (USA)
A. Korbut (USSR)
K. O. Kortanek (USA)
J. Krarup (Denmark)
H. W. Kuhn (USA)
A. Land (England)
E. L. Lawler (USA)
C. Lemarechal (France)
F. Lootsma (Netherlands)
L. Lovász (Hungary)
G. P. McCormick (USA)
O. L. Mangasarian (USA)
B. Mond (Australia)
G. L. Nemhauser (USA)
F. Nozicka (CSSR)
W. Oettli (W. Germany)
A. Orden (USA)
M. W. Padberg (USA)
B. T. Poljak (USSR)
M. J. D. Powell (England)
A. Prékopa (Hungary)
W. R. Pulleyblank (Canada)
M. R. Rao (India)
K. Ritter (W. Germany)
S. M. Robinson (USA)
R. T. Rockafellar (USA)
J. B. Rosen (USA)
H. E. Scarf (USA)
N. Z. Shor (USSR)
J. Stoer (W. Germany)
M. Todd (USA)

A. W. Tucker (USA)   L. A. Wolsey (Belgium)   M. Grötschel
H. Tuy (Vietnam)   M. Yue (China)   (W. Germany)
S. Walukiewicz (Poland)   Co-Chairman of the Or-
R. Wets (USA)   ex officio   ganizing Committee
A. Wierzbicki (Poland)   A. Bachem
P. Wolfe (USA)   (W. Germany)

Organizing Committee

A. Bachem   L. Butz   M. Hahmann
  (Co-Chairman)   W. H. Cunningham   H. Higgins
M. Grötschel   U. Derigs   M. Jünger
  (Co-Chairman)   W. Ehbrecht   R. v. Randow
J. Araoz   A. Frank   G. Reinelt
D. Bosma   F. Gotterbarm   R. Schrader

I. About the XIth International Symposium on Mathematical Programming

Welcoming Addresses

Welcoming Address of the President of the Federal Republic of Germany

The ever increasing complexity of our world necessitates the adoption of highly sophisticated planning methods, and I am fully aware of the essential role played by mathematics today in mastering the problems of our time. It was therefore with particular pleasure that I accepted the sponsorship of the XI. International Symposium on Mathematical Programming.

Many facets of our lives are governed and dominated by technology whose progress in turn depends widely on new developments in mathematics and its applicability. A technology based on mathematical programming methods can contribute substantially to the optimal use of our sparse resources and to the minimization of pollution.

I hope that this international congress will not only be a platform for the exchange of new ideas in scientific methodology, but that it will strengthen the mutual understanding between differing cultures and societies.

I wish you all a very successful conference and a very happy stay in our country.

(Prof. Dr. K. Carstens)

Welcoming Address of the President of the Deutsche Forschungsgemeinschaft

I welcome you cordially to the International Symposium of your Society and am glad that you have followed the organizers' invitation to the Federal Republic of Germany.

Mathematics and mathematicians play an important part in the scientific tradition of our country. They were, however, not always in complete accord with our poets and thinkers; Goethe for instance, whose death 150 years ago we recently commemorated, and who was also active scientifically, wrote: "Mathematicians are foolish fellows and so devoid of even a suspicion of what is essential that one must forgive them their conceit"; and Gauss's opinion of philosophers was no whit better: "Muddled concepts and definitions are nowhere more at home than among philosophers who are not mathematicians."

So a fruitful dialog between mathematicians and scientists applying mathematics in other disciplines is particularly commendable; and the German Research Society's Special Collaborative Program "Economic Prognosis, Decision and Equilibrium Models" which contributed to the organization of this Symposium, is an example of a very successful cooperation of this kind. Your discipline with its countless applications, even extending as far as geology, my own field of research, provides an excellent basis for such a dialog; and one of the first applications of Linear Programming, in the organization of the 1949 Berlin air lift, established a more than scientific link with Germany.

Nowadays our connections abroad are as wide and active as never before; but even though these connections may – sometimes rightly – be criticized as "scientific jet-setting", their scientific, human and even political advantages should not be underestimated. Young scientists, in particular, can profit from the unique opportunity of presenting and discussing their ideas at international meetings. How often has a conversation or a lecture at a congress provided the crucial impetus for a field or an individual scientist! Examples are Hilbert's lectures at the International Congress of Mathematicians in Paris in 1900, still figuring largely in present day mathematics; or the discussions between Niels Bohr, Albert Einstein, the 25 years old Werner Heisenberg and others at the physics meetings in Como and Solvay in 1927, when the uncertainty principle was born; or the by no means uncontroversial impact of the Asilomar Conference on gene technology.

I hope that your Symposium with its lectures, its discussions and its social events will also prove fruitful in this sense and that you will keep it in pleasant memory.

(Prof. Dr. E. Seibold)

Welcoming Address of the Rektor
of the Rheinische Friedrich-Wilhelms-Universität

It is a great pleasure for me to welcome you, the participants of the XI. International Symposium on Mathematical Programming, on behalf of the University of Bonn. I hope that your meeting will be most successful and that you will have a pleasant stay in Bonn.

The University of Bonn was founded in 1818 and after an eventful history grew into one of the largest universities in Germany. Today it has about 38,000 students and a large number of well-known research institutions covering a wide range of fields from the humanities to pure and applied sciences.

The University of Bonn has a long tradition in the area of pure mathematics documented by the names of many eminent mathematicians who have taught here.

In the last twenty years a strong emphasis has been placed on various subjects of applied mathematics to cope with the new and demanding problems arising in economics, engineering, social sciences etc. The University is proud that these efforts have fallen on fertile ground and created very lively research groups, in particular in the field of optimization.

I would like to thank the members of the Institut für Operations Research and of the Sonderforschungsbereich 21 of the University of Bonn for their substantial efforts in organizing this symposium, and I hope that these efforts will prove to be a lasting contribution to the advancement of science and to an international cooperation to solve the problems of our world.

(Prof. Dr. W. Besch)

Welcoming Address of the Lord Mayor of Bonn

I extend a hearty welcome to all participants of the XI. International Symposium on Mathematical Programming, and I am happy that this international exchange of scientific thought is taking place for the first time in Germany, and more particularly in Bonn.

You as scientists are working in a relatively young field. In contrast, it is the older cities, of which Bonn is one, that stand to benefit from your achievements. That the applications of mathematical programming to economic problems would one day become important for our city was certainly not dreamt of in its cradle days two thousand years ago. However, with the steady growth from a Roman garrison outpost through a small medieval town to the well-known university city and finally the capital of the Federal Republic of Germany, have come not only advantages but also problems. Bonn is today the amalgamation of more than thirty former villages, and has nearly 300,000 inhabitants. Its council and administration, together with its citizens, are taking great pains to preserve the historical village centres and the charming corners of our old city which is so rich in tradition. But as with every pulsating city, Bonn too has to labor under the strains of our time: traffic problems, environmental pollution and financial restrictions draw our attention to increasingly urgent problems. I am confident, however, that we will, with the help of some of the many successful mathematical programming methods, master these problems and be able to plan and work more rationally.

I wish you a successful conference and trust that you will return home much enriched. In particular I hope that you will find time to get to know our beautiful city and to discover some of its many charms.

(Dr. H. Daniels)

Welcoming Address of the President of the Mathematical Programming Society

In opening the Eleventh International Symposium of the Mathematical Programming Society, may I take this occasion, on behalf of the society, on behalf of the members who are here today, and on behalf of all the participants, to express our heart-felt thanks to the President of the Federal Republic of Germany, to the Mayor of Bonn, and to the Rector of the Rheinische Friedrich-Wilhelms-Universität Bonn, for the most gracious hospitality they have offered to us. It is a great pleasure for our Society to meet in Bonn, the capital of the Federal Republic of Germany, a city of upmost distinction, and the center of a large worldwide famous University. I also wish to extend our gratitude to the Director of the Institut für Ökonometrie und Operations Research, Professor Dr. Bernhard Korte, Chairman of the Program Committee, and to the Co-Chairmen of the Organizing Committee: this meeting could not have taken place without their dedicated and successful work.

The Mathematical Programming Society is an international organization dedicated to the promotion and the maintenance of high professional standards in the subject of mathematical programming. It publishes the journal Mathematical Programming, consisting of technical articals on all aspects of the subject, and the series Mathematical Programming Studies. Every three years the Society as a whole meets in its International Symposium on Mathematical Programming at some world center of research.

The roots of mathematical programming can be found in the applied mathematics of the nineteenth century, but its emergence as a discipline right awaited the great advances of the nineteen-forties and fifties in the application of mathematical methods to important problems of business, industry, and technology. In the nineteen-sixties it became clear that there was a true international community of researchers and practitioners of mathematical programming in 1970, under the chairmanship of Alex Orden, a founding committee, of which I am very proud to have been a member, proceeded to form a Society from this Community, with the aims of ensuring the continuity of the International Symposia of Mathematical Programming, and establishing a journal to serve the community's interests. In 1970 the Committee selected an Editor-in-Chief and a publisher for the journal Mathematical Programming. In 1971 it set up a secretariat for the Society, and in early 1972 enrolled 366 Charter Members, who adopted a Constitution and in 1973 elected the first officers. The Society became a "Kindred Society" of the International Federation of Operational Research Societies in 1975.

Our society is then, formally, very young, though its birth might be dated back to the Chicago Symposium and its 34 papers in 1949. It is now a well-established Society, which sponsors and supports activities in mathematical programming throughout the world, in addition to its Newsletters, publications, and six Standing Committees. The program of the present Sympsoium bears evidence of its continuing success and development.

(Prof. Dr. J. Abadie)

Welcoming Address of the Chairman of the International Program Committee

Welcome to Bonn! The XI. Symposium on Mathematical Programming will continue the fine tradition of a remarkable line of meetings which began in Chicago in 1949. There is no doubt that this triennial international scientific meeting is the most important conference in this area.

I am happy that the 1982 symposium is taking place in Bonn and at this university. We shall have the opportunity to welcome a large number of participants and lecturers from very many countries, and this time, in addition to the traditionally large contingent from North America many scholars from Western and also from Eastern Europe will participate. Furthermore larger delegations from Asia, Africa and Southamerica and in particular from developing countries are present at this meeting. This gives us particular pleasure.

It can be stated with great satisfaction that the field of mathematical programming plays an increasingly important role within the area of applied (or better: applicable) mathematics. Founded in the late forties, it is today very much used in engineering, natural sciences, economics and biological sciences. Investigations in computing centers have revealed that among all software packages computer codes on mathematical programming and in particular linear programming are most commonly used. It is well known that the economic effect of using linear programming is measured in very large numbers.

This tremendous success in the past however, should not distract our attention from trends that are developing in our discipline. In 1947 John von Neumann, whose scientific work bears certain relationship to mathematical programming stated his deep concern at the general development of mathematics away from its real sources in applications by asking if mathematics is in danger of becoming baroque or is perhaps already in high-baroque? At a meeting of experts several years ago the question was raised whether mathematical programming in particular is already becoming baroque? It was unanimously decided that apart from some baroque frills the area is still far from its classical era and in some parts is still in its stone-age.

This symposium will bear witness to the fact that mathematical programming is still very much alive and promises to remain so for a long time.

(Prof. Dr. B. Korte)

List of Sponsors

The local organizers of the XIth International Symposium on Mathematical Programming gratefully acknowledge the sponsorship and financial support received from the following institutions, whose generous and unselfish help has contributed substantially to the success of the conference.

Public Institutions

Deutsche Forschungsgemeinschaft, Bonn

Der Minister für Wissenschaft und Forschung des Landes Nordrhein-Westfalen, Düsseldorf

Deutscher Akademischer Austauschdienst, Bonn

Rheinische Friedrich-Wilhelms-Universität, Bonn

Gesellschaft von Freunden und Förderern der Rheinischen Friedrich-Wilhelms-Universität zu Bonn

Der Minister für Innerdeutsche Beziehungen, Bonn

Deutsche Stiftung für Internationale Entwicklung, Berlin

Stadt Bonn

The Mathematical Programming Society

European Research Office, London

Special Interest Group on Mathematical Programming (SIGMAP) of the Association for Computing Machineries (ACM)

Office of Naval Research, London

Stifterverband für die deutsche Wissenschaft, Essen

Rheinisches Landesmuseum, Bonn

Braunschweigisches Landesmuseum, Braunschweig

Private Firms

Robert Bosch GmbH, Stuttgart

Deutsche BP Aktiengesellschaft, Hamburg

IBM Deutschland GmbH, Stuttgart

Daimler-Benz Aktiengesellschaft, Stuttgart

Deutscher Herold Lebensversicherung-AG, Bonn

Henkel Kommanditgesellschaft auf Aktien, Düsseldorf

Kälte Hunke, Bonn

Springer-Verlag GmbH & CO. KG, Heidelberg

Volkswagenwerk Aktiengesellschaft, Wolfsburg

Ruhrgas Aktiengesellschaft, Essen

Westdeutsche Landesbank Girozentrale, Münster

Birkhäuser-Verlag AG, Basel

Universitätsbuchhandlung Bouvier, Bonn

Athenäum Verlag GmbH, Königstein/Ts.

The Fulkerson Prize and the Dantzig Prize

During the opening session of the XIth International Symposium on Mathematical Programming the two most prestigeous prizes in the field have been awarded.

The Fulkerson Prize

The Fulkerson Prize for outstanding papers in the area of discrete mathematics is sponsored jointly by the Mathematical Programming Society and the American Mathematical Society. Beginning in 1979, up to three awards are being presented at each (triennial) International Congress of the Mathematical Programming Society; they are paid out of a memorial fund administered by the American Mathematical Society that was established by friends of the late Delbert Ray Fulkerson to encourage mathematical excellence in the fields of research exemplified by his work. Thus, this is the second time the Fulkerson Prize has been awarded.

The specifications for the Fulkerson Prize read:

"Papers to be eligible for the Fulkerson Prize should have been published in a recognized journal during the six calendar years preceding the year of the Congress. This extended period is in recognition of the fact that the value of fundamental work cannot always be immediately assessed. The prizes will be given for single papers, not series of papers or books, and in the event of joint authorship the prize will be divided.

The term "discrete mathematics" is intended to include graph theory, networks, mathematical programming, applied combinatorics, and related subjects. While research work in these areas is usually not far removed from practical applications, the judging of papers will be based on their mathematical quality and significance."

The Selection Committee for the award has two members appointed by the chairman of the Mathematical Programming Society and one member appointed by the president of the American Mathematical Society. The committee members serve for at most two rounds of awards, with terms overlapping where possible for the sake of continuity. The Selection Committee devises its own procedures for acquiring nominations or otherwise searching out papers of interest, taking pains, however, not to overlook the work of young, relatively unknown mathematicians.

The Selection committee for the Fulkerson Prize of 1982 consisted of R. L. Graham (chairman), R. M. Karp and V. L. Klee. The unanimous recommendations of the committee were approved by the Mathematical Programming Society and the American Mathematical Society. On August 23, 1982 the following three Fulkerson Prizes were awarded as follows:

1. One award (jointly) to L. G. Khachiyan and to D. B. Iudin and A. S. Nemirovskii for their respective papers (Khachiyan) "A Polynomial Algorithm in Linear Programming", Doklady Akademiia Nauk SSSR 244 (1979), 1093–1096 (translated in Soviet Mathematics Doklady 20, 191–194, 1979) and (Iu-

din and Nemirovskii) "Informational Complexity and Effective Methods of Solution for Convex Extremal Problems", Ekonomika i Matematicheskie Metody 12 (1976), 357–369 (translated in Matekon: Translations of Russian and East European Math. Economics 13, 25–45, Spring '77).
2. One award (jointly) to M. Grötschel, L. Lovász and A. Schrijver for their paper "The Ellipsoid Method and Its Consequences in Combinatorial Optimization", Combinatorica 1 (1981), 169–197.
3. One award (jointly) to G. P. Egorychev and D. I. Falikman for their respective papers (Egorychev) "The solution of van der Waerden's problem for permanents", Dokl. Akad. Sci. SSSR 258 (1981), 1041–1044 (Russian); Advances in Math. 42 (1981), 299–305 and (Falikman) "A proof of the van der Waerden conjecture on the permanent of a double stochastic matrix", Mat. Zametki 29 (1981), 931–938.

The Dantzig Prize

The George B. Dantzig Prize in Mathematical Programming is jointly sponsored by the Society of Industrial and Applied Mathematics and the Mathematical Programming Society. The Dantzig Prize is being presented triennially, twice at the International Symposium on Mathematical Programming and every third time at the national SIAM meeting. It was awarded for the first time at the Symposium in Bonn.

The specifications of the George B. Dantzig Prize read:

"The prize is awarded for original work which by its breadth and scope constitutes an outstanding contribution to the field of mathematical programming. The contribution(s) for which the award is made must be publicly available and may belong to any aspect of Mathematical Programming in its broadest sense.

The contributions eligible for considerations are not restricted with respect to the age or number of their authors although preference should be given to the singly-authored work of "younger" people."

The members of the selection committee for the 1982 Dantzig prize were E. M. L. Beale, J. Dennis, R. J.-B. Wets (chairman) and Ph. Wolfe. The committee unanimously agreed to award the George B. Dantzig Prize jointly to Michael J. D. Powell and to R. Tyrrell Rockafellar for their work in nonlinear optimization and for the scientific leadership they have provided to the mathematical programming community. The citations for the two laureats read:

"The George B. Dantzig Prize in Mathematical Programming is presented to Michael J. D. Powell on this 23rd day of August 1982 for his pioneering work in the numerical optimization of nonlinear functions done at the outset in collaboration with R. Fletcher, and for his scholarship in recognizing, and leadership in championing the work of others."

"The George B. Dantzig Prize in Mathematical Programming is presented to R. Tyrrell Rockafellar on this 23rd day of August 1982 for his key contributions to the theory of nonlinear optimization, in particular for the development of the general duality framework and its applications, and for his work on subdifferentiability in the convex and nonconvex case."

I. About the XIth International Symposium on Mathematical Programming

As the geographical distribution of the 1982 prize winners shows high quality research in the area of mathematical programming is done not only in a few leading countries but in many parts of the world. Five of the awardees are from the Soviet Union and one awardee from each of the following countries: Hungary, Netherlands, Great Britain, USA, West Germany. Unfortunately none of the five prize winners from the Soviet Union was able to attend the ceremony in person despite considerable efforts of the organizing committee.

II. Mathematical Programming:
The State of the Art – Bonn 1982

Predictor-Corrector and Simplicial Methods for Approximating Fixed Points and Zero Points of Nonlinear Mappings

E. L. Allgower
Colorado State University, Department of Mathematics, Fort Collins, Colorado 80523, USA

K. Georg
Institut für Angewandte Mathematik, Universität Bonn, Wegelerstr. 6, 5300 Bonn 1

1. Introduction

Our aim here is to provide a brief introductory overview of numerical homotopy methods for approximating a solution of a system of nonlinear equations

(1) $$G(x)=0$$

where the map $G: \mathbb{R}^N \to \mathbb{R}^N$ satisfies some appropriate conditions.

Classically, the zero point problem has been frequently solved by iterative methods such as Newton's method and modifications thereof. For comprehensive discussions, see e.g. [101.8], [102], [126]. Among the difficulties and pitfalls which cause the classical methods frequently to perform inefficiently or even fail are:
1. The lack of good starting points.
2. Slow convergence.
3. The lack of smoothness or continuity of G.

In recent years, two general remedial techniques to overcome these difficulties have been devised:
1. Quasi-Newton and hybrid methods.
2. Continuation (homotopy, imbedding) methods.

We will not deal with Quasi-Newton methods here. However, the reader who is familiar with them may recognize that these two methods are not totally unrelated, and that some mutual borrowing or lending has historically occurred. For the reader who wishes to become informed on Quasi-Newton methods we suggest the introductory book [38.2] and the overview articles [37], [38].

In this limited space it will not be possible even to give a comprehensive survey of homotopy methods. We shall instead confine ourselves to offering a relatively comprehensive and up-to-date bibliography, and a brief description of the basic ideas of several relatively simple examples of continuation algorithms. Under the terminology we have chosen to adopt, the general scheme of a continuation algorithm involves two parts:

(2) **Construction of a homotopy.** A homotopy $H: \mathbb{R}^N \times \mathbb{R} \to \mathbb{R}^N$ has to be chosen which deforms the problem $G(x)=0$ into a trivial problem $G_0(x)=0$ by $H(x, 0) = G_0(x)$ and $H(x, 1) = G(x)$, $x \in \mathbb{R}^N$. This deformation has to furnish a continuum $c \subset H^{-1}(0)$ which links a known zero point x_0 of G_0 with a desired zero point x_1 of G, cf. figure (4).

(3) **Numerical curve tracing.** A numerical process for tracing the continuum c from x_0 to x_1.

(4) Figure

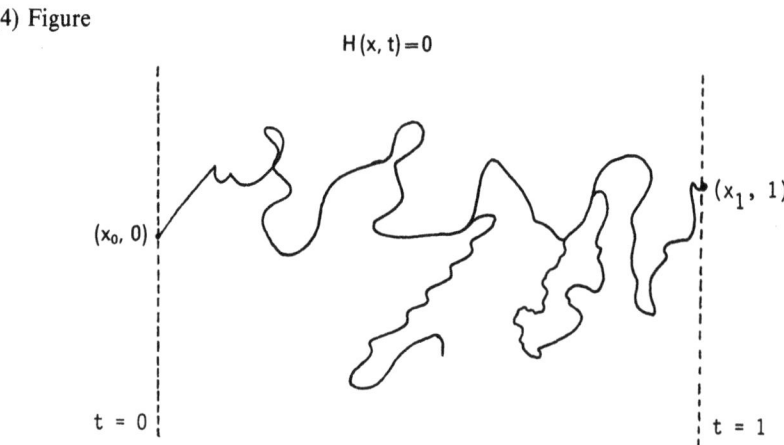

The two fundamental methods which we will discuss for implementing (3) are:
1. Predictor-corrector methods for numerically integrating Davidenko's equation.
2. Simplicial methods, wherein a polygonal path c_T is traced instead of c.

Although many of the algorithms cited in our bibliography do not involve an additional parameter, most of them also have a homotopy interpretation.

The idea of using deformations as a mathematical tool dates at least to Poincaré [105.3]. The use of deformations to solve nonlinear systems of equations may be traced back at least to Lahaye [90.44], [90.45]. Classical embedding methods may be regarded as a forerunner of the predictor-corrector (p.c.) methods which will be discussed here. Classical embedding methods have been a popular numerical device for a considerable number of years (for extensive surveys, see [48], [88.5], [102], [148], [149.5]). These methods are limited, however, by the fact that they cannot negotiate around turning points with respect to the underlying parameter. To our knowledge, Haselgrove [71] may be the first to have overcome this difficulty. Unfortunately, Haselgrove's paper contained no numerical examples and the idea was not soon pursued. In more recent years numerous authors have published similar ideas. Some of these are cited in Section 4.

II. Predictor-Corrector and Simplicial Methods for Approximating

Scarf's constructive proof of the Brouwer fixed point theorem (in the sense that it could be numerically implemented) [122] soon inspired other constructive approaches; analogous existence results of nonlinear analysis where obtained by similar approaches, classical homotopy arguments could be numerically implemented, new continuation arguments were developed to solve nonlinear problems.

Scarf's constructive approach and many of the subsequent simplicial algorithms are based upon an idea of complementary pivoting which originated in a paper by Lemke and Howson [92]. Simplicial algorithms using homotopies, and thereby allowing the possibility of automatically obtaining successively better approximations to fixed points or zero points were given by Eaves [43] and Merrill [96]. Improved algorithms have since been given by numerous authors. For references, see the bibliography and Section 6.

A constructive proof of the Brouwer fixed point theorem for smooth mappings was given by Kellogg, Li and Yorke [83]. The proof was based upon the non-retraction principle of Hirsch [72.8] and the implicit function theorem. The constructive aspect of the proof rested in the tracing of an implicitly defined curve via a numerical method for solving initial value problems. Subsequently, Chow, Mallet-Paret and Yorke [27], and Garcia and Gould [51] studied the homotopy approach for approximating fixed points or zero points via smooth homotopy curves. Certain of the smooth homotopy methods have been shown to be very closely related to the global Newton method introduced by Branin [21] and further studied by Smale and Hirsch [73], [129].

Since our mathematical discussions will be confined to describing the basic aspects of a few homotopy algorithms, we suggest the following articles and books for further reading: [3.5], [6], [44], [46], [54.3], [64.2], [67.5], [81], [87], [112], [113.15], [117.5], [124], [133], [137], [142.3], [146]. The titles of the above cited works generally indicate whether they deal exclusively with smooth (predictor-corrector) or simplicial methods.

The main ideas which will be developed here are the creation of curves in $\mathbb{R}^{N+1}$ which are implicitly defined by an equation e.g.

(5) $$H(u) = 0$$

where $H: \mathbb{R}^{N+1} \to \mathbb{R}^N$, and then the tracing of such curves by numerical integration techniques, by predictor-corrector methods and by simplicial methods.

Restrictions of space require that we shall discuss the details of only one application of homotopy continuation methods viz. unconstrained optimization. The following articles involve applications in which homotopy methods of the sort we discuss here have recently been applied. In a number of these cases numerical examples have also been calculated.

1. Complementary: [48.1], [48.5], [67.4], [84.8],]94.4], [94.8], [151.3].
2. Economics: [47.3], [94.5], [94.55], [94.56], [129], [139.2], [140.5], [142.4], [145.52], [153.54], [154.91].
3. Engineering: [25.5], [102.6], [110.5], [112], [113], [149.2], [149.3], [151.8], [152.2]–[152.55].

4. Differential equations: [8.1], [24.7], [27], [39], [39.8], [61], [64.2], [71], [103.1], [110.5], [111], [125.3], [150], [151.7].
5. Nonlinear programming: [12.4], [21.1], [44.2], [48.2], [87.1], [118], [129.4], [146.91], [154.8], [154.9], [154.91].
6. Nonlinear eigenvalue problems and bifurcation: [1.7], [6], [33.7], [75], [76], [79.1], [82], [102.75], [102.8], [103], [108.2].
7. Networks: [20.08], [25.6], [28.2], [78.5], [102.6].
8. Variational inequalities: [59], [61], [67.4], [74.5], [145.52].
9. Constructive proofs: [3], [8.2], [12.85], [12.9], [12.92], [26], [27], [41.5], [41.52], [42], [45], [48.75], [49.3], [49.7], [52], [53], [54], [73], [74.55], [83], [87.17], [87.2], [89], [89.5], [90], [92.3], [94.6], [96], [96.8], [103], [122], [129], [133], [142.8], [146], [153.9], [154.5].

Several computer codes have been published for continuation-type algorithms: [74.54], [76.5], [113.15], [116.6], [142.5], [152].

Articles in which computational results have been given are:
1. Simplicial: [12.4], [47.3], [48.1], [64.2], [75], [84.2], [85.3], [87.17], [87.9], [90], [96], [98.5], [102.8], [116.5]-[117], [118], [118.6], [121], [124], [126.92], [139.5], [142.4], [146.35], [146.5], [146.7], [153.71].
2. Predictor-corrector (numerical integration): [18.8], [20.08], [24.7], [25.5], [25.6], [28.2], [34], [64], [64.2], [73], [74.5], [79], [80], [83], [93.45], [94.9], [95], [97.5], [102.75], [105.5], [106], [111]-[113], [113.5], [125.3], [149.2]-[149.35], [150]-[152.55].

The remaining contents here are organized as follows. In Section 2 we give background discussions concerning the construction of homotopies and their regular points, the implicit definition of homotopy curves, and conditions under which homotopy curves will connect to solutions. In Section 3 we outline the details of some predictor-corrector algorithms for numerically tracing a smooth curve which is implicitly defined. In Section 4 an application of homotopy methods to unconstrained optimization is discussed. In Section 5 details of curve tracing by simplicial methods are given. In Section 6 a simplicial algorithm using a refining triangulation is sketched and applied to quasi-linear mappings in order to approximate a fixed point or zero point. Some important further aspects are briefly cited.

We apologize for the somewhat curious reference numbers. The explanation lies in the fact that our bibliography is stored on a computer tape. Instead of performing tedious renumberings when we update, we have chosen to exploit the denseness of the rational numbers.

2. Background: Homotopies, Regularity and Existence

In this section we will briefly discuss some of the underlying matters of homotopy methods which will enable us to conclude the existence of a smooth path connecting some chosen starting point to a zero point of the mapping $G: \mathbb{R}^N \to \mathbb{R}^N$. For reasons of simplicity of exposition we shall assume that all mappings are smooth $(:= C^\infty)$.

II. Predictor-Corrector and Simplicial Methods for Approximating

(1) **Homotopy.** Let our nonlinear problem be given in the form

(2) $$G(x) = 0$$

where $G: \mathbb{R}^N \to \mathbb{R}^N$ is smooth. We consider a smooth homotopy $H: \mathbb{R}^N \times \mathbb{R} \to \mathbb{R}^N$ such that

(3) $$H(x, 1) = G(x), \quad x \in \mathbb{R}^N$$

and such that the nonlinear system

(4) $$H(x, 0) = 0$$

has a known (or easily obtained) solution $x_0 \in \mathbb{R}^N$. Two typical examples are:

(5) $$H(x, t) = (1-t)A(x-x_0) + tG(x),$$

where A is some nonsingular (N, N)-matrix, and the global homotopy

(6) $$H(x, t) = G(x) + (t-1)G(x_0),$$

cf. [51.2], [81].

Let us assume that the zero point x_0 of (4) is regular, i.e., that the partial derivative H_x is nonsingular at $(x_0, 0)$. By the implicit function theorem, it follows that near $(x_0, 0)$ the solution set $H^{-1}(0)$ is given by a smooth curve $c(s)$. The aim is to trace this curve until the level $t = 1$ is encountered and hence a solution of (2) is found, cf. figure (7).

(7) Figure

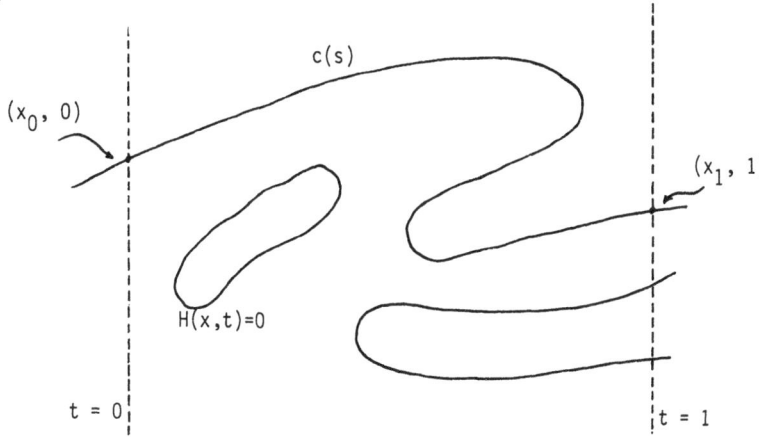

(8) **Regularity.** We first consider the question whether the solution curve $c(s) \in H^{-1}(0)$ may be extended for all $s \in \mathbb{R}$, and whether it consists of regular points of H.

(9) **Definition.** If a map $F: \mathbb{R}^m \to \mathbb{R}^p$ is smooth, a point $x \in \mathbb{R}^m$ is called regular if the Jacobian DF has maximal rank in x, and a value $y \in \mathbb{R}^p$ is called regular if $H^{-1}(y)$ is empty or consists only of regular points.

If zero is a regular value of H, a repeated application of the implicit function theorem shows that $H^{-1}(0)$ consists only of smooth 1-manifolds, each of which may be represented by a curve $c(s)$, see e.g. [97]. The answer to question (8) is provided by Sard's theorem, which we state in the following parametrized form, see e.g. [1], [27], [72.9]:

(10) **Theorem.** Let $F: \mathbb{R}^n \times \mathbb{R}^m \to \mathbb{R}^p$ be smooth and let zero be a regular value. Then for almost all "parameters" $d \in \mathbb{R}^m$, also the restricted map $F(\cdot, d)$ has zero as a regular value.

Using (10) as a perturbation technique, i.e., considering the starting point x_0 in (5) or (6) as the "parameters", a positive answer to question (8) may be given, cf. [3], [27], [105.2]. Let us therefore assume now that zero is a regular value of our homotopy (1). This implies the existence of the following curve:

(11) **Solution curve.** Let $c(s)$ be a smooth curve in $H^{-1}(0)$, defined for all $s \in \mathbb{R}$ and parametrized, say, according to arc length, such that the tangent $c(0)$ points into the half space $\mathbb{R}^N \times (0, \infty)$, cf. figure (7).

(12) **Existence.** The crucial question for the homotopy method is whether the solution curve $c(s)$ meets the level $\mathbb{R}^N \times \{1\}$ for some $s > 0$, and thus furnishes a solution of (2).

Sufficient conditions for existence (12) can be obtained from classical homotopy and degree theory, i.e., the continuation method of Leray and Schauder [93], see also [3], [8.7], [102.8], [103]. Let us study an example which will re-occur in Section 6 in the context of non-continuous mappings G.

(13) **Lemma.** Suppose that in the homotopy (5), A is the derivative of G at infinity, i.e., $\|G(x) - A(x)\| = 0(\|x\|)$ as $\|x\| \to \infty$. Then existence (12) holds.

One may regard (13) as a constructive version of Brouwer's fixed point theorem when smoothness and regularity are fulfilled (see e.g. [10], [27], [83]). For the global homotopy (6), we state the following result of Smale [129], see also [81]:

(14) **Theorem.** In case (6), let x_0 belong to the (smooth) boundary ∂U of an open bounded subset $U \subset \mathbb{R}^N$, and suppose that for all $x \in \partial U$, $DG(x)$ is nonsingular and the "Newton direction" $-DG(x)^{-1}G(x)$ points into U. Then existence (12) holds.

3. Initial Value Problems and a Predictor-Corrector Algorithm for Curve Following

In the preceding section, except for the fact that the homotopy parameter t may have been artificially introduced, and may be used to describe the initial point $(x_0, 0)$ and the terminal point $(x_1, 1)$, which we want to approximate, this addi-

II. Predictor-Corrector and Simplicial Methods for Approximating

tional parameter does not play any distinguished role in our discussions. Hence we will consider in the sequel the problem of numerically tracing a curve $c(s) \in H^{-1}(0)$ where $H: \mathbb{R}^{N+1} \to \mathbb{R}^N$ is a sufficiently smooth (e.g. C^∞) map and zero a regular value of H.

(1) Definition. Let A be an $(N, N+1)$-matrix with maximal rank N. By $T(A)$ we denote the vector in $\mathbb{R}^{N+1}$ which is uniquely defined by
1. $T(A) \in \ker(A)$;
2. $\|T(A)\| = 1$ (Euclidean norm);
3. $\det \begin{pmatrix} A \\ T(A)^T \end{pmatrix} > 0$ (orientation).

Let us consider the above-mentioned solution curve $c(s) \in H^{-1}(0)$. By assuming parametrization with respect to arc length, and differentiating the identity $H(c(s)) = 0$, one immediately sees that the solution curve $c(s)$ satisfies the following autonomous differential equation

$$\dot{u} = T(DH(u)). \tag{2}$$

Here $(\ )^{\cdot}$ denotes the differentiation with respect to arc length s, and DH denotes the Jacobian of H with respect to all variables. Clearly, condition (1)3 singles out one of the two possible orientations for $c(s)$. In this way $c(s)$ is uniquely defined by (2) and an initial value condition, say

$$c(0) = u_0 \in H^{-1}(0). \tag{3}$$

Since zero is a regular value of H, it is readily shown that $c(s) \in H^{-1}(0)$ is defined for all $s \in \mathbb{R}$. Equation (2) is basically due to Davidenko [33] and suggests a numerical curve tracing procedure for $c(s)$, namely to apply standard numerical integration methods for initial value problems. For references where this has been done, with numerous practical applications, see the works of L.T. Watson and colleagues [150]–[152.55].

Since the right-hand side of (2) is in the kernel of DH, the map H is constant on solutions of (2), and this implies that errors in the numerical integration tend to accumulate. Moreover, the primary task is generally not to approximate $c(s)$ with a high degree of accuracy, but rather merely to stay sufficiently near so as to be able to approximate any special point to an arbitrary (but practical) degree of accuracy, whenever it becomes desirable to do so, e.g. when level $t=1$ is reached. The fact that $c(s)$ is a zero set of H suggests using Newton's method in order to obtain a point which is nearer to the curve than the current point generated by integration steps, i.e., to use Newton's method as a corrector step.

(4) Definition. Let A be an $(N, N+1)$-matrix with maximal rank N. Then the Moore-Penrose inverse A^+ of A is the $(N+1, N)$-matrix obtained in the following way:

$$\begin{pmatrix} A \\ T(A)^T \end{pmatrix}^{-1} = (A^+, T(A)), \tag{5}$$

i.e., by inverting A "perpendicular to" the kernel of A, see e.g. [17], [109.7].

To illustrate the Newton step we have in mind, suppose that u is a given point near $H^{-1}(0)$. Define v to be a point in $H^{-1}(0)$ which has minimal distance to u in the Euclidean norm. It is an easy exercise to see that the Lagrange equations lead to the following characterization of v:

(6) $$H(v)=0, \quad (u-v)^T T(DH(v))=0.$$

From (4), it follows that Newton's method, applied to (6) for the initial point u, leads to the following approximation $\bar{u}$ of v:

(7) $$\bar{u}:=u-DH(u)^+ H(u),$$

cf. figure (8).

(8) Figure

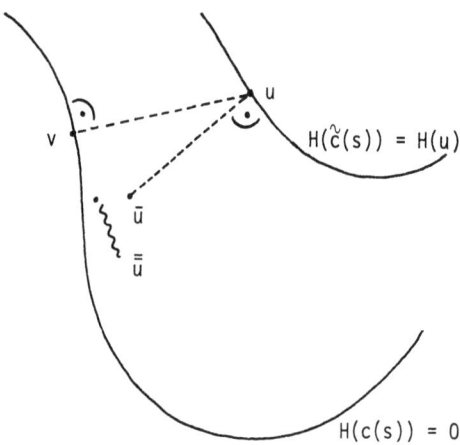

Equation (7) defines a well-known Newton's method for calculating solutions to underdetermined systems of nonlinear equations. It enjoys the usual Newton-Kantorovitch type properties [102], cf. [16]. In the context of curve following, it seems to have first been introduced by Haselgrove [71].

A continuous version of (7) is given by the following autonomous equation

(9) $$\dot{u}=-DH(u)^+ H(u),$$

i.e., (7) is obtained from (9) via Euler steps of length 1. The two flows (2) and (9) are orthogonal to each other, and the solutions of (9) tend to $H^{-1}(0)$, cf. [136]. Hence, the curve $c(s) \in H^{-1}(0)$ defined by (2)+(3) has better stability properties with respect to the following modification of Davidenko's equation

(10) $$\dot{u}=T(DH(u))-\alpha DH(u)^+ H(u),$$

which has been given by Georg [62]. Numerical tests [106] have shown that usual IVP-codes applied to (10) lead to satisfactory results as long as one does not want to follow the curve "too precisely". The choice of the damping pa-

II. Predictor-Corrector and Simplicial Methods for Approximating

rameter requires some care in order to avoid "artificial stiffness". An extensive discussion in [64.2] shows that $\alpha > 0$ should be chosen in such a way that the interval $[-2\alpha h, 0]$ is contained in the region of absolute stability (see e.g. [55.5], [67.8], [127]) of the adopted integration method, where h denotes the current step size.

Even though the numerical integration of Davidenko's equation (2) can be considerably improved by using the modification (10), such methods do not generally perform as well as the predictor-corrector methods which take explicit advantage of Newton's method (7) as a corrector and which we will now describe below. The primary advantage which the integration methods offer is that the determination of $T(DH(u))$, $DH(u)^+$ and the subsequent numerical integration can be performed via sophisticated computer library routines. For the former, see e.g. [41.3] and for the latter e.g. [127].

The predictor-corrector type of algorithms for curve following, which we now proceed to describe seem to date to Haselgrove [71]. For references we suggest [3.5], [6], [64], [64.2], [79], [80], [93.5], [95], [105.5], [111], [112], [125.9], [147.5].

Suppose the subsequent points $u_0, \ldots, u_n$ on a solution curve $c(s) \in H^{-1}(0)$ have been generated by some method, cf. figure (12). A predictor point $u_{n+1,0}$ is estimated by some method which has yet to be discussed. Then Newton's method (7) is started at this point and generates approximations

(11) $\qquad u_{n+1,q+1} := u_{n+1,q} - DH(u_{n+1,q})^+ H(u_{n+1,q}), \qquad q = 0, 1, \ldots$

of a new point u_{n+1} on the curve. The convergence $u_{n+1,q} \to u_{n+1}$ as $q \to \infty$ is quadratic.

(12) Figure

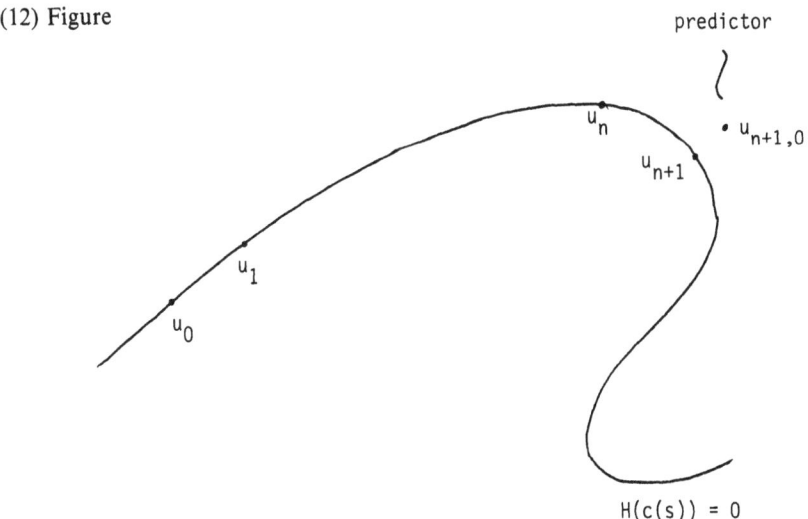

The corrector (11) gives the possibility of returning back to the solution curve as precisely as is wished, provided the starting (predictor) point $u_{n+1,0}$ is within the domain of attraction of Newton's method. This is an important difference from usual corrector steps in numerical integration. Consequently, the general opinion of workers in this field is that, because of the quality of Newton's method, it is unnecessary to perform sophisticated predictor steps. Generally, a tangential step is chosen:

(13) $$u_{n+1,0} = u_n + h_n T(DH(u_n)),$$

where h_n is some step length.

Because of the need for brevity here we shall merely give a general pattern of predictor-corrector algorithms and thereafter discuss a few aspects of the particular steps.

(14) **The general pattern of a predictor-corrector algorithm.**
 1. *Start.* Let u be an acceptable approximation to a point of $H^{-1}(0)$.
 2. *Jacobian and tangent.* Let A be equal to $DH(u)$ or an approximation thereof, and let $T(A)$ be given by (1).
 3. *Stepsize.* Choose the predictor step size $h > 0$ in a suitable way.
 4. *Predictor.* Set $v := u + hT(A)$.
 5. *Predictor test.* Is the predictor acceptable? If not, reduce h and go to 4.
 6. *Jacobian.* Obtain an approximation B to $DH(v)$.
 7. *Corrector.* Set $w := v - B^+ H(v)$, where B^+ is defined in (4), (5).
 8. *Corrector test.* Is the performance of the corrector acceptable. If not, reduce h and go to 4.
 9. *Accuracy test.* Does w approximate a new point on $H^{-1}(0)$ to within a desired tolerance?
   If yes, set $u := w$ and go to 2.
   If no, set $v := w$ and go to 6.

(15) **Step size control.** The strategy for selecting the step size h in (14)3 should be governed primarily by the particular implementation of the predictor-corrector procedure (see [64.2]), and by the particular type of nonlinear problem one wishes to solve.

The overriding criterion, however, should be that for whatever h is chosen, the corrector steps (14)7 should converge with a reasonable rate of contraction and should converge to a point further along the curve in the direction of traversing. With these considerations in mind, one then wants to trace a component of $H^{-1}(0)$ with a minimum of computational effort. Several authors have suggested strategies for particular types of problems, see e.g.: [35.5], [39], [47.6], [64.1], [68.1], [71], [83.4], [88.5], [95], [113.18], [148.4].

In making too large step sizes in the predictor steps (14)4, it may occur that v falls outside of the domain of convergence of the corrector process (14)7, or that v comes too close to some unwanted piece of $H^{-1}(0)$. Thus, in general, maximum tolerances should be maintained for such quantities as:

 1. Contraction, e.g. $\dfrac{\|H(w)\|}{\|H(v)\|}$ or $\dfrac{\|B^+ H(w)\|}{\|B^+ H(v)\|}$ in (14)8.
 2. Distance, e.g. $\|H(v)\|$ resp. $\|B^+ H(v)\|$ in (14)5
   or $\|H(w)\|$ resp. $\|B^+ H(w)\|$ in (14)8.

II. Predictor-Corrector and Simplicial Methods for Approximating 25

3. Angle, e. g. $\dfrac{\|w-v\|}{\|v-u\|}$ in (14)8.

4. Number of corrector steps necessary to attain the desired accuracy in (14)9.

5. In addition, it is necessary to specify a minimum step size $h_{\min}$ and stop the algorithm if $h \leqslant h_{\min}$.

(16) **Higher-order predictors.** In choosing predictor steps, the following two rules should be observed:

1. Use only information already obtained in order to reduce costly new calculations.

2. Use formulas which are not too sensitive to small variations of the u_i in directions perpendicular to the solution curve, since one may wish to stop the corrector (14)7 at an early stage in order to save computational costs.

A good choice are Adams-Bashforth predictors which are performed with respect to (2). They are discussed in [64.1], [64.2].

(17) **Solving the linear equations.** The predictor-corrector methods require for each step the numerical solution of linear equations where an $(N, N+1)$-matrix A is involved which equals or approximates the Jacobian DH at a point near $H^{-1}(0)$. In particular, problems of the following type have to be solved:

1. Calculate a nontrivial vector in ker A.

2. Calculate $A^+ u = v$ where u is given and v is sought.

Such solutions may be obtained by suitable decompositions of A, see e.g. [17], [109.7]. Let us briefly show that for a simple example the main numerical effort lies in decomposing A, and that subsequently the solutions to 1. and 2. are obtained very cheaply. Suppose Q is an orthogonal $(N+1, N+1)$-matrix such that $QA^T = R^T$ has upper triangular form (Householder transformation). Then the last column of Q^T is in ker A, and $A^+ = Q^T R^+$, where the operation R^+ is obtained by simple recursion formulas. Note also that it is not difficult to determine the "proper" sign in (1)3.

(18) **Remarks about evaluating approximate Jacobians.** It will have occured to the reader that the recalculation of the Jacobian in (14)2, (14)6 and its subsequent decomposition for each iterate would be excessively costly. Since the step size h is being adjusted to attain a reasonable rate of contraction, one may be satisfied with using modified Newton steps e. g. A in (14)2 or B in (14)6 is recalculated and decomposed only occasionally. An even better approach is to use Broyden-type low rank updates for the approximate Jacobian or some decomposition thereof. In [64], [64.2] adaptations are given for corresponding updates of underdetermined systems.

In many applications e. g. discretizations of various types of differential equations, network theory and structural mechanics, large sparse Jacobians arise which also have special structure. In the context of continuation methods, the corresponding Jacobians are again for underdetermined systems and hence some of the special structure (e. g. bandedness) is not exactly maintained. Nevertheless, in such cases special adaptations for approximating Jacobians or their decompositions may be very worthwhile. For large systems it may also be worthwhile to utilize iterative methods (e.g. SOR or conjugate gradients) to solve the linear systems. These are areas where for each application often a special technique should be developed. Examples where some adaptation of this nature has been made are [24.7], [25.5], [25.6], [28.2], [74.5], [110.5], [113].

(19) **Calculating special points.** In many applications there are certain points along $H^{-1}(0)$ which are of special interest, and hence it might be wished to approximate such points particularly well (e.g. when $t=1$ in a homotopy method (Section 2(1), or also "turning points"). We may generally formulate such points as points which satisfy an additional constraint, say
  1. $\phi(u(s))=0$ or
  2. $\phi(u(s))=\min!$

For example, in the case of the homotopy methods (Section 2(1)), if we want to approximate a point with $t=1$, we may set

$$\phi(u(s)) = e_{N+1}^T u(s) - 1.$$

If we formally apply Newton's method to solve 1., we have
  3. $s_{n+1} - s_n = -\phi(u(s))/\operatorname{grad}\phi(u(s_n))^T \dot{u}(s_n)$.

Interpreting 3. in the context of our pattern algorithm (14), we may set $s_{n+1} - s_n$ as an approximate guess for our step size h_n. In the case of 2., the situation is more complicated, but it may be treated in a similar way. Other approaches for calculating special points are discussed in [21.72], [94.9], [97.5], [102.75], [105.5], [105.6].

(20) **Bifurcation.** Under the regularity assumptions made at the beginning of this section, it was possible to conclude (see 1(3)) that

  1. $\det \begin{pmatrix} DH(c(s)) \\ \dot{c}(s)^T \end{pmatrix} > 0$

when following a curve $c(s)$ in $H^{-1}(0)$. If the regularity assumption is not fulfilled, and there is some point $u(s^*)$ on $c(s)$ at which the determinant in 1. changes sign, then it is known ([29], [88]) that an "odd" bifurcation occurs at $u(s^*)$. The checking of 1. can be very cheaply performed once the decomposition of the matrix A (approximating $DH(c(s))$) has been performed, see (17). Limitations of space prevent discussions of how a switching of branches near $u(s^*)$ can be effected. For some references, see [3.5], [18.4], [19.65], [21.72], [33.7], [64], [79.1], [80], [97.4], [103], [108.2], [111], [126.5], [152.9], [153.1].

(21) **Failure.** Basically, a predictor-corrector algorithm will fail when some step size h_n is attained with $h_n < h_{\min}$. This will generally mean that a point on $H^{-1}(0)$ has to be approximated where either the domain of contraction of the corrector process (14)7 is very small or the curvature is very high. Some possible remedies when this occurs are:

  (a) readjustment of the chosen step size strategy;
  (b) performing a perturbation on H;

II. Predictor-Corrector and Simplicial Methods for Approximating   27

(c) changing an underlying constraint in the corrector process (see [105.5]);
(d) resorting to a simplicial algorithm such as those which are discussed in Sections 5 and 6.

4. Monotonicity and Critical Points

We now briefly discuss an application of homotopy methods (Section 2(1)) to unconstrained optimization. Let us assume that $H: \mathbb{R}^N \times \mathbb{R} \to \mathbb{R}^N$ is a smooth homotopy with zero a regular value and such that

(1) $\qquad H(x, 0) = \operatorname{grad} e(x), \qquad H(x, 1) = \operatorname{grad} f(x), \qquad x \in \mathbb{R}^N$

for smooth functionals $e, f: \mathbb{R}^N \to \mathbb{R}$. In addition, it is assumed that x_0 is a known critical point of e, i.e.

$$H(x_0, 0) = \operatorname{grad} e(x_0) = 0.$$

(2) **Definition.** A solution curve

$$c(s) = (x(s), t(s)) \text{ in } H^{-1}(0) \text{ for } 0 \leqslant s \leqslant b$$

with $(x(0), t(0)) = (x_0, 0)$ and $(x(b), t(b)) = (x_1, 1)$ will be called monotone if $\dot{t}(s)$ does not change sign for $0 \leqslant s \leqslant b$. If $\dot{t}$ changes sign at s^*, we say that c has a turning point at s^*.

(3) **Monotonicity** implies invariance of the Morse index:
It is simple to prove [6] that if c is monotone, then the Hessian matrices

$$D^2 e(x_0) = D_x H(x, 0)|_{x=x_0}, \qquad D^2 f(x_1) = D_x H(x, 1)|_{x=x_1}$$

have the same number of negative eigenvalues. In particular, monotone solution curves connect minima to minima.

The preceding suggests that it would be very desirable to be able to construct homotopies H such that the curves c are monotone. It has been remarked [28] that for complex analytic maps $\tilde{H}: \mathbb{C}^N \times \mathbb{R} \to \mathbb{C}^N$, the solution curves $\tilde{c}$ in $\mathbb{C}^N \times \mathbb{R}$ are monotone if zero is a regular value of $\tilde{H}$.

It has been suggested [154.9] that a real homotopy $H: \mathbb{R}^N \times \mathbb{R} \to \mathbb{R}^N$ may be imbedded into a complex homotopy $\tilde{H}$ in order to make use of (3). This can, of course, only be of practical interest when the complex solution curve attains a real solution at the level $t = 1$. Although this idea is appealing and may be useful, there are some pitfalls which need to be recognized. We shall only briefly discuss them here.

Let $H: \mathbb{C}^N \times \mathbb{R} \to \mathbb{C}^N$ be a complex imbedding of a real analytic map, i.e., h is analytic and such that $\overline{h(z, t)} = h(\bar{z}, t)$ for $z \in \mathbb{C}^N$. Define the homotopy $H: \mathbb{C}^N \times \mathbb{R} \to \mathbb{C}^N$ by

(4) $\qquad\qquad\qquad H(z, t) = D_z h(z, t).$

Define $H^r, H^i: \mathbb{R}^N \times \mathbb{R}^N \times \mathbb{R} \to \mathbb{R}^N$ by

(5)
$$H^r(x, y, t) := \tfrac{1}{2}(H(x+iy, t) + H(x-iy, t))$$
$$H^i(x, y, t) := \frac{-i}{2}(H(x+iy, t) - H(x-iy, t)).$$

Finally define $\hat{H}: \mathbb{R}^{2N} \times \mathbb{R} \to \mathbb{R}^{2N}$ by

(6)
$$\hat{H}(x, y, t) := \begin{pmatrix} H^r(x, y, t) \\ -H^i(x, y, t) \end{pmatrix}.$$

The Davidenko equation corresponding to Section 3(2) for $\hat{H}(x, y, t) = 0$ assumes the form

(7)
$$\begin{pmatrix} D_t H^r & D_x H^r & D_y H^r \\ -D_t H^i & -D_x H^i & -D_y H^i \end{pmatrix} \begin{pmatrix} \dot{t} \\ \dot{x} \\ \dot{y} \end{pmatrix} = \begin{pmatrix} 1 \\ 0 \\ 0 \end{pmatrix}.$$

From (5) we have $D_y H^r = -D_x H^i$ and $D_y H^i = D_x H^r$ and hence (7) may be written

(8)
$$\begin{pmatrix} D_t H^r & D_x H^r & -D_x H^i \\ -D_t H^i & -D_x H^i & -D_x H^r \end{pmatrix} \begin{pmatrix} \dot{t} \\ \dot{x} \\ \dot{y} \end{pmatrix} = \begin{pmatrix} 1 \\ 0 \\ 0 \end{pmatrix}.$$

Hence from (7)

$$\begin{pmatrix} \dot{t} & \dot{x}^T & \dot{y}^T \\ D_t H^r & D_x H^r & D_y H^r \\ -D_t H^i & -D_x H^i & -D_y H^i \end{pmatrix} \begin{pmatrix} \dot{t} & 0^T \\ \dot{x} & I_{2N} \\ \dot{y} & \end{pmatrix} = \begin{pmatrix} 1 & \dot{x}^T & \dot{y}^T \\ 0 & D_x H^r & -D_x H^i \\ 0 & -D_x H^i & -D_x H^r \end{pmatrix}$$

and consequently,

(9)
$$\dot{t} \det \begin{pmatrix} \dot{t} & \dot{x}^T & \dot{y}^T \\ D_t \hat{H} & D_x \hat{H} & D_y \hat{H} \end{pmatrix} = \det \begin{pmatrix} D_x H^r & -D_x H^i \\ -D_x H^i & -D_x H^r \end{pmatrix} = \det D_{x,y} \hat{H}.$$

From (4), (5), (9) it follows that $D_{x,y} \hat{H}$ is a symmetric matrix and hence has only real eigenvalues. Furthermore, from the special form of $D_{x,y} \hat{H}$ (see (9)), it is easily seen that the eigenvalues occur in symmetric pairs about 0. Hence,

(10) $(-1)^N \det D_{x,y} \hat{H} \geq 0$ holds everywhere.

From (9), (10) it follows that if

$$\hat{c}(s) = (x(s), y(s), t(s)) \quad \text{for } 0 \leq s \leq b$$

is a smooth solution curve in $\hat{H}^{-1}(0)$ which consists of regular points $c(s)$ on some deleted neighborhood of s^*, then

(11) $\dot{t}$ changes sign at s^* if and only if $\det \begin{pmatrix} \dot{t} & \dot{x}^T & \dot{y}^T \\ D_t \hat{H} & D_x \hat{H} & D_y \hat{H} \end{pmatrix}$ does.

Hence:

(12) a turning point of $\hat{c}$ is also a bifurcation point.

Although it is possible (because of (10)) to extract a piecewise smooth solution curve $\hat{c}$ in $\hat{H}^{-1}(0)$ which is monotone, the crucial difficulty is that it can no longer be assured that the signs of the eigenvalues of $D_x H^r$ stay constant on $\hat{c}$. Indeed, it is possible to construct examples in which eigenvalues of $D_x H^r$ change sign on $\hat{c}$ when $y(s) \neq 0$, but no bifurcation is occurring. Thus in general, it is not assured that

$$D_x H^r(x_0, 0, 0) \quad \text{and} \quad D_x H^r(x_1, 0, 1)$$

have the same sign pattern of eigenvalues even if $(x_0, 0, 0)$ and $(x_1, 0, 1)$ are connected by a piecewise smooth and monotone solution curve.

5. Simplicial Continuation Methods

Consider again a map $H: \mathbb{R}^{N+1} \to \mathbb{R}^N$ and the equation $H(u) = 0$. In Section 3 we discussed predictor-corrector methods which approximately follow an "exact" solution curve $c(s) \in H^{-1}(0)$. Now we consider methods which "exactly" follow an approximate solution curve $c_T(s) \subset H_T^{-1}(0)$, where H_T is a piecewise linear (PL) approximation of H corresponding to a given triangulation T of $\mathbb{R}^{N+1}$. The polygonal path $c_T(s)$ is traced by techniques known from linear programming. We shall refer to such methods as simplicial continuation methods. We suggest the following articles for supplementary reading: [6], [44], [54.3], [75], [104], [124], [137], [142.3].

(1) Figure

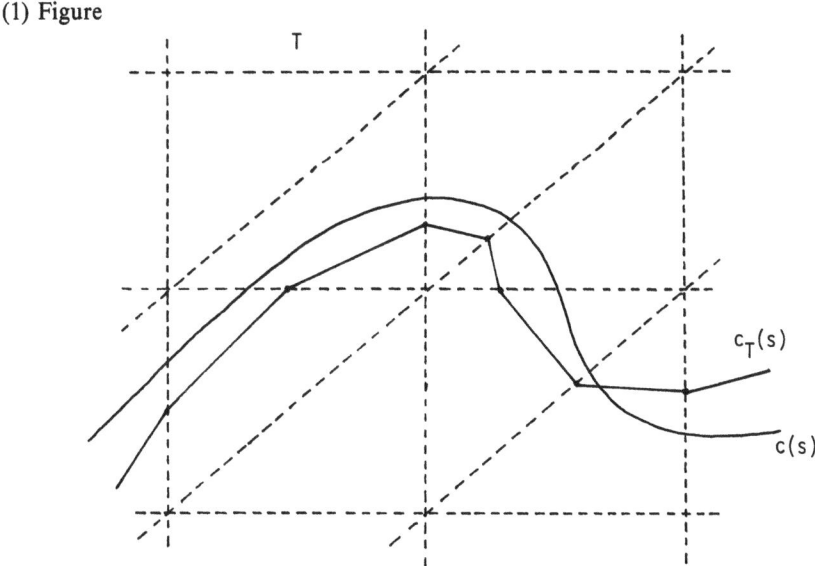

For intuitive purposes, the reader is invited to assume for the moment that H is again smooth and zero a regular value of H; thus, $c_T(s)$ approximates a solution curve $c(s) \subset H^{-1}(0)$ with order $O(h^2)$, where h denotes the mesh size of T. However, one advantage of simplicial methods is the fact that such smoothness assumptions are no longer necessary. This will be seen in Section 6. In this section we give an introduction into the general procedure of a simplicial algorithm.

(2) **Definition.** By a k-simplex $\sigma = [u_1, \ldots, u_{k+1}]$ we mean the convex hull of $k+1$ affinely independent points $u_1, \ldots, u_{k+1}$, called vertices of σ.

(3) **Definition.** If $\sigma = [u_1, \ldots, u_{k+1}]$ is a k-simplex and $\{v_1, \ldots, v_{m+1}\}$ a subset of its vertices, we call the m-simplex $\tau = [v_1, \ldots, v_{m+1}]$ an m-face of σ. Hence, 0-faces are the vertices and 1-faces are the edges of σ.

(4) **Definition.** A triangulation of $\mathbb{R}^{N+1}$ is a system $T = \{\sigma\}$ of $(N+1)$-simplices with the following properties:
1. The union of all $\sigma \in T$ is $\mathbb{R}^{N+1}$.
2. The intersection of two simplices of T is either void or a common face.
3. The covering T of $\mathbb{R}^{N+1}$ is locally finite, i.e., every compact subset of $\mathbb{R}^{N+1}$ meets only a finite number of simplices of T.

For our purposes, the important property of a triangulation is the fact that any vertex may be pivoted in exactly one way:

(5) **Lemma.** Let $\sigma = [u_1, \ldots, u_{N+2}]$ be an $(N+1)$-simplex of a triangulation T of $\mathbb{R}^{N+1}$, and u_i be a vertex of σ. Then there is exactly one vertex $u_i' \neq u_i$ such that σ' is again an $(N+1)$-simplex of T, where σ' is obtained from σ by replacing u_i with u_i'.

A proof may be found in [137]. Note that (5) can be expressed equivalently by saying that any N-face is contained in exactly two different $(N+1)$-simplices of T. We call the procedure of passing from σ to σ', respectively, from u_i to u_i' a *pivoting step*, cf. figure (6).

(6) Figure

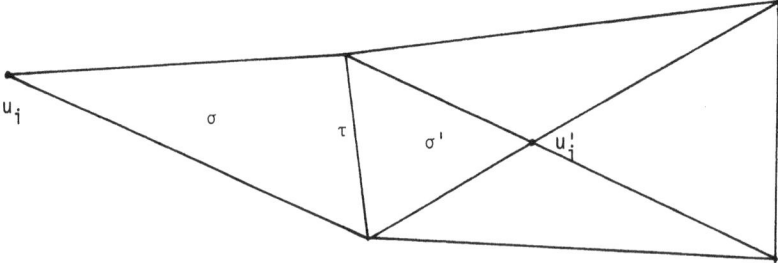

In numerical implementations of simplicial algorithms, only one (current) simplex is stored, and the triangulation is available via its pivoting steps which in all practical cases are simple and may be performed via a subroutine. The computational cost of one pivoting step is at most $O(N)$ arithmetic operations

II. Predictor-Corrector and Simplicial Methods for Approximating

in all such cases. The simplest triangulation is due to Freudenthal [49]. Its pivoting rules are given by an elementary reflection [5]:

We store a simplex $\sigma = [u_1, \ldots, u_{N+2}]$ together with a fixed cyclic ordering of its vertices, given for example by the permutation

$$\pi = \begin{pmatrix} 1 & \ldots & N+1 & N+2 \\ 2 & \ldots & N+2 & 1 \end{pmatrix}.$$ Then a vertex u_i is pivoted by

(7) $$u_i' := u_{\pi(i)} + u_{\pi^{-1}(i)} - u_i,$$

i.e., by reflecting u_i across the centroid of its two "neighbors" $u_{\pi(i)}, u_{\pi^{-1}(i)}$. Starting with any $(N+1)$-simplex σ in $\mathbb{R}^{N+1}$, this pivoting rule generates only simplices which belong to a triangulation T_σ of $\mathbb{R}^{N+1}$, where T_σ is obtained from Freudenthal's triangulation by some affine isomorphism.

(8) Figure. (Pivoting by reflextion)

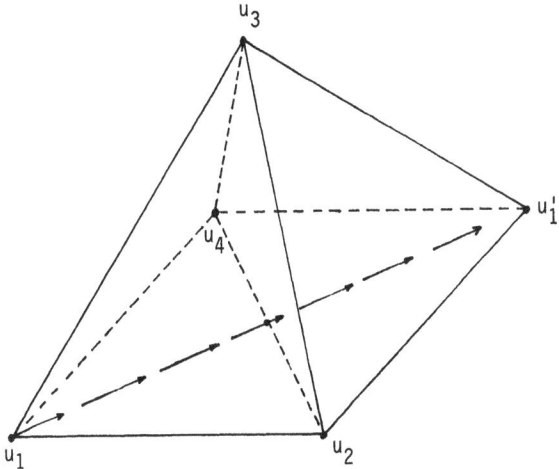

It may seem more natural to the reader to reflect u_i at the barycenter of the remaining N-face τ obtained from σ by omission of u_i. However, for $N \geq 2$ this rule does not lead to a triangulation, and for $N = 0, 1$ it coincides with (7).

(9) **Definition.** Let $\tau = [u_1, \ldots, u_{k+1}]$ be a k-simplex in $\mathbb{R}^{N+1}$, and denote by aff(τ) the affine hull of τ. Let $H: \mathbb{R}^{N+1} \to \mathbb{R}^N$, we interpolate H on the vertices of τ by an affine map $H_\tau: \text{aff}(\tau) \to \mathbb{R}^N$ as follows. Given a point $u \in \text{aff}(\tau)$, with its barycentric coordinates $a \in \mathbb{R}^{k+1}$:

(10) $$\begin{pmatrix} 1 & \ldots & 1 \\ u_1 & \ldots & u_{k+1} \end{pmatrix} a = \begin{pmatrix} 1 \\ u \end{pmatrix},$$

the affine interpolation is defined by

(11) $$\begin{pmatrix} 1 & \ldots & 1 \\ H(u_1) & \ldots & H(u_{k+1}) \end{pmatrix} a = \begin{pmatrix} 1 \\ H_\tau(u) \end{pmatrix}.$$

(12) **Definition.** Let T be a triangulation of $\mathbb{R}^{N+1}$ and $H: \mathbb{R}^{N+1} \to \mathbb{R}^N$ a map. The PL-approximation H_T of H (with respect to T) is given by

(13) $$u \in \sigma \in T \Rightarrow H_T(u) := H_\sigma(u).$$

Note that by (11), this definition does not depend on the special choice of σ.

In the literature, the map H is often referred to in this context as a "vector labeling". In order to discuss the solution set $H_T^{-1}(0)$, we need the crucial notion of a completely labeled simplex.

(14) **Definition.** A matrix is called lexicographically positive if the leading nonzero coefficient of each row is positive.

(15) **Definition.** Given a "labeling" $H: \mathbb{R}^{N+1} \to \mathbb{R}^N$ and an N-simplex $\tau: [u_1, \ldots, u_{N+1}]$, we call τ completely labeled (with respect to H) if the "labeling matrix"

(16) $$\begin{pmatrix} 1 & \cdots & 1 \\ H(u_1) & \cdots & H(u_{N+1}) \end{pmatrix}$$

is nonsingular and has a lexicographically positive inverse. Equivalently, τ is completely labeled if there exists a number $\delta > 0$ such that the equation

(17) $$H_\tau(u) = (\varepsilon, \varepsilon^2, \ldots, \varepsilon^N)^T$$

has solutions u_ε in τ for all $\varepsilon \in [0, \delta]$.

Hence, a completely labeled simplex σ not only contains a zero of the affine interpolation of H, but also a small solution arc u_ε. This perturbation technique reminds us of Sard's theorem and has been introduced by Charnes [25.7] in order to deal with degeneracies in linear programming. As a result, the following "door in-door out" principle is obtained, which is similar to the pivoting steps of linear programming:

(18) **Lemma.** Let σ be an $(N+1)$-simplex in $\mathbb{R}^{N+1}$ and $H: \mathbb{R}^{N+1} \to \mathbb{R}^N$ a labeling. Then σ either contains exactly two completely labeled N-faces or none at all.

We borrow from [43.5] a heuristic, but convincing description of a simplicial continuation algorithm. Think of the $(N+1)$-simplices of a triangulation T of $\mathbb{R}^{N+1}$ as "rooms" of the "building" $\mathbb{R}^{N+1}$ and of the completely labeled N-faces as "doors". By (5), any door belongs to exactly two rooms, and by (18), any room has either no doors or exactly two. We traverse the rooms by observing the following rule: if a room is entered by one door, it has to be left by the other.

(19) **Algorithm.** Assume a labeling $H: \mathbb{R}^{N+1} \to \mathbb{R}^N$ and a triangulation T of $\mathbb{R}^{N+1}$ are given.
  1. *Start.* Let σ_1 be an $(N+1)$-simplex of T and let τ_1 be a completely labeled N-face of σ_1. Set $i := 1$.
  2. *Linear equations.* Find the uniquely determined N-face τ_{i+1} of σ_i which is completely labeled and different from τ_i.

II. Predictor-Corrector and Simplicial Methods for Approximating

3. *Pivot.* Find the uniquely determined $(N+1)$-simplex σ_{i+1} of T which contains τ_{i+1} and is different from σ_i. Set $i:=i+1$ and go to 2.

The algorithm defines an infinite sequence of completely labeled N-faces $\tau_1, \tau_2, \ldots$ and an infinite sequence of corresponding $(N+1)$-simplices $\sigma_1, \sigma_2, \ldots$. Two different types of results are possible:

(20)
1. *No cycling.* All τ_i and all σ_i are pairwise different.
2. *Cycling.* There exists an integer $p > 1$ such that
$$\tau_i = \tau_j \Leftrightarrow \sigma_i = \sigma_j \Leftrightarrow i = j \text{ modulo } p.$$

As we will see later, solving (19)2 involves the numerical solution of linear equations, and as a consequence, the zero of H_{τ_i} is obtained without much additional cost.

(21) **Definition.** Given algorithm (19), for any completely labeled τ_i, we define $v_i \in \tau_i$ as the (uniquely determined) zero of the interpolation H_{τ_i}. Consider the polygonal path $c_T(s)$, parametrized according to arc length s, which encounters successively the zeros $v_1, v_2, \ldots$, where the parameter s is normalized, say, by $c_T(0) = v_1$. Clearly, $c_T(s) \in H_T^{-1}(0)$. We call $c_T(s)$ the PL-path generated by the simplicial algorithm (19). It is either homeomorphic to a straight line or a closed circle, cf. (20).

(22) Figure

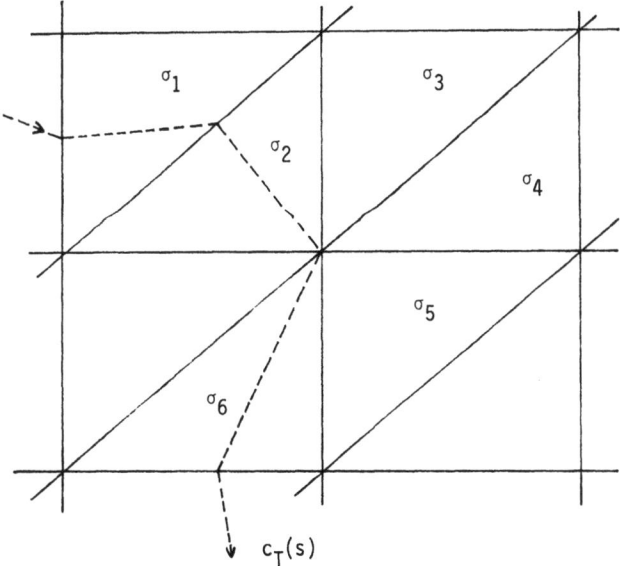

We give a sketch of the computational effort which is involved in performing (19). As has already been pointed out, step 3 requires at most $O(N)$ arithmetic operations in practical cases and hence may be neglected compared to the cost of step 2. As in linear programming, the perturbation technique (17) is mainly of theoretical interest. For numerical purposes, the necessary conditions of the following lemma may be considered to be also sufficient:

(23) **Lemma.** In algorithm (19), assume that the two completely labeled N-faces τ_i and τ_{i+1} are obtained from the $(N+1)$-simplex $\sigma_i = [u_i, \ldots, u_{N+2}]$ by omitting the vertex with index j^- (respectively j^+). Consider a non-zero solution $t = (t(1), \ldots, t(N+2))^T \in \mathbb{R}^{N+2}$ of the homogeneous equations

(24) $$\begin{pmatrix} 1 & \cdots & 1 \\ H(u_1) & \cdots & H(u_{N+2}) \end{pmatrix} t = 0$$

and a solution $c = (c(1), \ldots, c(N+2))^T \in \mathbb{R}^{N+2}$ of the inhomogeneous equations

(25) $$\begin{pmatrix} 1 & \cdots & 1 \\ H(u_1) & \cdots & H(u_{N+2}) \end{pmatrix} c = \begin{pmatrix} 1 \\ 0 \end{pmatrix}.$$

Then $t(j^-)t(j^+) < 0$, and we may assume $t(j^-) < 0 < t(j^+)$. Furthermore, the following assertions hold:

(26) j^- maximizes $j \to c(j)/t(j)$ for $t(j) < 0$.
(27) j^+ minimizes $j \to c(j)/t(j)$ for $t(j) > 0$.

The next step in algorithm (19) now consists of pivoting the vertex u_{j^+} into a new vertex u'_{j^+}. Thus, new linear equations (24), (25) are obtained which differ from the old ones only in column j^+. Consequently, updating techniques in the sense of Bartels and Golub [14], see also [65], are adequate numerical tools for solving these equations.

6. Simplicial Homotopy Methods

Simplicial continuation methods can be effectively used to calculate a zero of a map $G: \mathbb{R}^N \to \mathbb{R}^N$. The main idea of the simplicial algorithms using continuous refinement (due to Eaves [43]) is to set up a homotopy H and then construct a triangulation T of $\mathbb{R}^{N+1}$ in such a way that the mesh size shrinks in the direction of the homotopy parameter. Thus, the solution path $c_T(s) \in H_T^{-1}(0)$ approximates a zero of G with increasing accuracy, cf. figure (1).

II. Predictor-Corrector and Simplicial Methods for Approximating

(1) Figure

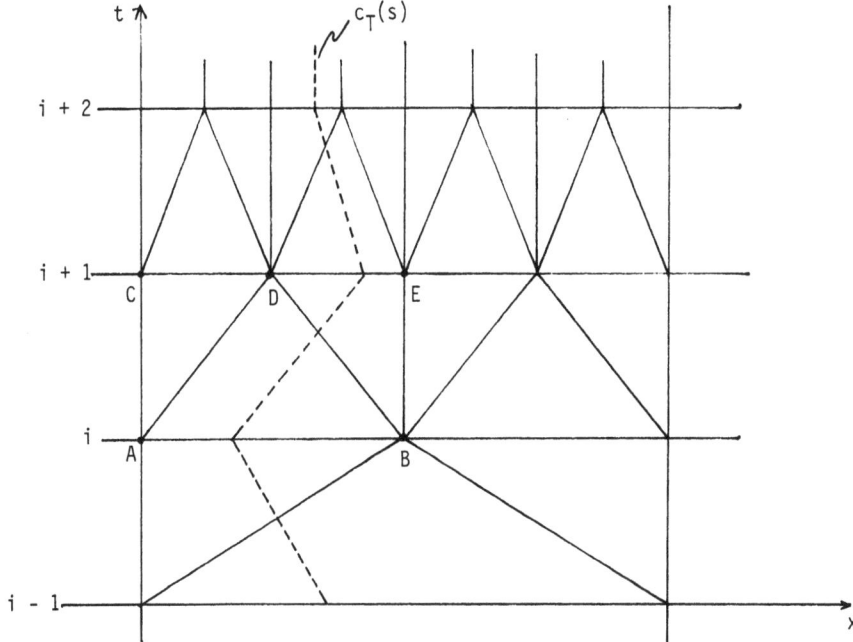

We now illustrate how simplicial algorithms may be used to calculate a solution of a nonlinear problem by sketching an approach due to Eaves and Saigal [45]. Let us first discuss the above-mentioned "shrinking" procedure in more detail.

(2) **Definition.** In $\mathbb{R}^{N+1} \triangleq \mathbb{R}^N \times \mathbb{R}$ we single out the last coordinate as a "homotopy parameter" and write a variable $u \in \mathbb{R}^{N+1}$ in the form, say, $u = (x, t)$ with $x \in \mathbb{R}^N$ and $t \in \mathbb{R}$. We shall occasionally make use of the canonical projection $P: \mathbb{R}^{N+1} \to \mathbb{R}^N$ given by $(x, t) \to x$. For $\sigma \subset \mathbb{R}^{N+1}$, we define the x-diameter of σ by

(3) $\qquad x\text{-diam}(\sigma) := \sup\{\|Pu - Pv\| : u, v \in \sigma\},$

where, for the sake of simplicity, $\|\cdot\|$ denotes the Euclidean norm.

(4) **Definition.** Let T be a triangulation of $\mathbb{R}^{N+1}$ and $C > 1$ a given "refining" factor. We call T a refining triangulation (with factor C) if the following conditions hold:
1. For any simplex $\sigma \in T$ there is a unique integer i, called the level of σ, such that $\sigma \subset \mathbb{R}^N \times [i, i+1]$.
2. If $\sigma \in T$ has level i, then any vertex of σ is either in $\mathbb{R}^N \times \{i\}$ or in $\mathbb{R}^N \times \{i+1\}$.
3. $C = \dfrac{\text{Max}\{x\text{-diam}(\sigma) : \sigma \in T \text{ has level } i+1\}}{\text{Max}\{x\text{-diam}(\sigma) : \sigma \in T \text{ has level } i\}}.$

The last condition describes the "shrinking" procedure, cf. figure (1). The first such refining triangulation, constructed by Eaves [43], had a factor $C<2$ and was modified by Eaves and Saigal [45] into one with factor $C=2$. Todd [137] gave an efficient refining triangulation with factor $C=2$, called J_3. Van der Laan and Talman [146.6] constructed a class of triangulations where the refining factor may be chosen to be any integer $C>1$, which may depend on the level i, see also [87.8]. Let us illustrate the applicability of the refining technique by discussing a homotopy for a zero problem which subsumes many important nonlinear problems, cf. [6], [42], [44], [61], [96], [118], [137]. Our presentation follows [64.2], where details of proofs are also given.

(5) **Definition.** We call a map $G: \mathbb{R}^N \to \mathbb{R}^N$ quasi-linear if the following two conditions hold:
1. G is locally bounded, i.e., the image of a compact set is bounded.
2. There exists a nonsingular (N, N)-matrix $DG(\infty)$ such that
$\|G(x) - DG(\infty)x\| = O(\|x\|)$ as $\|x\| \to \infty$.

$DG(\infty)$ is uniquely defined and called the derivative of G at ∞. Note that we do not assume any smoothness or even continuity of G. As a consequence, we have to be somewhat careful concerning the zero points, namely, we have to consider them with respect to a set-valued hull.

(6) **Definition.** Let $G: \mathbb{R}^N \to \mathbb{R}^N$ be locally bounded. We denote by $F(x)$ the filter of neighborhoods of a point $x \in \mathbb{R}^N$, and by $[\mathbb{R}^N]$ the system of all convex compact non-empty subsets of $\mathbb{R}^N$. Then the set-valued hull of G is defined to be the map $[G]: \mathbb{R}^N \to [\mathbb{R}^N]$ given by

(7) $$[G](x) := \cap \{\overline{\text{co}}(G(V)): V \in F(x)\}.$$

Here $\overline{\text{co}}(\cdot)$ denotes the closed convex hull.

It is easily seen that $[G]$ is the smallest set-valued map $\mathbb{R}^N \to [\mathbb{R}^N]$ which is upper semi-continuous and contains G, cf. [115]. Furthermore, G is continuous at a point $x \in \mathbb{R}^N$ if and only if $[G](x)$ is a singleton. For many applications, the following lemma shows how to construct the hull $[G]$:

(8) **Lemma.** Let $\{M_\alpha : \alpha \in I\}$ be a locally finite subdivision of $\mathbb{R}^N$, and let $G_\alpha: M_\alpha \to \mathbb{R}^N$ be continuous ($\alpha \in I$). We assume that G is obtained by the union $G := \cup G_\alpha|_{M_\alpha}$. Let us call an index $\alpha \in I$ active in $x \in \mathbb{R}^N$ if $x \in \overline{M_\alpha}$. Then

(9) $$[G](x) = \{G_\alpha(x): \alpha \text{ is active at } x\}.$$

It is an easy exercise to show that the following lemma is equivalent to Kakutani's fixed point theorem when considered in $\mathbb{R}^N$:

(10) **Lemma.** If $G: \mathbb{R}^N \to \mathbb{R}^N$ is quasi-linear, then there exists at least one zero of $[G]$, i.e., a point $x \in \mathbb{R}^N$ such that $0 \in [G](x)$.

We now sketch how such a zero point is approximated by a simplicial continuation method.

II. Predictor-Corrector and Simplicial Methods for Approximating

(11) **Algorithm.** Let $G: \mathbb{R}^N \to \mathbb{R}^N$ be quasi-linear and T a refining triangulation of $\mathbb{R}^{N+1}$. We define a homotopy $H: \mathbb{R}^{N+1} \to \mathbb{R}^N$ by

(12)
$$H(x, t) := \begin{cases} G(x) & \text{for } t > 0, \\ DG(\infty)(x - x_0) & \text{for } t \leq 0. \end{cases}$$

Here, $x_0 \in \mathbb{R}^N$ plays the role of a starting point. It is clear that the zeros of H_T in $\mathbb{R}^N \times (-\infty, 0]$ are all contained in the trivial solution path $c_T(s) = (x_0, s)$ for $s \leq 0$. We start the simplicial algorithm (19) of Section 5 on this path into $\mathbb{R}^N \times (0, \infty)$ and thus extend $c_T(s)$ for all $s > 0$, cf. figure (16). Let us denote the coordinate of c_T by $c_T(s) = (x_T(s), t_T(s))$. Using the asymptotic estimate (5)1., and the local finiteness of T, it is not difficult to show that $x_T(s)$ remains bounded and $t_T(s) \to \infty$ as $s \to \infty$. If $x_T(\infty)$ denotes the set of accumulation points of $x_T(s)$ as $s \to \infty$, the following two properties are readily shown:

(13) The set $x_T(\infty)$ is compact, non-empty and connected.

(14) Any accumulation point $\bar{x} \in x_T(\infty)$ is a zero of $[G]$.

As a corollary, we obtain:

(15) If $[G]$ has only isolated zeros, then $x_T(s)$ converges toward such a zero as $s \to \infty$.

The hypothesis in (15) is quite natural. If a nonlinear problem is formulated as a zero problem, find $0 \in [G](x)$, then the assumption means (in a heuristic way of speaking) that the problem is reasonably posed.

(16) Figure

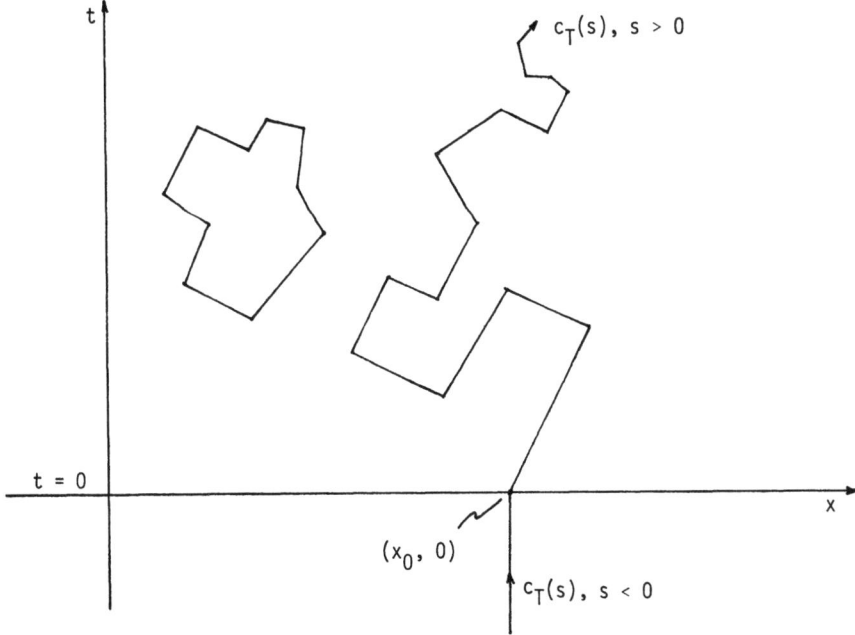

The following illustration of how quasi-linear maps may be applied was given in [61].

(17) **Example.** Let $F: \mathbb{R}^N \to \mathbb{R}^N$ be a continuous map and $\Omega \subset \mathbb{R}^N$ a bounded open zero neighborhood. Define

$$G(x) := \begin{cases} x & \text{for } x \notin \Omega, \\ x - F(x) & \text{for } x \in \Omega. \end{cases}$$

It is clear that G is quasi-linear, and hence algorithm (11) furnishes a zero point $\bar{x}$ of $[G]$. By using (8), it can be shown that only two cases are possible:
1. $\bar{x} \in \Omega$ and $F(\bar{x}) = \bar{x}$.
2. $\bar{x} \in \partial\Omega$ and $F(\bar{x}) = \lambda \bar{x}$ for some $\lambda \geq 1$.

In the first case, we found a fixed point of F in Ω. The theorem of Leray and Schauder [93] (applied in $\mathbb{R}^N$) ensures its existence, provided that case 2 does not hold. Hence, if algorithm (11) does not find a fixed point of F in Ω, i.e. in case 2, it gives us a good explanation for its failure. In fact, in many optimization problems, such a point indicates a direction of descent [61].

In the preceding discussion we sketched a particular homotopy algorithm using refining triangulations, because it is relatively easy to describe and comprehend. In recent years many types of simplicial algorithms have been given for solving a zero point problem such as: Find $0 \in [G](x)$ with G quasi-linear. Such algorithms are generally referred to as "simplicial fixed point algorithms". Our bibliography contains most of these works up to the present date. Limitations of space prevent further detailed discussions. We content ourselves therefore with briefly sketching several useful or important aspects.

(18) **Integer labeling.** Instead of using vector labeling, one may use an integer labeling $H: \mathbb{R}^{N+1} \to \mathbb{Z}$. This is connected with Sperner's lemma [131] and was one of the first approaches to fixed point algorithms, cf. [8.2], [28.4], [88.9], [89]. Since integer labeling leads to a very coarse approximation of a given nonlinear problem, it is usually not recommended for smooth problems, but may be interesting for problems of a more combinatorial nature. For additional reading, see [51.1], [54.3], [108.1], [124], [133], [137], [142.8], [146].

(19) **Restart algorithms.** Instead of traversing the whole PL-path of the homotopy algorithm (11), one may stop at a certain level and use the latest point $x_T(s)$ as a new starting point in (12). Clearly, the mesh size of the new triangulation T should be suitably reduced. For more details, see e.g. [87.8], [95.9], [96], [137], [142.3], [146.35], [146.9].

(20) **Variable dimension algorithms.** As one immediately sees from (12), the homotopy parameter t plays a much more artificial role in simplicial algorithms. Basically, it simply gives a reading of how much "shrinking" has been made in the refining process. In fact, it is possible to get rid of the t parameter altogether by constructing an algorithm which starts at a point (e.g. an approximate solution) and generates completely labeled simplices of varying dimensions. These techniques were begun by van der Laan and Talman [146.2] and are mainly used together with restart methods. It is possible to give an equivalent characterization of such algorithms in a homotopy formulation by con-

structing certain rather sophisticated triangulations. For details on these ideas see e.g. [87], [87.8], [87.9], [133], [139.4], [142.3], [142.8], [146], [146.35], [146.9], [154.5].

(21) **Sparseness.** For high-dimensional nonlinear problems (say $N \geqslant 50$) simplicial algorithms become too slow unless they use special structure such as sparseness or separability of maps. Such structure reflects itself in the fact that the PL-path $c_T(s)$ of the simplicial algorithm changes direction only occasionally, and not every time it crosses a completely labeled N-face. This observation is essentially due to Kojima [85.3] and has been effectively developed by Todd. By adapting the homotopy and triangulations in a suitable way, this effect may be improved. When the direction of the path does not change, a pivoting step is of very low computational cost. For more details, see e.g. [6], [118.2], [118.5], [118.6], [139.3], [141], [141.3], [142.1].

(22) **Acceleration.** Withour further assumptions on G, nothing can be said about the rate of convergence in (15), see e.g. [21.77]. To ensure linear convergence of algorithm (11), assumptions in the spirit of the Newton-Kantorovitch theorems [102] are necessary, cf. [30], [116.3], [116.6], [117]. Saigal was the first to see the close interrelationship between simplicial steps and Newton's method, see also [137], [139.7]. To accelerate a simplicial homotopy algorithm, several papers discuss techniques of mixing simplicial and Newton steps, see e.g. [64.2], [116.6], [119], [119.4]. In the context of simplicial continuation methods, i.e. when a whole curve $c(s)$ is to be approximated by a PL-path $c_T(s)$, an analogous argument leads to a mixing of simplicial and predictor-corrector steps. This has been recently investigated by Saupe [121], [121.3].

(23) **Adaptations to constraints.** It is possible (and very advantageous) to adapt the triangulations and the complementary pivoting procedure to constraints occurring in nonlinear optimization. This has been done by Todd [139.2], [140.5], [142.4].

Bibliography

0.7   Abbott, J. P. & Brent, R. P.: A note on continuation methods for the solution of nonlinear equations. J. Austral. Math. Soc. 20 (1978) 157-164

1.   Abraham, R. & Robbin, J.: Transversal mappings and flows. W. A. Benjamin, 1967

1.7   Alexander, J. C.: Bifurcation of zeroes of parametrized functions. J. Func. Anal. 29 (1978) 37-53

1.8   Alexander, J. C.: Numerical continuation methods and bifurcation. In: Functional differential equations and approximation of fixed points, H.-O. Peitgen, H.-O. Walther (eds), Lecture Notes in Math. 730, Springer Verlag (1979) 1-15

2.   Alexander, J. C. & Kellogg, R. B. & Li, T. Y. & Yorke, J. A.: Piecewise smooth continuation. Preprint, 1979

3. Alexander, J. C. & Yorke, J. A.: The homotopy continuation method: numerically implementable topological procedures. Trans. AMS 242 (1978) 271-284
3.4 Allgower, E. L.: Numerische Approximation von Lösungen nichtlinearer Randwertaufgaben mit mehreren Lösungen. ZAMM 54 (1974) T206-T207
3.46 Allgower, E. L.: Application of a fixed point search algorithm. In: Fixed points: Algorithms and applications, S. Kramardian (ed.), Academic Press, New York (1977) 87-111
3.5 Allgower, E. L.: A survey of homotopy methods for smooth mappings. In: Numerical solution of nonlinear equations, E. L. Allgower, K. Glashoff, H.-O. Peitgen (eds), Springer Lecture Notes in Math. 876 (1981) 1-29
4. Allgower, E. L. & Georg, K.: Triangulations by reflections with applications to approximation. In: Numerische Methoden der Approximationstheorie, L. Collatz, G. Meinardus, H. Werner (eds), Birkhäuser Verlag ISNM 42 (1978) 10-32
5. Allgower, E. L. & Georg, K.: Generation of triangulations by reflections. Utilitas Math. 16 (1979) 123-129
6. Allgower, E. L. & Georg, K.: Simplicial and continuation methods for approximating fixed points and solutions to systems of equations. SIAM Review 22 (1980) 28-85
7. Allgower, E. L. & Georg, K.: Homotopy methods for approximating several solutions to nonlinear systems of equations. In: Numerical solution of highly nonlinear problems, W. Förster (ed.), North-Holland Publ. Comp. (1980) 253-270
7.5 Allgower, E. L. & Georg, K.: Relationships between deflation and global methods in the problem of approximating additional zeroes of a system of nonlinear equations. To appear in the proc. of the Nato Advanced Research Institute on homotopy methods and global convergence, Plenum Press
8. Allgower, E. L. & Glashoff, K. & Peitgen, H.-O. (eds): Numerical solution of nonlinear equations. Springer Lecture Notes in Math. 876 (1981)
8.1 Allgower, E. L. & Jeppson, M. M.: The approximation of solutions of nonlinear elliptic boundary value problems with several solutions. In: Lecture Notes in Math. 333, Springer Verlag (1973) 1-20
8.2 Allgower, E. L. & Keller, C. L.: A search routine for a Sperner simplex. Computing 8 (1971) 157-165
8.7 Alligood, K.: Homological indices and homotopy continuation. Ph. D. Thesis, University of Maryland, 1979
10. Amann, H.: Lectures on some fixed point theorems. Monografias de Matematica. IMPA, Rio de Janeiro 1974
12.3 Avila, J. H.: The feasibility of continuation methods for nonlinear equations. SIAM J. Numer. Anal. 11 (1974) 102-122
12.4 Awoniyi, S. A. & Todd, M. J.: An efficient simplicial algorithm for computing a zero of a convex union of smooth functions. Preprint, 1981
12.6 Balinski, M. L. & Cottle, R. W. (eds): Complementarity and fixed point problems. Math. Programming Study 7 (1978)
12.85 Barany, I.: Borsuk's theorem through complementary pivoting. Math. Programming 18 (1980) 84-88
12.9 Barany, I.: Subdivisions and triangulations in fixed point algorithms. Preprint, International Research Institute for Management Science, Moscow, 1979
12.92 Baranov, A. V.: A method of calculating the complex roots of a system of nonlinear equations. Zh. Vyschist. Mat. Nat. 12 (1972) 199-203
14. Bartels, R. H. & Golub, G. H.: The simplex method of linear programming using LU-decompositions. CACM 12 (1969) 266-268

16. Ben-Israel, A.: A modified Newton-Raphson method for the solution of systems of equations. Israel J. Math. 3 (1965) 94–98
16.5 Ben-Israel, A.: A Newton-Raphson method for the solution of systems of equations. J. Math. Anal. Appl. 15 (1966) 243–252
17. Ben-Israel, A. & Greville, T. N. E.: Generalized inverses: Theory and applications. Wiley-Interscience Publ., 1974
18.4 Beyn, W.-J.: On discretizations of bifurcation problems. In: Bifurcation problems and their numerical solution, H. D. Mittelman, H. Weber (eds), ISNM 54, Birkhäuser Verlag Basel (1980), 46–73
18.8 Bissett, E. J. & Cavendish, J. C.: A numerical technique for solving nonlinear equations possessing multiple solutions. Techn. Report GMR-3057, MA-160, General Motors Research Laboratories, 1979
18.9 Bittner, L.: Einige kontinuierliche Analogien von Iterationsverfahren. In: Funktionalanalysis, Approximationstheorie, Numerische Mathematik, ISNM 7, Birkhäuser Verlag, Basel (1976) 114–135
19.5 Boggs, P. T.: The solution of nonlinear systems of equations by a-stable integration techniques. SIAM J. Numer. Anal. 8 (1971) 767–785
19.65 Bohl, E.: Chord techniques and Newton's method for discrete bifurcation problems. Numer. Math. 34 (1980), 111–124
19.75 Bosarge, W.: A continuous analog for some Newton's method convergence theorems. IBM Scientific Center Rep. # 320.2381 (1970)
19.85 Bowman, C. & Karamardian, S.: Error bounds for approximate fixed points. In: Fixed points: Algorithms and applications, S. Karamardian (ed.), Academic Press, New York (1977)
20.08 Branin, Jr., F. H.: Solution of nonlinear DC network problems via differential equations. Memoirs Mexico 1971 IEEE conference on systems, networks, and computers. Oaxtepee, Mexico
21. Branin, Jr., F. H.: Widely convergent methods for findings multiple solutions of simultaneous nonlinear equations. IBM J. Res. Develop. 16 (1972) 504–522
21.1 Branin, Jr., F. H. & Hoo, S. K.: A method for finding multiple extrema of a function of n variables. In: Numerical methods for nonlinear optimization, F. Lootsma (ed.), Academic Press (1972) 231–237
21.3 Brent, R. P.: On the Davidenko-Branin method for solving simultaneous nonlinear equations. IBM J. Res. Develop. (1972) 434–436
21.72 Brezzi, F. & Rapaz, J. & Raviart, P. A.: Finite dimensional approximation of nonlinear problems. Part 1: Branches of nonsingular solutions. Numer. Math. 36 (1980) 1–25. Part 2: Limit points. Numer. Math. 37 (1981) 1–28. Part 3: Simple bifurcation points. Numer. Math. 38 (1982) 1–30
21.77 Brooks, P. S.: Infinite regression in the Eaves-Saigal algorithm. Math. Programming 19 (1980) 313–327
21.8 Brouwer, L. E. J.: Über Abbildung von Mannigfaltigkeiten. Math. Ann. 71 (1912) 97–115
23.6 Browder, F. E.: On continuity of fixed points under deformations of continuous mappings. Summa Brasilia Mathematica 4,5 (1960) 183–191
24.1 Broyden, C. G.: A class of methods for solving nonlinear simultaneous equations. Math. Comp. 19 (1965) 577–593
24.2 Broyden, C. G.: A new method of solving nonlinear simultaneous equations. Computer Journal 12 (1969) 94–99
24.4 Carr, J. & Mallet-Paret, J.: Smooth homotopies for finding zeroes of entire functions. Preprint, 1982

24.5   Cellina, A. & Sartori, C.: The search for fixed points under perturbations. Rend. Sem. Mat. Univ. Padova 59 (1978) 199–208
24.7   Chan, T. F. & Keller, H. B.: Arclength continuation and multi-grid techniques for nonlinear eigenvalue problems. To appear in SIAM J. Sci. Stat. Comput. (1982)
25.5   Chao, K. S. & Liu, D. K. & Pan, C. T.: A systematic search method for obtaining multiple solutions of simultaneous nonlinear equations. IEEE Transactions on Circuits and Systems CAS-22 (1975) 748–753
25.6   Chao, K. S. & Saeks, R.: Continuation methods in circuit analysis. Proc. of the IEEE 65 (1977) 1187–1194
25.7   Charnes, A.: Optimality and degeneracy in Linear Programming. Econometrica 20 (1952) 160–170
26.   Charnes, A. & Garcia, C. B. & Lemke, C. E.: Constructive proofs of theorems relating to: $F(x)=y$, with applications. Math. Programming 12 (1977) 328–343
27.   Chow, S. N. & Mallet-Paret, J. & Yorke, J. A.: Finding zeroes of maps: Homotopy methods that are constructive with probability one. Math. Comput. 32 (1978) 887–899
28.   Chow, S. N. & Mallet-Paret, J. & Yorke, J. A.: A homotopy method for locating all zeroes of a system of polynomials. In: Functional differential equations and approximation of fixed points, H.-O. Peitgen, H.-O. Walther (eds), Springer Lecture Notes in Math. 730 (1979) 77–88
28.2   Chua, L. O. & Ushida, A.: A switching-parameter algorithm for finding multiple solutions of nonlinear resistive circuits. Circuit Theory and Appl. 4 (1976) 215–239
28.4   Cohen, D. I. A.: On the Sperner lemma. J. Comb. Theory 2 (1967) 585–587
29.   Crandall, M. G. & Rabinowitz, P. H.: Bifurcation from simple eigenvalues. J. Func. Anal. 8 (1971) 321–340
30.   Cromme, L.: Sind Fixpunktverfahren effektiv? NAM-Bericht # 25, Univ. Göttingen, 1980
33.   Davidenko, D.: On a new method of numerical solution of systems of nonlinear equations. Doklady Akad. Nauk USSR 88 (1953) 601–602 (in russian)
33.7   Decker, D. W. & Keller, H. B.: Solution branching – a constructive technique. In: New approaches to nonlinear problems in dynamics, P. Holmes (ed.), SIAM, Philadelphia (1980) 53–69
34.   Decker, D. W. & Keller, H. B.: Path following near bifurcation. Communications on Pure and Appl. Math. 34 (1981) 149–175
35.1   Deist, F. H. & Sefor, L.: Solution of systems of nonlinear equations by parameter variation. Computer Journal 10 (1967) 78–82
35.2   Den Heijer, C.: The numerical solution of nonlinear operator equations by imbedding methods. Ph. D. Thesis, Amsterdam 1979
35.5   Den Heijer, C. & Rheinboldt, W. C.: On steplength algorithms for a class of continuation methods. SIAM J. Numer. Anal. 18 (1981) 925–948
37.   Dennis, Jr., J. E. & Moré, J. J.: Quasi-Newton methods, motivation and theory. SIAM Review 19 (1977) 46–89
38.   Dennis, Jr., J. E. & Schnabel, R. B.: Least change secant updates for Quasi-Newton methods. SIAM Review 21 (1979) 443–459
38.2   Dennis, Jr., J. E. & Schnabel, R. B.: Numerical methods for unconstrained optimization and nonlinear equations. To appear
39.   Deuflhard, P.: A stepsize control for continuation methods and its special application to multiple shooting techniques. Numer. Math. 33 (1979) 115–146
39.8   Deuflhard, P. & Pesch, H. J. & Rentrop, P.: A modified continuation method for

	numerical solution of nonlinear two-point boundary value problems by shooting techniques. Numer. Math. 26 (1976) 327-343
40.	Doedel, E.: Auto: A program for the automatic bifurcation analysis of autonomous systems. Preprint
41.3	Dongarra, J. J. & Bunch, J. R. & Moler, C. B. & Stewart, G. W.: LINPAC user's guide. SIAM, Philadelphia (1979)
41.5	Drexler, F.-J.: Eine Methode zur Berechnung sämtlicher Lösungen von Polynomgleichungssystemen. Numer. Math. 29 (1977) 45-58
41.52	Drexler, F.-J.: A homotopy method for the calculation of all zeroes of zero-dimensional polynomial ideals. In: Continuation methods, H. Wacker (ed.), Academic Press, New York (1978) 69-94
42.	Eaves, B. C.: Computing Kakutani fixed points. SIAM J. Appl. Math. 21 (1971) 236-244
43.	Eaves, B. C.: Homotopies for computation of fixed points. Math. Programming 3 (1972) 1-22
43.4	Eaves, B. C.: Solving regular piecewise linear convex equations. Math. Programming Study 1 (1974) 96-119
43.5	Eaves, B. C.: Properly labeled simplexes. Studies in Optimization 10, MAA, Studies in Math., G. B. Dantzig, B. C. Eaves (eds), (1974) 71-93
44.	Eaves, B. C.: A short course in solving equations with PL homotopies. SIAM-AMS Proc. 9 (1976) 73-143
44.2	Eaves, B. C.: Computing stationary points. Mathematical Programming Study 7, M. L. Balinski, R. W. Cottle (eds), North Holland, Amsterdam (1978) 1-14
45.	Eaves, B. C. & Saigal, R.: Homotopies for computation of fixed points on unbounded regions. Math. Programming 3 (1972) 225-237
46.	Eaves, B. C. & Scarf, H.: The solution of systems of piecewise linear equations. Math. of Op. Res. 1 (1976) 1-27
47.3	Engels, C. R.: Economic equilibrium under deformation of the economy. In: Analysis and computation of fixed points, S. M. Robinson (ed.), Academic Press, New York (1980) 213-410
47.6	Engl, H. & Wacker, H.-J. & Zarzer, E.: Bemerkungen zur Aufwandsminimierung bei Stetigkeitsmethoden sowie Alternativen bei der Behandlung der singulären Situation. ISNM 38, Birkhäuser Verlag, Basel (1977) 175-193
47.8	Feilmeier, M.: Numerische Aspekte bei der Einbettung nichtlinearer Probleme. Computing 9 (1972) 355-364
48	Ficken, F.: The continuation method for functional equations. Comm. Pure and Appl. Math. 4 (1951) 435-456
48.1	Fisher, M. L. & Gould, F. J.: A simplicial algorithm for the nonlinear complementarity problem. Math. Programming 6 (1974) 281-300
48.2	Fisher, M. L. & Gould, F. J. & Tolle, J. W.: A simplicial algorithm for the mixed nonlinear complementary problem with applications to convex programming. Preprint, 1974
48.5	Fisher, M. L. & Tolle, J. W.: The nonlinear complementarity problem: existence and determination of solutions. SIAM J. Control and Optimization 612-623
48.7	Forster, W. (ed.): Numerical solution of highly nonlinear problems. North-Holland, Amsterdam 1980
48.75	Freidenfels, J.: Fixed-point algorithms and almost-complete sets. TR71-17, Operations Research House, Stanford Univ. 1971
48.8	Freudenstein, F. & Roth, B.: Numerical solution of systems of nonlinear equations. J. Assoc. Comput. Mech. 10 (1963) 550-556

49. Freudenthal, H.: Simplizialzerlegungen von beschränkter Flachheit. Annals of Math. 43 (1942) 580-582
49.3 Freund, R. M. & Todd, M. J.: A constructive proof of Tucker's combinatorial lemma. J. Comb. Theory (Series A) 30 (1981) 321-325
49.5 Fujisawa, T. & Kuh, E.: Piecewise linear theory of nonlinear networks. SIAM J. Appl. Math. 22 (1972) 307-328
49.7 Garcia, C. B.: A fixed point theorem including the last theorem of Poincaré. Math. Programming 8 (1975) 227-239
49.9 Garcia, C. B.: A hybrid algorithm for the computation of fixed points. Management Science 22 (1976) 606-613
49.95 Garcia, C. B.: Continuation methods for simplicial mappings. In: Fixed points: Algorithms and applications, S. Karamardian, Academic Press, New York (1977) 149-164
50. Garcia, C. B.: Computation of solutions to nonlinear equations under homotopy invariance. Math. of Op. Res. 2 (1977) 25-29
51. Garcia, C. B. & Gould, F. J.: A theorem on homotopy paths. Math. of Op. Res. 3 (1978) 282-289
51.1 Garcia, C. B. & Gould, F. J.: Scalar labelings for homotopy paths. Math. Programming 17 (1979) 184-197
51.2 Garcia, C. B. & Gould, F. J.: Relations between several path following algorithms and local and global Newton methods. SIAM Review 22 (1980) 263-274
51.5 Garcia, C. B. & Li, T. Y.: On the number of solutions to polynomial systems of equations. SIAM J. Numer. Anal. 17 (1980) 540-546
51.7 Garcia, C. B. & Zangwill, W. I.: Global continuation methods for finding all solutions to polynomial systems of equations in n variables. In: Symposium on Extremal Methods and Systems Analysis, Springer-Verlag 1978
52. Garcia, C. B. & Zangwill, W. I.: Determining all solutions to certain systems of nonlinear equations. Math. of Op. Res. 4 (1979) 1-14
53. Garcia, C. B. & Zangwill, W. I.: Finding all solutions to polynomial systems and other systems of equations. Math. Programming 16 (1979) 159-176
54. Garcia, C. B. & Zangwill, W. I.: An approach to homotopy and degree theory. Math. of Op. Res. 4 (1979) 390-405
54.1 Garcia, C. B. & Zangwill, W. I.: The flex simplicial algorithm. In: Numerical solution of highly nonlinear problems, W. Forster (ed.), North-Holland (1980) 71-92
54.3 Garcia, C. B. & Zangwill, W. I.: Pathways to solutions, fixed points, and equilibria. Prentice-Hall, Englewood Cliffs, N. J. 1981
54.7 Gavurin, M. K.: Nonlinear functional equations and continuous analoga of iterative methods. Isv. Vyssh. Uchebn. Zaved. Matematika 6 (1958) 18-31
55.5 Gear, C. W.: Numerical initial value problems in ordinary differential equations. Prentice-Hall, Englewood Cliffs 1971
57. Georg, K.: Algoritmi simpliziali come realizzazione numerica del grado di Brouwer. In: A survey on the theoretical and numerical trends in nonlinear analysis. Gius. Laterza & Figli, Bari (1979) 69-120
59. Georg, K.: An application of simplicial algorithms to variational inequalities. In: Functional differential equations and approximation of fixed points, H.-O. Peitgen, H.-O. Walther (eds), Springer Lecture Notes in Math. 730 (1979) 126-135
61. Georg, K.: A simplicial deformation algorithm with applications to optimization, variational inequalities and boundary value problems. In: Numerical solution of highly nonlinear problems, W. Förster (ed.), North-Holland Publ. Comp. (1980) 361-375

62. Georg, K.: Numerical integration of the Davidenko equation. In: Numerical solution of nonlinear equations, E. Allgower, K. Glashoff, H.-O. Peitgen (eds), Springer Lecture Notes in Math. 878 (1981) 128-161
63. Georg, K.: A numerically stable update for simplicial algorithms. In: Numerical solution of nonlinear equations, E. Allgower, K. Glashoff, H.-O. Peitgen (eds), Springer Lecture Notes in Math. 878 (1981) 117-127
64. Georg, K.: On tracing an implicitly defined curve by Quasi-Newton steps and calculating bifurcation by local perturbation. SIAM J. of SSC 2 (1981) 35-50
64.1 Georg, K.: A note on stepsize control for numerical curve following. To appear in the proc. of the Nato Advanced Research Institute on homotopy methods and global convergence, Plenum Press
64.2 Georg, K.: Zur numerischen Realisierung von Kontinuitätsmethoden mit Prädiktor-Korrektor- oder simplizialen Verfahren. Habilitationsschrift, Univ. Bonn (1982)
65. Gill, P. E. & Golub, G. H. & Murray, W. & Saunders, M. A.: Methods for modifying matrix factorizations. Math. of Comp. 28 (1974) 505-535
67.4 Gould, F. J. & Schmidt, C. P.: An existence result for the global Newton method. In: Variational inequalities and complementarity problems, [Proc. Internat. School, Erice 1978], Wiley, Chichester (1980) 187-194
67.5 Gould, F. J. & Tolle, J. W.: A unified approach to complementarity in optimization. Discrete Math. 7 (1974) 225-271
67.52 Gould, F. J. & Tolle, J. W.: An existence theorem for solutions to $f(x) = 0$. Math. Programming 11 (1976) 252-262.
67.8 Grigorieff, R. D.: Numerik gewöhnlicher Differentialgleichungen 1: Einschrittverfahren. B. G. Teubner, 1972
67.9 Grigorieff, R. D.: Numerik gewöhnlicher Differentialgleichungen 2: Mehrschrittverfahren. B. G. Teubner, 1977
68.1 Hackl, J. & Wacker, H. J. & Zulehner, W.: An efficient step size control for continuation methods. BIT 20 (1980) 475-485
71. Haselgrove, C. B.: The solution of non-linear equations and of differential equations with two-point boundary conditions. Comput. J. 4 (1961) 255-259
71.3 Hayes, L. & Wasserstrom, E.: Solution of non-linear eigenvalue problems by the continuation method. J. Inst. Math. Appl. 17 (1976) 5-14
72.8 Hirsch, M. W.: A proof of the nonretractibility of a cell onto its boundary. Proc. Amer. Math. Soc. 14 (1963) 364-365
72.9 Hirsch, M. W.: Differential topology. Springer Verlag, Berlin (1976)
73. Hirsch, M. W. & Smale, S.: On algorithms for solving $F(x) = 0$. Comm. Pure and Appl. Math. 32 (1979) 281-312
73.5 Hoang, T.: Pivotal methods for computing fixed points: a unified approach. Travaux Sem. Anal. Convexe 6 (1978), Exp. no. 20, 35 pp
74.5 Ikeno, E. & Ushida, A.: The arc-length method for the computation of characteristic curves. IEEE Transactions on Circuits and Systems
74.53 Jacovlev, M. N.: On the solution of nonlinear equations by differentiation with respect to a parameter. USSR Computational Math. and Mech. Phys. 4 (1964) 146-149
74.54 Jansen, R. & Louter, R.: An efficient way of programming Eaves' fixed point algorithm. In: Numerical solution of highly nonlinear problems, W. Forster (ed.), North-Holland (1980) 115-168
74.55 Jeppson, M.: A search for the fixed points of a continuous mapping. In: Mathematical topics in economic theory and computation, R. Day, S. Robinson SIAM, Philadelphia (1972) 122-129

75. Jürgens, H. & Peitgen, H.-O. & Saupe, D.: Topological perturbation in the numerical study of nonlinear eigenvalue- and bifurcation problems. Proc. symposium on analysis and computation of fixed points, S. M. Robinson (ed.), Academic Press, New York (1980) 139-181

76. Jürgens, H. & Saupe, D.: Methoden der simplizialen Topologie zur numerischen Behandlung von nichtlinearen Eigenwert- und Verzweigungsproblemen. Diplomarbeit, Bremen 1979

76.5 Jürgens, H. & Saupe, D.: SCOUT – Simplicial continuation utilities, user's manual and installation guide, in preparation

77. Kakutani, S.: A generalization of Brouwer's fixed point theorem. Duke Math. J. 8 (1941) 457-459

78. Karamardian, S. (ed.): Fixed points: Algorithms and applications. Proc. conf. on computing fixed points with applications, Academic Press, New York 1977

79. Kearfott, R. B.: A derivative-free arc continuation method and a bifurcation technique. In: Numerical solution of nonlinear equations, E. Allgower, K. Glashoff, H.-O. Peitgen (eds), Springer Lecture Notes in Math. 878 (1981) 182-198

79.1 Kearfott, R. B.: Some general derivative-free bifurcation techniques. Preprint

79.35 Keener, J. P. & Keller, H. B.: Perturbed bifurcation theory. Arch. Rational Mech. Anal. 50 (1974) 159-175

79.5 Keller, H. B.: Shooting and imbedding for two-point boundary value problems. J. Math. Anal. Appl. 36 (1971) 598-610

80. Keller, H. B.: Numerical solution of bifurcation and nonlinear eigenvalue problems. In: Applications of bifurcation theory, P. H. Rabinowitz (ed.), Academic Press (1977) 359-348

81. Keller, H. B.: Global homotopies and Newton methods. In: Recent advances in numerical analysis, C. De Boor, G. H. Golub (eds), Academic Press (1978) 73-94

82. Keller, H. B.: Constructive methods for bifurcation and nonlinear eigenvalue problems. Springer Lecture Notes in Math. 704 (1979) 241-251

82.5 Keller, H. B. & Langford, W. F.: Iterations, perturbations and multiplicities for nonlinear bifurcation problems. Arch. Rational Mech. Anal. 48 (1972) 83-108

83. Kellogg, R. B. & Li, T. Y. & Yorke, J. A.: A constructive proof of the Brouwer fixed point theorem and computational results. SIAM J. Numer. Anal. 4 (1976) 473-483

83.2 Kellogg, R. B. & Li, T. Y. & Yorke, J. A.: A method of continuation for calculating a Brouwer fixed point. In: Computing fixed points with applications, S. Karamardian (ed.), Academic Press (1977) 133-147

83.4 Kim, B. B.: On stepsize of parameter continuation method combined with Newton method. (Korean) Cho-son In-min Kong-hwa-kuk Kwa-hak-won, Tong-bo 4 (1980) 1-4

84.2 Köberl, D.: The solution of nonlinear equations by the computation of fixed points with a modification of the sandwich method. Computing 25 (1980) 175-179

84.8 Kojima, M.: Computational methods for solving the nonlinear complementarity problem. Keio Engineering Reports 27 (1974) 1-41

85.3 Kojima, M.: On the homotopic approach to systems of equations with separable mappings. Math. Programming Study 7 (1978), M. L. Balinski, R. W. Cottle (eds), 170-184

85.6 Kojima, M.: A modification of Todd's triangulation J3. Math. Programming 15 (1978) 223-227

86. Kojima, M.: Studies on piecewise-linear approximations of piecewise-C1-map-

pings in fixed points and complementarity theory. Math. Oper. Res. 3 (1978) 17–36

86.5 Kojima, M.: A note on a new algorithm for computing fixed points by van der Laan and Talman. In: Numerical solution of highly nonlinear problems, W. Forster (ed.), North-Holland, Amsterdam (1980) 37–42

87. Kojima, M.: An introduction to variable dimension algorithms for solving systems of equations. In: Numerical solution of nonlinear equations, E. L. Allgower, K. Glashoff, H.-O. Peitgen (eds), Springer Lecture Notes in Math. 878 (1981) 199–237

87.1 Kojima, M.: A complementary pivoting approach to parametric nonlinear programming. Preprint

87.15 Kojima, M. & Hirabayashi, R.: Continuous deformation of nonlinear programs. Preprint # B-101, Tokyo Inst. of Techn., Sept. 1981

87.17 Kojima, M. & Mizuno, S.: Computation of all solutions to a system of polynomial equations. Tokyo Inst. of Techn., Dec. 1981

87.2 Kojima, M. & Nishino, H. & Arima, N.: A PL homotopy for finding all the roots of a polynomial. Math. Programming 16 (1979) 37–62

87.8 Kojima, M. & Yamamoto, Y.: Variable dimension algorithms. Part I: Basic theory. Part II: Some new algorithms and triangulations with continuous refinement of mesh size. Preprints, Tokyo Inst. of Techn., 1980

87.9 Kojima, M. & Yamamoto, Y.: A unified approach to several restart fixed point algorithms for their implementation and a new variable dimension algorithm. Univ. of Tsukuba, Japan, Aprol 1982

88. Krasnosel'skii, M. A.: Topological methods in the theory of nonlinear integral equations. Pergamon Press 1964

88.2 Kubicek, M.: Algorithm 502, Dependence of solutions of nonlinear systems on a parameter. ACM-TOMS 2 (1976) 98–107

88.3 Kubicek, M. & Hlavacek, V.: One-parameter imbedding techniques for the solution of nonlinear boundary value problems. Appl. Math. Comput. 4 (1978) 317–357

88.4 Kubicek, M. & Holodniok, M. & Hlavacek, V.: Solution of nonlinear boundary value problems XI. One-parameter imbedding methods. Chem. Eng. Sci. 34 (1979) 645–650

88.5 Kubicek, M. & Holodniok, M. & Marek, I.: Numerical solution of nonlinear equations by one-parameter imbedding methods. Numer. Funkt. Anal. and Optimiz. 3 (1981) 223–264

88.9 Kuhn, H. W.: Simplicial approximation of fixed points. Proc. Nat. Acad. Sci. USA 61 (1968) 1238–1242

89. Kuhn, H. W.: Approximate search for fixed points. In: Computing methods in optimization problems 2, L. A. Zadek, L. W. Neustat, A. V. Balakrishnen (eds), Academic Press (1969) 199–211

89.5 Kuhn, H. W.: A new proof of the fundamental theorem of algebra. Math. Programming Study 1 (1974) 148–158

89.6 Kuhn, H. W.: Finding roots by pivoting. In: Fixed points: algorithms and applications, S. Karamardian (ed.), Academic Press (1977) 11–39

90. Kuhn, H. W. & Mackinnon, J. G.: Sandwich method for finding fixed points. J. Opt. Theory and Appl. 17 (1975) 189–204

90.2 Laasonen, P.: An imbedding method of iteration with global convergence. Computing 5 (1970) 253–258

90.44 Lahaye, E.: Une méthode de résolution d'une catégorie d'équations transcendantes. C. R. Acad. Sci. Paris 198 (1934) 1840–1842

90.45 Lahaye, E.: Solution of systems of transcendental equations. Acad. Roy. Belg. Bull. Cl. Sci. 5 (1948) 805–822

90.7 Leder, D.: Automatische Schrittweitensteuerung bei global konvergenten Einbettungsmethoden. ZAMM 54 (1970) 319–342

92. Lemke, C. E. & Howson, J. T.: Equilibrium points of bimatrix games. SIAM J. Appl. Math. 12 (1964) 413–423

93. Leray, J. & Schauder, J.: Topologie et équations fonctionelles. Ann. Sci. Ecole Norm. Sup. 51 (1934) 45–78

93.3 Li, T. Y.: Computing the Brouwer fixed point by following the continuation curve. Proc. Sem., Dalhousie Univ., Halifax, Academic Press, New York (1976) 131–135

93.4 Li, T. Y. & Mallet-Paret, J. & Yorke, J. A.: Regularity results for real analytic homotopies. To appear in: Numer. Math.

93.45 Li, T. Y. & Yorke, J. A.: Path following approach for solving nonlinear equations: homotopy, continuous Newton and projection. In: Functional differential equations and approximation of fixed points, H.-O. Peitgen, H.-O. Walther (eds), Springer Lecture Notes in Math. 730 (1979) 257–264

93.5 Li, T. Y. & Yorke, J. A.: A simple reliable numerical algorithm for following homotopy paths. In: Analysis and computation of fixed points, S. M. Robinson (ed.), Academic Press, New York (1980) 73–91

94.4 Lüthi, H.-J.: A simplicial approximation of a solution for the nonlinear complementarity problem. Math. Programming 9 (1975) 278–293

94.5 Lüthi, H.-J.: Komplementaritäts- und Fixpunktalgorithmen in der mathematischen Programmierung, Spieltheorie und Ökonomie. Lecture Notes in Economics and Mathematical Systems 129 (1976), Springer Verlag

94.55 Mackinnon, J. G.: Solving economic general equilibrium models by the sandwich method. In: Fixed points: Algorithms and applications, S. Karamardian (ed.), Academic Press, New York (1977)

94.56 Mackinnon, J. G.: Solving urban general equilibrium problems by fixed point methods. In: Analysis and computation of fixed points, S. M. Robinson (ed.), Academic Press, New York (1980) 197–212

94.6 Mallet-Paret, J. & Carr, J.: A homotopy method for transcendental equations. In: Proceedings of the Nato Institute on Homotopy Methods and Global Convergence, Sardinia 1981

94.78 Megiddo, N.: On the parametric nonlinear complementarity problem. Math. Programming Study 7 (1978) 142–150

94.8 Megiddo, N. & Kojima, M.: On the existence and uniqueness of solutions nonlinear complementarity theory. Math. Programming 12 (1977) 110–130

94.9 Melhem, R. G. & Rheinboldt, W. C.: A comparison of methods for determining turning points of nonlinear equations. Techn. Report ICMA-82-32, Univ. of Pittsburgh, Jan. 1982

94.99 Menzel, R. & Schwetlick, H.: Über einen Ordnungsbegriff bei Einbettungsalgorithmen zur Lösung nichtlinearer Gleichungen. Computing 16 (1976) 187–199

95. Menzel, R. & Schwetlick, H.: Zur Lösung parameterabhängiger nichtlinearer Gleichungen mit singulären Jacobi-Matrizen. Numer. Math. 30 (1978) 65–79

95.9 Merrill, O.: A summary of techniques for computing fixed points of continuous mappings. In: Mathematical topics in economic theory and computation, R. Day, S. Robinson (eds), SIAM, Philadelphia (1972) 130–149

96. Merrill, O. H.: Applications and extensions of an algorithm that computes fixed points of a certain upper semi-continuous point to set mapping. Ph. D. Thesis, Univ. of Michigan 1972

II. Predictor-Corrector and Simplicial Methods for Approximating

96.5   Meyer, G. H.: On solving nonlinear equations with a one-parameter operator imbedding. SIAM J. Numer. Anal. 5 (1968) 739–752

96.8   Meyerson, M. D. & Wright, A. H.: A new and constructive proof of the Borsuk-Ulam theorem. Proc. Amer. Math. Soc. 73 (1979) 134–136

97.   Milnor, J. W.: Topology from the differential viewpoint. Univ. Press of Virginia, Charlottesville, VA, 1969

97.3   Mittelmann, H. D. & Weber, H.: Numerical methods for bifurcation problems – A survey and classification. Preprint, Univ. of Dortmund, 1980

97.4   Mittelmann, H. D. & Weber, H.: A bibliography on numerical methods for bifurcation problems. Univ. of Dortmund, 1981

97.5   Moore, G. & Spence, A.: The calculation of turning points of nonlinear equations. SIAM J. Numer. Anal. 17 (1980) 567–576

98.5   Netravali, A. N. & Saigal, R.: Optimum quantizer design using a fixed-point algorithm. The Bell System Techn. J. 55 (1976) 1423–1435

98.7   Nirenberg, L.: Topics in nonlinear functional analysis. Courant Institute, New York 1974

100.8   Ohiwa, H.: A new algorithm for non-linear simultaneous equations. J. Inst. Math. Appl. 21 (1978) 189–196

101.8   Ortega, J. M. & Rheinboldt, W. C. (eds): Numerical solution of nonlinear problems. SIAM, Philadelphia (1970)

102.   Ortega, J. M. & Rheinboldt, W. C.: Iterative solution of nonlinear equations in several variables, Academic Press, New York 1970

102.6   Pan, C. T. & Chao, K. S.: Multiple solutions of nonlinear equations: roots of polynomials. IEEE Transactions on Circuits and Systems CAS-27 (1980) 825–832

102.75   Paumiers, J. C.: Une méthode numérique pour le calcul des points de retournement. Application a un problème aux limites non linéair. I: Etude théorique et expérimentation de la méthode. Numer. Math. 37 (1981) 433–444. II: Analyse numérique d'un problème aux limites non linéair. Numer. Math. 37 (1981) 445–452

102.8   Peitgen, H.-O.: Topologische Perturbationen beim globalen numerischen Studium nichtlinearer Eigenwert- und Verzweigungsprobleme. To appear in: Jahresbericht der DMV

103.   Peitgen, H.-O. & Prüfer, M.: The Leray-Schauder continuation method is a constructive element in the numerical study of nonlinear eigenvalue and bifurcation problems. In: Functional differential equations and approximation of fixed points, H.-O. Peitgen, H.-O. Walther (eds), Springer Lecture Notes in Math. 730 (1979) 326–409

103.1   Peitgen, H.-O. & Saupe, D. & Schmitt, K.: Nonlinear elliptic boundary value problem versus their finite difference approximation: Numerically irrelevant solutions. J. Reine Angew. Math. 322 (1981) 74–117

103.4   Peitgen, H.-O. & Schmitt, K.: Positive and spurious solutions of nonlinear eigenvalue problems. In: Numerical solution of nonlinear equations, E. L. Allgower, K. Glashoff, H.-O. Peitgen (eds), Springer Lecture Notes in Math. 878 (1981) 257–324

104.   Peitgen, H.-O. & Siegberg, H. W.: An epsilon-perturbation of Brouwer's definition of degree. In: Proc. Conf. Theorie des points fixes, E. Fadell, G. Fournier (eds), Springer Lecture Notes in Math.

105.   Peitgen, H.-O. & Walther, H.-O. (eds): Functional differential equations and approximation of fixed points. Springer Lecture Notes in Math. 730 (1979)

105.2   Percell, P.: Note on a global homotopy. Numer. Func. Anal. and Optim. 2 (1980) 99–106

105.3   Poincaré, H.: Sur les courbes définées par une équation differentielle. I (1881), II (1882), III (1885), IV (1886). Oeuvres I, Gauthier-Villars, Paris
105.5   Pönisch, G. & Schwetlick, H.: Computing turning points of curves implicitly defined by nonlinear equations depending on a parameter. Computing 26 (1981) 107–121
105.6   Pönisch, G. & Schwetlick, H.: Ein lokal überlinear konvergentes Verfahren zur Bestimmung von Rückkehrpunkten implizit definierter Raumkurven. Numer. Math. 38 (1982) 455–466
106.   Potthoff, M.: Zur numerischen Approximation von implizit gegebenen Kurven mit Anwendungen auf Homotopiemethoden. Diplomarbeit, Bonn 1981
106.3   Powell, M. J. D.: A hybrid method for nonlinear equations. In: Numerical methods for nonlinear algebraic equations, P. Rabinowitz (ed.), Gordon and Breach, 1970
106.4   Powell, M. J. D.: A Fortran subroutine for solving systems of nonlinear algebraic equations. In: Numerical methods for nonlinear algebraic equations, P. Rabinowitz (ed.), Gordon and Breach, 1970
108.1   Prüfer, M.: Sperner simplices and the topological fixed point index. Preprint # 134, SFB72, Univ. of Bonn, 1977
108.2   Prüfer, M.: Calculating global bifurcation. In: Continuation methods, H. J. Wacker (ed.), Academic Press (1978) 187–213
108.3   Prüfer, M. & Siegberg, H. W.: On computational aspects of topological degree in $\mathbb{R}^n$. In: Functional differential equations and approximation of fixed points, H.-O. Peitgen, H.-O. Walther (eds), Springer Lecture Notes in Math. 730 (1979) 410–433
108.4   Prüfer, M. & Siegberg, H. W.: Complementary pivoting and the Hopf degree theorem. J. Math. Anal. Appl. 84 (1981) 133–149
109.   Rabinowitz, P. H.: Some global results for nonlinear eigenvalue problems. J. Func. Anal. 7 (1971) 487–513
109.7   Rao, C. R. & Mitra, S. K.: Generalized inverses of matrices and its applications. Wiley, New York 1971
109.9   Reiser, P. M.: A modified integer labelling for complementarity algorithms. Math. Oper. Res. 6 (1981) 129–139
110.   Rheinboldt, W. C.: Methods for solving systems of nonlinear equations. SIAM, Regional Conf. Ser. in Appl. Math. 14 (1974)
110.5   Rheinboldt, W. C.: Numerical continuation methods for finite element applications. In: Formulation and algorithms in finite element analysis, K. J. Bathe, J. T. Oden, W. Wunderlich (eds), MIT Press (1977) 599–631
110.7   Rheinboldt, W. C.: An adaptive continuation Process for solving systems of nonlinear equations. Polish Academy of Science, Banach Ctr. Publ. 3 (1977) 129–142
111.   Rheinboldt, W. C.: Numerical methods for a class of finite dimensional bifurcation problems. SIAM J. Numer. Anal. 15 (1978) 1–11
112.   Rheinboldt, W. C.: Solution fields of nonlinear equations and continuation methods. SIAM J. Numer. Anal. 17 (1980) 221–237
113.   Rheinboldt, W. C.: Numerical analysis of continuation methods for nonlinear structural problems. Computer & Structures 13 (1981) 103–113
113.1   Rheinboldt, W. C.: Computation of critical boundaries on equilibrium manifolds. To appear in SIAM J. Numer. Anal. (1981)
113.15 Rheinboldt, W. C. & Burkardt, J. V.: A program for a locally-parametrized continuation process. To appear in ACM TOMS (1982)
113.18 Ribaric, M. & Seliskar, M.: On optimization of stepsize in the continuation method. Mathematica Balcanica 4.97 (1974) 517–521

113.3   Riks, E.: An incremental approach to the solution of snapping and buckling problems. Int. J. Solids and Structures 15 (1979) 524–551
113.5   Riks, E.: A unified method for the computation of critical equilibrium states of nonlinear elastic systems. Preprint
114.5   Robinson, S. M. (ed.): Analysis and computation of fixed points. Academic Press, New York 1980
115.   Rockafellar, R. T.: Convex analysis. Princeton Univ. Press 1970
116.2   Saari, D. G. & Saigal, R.: Some generic properties of paths generated by fixed point algorithms. To appear in: Proc. Sympos. on analysis and computation of fixed points, S. M. Robinson (ed.), Academic Press (1980) 57–72
116.3   Saigal, R.: Investigations into the efficiency of fixed point algorithms. In: Fixed points: algorithms and applications, S. Karamardian (ed.), Academic Press, New York 1977
116.5   Saigal, R.: On paths generated by fixed point algorithms. Math. of Op. Res. 1 (1976) 359–380
116.6   Saigal, R.: Accelerated Eaves-Saigal algorithm for computing fixed points. Preprint, North Western Univ. (1976)
117.   Saigal, R.: On the convergence rate of algorithms for solving equations that are based on methods of complementary pivoting. Math. of Op. Res. 2 (1977) 108–124
117.4   Saigal, R.: On piecewise linear approximations to smooth mappings. Math. Oper. Res. 4 (1979) 153–161
117.5   Saigal, R.: Fixed point computing methods. Encyclopedia of Computer Science and Technology, vol. 8 (1979) 545–566, Marcel Dekker Inc.
118.   Saigal, R.: The fixed point approach to nonlinear programming. Math. Programming Stud. 10 (1979) 142–157
118.2   Saigal, R.: Efficient algorithms for computing fixed points when mappings may be separable. Northwestern Univ., 1979
118.5   Saigal, R.: An efficient procedure for traversing large pieces in fixed point algorithms. Preprint, Northwestern Univ., 1981
118.6   Saigal, R.: A homotopy for solving large, sparse and structured fixed point problems. Preprint, Northwestern Univ. 1981
119.   Saigal, R. & Shin, Y. S.: Perturbations in fixed point algorithms. In: Functional differential equations and approximation of fixed points, H.-O. Peitgen, H.-O. Walther (eds), Springer Lecture Notes in Math. 730 (1979)
119.2   Saigal, R. & Solow, D. & Wolsey, L. A.: A comparative study of two algorithms to compute fixed points over unbounded regions. In: Proc. VII-th Math. Programming symposium, Stanford (1975)
119.4   Saigal, R. & Todd, M. J.: Efficient acceleration techniques for fixed point algorithms. SIAM J. Numer. Anal. 15 (1978) 997–1007
120.   Sard, A.: The measure of the critical values of differential maps. Bull. AMS 48 (1942) 883–890
121.   Saupe, D.: On accelerating PL continuation algorithms by predictor corrector methods. To appear in: Math. Programming
121.3   Saupe, D.: Beschleunigte PL-Kontinuitätsmethoden und periodische Lösungen parametrisierter Differentialgleichungen mit Zeitverzögerung. Ph. D. thesis, Univ. of Bremen, 1982
122.   Scarf, H. E.: The approximation of fixed points of a continuous mapping. SIAM J. Appl. Math. 15 (1967) 1328–1343
123.   Scarf, H. E.: The core of an n person game. Econometrica 35 (1967) 50–69

124. Scarf, H. E. & Hansen, T.: Computation of economic equilibria. Yale Univ. Press, New Haven 1973
124.8 Schidlovskaja, N. A.: Application of the method of differentiation by a parameter for solving nonlinear equations in Banach spaces. (Russian). Leningrad Gos. Univ. Uceu. Zap., Ser. Mat. Nauk. 33 (1958) 5-17
125.3 Schrauf, G.: Numerical investigation of Taylor-vortex flows in a spherical gap. Bonn, 1982
125.9 Schwetlick, H.: Ein neues Prinzip zur Konstruktion implementierbarer, global konvergenter Einbettungsalgorithmen. Beiträge Numer. Math. 4 (1975) 215-228 and 5 (1976) 201-206
126. Schwetlick, H.: Numerische Lösung nichtlinearer Gleichungen. VEB Deutscher Verlag der Wiss. 1979
126.1 Schwetlick, H.: Effective methods for computing turning points of curves implicitly defined by nonlinear equations. Preprint, Halle, 1981
126.5 Seydel, R.: Numerical computation of branch points in ordinary differential equations. Numer. Math. 32 (1979) 51-68
126.6 Seydel, R.: Numerical computation of branch points in nonlinear equations. Numer. Math. 33 (1979) 339-352
126.8 Shamir, S.: A homotopy fixed point algorithm with an arbitrary integer refinement factor. Preprint, Stanford Univ., 1979
126.9 Shamir, S.: The dynamic shift algorithm – a homotopy fixed point algorithm with a dynamic triangulation having an arbitrary grid refinement factor. Preprint, Stanford Univ., 1979
126.92 Shamir, S.: Two new triangulations for homotopy fixed point algorithms with an arbitrary grid refinement. In: Analysis and computation of fixed points, S. M. Robinson (ed.), Academic Press, New York (1980) 25-56
127. Shampine, L. F. & Gordon, M. K.: Computer solution of ordinary differential equations. The initial value problem. W. H. Freeman and Comp. 1975
127.4 Shapley, L. S.: A note on the Lemke-Howson algorithm. Math. Programming Study 1 (1974) 175-189
128.05 Shoven, J. B.: Applying fixed point algorithms to the analysis of tax policies. In: Fixed points: algorithms and applications, S. Karamardian (ed.), Academic Press, New York (1977) 403-434
129. Smale, S.: A convergent process of price adjustment and global Newton method. J. of Math. Economics 3 (1976) 1-14
129.4 Solow, D.: Comparative computer results of a new complementary pivot algorithm for solving equality and inequality constrained optimization problems. Math. Programming 18 (1981) 169-185
129.5 Solow, D.: Homeomorphisms of triangulations with applications to computing fixed points. Math. Programming 20 (1981) 213-224
131. Sperner, E.: Neuer Beweis über die Invarianz der Dimensionszahl und des Gebietes. Abh. Math. Sem. Univ. Hamburg 6 (1928) 265-272
133. Talman, A. J. J.: Variable dimension fixed point algorithms and triangulations. Ph. D. Thesis, Amsterdam (1980), Mathematical Centre Tracts 128
133.5 Talman, A. J. J. & Van der Heyden, L.: Algorithms for the linear complementarity problem which allow an arbitrary starting point. Preprint, 1981
136. Tanabe, K.: Continuous Newton-Raphson method for solving an underdetermined system of nonlinear equations. Nonlinear Analysis, Theory, Methods & Applications 3 (1979) 495-503
136.7 Thurston, G. A.: Continuation of Newton's method through bifurcation points. J. Appl. Mech. Techn. Phys. 36 (1969) 425-430

136.8 Todd, M. J.: A generalized complementary pivoting algorithm. Math. Programming 6 (1974) 243–263
137. Todd, M. J.: The computation of fixed points and applications. Springer Verlag, Lecture Notes in Econ. and Math. Systems 124 (1976)
138. Todd, M. J.: On triangulations for computing fixed points. Math. Programming 10 (1976) 322–346
139. Todd, M. J.: Orientation in complementary pivot algorithms. Math. of Op. Res. 1 (1976) 54–66
139.1 Todd, M. J.: Union Jack triangulations. In: Fixed points: algorithms and applications, S. Karamardian (ed.), Academic Press (1977) 315–336
139.2 Todd, M. J.: New fixed-point algorithms for economic equilibria and constrained optimization. Preprint, Cornell Univ., 1977
139.3 Todd, M. J.: Exploiting structure in fixed-point computation. Preprint, Univ. of Wisconsin-Madison, 1978
139.4 Todd, M. J.: Fixed-point algorithms that allow restarting without an extra dimension. Preprint, Cornell Univ., 1978
139.5 Todd, M. J.: Improving the convergence of fixed-point algorithms. Mathematical Programming Study 7 (1978), M. L. Balinski, R. W. Cottle (eds), North-Holland, Amsterdam, 151–169
139.7 Todd, M. J.: On the Jacobian of a function at a zero computed by a fixed point algorithm. Math. Oper. Res. 3 (1978) 126–132
139.9 Todd, M. J.: Piecewise linear paths to minimize convex functions may not be monotonic. Math. Programming 17 (1979) 106–108
140. Todd, M. J.: Global and local convergence and monotonicity results for a recent variable-dimension simplicial algorithm. In: Numerical solution of highly nonlinear problems, W. Forster (ed.), North-Holland, Amsterdam (1980) 43–76
140.5 Todd, M. J.: A quadratically convergent fixed point algorithm for economic equilibria and linearly constrained optimization. Math. Programming 18 (1980) 111–126
141. Todd, M. J.: Exploiting structure in piecewise-linear homotopy algorithms for solving equations. Math. Programming 18 (1980) 233–247
141.3 Todd, M. J.: Traversing large pieces of linearity in algorithms that solve equations by following piecewise-linear paths. Math. of Op. Res. 5 (1980) 242–257
142. Todd, M. J.: Numerical stability and sparsity in piecewise-linear algorithms. In: Proc. symposium on analysis and computation of fixed points, S. M. Robinson (ed.), Academic Press, New York (1980) 1–24
142.1 Todd, M. J.: Piecewise-linear homotopy algorithms for sparse systems of nonlinear equations. To appear in: SIAM J. on Control and Opt.
142.2 Todd, M. J.: On the computational complexity of piecewise-linear homotopy algorithms. Univ. of Cambridge, 1981
142.3 Todd, M. J.: An introduction to piecewise-linear homotopy algorithms for solving systems of equations. To appear in the proc. of a summer school on numerical analysis held at the Univ. of Lancester, July–August 1981
142.4 Todd, M. J.: Efficient methods of computing economic equilibria. To appear in the proc. conf. on applied general equilibrium analysis held in San Diego, 1981
142.5 Todd, M. J.: Plago: a Fortran implementation of a piecewise-linear homotopy algorithm for solving systems of nonlinear equations. Cornell Univ., 1981
142.8 Todd, M. J. & Wright, A. H.: A variable-dimension simplicial algorithm for antipodal fixed-point theorems. Numer. Funct. Anal. Optim. 2 (1980) 155–186

145.52 Tuy, H.: Pivotal methods for computing equilibrium points: unified approach and a new restart algorithm. Math. Programming 16 (1979) 210–227

145.53 Tuy, H. & Thoai, V. & Muu, L. D.: A modification of Scart's algorithm allowing restarting. Math. Operationsforsch. Statist., Ser. Optimization 9 (1978) 357–372

145.58 Van der Heyden, L.: A refinement procedure for computing fixed points. Math. Oper. Res. 7 (1982) 295–313

146. Van der Laan, G.: Simplicial fixed point algorithms. Ph. D. Thesis, Amsterdam (1980), Mathematical Centre Tracts 129

146.2 Van der Laan, G. & Talman, A. J. J.: A new algorithm for computing fixed points. Preprint, 1978

146.35 Van der Laan, G. & Talman, A. J. J.: A restart algorithm for computing fixed points without an extra dimension. Math. Programming 17 (1979) 74–84

146.5 Van der Laan, G. & Talman, A. J. J.: An improvement of fixed point algorithms by using a good triangulation. Math. Programming 18 (1980) 274–285

146.6 Van der Laan, G. & Talman, A. J. J.: A new subdivision for computing fixed points with a homotopy algorithm. Math. Programming 19 (1980) 78–91

146.7 Van der Laan, G. & Talman, A. J. J.: A restart algorithm without an artifical level for computing fixed points on unbounded regions. In: Functional differential equations and approximation of fixed points, H.-O. Peitgen, H.-O. Walther (eds), Springer Lecture Notes in Math. 730 (1979) 247–256

146.9 Van der Laan, G. & Talman, A. J. J.: A class of simplicial restart fixed point algorithms without an extra dimension. Math. Programming 20 (1981) 33–48

146.91 Van der Laan, G. & Talman, A. J. J.: Simplicial algorithms for finding stationary points, a unifying approach. Tilburg Univ., Jan. 1982

146.93 Van der Laan, G. & Talman, A. J. J.: On the computation of fixed points in the product space of unit simplices and an application to noncooperative n person games. Math. Oper. Res. 7 (1982) 1–13

146.95 Vertgeim, B. A.: On an approximate determination of the fixed points of continuous mappings. Soviet Math. Dokl. 11 (1970) 295–298

147.5 Wacker, H.-J. (ed.): Continuation methods. Academic Press, New York, 1978

148. Wacker, H.-J.: A summary of the developments on imbedding methods. In: Continuation methods, H.-J. Wacker (ed.), Academic Press (1978) 1–35

148.3 Wacker, H. J. & Engl, H. & Zarzer, E.: Bemerkungen zur Aufwandsminimierung sowie Alternativen bei der Behandlung der singulären Situationen. ISNM 38 (1977) 175–193, Birkhäuser Verlag

148.4 Wacker, H. J. & Zarzer, E. & Zulehner, W.: Optimal stepsize control for the globalized Newton method. In: Continuation methods, H. J. Wacker (ed.), Academic Press (1978) 249–276

149.2 Wang, C. Y. & Watson, L. T.: Squeezing of a viscous fluid between elliptic plates. Appl. Sci. Res. 35 (1979) 195–207

149.3 Wang, C. Y. & Watson, L. T.: Viscous flow between rotating discs with injection on the porous disc. Z. Angew. Math. Phys. 30 (1979) 773–787

149.35 Wang, C. Y. & Watson, L. T.: On the large deformations of C-shaped springs. Internat. J. Mech. Sci. 22 (1980) 395–400

149.5 Wasserstrom, E.: Numerical solutions by the continuation method. SIAM Review 15 (1973) 89–119

150. Watson, L. T.: An algorithm that is globally convergent with probability one for a class of nonlinear two-point boundary value problems. SIAM J. Numer. Anal. 16 (1979) 394–401

150.5 Watson, L. T.: Fixed points of C^2 maps. J. Comput. Appl. Meth. 5 (1979) 131–140

151. Watson, L. T.: A globally convergent algorithm for computing fixed points of C^2 maps. Appl. Math. Comput. 5 (1979) 297–311

151.3 Watson, L. T.: Solving the nonlinear complementarity problem by a homotopy method. SIAM J. Control and Optim. 17 (1979) 36–46

151.6 Watson, L. T.: Computational experience with the Chow-Yorke algorithm. Math. Programming 19 (1980) 92–101

151.7 Watson, L. T.: Numerical study of porous channel flow in a rotating system by a homotopy method. J. Comput. Appl. Math. 7 (1981)

151.8 Watson, L. T.: Solving finite difference approximations to nonlinear two-point boundary value problems by a homotopy method. SIAM J. Sci. Stat. Comput. 1 (1980) 467–480

152. Watson, L. T. & Fenner, D.: Algorithm 555: Chow-Yorke algorithm for fixed points or zeroes of $C \wedge 2$ maps. ACM Trans. on Math. Software 6 (1980) 252–260

152.3 Watson, L. T. & Li, T. Y. & Wang, C. Y.: The elliptic porous slider – a homotopy method. J. Appl. Mech. 45 (1978) 435–436

152.4 Watson, L. T. & Wang, C. Y.: Decleration of a rotating disc in a viscous fluid. 22 (1979) 2267–2269

152.4 Watson, L. T. & Wang, C. Y.: A homotopy method applied to elastica problems. Int. J. Solid Structures 17 (1981) 29–37

152.5 Watson, L. T. & Yang, W. H.: Optimal design by a homotopy method. Applicable Anal. 10 (1980) 275–284

152.55 Watson, L. T. & Yang, W. H.: Methods for optimal engineering design problems based on globally convergent methods. Computers and Structures 13 (1981) 115–119

152.9 Weber, H.: Numerische Behandlung von Verzweigungsproblemen bei gewöhnlichen Differentialsgleichungen. Numer. Math. 32 (1979) 17–29

153.05 Weber, H.: On the numerical solution of some finite dimensional bifurcation problems. Numer. Funct. Anal. Optimiz. 3 (1981) 341–366

153.1 Weber, H.: Zur Verzweigung bei einfachen Eigenwerten. Manuscripta Math. 38 (1982) 77–86

153.71 Wilmuth, R. S.: A computational comparison of fixed point algorithms which use complementary pivoting. In: Fixed points: algorithms and applications, S. Karamardian, Academic Press, New York (1977) 249–280

153.75 Wolsey, L. A.: Convergence, simplicial paths and acceleration methods for simplicial approximation algorithms for finding a zero of a system of nonlinear equations. CORE discussion paper # 7427, Univ. Cath. de Louvain, Belgium (1974)

153.8 Wright, A. H.: The octrahedal algorithm, a new simplicial fixed point algorithm. Math. Programming 21 (1981) 47–69

153.9 Wright, A.: Finding all solutions to a system of polynomial equations. Western Michigan Univ., May 1982

154.5 Yamamoto, Y.: A new variable dimension algorithm for the fixed point problem. To appear in Math. Programming

154.6 Yamamoto, Y.: Subdivisions and triangulations induced by a pair of subdivided manifolds. Preprint

154.7 Yamamoto, Y.: The 2-ray method: a new variable dimension fixed point algorithm with integer labelling. Univ. of Tsukuba, Japan, June 1982

154.8 Yamashita, H.: A continuous path method of optimization and its application to global optimization. In: Survey of mathematical programming (Proc. Ninth Intern. Math. Programming Sympos., Budapest, 1976), Vol. 1 (1979) 539–546, North-Holland

154.9 Zangwill, W. I.: Determing all minima of certain functions. Univ. of Chicago, August 1981

154.91 Zangwill, W. I. & Garcia, C. B.: Equilibrium Programming: the path following approach and dynamics. Math. Programming 21 (1981) 262–289

Polyhedral Theory and Commutative Algebra

L. J. Billera*
Cornell University, School of Operations Research, Upson Hall, Ithaca, NY 14853, USA

Abstract. An expository account is presented describing the use of methods of commutative algebra to solve problems concerning the enumeration of faces of convex polytopes. Assuming only basic knowledge of vector spaces and polynomial rings, the enumeration theory of Stanley is developed to the point where one can see how the Upper Bound Theorem for spheres is proved. A briefer account is then given of the extension of these techniques which yielded the proof of the necessity of McMullen's conjectured characterization of the f-vectors of convex polytopes. The latter account includes a glimpse of the application of these methods to the study of integer solutions to systems of linear inequalities.

Introduction and Summary

There has been much progress lately on questions concerning the enumeration of faces of various dimensions in convex polytopes. Specifically, one might be interested in bounds on the number of faces of one dimension (say, the number of vertices) given information on the number of faces of some other dimension (say, the number of facets). More generally, one might ask whether there exists a polytope having a predesignated number of faces of each dimension. Questions such as these go back to Euler, and they remain of fundamental interest in the context of mathematical programming today.

An important development in this field has been the introduction by Stanley of some methods from the field of commutative algebra. He used these techniques first to obtain an extension of the Upper Bound Theorem of McMullen from polytopes to general triangulated spheres. This theorem gives tight upper bounds on the number of faces in each dimension in terms of the number of vertices (and, by polarity in the case of polytopes, also bounds in terms of the number of facets). Subsequently, he combined these methods with some recent results in algebraic geometry to complete the proof of McMullen's conjectured characterization of the face-counting vectors of simplicial convex polytopes. The latter result applies equivalently to simple (i.e., nondegenerate) polytopes and provides a complete description of the relationships between the

---

* Prepared for the XI International Symposium on Mathematical Programming, Bonn, August 23–27, 1982. Supported in part by the National Science Foundation under Grant MCS81-02353.

numbers of faces of various dimensions for polytopes in either class. To date, there is no proof of this purely combinatorial result that avoids the algebraic machinery of Stanley.

This paper will describe this use of ring theoretic methods on problems of facial enumeration for convex polytopes. After introducing the problem and describing the major results for f-vectors of polytopes, some basic ideas from commutative algebra will be introduced and discussed. To each simplicial convex polytope P, we then associate a commutative ring, certain invariants of which are related to the number of faces of P in each dimension. We will see how classical algebraic results on these invariants together with some recent developments in commutative algebra related to invariant theory lead to a proof of the Upper Bound Theorem in a general setting provided by a certain class of simplicial complexes. A theorem of Reisner which characterizes complexes in this class is then discussed. Finally, after describing the construction used by the author and Lee to prove the sufficiency of McMullen's conditions, we will examine Stanley's proof of their necessity in enough detail to begin to see a striking connection between the numbers of faces of a polytope and the integer points in a related system of convex cones. The possibility of using these methods to shed light on general integer programming problems is one of the most exciting aspects of this area of research.

f-Vectors of Convex Polytopes

By a *convex polytope* P we mean the convex hull of a finite point set in a real Euclidean space. Equivalently, P can be defined as the bounded intersection of finitely many closed half-spaces. By a face F of P we mean the intersection of P with a hyperplane having the property that P is contained in one of its closed half-spaces. Thus, the empty set is always a face of P, and we call P a face of P (whether or not it arises in the above manner). All other faces will be called *proper* faces, and they are finite in number. Each face of a polytope P is again a polytope.

We define the *dimension* of a polytope P, $\dim P$, to be the dimension of $\mathrm{aff}(P)$, its affine hull, and say that P is a *d-polytope* if $\dim P = d$. In this case each face of P, except P itself, has dimension less than d. For each $i = -1, 0, 1, \ldots, d-1$, let $f_i(P)$ denote the number of i-dimensional faces of P. In particular, $f_{-1}(P) = 1$ counts the empty face, $f_0(P)$ is the number of *vertices*, $f_1(P)$ is the number of *edges* and $f_{d-1}(P)$ is the number of *facets* of P. We denote by $f(P)$ the vector $(f_{-1}(P), f_0(P), \ldots, f_{d-1}(P))$, called the *f-vector* of P. For a comprehensive treatment of the theory of convex polytopes and, in particular, of f-vectors see [11]. For survey of the latter topic which includes a discussion of the more recent results, see [19].

Let $f(P^d)$ denote the set of all f-vectors of d-polytopes. There is considerable interest in describing the set $f(P^d)$ exactly, but this remains an open problem. However, we can describe $\mathrm{aff}(f(P^d))$ and certain inequalities satisfied by

II. Polyhedral Theory and Commutative Algebra

each $f \in f(P^d)$. First, each $f \in f(P^d)$ satisfies the *Euler Equation*

$$f_0 - f_1 + f_2 - \ldots \pm f_{d-1} = 1 - (-1)^d,$$

and, further, this equation specifies the affine hull, namely

$$\text{aff}(f(P^d)) = \{(f_{-1}, f_0, \ldots, f_{d-1}): f_{-1} = 1, f_0 - f_1 + \ldots = 1 - (-1)^d\}.$$

The inequalities are somewhat harder to describe. To this end, consider the moment curve in $\mathbb{R}^d$ given by $x(t) = (t, t^2, t^3, \ldots, t^d)$ and choose real numbers $t_1 < t_2 < \ldots < t_n$, with $n > d$. Define $C(n, d)$ to be the convex hull of $V = \{x(t_1), x(t_2), \ldots, x(t_n)\}$. While the actual polytope obtained by this procedure depends on the choices of the t_i's, it is known that its combinatorial structure, in particular, its f-vector, is independent of the t_i's. We use the symbol $C(n, d)$ to refer to this combinatorial type. It is easily seen that $C(n, d)$ is a simplicial d-polytope, that is each facet (and thus, each proper face) is a simplex (an $(r-1)$-polytope having just r vertices).

One of the most remarkable properties of the polytope $C(n, d)$ is that it is neighborly, that is, each pair of vertices forms an edge of $C(n, d)$. In fact, for $k = 1, \ldots, [d/2]$, the convex hull of *any* k-subset of V is a face of $C(n, d)$. (Here $[x]$ denotes the largest integer less than or equal to x.) Thus among all d-polytopes with n vertices, $C(n, d)$ clearly has the maximum number of i-faces for $i = 0, 1, \ldots, [d/2] - 1$. That $C(n, d)$ has the maximum number of i-faces, among all d-polytopes with n vertices, for *all* i is the content of the *Upper Bound Theorem*, first formulated by Motzkin [27] and proved by McMullen [22]. Thus we have that for all d-polytopes P with n vertices

$$f(P) \leq f(C(n, d)).$$

Since the number of i-faces of $C(n, d)$ is known for each i as a function of n and d, this gives upper bounds for each $f_i(P)$ in terms of $n = f_0(P)$ (and d), and thus inequalities which must be satisfied for each $f \in f(P^d)$.

A particularly important component of $f(C(n, d))$ is

$$f_{d-1}(C(n, d)) = \binom{n - [(d+1)/2]}{n - d} + \binom{n - [(d+2)/2]}{n - d},$$

which is the maximum number of facets in a d-polytope with n vertices (or, by polarity, the maximum number of vertices in a d-polytope with n facets). Of course, by the above discussion we have

$$f_i(C(n, d)) = \binom{n}{i+1}$$

for $i = 0, 1, \ldots, [d/2] - 1$. See [11] or [25] for a general expression for the terms of $f(C(n, d))$ and its derivation.

A complete description of $f(P^d)$ has remained elusive for general d. There do not seem to be even reasonable conjectures as to a final set of conditions. However, if one restricts to the case of simplicial polytopes, then the situation is considerably better understood. In fact, the set $f(P_s^d)$ of all f-vectors of simplicial d-polytopes is completely known, being specified entirely by a list of linear equations, linear inequalities and nonlinear inequalities. We will describe these in turn. First note that by the usual polyhedral polarity [11], to each d-polytope P there corresponds another d-polytope P^* which has the property that $f_i(P) = f_{d-1-i}(P^*)$. In particular, when P is simplicial, then P^* is *simple* (or nondegenerate, in linear programming terms), that is, each vertex is on precisely d facets. Thus, describing $f(P_s^d)$ is equivalent to describing the f-vectors of all nondegenerate d-polytopes. Further, one proves the Upper Bound Theorem by first showing that the maximum numbers of faces must occur in simplicial polytopes, and then proving the Upper Bound Theorem for simplicial polytopes.

First, we note that each $f \in f(P_s^d)$ satisfies the *Dehn-Sommerville Equations*

$$E_k^d: \sum_{j=k}^{d-1} (-1)^j \binom{j+1}{k+1} f_j = (-1)^{d-1} f_k$$

for $k = -1, 0, 1, \ldots, d-1$. The equation E_{-1}^d is just the Euler equation. It is known that $[(d+1)/2]$ of these equations are independent, and they completely determine $\mathrm{aff}(f(P_s^d))$. Thus the dimension of $\mathrm{aff}(f(P_s^d))$ is $[d/2]$. See [11] or [25] for details.

To be able to describe the remaining conditions, we must apply a change of variables to the space of f-vectors, first suggested by Sommerville [32], which recasts the Dehn-Sommerville equations in a particularly simple form. To this end, take f in $f(P_s^d)$ and define the polynomial

$$f(t) = \sum_{j=-1}^{d-1} f_j t^{j+1}$$

and let

$$h(t) = (1-t)^d f(t/(1-t)).$$

Since f is of degree d, h is again a polynomial, say

$$h(t) = \sum_{i=0}^{d} h_i t^i.$$

If $f = f(P)$ then we define the h-vector of P to be the vector of coefficients of $h(t)$, that is $h(P) = (h_0, h_1, \ldots, h_d)$. Note that $h_0(P) = 1$ since $f_{-1}(P) = 1$.

The vector $h(P)$ can be obtained from $f(P)$ by means of an invertible linear transformation, whose inverse is given by

$$f_j = \sum_{i=0}^{j+1} \binom{d-i}{d-j-i} h_i.$$

Thus f_j is a *nonnegative* linear combination of $h_0, \ldots, h_{j+1}$, and so an inequality of the form $h(P) \le h(P')$ implies the corresponding inequality $f(P) \le f(P')$. In fact, the proof of the Upper Bound Theorem proceeds by showing $h(P) \le h(C(n,d))$ for the simplicial d-polytopes with n vertices. In terms of the h-vector of P, the Dehn-Sommerville equations become $h_i = h_{d-i}$, for $i = 0, 1, \ldots, [d/2]$. (See [26] or [25] where our h_{k+1} corresponds to their $g_k^d(P)$.)

II. Polyhedral Theory and Commutative Algebra

Let $h(P_s^d)$ denote the set of h-vectors of simplicial d-polytopes. By the above discussion, knowing $h(P_s^d)$ is equivalent to knowing $f(P_s^d)$. We can now describe the set of linear inequalities satisfied by all $f \in f(P_s^d)$. They were first proposed by McMullen and Walkup in the form of a *Generalized Lower Bound Conjecture*, which stated that $h_{i+1} \geq h_i$, for $i = 0, 1, \ldots, [d/2] - 1$. Thus, in light of the Dehn-Sommerville equations, these inequalities imply that the h-vector must be unimodal. In terms of the f_i's, $h_{i+1} \geq h_i$ implies a lower bound on f_i as a linear function of the f_j's for $j < i < [d/2]$; these lower bounds imply the lower bounds given in the so-called *Lower Bound Theorem* proved by Barnette [2] (see [26]).

To complete the description, we must establish a last bit of notation. For positive integers h and i, we note that h can always be written uniquely in the form

$$h = \binom{n_i}{i} + \binom{n_{i-1}}{i-1} + \ldots + \binom{n_j}{j},$$

where $n_i > n_{i-1} > \ldots > n_j \geq j \geq 1$. (Choose n_i to be the largest integer with $h \geq \binom{n_i}{i}$, etc.) Define the ith *pseudopower* of h to be

$$h^{(i)} = \binom{n_i + 1}{i + 1} + \binom{n_{i-1} + 1}{i} + \ldots + \binom{n_j + 1}{j + 1}.$$

Put $0^{(i)} = 0$ for all i.

We state the nonlinear inequalities on the components of the h-vector (and thus the f-vector) together with the earlier conditions in the form of a characterization of $h(P_s^d)$.

Theorem (McMullen's Conditions): An integer vector $h = (h_0, h_1, \ldots, h_d)$ is the h-vector of a simplicial convex d-polytope if and only if the following three conditions hold:

(i) $h_i = h_{d-i}$, $i = 0, 1, \ldots, [d/2]$,
(ii) $h_{i+1} \geq h_i$, $i = 0, 1, \ldots, [d/2] - 1$, and
(iii) $h_0 = 1$ and $h_{i+1} - h_i \leq (h_i - h_{i-1})^{(i)}$, $i = 1, \ldots, [d/2] - 1$

This characterization was conjectured in 1971 by McMullen [24], [25], and proved by him for $d \leq 5$ and for the case of d-polytopes having n vertices where $d < n < d + 3$. The sufficiency of these conditions was proved in 1979 by Billera and Lee [3], [4], followed almost immediately by a proof of necessity by Stanley [42]. The proof of sufficiency depends heavily on insights provided by earlier work of Stanley [36] in which the Upper Bound Theorem was extended to general triangulations of spheres by means of techniques of commutative algebra. The proof of necessity extends this earlier work, introducing powerful new techniques from algebraic geometry. In the rest of this paper, we will describe the algebraic techniques developed by Stanley and outline the steps leading to the proof of this result.

Note that the f-vectors of unbounded polyhedra are not covered by any of the results discussed here. For treatments of this topic, see [17], [6], [5] and [18].

As an interesting consequence of the sufficiency proof, Lee [18a] has shown that for every simple d-polytope P with n facets there is another such polytope P' with $f(P')=f(P)$ such that the diameter of P' is at most $n-d+1$, one more than the bound given by the Hirsch conjecture. Thus if there are examples where this conjecture fails badly, it will not be due solely to the *number* of edges (or to the number of faces of any other dimension).

The Stanley-Reisner Ring of a Simplicial Complex

By a simplicial complex on a vertex set $V=\{v_1, v_2, \ldots v_n\}$ we mean a subset Δ of 2^V such that $G \in \Delta, F \subset G$ implies $F \in \Delta$ and each $\{v_i\} \in \Delta$. In other words, Δ is an independence system for which each one-element set is independent. We define the dimension of Δ, dimΔ, to be $d-1$, where $d = \max_{F \in \Delta} |F|$, the maximum cardinality of an F in Δ. Our main example of a simplicial complex will be the boundary complex of a simplicial d-polytope P: the set V will be the vertex set of P and Δ will consist of all vertex sets of proper faces of P plus the empty set. In this case, the two uses of d are consistant.

An element F of Δ will be called a *face* (or *simplex*) of Δ, and if $|F|=i+1$, F will be called an i=face. We define $f_i(\Delta)$ to be the number of i-faces of Δ, and we call $f(\Delta)=(f_{-1}(\Delta), f_0(\Delta), \ldots, f_{d-1}(\Delta))$ the *f-vector* of Δ (again $f_{-1}(\Delta)=1$ counts the empty face). We define the quantities $h_i(\Delta)$, $0 \leqslant i \leqslant d$, by means of the same polynomial relation as in the case of polytopes, and let $h(\Delta)=(h_0(\Delta), \ldots, h_d(\Delta))$ be the *h-vector* of Δ.

Let k be a field, and let $R=k[X_1, \ldots, X_n]$ be the ring of polynomials in n indeterminates over k. If $X_1^{a_1} \ldots X_n^{a_n}$ is a monomial in R (where the a_i's are nonegative integers), we define its *support* to be the set $\{v_i: a_i \neq 0\}$. We define an ideal I_Δ in R by

$$I_\Delta = \langle X_{i_1} \ldots X_{i_k} : i_1 < i_2 < \ldots < i_k, \{v_{i_1}, \ldots, v_{i_k}\} \notin \Delta \rangle,$$

the ideal generated by all sqaure-free monomials whose support is *not* a face of Δ. Finally we define the ring $k[\Delta]=R/I_\Delta$. $k[\Delta]$ has come to be called the *Stanley-Reisner ring* of Δ. It was first considered in this form by Stanley [35], [36] and Reisner [29], although it is a special case of a more general notion considered earlier by Hochster [13].

In the case where Δ is the boundary complex of a triangle with vertices v_1, v_2, and v_3, i.e., $\Delta = \{\{v_1, v_2\}, \{v_1, v_3\}, \{v_2, v_3\}, \{v_1\}, \{v_2\}, \{v_3\}, 0\}$, I_Δ is the ideal generated by the monomial $X_1 X_2 X_3$. The ring $k[\Delta]$ consists essentially of all polynomials in X_1, X_2 and X_3 none of whose monomials is divisible by $X_1 X_2 X_3$.

In order to get information on the h-vector (and thus the f-vector) of the complex Δ from the ring $k[\Delta]$, we need to discuss the notion of a grading on $k[\Delta]$.

II. Polyhedral Theory and Commutative Algebra

Graded Algebras and Their Hilbert Functions

Let k be a field. By a *graded k-algebra* we mean a commutative, associative ring A such that $k \subset A$ (and so A is a vector space over k), together with a collection of subspaces A_i of A indexed by the nonnegative integers such that
  1) A is the vector space direct sum of the A_i,
     denoted $A = A_0 \oplus A_1 \oplus \ldots \oplus A_i \oplus \ldots$,
  2) $A_0 = k$,
  3) $A_i A_j \subset A_{i+j}$, i.e., the product of an element of A_i with an element of A_j is in A_{i+j},
  4) A is finitely generated as a k-algebra, i.e., there are finitely many elements $x_1, \ldots, x_r$ in A such that each element of A can be written as a polynomial in $x_1, \ldots, x_r$ with coefficients in k.

We note that the polynomial in (4) is not necessarily unique. If the x_i's can all be chosen from A_1, then A is called *standard*.

A simple example of a graded k-algebra is $A = k[X]$, the polynomial ring over k in one indeterminate. Here A_i is the set of all k-multiples of the monomial X^i. A bit more interesting is the ring $= k[X_1, \ldots, X_n]$, where A_i is the vector space generated by all monomials of degree i. Normally, one takes the *degree* of a monomial $X_1^{a_1}, \ldots, X_n^{a_n}$ to be the total degree, $a_1 + a_2 + \ldots + a_n$. Notice, however, that by defining degree by $w_1 a_1 + w_2 a_2 + \ldots + w_n a_n$, where the w_i's are arbitrary positive integers, one gets different *gradings* on A, i.e., different direct summands A_i. Choosing different gradings for an algebra can prove useful for some applications. For our puposes, however, we will always consider a polynomial ring to be graded by total degree, making it a standard k-algebra.

An element in a graded k-algebra A is said to be homogeneous (of degree i) if it belongs to A_i for some i. Since the A_i's are vector subspaces of A, any k-linear combination of homogeneous elements of the same degree is again homogeneous of that degree, and by (3) above, the product of homogeneous elements of any degree is homogeneous. By (1), each element a of A can be written uniquely as a finite sum of homogeneous elements, at most one of each degree. These are called the *homogeneous components* of a. It follows from this that the x_i's in (4) can always be chosen to be homogeneous. In a polynomial ring, the notion of homogeneous element coincides with the usual notion of homogeneous polynomial

An ideal I in A is said to be homogeneous if I can be generated by homogeneous elements or, equivalently, if for each a in I, all the homogeneous components of a are also in I. Each such ideal I, considered as a k-vector space, has a direct sum decomposition

$$I = I_0 \oplus I_1 \oplus I_2 \oplus \ldots$$

where $I_j = I \cap A_j$. Note that if I is a proper ideal, that is, $I \neq A$, then $I_0 = 0$. (From now on the term "ideal" will necessarily mean "proper ideal".) Thus each homogeneous ideal is contained in the maximal (homogeneous) ideal

$$A_+ = \cup \{A_i : i > 0\},$$

sometimes called the *irrelevant* maximal ideal. The most important fact to be noted about a homogeneous ideal I in a graded ring A is that the quotient ring A/I has a natural grading given by $(A/I)_j = A_j/I_j$. Thus the quotient of a (standard) graded k-algebra by a homogeneous ideal is again a (standard) graded k-algebra.

It is immediate that the standard graded k-algebras are precisely the quotients of polynomial rings (with the standard grading) by homogeneous ideals. If we let Δ be a simplicial complex and define the ideal I_Δ and the ring $k[\Delta]$ as before, then I_Δ is a homogeneous ideal in the polynomial ring R since it is generated by monomials, which are surely homogeneous (in any of the gradings mentioned for R). Thus $k[\Delta]$ is a standard graded k-algebra for any simplicial complex Δ.

Let A be any graded k-algebra. Since each A_i is a vector space over k, it makes sense to consider its dimension, denoted $\dim_k A_i$. It follows from the fact that A is finitely generated as an algebra that $\dim_k A_i$ is finite for each i. (A is in fact a Noetherian ring). We define the *Hilbert function* of A by

$$H(A, i) = \dim_k A_i, \quad i \geq 0.$$

The *Hilbert series* of A is defined to be the formal power series

$$F(A, t) = \sum_{i > 0} H(A, i) t^i.$$

Since $A_0 = k$, we always have $H(A, 0) = 1$, and so the constant term of $F(A, t)$ is always 1. For $A = k[X]$, each $H(A, i) = 1$ and so in this case

$$F(A, t) = 1/(1-t) = 1 + t + t^2 + t^3 + \dots.$$

When $A = k[X_1, \dots, X_n]$, a straightforward counting argument gives

$$F(A, t) = [1/(1-t)]^n$$

since in this case $H(A, i)$ is the number of monomials of degree i in n variables. Note that for any simplicial complex D, $H(k[\Delta], 1)$ is always the number of vertices in Δ.

In the case that A is a *standard* k-algebra, it is known that for i sufficiently large, $H(A, i)$ is a polynomial function of i. This polynomial is called the *Hilbert polynomial* of A, and its degree is $d-1$ where $d = \dim A$, the Krull dimension of A. In general, $\dim A$ is defined to be the longest length r of strictly increasing chain

$$\mathfrak{p}_0 \subset \mathfrak{p}_1 \subset \dots \subset \mathfrak{p}_r$$

of (homogeneous) prime ideals or, equivalently, the maximum number of elements of A which are algebraically independent over the field k (i.e., satisfy no nontrivial polynomial over k). See, for example, [1] for further details. We will use this observation about the degree of the Hilbert polynomial later to show that $\dim k[\Delta] = 1 + \dim \Delta$ for a simplicial complex Δ.

As an example, let A be the standard k-algebra $k[X, Y]/\langle XY \rangle$. A is the ring of a simplicial complex consisting merely of two points (and no other simplices). Then $H(A, 0) = 1$ and $H(A, i) = 2$ for $i > 0$. Thus the Hilbert polynomial is the constant polynomial 2 of degree 0 and so $\dim A = 1$. Finally, $F(A, t) = (1+t)/(1-t)$.

Let A be a graded k-algebra of Krull dimension d. A (homogeneous) *system of parameters* for A is a set $\theta_1, \theta_2, \ldots, \theta_d$ of homogeneous elements in A_+ (i.e., homogeneous elements of positive degree) such that A is a finitely generated module over the subalgebra $k[\theta_1, \ldots, \theta_d]$ (the algebra of all polynomials in the elements $\theta_1, \ldots, \theta_d$). A system of parameters is always algebraically independent over k (see [31]) and so $k[\theta_1, \ldots, \theta_d]$ is in fact a polynomial ring. An equivalent condition for $\theta_1, \ldots, \theta_d$ to be a system of parameters is that $A/\langle \theta_1, \ldots, \theta_d \rangle$ be a finite dimensional vector space over k, where $\langle \theta_1, \ldots, \theta_d \rangle$ denotes the ideal generated by the θ_i's.

In the case A is a polynomial ring in n indeterminates, the Krull dimension is clearly n, and the indeterminates themselves form a system of parameters, as will any nonsingular k-linear transformation of the indeterminates. Consider $A = k[X]$ and let $\theta = X^2$. Since for every $f \in A$, $f(X) = g(X^2) + Xh(X^2)$, with g and h polynomials, A is finitely generated over $k[X^2]$ with generators 1 and X, and so θ is a system of parameters for A. Equivalently, $A/\langle \theta \rangle$ is generated as a vector space over k by the images of 1 and X.

It is a consequence of the Noether Normalization Lemma that a system of parameters always exists for any graded k-algebra A (see, e.g., [1], [31] or [48]). Further, if A is standard and k is an infinite field (like the rationals), there exists a system of parameters each of whose elements is homogeneous of degree 1 (that is, each $\theta_i \in A_1$) [1]. The proof in this case is quite intuitive, proceeding somewhat like the process of obtaining a basis for a vector space from a set of generators. One starts with a set of algebra generators $x_1, \ldots, x_r$ for A (which are chosen from A_1 if A is standard). If these are algebraically independent, then they form a system of parameters, and we are done. If not, then there is a polynomial relation satisfied by $x_1, \ldots, x_r$. Assume for the moment that this relation is monic in x_r, that is, the coefficient of the highest power of x_r, say the mth, is 1. Then x_r^m (and all higher powers of x_r) can be expressed as a sum of lower powers of x_r multiplied by polynomials in $x_1, \ldots, x_{r-1}$. Thus A is a finitely generated module over $k[x_1, \ldots, x_{r-1}]$. Now continue the process with $k[x_1, \ldots, x_{r-1}]$; it will eventually stop with a system of parameters. While the relation obtained at each step will not always be monic, it is always possible, since the field is infinite, to replace the x_i's by linear combinations of the x_i's to make it monic.

To illustrate, consider $A = k[x, y]/\langle xy \rangle$. At the first step one gets the relation $xy = 0$, which is not monic in either variable. Making the linear change
$$x' = x - y$$
one gets the relation
$$0 = xy = (y + x')y = y^2 + x'y$$
which is monic in y. We get, then, that x' is a system of parameters for A, which is generated by 1 and y as a module over $k[x']$.

Systems of parameters which are homogeneous of degree 1 need not exist for standard graded k-algebras when k is a finite field. In fact when Δ is a 1-dimensional simplicial complex (i.e., a graph) and k is a field with q elements then one can show that $k[\Delta]$ has a system of parameters which is homogeneous of degree 1 if and only if Δ is $q+1$ colorable. This is a consequence of a result of Stanley, mentioned in [50], which states that (for a d-dimensional complex Δ) homogeneous elements of degree 1, $\theta_1, \ldots, \theta_d$, form a system of parameters for $k[\Delta]$ if and only if the d by n matrix of coefficients expressing the θ_i as linear combinations of the x_j's (generators of $k[\Delta]_1$ corresponding to vertices of Δ) has the property that, for each face of Δ, the set of columns corresponding to its vertices is linearly independent.

Cohen-Macaulay Rings: h-Vectors as Hilbert Functions

We turn now to the task of showing how the h-vector of a simplicial convex polytope can be viewed as the Hilbert function of a certain graded k-algebra [36]. We begin by considering a general simplicial complex D and restrict later to a special class of complexes which includes boundary complexes of simplicial convex polytopes as well as general simplicial balls and spheres.

Let Δ be a $(d-1)$-dimensional simplicial complex and $k[\Delta]$ be the associated k-algebra. Recall that the f-vector $f(\Delta) = (f_{-1}, f_0, \ldots, f_{d-1})$ and h-vector $h(\Delta) = (h_0, \ldots, h_d)$ are related by the polynomial equation

$$h(t) = (1-t)^d f(t/(1-t))$$

where

$$f(t) = f(\Delta, t) = \sum_{j=-1}^{d-1} f_j t^{j+1}$$

and

$$h(t) = h(\Delta, t) = \sum_{j=0}^{d} h_j t^j.$$

Now $H(k[\Delta], m)$ is the number of monomials of degree m with support in Δ, and a straightforward counting argument proves the following result.

Lemma 1: $H(k[\Delta], 0) = 1$ and for all $m > 0$,

$$H(k[\Delta], m) = \sum_{j=0}^{d-1} f_j \binom{m-1}{j}.$$

Corollary 2: $\dim k[\Delta] = d = 1 + \dim \Delta$

Proof: For $m > 0$, $H(k[\Delta], m)$ is given by a polynomial of degree $d-1$, and so $\dim k[\Delta] = d$ by earlier comments.

Lemma 3: $(1-t)^d F(k[\Delta], t) = h(\Delta, t)$.

Proof: It is sufficient to show that the Hilbert series of $k[\Delta]$ is given by $f(t/(1-t))$. But

$$f(t/(1-t)) = \sum_{j=-1}^{d-1} f_j t^{j+1} (1+t+t^2+\ldots)^{j+1},$$

II. Polyhedral Theory and Commutative Algebra

and it is again straightforward, using Lemma 1, to verify that the coefficient of t^m in this series is given by $H(k[\Delta], m)$.

We are now led to ask, for a standard graded k-algebra A of Krull dimension d, when $(1-t)^d F(A, t)$ is again the Hilbert series of a graded k-algebra. We have seen that this series is a polynomial when $A = k[\Delta]$ for some $(d-1)$-dimensional complex Δ. In fact, it is generally true that $(1-t)^r F(A, t)$ is a polynomial when A is a standard graded k-algebra which can be generated as an algebra by r elements of degree 1 (see, e.g. [45]). This can be proved using the following result [39, Theorem 3.1]. Recall that if I is homogeneous ideal in A, then I inherits a grading from A given by $I_j = I \cap A_j$. We can then define $H(I, j) = \dim_k I_j$ and $F(I, t)$ as before.

For $a \in A$, define $\operatorname{Ann} a = \{r \in A : ra = 0\}$. $\operatorname{Ann} a$ is an ideal which is homogeneous when a is.

Lemma 4: Let A be a graded k-algebra, and let θ be a homogeneous element of degree $m > 0$. Then

$$(1-t^m) F(A, t) = F(A/\langle\theta\rangle, t) - t^m F(\operatorname{Ann} \theta, t).$$

Proof: By linear algebra we get for each j and homogeneous ideal I

$$H(A, j) = H(I, j) + H(A/I, j)$$

and so

$$F(A, t) = F(I, t) + F(A/I, t).$$

Now let $I = \langle\theta\rangle$ and note that for each j we have

$$H(I, j+m) = \dim_k (\theta A_j)$$
$$= \dim_k A_j - \dim_k (A_j \cap \operatorname{Ann} \theta)$$
$$= H(A, j) - H(\operatorname{Ann} \theta, j),$$

and so

$$F(I, t) = t^m [F(A, t) - F(\operatorname{Ann} \theta, t)].$$

Eliminating $F(I, t)$ yields the desired result.

Lemma 4 is of special interest in the case where $\operatorname{Ann} \theta$ is the zero ideal, that is, where θ is not a *zero divisor*. For then $F(\operatorname{Ann} \theta, t) = 0$ and if, further, the degree of θ is 1, we have

$$(1-t) F(A, t) = F(A/\langle\theta\rangle, t),$$

the Hilbert series of the graded k-algebra $A/\langle\theta\rangle$. The problem is now whether we can choose θ so that $A/\langle\theta\rangle$ itself has a homogeneous non-zero divisor (of degree 1), and so on, continuing the process for d steps. We next describe the class of algebras for which this process can always be carried out.

Recall that a graded k-algebra A of Krull dimension d always has a system of parameters, that is, a set $\theta_1, \ldots, \theta_d$ of homogeneous elements in A_+ such that A is a finitely generated module over the subalgebra $k[\theta_1, \ldots, \theta_d]$. A is said to be a *Cohen-Macaulay* graded k-algebra if for some (equivalently, for each) system of parameters $\theta_1, \ldots, \theta_d$, A is a *free* module over $k[\theta_1, \ldots, \theta_d]$. This means that there are homogeneous elements $\eta_1, \ldots, \eta_r$ in A so that each a in A has a *unique* representation of the form

$$a = \sum_{i=1}^{r} \eta_i p_i(\theta_1, \ldots, \theta_d)$$

where the p_i's are polynomials in $\theta_1, \ldots, \theta_d$. It is a characterizing property of Cohen-Macaulay k-algebras that every system of parameters is a *regular sequence* (or an A-sequence), that is, each system of parameters $\theta_1, \ldots, \theta_d$ for A has the property that for $i = 1, \ldots, d$, θ_i is not a zero divisor on the ring $A/\langle\theta_1, \ldots, \theta_{i-1}\rangle$. (In fact, the existence of a sequence having this property is normally taken as the definition of a Cohen-Macaulay ring, the freeness being a consequence in this case. See [31] or [15] for details.)

We can now state a key result of Stanley [36], [39, Corollary 3.2], whose origins trace back to work of Macaulay [20].

Theorem 5: Let A be a standard graded k-algebra of Krull dimension d, and let $\theta_1, \ldots, \theta_d$ be a system of parameters for A which is homogeneous of degree 1. Suppose A is Cohen-Macaulay. Then

$$(1-t)^d F(A, t) = F(B, t)$$

where $B = A/\langle\theta_1, \ldots, \theta_d\rangle$.

Proof: Since A is Cohen-Macaulay, $\theta_1, \ldots, \theta_d$ is a regular sequence. The assertion now follows from Lemma 4 and the subsequent discussion.

We will say that a complex Δ is a *Cohen-Macaulay complex* (over k) if its associated k-algebra $k[\Delta]$ is Cohen-Macaulay. Later we will describe conditions on Δ which will insure that $k[\Delta]$ is Cohen-Macaulay. These conditions will depend on the choice of field k, but it suffices for now to note that if Δ is Cohen-Macaulay over any field, then it will be Cohen-Macaulay over the rationals. Thus if we omit mention of k when we specify that Δ is a Cohen-Macaulay complex, we will assume that k is the field of rational numbers, Q. Recall that for any complex Δ, $k[\Delta]$ is always a standard k-algebra, and thus, if k is an infinite field, there always exists a system of parameters which is homogeneous of degree 1.

When we refer to a vector of finite length as a Hilbert function, we mean that vector with a sequence of zeros appended to it. The following is a consequence of Theorem 5 and Lemma 3.

Corollary 6: The h-vector of a Cohen-Macaulay complex Δ (over k) is always the Hilbert function of a standard graded k-algebra B. In fact, we can always choose $k = Q$, and take B to be $k[\Delta]/\langle\theta_1, \ldots, \theta_d\rangle$, where $d = \dim \Delta + 1$, and the θ_i's are homogeneous elements of degree 1 which form a system of parameters for $k[\Delta]$. In this case, h_i counts the number of η_i of degree i, where the elements $\eta_1, \ldots, \eta_r$ are as specified as above.

II. Polyhedral Theory and Commutative Algebra

It follows from work of Hochster [13] that $k[\Delta]$ is Cohen-Macaulay whenever Δ is the boundary complex of a simplicial convex polytope. This uses the fact, shown by Bruggesser and Mani, that boundary complexes of polytopes are always shellable [7]. A complete characterization of when $k[\Delta]$ is Cohen-Macaulay was later supplied by Reisner [29], who showed that, in fact, $k[\Delta]$ is Cohen-Macaulay for any simplicial sphere or ball. We will discuss Reisner's result a bit later, after discussing the relevance of Corollary 6 to the Upper Bound Theorem.

O-Sequences and the Upper Bound Theorem

As with simplicial convex polytopes, one can ask for any simplicial complex Δ, having n vertices and dimension $d-1$, whether the inequality of f-vectors

$$f(\Delta) \leq f(C(n,d)) \tag{1}$$

holds where, as before, $C(n,d)$ is the boundary complex of the cyclic d-polytope with n vertices. As before, this inequality is implied by the inequality of h-vectors

$$h(\Delta) \leq h(C(n,d)). \tag{2}$$

In the presence of the Dehn-Sommerville equations

$$h_i(\Delta) = h_{d-i}(\Delta), \quad i = 0, \ldots, [d/2]$$

(2) is equivalent to

$$h_i(\Delta) \leq \binom{n-d+i-1}{i}, \quad i = 0, \ldots, [d/2]. \tag{3}$$

See [22] or [25] for details. We will show that (3) holds whenever Δ is a Cohen-Macaulay complex. In particular, since the Dehn-Sommerville equations hold for triangulations of $(d-1)$-spheres as well as for the boundaries of simplicial d-polytopes ([16], [11]), we get that (2) and hence (1) hold for these complexes, proving the Upper Bound Theorem for spheres as well as for polytopes.

The key to showing (3) for a Cohen-Macaulay complex Δ is Corollary 6, which says that the h-vector of such a complex is the Hilbert function of a standard graded k-algebra. Accordingly, we turn now to characterizing such Hilbert functions.

Let $X_1, \ldots, X_n$ be a list of indeterminates, and let m and m' be monomials in the X_i's. We denote $m|m'$ to mean m divides m', that is, the power of each X_i in m is at most that in m'. By an order ideal of monomials we mean a nonempty set M of monomials such that $m' \in M$, $m|m' \Rightarrow m \in M$, that is, all divisors of monomials in M are also in M. In particular, $1 \in M$ for each order ideal of monomials M. For a monomial m, we write $\deg m$ to denote the total degree of m computed in the standard fashion (with $\deg X_i = 1$). A sequence of integers $H_0, H_1, H_2, \ldots$ is called on *O-sequence* if there exists an order ideal of monomials M such that for each i, $H_i = |\{m \in M: \deg m = i\}|$.

We define an ordering on the set of all monomials in $X_1, \ldots, X_n$ by $m < m'$ if $\deg m < \deg m'$ or if $\deg m = \deg m'$ and m precedes m' in the lexicographic order induced by $X_1 < X_2 < \ldots X_n$ (so, for instance, $X_1 < X_1^2 < X_2 < X_1 X_2 < X_2^2$). The following theorem is due to Stanley [39], [36], who attributes it essentially to Macaulay [21].

Theorem 7: Let $H_0, H_1, H_2, \ldots$ be a sequence of integers. The following four conditions are equivalent.

(i) There exists a standard graded k-algebra A with Hilbert function $H(A, i) = H_i$ for $i \geq 0$.

(ii) $H_0, H_1, H_2, \ldots$ is an O-sequence.

(iii) $H_0 = 1$, $H_1 \geq 0$ and for $i \geq 1$

$$0 \leq H_{i+1} \leq H_i^{(i)}.$$

(iv) Let $n = H_1$, and for each $i \geq 0$, let M_i be the (lexicographically) first H_i monomials of degree i in n variables. Then $M = \bigcup_{i \geq 0} M_i$ is an order ideal of monomials.

The hard part of this theorem is the implication (ii) ⇒ (iv), which was proved by Macaulay. The proof of a more general version can be found in [8]. Condition (iii) was first considered by Stanley [35] [36]. A proof of the equivalence of (iii) and (iv) can be found in [4, Prop. 1]. The connection between the algebra and the combinatorics is provided by the equivalence of (i) and (ii), which we consider further.

To see (ii) ⇒ (i), suppose M is an order ideal of monomials in $X_1, \ldots, X_n$ counted by the O-sequence H_i. Let $A = k[X_1, \ldots, X_n]/I$, where I is the homogeneous ideal generated by those monomials not in M. Then $H(A, i) = H_i$ for each i. On the other hand, (i) ⇒ (ii) is a consequence of the following result [39, Theorem 2.1].

Lemma 8: Let A be a standard graded k-algebra and let $x_1, \ldots, x_n$ be a set of homogeneous elements of degree 1 which generate A as a k-algebra. Then there is an order ideal of monomials in $x_1, \ldots, x_n$ which forms a vector space basis for A over k.

The proof of Lemma 8 proceeds by a greedy algorithm. Inductively form a k-basis M for A by putting 1 in M, and continue at each step by choosing the first monomial (in the ordering defined above) which is linearly independent of those already in M. It follows easily that M is a k-basis which is an order ideal of monomials.

Theorem 7 allows us to prove Stanley's Upper Bound Theorem for Cohen-Macaulay complexes [36, Corollary 4.4].

Theorem 9: Let Δ be a $(d-1)$-dimensional Cohen-Macaulay complex with n vertices. Then for $i = 0, \ldots, d$,

$$h_i(\Delta) \leq \binom{n-d+i-1}{i}.$$

Proof: We know that $h(\Delta)$ is the Hilbert function of a standard graded k-algebra. One can check directly from the definition of h that $h_1(\Delta) = n - d$. Otherwise, using the notation of Corollary 6, note that

$$h_1(\Delta) = \dim_k B_1$$
$$= \dim_k k[\Delta]_1 / I_1$$
$$= H(k[\Delta], 1) - H(I, 1)$$

where $I = \langle \theta_1, \ldots, \theta_d \rangle$. We know that $H(k[\Delta], 1) = n$, and since the θ_i's are all homogeneous of degree 1 and linearly independent (the are *algebraically* independent), we have $H(I, 1) = d$, proving the assertion.

We complete the proof in two different ways. First, by (ii) of Theorem 7, there is an order ideal of monomials in $h_1(\Delta) = n - d$ variables having $h_i(\Delta)$ monomials of degree i. This can be no more than the *total* number of monomials of degree i with $n - d$ variables, which is $\binom{n-d+i-1}{i}$. Alternatively, we proceed by induction. Note that

$$h_1(\Delta) = n - d = \binom{n-d}{1},$$

and if

$$h_{i-1}(\Delta) \leq \binom{n-d+i-2}{i-1}$$

then by (iii) of Theorem 7

$$h_i(\Delta) \leq h_{i-1}^{\langle i-1 \rangle}$$
$$\leq \binom{n-d+i-2}{i-1}^{\langle i-1 \rangle}$$
$$= \binom{n-d+i-1}{i},$$

and again the result follows.

Note that (due to the presence of the Dehn-Sommerville equations) the Upper Bound Theorem for spheres and polytopes requires the inequalities in Theorem 9 only for i up to $[d/2]$. The rest are redundant in this case. For general Cohen Macaulay complexes, however, the later inequalities are not necessarily consequences of the earlier ones. For Cohen-Macaulay complexes that do not satisfy the Dehn-Sommerville equations (simplicial balls, for example), Theorem 9 does provide an upper bound on the f-vector, though it is not that given by (1). See [30] for a treatment of upper bounds in even more general complexes.

Finally, we have noted that the Cohen-Macaulay property, and thus the Upper Bound Theorem, for the boundary complex of a simplicial polytope follows from the shellability of polytopes. But this is precisely the property needed in McMullen's proof for this case (see [22] or [25]), although he requires that certain special shellings exist, and this may not be true for a general shellable sphere. Further, not all spheres are shellable.

Reisner's Theorem: When $k[\Delta]$ is Cohen-Macaulay

The question of whether one could characterize topologically those complexes Δ for which $k[\Delta]$ is Cohen-Macaulay was first raised by Hochster [13], who was originally motivated by questions in invariant theory. A complete answer was given by Reisner [29, Theorem 1], whose work was done independently of that of Stanley on upper bounds [35], and who was unaware of the implications of his work in enumeration theory.

To describe Reisner's conditions, we need the following notation. If F is a face in the simplicial complex Δ, the *link* of F in Δ, denoted $lk[F,\Delta]$, is defined by

$$lk[F,\Delta] = \{G \in \Delta : G \cap F = \emptyset, G \cup F \in \Delta\}.$$

For each F, $lk[F,\Delta]$ is a subcomplex of Δ. (It is the *contraction* of F in the case that Δ is the complex of independent sets in a matroid.) Note that $lk[\emptyset, \Delta] = \Delta$.

Theorem 10: For a simplicial complex Δ, $k[\Delta]$ is Cohen-Macaulay if and only if for each face F of Δ, including $F = \emptyset$,

$$\tilde{H}_i(lk[F,\Delta]; k) = 0 \quad \text{for } i < \dim lk[F,\Delta],$$

that is, the reduced homology of $lk[F,\Delta]$ with coefficients in k vanishes except possibly in the dimension of $lk[F,\Delta]$.

It is known that if Δ is a triangulation of a manifold or a manifold with boundary, then the homology conditions on the links of nonempty faces F automatically hold, so it reduces to a requirement that the reduced homology of Δ vanishes below $\dim \Delta$. In particular, this is true for triangulations of balls and spheres. (See, for example, [12].) It is easily checked that this result implies that for 1-dimensional complexes (i.e. graphs), being Cohen-Macaulay is equivalent to being connected. A triangulation of the projective plane shows that whether $k[\Delta]$ is Cohen-Macaulay for a given Δ may depend upon the choice of k [29, Remark 3]. It is relatively easy to show using these conditions that each shellable complex is Cohen-Macaulay. (See [9] for the definition of shellable complex.)

To help understand the meaning of this result, we will briefly describe the notion of the reduced homology groups of a simplicial complex with coefficients in a field k. It will be clear that these conditions are really combinatorial and algebraic in nature.

For a d-dimensional simplicial complex Δ, let C_i, $i = -1, 0, 1, \ldots, d$, be the k-vector space (formally) spanned by all i-faces of Δ, the so-called space of i-*chains* of Δ. C_i is merely a vector space whose basis elements correspond to the i-faces of Δ, or, equivalently, it is the subspace of $k[\Delta]_i$ spanned by all square-free monomials. Let $V = \{v_1, \ldots, v_n\}$ be the set of vertices of Δ, and for an r-face $F = \{v_{i_0}, v_{i_1}, \ldots, v_{i_r}\}$, where $i_0 < i_1 < \ldots < i_r$, we define F_j to be the $(r-1)$-face $F \setminus \{v_{i_j}\}$, for $j = 0, \ldots, r$. For $r = 0, \ldots, d$, we define the rth *boundary map*, a linear transformation

$$\partial_r : C_r \to C_{r-1},$$

II. Polyhedral Theory and Commutative Algebra

by specifying it on the basis of C_r and extending by linearity. For an r-face F, define

$$\partial_r(F) = \sum_{j=0}^{r} (-1)^j F_j$$

$$= F_0 - F_1 + F_2 - \ldots \pm F_r.$$

For convenience, we define $\partial_{d+1} = 0$.

It is not hard to check that with this definition of boundary map, the composition $\partial_r \partial_{r+1} = 0$, for $r = 0, \ldots, d$. Thus for each such r we have $\operatorname{Im} \partial_{r+1} \subset \operatorname{Ker} \partial_r$, where Im and Ker denote the image and the kernel (nullspace) respectively. We now can define the rth *reduced homology* of Δ with coefficients in k to be the quotient vector space

$$\tilde{H}_r(\Delta; k) = \operatorname{Ker} \partial_r / \operatorname{Im} \partial_{r+1}$$

for $r = 0, 1, \ldots, d$. (To define the regular (non-reduced) homology, one proceeds in the same way, except with $C_{-1} = 0$. The only difference is in the dimension of the 0th space, which is the number of connected components of Δ in the regular case, and one less in the reduced case. See, for example, [12] or [28] for details.)

A discussion of the proof of Reisner's Theorem is beyond the scope of this paper. It involves the use of a characterization of the Cohen-Macaulay property via the theory of local cohomology of rings. The homology of Δ arises out of the calculation of the cohomology of $k[\Delta]$. Other proofs of this result can be found in [14] and [49], the latter being the most elementary.

Necessity and Sufficiency of the McMullen Conditions

In light of Theorem 7, we can give the following restatement of McMullen's conditions.

Theorem 11: An integer vector $h = (h_0, h_1, \ldots, h_d)$ is the h-vector of a simplicial d-polytope if and only if it satisfies the Dehn-Sommerville equations ($h_i = h_{d-i}$) and the sequence of differences

$$H_0, H_1, \ldots, H_{[d/2]},$$

given by $H_0 = h_0$ and $H_i = h_i - h_{i-1}$ for $i = 1, \ldots, [d/2]$, is an O-sequence.

The proof of the sufficiency of these conditions makes use of the special order ideal of monomials provided by (iv) of Theorem 7, when the sequence of H_i's is an O-sequence, to construct a simplicial d-polytope P such that $h = h(P)$. The idea of this construction is to associate with each monomial in this order ideal a facet of the cyclic polytope $C(n, d+1)$, where $n = h_1 + d$. One shows that for suitable choice of $t_1 < t_2 < \ldots < t_n$, there will be a point z in R^{d+1} which is beyond precisely those facets of $C(n, d+1)$ corresponding to the order ideal and beneath the rest. The desired polytope P turns out to be the intersection of $P' = \operatorname{conv}(C(n, d+1) \cup \{z\})$ with a hyperplane strictly separating z from

$C(n, d+1)$. The idea of constructing polytopes in this way over cyclic polytopes was inspired by a construction of Klee [17].

The association of facets of $C(n, d+1)$ to monomials depends on a combinatorial characterization of the facets of cyclic polytopes known as *Gale's evenness criterion*, which states roughly that between any two $x(t_i)$ not in a given facet, there must be an even number of $x(t_i)$ in the facet (with between being in the sense of the ordering of the t_i's). This induces a pairing of some of the vertices in any facet, and in the facet associated to a given monomial, the exponent of X_i in the monomial denotes the number of pairs of vertices which are to be displaced i units to the right from the position of this pair in the facet $\{x(t_1), \ldots, x(t_{d+1})\}$. Thus, considering just the incidence vector of the $x(t_i)$'s in a facet, the monomial 1 corresponds to the facet $(1, 1, \ldots, 1, 0, \ldots, 0)$, X_1 to the facet $(1, \ldots, 1, 0, 1, 1, 0, \ldots, 0)$, X_2 to $(1, \ldots, 1, 0, 0, 1, 1, 0, \ldots, 0)$ and X_1^2 to $(1, \ldots, 1, 0, 1, 1, 1, 1, 0, \ldots, 0)$. See [3] for a longer sketch of the proof, and [4] for all the details.

To show that the conditions are necessary, it is enough, by (i) of Therorem 7, to find a standard graded k-algebra with Hilbert function $H(A, i) = H_i$ for $i = 0, 1, \ldots, [d/2]$. Recall that by Corollary 6, if $h = h(P)$ for a simplicial d-polytope P, then $h_i = H(B, i)$ for the graded k-algebra $B = k[P]/\langle \theta_1, \ldots, \theta_d \rangle$, where the θ_i's form a system of parameters for $k[P]$. Now

$$B = B_0 \oplus B_1 \oplus B_2 \oplus \ldots \oplus B_d$$

where each B_i is a k-vector space of dimension h_i. Stanley shows [42] that for a suitable choice of θ_i's there is an element $w \in B_1$ such that the linear map

$$T_i: B_i \to B_{i+1}$$

given by $T_i(x) = wx$ (recall that the B_i's form a grading) is injective (one-to-one) for $i < [d/2]$. Since w is homogeneous, the ring $A = B/\langle w \rangle$ is a graded k-algebra. Now take $i < [d/2]$. Since T_i is injective we have $\dim_k \text{Im } T_i = \dim_k B_i$ and so

$$H(A, i+1) = \dim_k B_{i+1}/\text{Im } T_i$$
$$= \dim_k B_{i+1} - \dim_k B_i$$
$$= h_{i+1} - h_i$$
$$= H_{i+1}$$

Thus the H_i's are seen to form an O-sequence.

To find the element w, Stanley constructs from P an abstract algebraic variety which turns out to be essentially a complex projetive variety. This *toric variety* X_P [10] has the property that for a suitably chosen system of parameters $\theta_1, \ldots, \theta_d$ for $k[P]$, the spaces B_i describe the cohomology of X_P in even dimensions. The existence of w then follows from the so-called "hard Lefschetz theorem" for X_P. While the definition of X_P is beyond the scope of this paper, it is particularly intriguing to note what information about P goes into the construction of X_P. Doing so allows us to briefly discuss an aspect of this subject with potential interest to mathematical programming far beyong the topic at hand.

Let L be a rational convex polyhedral cone (i.e., given by finitely many rational inequalities) in $\mathbb{R}^n$. We associate a graded k-algebra with L in the follow-

II. Polyhedral Theory and Commutative Algebra

ing way. For each $z \in \mathbb{Z}^n$, where $\mathbb{Z}$ is the ring of integers, let X^z denote the monomial $X_1^{z_1} \ldots X_n^{z_n}$, $z = (z_1, \ldots, z_n)$. Now define A_L to be the subring of $k[X_1, \ldots, X_n]$ spanned by $\{X^z : z \in L \cap \mathbb{Z}^n\}$. Since L is a convex cone (so is closed under addition), it is immediate that A_L consists of all k-linear combinations of monomials X^z where z is an integer point of L. Using the fact that L is finitely generated, it is an easy argument to show that A_L is finitely generated as a k-algebra. The Krull dimension of A_L is the same as the usual dimension of the cone L. It is a theorem of Hochster [13] that A_L is Cohen-Macaulay for each L. (See also [10].) In some sense, this result says something about the regularity of the lattice points in convex cones. In general, a study of these rings may prove useful in understanding the difficult problems involving integral solutions to linear systems. For work related to this topic, see [10], [13], [33], [34], [39], [37], [40], [44], [43] and [45], the last two of which are surveys.

To "describe" the variety X_P, let us assume that the origin is an interior point of P and that all vertices of P are rational vectors. (Since P is simplicial, a small perturbation of the vertices of P can assure this without changing the combinatorial structure of P.) For each face F of P let cone(F) be the (rational) cone spanned by F (and the origin), and let L be the polar cone to cone(F), i.e., $L^* = \text{cone}(F)$. Denote by $A(F)$ the k-algebra A_L defined above. The toric variety is made up of the so-called affine schemes $\text{Spec} A(F)$, called affine toric varieties. The recipe for constructing X_P from the $\text{Spec} A(F)$ is derived directly from how the various F's fit together in P. (See [10] or [46] for more details.) We should point out here that the crucial property of X_P for this proof, its projectivity, is equivalent to the convexity of P. Thus this proof does not work for non-convex spherical complexes, and so the necessity of McMullen's conditions (other than the Dehn-Sommerville equations) remains an open question for triangulations of general spheres, piecewise linear (i.e., combinatorial [9]) spheres or even shellable spheres. This situation should be contrasted with the case of upper bounds.

The important thing to note here is that the information used to specify X_P, and then prove the necessity of last two of McMullens's conditions, come from the location of integer points in the polars of the cones specified by the faces of P. This provides a rather surprising connection between two seemingly unrelated aspects of polyhedral theory: enumeration and integrality. Whether or not this indicates the existence of a deep interplay between these subjects which will aid our understanding and ability to deal with problems of integrality is a question that can only be answered by further research.

Bibliography

[1] M. F. Atiyah and I. G. Macdonald, *Introduction to Commutative Algebra*, Addison-Wesley, Reading, Massachusetts, 1969.

[2] D. W. Barnette, A proof of the lower bound conjecture for convex polytopes, *Pacific J. Math.* 46 (1973), 349–354.

[3] L. J. Billera and C. W. Lee, Sufficiency of McMullen's conditions for f-vectors of simplicial polytopes, *Bull. Amer. Math. Soc.* (New Series) 2 (1980), 181–185.

[4] L. J. Billera and C. W. Lee, A proof of the sufficiency of McMullen's conditions for f-vectors of simplicial convex polytopes, *Jour. Combinatorial Theory (A)* 31 (1981), 237-255.

[5] L. J. Billera and C. W. Lee, The numbers of faces of polytope pairs and unbounded polyhedra, *Europ. J. Combin.* 2 (1981), 307-332.

[6] A. Björner, The minimum number of faces of a simple polyhedron, *Europ. J. Combin.* 1 (1980), 27-31.

[7] H. Bruggesser and P. Mani, Shellable decompositions of cells and spheres, *Math. Scand.* 29 (1971), 197-205.

[8] G. Clements and B. Lindström, A generalization of a combinatorial theorem of Macauly, *J. Combinatorial Theory* 7 (1969), 230-238.

[9] G. Danaraj and V. Klee, Which spheres are shellable?, *Ann. Discrete Math.* 2 (1978), 32-52.

[10] V. I. Danilov, The geometry of toric varieties, *Russian Math. Surveys* 33:2 (1978), 97-154; translated from *Uspekhi Mat. Nauk.* 33:2 (1978), 85-134.

[11] B. Grünbaum, *Convex Polytopes*, Wiley Interscience, New York, 1967.

[12] P. J. Hilton and S. Wylie, *Homology Theory*, Cambridge Univ. Press, 1960.

[13] M. Hochster, Rings of invariants of tori, Cohen-Macaulay rings generated by monomials, and polytopes, *Annals of Math.* 96 (1972), 318-337.

[14] M. Hochster, Cohen-Macaulay rings, combinatorics, and simplicial complexes, in *Ring Theory II (Proc. Second Oklahoma Conference)*, B. R. McDonald and R. Morris, eds., Marcel Dekker, New York and Basel, 1977, 171-223.

[15] I. Kaplansky, *Commutative Rings*, rev. ed., Univ. of Chicago Press, 1974.

[16] V. Klee, A combinatorial analogue of Poincare's duality theorem, *Can. Jour. Math.* 16 (1964), 517-531.

[17] V. Klee, Polytope pairs and their relationship to linear programming, *Acta Math.* 133 (1974), 1-25.

[18] C. W. Lee, Bounding the numbers of faces of polytope pairs and simple polyhetra, *Ann. Discrete Math.* (to appear).

[18a] C. W. Lee, Two combinatorial properties of a class of simplicial polytopes, Research Report RC 8957, IBM, Yorktown Heights, New York, 1981.

[19] C. W. Lee, Characterizing the numbers of faces of a simplicial convex polytope, in *Convexity and Related Combinatorial Geometry*, D. C. Kay and M. Breen, eds., Marcel Dekker, New York and Basel, 1982.

[20] F. S. Macaulay, *The Algebraic Theory of Modular Systems*, Cambridge Tracts in Mathematics and Mathematical Physics, No. 19, Cambridge Univ. Press, London, 1916.

[21] F. S. Macaulay, Some properties of enumeration in the theory of modular systems, *Proc. London Math. Soc.* 26 (1927), 531-555.

[22] P. McMullen, The maximum numbers of faces of a convex polytope, *Mathematika* 17 (1970), 179-184.

[23] P. McMullen, The minimum number of faces of a convex polytope, *J. London Math. Soc. (2)* 3 (1971), 350-354.

[24] P. McMullen, The numbers of faces of simplicial polytopes, *Israel J. Math.* 9 (1971), 559-570.

[25] P. McMullen and G. C. Shephard, *Convex Polytopes and the Upper Bound Conjecture*, London Math. Soc. Lecture Notes Series, Vol. 3, 1971.

[26] P. McMullen and D. W. Walkup, A generalized lower-bound conjecture for simplicial polytopes, *Mathematika* 18 (1971), 264-273.

[27] T. S. Motzkin, Comonotone curves and polyhedra, Abstract 111, *Bull. Amer. Math. Soc.* 63 (1957), p. 35.

II. Polyhedral Theory and Commutative Algebra

[28] L. S. Pontryagin, *Foundations of Combinatorial Topology*, Graylock Press, Rochester, N. Y., 1952.

[29] G. Reisner, Cohen-Macaulay quotients of polynomial rings, *Advances in Math.* 21 (1976), 30-49.

[30] P. Schenzel, On the number of faces of simplicial complexes and the purity of Frobenius, *Math. Z.* 178 (1981), 125-142.

[31] W. Smoke, Dimension and multiplicity for graded algebras, *J. Algebra* 21 (1972), 149-173.

[32] D. M. Y. Sommerville, The relations connecting the angle-sums and volume of a polytope in space of n dimensions, *Proc. Roy. Soc. London,* ser. A 115 (1927), 103-119.

[33] R. Stanley, Linear homogeneous diophantine equations and magic labelings of graphs, *Duke Math. J.* 40 (1973), 607-632.

[34] R. Stanley, Combinatorial reciprocity theorems, *Advances in Math.* 14 (1974), 194-253.

[35] R. Stanley, Cohen-Macaulay rings and constructible polytopes, *Bull. Am. Math. Soc.* 81 (1975) 133-135.

[36] R. Stanley, The upper bound conjecture and Cohen-Macaulay rings, *Studies in Applied Math.* 54 (1975), 135-142.

[37] R. Stanley, Magic labelings of graphs, symmetric magic squares, systems of parameters, and Cohen-Macaulay rings, *Duke Math. J.* 43 (1976), 511-531.

[38] R. Stanley, Cohen-Macaulay complexes, in *Higher Combinatorics,* M. Aiger, ed., Reidel, Dordrecht and Boston, 1977, 51-62.

[39] R. Stanley, Hilbert functions of graded algebras, *Advances in Math.* 28 (1978), 57-83.

[40] R. Stanley, Combinatorics and invariant theory, in *Relations between Combinatorics and Other Parts of Mathematics,* D. K. Ray-Chaudhuri, ed., Proceedings of Symposia in Pure Mathematics, Vol. XXXIV, American Mathematical Society, Providence, 1979, 345-355.

[41] R. Stanley, Decompositions of rational convex polytopes, *Annals of Discrete Math.* 6 (1980), 333-342.

[42] R. Stanley, The number of faces of a simplicial convex polytope, *Advances in Math.* 35 (1980), 236-238.

[43] R. Stanley, Interactions between commutative algebra and combinatorics, Report No. 4, Mathematiska Institutionen, Stockholms Universitet, Sweden, 1982.

[44] R. Stanley, Linear diophantine equations and local cohomology, *Inv. Math.* 68 (1982), 175-193.

[45] R. Stanley, An introduction to combinatorial algebra, *Proceedings of the Silver Jubilee Conference on Combinatorics,* University of Waterloo (to appear).

[46] B. Tessier, Varietes toriques et polytopes, *Séminaire Bourbaki, 1980/81,* no. 565, Lecture Notes in Mathematics, Vol. 901, Springer Verlag, Berlin, Heidelberg, 1981.

[47] D. W. Walkup, The lower bound conjecture for 3- and 4-manifolds, *Acta Math.* 125 (1970), 75-107.

[48] O. Zariski and P. Samuel, *Commutative Algebra,* Vol. II, van Nostrand, Princeton, N.J., 1960.

[49] K. Baclawski and A. M. Garsia, Combinatorial decompositions of a class of rings, *Advances in Math.* 39 (1981), 155-184.

[50] R. Stanley, Balanced Cohen-Macaulay complexes, *Trans. Amer. Math. Soc.* 249 (1979), 139-157.

Reminiscences About the Origins of Linear Programming

G. B. Dantzig[1]
Stanford University, Department of Operations Research, Stanford, CA 94305, USA

Abstract: The author recalls the early days of linear programming, the contributions of von Neumann, Leontief, Koopmans and others.

Linear Programming is viewed as a revolutionary development giving man the ability to state general objectives and to find, by means of the simplex method, optimal policy decisions for a broad class of practical decision problems of great complexity.

In the real world, planning tends to be ad hoc because of the many special-interest groups with their multiple objectives. Much work remains to develop a more disciplined infrastructure for decision making in which the full potential of mathematical programming models could be realized.

Since its conception in 1947 in connection with the planning activities of the military, linear programming has come into wide use. In academic circles mathematicians, economists, and those who go by the name of Operations Researchers or Management Scientists, have written hundreds of books on the subject and, of course, an *unaccountable* number of articles.

Interestingly enough, in spite of its wide applicability to everyday problems, linear programming was unknown prior to 1947. It is true that two or three individuals may have been aware of its potential – for example Fourier in 1823 and de la Vallee Poussin in 1911. But these were isolated cases. Their works were soon forgotten. L. Kantorovich in 1939 made an extensive proposal that was neglected by the U.S.S.R. It was only after the major developments in mathematical programming had taken place in the West that Kantorovich's paper became known around 1959. To give some idea of how meager the research effort was: T. Motzkin in his Ph.D. thesis lists only 42 papers before 1936 on linear inequality systems authored by such names as Stokes, Dines, McCoy, Farkas.

My own contributions to the field grew out of my World War II experience. I had become an expert on programming planning methods using desk calculators. In 1946 I was the Mathematical Advisor to the U.S. Air Force Comptroller. I had just formally completed my Ph.D. and was looking for an academic position. In order to entice me into *not* taking another job, my colleagues D.

---

[1] This research was paritally supported by the Department of Energy Contract DE-AM03-76SF00326, PA No. DE-AT03-76ER72018; Office of Naval Research Contract N00014-75-C-0267; National Science Foundation Grants MCS-7681259, MCS-7926009 and ECS-8012974.

II. Reminiscences About the Origins of Linear Programming

Hitchcock and M. Wood challenged me to see what could be done to mechanize the planning process, i.e., to find a way to more rapidly compute a time-staged deployment, training and logistical supply *program*. In those days mechanization meant using analog devices or punch card equipment.

Consistent with my training as a mathematician, I set out to formulate a model. I was fascinated by the work of Wassily Leontief who proposed in 1932 a simple matrix structure which he called the *Interindustry Input-Output Model* of the American Economy. It was simple in concept and could be implemented in sufficient detail to be useful for practical planning. I soon saw that it had to be generalized. Leontief's was a steady-state model and what was needed was a highly *dynamic model*, one that could change over time. In Leontief's model there was a one-to-one correspondence between the production processes and the items produced by these processes. What was needed was a model with many *alternative activities*. The application was to be *large scale*, hundreds of items and activities. Finally it had to be *computable*. Once the model was formulated, there had to be a practical way to compute what quantities of these activities to engage in that was consistent with their respective input-output characteristics and with given resources. The model I formulated would be described today as a time-staged, dynamic linear program with a staircase matrix structure. Initially there was no objective function; explicit goals did not exist because practical planners simply had no way to implement such a concept.

A simple example illustrates the fundamental difficulty of formulating a planning problem using an activity analysis approach. Consider the problem of assigning 70 men to 70 jobs. An "activity" consists in assigning the i-th man to the j-th job. The restrictions are: (1) each man must be assigned, there are 70 such, and (2) each job must be filled, also 70. The level of an activity is either 1, meaning it *will* be used or 0, meaning it will *not*. Thus are 2×70 or 140 restrictions, 70×70 or 4900 activities with 4900 corresponding zero-one decision variables. *Unfortunately* there are 70! different possible solutions or ways to make the assignments. The problem is to compare one with another and select one which is "best" by some criterion.

Now 70! is a big number, greater than 10^{100}. Suppose we had an IBM 370-168 available at the time of the big bang 15 billion years ago. Would it have been able to look at all the 70! combinations by the year 1981? No! Suppose instead it could examine 1 billion assignments per second? The answer is still no. Even if the Earth were filled with such computers all working in parallel, the answer would still be no. If, however, there were 10^{50} earths or 10^{44} suns all filled with nano-second speed computers all programmed in parallel from the time of the big bang until sun grows cold, then perhaps the answer is *yes*.

This simple example illustrates why up to 1947, and for the most part to this day, a great gulf exists between man's aspirations and his actions. Man may wish to state his wants *in terms of an objective to be extremized* but there are so many different ways to go about it, each with its advantages and disadvantages, that it was impossible to compare them all and choose which among them is the best. Invariably man has turned to a leader whose "experience" and "mature judgment" would guide the way. Those in charge like to do this by simply issuing a series of ground rules or edicts to be executed by those developing the program. This was the situation in late 1946. I had formulated a model that satisfactorily represented the technological relations usually encountered in practice. In place of an explicit goal or objective function were a large number of ad-hoc ground rules issued by those in authority to aid the selection. Without the latter, there would be, in most cases, an astronomical number of feasible solutions to choose from.

All that I have related up to now in the early development took place before the advent of the computer, more precisely, before in late 1946 we were aware that it was going to exist.

To digress for a moment, I would like to say a few words about the electronic computer. To me, and I suppose to all of us, one of the most startling developments of all time is the penetration of the computer into almost every phase of human activity. Before a computer can be intelligently used, however, a model must be formulated and good algorithms developed. To build a model requires the *axiomatization* of a subject matter field. In time this gives rise to a whole new mathematical discipline which is then studied for its own sake. Thus, with each new penetration of the computer, a new science is born.

Von Neumann notes this tendency to axiomatize in his paper on *The General and Logical Theory of Automata*. In it he states that "automata have been playing a continuously increasing role in the natural sciences. Automata have begun to invade certain parts of mathematics too, particularly but not exclusively mathematical physics or applied mathematics. Their role in mathematics presents an interesting counterpart to certain functional aspects of organization in nature. For example the natural systems such as the central nervous system are of enormous complexity and it is clearly necessary (in order to understand them) first to subdivide what they represent into several parts which to a certain extent are independent, elementary units. The problem then consists of understanding how these elements are organized as a whole. It is the latter problem which is likely to attract those who have the background and tastes of the mathematician. With this attitude, he will be inclined to forget the origins and then, after the process of axiomatization is complete, concentrate on the mathematical aspects. (End of the von Neumann paraphrase.)

By mid-1947 I decided that the *objective* had to be made explicit. I formulated the planning problem as a set of axioms. The axioms concerned the relations between two kinds of sets: the first were the set of items being produced or consumed and the second the set of activities or production processes in which items could be inputed or outputed in fix proportions as long as the proportions were non-negative multiples of each other. The resulting mathematical system to be solved was the minimization of a linear from subject to linear equations and inequalities. The use of the linear from as the objective function to be extremized was the *novel* feature.

Now came the non-trivial question: Can one solve such systems? At first I assumed the Economists had worked on this problem. So I visited T. C. Koopmans in June 1947 at the Cowles Foundation at the University of Chicago to learn what I could from mathematical economists. Koopmans became quite excited. During World War II he worked for the Allied Shipping Board on a transportation model and so he had the theoretical as well as the practical planning background necessary to appreciate what I was presenting. He saw immediately the implications for general economic planning. From that time on, Koopmans took the lead in bringing the potentialities of linear programming models to the attention of young economists like K. Arrow, P. Samuelson, H. Simon, R. Dorfman, L. Hurwicz to name but a few. Their research led to several Nobel Prizes.

Seeing that economists did not have a method of solution, I next decided to try my own luck at finding an algorithm. I owe a great debt to Jerzy Neyman, the mathematical statistician, who guided my graduate work at Berkeley. My thesis was on two famous unsolved problems in mathematical statistics which I mistakenly thought was a homework assignment and solved. One of them, later published joint with A. Wald, was on the Neyman-Pearson Lemma. In today's terminology, my thesis was on the existence of Lagrange multipliers (or dual variables) for a general linear program over a continuum of variables each bounded between zero and one and satisfying linear constraints expressed in the form of Lebesgue integrals. There was also a linear objective to be extremized. The particular geometry used in my thesis was in the dimension of the columns instead of the rows. This column geometry gave me the insight that made me believe the Simplex Method would be a very efficient solution technique for solving linear programs. This I proposed in the summer of 1947 and by good luck it *worked!*

It was nearly a year later however in 1948, before we realized just how powerful the Simplex Method really was. In the meantime, I decided to consult with the "great" Johnny von Neumann to see what he could suggest in the way of solution techniques. He was considered by many as the leading mathematician in the world. On October 3, 1947, I visited him for the first time at the Institute for Advanced Study at Princeton. I remember trying to describe to him, as I would to an ordinary mortal, the Air Force problem. I began with the formulation of the linear programming model in terms of activities and items, etc. Von Neumann did something which I believe was uncharacteristic of him. "Get to the point," he said impatiently. Having at times a somewhat low kindling-point, I said to myself "O. K., if he wants a quicky, then that's what he'll get." In under one minute I slapped the geometric and the algebraic version of the problem on the blackboard. Von Neumann stood up and said "Oh that!" Then for the next hour and a half, he proceeded to give me a lecture on the mathematical theory of linear programs.

At one point seeing me sitting there with my eyes popping and my mouth open (after all I had searched the literature and found nothing), von Neumann said: "I don't want you to think I am pulling all this out of my sleeve on the spur of the moment like a magician. I have just recently completed a book with Oscar Morgenstern on the Theory of Games. What I am doing is conjecturing that the two problems are equivalent. The theory that I am outlining for your problem is an analogue to the one we have developed for games. Thus I learned about Farkas' Lemma, and about duality for the first time. Von Neumann promised to give my problem some thought and to contact me in a few weeks. He did write to me proposing an iterative scheme which Alan Hoffman and his group at the Bureau of Standards around 1952 compared with the Simplex Method and also with proposals of Motzkin. The Simplex Method came out a clear winner.

As a result of another visit to Princeton in June 1948, I met Al Tucker. Soon Tucker and his students H. Kuhn and D. Gale began their historic work on game theory, nonlinear programming and duality theory. The Princeton group became the focal point among mathematicians doing research in these fields. Twelve years later I remember a conversation with Professor Tucker, who had been reading the manuscript of my book *Linear Programming and Extensions.* Our conversation went like this: "Why", he asked, "do you ascribe duality to von Neumann and not to my group?" "Because he was the first to show it to me." He said, "that is strange for we have found nothing in writing about what von Neumann has done. What we have is his paper *On a Maximizing Problem.* "True," I said, "but let me send you a paper I wrote as a result of my first meeting with von Neumann." I sent him my report *A Theorem on Linear Inequalities,* dated 5 January 1948, which contained (as far as I know) the first rigorous proof of duality. Later Tucker asked me, "Why didn't you publish it?", to which I replied: "Because it was not my result – it was von Neumann's. All I did was write up, for internal circulation, my own proof of what von Neumann outlined. It was my way of educating the people in my office in the Pentagon." Today everyone cites von Neumann as the originator of the duality theorem and credits Tucker, Kuhn and Gale as the publishers of the first rigorous proof.

Not too long after my first meeting with Tucker there was a meeting of the Econometric Society in Wisconsin attended by well-known statisticians, mathematicians and economists like H. Hotelling, T. Koopmans, J. von Neumann, and many others all well known today who were then just starting their careers. I was a young unknown. I remember being quite frightened with the idea of presenting for the first time to such a distinguished audience the concept of linear programming.

After my talk, the chairman called for discussion. For a moment there was *silence;* then a hand raised. It was Hotelling's. I must hasten to explain that Hotelling was *huge.* He used to love to swim in the ocean and when he did, it is said that the level of the ocean rose perceptively. This huge whale of a man stood up in the back of the room. His expressive face took on one of those all-knowing smiles we all know so well. He said devastatingly: *"But we all know the world is non-linear".* Then he majestically sat down. And there I was, a virtual unknown, frantically trying to compose the proper reply to the great Hotelling.

Suddenly another hand in the audience was raised. It was von Neumann. "Mr. Chairman, Mr. Chairman," he said, "If the Speaker does not mind, I would like to reply for him." Naturally I agreed. Von Neumann said: "The speaker titled his talk 'Linear Programming' and he carefully stated his axioms. If you have an application that satisfies the axioms, use it. If it does not, then *don't,*" and he sat down. In the final analysis, of course, Hotelling was right. The world is highly non-linear. Fortunately systems of linear inequalities (as opposed to equalities) permits us to approximate most of the kinds of non-linear relations encountered in practical planning.

II. Reminiscences About the Origins of Linear Programming

In 1949, exactly two years from the time the linear programming was first started, the first conference on mathematical programming (sometimes referred to as the 0 Symposium on Mathematical Programming) was held at the University of Chicago. Koopmans, the organizer, later titled the proceedings of the conference "Activity Analysis of Production and Allocation." Economists like T. Koopmans, K. Arrow, P. Samuelson, L. Hurwitz, D. Dorfman, Georgescu-Roegen, and H. Simon, mathematicians like A. Tucker, H. Kuhn, and D. Gale and Air Force types like Marshall Wood, Murray Geisler, and myself all made contributions.

The advent or *rather the promise* of the electronic computer, the exposure of theoretical mathematicians and economists to real problems during the war, the interest in mechanizing the planning process, and the availability of money for such applied research, all converged during the period 1947–1949. The time was ripe. The research accomplished in these two short years, in my opinion, is one of the remarkable events of history. The Proceedings of the Conference remains to this very day an important basic reference, a classic!

While editing the proceedings, Koopmans asked me to do something to get rid of a condition I assumed to prove the Simplex Method. He wanted me to try to prove that the algorithm would converge without a nondegeneracy assumption, an assumption which I felt initially was reasonable. After all what was the probability of four planes in three space meeting in a point (for example)! But then something unexpected happened. It turned out that although the probability of a L.P. being degenerate was zero, every practical problem tested by my branch in the Air Force turned out to be so. Degeneracy couldn't happen but it did. It was the rule *not* the exception!

I proposed a method of *perturbation* of the RHS as a way of avoiding degeneracy when using the simplex method. The proofs I outlined and gave as homework exercises to classes that I was teaching at the time. Edmondston and others turned in proofs (March 1951). In the summer of 1951, Philip Wolfe, then a student at Berkeley, spent the Summer with me at the Pentagon and proposed a lexicographic interpretation of the perturbation idea which P. Wolfe, A. Orden and I published as a joint paper. A. Charnes independently developed a different perturbation scheme. Years later, Wolfe proposed a third way (based on my inductive proof of the simplex method) that is, in my opinion, the best one because it resolves degeneracy using only one extra column of information. Whether or not such a scheme is needed in practice has never been settled. It has been observed recently (1981) that even when there is no degeneracy, there is a high probability of *near* degeneracy. This suggests that pivot selection criteria should be designed to seek feasible solutions in directions away from degenerate and near degenerate basic feasible solutions. Doing so should reduce the total number of iterations.

The simplex method is also a powerful theoretical tool for proving theorems. To prove theorems it is essential that the algorithm include a way of avoiding degeneracy.

In the 1950's and 1960's many new subfields began to emerge. I have only time to say a few words about each.

Non-linear programming began around 1951 with the famous Kuhn-Tucker Conditions which are related to the Fritz-John Conditions (1948). Later Terry Rockafeller, P. Wolfe, R. Cottle, and others developed the theory of non-linear programming and extended the notions of duality.

Commercial Applications were begun in 1951 by Charnes and Cooper. Soon thereafter, practical applications began to dominate.

Network Flow Theory began to evolve around 1954. Ford, Fulkerson, and Hoffman showed the connections to graph theory. Recent research on combinatorial optimization is an outgrowth of their research.

Large-Scale Methods (my field) began in 1955 with my paper "Upper Bounds, Block Triangular Systems, and Secondary Constraints". In 1959–60 Wolfe and I published our papers on the Decomposition Principle.

Stochastic Programming began in 1955 with my paper "Linear Programming under Uncertainty", an approach which has been greatly extended by R. Wets in the 1960's. Important contributions to this field have been made by A. Charnes. Stochastic Programming is, in my opinion, one of the most promising fields for future research, one closely tied to large-scale methods.

Integer Programming began in 1958 by R. Gomory. Unlike the earlier work on the Travelling Salesman Problem by Fulkerson, Johnson and myself, Gomory showed how to systematically generate the cutting planes. Branch and bound techniques, which we also used in our paper, have been studied by E. Balas and many others. Branch and bound has turned out to be the most successful approach in practice for solving Integer Programs.

Complementary Pivot Theory was started around 1962–63 by Richard Cottle and myself and greatly extended by Cottle. It was an outgrowth of Wolfe's method for solving quadratic programs. In 1964 Lemke and Howson applied the algorithm to bimatrix games. In 1965 C. Lemke extended the approach to other non-convex programs. In the 1970's, H. Scarf, H. Kuhn, and B. Eaves extended it to solving fixed point problems. Lemke's results represent a historic breakthrough into the non-convex domain.

Polynomial Time Algorithms. In 1978 L. G. Khachian showed that an ellipsoidal type algorithm could solve all linear programs in polynomial time. An important theoretical breakthrough, but so far not one that can be used to solve practical problems. Klee and Minty have shown that one variant of the simplex method is not polynomial. This leaves open the related question why does the simplex method solve the wide class of practical linear programs encountered in approximately linear time?

In the late 1960's and 1970's the various subfields of the mathematical programming that I just outlined have each, in turn, grown exponentially. It is impossible for me in this short presentation to sketch these developments.

Before closing let me tell some stories about how various linear programming terms arose. The military refer to their various plans or proposed schedules of training, logical supply and deployment of combat units as a *program*. When I had first analyzed the Air Force planning problem and saw that it could be formulated as a system of linear inequalities, I called my first paper: *Programming in a Linear Structure*. In the summer of 1948, Koopmans and I visited the RAND Corporation. One day we took a walk near the Santa Monica beach. Koopmans said: "Why not shorten Programming in a Linear Structure to Linear Programming?" I replied: "That's it! From now on that will be its name." Later that day I gave a talk at RAND with that title. The term Mathematical Programming is due to Robert Dorfman who felt as early as 1949 that the term linear programming was too restrictive. The term Simplex Method arose out of a discussion with T. Motzkin who felt that the approach that I was using in the geometry of the columns was best described as a movement from one simplex to a neighboring one. Mathematical Programming is also responsible for many terms which are now standard in mathematical literature. Terms like Arg Min, Arg Max, Lexico-Max, Lexico-Min. The term Dual is not new. But surprisingly the term primal, introduced around 1954, is. It came about this way: W. Orchard-Hays, who is responsible for the first commercial grade L.P. software, said to me at RAND one day around 1954: "We need a word that stands for 'the original problem of which this is the dual'." I, in turn, asked my father, Tobias Dantzig, mathematician and author, well-known for his books popularizing the history of mathematics. He knew his Greek and Latin. Whenever I tried to bring up the subject of linear programming, Toby (as he was affectionately known), would become bored, but on this occasion he did give the matter some thought and suggested Primal as the natural antonym since both primal and dual derive from the Latin. It was Toby's one and only contribution to linear programming; his sole contribution unless, of course, you want to count the training he gave me in classical mathematics or his part in my conception.

If I were asked to summarize my early and perhaps my most important contributions to linear programming, I would say they are three:
(1) Recognizing (as a result of five war-time years as a practical program planner) that most practical planning relations could be reformulated as a system of linear inequalities.
(2) Expressing criteria for selection of good or best plans in terms of explicit goals (e.g., linear objective forms) and not in terms of ground rules which are at best only a means for carrying out the objective not the objective itself.
(3) Inventing the simplex method which transformed a rather simple, possibly interesting approach to economic theory into a basic tool for practical planning of large complex systems.

The tremendous power of the simplex method is difficult to realize. To solve by brute force the Assignment Problem which I mentioned earlier would require a solar system full of nano-second electronic computers running from the time of the big bang until the time the Universe grows cold to scan all the permutations in order to be certain to find the one which is best. Yet it takes only a second to find the Optimum using an IBM 370-168 and standard simplex-method software.

In retrospect it is interesting to note that the original problem that started my research is still outstanding – namely the problem of planning or scheduling dynamically over time. Many proposals have been made on ways to solve large-scale systems of this type such as the Nested Decomposition Principle.

Today this is an active, exciting and difficult field having important long-term planning applications that could contribute to the well-being and stability of the world.

Prior to linear programming it was not meaningful to explicitly state general goals and so objectives were confused with the ground rules for solution. Ask a military commander what the goal is and he will say "The goal is to win the war." Upon being pressed to be more explicit, a Navy man will say "The way to win the war is to build battleships," or, if he is an Air Force general, he will say "The way to win is to build a great fleet of bombers." Thus the means becomes the objective and this in turn spawn new ground rules as to how to go about building bombers or space shuttles that again become confused with the goals, etc., down the line.

The ability to state general objectives and then find optimal policy solutions to practical decision problems of great complexity is a revolutionary development.

In closing, therefore, I would like to reminisce about the future. Modern economies are complex. They are subject to dangerous dynamic forces. Growing population, decreasing resource base, growing gap between rich and poor countries, and political trouble-making on an international scale.

Despite these dynamic forces and the complexity of world economics, policy makers continue to make decisions without the benefit of the powerful analytical tools that are available to them. That their decisions are bad, is self evident.

For this reason, I have become more and more concerned that models are not becoming part of the infra-structure of the decision-making process. I have asked myself why. In the real world, there are many special-interest groups who influence the decision process both positively and negatively. Indeed their input is essential because of their special responsibilies and knowledge. There are multi objectives and these are often vaguely stated. In brief, decision making is being done in too ad hoc a way for mathematical programming models to be relevant. This suggests that ways must be found to make the planning process more disciplined in order that these groups can better evaluate the alternatives and reach a common understanding. In such a framework, I believe, the full potential of mathematical programming could be brought to bear on the critical issues facing the nation and the world.

Penalty Functions

R. Fletcher
University of Dundee, Mathematics Department, Dundee DD1 4HN, Scotland, United Kingdom

Abstract. This state of the art review starts with a discussion of the classical (Courant) penalty function and of the various theoretical results which can be proved. The function is used sequentially and numerical results are disappointing; the reasons for this are explained, and cannot be alleviated by extrapolation. These difficulties are apparently overcome by using the multiplier (augmented Lagrangian) penalty function, and the theoretical background and practical possibilities are described. However the current approach for forcing convergence has its disadvantages and there is scope for more research.

Most interest currently centres on *exact* penalty functions, and in particular the l_1 exact penalty function. This is a nonsmooth function so cannot be minimized adequately by current techniques for smooth functions. However it is very useful as a criterion function in association with other techniques such as sequential QP. A thorough description of the l_1 penalty function is given, which aims to clarify the first and second order conditions associated with the minimizing point. There are some disadvantages of using nonsmooth penalty functions including the existence of curved grooves which can be difficult to follow, and the possibility of the Maratos effect occurring. Suggestions for alleviating these difficulties are discussed.

These effects have recently caused more emphasis in the search for suitable smooth exact penalty functions, and a survey of research in this area, and the theoretical and practical possibilities, is given.

Finally some of the many other ideas for penalty functions are discussed briefly.

1. Introduction

When solving an optimization problem in which the constraints cannot easily be eliminated, it is necessary to balance the aims of reducing the objective function and staying inside the feasible region. This invariably leads to the idea of a penalty function which is some combination of the objective and constraint functions which enables the former to be minimized whilst controlling constraint violations by penalizing them. Penalty functions are most frequently used in the context of nonlinear programming, either as a means of reducing the problem to a more simple unconstrained problem, or to provide a criterion function for inducing global convergence in other methods. To simplify the presentation, two prototype situations are considered: either the equality constraint problem

$$\text{minimize } f(x), \quad x \in \mathbb{R}^n$$
$$\text{subject to } c(x) = 0 \quad (1.1)$$

where $c(x)$ is $\mathbb{R}^n \to \mathbb{R}^m$, $m \leq n$, or the inequality constraint problem

$$\text{minimize } f(x), \quad x \in \mathbb{R}^n$$
$$\text{subject to } c_i(x) \geq 0 \quad i = 1, 2, \ldots, m. \quad (1.2)$$

Usually the generalization to solve mixed problems is straightforward. In each case a local solution x^* is sought: indeed all the references to 'solution' or 'minimizer' in the paper pertain to local results. It is assumed that the constraint functions are nonlinear: if some constraints exist which are linear (especially simple bounds on x) then special treatment for these constraints is desirable. Essentially it is more efficient to use active linear constraints (especially bounds) to eliminate variables and so reduce the problem size, rather than to include the linear constraints in the penalty term.

It is assumed that the reader is familiar with standard optimality conditions for nonlinear programming. There are many good references but I shall use the notation and style of Fletcher (1981a). It is assumed that the functions $f(x)$ and $c_i(x)$ are smooth (at least $\mathbb{C}^1$ and usually $\mathbb{C}^2$) and the notation $\nabla f (=g)$ and $\nabla^2 f$ is used for the gradient and Hessian matrices of f. Likewise the notation $\nabla c_i (=a_i)$ is used for the normal vector of a constraint function $c_i(x)$. Often the column vectors a_i are collected into a Jacobian matrix A, either for all the constraints or just the active constraints, according to context. Active constraints are denoted by the index set

$$\mathscr{A} = \mathscr{A}(x) = \{i : c_i(x) = 0\} \quad (1.3)$$

where x is a feasible point. Methods for nonlinear problems are usually iterative and the notation $x^{(k)} k = 1, 2, \ldots$ is used for the sequence of iterates, where $x^{(1)}$ is user supplied. Quantities derived from $x^{(k)}$ are denoted interchangeably by $c(x^{(k)}) = c^{(k)}$ etc. and the same convention is used for quantities derived from x^*, x' etc. To state necessary optimality conditions requires regularity assumptions which I shall invariably assume are satisfied: often the sufficient assumption that the vectors a_i^* $i \in \mathscr{A}^*$ are linearly independent is made. It is convenient to introduce the Lagrangian function

$$\mathscr{L}(x, \lambda) = f(x) - \lambda^T c(x). \quad (1.4)$$

First order necessary conditions or *Kuhn Tucker (KT) conditions* are that x^* is a feasible point ($c^* = 0$ for (1.1) or $c^* \geq 0$ for (1.2)), and there exist Lagrange multipliers λ^* such that

$$\nabla \mathscr{L}(x^*, \lambda^*) = g^* - A^* \lambda^* = 0. \quad (1.5)$$

In addition for (1.2), $\lambda^* \geq 0$ must hold and $\lambda^{*T} c^* = 0$ (the complementarily condition: essentially inactive constraints have zero multipliers). Second order necessary conditions are that the Lagrangian function has nonnegative curvature for all feasible directions of zero slope: this implies that

$$s^T W^* s \geq 0 \quad \text{for all } s : a_i^{*T} s = 0, \quad i \in \mathscr{A}^* \quad (1.6)$$

II. Penalty Functions

where $W^* = \nabla_x^2 \mathscr{L}(x^*, \lambda^*)$. Sufficient conditions for x^* to be an isolated local solution of (1.2) are that x^* is a KT point (i.e. x^* and some λ^* satisfy first order conditions), strict complementarity holds ($c_i = 0 \Rightarrow \lambda_i > 0$), and

$$s^T W^* s > 0 \quad \text{for all } s: s \neq 0, a_i^{*T} s = 0, \quad i \in \mathscr{A}^*. \tag{1.7}$$

Tighter conditions can be determined but I wish to give an insight in what follows into the effect that second order conditions play in ensuring that penalty functions have local minima. This simple form of the conditions enables this to be done briefly, yet loses very little generality. I shall also make the assumption that the vectors $a_i^* i \in \mathscr{A}^*$ are linearly independent and refer to these sufficient conditions collectively as 'standard conditions'.

A wide variety of penalty functions has been suggested, often with great ingenuity. There are two main types: *seqential,* in which x^* is located as the limit point of a sequence of unconstrained penalty function minimizations, and *exact,* in which x^* is the minimizer of a single penalty function. Exact penalty functions are potentially the most efficient, since they avoid the sequential aspect, and they are currently attracting the most research interest. They can be used in two ways; either to provide a transformation of the nonlinear programming problem to an unconstrained minimization problem, or as a criterion function for use with other direct methods for nonlinear programming. Early sequential penalty functions, described in section 2, have serious numerical disadvantages but they are included for completeness and because the theory underpins many subsequent developments. These numerical disadvantages are largely avoided in the augmented Lagrangian penalty functions described in section 3. Being sequential, these are currently unfashionable, but are still useful in providing an easily used approach to nonlinear programming. Turning to exact penalty functions, the most important is arguably the nonsmooth l_1 exact penalty function. Since optimality conditions for such a function are not as well known as for smooth functions, a quite detailed treatment is given in section 4, together with a discussion of various ways in which the function is used. This includes applications not only to nonlinear programming but also to linear and quadratic programming. The nonsmooth nature of the l_1 penalty function does have some attendant disadvantages, but these can largely be overcome and currently I regard this as the most promising penalty function approach to nonlinear programming.

In view of these disadvantages there has been a resurgence of interest recently into smooth exact penalty functions. Whilst I do not fully share this enthusiasm, I have tried in section 5 to review recent developments and assess their merits and demerits. Finally in section 6 a brief review of some of the many other ideas for penalty functions is given.

As well as comparing and contrasting these penalty functions in what follows, I have also tried to give some insight into how different optimality conditions come into play. One of the difficulties is the choice of the penalty parameters which arise, and it is not easy to choose these without an understanding of the effect they play in creating minima in the penalty functions. For instance it is important to appreciate that in the l_1 penalty function the choice of parameter (v) is determined by the Lagrange multipliers λ^*, whereas in smooth penalty functions it is determined by the curvature of the Lagrangian matrix W^*. An important question is whether such parameters should be fixed by the user or automatically adjusted by software. There is currently no agreement on this and for reasons of space I refer the reader to the discussion in Powell (1982) chapter 4, and that in Powell's paper in this volume. Another consideration of a similar nature is that it might be advantageous to scale either the variables or the constraint functions in the original problem; for example it is desirable that the contribution of each of the individual terms in forming say $c^T c$ or $\|c\|$ is roughly comparable. Again there is the question of whether this should be user supplied or automatically adjusted; see for example the discussion in Powell (1982), chapter 6.

2. Early penalty functions

Early penalty functions were generally sequential in nature and for the equality problem (1.1) the simplest is the function

$$\phi(x, \sigma) = f(x) + \tfrac{1}{2}\sigma c(x)^T c(x), \qquad (2.1)$$

usually attributed to Courant (1943). The penalty is a sum of squares of constraint terms, weighted by a parameter σ. However the solution x^* of (1.1) is not in general a minimizer of (2.1) for any finite σ but only in the limit $\sigma \to \infty$. Thus the technique of solving a sequence of minimization problems is suggested. This is traditionally implemented as follows

(i) choose a fixed sequence $\{\sigma^{(k)}\} \to \infty$, typically $\{1, 10, 10^2, 10^3 \ldots\}$
(ii) for each $\sigma^{(k)}$ find a local minimizer, $x(\sigma^{(k)})$ say, to

$$\min_x \phi(x, \sigma^{(k)}) \qquad (2.2)$$

(iii) terminate when $c(x(\sigma^{(k)}))$ is sufficiently small.

It is important to emphasize that in practice step (ii) in (2.2) is done *numerically*, that is by the application of an unconstrained minimization method. The choice of this method will depend on whether or not derivatives are available and on the size of the problem. Often $x(\sigma^{(k)})$ is used as the initial approximation in minimizing $\phi(x, \sigma^{(k+1)})$, and other information such as inverse Hessian approximations can also be passed forward from one iteration to the next. In fact algorithm (2.2) is idealized in that step (ii) cannot be solved exactly in a finite number of operations, and it is assumed that $x(\sigma^{(k)})$ is obtained as accurately as possible.

A variety of results relating to the convergence of this sequential penalty function can be given. In doing this, quantities derived from $\sigma^{(k)}$, such as $x(\sigma^{(k)})$, $f(x(\sigma^{(k)}))$ etc. are denoted by $x^{(k)}$, $f^{(k)}$ etc. Mostly just an outline of these results is given and the details are filled in, for example, by Fletcher (1981a). If one is prepared to assume that global solutions are computed in (2.2), step (ii), and $f(x)$ is bounded below on the feasible region, then a direct result can be obtained that if $\sigma^{(k)} \to \infty$ then $c^{(k)} \to 0$ and any accumulation point

II. Penalty Functions

of the sequence $\{x^{(k)}\}$ solves (1.1). This result does not require differentiability or the existence of a KT point for (1.1). In practice only local minima may be computed and it is possible to prove somewhat weaker results under this assumption, and also to get asymptotic estimates of the rate of convergence. In doing this the vector

$$\lambda^{(k)} = -\sigma^{(k)} c^{(k)} \quad (2.3)$$

is defined, and can be regarded as a Lagrange multiplier estimate by virtue of (2.5). Again let $\sigma^{(k)} \to \infty$, let x^* be any accumulation point of the sequence $\{x^{(k)}\}$, and assume that rank $A^* = m$. The minimality of $x^{(k)}$ implies that

$$\nabla \phi(x^{(k)}, \sigma^{(k)}) = g^{(k)} + \sigma^{(k)} A^{(k)} c^{(k)} = 0$$

and hence

$$g^{(k)} = A^{(k)} \lambda^{(k)} \quad (2.4)$$

Since rank $A^* = m$ it follows that

$$\lambda^{(k)} = A^{(k)+} g^{(k)} = \lambda^* + o(1) \quad (2.5)$$

where $\lambda^* = A^{*+} g^*$ is the Lagrange multiplier vector for (1.1) at x^*. It then follows by continuity from (2.4) that $g^* = A^* \lambda^*$ and from (2.3) that $c^{(k)} = -\lambda^*/\sigma^{(k)} + o(1/\sigma^{(k)})$ so that $c^* = 0$. Thus x^* is a KT point for (1.1). It also follows using first order Taylor series for f and c about $x^{(k)}$ that

$$f^* = f^{(k)} - c^{(k)T} \lambda^{(k)} + o(x^{(k)} - x^*)$$

and hence that

$$f^* = \phi^* = \phi^{(k)} + \tfrac{1}{2} \sigma^{(k)} c^{(k)T} c^{(k)} + o(x^{(k)} - x^*) \quad (2.6)$$

which provides a useful way of estimating f^*. Finally it can be shown with a further assumption that

$$x^{(k)} - x^* = -T^* \lambda^*/\sigma^{(k)} + o(1/\sigma^{(k)}) \quad (2.7)$$

where T^* is some matrix, which indicates that the rate of convergence of algorithm (2.2) is linear and one extra decimal place is obtained in the solution on each iteration, which is in agreement with practical experience. The asymptotic expression (2.7) has also been used in an extrapolation scheme (SUMT, Fiacco and McCormick, 1968 in the context of barrier functions). Estimates of $x^{(k)} - x^*$ can be obtained which may be used in a termination criterion, and as a means to provide better initial estimates when minimizing $\phi(x, \sigma^{(k)})$.

This well developed theoretical background may make it appear that, apart from the inefficiency of sequential minimization, the method is a robust one which can be used with confidence. In fact this is not true at all and there are severe numerical difficulties which arise when the method is used in practice. These are caused by the fact that as $\sigma^{(k)} \to \infty$, it is increasingly difficult to solve the minimization problem (2.2) step (ii). In particular the location of the solution in the space of directions orthogonal to the vectors a_i^* is difficult. Thus although there is no difficulty in making c^* small, there is considerable difficulty in satisfying the first order condition $g^* = A^* \lambda^*$. This difficulty is associated with the fact that (for $0 < m < n$) the Hessian matrix $\nabla^2 \phi^{(k)}$ becomes increasing ill-conditioned as $\sigma^{(k)} \to \infty$. Since

$$\nabla^2 \phi^{(k)} = \nabla^2 \mathscr{L}^{(k)} + \sigma^{(k)} A^{(k)} A^{(k)T}$$

and since the matrix AA^T has rank m, it is easy to see that m eigenvalues of $\nabla^2 \phi^{(k)}$ approach ∞ as $\sigma^{(k)} \to \infty$, whereas the remainder are bounded. Thus the condition number of $\nabla^2 \phi^{(k)}$ approaches ∞. In practice this shows up in that large values of $\nabla \phi$ arise whilst the minimization method is unable to make progress in reducing ϕ.

These remarks have implications for the choice of the sequence $\{\sigma^{(k)}\}$. Choosing a very large $\sigma^{(1)}$ gives a minimization problem which is difficult to solve accurately. Alternatively choosing a small $\sigma^{(1)}$ and increasing $\sigma^{(k)}$ gradually, makes $x^{(k)}$ a good initial approximation for minimizing $\phi(x, \sigma^{(k+1)})$ and hence enables a more accurate solution to be obtained, but is very inefficient. The typical sequence in (2.2) step (ii) is a trade off between these extremes. In practice a value of approximately $\varepsilon^{-1/2}$ is an upper limit to $\sigma^{(k)}$ were ε is the relative precision of the computation. At this stage it is necessary to observe the errors in the KT conditions $\underline{c}^* = \underline{0}$ and especially $g^* = A^* \lambda^*$ to assess how accurately $\underline{x}^*$ is approximated. Also (2.6) can be used to estimate the best value of f^*. In fact many users do not carry out the sequential technique at all but carry out a single minimization with a moderately large value of σ. The above advice about assessing the size of the errors is also appropriate to this situation.

As mentioned above, algorithm (2.2) and the consequent theory are idealized in that exact minimization of the penalty function (2.1) is assumed. In fact it is straightforward to give similar results when approximate minimization is allowed. Let the test for terminating step (ii) in (2.2) with an approximate minimizer $\underline{x}^{(k)}$ be

$$\|\nabla \phi(\underline{x}^{(k)}, \sigma^{(k)})\| \leq v \|\underline{c}^{(k)}\| \tag{2.8}$$

for some fixed $v > 0$. Following the argument from (2.3) onwards, the assumption that rank $A^* = m$ ensures that there exists a constant $\alpha > 0$ such that $\|A^{(k)} \underline{c}^{(k)}\| \geq \alpha \|\underline{c}^{(k)}\|$. Consequently by definition of $\nabla \phi$

$$v \|\underline{c}^{(k)}\| \geq \|g^{(k)} + \sigma^{(k)} A^{(k)} \underline{c}^{(k)}\| \geq \sigma^{(k)} \alpha \|\underline{c}^{(k)}\| - \|g^{(k)}\| \tag{2.9}$$

and hence as $\sigma^{(k)} \to \infty$ it follows that $\underline{c}^{(k)} \to \underline{0}$. Then (2.8) shows that

$$g^{(k)} = A^{(k)} \lambda^{(k)} + o(1)$$

and the results about convergence to a KT point follow as above. It is interesting to relate (2.8) to the practical difficulties caused by ill-conditioning in that it excludes the possibilities that large gradients can be accepted in a minimizer, and can be seen as an attempt to force the KT condition $g^{(k)} = A^{(k)} \lambda^{(k)}$ to hold. Unfortunately when $\sigma^{(k)}$ becomes large the minimization method is not able to find a point which satisfies (2.8). Nonetheless (2.8) is potentially useful in indicating that no further progress can be made in this situation, as well as reducing somewhat the effort required to find $\underline{x}^{(k)}$ for small values of $\sigma^{(k)}$.

A penalty function for the inequality constraint problem (1.2) can also be given by following the idea that the penalty in (2.1) is a sum of squares of con-

II. Penalty Functions

straint violations. Thus the penalty function

$$\phi(x, \sigma) = f(x) + \tfrac{1}{2} \sigma \sum_{i=1}^{m} [\min(c_i, 0)]^2 \qquad (2.10)$$

is suggested. There is no difficulty in extending the above results to this case (or to a penalty function which arises from a mixture of equations and inequalities). The result which assumes that global solutions are computed is easily obtained by reducing the inequality $c_i(x) \geq 0$ to the equation $\min(c_i(x), 0) = 0$ and using the previous result, since differentiability is not required. The result which assumes local solutions are computed is replicated by defining

$$\lambda_i^{(k)} = -\sigma^{(k)} \min(c_i^{(k)}, 0) \quad i = 1, 2, \ldots, m \qquad (2.11)$$

in place of (2.3), and assuming that the vectors a_i^* $i \in \mathscr{A}^*$ are linearly independent (full rank) at x^*. The numerical difficulties carry over to the inequality case, and in fact are made worse because the penalty function (2.10) has a jump discontinuity in second derivative whenever $c = 0$, for example at x^*. This jump discontinuity tends to ∞ as $\sigma^{(k)} \to \infty$. Thus additional difficulties are put in the way of a Newton like method which attempts to estimate the second derivative matrix $\nabla^2 \phi$.

Another class of sequential minimization techniques is available to solve the inequality constraint problem (1.2), known as barrier function methods. These are characterized by their property of preserving strict constraint feasibility at all times by augmenting $f(x)$ with a barrier term which is infinite on the constraint boundaries. The two most important examples are the inverse barrier function

$$\phi(x, r) = f(x) + r \sum_i [c_i(x)]^{-1} \qquad (2.12)$$

(Carroll, 1961) and the logarithmic barrier function

$$\phi(x, r) = f(x) - r \sum_i \log(c_i(x)) \qquad (2.13)$$

(Frisch, 1955). As with σ in algorithm (2.2), the parameter r is used to control the barrier function iteration. In this case a sequence $\{r^{(k)}\} \to 0$ is chosen which successively reduces the effect of the barrier term, except close to the boundary. If $x(r^{(k)})$ is defined as the minimizer of $\phi(x, r^{(k)})$ then $x(r^{(k)}) \to x^*$ as $r^{(k)} \to 0$ and this result can be established in both a global sense and a local sense, by following similar arguments as for the Courant penalty function (2.1). Likewise estimates of Lagrange multipliers and extrapolation techniques can also be developed.

Unfortunately barrier functions suffer from the same numerical difficulties in the limit as for the penalty function algorithm, in particular the badly determined nature of $x(r^{(k)})$ due to ill-conditioning in the Hessian matrix $\nabla^2 \phi^{(k)}$ and large gradients $\nabla \phi$. Moreover there are additional problems which arise. The barrier function is undefined for infeasible points and the simple expedient of setting it to infinity can make the line search inefficient. Also the singularity makes the conventional quadratic or cubic interpolations in the line search work less efficiently. Thus a special purpose line search is required and the aim of a simple to use algorithm is lost. Another difficulty is that an initial interior feasible point is required and this itself is a non trivial problem involving the strict solution of a set of nonlinear inequalities. There are also difficulties in extending the method to handle mixed problems which include equations. In view of these difficulties the sequential application of barrier functions attracts little interest. However hybrid methods using a barrier function to maintain feasibility have been investigated with more success (Osborne and Ryan, 1972, Murray and Wright, 1978).

3. Augmented Lagrangian Penalty Functions

The way in which the penalty function (2.1) is used to solve (1.1) can be envisaged as an attempt to create a local minimizer at x^* in the limit $\sigma^{(k)} \to \infty$. However x^* can be made to minimize ϕ for finite σ by changing the origin of the penalty term. This suggests using the function

$$\phi(x, \theta, \sigma) = f(x) + \tfrac{1}{2} \Sigma \sigma_i (c_i(x) - \theta_i)^2$$
$$= f(x) + \tfrac{1}{2} (\underline{c}(x) - \underline{\theta})^T S (\underline{c}(x) - \underline{\theta}) \quad (3.1)$$

(Powell, 1969), where $\underline{\theta}, \underline{\sigma} \in \mathbb{R}^m$ and $S = \operatorname{diag} \sigma_i$. The parameters θ_i correspond to shifts of origin and the $\sigma_i \geq 0$ control the size of the penalty, like σ in (2.1). For a given problem there is usually a particular value of $\underline{\theta}$, which depends on $\underline{\sigma}$, such that x^* minimizes $\phi(x, \underline{\theta}, \underline{\sigma})$. This suggests an algorithm which attempts to locate the optimum value of $\underline{\theta}$, whilst keeping $\underline{\sigma}$ finite and so *avoids the ill conditioning in the limit* $\underline{\sigma} \to \infty$.

In fact it is more convenient to define

$$\lambda_i = \theta_i \sigma_i \quad i = 1, 2, \ldots, m \quad (3.2)$$

and ignore the term $\tfrac{1}{2} \Sigma \sigma_i \theta_i^2$ (independent of x) giving

$$\phi(x, \underline{\lambda}, \underline{\sigma}) = f(x) - \underline{\lambda}^T \underline{c}(x) + \tfrac{1}{2} \underline{c}(x)^T S \underline{c}(x) \quad (3.3)$$

(Hestenes, 1969). There exists a corresponding optimum value of $\underline{\lambda}$ for which x^* minimizes $\phi(x, \underline{\lambda}, \underline{\sigma})$, which in fact is the Lagrange multiplier $\underline{\lambda}^*$ at the solution to (1.1). This result is now true independent of $\underline{\sigma}$, so it is usually convenient to ignore the dependence on $\underline{\sigma}$ and write $\phi(x, \underline{\lambda})$ in (3.3), and to use $\underline{\lambda}$ as the control parameter in a sequential minimization algorithm as follows
 (i) determine a sequence $\{\underline{\lambda}^{(k)}\} \to \underline{\lambda}^*$
 (ii) for each $\underline{\lambda}^{(k)}$ find a local minimizer, $x(\underline{\lambda}^{(k)})$ say, to $\min_x \phi(x, \underline{\lambda})$
 (iii) terminate when $\underline{c}(x(\underline{\lambda}^{(k)}))$ is sufficiently small.
The main difference between this algorithm and (2.2) is that $\underline{\lambda}^*$ is not known in advance so the sequence in step (i) cannot be predetermined. However it is shown below how such a sequence can be constructed. Because (3.3) is obtained from (2.1) by adding a multiplier term $-\underline{\lambda}^T \underline{c}$, (3.3) is often referred to as a *multiplier penalty function*. Alternatively (3.3) is a Lagrangian function in

which the objective f is augmented by the term $\frac{1}{2} c^T S c$. Hence the term *augmented Lagrangian function* is also used to describe (3.3).

The result that λ^* is the optimum choice of the control parameter vector in (3.3) can be derived by showing that x^* minimizes $\phi(x, \lambda^*)$ if the elements of σ are sufficiently large. Differentiating (3.3) gives

$$\nabla \phi(x^*, \lambda) = g^* - A^*\lambda^* + A^* S c^* = 0$$

if x^*, λ^* is a KT point. Differentiating again gives

$$\nabla^2 \phi(x^*, \lambda^*) = W^* + A^* S A^{*T}$$

If standard conditions hold, W^* is strictly positive definite on the null space of A^* (see (1.7)); also S can be chosen sufficiently large so that the term $A^* S A^{*T}$ dominates the contribution from W^* on the range space of A^*. Hence $\nabla^2 \phi(x^*, \lambda^*)$ is positive definite and so the result is proved. A less informal proof of this result is given by Fletcher (1981a), extending also to the case that A^* is rank deficient. The above proof has important practical implications in that it indicates exactly a threshold value for the penalty parameters σ_i (that is S). In the proof, the need for x^*, λ^* to satisfy strict second order conditions is not easily relaxed. For example if $f = x_1^4 + x_1 x_2$ and $c = x_2$ then $x^* = 0$ solves (1.1) with unique multiplier $\lambda^* = 0$. However second order conditions are not satisfied, and in fact $x^* = 0$ does not minimize $\phi(x, 0, \sigma) = x_1^4 + x_1 x_2 + \frac{1}{2} \sigma x_2^2$ for any value of σ.

To make further progress, quantities derived from a fixed value of λ in (3.3) are examined. The minimizer $x(\lambda)$ can be regarded as having been determined by solving the nonlinear equations (in x)

$$\nabla \phi(x, \lambda) = 0. \tag{3.5}$$

Because $\nabla^2 \phi$ is the Jacobian of this system and is positive definite at x^*, λ^* it follows from the implicit function theorem that $x(\lambda)$ is a $\mathbb{C}^1$ function defined in a neighbourhood of x^*, λ^*. It is also important to examine the function

$$\psi(\lambda) \triangleq \phi(x(\lambda), \lambda) \tag{3.6}$$

which is derived from λ by first finding $x(\lambda)$. By the local optimality of $x(\lambda)$ for x, λ in a neighbourhood of x^*, λ^*, it follows that

$$\psi(\lambda) = \phi(x(\lambda), \lambda) \leq \phi(x^*, \lambda) = \phi(x^*, \lambda^*) = \psi(\lambda^*) \tag{3.7}$$

(using $c^* = 0$). Thus λ^* *is a local unconstrained maximizer of* $\psi(\lambda)$. This result is also true globally if $x(\lambda)$ is a global maximizer of $\phi(x, \lambda)$. Thus methods for generating a sequence $\lambda^{(k)} \to \lambda^*$ can be derived by applying unconstrained minimization methods to $-\psi(\lambda)$.

It follows from (3.6) and (3.5) that

$$\nabla_\lambda \psi(\lambda) = -c(x(\lambda)) \tag{3.8}$$

$$\nabla_\lambda^2 \psi(\lambda) = -A^T W_\sigma A|_{x(\lambda)} \tag{3.9}$$

where $W_\sigma = \nabla_x^2 \phi(x(\lambda), \lambda)$ (see Fletcher (1981a) for example). Thus the most obvious minimization method to use is Newton's method. Given an initial estimate $\lambda^{(1)}$ the iteration formula is

$$\lambda^{(k+1)} = \lambda^{(k)} - (A^T W_\sigma^{-1} A)^{-1} \mathcal{L}|_{\underline{x}(\underline{\lambda}^{(k)})}. \quad (3.10)$$

This method requires W_σ, and hence explicit formulae for second derivatives, which is disadvantageous. However when only first derivatives are available, and a quasi-Newton method is used to find $\underline{x}(\underline{\lambda}^{(k)})$, then the resulting inverse Hessian approximation can be used to estimate W_σ^{-1}. Using this matrix in (3.10) (Fletcher, 1975) enables the advantages of Newton's method to be obtained whilst only requiring first derivatives. A different method is suggested by Powell (1969) and by Hestenes (1969), and is best motivated by the fact that for large σ,

$$(A^T W_\sigma^{-1} A)^{-1} \approx S. \quad (3.11)$$

When this approximation is used in (3.10) the iteration

$$\underline{\lambda}^{(k+1)} = \underline{\lambda}^{(k)} - S\underline{c}^{(k)} \quad (3.12)$$

is obtained. No derivatives are required by this formula so it is particularly convenient when the routine which minimizes $\phi(\underline{x}, \underline{\lambda})$ does not calculate or estimate derivatives. Furthermore by making S sufficiently large, arbitrarily good relative agreement in (3.11) can be obtained, and hence an arbitrarily fast rate of linear convergence of $\underline{\lambda}^{(k)}$ to $\underline{\lambda}^*$.

Although these methods have good local convergence properties, they must be supplemented in a general algorithm to ensure global convergence. This can be done by an algorithm due to Powell (1969).
  (i) Initially set $\underline{\lambda} = \underline{\lambda}^{(1)}$, $\underline{\sigma} = \underline{\sigma}^{(1)}$, $k = 0$, $\|\underline{c}^{(0)}\|_\infty = \infty$
  (ii) find the minimizer $\underline{x}(\underline{\lambda}, \underline{\sigma})$ of $\phi(\underline{x}, \underline{\lambda}, \underline{\sigma})$ and denote $\underline{c} = \underline{c}(\underline{x}(\underline{\lambda}, \underline{\sigma}))$.
  (iii) If $\|\underline{c}\|_\infty > \frac{1}{4}\|\underline{c}^{(k)}\|_\infty$ set $\sigma_i = 10\sigma_i \, \forall i : |c_i| > \frac{1}{4}\|c^{(k)}\|_\infty$ and goto (ii)   (3.13)
  (iv) Set $k = k+1$, $\underline{\lambda}^{(k)} = \underline{\lambda}$, $\underline{\sigma}^{(k)} = \underline{\sigma}$, $\underline{c}^{(k)} = \underline{c}$
  (v) Set $\underline{\lambda} = \underline{\lambda}^{(k)} - S^{(k)}\underline{c}^{(k)}$ and goto (ii)

The aim of the algorithm is to achieve linear convergence at a rate $\frac{1}{4}$ or better. If any component c_i is not reduced at this rate, the corresponding penalty σ_i is increased tenfold which induces more rapid convergence. A simple proof that $\underline{c}^{(k)} \to \underline{0}$ can be given along the following lines. Clearly this happens unless the inner iteration ((iii)→(ii)) fails to terminate. In this iteration $\underline{\lambda}$ is fixed, and if for any i, $|c_i| > \frac{1}{4}\|\underline{c}^{(k)}\|_\infty$ occurs infinitely often, then $\sigma_i \to \infty$. As in section 2 this implies that $c_i \to 0$ which contradicts the infinitely often assumption, and proves termination. This convergence result is true whatever formula is used in step (v) of (3.13). It follows as in section 2 that any limit point is a KT point of (1.1) with $\underline{x}^{(k)} \to \underline{x}^*$ and $\underline{\lambda}^{(k)} - S^{(k)}\underline{c}^{(k)} \to \underline{\lambda}^*$. For formulae (3.10) and (3.12), the required rate of convergence is obtained when $\underline{\sigma}$ is sufficiently large, and then the basic iteration takes over in which $\underline{\sigma}$ stays constant and only the $\underline{\lambda}$ paramteres are changed.

In practice this proof is not as powerful as it might seem. Unfortunately increasing $\underline{\sigma}$ can lead to difficulties caused by illconditioning as described in section 2, in which case accuracy in the solution is lost. Furthermore when no feasible point exists, this situation is not detected, and $\underline{\sigma}$ is increased without

II. Penalty Functions

bound which is an unsatisfactory situation. I think there is scope for other algorithms to be determined which keep σ fixed at all times and induce convergence by ensuring that $\psi(\lambda)$ is increased sufficiently at each iteration, for example by using a restricted step modification of Newton's method.

There is no difficulty in modifying these penalty functions to handle inequality constraints. For the inequality constraint problem (1.2) a suitable modification of (3.1) (Fletcher, 1975) is

$$\phi(x, \theta, \sigma) = f(x) + \tfrac{1}{2} \sum_i \sigma_i (c_i(x) - \theta_i)_-^2 \qquad (3.14)$$

where $a_- = \min(a, 0)$. Although $\phi(x, \theta, \sigma)$ has second derivative jump discontinuities at $c_i(x) = \theta_i$, these are finite and are usually remote from the solution and in practice do not appear to affect the performance of the unconstrained minimization routine adversely. Although smoother augmented Lagrangians can be constructed (Mangasarian, 1973) I do not think that there is any evidence that they are preferable. As with (3.1) it is possible to rearrange the function, using (3.2) and omitting terms independent of x, to give the multiplier penalty function

$$\phi(x, \lambda, \sigma) = f(x) + \sum_i \begin{cases} -\lambda_i c_i + \tfrac{1}{2} \sigma_i c_i^2 & \text{if } c_i \leq \lambda_i/\sigma_i \\ -\tfrac{1}{2} \lambda_i^2/\sigma_i & \text{if } c_i \geq \lambda_i/\sigma_i \end{cases} \qquad (3.15)$$

where $c_i = c_i(x)$. Rockafellar (1974) first suggested this type of function, and as with (3.3), it is the most convenient form for developing the theory.

Most of the theoretical results can be extended to the inequality case. The result that x^* minimizes $\phi(x, \lambda^*, \sigma)$ for sufficiently large σ follows directly from the equality result if strict complementarity holds, although it can also be proved in the absence of this condition (Fletcher, 1975). The dual function $\psi(\lambda)$ is again defined by (3.6) and an analogous global result to (3.7) is

$$\psi(\lambda) = \phi(x(\lambda), \lambda) \leq \phi(x^*, \lambda)$$

$$= f^* + \sum_i \begin{cases} -c_i^* \lambda_i + \tfrac{1}{2} \sigma_i c_i^{*2} & c_i^* \leq \lambda_i/\sigma_i \\ -\tfrac{1}{2} \lambda_i^2/\sigma_i & c_i^* \geq \lambda_i/\sigma_i \end{cases}$$

$$\leq f^* + \sum_i \begin{cases} -\tfrac{1}{2} \sigma_i c_i^{*2} \\ -\tfrac{1}{2} \lambda_i^2/\sigma_i \end{cases}$$

$$\leq f^* = \phi(x^*, \lambda^*) = \psi(\lambda^*) \qquad (3.16)$$

This result is also true locally if strict complementarity holds and can probably be extended in absence of this condition, again by following Fletcher (1975). Derivative expressions analogous to (3.8) and (3.9) are easily obtained as

$$d\psi(\lambda)/d\lambda_i = -\min(c_i, \theta_i) \quad i = 1, 2, \ldots, m \qquad (3.17)$$

where $c_i = c_i(x(\lambda))$ and $\theta_i = \lambda_i/\sigma_i$, and

$$\nabla^2 \psi(\lambda) = \begin{bmatrix} -A^T W_\sigma^{-1} A & 0 \\ 0 & -S^{-1} \end{bmatrix}_{x(\lambda)} \qquad (3.18)$$

where the columns of A correspond to indices $i: c_i < \theta_i$ and those of S^{-1} to indices $i: c_i \geq \theta_i$. Algorithms for determining the sequence $\{\lambda^{(k)}\}$ in step (i) of algorithm (3.4) can be determined using these derivative expressions. An equivalent form of Newton's method (3.10) is possible, although in view of the implicit inequalities $\lambda \geq 0$ it is probably preferable to choose $\lambda^{(k+1)}$ to solve a subproblem

$$\begin{array}{c} \text{maximize } q^{(k)}(\lambda) \\ \lambda \\ \text{subject to } \lambda \geq 0 \end{array} \qquad (3.19)$$

where $q^{(k)}(\lambda)$ is obtained by truncating a Taylor series for $\psi(\lambda)$ about $\lambda^{(k)}$ after the quadratic term. In fact the simple structure of (3.19) enables a more simple problem to be solved in terms of just the indices $i: c_i < \theta_i$. The result in (3.11) can also be used to determine an extension of (3.12), that is

$$\lambda_i^{(k+1)} = \lambda_i^{(k)} - \min(\sigma_i c_i^{(k)}, \lambda_i^{(k)}) \quad i = 1, 2, \ldots, m \qquad (3.20)$$

These formulae can be incorporated in a globally convergent algorithm like (3.13) by using $\|\nabla \psi\|_\infty$ in place of $\|c\|_\infty$ to monitor the rate of convergence.

A selection of numerical experiments is given by Fletcher (1975), which seems to indicate that whilst both the Newton-like formulae or the Powell-Hestenes formulae for updating $\lambda^{(k)}$ are effective, the Newton-like method is somewhat more efficient. Local convergence is rapid and high accuracy can usually be achieved in about 4-6 minimizations. When this occurs with modest values of σ then no difficulties due to ill-conditioning and loss of accuracy are observed. Furthermore the Hessian matrix can be carried forward, and updated if σ_i is increased, so the computational effort for the successive minimizations goes down rapidly. Since $\phi(x, \lambda, \sigma)$ is always well defined, there is no difficulty in coping with infeasible points, and it is easy to program the method using an existing quasi-Newton subroutine. The main disadvantage is that the sequential nature of method is less efficient than the more direct approach of the next section, and this is supported by the more recent numerical results of Schittkowski (1980). Also the global convergence result based on increasing σ, whilst powerful in theory, does not always work well in practice, and there are practical applications in which it has caused ill-conditioning and low accuracy.

Two final points are worthy of note. Firstly if the problem is a mixture of linear and nonlinear constraints, then it may be worth incorporating only the nonlinear constraints into the penalty function, and the linear constraints can be included when $\phi(x, \lambda, \sigma)$ is minimized, for example at step (ii) in algorithm (3.4). This is especially true for bounds on the variables, since minimization with bounds is not a significant complication on an unconstrained minimization routine. Another point is that approximate minimization of the penalty function (see (2.8) for example) can also be considered, and a review of recent work in this area is given by Coope and Fletcher (1980). In this case the algorithms start to become more like the direct methods and Coope and Fletcher give an algorithm which incorporates the SQP correction defined in (4.28).

II. Penalty Functions

4. Nonsmooth exact penalty functions

A general class of nonsmooth exact penalty functions can be constructed using functions like

$$\phi(x) = vf(x) + \|\underline{c}(x)\| \qquad (4.1)$$

for (1.1) or

$$\phi(x) = vf(x) + \|\underline{c}(x)_-\| \qquad (4.2)$$

for (1.2), where v is some weighting parameter and the norm in (4.2) is monotonic but otherwise arbitrary. This theory is set out in Fletcher (1981), although it is rather more general than is required for most practical applications.

This chapter is concerned with the choice of the l_1 norm in constructing these functions, since it is used in the large majority of applications, and a more simple format than the above will be used. Optimality conditions for nonsmooth functions are currently not as well known as for smooth functions and the first part of the chapter studies optimality conditions for the function

$$\phi(x) = f(x) + \sum_{i \in E} |c_i(x)| + \sum_{i \in I} c_i(x)_+ \qquad (4.3)$$

where $a_+ = \max(a, 0)$, which contains l_1 terms. E and I are finite disjoint index sets which can be associated with the equations and inequalities in a nonlinear programming problem. Next the application of (4.3) to a nonlinear programming problem is considered, and in particular the extent to which a minimizer of (4.3) corresponds to a solution of the latter. Finally algorithms for minimizing (4.3) are studied; in particular the nonsmooth nature of (4.3) causes some difficulties, and of course conventional techniques for smooth problems are inadequate. Also the use of (4.3) in linear and quadratic programming is discussed.

Optimality conditions for nonsmooth functions involve the investigation of directional derivatives. Define the notation

$$D_s \phi(x') = \lim_{\theta \downarrow 0} \frac{\phi(x(\theta)) - \phi(x')}{\theta} \qquad (4.4)$$

where $x(\theta) = x' + \theta \underline{s}$ is a ray from x' in the direction $\underline{s}$, or more generally may be an arc or a directional sequence. This provides a basis for deriving optimality conditions for (4.3): first a brief sketch of a general approach is given using convex analysis. An alternative, more direct, approach is then described which gives more insight into the role played by the different elements.

A general approach to the determination of optimality conditions is to regard (4.3) as a special case of the function

$$\phi(x) = f(x) + h(\underline{c}(x)) \qquad (4.5)$$

where $f(x) (\mathbb{R}^n \to \mathbb{R})$ and $\underline{c}(x) (\mathbb{R}^n \to \mathbb{R}^m)$ are smooth but otherwise arbitrary and $h(\underline{c}) (\mathbb{R}^m \to \mathbb{R})$ is nonsmooth but convex. For the convex function $h(\underline{c})$ the subdifferential $\partial h(\underline{c})$ is the set of gradients of all supporting linear approximations at $\mathfrak{c}$ to be the graph of $h(\mathfrak{c})$. It is not difficult to see that the directional derivative along $\mathfrak{s}$ is determined by the supporting linear approximation that most closely approximates $h(\mathfrak{c})$ along $\mathfrak{s}$, that is

$$D_s h(\mathfrak{c}) = \max_{\lambda \in \partial h} \mathfrak{s}^T \lambda \tag{4.6}$$

For the composite term $h(\mathfrak{c}(x))$, a direction $\mathfrak{s}$ in x-space gives rise to a direction $A^T \mathfrak{s}$ in $\mathfrak{c}$-space (chain rule), so $D_s h(\mathfrak{c}(x')) = \max_{\lambda \in \partial h'} \mathfrak{s}^T A \lambda$. Thus the directional derivative of (4.5) can be written

$$D_s \phi(x') = \max_{\gamma \in \partial \phi'} \gamma^T \mathfrak{s} \tag{4.7}$$

where

$$\partial \phi(x') = \{\gamma : \gamma = g' + A'\lambda, \lambda \in \partial h'\}. \tag{4.8}$$

If x^* is a local minimizer of $\phi(x)$ then

$$D_s \phi(x^*) \geq 0 \quad \forall \mathfrak{s}. \tag{4.9}$$

Using a separating hyperplane argument it can be shown that (4.9) is equivalent to

$$\mathfrak{0} \in \partial \phi(x^*). \tag{4.10}$$

which is therefore a necessary condition for a local minimizer. Another way to state (4.10) is to introduce the Lagrangian function

$$\mathcal{L}(x, \lambda) = f(x) + \lambda^T \mathfrak{c}(x). \tag{4.11}$$

Thus a *first order necessary condition* is that if x^* minimizes $\phi(x)$ in (4.5) then there exists a vector $\lambda^* \in \partial h^*$ such that

$$\nabla \mathcal{L}(x^*, \lambda^*) = g^* + A^* \lambda^* = \mathfrak{0}. \tag{4.12}$$

In the case that $h(\mathfrak{c}) = \sum_{i \in E} |c_i| + \sum_{i \in I} (c_i)_+$ it is easily seen from the graph of $|\mathfrak{c}|$ or $\mathfrak{c}_+$ that

$$\partial h(\mathfrak{c}^*) = \{\lambda : \begin{matrix} -1 \leq \lambda_i \leq 1 \\ c_i^* \neq 0 \Rightarrow \lambda_i = \operatorname{sign} c_i^* \end{matrix} \bigg\rangle i \in E \tag{4.13}$$

$$\begin{matrix} 0 \leq \lambda_i \leq 1 \\ c_i^* > 0 \Rightarrow \lambda_i = 1 \\ c_i^* < 0 \Rightarrow \lambda_i = 0 \end{matrix} \bigg\rangle i \in I \tag{4.14}$$

which completes the description of the first order conditions for (4.3). It is possible to continue the analysis to derive second order conditions (also see below) and a complete description is given by Fletcher (1981a).

An alternative more direct proof of these conditions can also be given; to shorten the presentation it is assumed that the set E is empty. Define the set $Z = Z(x) = \{i : c_i(x) = 0\}$. The proof requires that the vectors g_i^* $i \in Z^*$ are linearly independent: in this respect it is weaker than the above proof, but it does have some useful constructive aspects. From (4.3) it follows that

$$D_s \phi(x) = \mathfrak{s}^T g + \sum_{i \notin Z} \sigma_i \mathfrak{s}^T g_i + \sum_{i \in Z} (\mathfrak{s}^T g_i)_+ \tag{4.15}$$

II. Penalty Functions

where $\sigma_i = 1$ if $c_i > 0$ or $\sigma_i = 0$ if $c_i < 0$. The proposition is that (4.9) is true iff there exists $\lambda^* \in (4.14)$ such that (4.12) holds. The 'if' result is straightforward since substitution of (4.12) and (4.14) into (4.15) yields

$$D_s \phi(x^*) = \sum_{i \in Z^*} ((s^T q_i^*)_+ - s^T q_i^* \lambda_i^*) \geq 0. \tag{4.16}$$

For the converse result, first observe that the existence of λ^* in (4.14) such that (4.12) holds is equivalent to

$$\exists \lambda^* \text{ such that } g^* + \sum_{i \notin Z^*} \sigma_i^* q_i^* + \sum_{i \in Z^*} \lambda_i^* q_i^* = 0 \tag{4.17}$$

$$0 \leq \lambda_i^* \leq 1 \quad i \in Z^*. \tag{4.18}$$

This is assumed to be false, and a direction is constructed for which $D_s \phi^* < 0$, thus providing a contradiction. One possibility is that (4.16) is false, in which case there exists vectors λ^* and $\mu \neq 0$ such that both $\bar{g}^* + A^* \lambda^* = \mu$ and $A^{*T} \mu = 0$, where $\bar{g}$ denotes $g + \sum_{i \notin Z} \sigma_i q_i$ and A collects the columns q_i, $i \in Z$. Then $s = -\mu$ gives rise to

$$D_s \phi^* = s^T \bar{g}^* + \sum_{i \in Z^*} (s^T q_i^*)_+ = -\mu^T \mu < 0$$

and hence s is the required descent direction. Alternatively (4.17) may be true but (4.18) false for some index $p \in Z^*$. Consider the vector $s = A^{*+T} e_p$ where $A^+ = (A^T A)^{-1} A^T$ exists by the independence assumption. It follows that $s^T q_p^* = 1$ and $s^T q_i^* = 0$, $i \in Z^*$, $i \neq p$. Then as in (4.16)

$$D_s \phi^* = \sum_{i \in Z^*} ((s^T q_i^*)_+ - s^T q_i^* \lambda_i^*) = 1 - \lambda_p^*.$$

Thus if (4.18) is false because $\lambda_p^* > 1$, this construction has given a descent direction. Finally if (4.18) is false because $\lambda_p^* < 0$ then the vector $s = -A^{*+T} e_p$ gives $s^T q_p^* = -1$ and hence $D_s \phi^* = \lambda_p^* < 0$, and again s is a descent direction.

It is valuable to summarize the content of this result when x^* is not a minimizer as follows. If $\bar{g}^*$ is not a linear combination of the vectors q_i^* $i \in Z^*$ then a descent direction exists which is orthogonal to these vectors. If $\bar{g}^* + A^* \lambda^* = 0$, but $0 \leq \lambda_p^* \leq 1$ is false for some $p \in Z^*$ then $\phi(x)$ can be reduced by *moving away from the surface* $c_p(x) = 0$. If $\lambda_p^* > 1$ then a direction which increases $c_p(x)$ is chosen, or if $\lambda_p^* < 0$ then a direction which decreases $c_p(x)$. It is straightforward to extend the above analysis to indices in E; in fact the result in (4.13) follows directly from the above since $|c_i| = (c_i)_+ + (-c_i)_+$.

The same approach can be used to obtain insight into second order conditions. Let the first order conditions (4.12)–(4.14) hold at x^* and in addition let $0 < \lambda_i^* < 1$ $i \in Z^*$ (essentially strict complementarity). Then the above analysis shows that directions for which $s^T q_i^* \neq 0$ for any $i \in Z^*$ are strict ascent directions $(D_s \phi^* > 0)$ for which second order effects do not operate. However for any s such that $s^T q_i^* = 0$, $i \in Z^*$, then $D_s \phi^* = 0$ (consider s and $-s$) and second order effects are important. For any such s, an arc $x(\theta)$ $\theta \in [0, \bar{\theta})$ can be constructed such that $x(0) = x^*$, $\dot{x}(0) = s$ and $c_i(x(\theta)) = 0$, $i \in Z^*$, (using the independence assumption). It follows from (4.11) that

$$\phi(x(\theta)) = \mathcal{L}(x(\theta), \lambda^*) = \phi^* + \tfrac{1}{2} \delta^T W^* \delta + o(\delta^T \delta) \tag{4.19}$$

by using a Taylor expansion, where $\delta = x(\theta) - x^*$ and $W^* = \nabla^2 \mathscr{L}(x^*, \lambda^*)$. By minimality of ϕ^* and taking the limit $\theta \downarrow 0$ it follows that

$$s^T W^* s \geq 0 \quad \forall s: s^T a_i^* = 0, \quad i \in Z^* \tag{4.20}$$

so that this is a second order necessary condition for x^* to minimize $\phi(x)$. It can be shown that second order sufficient conditions are those obtained if strict inequality holds in (4.20) (in addition to (4.12)–(4.14) and strict complementarity). This result can be extended to exclude the assumption of strict complementarity.

Another useful set of first order conditions arises when (4.3) is minimized subject to simple bounds

$$l \leq x \leq u. \tag{4.21}$$

and is readily obtained by extending the above argument. Assuming $l < u$ without loss of generality, a single multiplier π_i^* can be assigned to each pair of inequalities in (4.21), and the necessary condition is that there exist multipliers $\lambda^* \in \partial h(c^*)$ and π^* such that

$$g^* + A^* \lambda^* - \pi^* = 0, \tag{4.22}$$

$$\begin{matrix} \pi_i^* \geq 0 \text{ if } x_i^* = l_i \\ \pi_i^* \leq 0 \text{ if } x_i^* = u_i \\ \pi_i^* = 0 \text{ otherwise.} \end{matrix} \tag{4.23}$$

Second order conditions analogous to (4.20) are also readily obtained.

It is now possible to investigate the use of (4.3) as an exact penalty function. Consider a nonlinear programming problem

$$\begin{aligned} \text{minimize} \quad & f(x) & & x \in \mathbb{R}^n \\ \text{subject to} \quad & c_i(x) = 0 & & i \in E \\ & c_i(x) \leq 0 & & i \in I. \end{aligned} \tag{4.24}$$

Note that the direction of the inequality is opposite to that in (1.2) for convenience. Let x^*, λ^* be a KT point for (4.24). Consider whether x^*, λ^* also satisfies first order conditions for (4.3). It is easily seen that (4.12) is satisfied, but unless $\|\lambda^*\|_\infty \leq 1$, the condition $\lambda^* \in \partial h(c^*)$ (4.12, 4.13) is not satisfied. To avoid this possibility the objective function can be scaled as $f \to vf$ before use. Then the Lagrange multipliers for (4.24) are scaled as $\lambda^* \to v\lambda^*$, and clearly if $v \leq 1/\|\lambda^*\|_\infty$ is chosen then $v\lambda^* \in \partial h(c^*)$. Hence a KT point for the scaled problem also satisfies first order conditions for (4.3). The converse is true if in addition x^* is feasible in (4.24). Turning to second order conditions, if x^* is feasible then the index sets Z^* and $\mathscr{A}^*$ are identical and it can be seen by comparing (1.6) and (4.20) that the second order conditions are the same if $\lambda_i^* < 1$ is assumed. This result is not true in absence of this condition (e.g. $\min x - \frac{1}{2} x^2$ subject to $0 \leq x \leq 1$, with $v = 1$ and $x^* = 0$) and so the stronger assumption $v < 1/\|\lambda^*\|_\infty$ is made. Of course scaling $f \to vf$ does not essentially change (4.24) but does change (4.3) to give

$$\phi(x) = vf(x) + \sum_{i \in E} |c_i(x)| + \sum_{i \in I} c_i(x)_+ \tag{4.25}$$

II. Penalty Functions

and it is this function that is used as the l_1 exact penalty function. Hence it follows that *if $v < 1/\|\lambda^*\|_\infty$ and x^* is feasible in (4.24) then first order and second order sufficient conditions for (4.24) and (4.25) are equivalent*, and if these conditions hold then x^* solves (4.24) iff x^* minimizes (4.25).

Attention is now given to the problem of minimizing the l_1 exact penalty function (4.25). Although it is a nonsmooth function, general methods (e.g. bundle methods) for nonsmooth problems are inefficient since they do not accound for the special structure of (4.25). Also methods which do not attempt to measure second order effects are disregarded since in general they do not converge superlinearly. One possibility is to use a method for minimizing (4.24) as a means of generating search directions for use with (4.25). Most important in this respect is the widely used sequential quadratic programming (SQP) method originated by Wilson (1963). Briefly, at a current iterate $x^{(k)}, \lambda^{(k)}$ the constraint functions in (4.24) are approximated by a first order Taylor series

$$c_i(x^{(k)}+\delta) \simeq l_i^{(k)}(\delta) = c_i^{(k)} + a_i^{(k)T}\delta, \tag{4.26}$$

and the objective function in (4.24) is approximated by a quadratic function

$$f(x^{(k)}+\delta) \simeq q^{(k)}(\delta) = f^{(k)} + g^{(k)T}\delta + \tfrac{1}{2}\delta^T W^{(k)}\delta. \tag{4.27}$$

where $W^{(k)} = \nabla_x^2 \mathcal{L}(x^{(k)}, \lambda^{(k)})$. This is a Taylor series together with terms involving $\nabla^2 c_i$ in W which account for constraint curvature. The functions f and c_i in (4.24) are replaced by $q^{(k)}$ and $l_i^{(k)}$ to give the quadratic programming subproblem

$$\begin{aligned}\underset{\delta}{\text{minimize}} \quad & q^{(k)}(\delta) \\ \text{subject to} \quad & l_i^{(k)}(\delta) = 0 \quad i \in E \\ & l_i^{(k)}(\delta) \leq 0 \quad i \in I\end{aligned} \tag{4.28}$$

which is solved to determine a correction $\delta^{(k)}$. Then $x^{(k+1)} = x^{(k)} + \delta^{(k)}$ and $\lambda^{(k+1)}$ is taken as the Lagrange multiplier vector for (4.28). This algorithm can be regarded as a realization of Newton's method applied to solve the nonlinear equations $\nabla_x \mathcal{L}(x, \lambda) = 0$ and $c_i(x) = 0$ for the active constants in (4.28). A useful variation of this method is suggested by Han (1977) in which $W^{(k)}$ is approximated by a positive definite matrix $B^{(k)}$ by using some variable metric updating formula which requires only first derivatives (VM-SQP say). In particular Powell (1978) suggests a suitable variation of the BFGS formula. In an attempt to show that this method is consistent with the penalty function (4.25), Han shows that if $v < 1/\|\lambda^{(k+1)}\|_\infty$ then the search direction resulting from (4.28) (with $W^{(k)} = B^{(k)}$) is a descent direction. By making various assumptions about the problem he also shows that if v is sufficiently small, then this result enables a global convergence proof to be established. In practice (e.g. Schittkowski, 1980), good numerical results are often obtained, especially in terms of the numbers of function and gradient calls required. However the solution of a QP subproblem on each iteration is an expensive step in comparison with the augmented Lagrangian methods of section 3.

To some extent however Han's convergence theorem obscures the fact that this approach does have some disadvantages. For instance the subproblem

(4.28) may be infeasible or (if $W^{(k)}$ is used) unbounded. Also the algorithm can fail by converging to a limit point at which vectors q_i $i \in \mathscr{A}$ are dependent, which is not a minimizer or even a stationary point (Fletcher, 1981b). In these situations there is no a-priori bound on $\lambda^{(k+1)}$ arising from (4.28), so the descent property may be violated and $B^{(k)}$ may be unbounded. These difficulties are all overcome (except possibly the last one) in an algorithm due to Fletcher (1981a). This algorithm is very simply explained as the replacement of f and c_i by $q^{(k)}$ and $l_i^{(k)}$, not into (4.24) giving SQP, but into the penalty function (4.25) itself. In addition a step restriction (trust region) strategy is used, so the subproblem which is solved on iteration k is

$$\text{minimize}_\delta \psi(\delta) \triangleq vq^{(k)}(\delta) + \sum_{i \in E} |l_i^{(k)}(\delta)| + \sum_{i \in I} (l_i^{(k)}(\delta))_+$$
$$\text{subject to } \|\delta\|_\infty \leq \rho^{(k)}. \qquad (4.29)$$

The solution of this problem $\delta^{(k)}$ and the Lagrange multiplier vector $\lambda^{(k+1)}$ for the l_1 terms then define the next iterate, as for SQP. The radius $\rho^{(k)}$ of the trust region is varied adaptively in a conventional way by comparing the actual and predicted reductions in ϕ. Close to a solution of (4.24), (4.28) and (4.29) yield identical results so the main feature of (4.29) is that it improves the global properties of SQP. Thus (4.29) shares the good practical performance of SQP but also solves a number of problems on which SQP fails (Fletcher, 1981b). A nice feature of the trust region in (4.29), which to some extent explains the good global properties, is that only a subset of 'locally active constraints' (which does not necessarily include all the equations) are zeroed by the solution of (4.29). Only the locally active constraint normals need be independent, which is automatically achieved by solving (4.29). On the other hand the solution of (4.28) attempts to maintain feasibility with respect to all the constraints. In dependent or nearly dependent situations remote from the solution this can be disadvantageous. It might appear that (4.29) is a substantially more complicated subproblem than (4.28). In fact this is not so, and (4.29) is very similar to a QP problem in structure and complexity, and can be solved by similar methods. It might be called an $l_1 QP$ problem. When software for $l_1 QP$ becomes more available, I think that the algorithm based on (4.29) ($Sl_1 QP$ say) will invariably be chosen in preference to SQP. As it stands $Sl_1 QP$ requires second derivatives but a similar method can be obtained in which $W^{(k)}$ is approximated by $B^{(k)}$ which is updated as above. Since there is an a-priori bound on $\lambda^{(k+1)}$ when solving (4.29) it seems much less likely that $B^{(k)}$ will become unbounded with $Sl_1 QP$, although this is currently an open question.

One adverse feature of an l_1 exact penalty function is the lack of smoothness caused by the l_1 terms, which gives rise to 'curved grooves' which are difficult for an algorithm to follow. Another manifestation of this is the 'Maratos effect' in which a unit step of the SQP method may *increase* $\phi(x)$ even though $x^{(k)}$ and $\lambda^{(k)}$ are arbitrarily close to x^*, λ^*. The reason for this is that although $x^{(k+1)} - x^* = o(x^{(k)} - x^*)$ and $\lambda^{(k+1)} - \lambda^* = o(\lambda^{(k)} - \lambda^*)$, it does not follow that $\Delta \phi^{(k)} / \Delta \psi^{(k)} = 1 + o(1)$ as would be the case for smooth optimization. (Here $\Delta \phi^{(k)} = \phi^{(k)} - \phi^{(k+1)}$ and $\Delta \psi^{(k)} = \psi(\delta^{(k)})$ are the actual and predicted reductions in ϕ.) In fact if first order terms are negligible at $x^{(k)}$ but not at $x^{(k+1)}$, as when

$x^{(k)}$ is close to an active constraint boundary, then $\Delta\phi^{(k)}/\Delta\psi^{(k)}<0$ can occur, which is the Maratos effect. A modification which avoids the effect is to allow a single additional 'second order correction' step after the basic step, similar to the projection step in a feasible direction algorithm (Coleman and Conn (1980), Fletcher 1982)). This only requires one additional evaluation of f and c and ensures that $\Delta\phi^{(k)}/\Delta\psi^{(k)} = 1+o(1)$. There is no difficulty in modifying the SQP or Sl_1QP algorithms and Fletcher (1982) shows that the latter is capable of substantially faster progress in following a curved groove. Similar ideas are also given in Mayne (1980) and in the 'watchdog technique' of Chamberlain et. al. (1982) which allows iterates that do not reduce $\phi(x)$ to be accepted by the SQP method in certain circumstances.

Finally the use of l_1 exact penalty functions in linear and quadratic programming is mentioned. In linear programming a cost function like (4.3) with $f=0$ and c linear is often used in finding feasible points, for example in the 'composite simplex method' of Wolfe (1965). This method incorporates the logic of simplex basis changes together with a line search to optimize (4.3) on each iteration. The idea can be generalized to solve the full problem by solving (4.3) with f as the linear cost function. As above, if $v<1/\|\lambda^*\|_\infty$ is chosen then the whole problem is solved in one pass, thus combining the more conventionally separate Phase I (feasibility) and Phase II (optimality) stages. The limit $v\downarrow 0$ is essentially the early 'big M' method ($M=1/v$). The same idea can be used in quadratic programming (l_1QP) and the basis of an algorithm is sketched out in Fletcher (1981a). There are some indications that these algorithms are more suitable for handling difficulties over near degeneracy, as well as being potentially more efficient in combining the separate Phase I and Phase II stages.

5. Smooth exact penalty functions

Many of the difficulties involved with nonsmooth exact penalty functions can be avoided if smooth exact penalty functions are used. However there are some attendant difficulties with the latter which have inhibited their widespread use. Nonetheless there has recently been renewed interest in this approach and a number of different approaches to the construction of a smooth exact penalty function have been suggested. My early ideas on the subject were applicable to the equality problem (1.1) and involved an augmented Lagrangian type of function

$$\phi(x) = f(x) - \lambda(x)^T c(x) + c(x)^T S c(x) \tag{5.1}$$

in which $\lambda(x)$ is determined by

$$\lambda(x) = A^+ g|_x \tag{5.2}$$

which is the solution of the overdetermined least squares problem $A\lambda = g$. In Fletcher (1970) two possibilities for S are considered; the obvious choice $S = \frac{1}{2}\sigma I$, and the choice $S = \sigma A^+ A^{+T}$. In both cases there exists a threshold value $\bar\sigma$ such that if $\sigma > \bar\sigma$, and standard conditions hold at a local solution x^* to (1.1), then x^* minimizes (5.1). In particular the choice $S = \sigma A^+ A^{+T}$ has some convenient properties in regard to the computation of $\nabla^2 \phi$, which is also an observation of Boggs and Tolle (1980). To see the mechanism by which a minimizer of (5.1) is created, it follows at x^* that $\lambda(x^*) = A^{*+} g^* = \lambda^*$ and hence

$$\nabla \phi^* = g^* - A^* \lambda^* + M^* c^* = 0$$

by the KT conditions, where M^* is $2A^* S^*$ plus terms involving $\nabla \lambda^T$ and ∇S. Furthermore, using $c^* = 0$ and defining $\Lambda = \nabla \lambda^T$,

$$\nabla^2 \phi^* = W^* - \Lambda^* A^{*T} - A^* \Lambda^{*T} + 2A^* S A^{*T}$$
$$= W^* - \Lambda^* A^{*T} - A^* \Lambda^{*T} + 2\sigma P^*$$

where $P = AA^+$ is a projection matrix on to range (A). It follows from standard conditions that W^* is positive definite on the null space of A^*, and σ can be chosen sufficiently large so that the term σP^* is dominant on the range space. Thus $\nabla \phi^* = 0$ and a threshold value for σ can be determined such that $\nabla^2 \phi^*$ is positive definite, which implies that x^* minimizes $\phi(x)$; the formal proof is very similar to that for augmented Lagrangians in section 3, but includes provision for the terms in Λ^*.

When attention is switched to the inequality problem (1.2) the choice $S = \sigma A^+ A^{+T}$ turns out to be even more significant. In the equality case it is possible to rearrange (5.1) to give

$$\phi(x) = f(x) - \pi(x)^T c(x) \tag{5.3}$$

where

$$\pi(x) = A^+ (g - \sigma A^{+T} c)|_x , \tag{5.4}$$

and $\pi(x)$ turns out to be the Lagrange multiplier vector of the subproblem

$$\begin{aligned} \underset{\delta}{\text{minimize}} \quad & \tfrac{1}{2}\sigma \delta^T \delta + g^T \delta \\ \text{subject to} \quad & A^T \delta + c = 0 \end{aligned} \tag{5.5}$$

where g, A and c are evaluated at x. The constraints in (5.5) are linearizations of the nonlinear constraints in (1.1) about x. This immediately suggests that for the inequality problem (1.2), (5.3) is again an appropriate penalty function (Fletcher, 1973), in which $\pi(x)$ is determined as the Lagrange multiplier vector of the subproblem

$$\begin{aligned} \underset{\delta}{\text{minimize}} \quad & \tfrac{1}{2}\sigma \delta^T \delta + g^T \delta \\ \text{subject to} \quad & A^T \delta + c \geq 0. \end{aligned} \tag{5.6}$$

A direct way of obtaining $\pi(x)$ is to solve the Wolfe dual of (5.6), which after some rearrangement becomes

$$\underset{\pi: \pi \geq 0}{\text{minimize}} \ \tfrac{1}{2}(A\pi - g)^T (A\pi - g) + \sigma \pi^T c \tag{5.7}$$

The merits and demerits of this approach are discussed in more detail later in this section.

More recent research has investigated the possibility of determining a smooth exact penalty function which is defined on the space of vectors

II. Penalty Functions

$(x, \lambda) \in \mathbb{R}^{n+m}$. The idea of Di Pillo and Grippo (1979) for the equality problem (1.1) is to augment the Lagrangian not only with a quadratic term in c, but also with a term which is quadratic in $\nabla_x \mathscr{L}(x, \lambda)$, giving the function

$$\phi(x, \lambda) = f(x) - c(x)^T \lambda + \tfrac{1}{2}\sigma \|c(x)\|_2^2 + \tfrac{1}{2}\rho \|M(x)(g(x) - A(x)\lambda)\|_2^2 \quad (5.8)$$

Various choices for the matrix M are proposed, such as $M = A^T$ or $M = A^+$ or $M = I$. For sufficiently large σ and $\rho (>0)$, a KT pair x^*, λ^* which solves (1.1) and satisfies standard conditions is also a local minimizer of (5.8). Bertsekas (1982) points out that if the problem of minimizing $\phi(x, \lambda)$ is decomposed as

$$\underset{x, \lambda}{\text{minimize}}\, \phi(x, \lambda) = \min_x \hat{\phi}(x) \triangleq \min_\lambda \phi(x, \lambda), \quad (5.9)$$

and if $M = A^+$, then the function $\hat{\phi}(x)$ can be interpreted in terms of Fletcher's function (5.1). This is true but somewhat misleading since it does not emerge in the desirable form in which $S = \sigma A^+ A^{+T}$. If $M = I$ is chosen, and (5.9) is used, then $\hat{\phi}(x)$ can be expressed as in (5.3) with

$$\bar{\lambda}(x) = A^+ \left(g + \frac{1}{\rho} A^{+T} c \right)\bigg|_x \quad (5.10)$$

in which there is a significant difference in sign (c.f. (5.4)). This seems to explain why no simple generalization of (5.8) (analogous to (5.7)) has been suggested for solving inequality problems. Nonetheless Di Pillo, Grippo and Lampariello (1981) do suggest a penalty function for inequalities which in its simplest form is

$$\phi(x, \lambda) = f(x) - \lambda^T u(x, \lambda) + \tfrac{1}{2}\sigma \|u(x, \lambda)\|_2^2 + \tfrac{1}{2}\rho \|v(x, \lambda)\|_2^2 \quad (5.11)$$

where

$$u(x, \lambda) = c(x) - w(x, \lambda) \quad (5.12)$$

$$v(x, \lambda) = A(x)^T \nabla_x \mathscr{L}(x, \lambda) + 4\Lambda w(x, \lambda) \quad (5.13)$$

$$w_i(x, \lambda) = \frac{(\sigma c_i(x) - \lambda_i - 4\rho \lambda_i a_i(x)^T \nabla_x \mathscr{L}(x, \lambda))_+}{\sigma + 16\rho \lambda_i^2} \quad (5.14)$$

($\Lambda = \text{diag}\, \lambda_i$, $a_+ = \max(a, 0)$). This has the same equivalence property as (5.8) in that a KT point which solves (1.2) is also minimizer of (5.11) (note that $\lambda \geqslant 0$ need not be imposed directly). However it is not obviously an immediate generalization of (5.8), since replacing the equations $c_i(x) = 0$ in (1.1) by two inequalities $c_i(x) \geqslant 0$ and $c_i(x) \leqslant 0$ and using (5.11)-(5.14) does not appear to reduce to (5.8). Also (5.11)-(5.14) is very complicated and contains non-quadratic terms in λ. The numerical results given by the authors refer to a still more complicated three parameter (σ, ρ, τ) function. Acceptable results are given for some parameter selections, but not all, and the difficulties in choosing parameters may be disadvantageous.

Another smooth penalty function in $\mathbb{R}^{n+m}$ is recently given by Han and Mangasarian (1981) and is both simple, and also applicable to the inequality problem (1.2). The function is constructed from the Wolfe dual of (1.2) by replacing the equation $\nabla_x \mathscr{L}(x, \lambda) = 0$ by a Courant penalty term (section 2). This defines the penalty problem

$$\underset{\substack{x,\lambda \\ \lambda > 0}}{\text{minimize}} \; \phi(x,\lambda) \triangleq -f(x) + \lambda^T c(x) + \tfrac{1}{2}\rho \|\nabla_x \mathscr{L}(x,\lambda)\|_2^2. \qquad (5.15)$$

To some extent there is a relationship between this idea and those in Fletcher (1973) in that decomposing (5.15) in a similar way to (5.9) would lead to a definition of $\lambda(x)$ by solving

$$\underset{\lambda : \lambda > 0}{\text{minimize}} \; \tfrac{1}{2}\rho (A\lambda - g)^T (A\lambda - g) + \lambda^T c \qquad (5.16)$$

which is identical to (5.7) if $\rho = 1/\sigma$. However the function $\phi(x, \lambda(x))$ which would result from (5.15) is not the same as the function $\phi(x)$ defined by Fletcher's method (5.3) and (5.7). In fact Han and Mangasarian do not use this decomposition but suggest that some sort of gradient projection type algorithm, applied in $\mathbb{R}^{n+m}$, is used. This approach requires second derivatives of the problem functions to be available. An apparent disadvantage of the penalty function (5.15) is that it can only be applied to problems for which $\nabla_x^2 \mathscr{L}$ is positive definite at the solution. The example of Fletcher (1981a), Question 9.23, illustrates the need for this condition. However when this condition does not hold the objective function may be augmented by a Courant penalty term (see (2.1)) in order to make $\nabla_x^2 \mathscr{L}(x^*, \lambda^*)$ positive definite. This device enables the penalty function (5.15) to be applied whenever standard conditions hold, albeit at the expense of introducing an additional parameter.

So far the derivation of various exact penalty functions has been discussed. However it is of prime importance to consider how readily these functions can be incorporated into an algorithm for nonlinear programming, and it is at this stage that some of the disadvantages of smooth exact penalty functions become apparent. As usual, considerations such as global convergence, superlinear rate of convergence, ease of use, numerical stability etc. are foremost. There are two main ways of applying a smooth exact penalty function,

(i) by direct application of a standard minimization method which has good convergence properties, and

(ii) as a criterion function in association with some auxiliary method of generating search directions or trial points.

Both methods have some disadvantages. In case (i) these stem from the fact that the penalty functions $\phi(x)$ or $\phi(x, \lambda)$ are defined in terms of derivatives ∇f and ∇c^T (g and A) of the problem functions. Hence to calculate $\nabla \phi$ for use in a convergent minimization method requires the Hessian matrices $\nabla^2 f$ and $\nabla^2 c_i$ $i = 1, 2, \ldots, m$ to be available (Fletcher (1973), Han and Mangasarian (1981)), which puts the method at a disadvantage vis-a-vis first derivative methods. An attempt to estimate $\nabla \phi$ using only first derivatives in conjunction with updating techniques is made by Fletcher and Lill (1970) but with only limited success. However, if second derivatives of f and c are available then it is possible to compute an $O(h)$ estimate of $\nabla^2 \phi(x^*)$ (Fletcher, 1972), which enables a Newton-type algorithm with second order convergence to be derived. This is preferable to using $\nabla \phi$ in a quasi-Newton method since the latter in general requires $\sim n$ iterations before a good estimate of $\nabla^2 \phi$ is built up.

II. Penalty Functions

More recently research has focussed on the case (ii) possibility above (Bertsekas (1982), Boggs and Tolle (1981), Di Pillo, Grippo and Lampariello (1981)). This avoids the need to compute derivatives of $\phi(x)$ or $\phi(x,\lambda)$ but introduces a consequant requirement that the penalty function and the auxiliary method are in some sense compatible. Ultimately it is desirable to prove that the resulting algorithm has all the usual properties, but a first step in this respect is to investigate whether or not the auxiliary method generates descent directions. The usual method for choosing search directions is the variable metric sequential quadratic programming method (*VM-SQP*: section 4). Bertsekas (1982) observes that if standard conditions hold and if x, λ is sufficiently close to x^*, λ^* then the *VM-SQP* direction is a descent direction for the Di Pillo-Grippo function (5.8). He also observes that for the function $\hat\phi(x)$ defined in (5.9), the *SQP* direction is a descent direction globally if σ in (5.8) is sufficiently large. Similar results are observed by Boggs and Tolle (1981) and Di Pillo, Grippo and Lampariello (1981). These results form the basis of most algorithms suggested in category (ii) and enable some sort of global convergence and superlinear rate of convergence results to be proved. I personally do not find this approach to be entirely satisfactory, and I think it may lead to the usual numerical difficulties associated with large penalty parameters. In this respect it is similar to Powell's (1969) global convergence strategy for augmented Lagrangian penalty functions (see (3.13)). Changing the parameters also changes the definition of $\phi(x)$ or $\phi(x,\lambda)$, so the advantage of having a fixed criterion function is lost.

The choice of a smooth exact penalty function from the various possibilities described above is also important. In particular the alternatives of minimizing a function either on $\mathbb{R}^n$ or on $\mathbb{R}^{n+m}$ merit some discussion. Although future research may prove otherwise, I think that currently there are a number of indications which favour a function defined on $\mathbb{R}^n$. Other things being equal, it is generally most efficient to eliminate variables from a minimization problem if this can be done easily, so as to solve as small a nonlinear problem as possible. This suggests for instance that an approach which explicitly retains the conditions $\lambda \geqslant 0$ is superior to one which does not, since it essentially eliminates the multipliers of the inactive constraints. However the approach of eliminating all the multipliers by solving (5.7) leads to an even smaller nonlinear problem in $\mathbb{R}^n$ so would appear to be preferable. Moreover many nonlinear programming problems have simple bounds

$$\underline{l} \leqslant \underline{x} \leqslant \underline{u} \tag{5.17}$$

on the variables and a good method should essentially be able to eliminate active bounds and solve a smaller nonlinear problem. The approach of Fletcher (1973) enables this to be done readily by adding (5.17) into (5.6) giving

$$\text{minimize}_{\underline{\delta}} \tfrac{1}{2}\sigma\underline{\delta}^T\underline{\delta} + \underline{g}^T\underline{\delta} \tag{5.18}$$

$$\text{subject to } A^T\underline{\delta} + \underline{c} \geqslant \underline{0}$$

$$\underline{l} \leqslant \underline{x} + \underline{\delta} \leqslant \underline{u}$$

where x is the current estimate of x^*. This problem is again solved to determine multipliers $\pi(x)$ for the constraints $A^T\delta+c\geqslant 0$, to be substituted into (5.3). The resulting function $\phi(x)$ is minimized subject to (5.17) which enables the active bounds to reduce the dimension of the nonlinear problem. It is not yet clear that other methods are able to take advantage of bounds in a similar way to reduce the problem size. Yet another advantage of solving (5.18) is that it avoids numerical problems associated with 'squaring' in the matrix A^TA which can potentially cause loss of precision. Also for large problems in which A is sparse, much sparsity is lost in forming A^TA.

A possible difficulty of using (5.3) and (5.6) (or (5.18)) is that $\phi(x)$ is nondifferentiable on surfaces across which the effective active set changes (as determined by $\pi(x)$). However $\phi(x)$ is always differentiable at x^* (even if x^*, π^* does not satisfy strict complementarity), and I think it is an open question as to whether the nonsmooth nature of $\phi(x)$ remote from the solution is a serious practical disadvantage. Another possible difficulty concerns the example given by Fletcher (1973) that illustrates a situation in which $\phi(x)$ can be discontinuous when x is not a feasible point. This is caused by a transitional degenerate situation: the problem can be perturbed to give either a problem with a single global solution, or a problem with two local minima separated by a surface on which $\phi(x)=\infty$. The situation is closely associated with loss of rank in the Jacobian matrix and may not prove to be serious in practice. In fact very little is known about the effect of rank deficiency in any of the smooth exact penalty functions. Another difficulty with using (5.6) or (5.18) is that if x is not a feasible point then it is possible that no feasible solution to (5.6) or (5.18) exists. In this case the dual (5.7) is unbounded, so presumably algorithms based on minimizing in $\mathbb{R}^{n+m}$ will also have difficulties in this situation. This situation also causes the SQP subproblem to be infeasible so some algorithms also fail on this account. A possible solution might be to include l_1 terms in (5.18) (this is equivalent to bounding the multipliers from above in a dual approach) and solve

$$\underset{\delta}{\text{minimize }} \tfrac{1}{2}\sigma\delta^T\delta+g^T\delta+\|(A^T\delta+c)_-\|_1$$
$$\text{subject to } l\leqslant x+\delta\leqslant u. \tag{5.19}$$

The multipliers of the l_1 term are assigned to $\pi(x)$ which is used in the usual way. The subproblem (5.19) is an l_1QP problem which can be solved as referred to in section 4.

6. Other penalty functions

Many other penalty functions have been suggested for nonlinear programming which are currently not as popular as some of those described above. However they often exhibit considerable ingenuity and contain interesting ideas which are worthy of note. Examples include the 'method of centres', the 'method of moving truncations', 'exponential penalty functions' and many others. A vir-

II. Penalty Functions

tual treasure house of references to this early work is the review of Lootsma (1972). In respect of the equality problem (1.1), one idea in particular which merits explicit mention arises from the observation that the function

$$\phi(x, f) = \left\| \begin{array}{c} f(x) - f \\ c_1(x) \\ c_2(x) \\ \vdots \\ c_m(x) \end{array} \right\|_p \tag{6.1}$$

is minimized by x^* if the control parameter $f = f(x^*)$. Thus a sequential penalty function can be envisaged in which a sequence of control parameters $\{f^{(k)}\} \to f^*$ is generated with the property that the minimizers $x(f^{(k)}) \to x^*$. Morrison (1968) suggests a method with $p = 2$ in which the sum of squares function $[\phi(x, f)]^2$ is minimized sequentially, but there is a potential difficulty that if $m < n$ then $\nabla^2 \phi(x^*, f)$ tends to a singular matrix as $f \to f^*$. This can slow down the rate of convergence of conventional algorithms. Gill and Murray (1976) introduce two new features, one of which is a special purpose method for solving ill-posed least squares problems. They also give a new formula for updating the control parameter $f^{(k)}$ which has a second order rate of convergence and controls cancellation errors.

Another penalty function which relates the ideas in (6.1) to multiplier penalty functions (section 3) and l_∞ exact penalty functions (e.g. (4.1)) is given by Bandler and Charalambous (1972). The function

$$\phi(x, p, \alpha, f) = \sum_{i=0}^{m} (f_i^p)^{1/p}$$
$$f_0 = f(x) - f \tag{6.2}$$
$$f_i = f(x) - f - \alpha_i c_i(x) \quad i \geq 1$$

is defined where $f < f(x^*)$ and $f_i > 0$ for all $i \geq 0$. There are many ways to force the minimizer of this function to converge to x^*. In a neighbourhood of x^* and in the limit $p \to \infty$ the function is equivalent to an exact l_∞ penalty function. Thus one possible mode of application is a Polya-type algorithm in which f and α are fixed and a control sequence $\{p^{(k)}\} \to \infty$ is used. Another possibility is to fix p and α and choose a sequence of control parameters $f^{(k)} \to f(x^*)$ as in (6.1). Finally p and f can remain fixed in which case a sequence $\alpha^{(k)} \to \lambda^*/(m+1)$ of scaled Lagrange multiplier estimates can be used as control parameters. Charalambous (1977) shows how a suitable updating formula for the $\alpha^{(k)}$ parameters can be determined.

An observation of a similar nature, which I include for what it is worth, is that an l_p Lagrangian exact penalty function may also merit consideration, which is

$$\phi(x, \lambda, \sigma) = f(x) - \lambda^T c(x) + \sigma \|c(x)\|_p \tag{6.3}$$

for an equality problem, in particular the case $p = 1$. To compare this with (4.1) let $\sigma = 1/\nu$. This is an exact penalty function for any fixed λ but the choice

$\lambda = \lambda^*$ is special in that the threshold value of σ is zero, so that difficulties due to selecting an appropriate parameter are removed. An exact penalty algorithm might be envisaged in which λ is updated from time to time, perhaps along the lines of section 3.

7. References

Bandler J. W. and Charalambous C. (1972), "Practical least p-th optimization of networks", IEEE Trans. Microwave Theo. Tech. (1972 Symposium Issue), 20, 834–840.

Bertsekas D. P. (1982), "Augmented Lagrangian and differentiable exact penalty methods", in "Nonlinear Optimization 1981" ed. M. J. D. Powell, Academic Press, London.

Boggs P. T. and Tolle J. W. (1980), "Augmented Lagrangians which are quadratic in the multiplier", J. Opt. Theo. Applns., 31, 17–26.

Boggs P. T. and Tolle J. W. (1981), "An implementation of a quasi-Newton method for constrained optimization", University of N. Carolina at Chapel Hill, O. R. and Systems Analysis Report 81-3.

Carroll C. W. (1961), "The created response surface technique for optimizing nonlinear restrained systems", Operations Res., 9, 169–184.

Chamberlain R. M., Lemarechal C., Pedersen H. C. and Powell M. J. D. (1982), "The watchdog technique for forcing convergence in algorithms for constrained optimization" in "Mathematical Programming Study 16, Algorithms for Constrained Minimization of Smooth Nonlinear Functions" eds. A. G. Buckley and J.-L. Goffin, North Holland, Amsterdam.

Charalambous C. (1977), "Nonlinear least p-th optimization and nonlinear programming", Math. Prog., 12, 195–225.

Coleman T. F. and Conn A. R. (1980), "Nonlinear programming via an exact penalty function: Asymptotic analysis", Univ. of Waterloo, Dept. of Comp. Sci. Report CS-80-30.

Coope I. D. and Fletcher R. (1980), "Some numerical experience with a globally convergent algorithm for nonlinearly constrained optimization", J. Opt. Theo. Applns., 32, 1–16.

Courant R. (1943), "Variational methods for the solution of problems of equilibrium and vibration", Bull. Amer. Math. Soc., 49, 1–23.

Di Pillo G. and Grippo L. (1979), "A new class of augmented Lagrangians in nonlinear programming" SIAM J. Control Opt., 17, 618–628.

Di Pillo G., Grippo L. and Lampariello F. (1981), "A class of algorithms for the solution of optimization problems with inequalities", CNR Inst. di Anal. dei Sistemi ed Inf. Report R18.

Fiacco A. V. and McCormick G. P. (1968), "Nonlinear Programming", John Wiley, New York.

Fletcher R. (1970), "A class of methods for nonlinear programming with termination and convergence properties", in "Integer and Nonlinear Programming" ed. J. Abadie, North Holland, Amsterdam.

Fletcher R. (1972), "A class of methods for nonlinear programming, III Rates of convergence" in "Numerical Methods for Nonlinear Optimization", ed. F. A. Lootsma, Academic Press, London.

Fletcher R. (1973), "An exact penalty function for nonlinear programming with inequalities", Math. Progr. 5, 129-150.

Fletcher R. (1975), "An ideal penalty function for constrained optimization", J. Inst. Maths. Applns., 7, 76-91.

Fletcher R. (1981a), "Practical Methods of Optimization, Vol. 2, Constrained Optimization", John Wiley, Chichester.

Fletcher R. (1981b), "Numerical experiments with an exact l_1 penalty function method", in "Nonlinear Programming 4", eds. O. L. Mangasarian, R. R. Meyer and S. M. Robinson, Academic Press, New York.

Fletcher R. (1982), "Second order corrections for nondifferentiable optimization", in "Numerical Analysis Proceedings, Dundee 1981", ed. G. A. Watson, Lecture Notes in Mathematics 912, Springer Verlag, Berlin.

Fletcher R. and Lill S. A. (1972), "A class of methods for nonlinear programming, II Computational experience", in "Nonlinear Programming", eds. J. B. Rosen, O. L. Mangasarian and K. Ritter, Academic Press, New York.

Frisch K. R. (1955), "The logarithmic potential method of convex programming", Memo., Univ. Inst. of Economics, Oslo, May 1955.

Gill P. E. and Murray W. (1976), "Nonlinear least squares and nonlinearly constrained optimization", in "Numerical Analysis, Dundee 1975", ed. G. A. Watson, Lecture Notes in Mathematics 506, Springer Verlag, Berlin.

Han S. P. (1977), "A globally convergent method for nonlinear programming", J. Opt. Theo. Applns., 22, 297-309.

Han S. P. and Mangasarian O. L. (1981), "A dual differentiable exact penalty function", Univ. of Wisconsin, Dept of Comp. Sci. Report 434, and in Math. Progr., 25, 293-306, (1983).

Hestenes M. R. (1969), "Multiplier and gradient methods", J. Opt. Theo. Applns., 4, 303-320.

Lootsma F. A. (1972), "A survey of methods for solving constrained minimization problems via unconstrained minimization" in "Numerical Methods for Nonlinear Optimization", ed. F. A. Lootsma, Academic Press, London.

Mangasarian O. L. (1973), "Unconstrained Lagrangians in nonlinear programming", Univ. of Wisconsin, Dept. of Comp. Sci. Report 174.

Mayne D. Q. (1980), "On the use of exact penalty functions to determine step length in optimization algorithms" in "Numerical Analysis, Dundee 1979", ed. G. A. Watson, Lecture Notes in Mathematics 773, Springer-Verlag, Berlin.

Morrison D. D. (1968), "Optimization by least squares", SIAM J. Num. Anal., 5, 83-88.

Murray W. and Wright M. H. (1978), "Projected Lagrangian methods based on the trajectories of penalty and barrier functions", Stanford Univ. Dept. of O. R. Report SOL 78-23.

Osborne M. R. and Ryan D. M. (1972), "A hybrid algorithm for nonlinear programming" in "Numerical Methods for Nonlinear Optimization", ed. F. A. Lootsma, Academic Press, London.

Powell M. J. D. (1969), "A method for nonlinear constraints in minimization problems", in "Optimization", ed. R. Fletcher, Academic Press, London.

Powell M. J. D. (1978), "A fast algorithm for nonlinearly constained optimization calculations", in "Numerical Analysis, Dundee 1977", ed. G. A. Watson, Lecture Notes in Mathematics 630, Springer Verlag, Berlin.

Powell M. J. D. (ed.) (1982), "Nonlinear Optimization 1981", Academic Press, London.

Rockafellar R. T. (1974), "Augmented Lagrange multiplier functions and duality in nonconvex programming", SIAM J. Control, *12*, 268–285.

Schittkowski K. (1980), "Nonlinear Programming Codes", Lecture Notes in Economics and Mathematical Systems 183, Springer-Verlag, Berlin.

Wilson R. B. (1963), "A simplicial algorithm for concave programming", PhD dissertation, Harvard Univ. Grad. School of Business Admin.

Wolfe P. (1965), "The composite simplex algorithm", SIAM Rev., *7*, 42–54.

Applications of the FKG Inequality and Its Relatives

R. L. Graham
Bell Laboratories, Department of Discrete Mathematics, 600 Mountain Avenue, Murray Hill, New Jersey 07974, USA

Introduction

In 1971, C. M. Fortuin, P. W. Kasteleyn and J. Ginibre [FKG] published a remarkable inequality relating certain real functions defined on a finite distributive lattice. This inequality, now generally known as the FKG inequality, arose in connection with these authors' investigations into correlation properties of Ising ferromagnet spin systems and generalized earlier results of Griffiths [Gri] and Harris [Har] (who was studying percolation models). The FKG inequality in turn has stimulated further research in a number of directions, including a variety of interesting generalizations and applications, particularly to statistics, computer science and the theory of partially ordered sets. It turns out that special cases of the FKG inequality can be found in the literature of at least a half dozen different fields, and in some sense can be traced all the way back to work of Chebyshev.

In this paper, I will survey some of this history as well as the more recent extensions and applications. I will also discuss various open problems along the way which I hope will convince the reader that this exciting area is still fertile ground for further research.

Background

We begin with an old result of Chebyshev (see [HLP]) which asserts that if f and g are both increasing (or both decreasing) functions on [0, 1] then the average value of the product fg is at least as large as the product of the average values of f and g, where the average is taken with respect to some measure μ on [0, 1].

In symbols, this is just

(1) $$\int_0^1 fg\,d\mu \geq \int_0^1 f\,d\mu \int_0^1 g\,d\mu$$

In the case that μ is a discrete measure we can restate (1) as follows: If $f(k)$ and $g(k)$ are both increasing (or both decreasing) and $\mu(k) \geq 0$ for $k = 1, 2, 3, \ldots$, then

$$\frac{\sum_k f(k)g(k)\mu(k)}{\sum_k \mu(k)} \geq \frac{\sum_k f(k)\mu(k)}{\sum_k \mu(k)} \cdot \frac{\sum_k g(k)\mu(k)}{\sum_k \mu(k)}$$

i.e.,

(2) $$\sum_k f(k)g(k)\mu(k) \sum_k \mu(k) \geq \sum_k f(k)\mu(k) \sum_k g(k)\mu(k).$$

The proofs of (1) and (2) follow immediately by expanding the inequality

$$\sum_{i,j} (f(i)-f(j))(g(i)-g(j))\mu(i)\mu(j) \geq 0.$$

Basically, the FKG inequality represents a way of extending (2) to the situation in which the underlying index set is only *partially* ordered, as opposed to the *totally* ordered index set of integers occurring in (2). The setting is as follows. Let $(\Gamma, <)$ be a finite distributive lattice, i.e., Γ is a finite set, partially ordered by $<$, for which the two functions $\wedge$ (meet or greatest lower bound) and $\vee$ (join or least upper bound) defined by:

$$x \wedge y := \max\{z \in \Gamma : z \leq x, z \leq y\},$$
$$x \vee y := \min\{z \in \Gamma : z \geq x, z \geq y\}$$

are well-defined and satisfy the distributive laws:

$$x \wedge (y \vee z) = (x \wedge y) \vee (x \wedge z)$$
$$x \vee (y \wedge z) = (x \vee y) \wedge (x \vee z)$$

for all $x, y, z \in \Gamma$.

It is well known that any such lattice can be realized as a sublattice of the lattice of all subsets of some finite set partially ordered by inclusion and with $x \wedge y = x \cap y$ and $x \vee y = x \cup y$.

We now suppose $\mu: \Gamma \to \mathbb{R}_0$, the nonnegative reals, satisfies

(*) $$\mu(x \wedge y)\mu(x \vee y) \geq \mu(x)\mu(y) \quad \text{for all } x, y \in \Gamma.$$

For reasons we shall mention later, a function μ satisfying (*) is often called *log supermodular*. Finally, a function $f: \Gamma \to \mathbb{R}$ is called *increasing* if

$$x \leq y \Rightarrow f(x) \leq f(g) \quad \text{for } x, y \in \Gamma$$

(with *decreasing* defined similarly).

The FKG inequality states:

If f and g are both increasing (or both decreasing) real functions on a finite distributive lattice Γ and $\mu: \Gamma \to \mathbb{R}_0$ is log supermodular then

(3) $$\sum_{x \in \Gamma} f(x)g(x)\mu(x) \sum_{x \in \Gamma} \mu(x) \geq \sum_{x \in \Gamma} f(x)\mu(x) \sum_{x \in \Gamma} g(x)\mu(x).$$

The original proof of (3) was somewhat complicated [FKG]. Several years after (3) appeared, Holley found the following beautiful generalization:

II. Applications of the FKG Inequality and Its Relatives

Theorem (Holley [Hol])

Suppose $\alpha, \beta: \Gamma \to \mathbb{R}_0$ satisfy

$$\alpha(x \wedge y)\beta(x \vee y) \geq \alpha(x)\beta(y) \quad \text{for all } x, y \in \Gamma.$$

Then for any increasing function $\theta: \Gamma \to \mathbb{R}_0$

(4) $$\sum_{x \in \Gamma} \alpha(x)\beta(x) \geq \sum_{x \in \Gamma} \beta(x)\theta(x).$$

However, this result was itself soon superseded by a striking result of Ahlswede and Daykin [AD 1] which we now describe.

The Ahlswede-Daykin Inequality

To state the inequality of Ahlswede and Daykin, we need the following simplifying notation. For subsets X and Y of Γ, define

$$X \wedge Y = \{x \wedge y : x \in X, y \in Y\},$$
$$X \vee Y = \{x \vee y : x \in X, y \in Y\}$$

and, for a function of $f: \Gamma \to \mathbb{R}$, define

$$f(X) = \sum_{x \in X} f(x).$$

As before, Γ denotes a finite distributive lattice.

Theorem [AD]

Suppose $\alpha, \beta, \gamma, \delta: \Gamma \to \mathbb{R}_0$ satisfy

(5) $$\alpha(x)\beta(y) \leq \gamma(x \vee y)\delta(x \wedge y) \quad \text{for all } x, y \in \Gamma.$$

Then

(5′) $$\alpha(X)\beta(Y) \leq \gamma(X \vee Y)\delta(X \wedge Y) \quad \text{for all } X, Y \subseteq \Gamma.$$

Note the attractive similarity between the hypothesis (5) and the conclusion (5′). This certainly contributes to the relative simplicity of the proof of the theorem, which we now give (also, see [Kle 2]).

Proof: It follows from our previous remarks that it suffices to prove the theorem for the case that $\Gamma = 2^{[N]}$ is the lattice of subsets of $[N] = \{1, 2, \ldots, N\}$ partially ordered by inclusion. In this case, the hypothesis is

(6) $$\alpha(x)\beta(y) \leq \gamma(x \cup y)\delta(x \cap y) \quad \text{for all } x, y \in 2^{[N]}$$

and the desired conclusion is

(6′) $$\alpha(X)\beta(Y) \leq \gamma(X \vee Y)\delta(X \wedge Y) \quad \text{for all } X, Y \subseteq 2^{[N]}.$$

The proof proceeds by induction on N. We first consider the case $N=1$, in which case $2^{[N]} = 2^{[1]} = \{\phi, \{1\}\}$. For $\sigma = \alpha, \beta, \gamma$ or δ, let σ_0 denote $\sigma(\phi)$ and σ_1 denote $\sigma(\{1\})$. Then (6) becomes

(7)
$$\alpha_0 \beta_0 \leqslant \gamma_0 \delta_0,$$
$$\alpha_1 \beta_0 \leqslant \gamma_1 \delta_0,$$
$$\alpha_0 \beta_1 \leqslant \gamma_1 \delta_0,$$
$$\alpha_1 \beta_1 \leqslant \gamma_1 \delta_1.$$

It is easy to check that (6') holds if either X or Y consists of a single element. This leaves only the case $X = \{\phi, \{1\}\} = Y$ to deal with. In this case, the inequality we must prove is

(7') $$(\alpha_0 + \alpha_1)(\beta_0 + \beta_1) \leqslant (\gamma_0 + \gamma_1)(\delta_0 + \delta_1).$$

Note that (7') would follow at once from (7) if one of the occurrences of $\gamma_1 \delta_0$ in (7') were $\gamma_0 \delta_1$ instead (by summing). As it is we have to work slightly harder. If any of $\alpha_0, \beta_0, \gamma_0, \delta_0$ is zero then (7') follows at once. It follows from this (and a little computation) that it is enough to consider the special case $\alpha_0 = \beta_0 = \gamma_0 = \delta_0 = 1$. Now (7) becomes

(8) $$\alpha_1 \leqslant \gamma_1, \quad \beta_1 \leqslant \gamma_1, \quad \alpha_1 \beta_1 \leqslant \gamma_1 \delta_1$$

and (7') becomes

(8') $$(1+\alpha_1)(1+\beta_1) \leqslant (1+\gamma_1)(1+\delta_1).$$

Again, (8') is immediate if $\gamma_1 = 0$ so we may assume $\gamma_1 > 0$. Since (8') becomes harder to satisfy as δ_1 decreases, it suffices to prove (8') when δ_1 is as small as possible, i.e., (by (8)) when $\delta_1 = \alpha_1 \beta_1 / \gamma_1$. In this case (8') becomes

$$(1+\alpha_1)(1+\beta_1) \leqslant (1+\gamma_1)\left(1 + \frac{\alpha_1 \beta_1}{\gamma_1}\right)$$

i.e.,

$$\alpha_1 + \beta_1 \leqslant \gamma_1 + \frac{\alpha_1 \beta_1}{\gamma_1}.$$

However, this last inequality is an immediate consequence of

$$(\gamma_1 - \alpha_1)(\gamma_1 - \beta_1) \geqslant 0$$

which is implied by (8). This proves the result for the case $N=1$.

Assume now that the assertion holds for $N = n-1$ for some $n \geqslant 2$. Suppose $\alpha, \beta, \gamma, \delta: 2^{[n]} \to \mathbb{R}_0$ satisfy the hypothesis (6) with $N = n$ and let $X, Y \subseteq 2^{[n]}$ be given. We will define new functions $\alpha', \beta', \gamma', \delta'$ mapping $2^{[n-1]} := T'$ into $\mathbb{R}_0$ as follows:

$$\alpha'(x') = \sum_{\substack{x \in X \\ x' = x \setminus \{n\}}} \alpha(x),$$

$$\beta'(y') = \sum_{\substack{y \in Y \\ y' = y \setminus \{n\}}} \beta(y),$$

II. Applications of the FKG Inequality and Its Relatives

$$\gamma'(z') = \sum_{\substack{z \in X \cup Y \\ z' = z \setminus \{n\}}} \gamma(z),$$

$$\delta'(w') = \sum_{\substack{w \in X \cap Y \\ w' = w \setminus \{n\}}} \delta(w).$$

Thus, for $x' \in T'$,

$$\alpha'(x') = \begin{cases} \alpha(x') + \alpha(x' \cup \{n\}) & \text{if } x' \in X, \ x' \cup \{n\} \in X, \\ \alpha(x') & \text{if } x' \in X, \ x' \cup \{n\} \notin X, \\ \alpha(x' \cup \{n\}) & \text{if } x' \notin X, \ x' \cup \{n\} \in X, \\ 0 & \text{if } x' \notin X, \ x' \cup \{n\} \notin X. \end{cases}$$

Observe that with these definitions

$$\alpha(X) = \sum_{x \in X} \alpha(x) = \sum_{x' \in T'} \alpha'(x') = \alpha'(T')$$

and, similarly,

$$\beta(Y) = \beta'(T'), \quad \gamma(X \vee Y) = \gamma'(T'), \quad \delta(X \wedge Y) = \delta'(T').$$

Therefore, *if*

(10) $\quad \alpha'(x')\beta'(y') \leq \gamma'(x' \cup y')\delta'(x' \cap y') \quad \text{for all } x', y' \in T'$

holds, then by the induction hypotheses we would have

$$\alpha(X)\beta(Y) = \alpha'(T')\beta'(T') \leq \gamma'(T')\delta'(T') = \gamma(X \vee Y)\delta(X \wedge Y)$$

since $T' \vee T' = T'$, $T' \wedge T' = T'$, which is just (6'), the desired conclusion.

So, it remains to prove (10). However, note that by (9) this is exactly the same computation as that performed for the case $N = 1$ with $x' \leftrightarrow \phi$ and $x' \cup \{n\} \leftrightarrow \{1\}$. Since we have already treated this case then the proof of the induction step is completed. This proves the theorem of Ahlswede and Daykin. □

We next indicate how the FKG inequality follows from the AD (= Ahlswede-Daykin) inequality. As usual it suffices to prove this in the case that $\Gamma = 2^{[N]} := T$ partially ordered by inclusion. Note that if A and B are upper ideals in T (i.e., $x, y \in A \Rightarrow x \cup y \in A$) then the indicator functions $f = I_A$ and $g = I_B$ (where $I_A(x) = 1$ if and only if $x \in A$) are increasing. Taking $\alpha = \beta = \gamma = \delta = \mu$ in (5) and $X = A$, $Y = B$ in (5') we have

(11) $\quad \mu(A)\mu(B) \leq \mu(A \vee B)\mu(A \wedge B).$

But

$$\mu(A) = \sum_{x \in A} \mu(x) = \sum_{x \in T} f(x)\mu(x),$$

$$\mu(B) = \sum_{x \in T} g(x)\mu(x),$$

$$\mu(A \wedge B) = \sum_{z \in A \wedge B} \mu(z) = \sum_{z \in T} f(z)g(z)\mu(z),$$

$$\mu(A \vee B) = \sum_{z \in A \vee B} \mu(z) \leq \sum_{z \in T} \mu(z).$$

Thus, (11) implies

$$\sum_z f(z)g(z)\mu(z) \sum_z \mu(z) \geq \sum_z f(z)\mu(z) \sum_z g(z)\mu(z)$$

which is just the FKG inequality for this case. The general FKG inequality is proved in just this way by first writing an arbitrary increasing function f on T as $f = \sum_i \lambda_i I_{A_i}$ where $\lambda_i \geq 0$ and the A_i are suitable upper ideals in T. That is, for

$$f = \sum_i \lambda_i I_{A_i}, \quad \lambda_i \geq 0,$$

we have

$$\sum_{z \in T} f(z)\mu(z) = \sum_{z \in T} \sum_i \lambda_i I_{A_i}(z)\mu(z)$$

$$= \sum_i \lambda_i \sum_{z \in T} I_{A_i}(z)\mu(z), \quad \text{etc.,}$$

and we can now apply the preceding inequality.

Some Consequences of the AD Inequality

By specializing the choices of α, β, γ and δ in (5), many results which have appeared in various places and times in the literature can be obtained. We now describe some of these.

To begin with, setting $\alpha = \beta = \gamma = \delta = 1$ we have $\alpha(T) = |T|$ for $T \subseteq [N]$ and we obtain

(12) $\qquad |X||Y| \leq |X \vee Y||X \wedge Y| \quad \text{for all } X, Y \subseteq 2^{[N]}$

This was first proved by Daykin [Day 1] (who also showed that this implies a lattice is distributive) and has as immediate corrollaries:

(a) (Seymour [Sey]). For any two upper ideals U, U' of $2^{[N]}$,

$$|U||U'| \leq |U \cap U'| \cdot 2^N.$$

(b) (Kleitman [Kle 1]). For any upper ideal U and lower ideal L of $2^{[N]}$,

$$|U||L| \geq |U \cap L| \cdot 2^N.$$

(c) (Marica-Schönheim [MS])

$$|A| \leq |A \setminus A| \quad \text{for all } A \subseteq 2^{[N]}.$$

Kleitman's result first appeared in 1966 and, in fact, directly implies (a). The 1969 result (c) of Marica and Schönheim arose in connection with the following (still unresolved) number-theoretic conjecture of the author.

Conjecture: If $0 < a_1 < a_2 < \ldots < a_n$ are integers then

$$\max_{i,j} \frac{a_i}{gcd(a_i, a_j)} \geq n.$$

The Marica-Schönheim inequality is equivalent to the validity of the conjecture when all the a_k are squarefree. To see this, suppose

$$a_i = \prod_k p_k^{\varepsilon_{ik}}, \quad \varepsilon_{ik} = 0 \text{ or } 1,$$

where p_k denotes the kth prime. To each a_i associate the set $S_i \subseteq \mathbb{Z}^+$ defined by

$$S_k = \{k : \varepsilon_{ik} = 1\}.$$

Then

$$a_i = \prod_{k \in S_i} p_k$$

and

$$\frac{a_i}{gcd(a_i, a_j)} = \prod_k p_k^{\varepsilon_{ik} - \min(\varepsilon_{ik}, \varepsilon_{jk})}$$

$$= \prod_{k \in S_i \setminus S_j} p_k.$$

By (c) there must be at least n *different* sets $S_i \setminus S_j$ and so, at least n different integers of the form $\dfrac{a_i}{gcd(a_i, a_j)}$. Thus, the largest such value must be at least as large as n, which is the desired conclusion. (Further generalizations can be found in [Mar] and [DL].)

The implication (12) ⇒ (c) is short and sweet – simply note that for $A, B \subseteq 2^{[N]}$,

$$|A||B| = |A||\bar{B}| \leq |A \vee \bar{B}||A \wedge \bar{B}|$$

$$= \overline{|A \vee \bar{B}|}|A \wedge \bar{B}|$$

$$= |\bar{A} \wedge B||A \wedge \bar{B}|$$

$$= |B \setminus A||A \setminus B|$$

and set $A = B$.

Partial results currently available for the conjecture can be found in the surveys [Won] and [EG].

Linear Extensions of Partial Orders

An important area in which the FKG and related inequalities have had important applications has been in the theory of partially ordered sets and the analysis of related sorting algorithms in computer science. Many algorithms for sorting n numbers $\{a_1, a_2, \ldots, a_n\}$ proceed by using binary comparisons $a_i : a_j$ to construct successively stronger partial orders P until a linear order finally emerges (e.g., see Knuth [Knu]). A fundamental quantity in analyzing the expected behavior of such algorithms is $Pr(a_i < a_j | P)$, the probability that the result of $a_i : a_j$ is $a_i < a_j$ when all linear orders consistent with P are considered equally likely. In this section we describe a number of recent results of this type.

First, we need some notation. Let $(P, <)$ be a finite partially ordered set and for $n = |P|$, let Λ denote the set of all $1-1$ mappings of P onto $[n]$. We will assign a uniform probability distribution on Λ so that each $\lambda \in \Lambda$ has the probability $\frac{1}{n!}$ of occurring. A map $\lambda \in \Lambda$ is said to be a *linear extension* of P if

$$x < y \text{ in } P \Rightarrow \lambda(x) < \lambda(y).$$

We denote the set of linear extensions of P by $\Lambda(P)$. We are going to consider the probabilities of certain "events" occuring where we will think of an event as some other partial order on the elements of P which is preserved by various $\lambda \in \Lambda$. Thus, $Pr(Q) = \frac{|\Lambda(Q)|}{n!}$ is just the fraction of $\lambda \in \Lambda$ which are linear extensions of Q, i.e., $u < v$ in Q implies $\lambda(u) < \lambda(v)$. Similarly, $Pr(P \text{ and } Q)$ is the fraction of $\lambda \in \Lambda$ which are linear extensions of both P and Q, while $Pr(P|Q)$ is defined as usual to mean $Pr(P \text{ and } Q)/Pr(Q)$, provided the denominator does not vanish.

To get into the spirit of this topic we first give a relatively simple result. Suppose $(P, <)$ consists of two *chains* $A = \{a_1 < \ldots < a_m\}$ and $B = \{b_1 < \ldots < b_n\}$. Of course, relations of the form $a_i < b_j$ and $b_r < a_s$ are also allowed. Suppose Q and Q' are two events both of which are unions of sets of the form $\{a_{i_1} < b_{j_1}, a_{i_2} < b_{j_2}, \ldots\}$, i.e., such that a's are less than b's. It is quite natural to believe that the events Q and Q' are *positively correlated* when considering linear extensions of P, i.e.,

$$Pr(Q|P \text{ and } Q') \geq Pr(Q|P)$$

or, more symmetrically,

(13) $\qquad Pr(Q \text{ and } Q'|P) \geq Pr(Q|P) Pr(Q'|P).$

In fact, the inequality in (13) is a theorem of Graham, Yao and Yao [GYY]. The first proof of (13) was a rather complicated combinatorial proof which used, among other things, the Marica-Schönheim inequality. Shortly after [GYY] appeared, Shepp [She 1] (and independently Kleitman and Shearer [KS]) provided rather short proofs based on the FRG inequality. Before sketching Shepp's proof, we give an example which shows that if the hypothesis that P consists of two chains is weakened even slightly then (13) no longer remains valid.

Example: Let $(P, <)$ consist of the set $\{a_1, a_2, b_1, b_2, b_3\}$ with the following partial order:

$$\{a_1 < a_2, a_1 < b_2, a_2 < b_1, b_1 < b_3, b_2 < b_3\}.$$

Let Q denote the event $\{a_1 < b_1\}$ and let Q' denote the event $\{a_2 < b_3\}$. A straightforward calculation shows that

$$Pr(Q|P \text{ and } Q') = \tfrac{3}{5} < \tfrac{5}{8} = Pr(Q|P),$$

which violates (13).

II. Applications of the FKG Inequality and Its Relatives

Proof of (13). (Shepp [She 1]). Define a lattice Γ with elements of the form $\bar{x}=(x_1,\ldots,x_m)$ where $1 \leqslant x_1 < x_2 < \ldots < x_m \leqslant m+n$. Let us define $\bar{x} \leqslant \bar{x}'$ to mean $x_i \leqslant x_i'$ for $1 \leqslant i \leqslant m$. Thus, we have

$$\bar{x} \vee \bar{x}' = (\ldots, \max(x_i, x_i'), \ldots)$$

$$\bar{x} \wedge \bar{x}' = (\ldots, \min(x_i, x_i'), \ldots).$$

It is easily checked that with these definitions Γ is a distributive lattice. For each $\bar{x} \in \Gamma$ we can associate a unique mapping $\lambda_{\bar{x}} \in \Lambda$ by setting:

$$\lambda_{\bar{x}}(a_i) = x_i, \quad \lambda_{\bar{x}}(b_j) = y_j$$

where $[m+n] \setminus \{x_1 < \ldots < x_m\} = \{y_1 < \ldots < y_n\}$.

Finally, define

$$\mu(\bar{x}) = \begin{cases} 1 & \text{if } \lambda_{\bar{x}} \in \Lambda(P) \\ 0 & \text{otherwise} \end{cases}$$

$$f(\bar{x}) = \begin{cases} 1 & \text{if } \lambda_{\bar{x}} \in \Lambda(Q) \\ 0 & \text{otherwise} \end{cases}$$

$$f'(\bar{x}) = \begin{cases} 1 & \text{if } \lambda_{\bar{x}} \in \Lambda(Q') \\ 0 & \text{otherwise} \end{cases}$$

To apply the FKG inequality, we must first verify the log supermodularity of μ:

(14) $\qquad \mu(\bar{x})\mu(\bar{x}') \leqslant \mu(\bar{x} \vee \bar{x}')\mu(\bar{x} \wedge \bar{x}') \quad \text{for all } \bar{x}, \bar{x}' \in \Gamma.$

Suppose $\mu(\bar{x})\mu(\bar{x}') = 1$. Then $\lambda_{\bar{x}} \in \Lambda(P)$, $\lambda_{\bar{x}'} \in \Lambda(P)$. If $a_i < a_j$ in P then

$$\lambda_{\bar{x}}(a_i) = x_i < x_j = \lambda_{\bar{x}}(a_j)$$
$$\lambda_{\bar{x}'}(a_i) = x_i' < x_j' = \lambda_{\bar{x}'}(a_j)$$

and so,

$$\lambda_{\bar{x} \vee \bar{x}'}(a_i) = \max(x_i, x_i') < \max(x_j, x_j') = \lambda_{\bar{x} \vee \bar{x}'}(a_j).$$

Similarly, if $b_i < b_j$ in P then

$$\lambda_{\bar{x} \vee \bar{x}'}(b_i) < \lambda_{\bar{x} \vee \bar{x}'}(b_j).$$

More interesting is the case that $a_i < b_j$ in P. Then

$$\lambda_{\bar{x}}(a_i) = x_i < y_j = \lambda_{\bar{x}}(b_j),$$
$$\lambda_{\bar{x}'}(a_i) = x_i' < y_j' = \lambda_{\bar{x}'}(b_j).$$

But this implies

$$x_i \leqslant i+j-1, \quad x_i' \leqslant i+j-1$$

and consequently,

$$\lambda_{\bar{x} \vee \bar{x}'}(a_i) = \max(x_i, x_i') \leqslant i+j-1,$$

i.e.,

$$\lambda_{\bar{x} \vee \bar{x}'}(a_i) < \lambda_{\bar{x} \vee \bar{x}'}(b_j).$$

The argument for $b_i < a_j$ is similar. This shows that $\lambda_{\bar{x} \vee \bar{x}'} \in \Lambda(P)$, i.e., $\mu(\bar{x} \vee \bar{x}') = 1$. In basically the same way it follows that $\mu(\bar{x} \wedge \bar{x}') = 1$. Therefore, we have shown that

$$\mu(\bar{x})\mu(\bar{x}') = 1 \Rightarrow \mu(\bar{x} \vee \bar{x}')\mu(\bar{x} \wedge \bar{x}') = 1$$

and so (14) always holds.

The final hypothesis to check before we can apply the FKG inequality is that f and f' are both decreasing. To see this, suppose $\bar{x} \leq \bar{x}'$ and $f(\bar{x}') = 1$. Then, by definition,

$$\lambda_{\bar{x}'} \in \Lambda(Q) = \bigcup_k \Lambda(Q_k)$$

where

$$Q_k = \{a_{i_1} < b_{j_1}, a_{i_2} < b_{j_2}, \ldots\}.$$

Thus, for some k, the elements x'_i of $\bar{x}'$ satisfy all the constraints $x'_{i_1} \leq i_1 + j_1 - 1$, $x'_{i_2} \leq i_2 + j_2 - 1$, ..., imposed by $\Lambda(Q_k)$. However, since $\bar{x} \leq \bar{x}'$ then $x_i \leq x'_i$ for all i, and in particular, $x_{i_1} \leq x'_{i_1} \leq i_1 + j_1 - 1$, etc. This implies that $\lambda_{\bar{x}} \in \Lambda(Q_k) \subseteq \Lambda(Q)$ and $f(\bar{x}) = 1$, and consequently, f is decreasing.

The FKG inequality can now be applied to the functions we have defined, yielding

$$\sum_{\bar{x} \in \Gamma} f(\bar{x})f'(\bar{x})\mu(\bar{x}) \sum_{\bar{x} \in \Gamma} \mu(\bar{x}) \geq \sum_{\bar{x} \in \Gamma} f(\bar{x})\mu(\bar{x}) \sum_{\bar{x} \in \Gamma} f'(\bar{x})\mu(\bar{x}).$$

Interpreting this in terms of P, Q and Q', we obtain

$$|Q \cap Q' \cap P||Q| \geq |Q \cap P||Q' \cap P|$$

i.e.,

$$Pr(Q \text{ and } Q'|P) \geq Pr(Q|P)Pr(Q'|P)$$

which is just (13). □

It is natural to try and extend (13) to more general partial orders P than just those which can be covered by two chains. However, examples such as the one previously given show that *some* additional hypotheses must be assumed in order for the desired positive correlation to hold. One such extension was provided by Shepp in the following result.

Theorem (Shepp [She 1]). Suppose $(P, <)$ is a union of two *disjoint* partial orders, i.e., $P = A \cup B$ where each pair a, b with $a \in A$ and $b \in B$ are incomparable. Then any two events Q and Q', each being the union of sets of the form $\{a_{i_1} < b_{j_1}, a_{i_2} < b_{j_2}, \ldots\}$ are positively correlated, i.e.,

$$Pr(Q \text{ and } Q'|P) \geq Pr(Q|P)Pr(Q'|P).$$

The only proof known for this result uses the FKG inequality in a somewhat more subtle way than in the preceding theorem (a similar use of FKG will be given in the next section).

Universal correlation. Suppose Q and Q' are a pair of partial orders on a common underlying set S. Following Winkler [Win 1], let us call Q and Q' *univer-*

sally correlated if for all possible partial orders P on S,

(15) $$Pr(Q \text{ and } Q'|P) \geq Pr(Q|P)Pr(Q'|P)$$

As an example, it had been a long-standing conjecture until quite recently that the pair $Q=\{x<y\}$ and $Q'=\{x<z\}$ were universally correlated. This conjecture, due to I. Rival and W. Sands and known as the XYZ conjecture, also has finally been settled (affirmatively) by Shepp, not surprisingly (by now) using the FKG inequality. An interesting application of this result can be found in [Win 2]. We give an outline of Shepp's snappy proof.

Theorem (Shepp [She 2]). *For any partial order P on the set $\{x_1, x_2, \ldots, x_n\}$, the sets $Q=\{x_1<x_2\}$ and $Q'=\{x_1<x_3\}$ are positively correlated, i.e.,*

$$Pr(Q \text{ and } Q'|P) \geq Pr(Q|P)Pr(Q'|P).$$

Proof (outline). Choose a large $N \gg n$ and let $\Omega = [N]^n = \{\bar{x} = (x_1, \ldots, x_m) : x_i \in [N]\}$. Define a partial order on Ω by:

$$\bar{x} \leq \bar{y} \text{ if and only if } x_1 \geq y_1, \; x_i - x_1 \leq y_i - y_1, \quad i = 2, 3, \ldots, N.$$

It is easy to verify that this indeed does define a partial order on Ω and further that

$$(\bar{x} \wedge \bar{y})_i = \min(x_i - x_1, y_i - y_1) + \max(x_1, y_1), \quad i \in [n]$$
$$(\bar{x} \vee \bar{y})_i = \max(x_i - x_1, y_i - y_1) + \min(x_1, y_1), \quad i \in [n].$$

Using these expressions together with the fact that the reals partially ordered by magnitude form a distributive lattice (so that $\min(a, \max(b, c)) = \max(\min(a, b), \min(a, c))$, etc.), it follows (after a little computation) that Γ also forms a distributive lattice.

Let f be the indicator function of the event $\{x_1 \leq x_2\}$, i.e.,

$$f(\bar{x}) = \begin{cases} 1 & \text{if } x_1 \leq x_2, \\ 0 & \text{otherwise} \end{cases}$$

and let g be the indicator function of the event $\{x_1 \leq x_3\}$. Again, an easy calculation shows that f and g are both increasing.

Finally, take μ to be the indicator function of P, i.e.,

$$\mu(\bar{x}) = \begin{cases} 1 & \text{if } \bar{x} \text{ satisfies the inequalities in } P, \\ 0 & \text{otherwise} \end{cases}$$

In order to apply the FKG inequality it remains to verify that $\mu(\bar{x}) \geq 0$ (obvious) and that μ is log supermodular, i.e.,

(16) $$\mu(\bar{x} \wedge \bar{y})\mu(\bar{x} \vee \bar{y}) \geq \mu(\bar{x})\mu(\bar{y}) \quad \text{for all } \bar{x}, \bar{y} \in \Omega.$$

Suppose $\mu(\bar{x}) = \mu(\bar{y}) = 1$. If $x_i < x_j$ in P then $x_i < x_j$ and $y_i < y_j$ and consequently

$$(\bar{x} \wedge \bar{y})_i \leq \min(x_j - x_1, y_j - y_1) + \max(x_1, y_1) = (\bar{x} \wedge \bar{y})_j$$

and, in a similar way,

$$(\bar{x} \vee \bar{y})_i \leq (\bar{x} \vee \bar{y})_j.$$

Thus, since $\bar{x} \wedge \bar{y}$ and $\bar{x} \vee \bar{y}$ both satisfy each inequality in P, $\mu(\bar{x} \wedge \bar{y}) = \mu(\bar{x} \vee \bar{y}) = 1$ and so, (16) holds.

Having verified the hypotheses of the FKG inequality, we can apply its conclusion. This yields

(17) $\quad Pr(x_1 \leq x_2, x_1 \leq x_3, P) Pr(P) \geq Pr(x_1 \leq x_2, P) Pr(x_1 \leq x_3, P).$

Letting $N \to \infty$, the probability that $x_i = x_j$ tends to zero and so, it follows that (17) also holds for the permutations (i.e., $1-1$ maps) induced by $x_1, \ldots, x_n$. Finally, dividing by $Pr(P) Pr(x_1 < x_3, P)$ we obtain the desired inequality and the theorem is proved. □

It is natural to ask for other examples of universally correlated partial orders, or even more ambitiously, to ask for a characterization of all such pairs. The apparent difficulty of the second task is increased when one notes that these are easy examples which show that such "reasonable" sets as $\{x_1 < x_2 < x_4\}$ and $\{x_1 < x_3 < x_4\}$ are *not* universally correlated. It is therefore rather surprising that there is in fact a striking characterization for such pairs.

In order to state this result (due to Winkler [Win 1]) we first need a definition. For a partial order P, let us call an inequality $x < y$ in P *minimal* if there is no z such that $x < z < y$ in P. We denote the set of minimal inequalities in P by $\Delta(P)$.

Theorem (Winkler [Win 1]). Two partial orders Q and Q' (on the same underlying set) are universally correlated if and only if
  (i) Q and Q' are consistent (i.e., for no x and y do we have $x < y$ in Q and $y < x$ in Q')
  (ii) $u < v$ in $\Delta(Q \cup Q') - \Delta(Q)$ and
       $x < y$ in $\Delta(Q \cup Q') - \Delta(Q')$
       $\Rightarrow$
       $u = x$ or $v = y$.

In particular, Winkler shows that whenever Q and Q' are universally correlated, it can in fact be proved by repeated applications of the XYZ theorem and so, ultimately rests on the FKG inequality. As pointed out in [Win 1], it follows from the theorem that the pairs

$$Q = \{x < y < z\}, \quad Q' = \{x < z\}$$
$$Q = \{x < y, x < z\}, \quad Q' = \{x < u, x < v\}$$
$$Q = \{u < v, x < y, x < z\}, \quad Q' = \{x < v\}$$
$$Q = \{u < v, x < y\}, \quad Q' = \{x < v, u < y\}$$

are universally correlated, whereas the (nearly) equally plausible pairs

$$Q = \{x < y\}, \quad Q' = \{x < u < v\}$$
$$Q = \{x < y < z, u < w\}, \quad Q' = \{x < z, u < v < w\}$$
$$Q = \{x < u, y < v\}, \quad Q' = \{x < v, y < v\}$$

are *not* universally correlated.

II. Applications of the FKG Inequality and Its Relatives

In fact, Winkler shows that if Q and Q' are partial orders on a common n-element set which are not universally correlated then there is a partially ordered set on at most $n+1$ elements on which Q and Q' are negatively correlated.

More Applications of FKG

Consider the set of all $2^{\binom{n}{2}}$ labelled graphs on the vertex set $[n]$ with the uniform probability of $2^{-\binom{n}{2}}$ assigned to each graph. Let us call a property P of a graph *increasing* if any graph formed by adjoining additional edges to a graph with property P also has property P. Examples of increasing properties are:
(a) G is Hamiltonian;
(b) G has chromatic number at least k;
(c) G has independence number at most k;
(d) G is connected (or, more generally, k-connected);
(e) G has girth at most k;
(f) G contains a clique of size k.

Using the FKG inequality, it is almost trivial to show that any pair of increasing (or any pair of decreasing) properties are positively correlated. To see this, simply define:

$\Gamma^n :=$ set of all graphs with vertex set $[n]$ partially ordered by inclusion of the edge sets.
$f = I_P$ - the indicator function of property P,
$f' = I_{P'}$ - the indicator function of property P',
$\mu = 1$.

Thus, μ is automatically log supermodular and $\mu(X) = |X|$ for $X \subseteq \Gamma_n$. The FKG inequality therefore applies to this situation and implies at once that P and P' are positively correlated.

Another area in which the FKG inequality has been applied effectively is that of unimodality and log convexity of sequences. Recall that a real sequence $(a_0, a_1, \ldots, a_n)$ is *log convex* if

$$a_k^2 \leq a_{k-1} a_{k+1} \quad \text{for } 1 \leq k \leq n-1.$$

An important property of log convex sequences is that they are unimodal. A typical result of this type (due to Seymour and Welsh [SW]) is the following.

Theorem [SW]. If $(a_0, a_1, \ldots, a_n)$ is log convex and positive and $(b_0, b_1, \ldots, b_n)$ and $(c_0, c_1, \ldots, c_n)$ are both increasing (or both decreasing) then

(18) $$\sum_{k=0}^{n} a_k b_k c_k \sum_{k=0}^{n} a_k \geq \sum_{k=0}^{n} a_k b_k \sum_{k=0}^{n} a_k c_k$$

Note the similarity in form of (18) to the FKG inequality (which is used in [SW] to prove (18)). If all a_k are set equal to 1 then (18) reduces to the previously mentioned result (2) of Chebyshev.

Order Preserving Maps of Partial Orders

Given a finite partially ordered set $(P, <)$ one can weaken the notion of a linear extension and require that a map $\rho: P \to [n]$ only satisfy

$$x < y \text{ in } P \Rightarrow \rho(x) < \rho(y)$$

We will call such a ρ *order-preserving*. Note that in particular ρ need not be $1-1$. It is only natural to expect that many of the results which hold for linear extensions also hold for order-preserving maps. While this in fact may well be true, there are still relatively few theorems available for this class of maps (no doubt, due in part to the fact that they have not been studied as much).

As an example of such an analogue, we mention the following (due to J. W. Daykin and the author).

For a partially ordered set P, define $R(P, n)$ to be the set of all order-preserving maps of P into $[n]$, and for $x \in P$, let

$$\text{range}(x) := \{\rho(x) : \rho \in R(P, n)\}.$$

Theorem: Suppose P is covered by two disjoint *chains* A and B. Let Q and Q' be partial orders on $A \cup B$ both being unions of sets of the form $\{a_{i_1} < b_{i_1}, a_{i_2} < b_{i_2}, \ldots\}$. Furthermore, assume for all pairs $a \in A$, $b \in B$ which are comparable in P that

(19) $$\text{range}(a) \cap \text{range}(b) = \phi.$$

Then Q and Q' are positively correlated in P, i.e.,

$$Pr(Q \text{ and } Q'|P) \geq Pr(Q|P) Pr(Q'|P)$$

where we assume that all maps of P into $[n]$ are equally likely.

The proof is a modification of that used for (13), complicated by the fact that ρ need not be $1-1$. The key new ingredient needed is the result (pointed out by J. W. Daykin [JWD]) that range(a) is always convex, i.e., an interval in $[n]$.

A related result for linear extensions which is not yet known to hold for order-preserving maps is the following beautiful result of Stanley.

For a finite partially ordered set P, an arbitrary element $x \in P$, and an arbitrary positive integer n, let $N_i(P, x, n)$ denote the number of linear extensions $\lambda: P \to [n]$ with $\lambda(x) = i$.

Theorem (Stanley [Sta]). For any P, $x \in P$ and $n \in \mathbb{Z}^+$, the sequence $N_i(P, x, n)$, $1 \leq i \leq n$, is log concave.

II. Applications of the FKG Inequality and Its Relatives

This was conjectured by Chung, Fishburn and Graham (strengthening an earlier conjecture of Rivest [Riv] that the $N_i(p, x, n)$ were always unimodal) who proved it when P could be covered by two disjoint chains (see [CFG]). Stanley's proof uses the Aleksandrov-Fenchel inequalities from the theory of mixed volumes (see [Bus], [Fen]).

So far, no one has been able to establish the corresponding result for order-preserving maps of P into $[n]$ although it must certainly be true.

It would seem that Stanley's theorem (and the analogue for order-preserving maps) should have a proof based on the FKG or AD inequalities. However, such a proof has up to now successfully eluded all attempts to find it.

Concluding Remarks

It is not possible, of course, because of space limitations to explore fully all the recent developments concerning the FKG inequality and its various generalizations and applications. We mention here several sources where the interested reader can find additional material on these topics.

To begin with, a wide variety of FKG-like inequalities have been investigated by Ahlswede and Daykin [AD 1], [AD 2], [AD 3] and Daykin [Day 2], [Day 4]. Kemperman [Kem 1] has given some very pretty extensions of work of Holley [Hol], Preston [Pre] and others [Car], [Bru], [KS] which consider the FKG inequality for measures on partially ordered measure spaces. A number of inequalities related to the FKG inequality have been developed in connection with certain concepts in the statistical theory of reliability (going back at least to Esary, Proschan and Walkup [EPW] in 1967 and Sarkar [Sar] in 1969). The interested reader will find many of these in the book of Barlow and Proschan [BP]. In fact, the FKG inequality is just one among a large class of statistical multivariate inequalities about which a number of survey papers and books have recently appeared (e.g., see [Eat], [Jog], [Ton], [MO], [Kem 2]). An interesting connection between the FKG inequality and majorization on partially ordered sets is given in [Lih]. Also, recent applications of various forms of the FKG inequality to modern theoretical physics can be found in [BR], which in particular contains the following (perhaps unfamiliar) version of FKG:

Theorem: Let $dv = e^w d^n q$ be a probability measure on $\mathbb{R}^n$ with $w \in C^2(\mathbb{R}^n)$. Suppose

$$\partial^2 w / \partial q_i \partial q_j \geq 0, \quad i \neq j.$$

Then

$$\int fg\, dv \geq \int f\, dv \int g\, dv$$

for all increasing functions of $\mathbb{R}^n$ (such that f, g and fg are dv-integrable).

We remark in closing that because of the intimate relation between log supermodular functions and ordinary sub- and supermodular functions (f is supermodular iff $\exp f$ is log supermodular), one suspects that there are in fact deeper connections between inequalities such as FKG and the many other striking properties enjoyed by such functions than are currently known. There is every reason to believe that we have yet to realize the full potential such a more complete understanding could provide.

References

[AD 1] R. Ahlswede and D. E. Daykin, An inequality for the weights of two families of sets, their unions and intersections. Z. Wahrscheinlichkeitstheorie and Verw. Gebiete 43 (1978), 183–185.

[AD 2] –, –, Inequalities for a pair of maps $S \times S \to S$ with S a finite set. Math. Z. 165 (1979), 267–289.

[AD 3] –, –, Integral inequalities for increasing functions, Math. Proc. Camb. Phil. Soc. 86 (1979), 391–394.

[BP] R. E. Barlow and F. Proschan, Statistical Theory of Reliability and Life Testing, Holt, New York, 1975.

[BR] G. A. Battle and L. Rosen, The FKG inequality for the Yukawa$_2$ quantum field theory, J. Statist. Physics 2 (1980), 123–193.

[Bru] Brunel, A., Une preuve elementaire du lemme de Holley. Théorie Ergodique, Lecture Notes in Mathematics, no. 532, Springer-Verlag (1976), 15–18.

[Bus] H. Busemann, Convex Surfaces, Interscience, New York, 1958.

[Car] Cartier, P., Inégalités de corrélation en mécanique statistique. Seminaire Bourbaki 25 (1973), 431–(01-23).

[CFG] F. R. K. Chung, P. C. Fishburn and R. L. Graham, On unimodality for linear extensions of partial orders, SIAM J. Algebraic and Discrete Methods 1 (1980), 405–410.

[Day 1] D. E. Daykin, A lattice is distributive iff $|A||B| \le |A \vee B||A \wedge B|$, Nanta Math., 10 (1977), 58–60.

[Day 2] –, Poset functions commuting with the product and yielding Chebychev type inequalities. In: Problèmes Combinatoires et Théorie des Graphs (Paris 1976), 93–98. Colloques Internationaux du C.N.R.S., No. 260, Paris: C.N.R.S. 1978.

[Day 3] –, Inequalities among the subsets of a set, Nanta Math., 12 (1980), 137–145.

[Day 4] –, A hierarchy of inequalities, Studies in Appl. Math., 63 (1980), 263–274.

[DL] – and L. Lovász, The number of values of Boolean functions, J. London Math. Soc. ser. 2, 12 (1976), 225–230.

[JWD] J. W. Daykin (personal communication).

[Eat] M. L. Eaton, A review of selected topics in multivariate probability inequalities, Ann. Statist. 10 (1982), 11–43.

[EG] P. Erdös and R. L. Graham, Old and new problems and results in combinatorial number theory, Monographie no. 28, l'Enseignement Math., Genève, 1980.

[EPW] J. D. Esary, F. Proschan and D. W. Walkup, Association of random variables with applications, Ann. Math. Statist., 38 (1967), 1466–1474.

[Fen] W. Fenchel, Inégalités quadratique entre les volumes mixtes des corps convexes, C. R. Acad. Sci., Paris, 203 (1936), 647–650.

[FKG] C. M. Fortuin, J. Ginibre and P. N. Kasteleyn, Correlation inequalities on some partially ordered sets, Commun. Math. Phys. 22 (1971), 89–103.

[Gra] R. L. Graham, Linear extensions of partial orders and the FKG inequality, in Ordered Sets, I. Rival, ed., 213–236, D. Reidel, Dordrecht, Holland, 1982.

[GYY] –, A. C. Yao and F. F. Yao, Some monotonicity properties of partial orders, SIAM J. Algebraic and Discrete Methods 1 (1980), 251-258.
[Gri]   R. B. Griffiths, Correlations in Ising Ferromagnets, J. Math. Phys., 8 (1967), 478-483.
[HLP] G. H. Hardy, J. E. Littlewood, and G. Pólya, Inequalities, 2nd ed., Cambridge University Press, New York, 1959.
[Har]   T. E. Harris, A lower bound for the critical probability in a certain percolation process, Proc. Camb. Phil. Soc., 56 (1960), 13-20.
[Hol]   R. Holley, Remarks on the FKG inequalities, Comm. Math. Phys., 36 (1974), 227-231.
[Jog]   K. Jogdeo, Association and probability inequalities, Ann. Statist. 5 (1977), 495-504.
[KS]   D. G. Kelley and S. Sherman, General Griffith's inequalities on correlations in Ising ferromagnets, J. Math. Phys. 9 (1968), 466-484.
[Kem 1] J. H. B. Kemperman, On the FKG-inequality for measures on a partially ordered space, Indag. Math. 39 (1977), 313-331.
[Kem 2] –, Book Review, Bull. Amer. Math. Soc. (New Series) 5 (1981), 319-324.
[Kle 1] D. J. Kleitman, Families of non-disjoint sets, J. Combinatorial Th., 1 (1966), 153-155.
[Kle 2] –, Extremal hypergraph problems, Surveys in Combinatorics, B. Bollobás, ed., London Math. Soc., Lecture Note Series 38, Cambridge Univ. Press (1979), 44-65.
[KS]   – and J. B. Shearer, Some monotonicity properties of partial orders, Studies in Appl. Math. 65 (1981), 81-83.
[Knu]   D. E. Knuth, The Art of Computer Programming, Vol. 3, Sorting and Searching, Addison-Wesley, Reading, Mass., 1973.
[Lith]   K.-W. Lih, Majorization on finite partially ordered sets. (To appear)
[Mar]   J. Marica, Orthogonal families of sets, Canad. Math. Bull. 14 (1971), 573.
[MS]   – and J. Schönheim, Differences of sets and a problem of Graham, Canad. Math. Bull., 12 (1969), 635-637.
[MO]   A. W. Marshall and I. Olkin, Inequalities: Theory of Majorization and Its Applications, Academic Press, New York, 1979.
[Pre]   C. J. Preston, A generalization of the FKG inequalities, Comm. Math. Phys. 36 (1974), 233-241.
[Riv]   R. Rivest (personal communication).
[Sar]   T. K. Sarkar, Some lower bounds of reliability, Tech. Rep. No. 124, Depts. of Op. Res. and Stat., Stanford Univ., 1969.
[Sey]   P. D. Seymour, On incomparable collections of sets, Mathematika, 20 (1973), 208-209.
[SW]   – and D. J. A. Welsh, Combinatorial applications of an inequality from statistical mechanics, Math. Proc. Camb. Phil. Soc. 77 (1975), 485-495.
[She 1] L. A. Shepp, The FKG inequality and some monotonicity properties of partial orders, SIAM J. Algebraic and Discrete Methods 1 (1980), 295-299.
[She 2] –, The XYZ conjecture and the FKG inequality, Ann. of Prob. 10 (1982), 824-827.
[Sta]   R. P. Stanley, Two combinatorial applications of the Aleksandrov-Fenchel inequalities, J. Combinatorial Theory (Ser. A) 31 (1981), 56-65.
[Ton]   Y. L. Tong, Probability Inequalities in Multivariate Distributions, Academic Press, New York, 1980.
[Win 1] P. Winkler, Correlation among partial orders. (To appear)
[Win 2] –, Average height in a partially ordered set, Dis. Math. 39 (1982), 337-341.
[Won]   W. Wong, On a number theoretic conjecture of Graham, Ph. D. Dissertation, Dept. of Math., Columbia Univ., 1981.

Semi-Infinite Programming and Applications*

S.-Å. Gustafson
Department of Numerical Analysis and Computing Science, Royal Institute of Technology, S-10044 Stockholm 70, Sweden.

K. O. Kortanek**
Carnegie-Mellon University, Department of Mathematics, Schenley Park, Pittsburgh, PA 15213, USA

Abstract. An important list of topics in the physical and social sciences involves continuum concepts and modelling with infinite sets of inequalities in a finite number of variables. Topics include: engineering design, variational inequalities and saddle value problems, nonlinear parabolic and bang-bang control, experimental regression design and the theory of moments, continuous linear programming, geometric programming, sequential decision theory, and fuzzy set theory. As an optimization involving only finitely many variables, semi-infinite programming can be studied with various reductions to finiteness, such as finite subsystems of the infinite inequality system or finite probability measures.

This survey develops the theme of finiteness in three main directions: (1) a *duality theory* emphasizing a perfect duality and classification analogous to finite linear programming, (2) a *numerical treatment* emphasizing discretizations, cutting plane methods, and nonlinear systems of duality equations, and (3) *separably-infinite programming* emphasizing its uniextremal duality as an equivalent to saddle value, biextremal duality. The focus throughout is on the fruitful interaction between continuum concepts and a variety of finite constructs.

1. Introduction: Consequences of the Finite in Semi-Infinite Programming

Semi-infinite programming is a next level of extension of elementary finite linear programming where now finitely many variables occur in infinitely many constraints. The work of many people over at least 20 years has shown the relative ease of formulating models of this type, while significant computational advances were made in solving such problems during the 70's.

As a typical problem of this type we take the liberty of quoting the concise description of the air pollution problem appearing in the [Chui and Ward, 1981] review of [Krabs, 1979]. The formulation is a one-sided approximation problem.

---

* This paper is dedicated to Professor Abraham Charnes on his 65th birthday.
** Preparation of this paper was supported in part by National Science Foundation Grant ECS-8209951.

II. Semi-Infinite Programming and Applications

"In a given two-dimensional control region S, a certain air quality is to be guaranteed. At the same time, the yearly average concentration of a pollutant (e.g. sulphur dioxide or carbon monoxide) is to be kept below a prescribed standard which is described by a real-valued function φ on S. The concentration observed in S is assumed to come from two kinds of sources: (a) sources which can be controlled and hence regulated, and (b) sources which cannot be controlled.

If, say, n controllable sources are present, then it is assumed that these contribute a yearly average $u_1, \ldots, u_n$ to the air pollution. We denote by u_0 the contribution from the uncontrollable sources. Then $u_0, u_1, \ldots, u_n$ are again real-valued functions on S whose actual determination as a rule is naturally a very difficult problem. From the requirement that the concentration φ is not to be exceeded, we are first led to the side conditions $\sum_{j=1}^{n} u_j(s) + u_0(s) \leqslant \varphi(s)$ for all $s \in S$. If any of these conditions is not satisfied, then the contributions of the controllable sources are reduced and, in fact, the jth source is reduced by a factor x_j, $0 \leqslant x_j \leqslant 1$ for $j = 1, \ldots, n$, so that the side conditions

$$\sum_{j=1}^{n} (1-x_j) u_j(s) + u_0(s) \leqslant \varphi(s) \quad \text{for all } s \in S \qquad (*)$$

are satisfied. Obviously these will be satisfied, if, for all $s \in S$, the condition $u_0(s) \leqslant \varphi(s)$ is satisfied.

If any $x_j \neq 0$ must be chosen, then, in general, costs naturally arise (e.g. by changing production plans, introducing air purifiers, etc.). In the simplest case these costs are assumed to be proportional to x_j. Let c_j be the proportionality constant, so that with the choice of reduction factors $x_1, \ldots, x_n \in [0, 1]$ the total cost is

$$c(x_1, \ldots, x_n) = \sum_{j=1}^{n} c_j x_j. \qquad (**)$$

One now tries to choose the factors $x_1, \ldots, x_n$ subject to the side conditions (*) so that the cost $c(x_1, \ldots, x_n)$ is as small as possible."

A numerical treatment of a semi-infinite program usually employs duality results patterned after the duality theory of finite linear programming. One readily sees what duality to expect under the best of circumstances if one arbitrarily chooses a finite number of constraints to observe while ignoring all others. For example, when all constraints are linear as in (*), then this choice results in a finite list of dual variables appearing in the associated dual pair of reduced linear programs. This pair strongly suggests what a dual program to the original infinitely constrained linear program should be, but one needs a structure to include all such choices and their associated sets of dual variables as part and parcel of a stand-alone optimization. The construction involves an infinite number of variables, as it must, but only a finite number of them ever need be different from zero. Roughly speaking, the dual program must allow finite sets of constraints to be freely chosen while for any such choice the associated finite list of dual variables must obey the duality laws of ordinary linear programming. The construction applies to certain nonlinear programs as well.

Let us first rewrite (*) and (**) in standard form and proceed straightaway to the construction of its dual program.

Seek
$$\min \sum_{j=1}^{n} c_j x_j \tag{1a}$$

subject to
$$\sum_{j=1}^{n} x_j u_j(s) \geq u_{n+1}(s), \quad \text{for all } s \in S \tag{1b}$$

$$x_j \geq 0 \tag{1c}$$

and
$$x_j \leq 1 \quad \text{for } j=1,\ldots,n, \tag{1d}$$

where $u_{n+1}(s) = \sum_{j=0}^{n} u_j(s) - \varphi(s)$. The value $u_{n+1}(s)$ is the excess pollution above standard at point s before emission reductions. Choose now a finite subset of S, say $s_1, \ldots, s_\ell$, and let the list of dual variables associated with the reduced system of inequalities be denoted by $\{\lambda_1, \ldots, \lambda_\ell\}$. Let y_{jL} and y_{jU} denote the dual variables respectively for (1c) and (1d), $j=1, \ldots, n$. Then the linear programming dual to the reduced problem (1) is:

Seek
$$\max \sum_{r=1}^{\ell} u_{n+1}(s_r) \lambda_r - \sum_{j=1}^{n} y_{jU} \tag{2a}$$

subject to
$$\sum_{r=1}^{\ell} u_j(s_r) \lambda_r + y_{jL} - y_{jU} = c_j \quad \text{for } j=1,\ldots,n \tag{2b}$$

and
$$\lambda_r \geq 0, \ r=1,\ldots,\ell; \quad y_{jL}, y_{jU} \geq 0 \quad \text{for } j=1,\ldots,n. \tag{2c}$$

Obviously program (2) depends on the choice of the finite subset of S. The structure which allows finite subsets to be freely chosen is what has been termed the space of *generalized finite sequences* over S with respect to $\mathbb{R}$, denoted $\mathbb{R}^{(S)}$, introduced in [Charnes, Cooper, and Kortanek, 1962, 1963].

Definition 1. Let S be a non-empty set. Let $\mathbb{R}^{(S)}$ be the set of all real valued functions on S having only finitely many non-zero values. Algebraic operations on the set $\mathbb{R}^{(S)}$ are as follows. For any λ and $\mu \in \mathbb{R}^{(S)}$ and real number r, the sum $\lambda + \mu \in \mathbb{R}^{(S)}$ is defined point-wise by $(\lambda + \mu)(s) = \lambda(s) + \mu(s)$, while the scalar multiplication $r\lambda$ is also determined by pointwise multiplication, i.e. $(r\lambda)(s) = r\lambda(s)$ for all $s \in S$. For any $\lambda \in \mathbb{R}^{(S)}$ define its *support* by $\text{supp}\lambda = \{s \in S | \lambda(s) \neq 0\}$. □

Employing this construction yields the desired dual program to (1).

Seek
$$\max \sum_{s \in \text{supp}\lambda} u_{n+1}(s) \lambda(s) - \sum_{j=1}^{n} y_{jU} \tag{3a}$$

from among those $\lambda \in \mathbb{R}^{(S)}$ and $y_{jL}, y_{jU} \in \mathbb{R}, j=1, \ldots, n$ satisfying

$$\sum_{s \in \text{supp}\lambda} u_j(s) \lambda(s) + y_{jL} - y_{jU} = c_j, \quad j=1,\ldots,n \tag{3b}$$

and $\lambda(s) \geq 0$ for all $s \in S$ and

$$y_{jL}, y_{jU} \geq 0 \quad \text{for } j=1, \ldots, n. \tag{3c}$$

Let us review some interpretations of program (3) following [Gorr and Kortanek, 1971] and [Gorr, Gustafson and Kortanek, 1972] and [Samuelsson, 1972].

The variable $\lambda(s)$ is the imputed value of a unit reduction in pollution at point $s \in S$. For simplicity let us disregard the variables y_{jL} and y_{jU}. Then program (3) seeks to maximize the imputed value of total reduction of excess pollution over the entire region S, subject to each pollutor's contribution to this value being equal to his incurred cost. The numbers $\{\lambda(s)\}$ act as *emission charges or prices* to provide incentives to pollutors to reduce emissions. For example, if pollutor j is offered the amount $\left[\sum_{s \in \text{supp}\lambda} u_j(s)\lambda(s)\right] x_j$ to reduce emissions by the fraction x_j, then at optimum this should equal his incurred cost in doing so according to (3b). The interpretation can be extended when the y_{jL} and y_{jU} variables are considered and also when the cost function (**) is nonlinear, see [Gustafson and Kortanek, 1973c] and [Gribik, 1979b].

When the coefficient functions $\{u_j\}_{j=1}^n$ are convex and u_{n+1} is concave, then program (1) is in duality with a one-variable linear program with variable coefficients of Dantzig and Wolfe's generalized linear programming see [Dantzig, 1963] and [Kortanek, 1976]. Programs of this type have been termed "C/C semi-infinite linear programs" in [Parks and Soyster, 1983], where it is shown that they are related to certain problems in set-inclusive programming [Soyster, 1973] and fuzzy set programming [Negoita and Sularia, 1976].

In this paper we have carried the introductory theme on the consequences of finiteness into three main sections: 2. Duality Theory; 3. Numerical Treatment; and 4. Saddle Value Problems and Separably-Infinite Programs. The outcome is a collection of 9 theorems, with 4 in Section 2, 4 in Section 3, and 1 in Section 4. Each theorem involves finiteness in some form or another such as conditions on finite subsystems of inequalities, program value attainment or program feasibility with a finite set of variables, or simply a discretization of an index set.

We have focused our review primarily on the 20-year period, 1962-1982.

2. Duality Theory

2.1. General Linear Inequality Systems

In finite linear programming (2) is usually referred to as the primal program. It will be convenient to maintain this tradition upon the introduction of generalized finite sequences. We restate program (3) with the set S now being arbitrary.

Program P. Let S be an arbitrary non-empty set and $u = (u_1, \ldots, u_n)$ be a finite list of real valued functions on S. Let u_{n+1} also be real valued on S, and $b \in \mathbb{R}^n$.

Find $\qquad v_P = \sup_s \sum_s u_{n+1}(s) \lambda(s)$

from among $\qquad \lambda \in \mathbb{R}^{(S)}, \lambda \geq 0$ satisfying

$$\sum_s u(s) \lambda(s) = b, \qquad (4)$$

where $\qquad \sum_s u(s)\lambda(s)$ is defined to be $\sum_{s \in \text{supp} \lambda} u(s)\lambda(s)$.

Program D. Same assumptions as in P.

Find $\qquad v_D = \inf y \cdot b$

from among $\quad y \in \mathbb{R}^n$ satisfying

$$y \cdot u(s) \geq u_{n+1}(s), \quad \text{for every } s \in S. \qquad (5)$$

When both programs are consistent, then the well known *duality inequality* holds for their values:

$$v_D \geq v_P. \qquad (6)$$

It is because of the duality inequality (6) that the program pair (P, D) is termed *dual*. In general, of course, one seeks conditions to guarantee the *duality equality* $v_D = v_P$.

Let us examine more carefully what happens when we consider dual pairs of finite linear programs obtained by observing only finitely many of the linear inequalities in (5). After all, these reductions have provided a motivation for introducing generalized finite sequences in Program P.

An example provides some insights.

Example 1. Let $S = \{s | 0 \leq s \leq 1\}$, $u_1(s) = 1$, $u_2(s) = s$, and $u_3(s) = \sqrt{s}$ for each $s \in S$, while $b = (1, 0)$. Program D seeks a one-sided affine approximation to u_3 having smallest intercept, i.e.,

$$v_D = \inf y_1$$

subject to $\quad y_1 + y_2 s \geq \sqrt{s}$ for all s, $0 \leq s \leq 1$. $\qquad (7)$

It is easy to verify that (7) is consistent, and that if (y_1, y_2) is feasible, then $y_1 > 0$. One can also see that the infimum value v_D is zero but not attained. The primal program has a unique optimal solution obtained by assigning "mass" one at the point $0 \in S$, namely define $\lambda^* \in \mathbb{R}^{(S)}$ by

$$\lambda^*(s) = \begin{cases} 1 & \text{if } s = 0 \\ 0 & \text{if } s \neq 0. \end{cases}$$

Example 1 illustrates the duality equality $v_P = v_D$ with attainment in P but non-attainment in D. Example 1 also illustrates another fact. If we adjoin $-y_1 \geq 0$

II. Semi-Infinite Programming and Applications

to (7), then every finite subsystem is consistent, but the augmented infinite inequality system is not.

"Finite subsystem consistency", namely where every finite subsystem of (5) is consistent has, nevertheless, certain implications for the entire inequality system. In this situation (5) is consistent in an extended number system defined as follows.

Definition 2. Let $\mathbb{R}[\theta]$ denote the polynomial ring obtained by adjoining a transcendental θ to the real number field $\mathbb{R}$, where the non-Archimedean ordering is derived from the infinite valuation in which $a < \theta$ for all $a \in \mathbb{R}$. For any polynomial $\sum_{r=0}^{q} a_r \theta^r \in \mathbb{R}[\theta]$, the polynomial part $\sum_{r=1}^{q} a_r \theta^r$ is termed its *infinite part* and a_0 is termed its *real part*. See [Zariski and Samuel, 1950], Chapter 1. $(\mathbb{R}[\theta])^p$ shall denote p copies of $\mathbb{R}[\theta]$. Corresponding to $y(\theta) = \sum_{r=0}^{q} a_r \theta^r$, define $y(k) = \sum_{r=0}^{q} a_r k^r$ for $k \in \mathbb{R}$.

They key tool is the following non-Archimedean Helley theorem, [Jeroslow and Kortanek, 1971].

Theorem 1. Assume that for every finite subset $T = \{s_r | r = 1, \ldots, \ell\}$ of S, the finite subsystem of (5),

$$y \cdot u(s_r) \geq u_{n+1}(s_r), \quad r = 1, \ldots, \ell$$

is consistent over $\mathbb{R}^n$. Then there is a solution $y(\theta) \in (\mathbb{R}[\theta])^n$ to (5) itself such that each of its components $y_j(\theta), j = 1, \ldots, n$, is a polynomial in θ of degree at most n.

The following example shows that polynomials of full degree n may be necessary:

$$
\begin{aligned}
y_1 &\geq k \\
-ky_1 + y_2 &\geq 0 \\
&\vdots \\
-ky_{n-1} + y_n &\geq 0,
\end{aligned}
$$

for all positive integers k.

Theorem 1 can be used to prove Blair's result on existence of ascent vectors. If (5) is inconsistent, then there exists $a \in \mathbb{R}^n$ having the following property:

for every positive integer n, there exists a finite subset $T = \{s_r | r = 1, \ldots, \ell\}$ such that $y \cdot u(s_r) \geq u_{n+1}(s_r), r = 1, \ldots, \ell$ implies $y \cdot a \geq n$, see [Blair, 1974].

Such vectors a are termed *ascent vectors*, and they can be used to construct asymptotic solutions to (4) for purposes of determining Duffin's concept of the *subvalue* of a linear program [Duffin, 1956]. Theorem 1 can also be used to characterize the set of ascent vectors, see [Kortanek, 1977a] and another alternative characterization [Blair, 1983].

In the extended number system we see now that the minimum in Example 1 is attained: set $\bar{y}_1 = 0$ and $\bar{y}_2(\theta) = r\theta$ for any fixed positive real r. Actually, Theorem 1 suggests an expansion of Program D to the following one, and a corollary on perfect duality, a term introduced in [Duffin, 1973].

Program D*. Same assumptions as P.

Find $\quad v_{D^*} = \inf t$

from among $\quad t \in \mathbb{R}$ and $y(\theta) \in (\mathbb{R}[\theta])^n$ which satisfy

$$t - y(\theta) \cdot b \geq 0$$

and $\quad y(\theta) \cdot u(s) \geq u_{n+1}(s)$ for every $s \in S$.

Corollary 1 [to Theorem 1]. Programs P and D* are in perfect duality in the sense that the following two properties hold:
  (p1) if one program is consistent and has a finite value, then the other program is consistent, and
  (p2) if both programs are consistent, then they have equal finite values.
Moreover, when v_{D^*} is finite, it is attained.

Theorem 1 may be used to introduce polynomial subgradients in convex analysis to assure the duality equality for finite valued programs, see [Kortanek and Soyster, 1981], and [Borwein, 1980] where a new polynomial subgradient calculus is developed.

Example 2 *[Polynomial Subgradients]*. Let E be any real linear space and let G be a bifunction from $\mathbb{R}^n$ to E, that is, to each $\tau \in \mathbb{R}^n$ (perturbation space) is associated an extended real-valued function on E, $G\tau: E \to [-\infty, \infty]$. Assume that G is a proper closed bifunction, which means that the graph function defined by $g(\tau, x) = (G\tau)(x)$ is proper, closed, and convex on $\mathbb{R}^n \times E$, see [Rockafellar, 1970]. In this case Program P becomes:

Seek $\quad (\inf G)(0) = \inf_{x \in E} (G0)(x)$.

Because $g(\tau, x)$ is convex, Program P is equivalent to its *convexification*:

P_C: Find $\quad (\inf G)(0) = \inf \sum_{(\tau, x)} g(\tau, x) \lambda(\tau, x)$

from among $\quad \lambda \in \mathbb{R}^{(\mathbb{R}^n \times E)}$ satisfying $\lambda \geq 0$

$$\sum_{(\tau, x)} \lambda(\tau, x) = 1, \quad \text{and}$$

$$\sum_{(\tau, x)} \tau_r \lambda(\tau, x) = 0, \quad \text{for } r = 1, \ldots, n.$$

However, Program P_C is a primal semi-infinite program with finite probability measures as variables, and with the appropriate identifications we obtain its polynomial (perfect) dual:

Program D_C^*.

Find $\quad \sup t$

from among $\quad t \in \mathbb{R}$, $y_0(\theta) \in \mathbb{R}[\theta]$, and $y(\theta) \in (\mathbb{R}[\theta])^n$ which satisfy

$$t - y_0(\theta) \leq 0$$

and
$$y_0(\theta)+y(\theta)\cdot\tau\leqslant g(\tau,x) \tag{8}$$
for every $(\tau,x)\in\mathbb{R}^n\times E$.

When $(\inf G)(0)$ is finite, then by perfect duality $y_0(\theta)=(\inf G)(0)$ and (8) implies
$$(\inf G)(0)+y(\theta)\cdot\tau\leqslant(\inf G)(\tau) \tag{9}$$
for every $\tau\in\mathbb{R}^n$. We summarize this result as follows.

Corollary 2 [to Theorem 1]. Assume that a proper closed bifunction G defined on $\mathbb{R}^n\times E$ has $(\inf G)(0)$ finite. Then the infimal function $\inf G$ has a polynomial subgradient in $(\mathbb{R}[\theta])^n$ at $\tau=0$.

Observe that without polynomial "multipliers" the dual program would be the Lagrangian dual:

Seek $\qquad LD = \sup\limits_{y\in\mathbb{R}^n}\ \inf\limits_{(\tau,x)} \{g(\tau,x)-y\cdot\tau\}$

from among those y for which the inner infimum is finite,

see [Luenberger, 1968] or [Rockafellar, 1970].

As an illustration consider
$$g(\tau,x)=\begin{cases} e^{-x_2} & \text{if } (x_1^2+x_2^2)^{1/2}-x_1\leqslant\tau \\ +\infty & \text{otherwise} \end{cases}$$
where $\tau\in\mathbb{R}$ and $x\in\mathbb{R}^2$. Here $(\inf G)(0)=1$ and $y_1(\theta)=\theta$ is a particular polynomial subgradient of the infimal function at $\tau=0$. This is a classical example in convex analysis of the occurance of a duality gap, for here the Lagrangian dual has value 0, i.e., $LD=0$. Polynomial subgradients are one way of removing duality gaps for finite-valued convex programs without imposing any additional assumptions, see [Kortanek, 1976].

The non-Archimedean Helley theorem and ascent ray theorem are types of direct reductions, where an infinite system or construction is replaced by a finite one. Further results involving either primal or dual program reductions appear in generality in [Ben-Tal, Ben-Israel, and Rosinger, 1979] and [Borwein, 1981a, 1981b], and are pursued in the next section. Other methods of removing duality gaps through approximation are given in [Duffin and Karlovitz, 1965] and [Karney, 1981a].

2.2 Semi-Infinite Programs with Linear Programming Duality Features

Duality relationships between Programs $P(4)$ and $D(5)$ are readily displayed in tabular form according to the following states for each program.

A *program is:*
(i) consistent and bounded (CONS, BD) if and only its value is finite,
(ii) consistent and unbounded (CONS, UBD) if and only if it is consistent without finite value,
(iii) inconsistent (INC) if and only if its constraints cannot be satisfied.

A *duality state* of the pair (P, D) is a combination of the states of P and D. The permissible and impossible duality states are conveniently presented in classification tables. In general, many more duality states arise in semi-infinite programming than in finite linear programming. The study can be approached by examining the convex cone spanned by the coefficients of all of the variables of Program P. In finite linear programming this cone is closed, while in general it is not.

Definition 3. The *moment cone* C associated with Program P is specified by

$$C_P = \left\{ \omega \in \mathbb{R}^{n+1} \Big| \omega = \sum_s \binom{u(s)}{u_{n+1}(s)} \lambda(s), \text{ for every } \lambda \in \mathbb{R}^{(S)}, \lambda \geq 0 \right\}. \quad \square$$

C_P is a convex cone containing 0. In [Karlin and Studden, 1966] it is termed the *moment space with respect to* $\{u:u_{n+1}\}$. Another important convex cone arises out of examining a homogenization of Program P, which itself gives information about boundedness of P, see [Duffin, 1956].

Definition 4. The *homogeneous derivant* or Program P is the following.

Program HP: Seek $\lambda \in \mathbb{R}^{(S)}$ such that $\sum_s u(s)\lambda(s) = 0$ and $-\sum_s u_{n+1}(s)\lambda(s) \leq -1$.
$\square$

The finite linear programming classification table is obtained when the moment cone C_{HP} of Program P is closed. A proof can be constructed from elementary separation arguments, see [Gustafson 1973b] and [Gustafson, Kortanek, and Samuelsson, 1974].

Theorem 2. Assume C_{HP} is closed. Then out of 9 conceivable duality states indicated in the classification table below, only 4 are permissible and so numbered. The remaining 5 impossible duality states are indicated with the letter "I". Moreover, in state 1, $v_P = v_D$ and v_P is attained.

Table 1. Permissible and impossible duality states for programs P and D

D \ P		CONS		INC
		BD	UBD	
CONS	BD	1	I	I
	UBD	I	I	2
INC		I	3	4

For consistent programs D the closure of the moment cone of the homogeneous derivant C_{HP} is weaker than the closure of C_P. For this case, the closure

II. Semi-Infinite Programming and Applications

of C_{HP} characterizes the "uniform LP duality" defined by the first two rows of Table 1 for any $b \in \mathbb{R}^n$, see [Duffin, Jeroslow and Karlovitz, 1983].

There are two conditions known for some time which when taken together are sufficient for C_P and C_{HP} to be closed.

(a) $\left\{ \begin{pmatrix} u(s) \\ u_{n+1}(s) \end{pmatrix} \middle| \text{ every } s \in S \right\}$ is compact in $\mathbb{R}^{n+1}$

and

(b) there exists $y^0 \in \mathbb{R}^n$ such that
$$y^0 \cdot u(s) > u_{n+1}(s) \quad \text{for every } s \in S.$$

The linear inequality system is said to be *canonically closed* if and only if (a) and (b) hold, see [Charnes, Cooper and Kortanek, 1969] and [Glashoff, 1979]. Within the last year a question that has been open for at least 20 years has been resolved in [Duffin, Jeroslow and Karlovitz, 1983]. They characterized the closure of C_{HP} when $D(5)$ is consistent by means of compactness (a), and interiority (b), except for at most a finite number of inequalities.

A characterization of the closure of C_P stated in terms of classification tables is given in [Gustafson, Kortanek and Samuelsson, 1974].

Analogous to Definition 4 one can define the homogeneous derivant for D and investigate more general duality relationships. When the duality states (and their asymptotic extensions) are considered simultaneously for both programs and their homogeneous derivants, then out of 121 mutually exclusive and collectively exhaustive duality states only 11 are possible, see [Kortanek, 1974] and [Kallina and Williams, 1971].

It is an easy consequence of linear programming theory, or alternatively Caratheodory's lemma, that when P is consistent it can be written in the following equivalent form.

Program P_n.

Find $\quad v_P = \sup \sum_{i=1}^{n} u_{n+1}(s_i) \lambda_i$

from among non-negative reals λ_i and points $s_i \in S$, $i = 1, 2, \ldots, n$ which satisfy

$$\sum_{i=1}^{n} u(s_i) \lambda_i = b. \quad \square$$

This program indicates that we need only consider solutions to P with at most n variables taking non-zero values.

For numerical purposes we impose continuity requirements on the coefficient functions in P over a subset S of $\mathbb{R}^n$. Towards this study we give now necessary and sufficient conditions for primal and dual feasible solutions to be optimal.

Theorem 3. *Let $\{\lambda_i\}_{i=1}^n$, $\{s_i\}_{i=1}^n$ be feasible for P_n and y feasible for D. Then these are optimal solutions if and only if the following two conditions hold.*

Complementary Slackness:

$$\lambda_i [y \cdot u(s_i) - u_{n+1}(s_i)] = 0, \quad i = 1, \ldots, n \quad (10\,\text{a})$$

and

Local Minimality:

$$y \cdot u(s) - u_{n+1}(s) \text{ has a local minimum} \quad (10\,\text{b})$$

at s_i whenever $\lambda_i > 0$. □

The proof of this theorem parallels finite linear programming with some subtlety arising from (10b). But y-feasibility requires $y \cdot u(s) - u_{n+1}(s)$, a continuous function of s, to be non-negative over S with a zero at s_i. Therefore, s_i is a local minimum of this continuous function.

Equations (10a) and (10b) represent equations to be solved for with respect to $\lambda_i \in \mathbb{R}$, $s_i \in S$ for $i = 1, 2, \ldots, n$ and $y \in \mathbb{R}^n$. By *system NL* (nonlinear) we shall mean (10a), (10b), and (10c) where the latter is:

$$\sum_{i=1}^{n} u(s_i) \lambda_i = b. \quad (10\,\text{c})$$

Observe that system NL provides necessary conditions for optimality. Any solution to system NL which also satisfies primal non-negativity

$$\lambda_i \geq 0, \quad i = 1, 2, \ldots, n$$

and dual feasibility

$$y \cdot u(s) \geq u_{n+1}(s) \quad \text{for every } s \in S$$

is respectively optimal for P and D.

System NL was first developed in [Gustafson, 1970] and extended to higher dimensional examples in [Gustafson and Kortanek, 1973a]. System NL has also been developed for a class of problems involving an infinite number of variables and an infinite number of constraints, see [Charnes, Gribik and Kortanek, 1980a].

System NL gives necessary conditions for solution of a more general semi-infinite programming formulation of [John, 1948] and also studied extensively in [Hettich and Zencke, 1982].

Program G. Let S be compact in $\mathbb{R}^m$, $Z_0 \subset \mathbb{R}^n$. Let $f: Z_0 \to \mathbb{R}$ and $g: Z_0 \times S \to \mathbb{R}$ be continuously differentiable. Find

$$v_G = \inf f(y) \quad (11\,\text{a})$$

from among those $y \in Z_0$ which satisfy

$$g(y, s) \leq 0 \quad (11\,\text{b})$$

for every $s \in S$.

The following result is proved in [Hettich and Zencke, 1982] and is an extension of Theorem 3.

Theorem 4. *Let y^* be an optimal solution to Program G. Assume that there exists $\xi \in \mathbb{R}^n$ such that*

$$\xi \cdot g_y(y^*, s) < 0 \quad \text{whenever} \quad g(y^*, s) = 0,$$

where
$$g_y(y^*, s) = \left(\frac{\partial g}{\partial y_1}(y^*, s), \ldots, \frac{\partial g}{\partial y_n}(y^*, s)\right).$$

Then there are q positive numbers, x_i, $i = 1, \ldots, q$ such that

$$f'(y^*) + \sum_{i=1}^{q} x_i g_y(y^*, s_i) = 0, \tag{12a}$$

$$x_i g(y^*, s_i) = 0, \quad i = 1, 2, \ldots, q, \tag{12b}$$

and

$$g(y^*, \cdot) \text{ has a local maximum at each } s_i. \tag{12c}$$

3. Numerical Treatment

3.1. Algebraic Representation Theorems

Two basic representation theorems of linear programming extend to semi-infinite programming: the Linear Independence with Extreme Points [LIEP] Theorem and the Opposite Sign [OSP] Theorem. The significance of these theorems for numerical treatment is that the *purification algorithm* [Charnes, Kortanek and Raike, 1965], which transforms any non-extreme point to an extreme point having objective function value at least as great, extends in a constructive way to semi-infinite programming.

Definition 5. The *feasible set* Λ of Program $P(4)$ is

$$\Lambda = \{\lambda \in \mathbb{R}^{(S)} | \sum_s u(s)\lambda(s) = b, \lambda \geq 0\}.$$

The subset $\{u(s)|s \in S\}$ of $\mathbb{R}^n$ is said to have the *opposite sign property* if and only if:

$$\alpha \in \mathbb{R}^{(S)}, \alpha \neq 0, \text{ and } \sum_s u(s)\alpha(s) = 0 \text{ implies}$$

$$\alpha(s)\alpha(s') < 0 \text{ for some } s \text{ and } s' \text{ in } S.$$

Theorem 5 [LIEP]. Assume Λ is non-empty. Then, $\lambda \in \Lambda$ is an extreme point of Λ if and only if $\{u(s)|s \in \text{supp}\,\lambda\}$ is a linearly independent set.

Theorem 6 [OSP]. A non-empty feasible set Λ is spanned by its extreme points if and only if $\{u(s)|s \in S\}$ has the opposite sign property.

These theorems appear in [Charnes, Cooper and Kortanek, 1963].

Example 3. Let S be the set of positive integers and define u on S by $u(n) = 2^{-n}$, for every $n \in S$. Let $b = 1$. Then the list of extreme points of Λ is specified by

$$\lambda^{(k)}(n) = \begin{cases} 2^n & \text{if } k = n \\ 0 & \text{if } k \neq n, \text{ for every } k \in S. \end{cases}$$

Theorems 5 and 6 hold if the real numbers $\mathbb{R}$ are replaced by any ordered field. These results are essentially of a finite, algebraic character in the sense that conditions are stated in terms of finite sets of vectors or a finite number of non-zero coordinates in an otherwise infinite sequence.

[Perold, 1981] obtains characterizations of extreme points arising in continuous time programming, see also [Levinson, 1966], and [Perold, 1978], where in the latter report a continuous time simplex method is presented.

Let us continue the approach to solving problems involving the continuum with yet other "finiteness" concepts.

3.2. Finite Discretizations and Linear Programming Approximations

In the introductory section the primal program P was constructed by a finiteness argument beginning with the infinitely constrained linear program D. On the other hand, choosing finite subsystems of the constraints (6) of D easily yields approximate solutions to D and P themselves. Each choice determines a *finite discretization* of the index set S, and the approximate solutions are obtained by solving the dual pair of reduced linear programs. At this point, however, the infinite set of linear inequalities may not yet be satisfied. Uniform discretizations may result in very large linear programs, but often a good approximation may be obtained through an appropriate discretization of S, see [Glashoff and Roleff, 1981] and [Streit and Nuttall, 1981]. These papers treat Chebysev approximation in complex variables which has application to the problem of antenna design.

There is another important use which a discretization serves. The dual pair of linear programming solutions provides a starting solution to system NL, (10). When the coefficient functions have derivatives, then the local minimality conditions (10b) become derivative conditions at points which are interior to S. Additional derivative equations arise from boundary points of S, and these must also be included in the gradient equivalent to (10b).

Having obtained a starting solution Newton-Raphson iterations can now be used to solve system NL. If the Newton-Raphson method converges, one need only check $\lambda_i \geqslant 0$ in (10c) for primal feasiblity, but the test for dual feasibility (5) can be difficult. If both feasibility tests are passed, then the results are indeed optimal for programs P and D by Theorem 3. If the Newton-Raphson method fails to converge, or if it converges to a point for which primal and dual feasibility is not attained, then the discretization is refined and the process repeated. The approach is based on [Gustafson, 1970], and a computer code has been developed in [Fahlander, 1973]. Convergence of optimal solutions to discretizations as the mesh of the discretization converges to 0 is established in [Gustafson and Kortanek, 1973a]. More general discretization methods are presented in [Hettich and Zencke, 1982].

The actual choice of discretization may be guided by other solution methods which generate sequences whose limit points solve the semi-infinite dual pair. The class of cutting plane methods is of this type. A cutting plane method may be stopped at any point during the course of iterations, and the resulting trial solution reduced by the purification algorithm to obtain a dual pair of basic optimal solutions. This pair is then the next starting solution for the Newton-Raphson method applied to system NL. We examine two of these cutting plane methods next. Clustering procedures for reducing the number of mass points are described in [Gustafson and Kortanek, 1973a], and by [Federov, 1972] in the context of statistical experimental design.

3.3. Cutting Plane Methods

We consider the following specialized version of Program D.

Program $\bar{D}$.

Find $\quad v_{\bar{D}} = \min y \cdot b$

from among $\quad y \in \mathbb{R}^n$ satisfying

$$y \cdot u(s) \geq u_{n+1}(s), \quad \text{for every } s \in S \tag{13a}$$

and

$$y \cdot a_i \geq d_i, \quad i = 1, \ldots, m \tag{13b}$$

under the following assumptions:
(i) $b \neq 0$, (ii) S is compact in $\mathbb{R}^m$, (iii) the polyhedral set defined by $a_i \in \mathbb{R}^n$, $d_i \in \mathbb{R}, i = 1, \ldots, m$ is non-empty and compact, (iv) the coefficient functions are continuous.

The following procedure is related to the classical cutting plane method for convex programs. It has no provision for deleting any of the cuts which are generated during the course of the iterations.

Alternating Algorithm for Program $\bar{D}$ (13) [Gustafson and Kortanek, 1973a].

Step 0: Let D_0 be the program

$$\min y \cdot b$$
$$\text{subject to} \quad y \cdot a_i \geq d_i, \quad i = 1, \ldots, m.$$

Let $k = 1$.

Step 1: Let y_k be a solution to Program D_{k-1}.

Step 2: Let s_k be a member of S for which $y_k \cdot u(s_k) - u_{n+1}(s_k) = \delta(y_k)$, where

$$\delta(y_k) = \min_{s \in S} [y_k \cdot u(s) - u_{n+1}(s)], \text{ the } \textit{discrepancy}.$$

If $\delta(y_k) \geq 0$, stop; y_k is optimal for $\bar{D}$. Otherwise, continue.

Step 3: Form Program D_k by adding the constraint

$$y \cdot u(s_k) \geq u_{n+1}(s_k)$$

to Program D_{k-1}. Set $k := k+1$ and return to Step 1.

It has been shown that if the method does not terminate, then limit points of the sequence $\{y_k\}_{k=1}^{\infty}$ are optimal for $\bar{D}$.

The next method does not require the most violated constraint (Step 2) and has provisions for dropping constraints. It is an extension of the [Elzinga and Moore, 1975] central cutting plane method, but it requires an interiority assumption.

Semi-Infinite Central Cutting Plane Algorithm [Gribik, 1979 a]

Adjoin the following assumption to Program $\bar{D}$: (v) there exists a feasible $\hat{y} \in \mathbb{R}^n$ for which $\hat{y} \cdot u(s) > u_{n+1}(s)$ for every $s \in S$.

Step 0: Let $\bar{v} > v_D$ and let SD_0 be the program

$$\max \sigma$$

subject to $\quad y \cdot b + \|b\| \sigma \leq \bar{v}$

and $\quad\quad y \cdot a_i \quad\quad \geq d_i, \quad i = 1, \ldots, m.$

Choose y_0 satisfying $y_0 \cdot a_i \geq d_i$, $i = 1, \ldots, m$ and let $k = 1$.

Step 1: Let $(y'_k, \sigma_k) \in \mathbb{R}^n \times \mathbb{R}$ be a solution to SD_{k-1}. If $\sigma_k = 0$, stop; otherwise go to Step 2.

Step 2: Delete constraints from SD_{k-1} according to either or both of the deletion rules or do not delete constraints. Call the resulting program also SD_{k-1}.

Step 3: (i) If $\delta(y'_k) \geq 0$, then add the constraint $y \cdot b + \|b\| \sigma \leq y'_k \cdot b$ to Program SD_{k-1}. Set $y_k = y'_k$.
(ii) Otherwise, find $s_k \in S$ such that

$$y'_k \cdot u(s_k) - u_{n+1}(s_k) \leq 0,$$

and add the constraint

$$y \cdot u(s_k) - \|u(s_k)\| \sigma \geq u_{n+1}(s_k)$$

to Program SD_{k-1}. Set $y_k = y_{k-1}$.

In either case, call the resulting program SD_k. Set $k := k+1$ and return to Step 1.

The deletion rules are more technical and are omitted for our purposes. The analogous convergence result maintains for the central cutting plane algorithm as for the alternating algorithm, but is has been proved that the former has a linear convergence rate between feasible points. Computational experience for the special case of geometric programming indicates that at the outset of the central cutting plane method, the convergence is superlinear, see [Gribik and Lee, 1978].

Let $\bar{P}$ denote the primal program associated with Program $\bar{D}$ (not specified here). If the central cutting plane method terminates, then it provides a dual optimal solution to the pair $(\bar{P}, \bar{D})$. If the algorithm does not terminate, then at some stage $\hat{k}$ it provides feasible points of $\bar{P}$ and $\bar{D}$ respectively. These feasible points may be purified [Theorems 5 and 6] so that at most n variables have non-zero values and such that the primal objective value is at least as great as stage $\hat{k}$. In this way the central cutting plane algorithm automatically provides initial points for Newton-Raphson iterations on system NL. If the Newton-Raphson should fail at this point, we may continue with the cutting plane algorithm.

Both cutting plane algorithms have been applied in geometric programming [Duffin, Peterson and Zener, 1967]. In [Gochet and Smeers, 1972] the alternating algorithm is applied to the semi-infinite programming equivalent of the primal geometric program. Step 2 is done by inspection in this case. They compared this with the use of the [Kelley, 1960] cutting plane algorithm on the associated convex program, and showed that the alternating algorithm generates deeper cuts than the Kelley algorithm.

In a similar manner [Gribik and Lee, 1978] showed that the semi-infinite central cutting plane algorithm can easily generate deeper cuts than the [Elzinga and Moore, 1975] algorithm applied to the convex program. Computational experience is also reported. Applications to experimental design appear in [Gribik and Kortanek, 1977] and [Gribik, 1979a].

[Blum and Oettli, 1975] have developed a cutting plane algorithm for linear semi-infinite programming satisfying the Haar condition which is often present in polynomial approximation problems. Their algorithm has a linear convergence rate and a uniform bound on the size of the linear programming subproblems. A recent cutting plane algorithm of [Topkis, 1982] for convex programming also has good rates of convergence and an effective cut-dropping procedure.

4. Saddle Value Problems and Separably-Infinite Programs

We review next the relationships between the conjugate function duality and semi-infinite programming duality theories. For this purpose let us consider a class of saddle value problems for concave-convex saddle functions $K(p,q)$, $p \in \mathbb{R}^m$, $q \in \mathbb{R}^n$ of the form

$$V_M = \sup_p \inf_q K(p,q) \quad \text{and} \quad V_N = \inf_q \sup_p K(p,q)$$

where

$$K(p,q) = g(p) + p \cdot A q + h(q),$$

where g is a proper closed concave function, h is a proper closed convex function, and A is an $m \times n$ matrix. It is well known that the study of such concave-convex saddle functions can be reduced to study pairs of dual convex programming problems. But each of these problems involves internal conjugate function optimizations as seen in the following dual pair of conjugate function programs corresponding to V_M and V_N, respectively.

Dual Conjugate Function Programs. Let g^*, h^* denote the conjugate functions of g and h respectively and assume that $h^*(-A)(\cdot)$ is not identically $+\infty$ on domain g and that $g^*(-A)(\cdot)$ is not identically $-\infty$ in domain h. Find

$$V_M = \sup_p \{g(p) - h^*(-A^T p)\}$$

and

$$V_N = \sup_q \{h(q) - g^*(-Aq)\},$$

see [Rockafellar, 1970].

Recently, [Charnes, Gribik and Kortanek, 1980b] developed a dual uniextremal principle equivalent to the saddle value problem which does not involve any internal optimizations. The construction employs supporting hyperplane representations which can always be taken to be canonically closed, see also [Lindberg, 1980], and an extension in [Borwein and Kortanek, 1981].

Hypograph of g

The *hypograph* of the closed proper concave function g is the set $\{(p_0, p) \in \mathbb{R} \times \mathbb{R}^m | p_0 \leq g(p)\}$ and a supporting hyperplane representation exists for this set of the form

$$v_0(t) p_0 + v(t) \cdot p \leq v_{m+1}(t) \quad \text{for every } t \in T$$

where T is an index set and where $v_0, v_{m+1}: T \to \mathbb{R}$, $v: T \to \mathbb{R}^m$. As a function v_0 must be non-negative, for otherwise if for some $\hat{t}$, $v_0(\hat{t}) < 0$, then the inequality indexed by $\hat{t}$ would be violated as $p_0 \downarrow -\infty$. Since g is proper, it cannot assume the value $+\infty$, and so v_0 cannot be identically zero, i.e., $\mathrm{supp}\, v_0 \neq \emptyset$.

Epigraph of h

The epigraph of the closed proper convex function h is the set $\{(q_0, q) \in \mathbb{R} \times \mathbb{R}^n | h(q) \leq q_0\}$, and in this case we write a supporting hyperplane system in the following form

$$u_0(s) q_0 + u(s) \cdot q \geq u_{n+1}(s) \quad \text{for every } s \in S$$

where S is an index set, $u_0, u_{n+1}: S \to \mathbb{R}$ and $u: S \to \mathbb{R}^n$. As before we find $u_0 \geq 0$, not identically zero.

Introducing appropriate generalized finite sequences we obtain the following pair of separably-infinite programs.

I. Find $\quad V_I = \sup p_0 + \sum_s u_{n+1}(s) \lambda(s)$

from among $\quad p_0 \in \mathbb{R}, \, p \in \mathbb{R}^m, \, \lambda \in \mathbb{R}^{(S)}$

subject to $\quad v_0(t) p_0 + v(t) \cdot p \leq v_{m+1}(t), \quad \text{every } t \in T$ \hfill (14)

$$\sum_s u_0(s) \lambda(s) = 1$$

$$-A^T p + \sum_s u(s) \lambda(s) = 0$$

and $\quad \lambda \geq 0.$

II. *Find* $V_{II} = \inf q_0 + \sum_{t} v_{m+1}(t)\eta(t)$

from among $q_0 \in \mathbb{R}, q \in \mathbb{R}^n, \eta \in \mathbb{R}^{(T)}$

subject to $u_0(s)q_0 + u(s) \cdot q \geqslant u_{n+1}(s)$ *every* $s \in S$ (15)

$$\sum_{t} v_0(t)\eta(t) = 1$$

$$-Aq + \sum_{t} v(t)\eta(t) = 0$$

and $\eta \geqslant 0.$

The following theorem can be proved using proper separation arguments.

Theorem 9 [Borwein and Kortanek, 1981]. Let $\mathscr{C}_S$ and $\mathscr{C}_T$ denote the moment cones associated with (15) and (14) respectively. Assume that both of these cones are closed. Assume further that

$$(g0^+)(p) + p \cdot Aq \geqslant 0 \quad \text{for every } q \in \text{domain } h$$

implies

$$(g0^+)(-p) - p \cdot Aq \geqslant 0 \quad \text{for every } q \in \text{domain } h$$

and

$$(h0^+)(q) + p \cdot Aq \leqslant 0 \quad \text{for every } p \in \text{domain } g$$

implies

$$(h0^+)(-q) - p \cdot Aq \leqslant 0 \quad \text{for every } p \in \text{domain } g.$$

Then $V_I = V_M = V_N = V_{II}$ with attainment in both Programs I and II. □

These conditions are related to those introduced in [Fan, 1959] and [Rockafellar, 1970]. Recent new developments on more general inequality systems and minimax theorems appear in [Pomerol, 1980]. Semi-infinite constructions in game theory appear in [Soyster, 1975] and [Tijs, 1979].

The topic of parametric and sensitivity analysis of semi-infinite programs usually involves some choices on how to model a differentiable program. Different choices of topologies can lead to different continuity or differentiable properties of the value function itself, see [Greenberg and Pierskalla, 1975], [Borwein, 1983b] and [Brosowski, 1981]. Coordinated with parametric programming studies are recent developments in second order differential conditions in semi-infinite programming see [Ben-Tal, 1980], [Ben-Tal, Teboulle and Zowe, 1979], [Ioffe, 1983], and [Borwein, 1983b]. These parallel some of the classical second order conditions of nonlinear programming [Fiacco and McCormick, 1968].

Acknowledgments. This paper has benefited greatly from discussions with many colleagues who participated in the 1978 Bad Honnef and 1981 Austin semi-infinite programming symposia. We are also indebted to Halina M. Strojwas for her painstaking review of this manuscript and its earlier versions.

References

Andreasson, D. O., and G. A. Watson: Linear Chebyshev Approximation Without Chebyshev Sets. BIT *16* (1970), 349–362.

Atwood, C. L.: Optimal and Efficient Designs of Experiments. Ann. Math. Statist. *40* (1969), 1570–1602.

Ben-Tal, A.: Second Order Theory of Extremum Problems. Extremal Methods and Systems Analysis (edited by A. V. Fiacco and K. O. Kortanek) 174. Berlin–Heidelberg–New York: Springer-Verlag, 1980, 336–356.

Ben-Tal, A., A. Ben-Israel, and E. Rosinger: A Helley-Type Theorem and Semi-Infinite Programming. Constructive Approaches to Mathematical Models (edited by C. V. Coffman and G. J. Fix). New York: Academic Press, 1979, 127–135.

Ben-Tal, A., M. Teboulle, and J. Zowe: Second Order Necessary Optimality Conditions for Semi-Infinite Programming Problems. Semi-Infinite Programming (edited by R. Hettich). Berlin–Heidelberg–New York: Springer-Verlag, 1979.

Blair, C. E.: A Note on Infinite Systems of Linear Inequalities in R^n. J. Math. Anal. and Appl. *1* (48) (1974), 150–154.

Blair, C. E.: A Survey of Ascent Ray Results and Their Applications. To appear (1983).

Blum, E. and W. Oettli: Mathematische Optimierung. Berlin–Heidelberg–New York: Springer-Verlag, 1975.

Böhning, D.: Numerische Methoden in der Versuchsplanung. Ph. D. Thesis, Dept. of Mathematics, Freie Universität. Berlin, 1980.

Bojanic, R. and R. DeVore: On Polynomials of Best One-Sided Approximation. L'Enseignement Mathematique *XII* (3) (1966).

Borwein, J. M.: Adjoint Process Duality. To appear in Math. of Operations Research.

Borwein, J. M.: Direct Theorems in Semi-Infinite Convex Programming. Math. Programming *21* (1981a), 301–318.

Borwein, J. M.: Convex Relations in Analysis and Optimization. Generalized Concavity in Optimization and Economics. New York: Academic Press, 1981b, 335–377.

Borwein, J. M.: The Limiting Lagrangian as a Consequence of Helley's Theorem. J.O.T.A. *33* (4) (1981c), 497–513.

Borwein, J. M.: Lexicographic multipliers. J. Math. Anal. Appl. 78 (1980), 309–327.

Borwein, J. M.: Semi-Infinite Programming: How Special Is It? To appear (1983).

Borwein, J. M. and K. O. Kortanek: Fenchel Duality and Separably-Infinite Programs. Optimization; Math. OP. & Stat. (Ilmenau), 14 (1983).

Brosowski, B.: Parametric Semi-Infinite Programming. Methoden und Verfahren der Mathematischen Physik. Bel. 22, Frankfurt/M., Bern, 1981.

Charnes, A.: Constrained Games and Linear Programming. Proc. Nat. Acad. Sci., USA *38* (1953), 639–641.

Charnes, A. and W. W. Cooper: Management Models and Industrial Applications of Linear Programming, Vols. I and II. New York: J. Wiley & Sons, 1961.

Charnes, A., W. W. Cooper and K. O. Kortanek: Duality, Haar Programs and Finite Sequence Spaces. Proc. Nat. Acad. Sci., USA *48* (1962a), 783–786.

Charnes, A., W. W. Cooper and K. O. Kortanek: A Duality Theory for Convex Programs with Convex Constraints. Bull. of the American Math. Soc. *68* (6) (1962b), 605–608.

Charnes, A., W. W. Cooper and K. O. Kortanek: Duality in Semi-Infinite Programs and Some Works of Haar and Caratheodory. Management Sci. *9* (1963), 209–228.

Charnes, A., W. W. Cooper and K. O. Kortanek: On Representations of Semi-Infinite Programs Which Have No Duality Gaps. Management Sci. *12* (1965), 113–121.

Charnes, A., W. W. Cooper and K. O. Kortanek: On the Theory of Semi-Infinite Programming and Some Generalizations of Kuhn-Tucker Saddle Point Theorems for Arbitrary Convex Functions. Nav. Res. Log. Quart. *16* (1969), 41–51.

Charnes, A., P. R. Gribik and K. O. Kortanek: Separably-Infinite Programming. Z. Op. Res. *24* (1980a), 33–45.

Charnes, A., P. R. Gribik and K. O. Kortanek: Polyextremal Principles and Separably-Infinite Programs. Z. Op. Res. *25* (1980b), 211–234.

Charnes, A., K. O. Kortanek and V. Lovegren: A Saddle Value Characterization of Fan's Equilibrium Points. To appear (1983).

Charnes, A., K. O. Kortanek and W. Raike: Extreme Point Solutions in Mathematical Programming: An Opposite Sign Algorithm. SRM No. 129, Northwestern University, June, 1965.

Charnes, A., K. O. Kortanek and S. Thore: An Infinite Constrained Game Duality Characterizing Economic Equilibrium. Dept. Mathematics, Carnegie-Mellon University, Pittsburgh, PA 15213, March, 1981.

Cheney, E. W.: Introduction to Approximation Theory. New York–St. Louis–San Francisco–Toronto–London–Sydney: McGraw-Hill, 1966.

Chui, C. K. and J. D. Ward: Book Review of Optimization and Approximation by W. Krabs, J. Wiley and Sons, 1979. Bull. AMS *3* (1980), 1065–1069.

Collatz, L. and W. Krabs: Approximationstheorie. Stuttgart: B. G. Teubner, 1973.

Collatz, L. and W. Wetterling: Optimierungsaufgaben, Zweite Auflage. Berlin–Heidelberg–New York: Springer-Verlag, 1971.

Cromme, L.: Eine Klasse von Verfahren zur Ermittlung bester nicht-linearer Tschebyscheff-Approximationen. Numer. Math. *25* (1976), 447–459.

Cybenko, G.: Affine Minimax Problems and Semi-Infinite Programming. Dept. of Mathematics, Tufts University, Medford, MA 02155, 1981.

Dahlquist, G. and Å. Björck: Numerical Methods. Englewood Cliffs, New Jersey: Prentice-Hall, 1974.

Dahlquist, G., S. Eisenstat and G. H. Golub: Bounds for the Error of Linear Systems of Equations Using the Theory of Moments. J. Math. Anal. Appl. *37* (1972), 151–166.

Dahlquist, G., S.-Å. Gustafson and K. Siklo'si: Convergence Acceleration from the Point of View of Linear Programming. BIT *1* (1965), 1–16.

Dantzig, G. B.: Linear Programming and Extensions. Princeton, New Jersey: Princeton University Press, 1963.

DeVore, R.: One-Sided Approximation of Functions. J. Approx. Theory *1* (1968), 11–25.

Dinkel, J. J., W. H. Elliot and Gary A. Kochenberger: Computational Aspects of Cutting-Plane Algorithms for Geometric Programming Problems. Mathematical Programming *13* (1977), 200–220.

Duffin, R. J.: Infinite Programs. In: Linear Inequalities and Related Systems (edited by H. W. Kuhn and A. W. Tucker). Princeton, New Jersey: Princeton University Press, 1965, 157–170.

Duffin, R. J.: Convex Analysis Treated by Linear Programming. Math. Prog. *4* (1973), 125–143.

Duffin, R. J., R. G. Jeroslow and L. A. Karlovitz: Duality in Semi-Infinite Linear Programming. To appear (1983).

Duffin, R. J. and L. A. Karlovitz: An Infinite Linear Program with a Duality Gap. Management Sci. *12* (1965), 122–134.

Duffin, R. J., E. L. Peterson and C. Zener: Geometric Programming. New York: J. Wiley & Sons, 1967.

Eckhardt, U.: Theorems on the Dimension of Convex Sets. Linear Algebra and Its Applications *12* (1975), 63–76.

Eckhardt, U.: Semi-Infinite Quadratic Programming. OR-Spek. *1* (1979), 51–55.
Eckhardt, U.: Representations of Convex Sets. Extremal Methods and Systems Analysis (edited by A. V. Fiacco and K. O. Kortanek) 174. Berlin–Heidelberg–New York: Springer-Verlag, 1980, 374–383.
Elzinga, J. and Th. G. Moore: A Central Cutting Plane Algorithm for the Convex Programming Problem. Math. Programming *8* (1975), 134–145.
Evans, J. P.: Duality in Mathematical Programming Over Markov Chains and Countable Action Spaces in Finite Horizon Problems. Management Sci. *15* (1969), 626–638.
Fahlander, K.: Computer Programs for Semi-Infinite Optimization. TRITA NA-7312, Department of Numerical Analysis and Computing Science, Royal Institute of Technology, S-10044 Stockholm 70, Sweden.
Fahlander, K., S.-Å. Gustafson and L. E. Olsson: Computing Optimal Air Pollution Abatement Strategies – Some Numerical Experiments on Field Data. Proceedings of the Fifth Meeting of the NATO/CCMS Expert Panel on Air Pollution Modeling, 1974.
Fan, K.: Convex Sets and Their Applications. Argonne National Laboratory, Summer, 1959.
Fan, K.: Asymptotic Cones and Duality of Linear Relations. J. Appr. Theory *2* (1969), 152–159.
Fedorov, V. V.: Theory of Optimal Experiments. New York: Academic Press, 1972.
Fiacco, A. V. and G. P. McCormick: Nonlinear Programming: Sequential Unconstrained Minimization Techniques. New York: John Wiley, 1968.
Flachs, J.: Saddle-Point Theorems for Rational Approximation. CSIR Technical Report, TWISK 219 (1981).
Glashoff, K.: Duality Theory of Semi-Infinite Programming. Semi-Infinite Programming (edited by R. Hettich). Berlin–Heidelberg–New York: Springer-Verlag, 1979.
Glashoff, K. and S.-Å. Gustafson: Numerical Treatment of a Parabolic Boundary Value Control Problem. J. Opt. Th. Appl. *19* (1976), 645–663.
Glashoff, K. and S.-Å. Gustafson: Einführung in die Lineare Optimierung. Darmstadt: Buchgesellschaft, 1978.
Glashoff, K. and S.-Å. Gustafson: Linear Optimization and Approximation. To appear Springer.
Glashoff, K. and K. Roleff: A New Method for Chebyshev Approximation of Complex-Valued Functions. Math. Comp. *36* (1981), 233–239.
Gochet, W. and Y. Smeers: On the Use of Linear Programs to Solve Prototype Geometric Programs. CORE Discussion Paper No. 7229, 1972.
Gochet, W., Y. Smeers and K. O. Kortanek: On a Classification Scheme for Geometric Programming and Complementarity Theorems. Applicable Analysis *6* (1976), 47–59.
Gonzaga, C. and E. Polak: On Constraint Dropping Schemes and Optimality Functions for a Class of Outer Approximation Algorithms. SIAM J. Contr. Opt. *17* (1979), 477–493.
Gonzaga, C., E. Polak and R. Trahan: An Improved Algorithm for Optimization Problems with Functional Inequality Constraints. IEEE Trans. Autom. Contr. *25* (1980), 49–54.
Gorr, W. L. and K. O. Kortanek: Numerical Aspects of Pollution Abatement Problems: Optimal Control Strategies for Air Quality Standards. Proceedings 10th Annual DGU Meetings, Bochum, 1971, 34–58.
Gorr, W., S.-Å. Gustafson and K. O. Kortanek: Optimal Control Strategies for Air Quality Standards and Regulatory Policies. Environment and Planning *4* (1972), 183–192.

Greenberg, H. J. and W. P. Pierskalla: Stability Theorems for Infinitely Constrained Mathematical Programs. J.O.T.A. *16* (1975), 409–428.

Gribik, P. R.: A Central Cutting Plane Algorithm for Semi-Infinite Programming Problems. In: Semi-Infinite Programming (edited by R. Hettich). Berlin–Heidelberg–New York: Springer-Verlag, 1979 a, 66–82.

Gribik, P. R.: Selected Applications of Semi-Infinite Programming. Constructive Approaches to Mathematical Models (edited by C. V. Coffman and G. J. Fix). New York: Academic Press, 1979 b, 171–188.

Gribik, P. R. and K. O. Kortanek: Equivalence Theorems and Cutting Plane Algorithms for a Class of Experimental Design Problems. SIAM J. Appl. Math. *32* (1977), 232–259.

Gribik, P. R. and D. N. Lee: A Comparison of Two Central Cutting Plane Algorithms for Prototype Geometric Programming Problems. Methods of Opns. Res. 31 (edited by W. Oettli and F. Steffens), 1978, 275–287.

Gustafson, S.-Å.: On the Computational Solution of a Class of Generalized Moment Problems. SIAM J. Numer. Anal. 7 (1970), 343–357.

Gustafson, S.-Å.: Die Berechnung von verallgemeinerten Quadraturformeln vom Gaußschen Typus, eine Optimierungsaufgabe. In: Numerische Methoden bei Optimierungsaufgaben (edited by L. Collatz and W. Wetterling), ISNM *17*. Basel: Birkhäuser, 1973 a, 59–71.

Gustafson, S.-Å.: Nonlinear Systems in Semi-Infinite Programming. In: Numerical Solutions of Nonlinear Algebraic Systems (edited by G. B. Byrnes and C. A. Hall). New York: Academic Press, 1973 b, 63–99.

Gustafson, S.-Å.: On Computational Applications of the Theory of the Moment Problem. Rocky Mountain J. on Math. *4* (1974), 227–240.

Gustafson, S.-Å.: On the Numerical Treatment of a Multi-Dimensional Parabolic Boundary-Value Control Problem. In: Optimization and Optimal Control, Lecture Notes in Mathematics. 477 (edited by R. Bulirsch, W. Oettli and J. Stoer). Berlin–Heidelberg–New York: Springer-Verlag, 1975.

Gustafson, S.-Å.: Stability Aspects on the Numerical Treatment of Linear Semi-Infinite Programs. TRITA-NA-7604, Dept. of Numerical Analysis and Computing Science, Royal Institute of Technology, S-10044, Stockholm 70, Sweden.

Gustafson, S.-Å.: On Numerical Analysis in Semi-Infinite Programming. In: Semi-Infinite Programming, Lecture Notes in Control and Information Sciences 15 (edited by R. Hettich). Berlin–Heidelberg–New York: Springer-Verlag, 1979.

Gustafson, S.-Å.: A Computational Scheme for Exponential Approximation. ZAMM *61* (1981), T284–T287.

Gustafson, S.-Å.: A General Three-Phase Algorithm for Nonlinear Semi-Infinite Programming Problems. To appear, (1983).

Gustafson, S.-Å. and K. O. Kortanek: Numerical Treatment of a Class of Semi-Infinite Programming Problems. NRLQ *20* (1973 a), 477–504.

Gustafson, S.-Å. and K. O. Kortanek: Numerical Solution of a Class of Convex Programs. Methods of Operations Research *16* (1973 b), 138–149.

Gustafson, S.-Å. and K. O. Kortanek: Mathematical Models for Air Pollution Control: Numerical Determination of Optimizing Policies. In: Models for Environmental Pollution Control (edited by R. A. Deininger). Ann Arbor: Ann Arbor Science Publishers, Inc., 1973 c, 251–265.

Gustafson, S.-Å. and K. O. Kortanek: Mathematical Models for Optimizing Air Pollution Abatement Policies: Numerical Treatment. Proceedings of the Bilateral U.S.-Czechoslovakia Environmental Protection Seminar, Pilsen, Czechoslovakia, 1973 d.

Gustafson, S.-Å. and K. O. Kortanek: Determining Sampling Equipment Locations by

Optimal Experimental Design with Applications to Environmental Protection and Acoustics. Proceedings of the 7th Symposium on the Interface, Iowa State University, Ames, Iowa, 1973 e.

Gustafson, S.-Å., K. O. Kortanek, and Samuelsson, H. M.: On dual programs and finite dimensional moment cones. Series in Numerical Optimization and Pollution Abatement, No. 8, Carnegie-Mellon University, Pittsburgh PA 1974.

Gustafson, S.-Å., K. O. Kortanek, and W. Rom: Non-Čebyševian Moment Problems. SIAM J. on Numer. Anal. 7 (1970), 335–342.

Gutknecht, M.: Ein Abstiegsverfahren für nicht-diskrete Tschebyscheff-Approximationsprobleme. In: Numer. Meth. der Approximationstheorie (edited by Collatz et al.) ISNM 42 (1978), 154–171.

Hettich, R.: Kriterien zweiter Ordnung für lokal beste Approximation. Numer. Math. 22 (1974), 409–417.

Hettich, R.: A Newton-Method for Nonlinear Chebyshev Approximation. Approx. Theory. Proc. Int. Colloqu. Bonn, Lecture Notes in Math. 556. Berlin–Heidelberg–New York: Springer-Verlag, 1976, 222–236.

Hettich, R. (editor): Semi-Infinite Programming, Lecture Notes in Control and Information No. 15. Berlin–Heidelberg–New York: Springer-Verlag, 1979.

Hettich, R. and H. Th. Jongen: On First and Second Order Conditions for Local Optima for Optimization Problems in Finite Dimensions. Methods of Oper. Res. 23 (1977), 82–97.

Hettich, R. and H. Th. Jongen: Semi-Infinite Programming: Conditions of Optimality and Applications. In: Optimization Techniques 2 (edited by J. Stoer). Berlin–Heidelberg–New York: Springer-Verlag, 1978, 1–11.

Hettich, R. and W. Van Honstede: On Quadratically Convergent Methods for Semi-Infinite Programming. In: Semi-Infinite Programming (edited by R. Hettich). Berlin–Heidelberg–New York: Springer-Verlag, 1979, 97–111.

Hettich, R. and W. Wetterling: Nonlinear Chebyshev Approximation by H-Polynomials. J. Approx. Theory 7 (1973), 198–211.

Hettich, R. and P. Zencke: Superlinear konvergente Verfahren für semi-infinite Optimierungsprobleme mit stark eindeutigen Fall. Universität Bonn, Preprint No. 354, SFB 72, (1980).

Hettich, R. and P. Zencke: Numerische Methoden der Approximation und semi-infiniten Optimierung. Stuttgart: Teubner, 1982.

Hoffman, K. H. and A. Klostermaier: A Semi-Infinite Programming Procedure. In: Approximation Theory II (edited by Lorentz et al.). New York–San Francisco–London. McGraw-Hill, 1976, 379–389.

Ioffe, A. D.: Second-Order Conditions in Nonlinear Nonsmooth Problems of Semi-Infinite Programming. To appear (1983).

Jeroslow, R. G.: A Limiting Lagrangian for Infinitely-Constrained Convex Optimization in R^n. J.O.T.A. 33 (3) (1981).

Jeroslow, R. G. and K. O. Kortanek: On Semi-Infinite Systems of Linear Inequalities. Israel J. Math. 10 (1971), 252–258.

John, F.: Extremum Problems with Inequalities as Subsidiary Conditions. In: Studies and Essays, Courant Anniversary Volume. New York: Interscience, 1948, 187–204.

Kallina, C. and A. C. Williams: Linear programming in reflexive spaces. SIAM Rev. (1971), 350–356.

Karlin, S. and W. J. Studden: Tchebycheff Systems: With Applications in Analysis and Statistics. New York–London–Sydney: Interscience Publishers, 1966.

Karney, D. F.: Duality Gaps in Semi-Infinite Linear Programming – An Approximation Problem. Math. Programming 20 (1981a), 129–143.

Karney, D. F.: Clark's Theorem for Semi-Infinite Convex Programs. Advances in Appl. Math. *2* (1981b), 7–13.

Kelley, J. E., Jr.: The Cutting Plane Method for Solving Convex Programs. SIAM *8* (1960), 703–712.

Kemperman, J. H. B.: On the Role of Duality in the Theory of Moments. To appear (1983).

Kiefer, J.: Optimum Experimental Designs. J. Roy. Statist. Soc. Ser. B. *21* (1959), 272–304.

Kiefer, J. and J. Wolfowitz: The Equivalence of Two Extremum Problems. Canadian J. Math. *12* (1960), 363–366.

Kortanek, K. O.: Classifying convex extremum problems over linear topologies having separation properties. J. Math. Anal. Appl. 46 (1974), 725–755.

Kortanek, K. O.: Extended Abstract of Classifying Convex Extremum Problems. Zentralblatt für Mathematik *283* (1975), 491–496.

Kortanek, K. O.: Perfect Duality in Generalized Linear Programming. In: Proc. 9th International Symp. Math. Prog. (edited by A. Prékopa), Hungarian Acad. Sci., Budapest. Amsterdam: North-Holland, 1976, 43–58.

Kortanek, K. O.: Constructing a Perfect Duality in Infinite Programming. Appl. Math. Opt. *3* (1977a), 357–372.

Kortanek, K. O.: Perfect Duality in Semi-Infinite and Generalized Convex Programming. In: Methods of Operations Research 25 (edited by A. Angermann, R. Kaerkes, K.-P. Kistner, K. Neumann, B. Rauhut, and F. Steffens). Meisenheim am Glan: Verlag Anton Hain, 1977b, 79–87.

Kortanek, K. O. and R. W. Pfouts: A Biextremal Principle for a Behavioral Theory of the Firm. J. Math. Modeling, *3* (1982), 573–590.

Kortanek, K. O. and A. L. Soyster: On Equating the Difference Between Optimal and Marginal Values of General Convex Programs. J. Opt. Theory and Appl. *33* (1981), 57–68.

Kortanek, K. O. and M. Yamasaki: Semi-Infinite Transportation Problems. J. Math. Anal. Appl. *88* (1982), 555–565.

Krabs, W.: Optimierung und Approximation. Stuttgart: B. G. Teubner, 1975.

Lehman, R. and W. Oettli: The Theorem of the Alternative, the Key Theorem, and the Vector-Maximum Problem. Math. Prog. *8* (1975), 332–344.

Levinson, N.: A Class of Continuous Linear Programming Problems. J. Math. Anal. Appl. *16* (1966), 73–83.

Lindberg, P. O.: Fenchel duality from LP duality, Optimization. Math. OP & Stat. (Ilmenau) *11* (1980), 171–180.

Luenberger, D. G.: Optimization by Vector Space Methods. New York: John Wiley, 1969.

Negoita, C. V. and M. Sularia: On Fuzzy Mathematical Programming and Tolerances in Planning. Institute of Management and Informatics, Bucharest, Economic Computation and Economic Cybernetics Studies and Research, 1976.

Osborne, M. R. and G. A. Watson: An Algorithm for Minimax Approximation in the Nonlinear Case. Computer J. *12* (1969), 63–68.

Parks, M. L. and A. L. Soyster: Semi-Infinite Programming and Fuzzy Set Programming, to appear (1983).

Perold, A. F.: Fundamentals of a Continuous Time Simplex Method. Technical Report SOL 78-26, Dept. Opns. Res., Stanford University, Stanford, CA 94305.

Perold, A. F.: Extreme Points and Basic Feasible Solutions in Continuous Time Linear Programming. SIAM J. Contr. Opt. *19* (1981), 52–63.

Pomerol, J. Ch.: About a minimax theorem of Matthies, Strang and Christiansen, Math. Prog. 19 (1980), 352–355.

Pukelsheim, F.: On Linear Regression Designs Which Maximize Information. J. Stat. Planning and Inference *4* (1980), 339–364.

Rockafellar, R. T.: Convex Analysis. Princeton University Press, Princeton, New Jersey, 1970.

Samuelsson, H. M.: A Note on the Duality Interpretation of an Air Pollution Abatement Model. Series in Numerical Optimization and Pollution Abatement, School of Urban and Public Affairs, Carnegie-Mellon University, Pittsburgh, PA, 1972.

Sibson, R.: Cutting-Plane Algorithms for D-Optimal Design. Dept. of Statistics, University of Cambridge, Cambridge, England, 1973.

Silvey, S. D.: Optimal Design. London: Chapman & Hall, 1980.

Silvey, S. D. and D. M. Titterington: A Geometic Approach to Optimal Design Theory. Biometrika *60* (1973), 21–32.

Soyster, A. L.: Convex Programming with Set Inclusive Constraints and Applications to Inexact Linear Programming. Operations Research *21* (1973), 1154–1157.

Soyster, A. L.: A Semi-Infinite Game. Mgt. Sci. *21* (1975), 806–812.

Soyster, A. L.: Inexact Linear Programming with Generalized Resource Sets. European J. of Opnl. Res. *3* (1979), 316–321.

Stoer, J.: Einführung in die Numerische Mathematik 2. Auflage. Berlin–Heidelberg–New York: Springer-Verlag, 1976.

Streit, R. L. and A. H. Nuttall: Linear Chebyshev Complex Function Approximation. Naval Underwater Systems Center, Newport, Rhode Island, USA (1981).

Tijs, S. H.: Semi-Infinite Linear Programs and Semi-Infinite Matrix Games. Nieuw Archif voor Wiskunde *27* (1979), 197–214.

Titterington, D. M.: Geometric Approaches to Design of Experiment. Math. Operationsforsch. Statist. Ser. Statist. *11* (1980), 151–163.

Topkis, D. M.: A Cutting Plane Algorithm with Linear and Geometric Rates of Convergence. J.O.T.A. *36* (1982), 1–22.

Tröltzsch, F.: Duality Theorems for a Class of Continuous Linear Programming Problems in a Space of Bochner-Integrable Abstract Functions. Math. Operationsforsch. Statist. Ser. Opt. *11* (1980), 375–388.

Tsay, J.-Y.: The Iterative Methods for Calculating Optimal Experimental Designs. Ph. D. Thesis, Purdue University, Lafayette, Indiana, 1974.

Van Honstede, W.: An Approximation Method for Semi-Infinite Problems. In: Semi-Infinite Programming (edited by R. Hettich). Berlin–Heidelberg–New York: Springer-Verlag, 1979, 126–136.

Veidinger, L.: On the Numerical Determination of the Best Approximation in the Chebyshev Sense. Numer. Math. *2* (1960), 99–105.

Watson, G. A.: The Calculation of Best One-Sided Lp-Approximations. Math. Comp. *27* (1973a), 607–620.

Watson, G. A.: One-Sided Approximation and Operator Equations. J. Inst. Maths. Applic. *12* (1973b), 197–208.

Watson, G. A.: On the Best Linear One-Sided Chebyshev Approximation. J. Approx. Theory *7* (1973c), 48–58.

Watson, G. A.: Globally Convergent Methods for Semi-Infinite Programming. Dept. of Mathematics, University of Dundee (1981a).

Watson, G. A.: Numerical Experiments with Globally Convergent Methods for Semi-Infinite Programming Problems. Dept. of Mathematics, University of Dundee (1981b).

Wetterling, W.: Anwendung des Newtonschen Iterationsverfahrens bei der Tschebyscheff-Approximation, insbesondere mit nichtlinear auftretenden Parametern. MTW Teil I: 61–63, Teil II: 112–115 (1963).

Whittle, P.: Some General Points in the Theory of Optimal Experimental Design. J. Roy. Statist. Soc. Ser. B. *35* (1973), 123-130.

Wynn, H. P.: The Sequential Generation on D-Optimal Designs. Ann. Math. Statist. *41* (1970), 1655-1664.

Wynn, H. P.: Results in the Theory and Construction of D-Optimal Experimental Designs (with Discussions). J. Roy. Statist. Soc. Ser. B. *34* (1972), 133-147.

Yamasaki, M.: Semi-Infinite Programs and Conditional Gauss Variational Problems. Hiroshima Math. J. *1* (1971), 177-226 and Corrections *2* (1972), 547.

Yamasaki, M.: Duality Theorems in Mathematical Programming and Their Applications. J. Sci. Hiroshima Univ. *30* (1968), 331-356.

Zariski, O. and P. Samuel: Commutative Algebra. New York: D. Van Nostrand Co., 1950.

Added in proof

The references indicated as "to appear (1983)" will appear in Semi-Infinite Programming and Applications: An International Symposium, Austin, Texas 1981 (Edited by A. V. Fiacco and K. O. Kortanek), Lecture Notes in Economics and Mathematical Systems, Vol. 215, Springer-Verlag, Berlin-Heidelberg-New York-Tokyo.

Applications of Matroid Theory

M. Iri
University of Tokyo, Faculty of Engineering, Dept. of Mathematical Engineering,
7-3-1 Hongo, Bunkyo-ku, Tokyo 113, Japan

Abstract. It will be shown that looking at a problem from the viewpoint of matroids enables us to understand the essence of the problem as well as its relations to other problems, clearly, preventing us from probable confusion into which we might have been involved without matroids, and that mathematical techniques developed in matroid theory are powerful for manipulating and solving the mathematical model which would otherwise have been impossible, or at best prohibitively complicated.
  Examples of problems to be discussed:
1. Topological, geometrical and physical matroids, or faithful and unfaithful representations in terms of matroids
2. Elements and their interconnections
3. Minimum-size systems of equations
4. Structural solvability of systems of equations
5. Two kinds of dualities

1. Introduction

Matroid theory, and related fields of mathematics such as theories concerning polymatroids, submodular functions, etc., have already been applied to many practical problems in engineering as well as to rather theoretical problems in other branches of combinatorial mathematics itself. However, the level of abstraction, the scale and the effectiveness of those applications are various.
  Speaking about applications of a branch of mathematics to practical problems, we should note two directions of link between mathematics and a more practical field. That is, we sometimes make use of known results in a branch of mathematics to get a new meaningful result in a practical field, where specializations and/or modifications of the mathematical results are often needed; and sometimes, we call it also an application to find out a new mathematical problem in a practical field and to analyse that mathematical problem theoretically. (It may happen in the latter case that the results of the analysis have no practical significance.) A "good" application should combine these two directions with each other, i.e., it should (i) start from finding a problem in a practical field which is important in the field per se but which has not been dealt with adequately because one has not been aware of an appropriate mathematical technique, (ii) construct a proper mathematical model of the problem, (iii) bestow mathematical considerations on the model by taking advantage of known mathematical results as well as by doing some new mathematical re-

II. Applications of Matroid Theory

search if already known results are not sufficient, and then (iv) bring about an innovation to the technology in the practical field.

In order to realize this "healthy cycle" in application, one must be versed, to a certain extent, not only in the relevant mathematical techniques but also in the state of the art of the object field. But for sufficient knowledge of the latter, one is apt to "invent" a theoretically handsome but practically ridiculous mathematics. Of course, we must not be too hasty in evaluating the practical usefulness of a mathematical result in application. It often takes a long time for us to become able to recognize a true significance of a novel work. But, this never means that one may think little or lightly of deepening one's understanding of the object field. It is only if one has a good understanding of the object field, or even a better understanding than the experts in the field, that one can break down a traditional prejudice on the basis of one's own mathematical thinking.

Moreover, the more systematic a mathematical approach, i.e., application of mathematics, to a practical field is, the happier shall we be. In other words, although fragmentary application of a mathematical technique to a problem may be of some significance in its own way, we should desirably aim at constructing a systematic mathematical theory which enables us to better recognize the essential structure of the field, to solve various important problems ever regarded as being difficult there, and to open a new vista of the field.

In the present essay, problems mainly from electric network theory will be taken up for the concrete examples by which to illustrate various methodological viewpoints connected with applications of matroid theory. Problems in elastic structure, signal-flow graphs, systems of equations, etc. will also be mentioned.

Since emphasis is laid on the relations between electric-network problems and matroids and the mathematical audience is anticipated, knowledge in matroid theory (and related mathematical fields) is largely assumed whereas motivations from the standpoint of applications will be described in considerable detail. So, the reader is requested to consult the mathematical literatures (e.g., [Bruter 1974], [Wilson 1973], [Welsh 1976]) in the reference list for mathematical preliminaries on matroids, polymatroids and submodular functions.

2. Equations Governing the Performance of an Electric Network

We shall consider a finite, lumped-constant, time-invariant electric network, as it is called in electric network theory. Such an electric network, when regarded as a system, consists of a finite number of "two-terminal" elements called "branches". They are, for example, voltage sources, current sources, (linear or nonlinear) resistors, inductors, capacitors, etc. The set of all elements of the network will be denoted by E. To each element $e_\kappa \in E$ ($\kappa = 1, \ldots, n$; $n = |E|$) are associated two physical quantities $\zeta^\kappa \equiv \zeta(e_\kappa)$ and $\eta_\kappa \equiv \eta(e_\kappa)$, called the *current in* branch e_κ and the *voltage across* branch e_κ, respectively. Mathematically, we need to prescribe an algebraic structure to which currents and voltages belong.

Usually, they are regarded as elements of a field; the field of real numbers $\mathbb{R}$ when we deal with a direct-current problem, i.e. the equilibrium state, and the field composed of the Laplace- or Fourier-transforms of real-valued functions of time (i.e. "wave forms"), or, of some kind of distributions or hyperfunctions such as the Mikusiński operators, when we consider dynamics.

An electric network is not merely a set of separate branches but the manner of their interconnection is also essential. The interconnection of branches is represented by a graph $G = (E, V, \partial^+, \partial^-)$ with E as the edge set, where V is the vertex set, and ∂^+ and $\partial^-: E \to V$ are the incidence functions, respectively, meaning "the initial vertex of" and "the terminal vertex of".

The equations which govern the performance of an electric network, i.e., which are to be satisfied by branch currents and voltages are classified into two sorts. The equations of the first sort, which are the *structural equations,* are purely graphical (or, we may say, *topological* or *combinatorial*), and are traditionally called "Kirchhoff's laws", and the equations of the other sort represent the *physical* characteristics of branches.

Specifically, if we denote a fundamental circuit matrix of G by R_p^κ ($\kappa = 1, \ldots, n$; $p = 1, \ldots, k$) and a fundamental cocircuit matrix of G by D_κ^a ($\kappa = 1, \ldots, n$; $a = 1, \ldots, m$), Kirchhoff's laws are expressed as

$$\text{Kirchhoff's current law:} \quad \sum_{\kappa=1}^{n} D_\kappa^a \xi^\kappa = 0 \quad (a = 1, \ldots, m), \tag{2.1}$$

$$\text{Kirchhoff's voltage law:} \quad \sum_{\kappa=1}^{n} R_p^\kappa \eta_\kappa = 0 \quad (p = 1, \ldots, k), \tag{2.2}$$

where m is the rank and k the nullity of G, and the entries of D_κ^a and R_p^κ are 0, 1 or -1. If we regard κ as the column index of the matrix D_κ^a and of R_p^κ, the matroid defined on the column set (which is identified with E in a natural way) of D_κ^a with respect to the linear dependence of column vectors is no other than the circuit matroid $M(G)$ of G and that defined on the column set (which is also identified with E) of R_p^κ is the cocircuit matroid $M^*(G)$ of G, and they are the dual of each other as matroids on E.

The physical characteristics of branches determine, in general, n independent nonlinear equations among the $2n$ variables ξ^κ's and η_κ's ($\kappa = 1, \ldots, n$):

$$f_i(\xi^1, \ldots, \xi^n; \eta_1, \ldots, \eta_n) = 0 \quad (i = 1, \ldots, n), \tag{2.3}$$

which are sometimes called the *constitutive equations*. We assume that, for meaningful values of ξ^κ's and η_κ's, we have

$$\text{the rank of the Jacobian matrix of } f_i = n. \tag{2.4}$$

That is, defining two matrices:

$$F^{(1)} = [f_{i\kappa}^{(1)}], \quad f_{i\kappa}^{(1)} = \frac{\partial f_i}{\partial \xi^\kappa},$$

$$F^{(2)} = [f_i^{(2)\kappa}], \quad f_i^{(2)\kappa} = \frac{\partial f_i}{\partial \eta_\kappa} \tag{2.5}$$

II. Applications of Matroid Theory

with κ as their column indices, we assume

$$\text{rank}[F^{(1)}|F^{(2)}]=n. \tag{2.6}$$

Everything is easy if the functions f_i are linear (not necessarily homogeneous) in ξ^κ's and η_κ's, but, even if some of them are nonlinear, we may consider the local problem, i.e. a perturbation, around a certain "normal" state $(\hat{\xi}^\kappa, \hat{\eta}_\kappa)$. So, without substantial loss in generality, we shall consider the equations:

$$\sum_{\kappa=1}^{n} f_{i\kappa}^{(1)} \xi^\kappa + \sum_{\kappa=1}^{n} f_i^{(2)\kappa} \eta_\kappa = c_i \quad (i=1,\ldots,n), \tag{2.7}$$

or

$$\sum_{\kappa=1}^{n} f_{i\kappa}^{(1)} (\xi^\kappa - \hat{\xi}^\kappa) + \sum_{\kappa=1}^{n} f_i^{(2)\kappa} (\eta_\kappa - \hat{\eta}_\kappa) = 0 \quad (i=1,\ldots,n), \tag{2.7'}$$

with the rank condition (2.6), for the physical characteristics of branches.

In connection with the constitutive equations, we have had the matrix

$$F=[F^{(1)}|F^{(2)}] \tag{2.8}$$

with $2n$ columns and n rows and of rank n. Since the column set of $F^{(1)}$ as well as that of $F^{(2)}$ can be identified with E, the column set of F can be identified with the disjoint union $E^{(1)} \cup E^{(2)}$ of two replicas of the set E. Thus, there appears another matroid on $E^{(1)} \cup E^{(2)}$ represented in terms of the matrix F.

Roughly speaking, if there are n branches, i.e., if $|E|=n$, there are $2n$ variables (among which n are currents in branches and n are voltages across branches), Kirchhoff's laws impose constraints of rank n on those variables (since the rank of $M(G)$ and that of $M^*(G)$ sum to n), and the constitutive equations also impose constraints of rank n (due to the assumption (2.6)), so that we have $2n$ conditions upon $2n$ variables. It is interesting to observe that the topological conditions constrain exactly half of the degrees of freedom of the variables and the physical characteristics kill the remaining half of the degrees of freedom.

From a more general standpoint of linear systems, an electric network is a typical example of a class of systems such as follows. A system is made of a set of n elements; with each of the elements are associated two kinds of variables, the variables of the first kind ξ^κ's called *intensive variables* and the other η_κ's *extensive variables* (in general, more than one intensive variable and the same number of extensive variables are associated with an element, and there is a one-to-one correspondence, or a bijection, between the set of intensive variables and that of extensive variables, so that we denote by $E(|E|=n)$ a set bijectively corresponding to each of them); there are *structural equations* of rank n, i.e., the intensive variables are subject to a system of linear homogeneous equations of the form (2.1), and the extensive variables to another system of equations of the form (2.2), where the matrices D and R are not necessarily connected with a graph but they must define a dual pair of matroids on their column sets; and the $2n$ (intensive and extensive) variables are subject also to *constitutive equations* of the form (2.3) and (2.4) (or, (2.7)/(2.7') and (2.6)).

An elastic structure, in the one-, two- or three-dimensional space, is another typical example of such a system, where "elements" are rods or bars connected with one another (and with a rigid base) at their ends, rigidly, by a hinge, or by a pin joint, or the like. Intensive variables ξ^κ's are forces (either tensions or compressions) and moments (bending and torsional) of rods, whereas extensive variables η_κ's are deformations, i.e. elongations (from normal lengths), bendings and torsions (and, sometimes, displacements and deflections) of rods. (In the following, we shall confine ourselves to the case where deformations are sufficiently small.) The structural equations (2.1) for the intensive variables are the conditions of "equilibrium of forces and moments", whereas those (2.2) for the extensive variables are the so-called conditions of "compatibility of deformations". Those structural equations are purely geometrical, i.e., the matrices in (2.1) and (2.2) are determined only by the geometrical configuration of the elastic structure and do not depend on what kinds of materials rods are made from nor on what the shape of the section of a rod is. The constitutive equations represent the elasticity properties of rods.

Although we shall not get in touch with elastic-structure problems in detail, some comparisons of electric-network problems with elastic-structure problems may be worth adding. The most remarkable contrast between the two will consist in that the constitutive equations are simpler for elastic structures whereas the structural equations are simpler for electric networks. In fact, Kirchhoff's laws are purely topological depending on the graphical structure of the network alone. The equilibrium or compatibility equations are more "quantitative"; they depend on the "metric" geometry of the structure, such as angles, lengths or relative positions. In an elastic structure, distinct rods seldom interact one another in the constitutive equations. In contrast, there are many electric devices now in use which are modelled as an assemblage of two-terminal elements, i.e. of branches, interacting ("mutually coupled" in electrical terminology) in the constitutive equations.

We have so far considered "closed" systems. It may seem that we need to consider also "open" systems since we have usually "external" or "applied" quantities (of currents, voltages, forces, moments, displacements, etc.). However, even when "external", say, voltages and currents exist, we can have a closed system by introducing branches called "voltage sources" and "current sources". A voltage (or current) source is a branch whose voltage (or current) has a prescribed value, and whose current (or voltage) does not appear in the constitutive equations. Therefore, the physical characteristics of "sources" can be expressed as part of equations of the form of (2.3) ((2.7) or (2.7')) with the rank condition (2.6). Another formulation of an open system into a closed one will be taken up in section 6.

Since the matroids determined by the structural equations for the intensive variables and for the extensive variables are mutually dual, it is natural to consider a dual pair of vector spaces to one of which the vectors with intensive variables ξ^κ as the components belong and to the other of which the vectors with extensive variables η_κ as the components belong. Usually the scalar product $\sum_{\kappa=1}^{n} \xi^\kappa \eta_\kappa$ makes sense, and has a physical dimension of power (as is the case

II. Applications of Matroid Theory

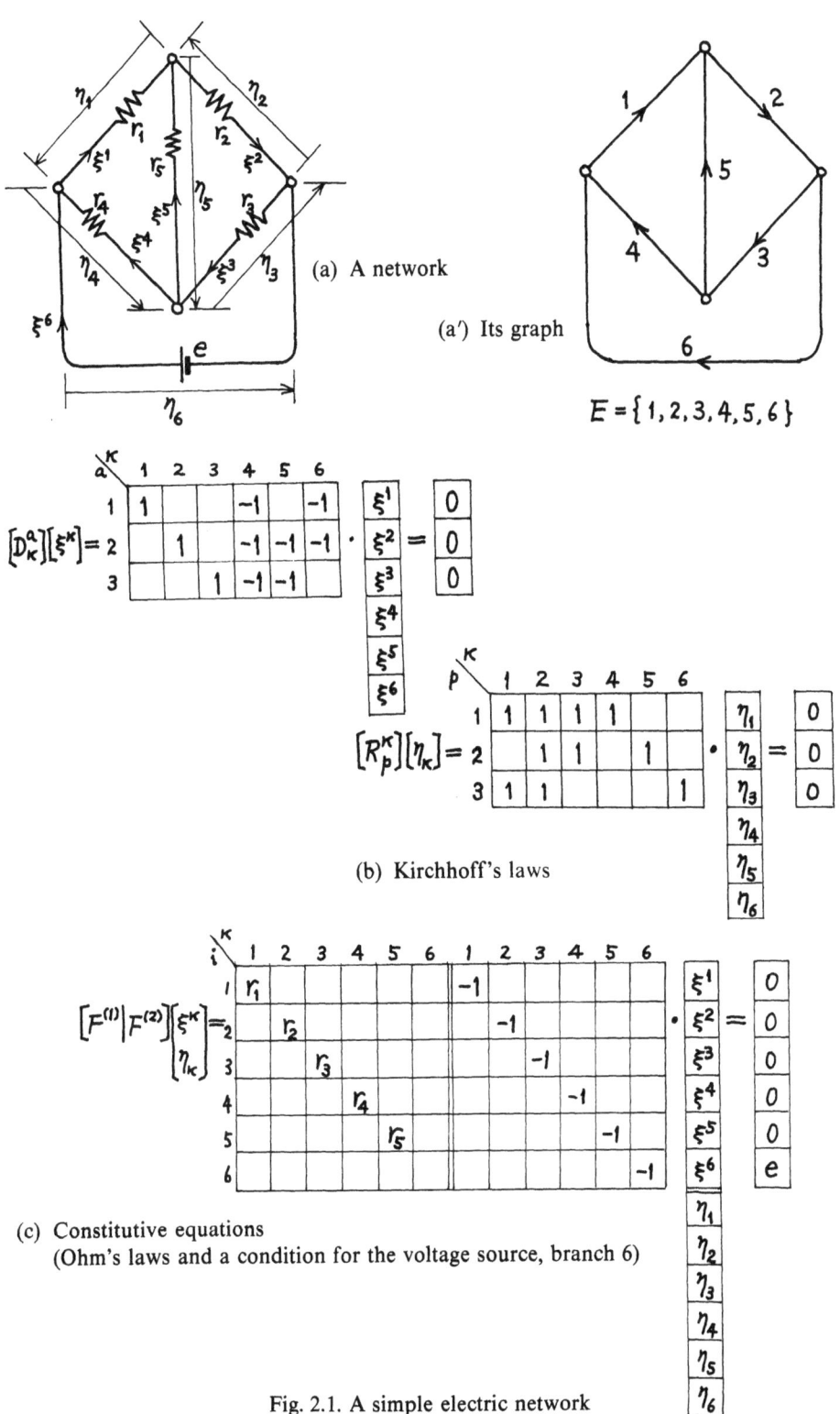

Fig. 2.1. A simple electric network

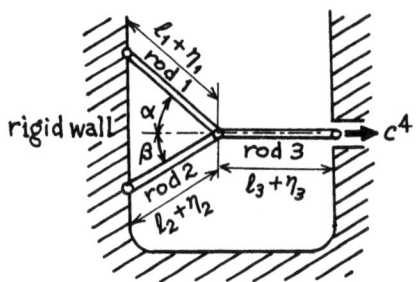

l_κ : normal length of rod κ
η_κ : infinitesimal elongation of rod κ
ξ^κ : tension of rod κ
c^4 : force applied to rod 3
α, β : angles between rods

(a) A two-dimensional elastic structure (rods are connected by pin joints)

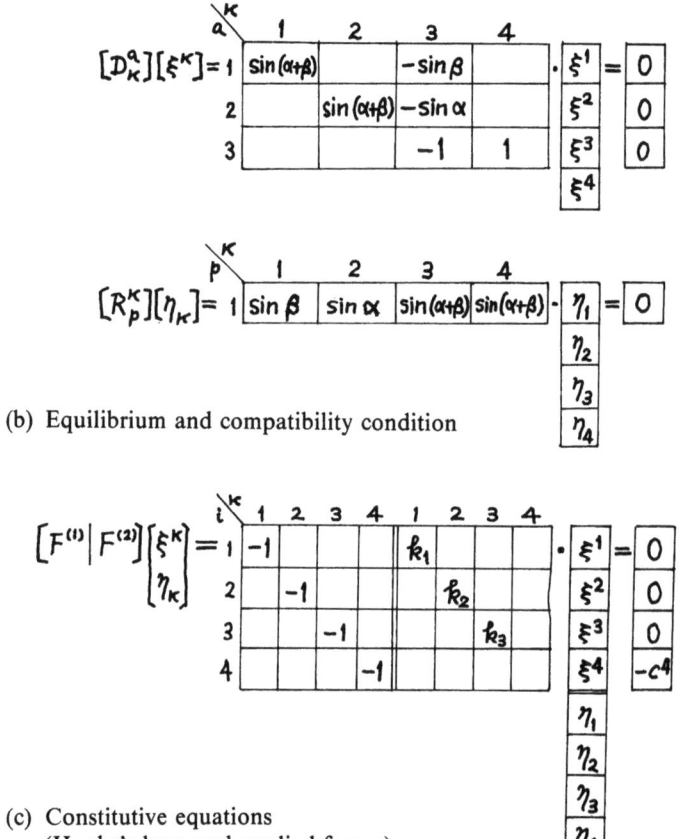

(b) Equilibrium and compatibility condition

(c) Constitutive equations
(Hooke's laws and applied forces)

Fig. 2.2. A simple elastic structure

with an electric network) or energy (as is the case with an elastic structure). Furthermore, since the assumption of the closedness of the system makes the structural equations homogeneous (i.e., makes the right-hand side vanish), the scalar product identically vanishes due to the duality of the two parts of the structural equations. This allures us to consider that the law of conservation of energy is not so much physical as topological or geometrical, but, to be more

II. Applications of Matroid Theory

conservative, we might consider that the fact that the energy conservation law takes a purely topological/geometrical form is itself a physics.

In Fig. 2.1 and Fig. 2.2, very small examples of an electric network and an elastic structure are illustrated.

A further remark should be made in order to render the mathematical model of a system so far described applicable to a wider class of electric networks. That is, when we deal with electric or electronic devices with more than two terminals, such as a three-terminal vacuum tube or transistor, it is an established trick to regard them as a set of two-terminal elements which are mutually coupled electrically and which are already partially connected topologically. For example, a tube and a transistor in Fig. 2.3(a) are regarded as a pair of branches connected at their ends as shown in Fig. 2.3(b) whose characteristics (with respect to a perturbation around a normal state; i.e. small-signal characteristics) are given in Fig. 2.3(c).

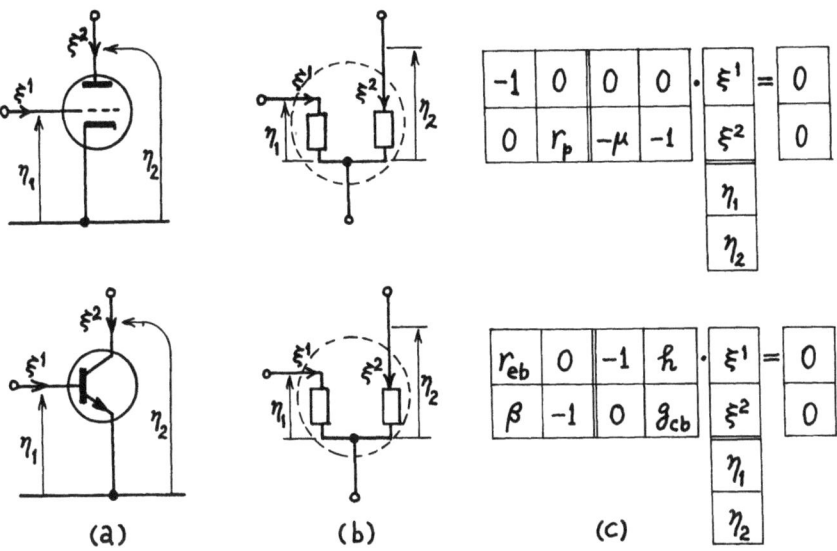

Fig. 2.3. Modelling a three-terminal element as partially connected two two-terminal elements

There is an entirely different expression of mathematical relations among relevant quantities of a system, called a *signal-flow graph*. It is regarded as an appropriate expression, especially, for a control system. However, a signal-flow graph is merely the sparsity structure of a system of equations disguised in a graph. As such, it has a wide application, but at the sacrifice of deeper insight into different properties of a system.

A signal-flow graph is a graph $G=(V, E, \partial^+, \partial^-)$ which has vertices ($\in V$) corresponding to variables (chosen so as to determine the state of the system completely, and consequently, called "state variables"), and has an edge e_κ ($\in E$) iff the variable corresponding to the vertex $\partial^+ e_\kappa$ appears explicitly in the function determining the value of the variable corresponding to the vertex

$\partial^- e_\kappa$. For the sake of comparison, the mathematical relations among the currents and voltages in the electric network of Fig. 2.1 are expressed in the form of a signal-flow graph in Fig. 2.4.

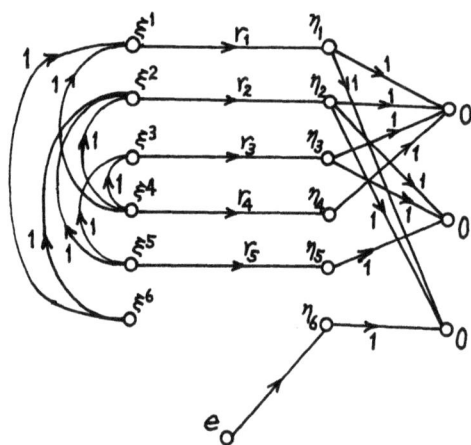

Fig. 2.4. A signal-flow graph for the network of Fig. 2.1.

The graphical or matroidal properties of the graph which are important in the analysis of a system represented by a signal-flow graph are somewhat different from those of the graph representing the topology of an electric network. In the latter case, the circuit matroid and the cocircuit matroid of the graph are of primary importance, and sometimes the matroid connected with the physical characteristics of branches also plays a substantial role, whereas, in the former case, the Menger-linkings on the graph and the matroidal properties connected with them are of our primary concern.

3. Topological, Geometrical and Physical Matroids, or Faithful and Unfaithful Representations

The observations to be presented in this section will not be rigorous mathematical ones, but they are intended to give a physical or engineering classification of matroids appearing in connection with systems analysis.

In the main part of section 2, we mentioned the following three kinds of matroids.

(1) Matroids which reflect the topological structure of an electric network, i.e., Kirchhoff's laws for currents and voltages.
(2) Matroids which reflect the metric-geometrical structure of an elastic structure, i.e., the equilibrium of forces and moments and the compatibility of deformations.
(3) Matroids which are concerned with constitutive equations, i.e. the physical characteristics of elements (branches of an electric network or rods of an elastic structure).

II. Applications of Matroid Theory

We shall figuratively call matroids of the first kind *topological* matroids, those of the second kind *geometrical,* and those of the third kind *physical,* respectively.

Topological matroids are distinguished from matroids of the other kinds by the following distinctive features.

The circuit matroid as well as the cocircuit matroid on the edge set of a network graph is regular and oriented. Hence, the matrix representing it (over any big field such as the field of real numbers) is totally unimodular, so that there is no "numerical" ambiguity in defining or "inputting" it, nor is there any difficulty in manipulating the matroid computationally. Furthermore, the representation over the smallest field, GF(2), preserves all essential informations. In other words, although the representation over GF(2) does not carry information about orientation, the orientation of a branch of an electric network is not so essential but it is a mere convention regarding in which direction we shall look at the branch, i.e., measure the current in and the voltage across it. Thus, the matroid representation of the topological structure of an electric network may be said to be "faithful". It should also be noted that the mathematical theory and computational techniques of electric networks can be extended in a natural way to the case where the circuit and cocircuit matroids of a graph are replaced by a dual pair of oriented regular matroids [Bruno-Weinberg 1976], [Bruter 1970], [Minty 1966], and, moreover, that an efficient procedure for reconstructing the original graph from its circuit or cocircuit matroid represented over GF(2) is now known [Bixby 1981], [Fujishige 1980b], [Iri 1962], [Iri 1968a].

In contrast with topological matroids, physical matroids are highly "quantitative". If we want to characterize faithfully the physical characteristics of branches of an electric network or of rods of an elastic structure, their interdependence, or the matroidal expression, is not enough. We usually need more information than a qualitative statement such as "this branch is a linear resistor, the voltage across and the current in it are proportional to each other, and they have nothing to do with the voltages and currents of the other branches"; i.e. we usually need to know "what the coefficient of proportionality, i.e., the resistance (or conductance), is". Thus, the matroid plays only partial rôle for the physical characteristics. However, in practical situations, being quantitative often means being uncertain. For example, a resistor of resistance 1Ω means actually a resistor of resistance "approximately equal to" 1Ω, i.e., a certain resistance, say, between $0.97\,\Omega$ and $1.03\,\Omega$. Hence, it is not only allowable but also important for us to investigate the properties of a system which are invariant under perturbations of physical characteristics of its elements. In fact, such a property is easy to technically realize, whereas a property which is not invariant under the perturbations will not be so easy. Algebraically, numerical values of parameters in the constitutive equations are assumed to be algebraically independent transcendentals in order to reflect these situations (see section 6).

Geometrical matroids, which appear in connection with elastic structures, share common properties partly with topological matroids and partly with physical matroids. In fact, they are similar to topological matroids in that they characterize the structural equations, whereas they are similar to physical matroids in that they are quantitative. Furthermore, unlike the case of an electric network, several intensive as well as extensive variables are associated with a single element (i.e., a rod) of an elastic structure, so that the ground set E (of intensive variables, or of extensive variables) is partitioned into subsets, each being associated with an element. This makes the independence or interdependence problem for forces/moments and deformations much more complicated than that for voltages and currents, and gives rise to matroid "parity" problems. The so-called "rigidity problems" can be regarded as the problems concerning the geometrical matroids. (We shall not discuss the rigidity problems in the present essay. See [Asimov-Roth 1978, 1979], [Crapo 1979], [Laman 1970], [Lovász-Yemini 1982], [Sugihara 1980].)

4. Minimum Fundamental Equations for an Electric Network as a Problem concerning the Topological Matroid

Let us consider an electric network with the set E of branches whose topological structure is given by the graph $G=(V, E, \partial^+, \partial^-)$. Except for pathological cases such as those considered in section 5, there is a dissection $E = E_C \cup E_V$ ($E_C \cap E_V = \emptyset$) of E such that the constitutive equations (2.7) can be solved for the voltages across branches of E_C and the currents in E_V, i.e., the voltages across E_C and the currents in E_V can be expressed as functions of the currents in E_C and the voltages across E_V. In terms of the matroid on $E^{(1)} \cup E^{(2)}$ represented by the matrix of (2.8), this means that the union $E_C^{(1)} \cup E_V^{(2)}$ of the image $E_C^{(1)}$ of E_C in $E^{(1)}$ and that $E_V^{(2)}$ of E_V in $E^{(2)}$ constitute a cobase. In the linear case, the functions will look like (4.1).

$$\eta_\lambda = \sum_{e_\mu \in E_C} z_{\lambda\mu} \xi^\mu + \sum_{e_\nu \in E_V} g_\lambda^\nu \eta_\nu + \tilde{v}_\lambda \quad (e_\lambda \in E_C),$$
$$\xi^\kappa = \sum_{e_\mu \in E_C} h_\mu^\kappa \xi^\mu + \sum_{e_\nu \in E_V} y^{\kappa\nu} \eta_\nu + \tilde{c}^\kappa \quad (e_\kappa \in E_V). \tag{4.1}$$

The matrix $\begin{bmatrix} z & g \\ h & y \end{bmatrix}$ is sometimes called the "branch immittance matrix" with respect to the dissection $E = E_C \cup E_V$ ($z_{\lambda\mu}$ = impedance from branch e_μ to branch e_λ, $y^{\kappa\nu}$ = admittance from branch e_ν to branch e_κ, g_λ^ν = voltage transfer ratio from branch e_ν to branch e_λ, h_μ^κ = current transfer ratio from branch e_μ to branch e_κ). Throughout the following arguments, we shall assume, for simplicity's sake, that the constitutive equations can be solved for any dissection of E into E_C and E_V. (If there are subsets of branches $\hat{E}_C$ and $\hat{E}_V$ which must be included in E_C and E_V, respectively, then we may modify the network in advance by opencircuiting all the branches of $\hat{E}_C$ and shortcircuiting all the branches of $\hat{E}_V$ in the following treatment.)

The so-called *fundamental system of equations* for an electric network is a system of equations which is obtained from the structural equations (2.1), (2.2) and the constitutive equations (2.7) by eliminating variables using the former

II. Applications of Matroid Theory

equations. There are two classical types of fundamental equations, the fundamental system of current-variable type and that of voltage-variable type. A fundamental system of current-variable type is constructed as follows.

(i) Choose an arbitrary base $\bar{B}(\subseteq E)$ of the cocircuit matroid $M^*(G)$ of G (i.e., a cobase of the circuit matroid $M(G)$). Then, by means of Kirchhoff's current law (2.1), all the other currents ξ^μ ($e_\mu \in E - \bar{B} \equiv B$) are expressed as linear combinations of the currents in $\bar{B}$. Let the expressions be

$$\xi^\mu = - \sum_{e_\kappa \in \bar{B}} N_\kappa^\mu \xi^\kappa \quad (e_\mu \in B), \tag{4.2}$$

where N_κ^μ is the non-unit part of the coefficient matrix of the basic form of (2.1) with respect to the base B of the circuit matroid $M(G)$, i.e., $[I|N] = [D]$ up to a permutation of columns and some row operations.

(ii) Assuming $E_C = E$ and $E_V = \emptyset$, express all the voltages η_λ ($e_\lambda \in E$) in terms of the currents ξ^κ in $\bar{B}$. If the voltage-current relations of the form (4.1) are assumed, the expressions will be of the form:

$$\eta_\lambda = \sum_{e_\mu \in B} z_{\lambda\mu} \xi^\mu + \sum_{e_\kappa \in \bar{B}} z_{\lambda\kappa} \xi^\kappa + \tilde{v}_\lambda$$

$$= \sum_{e_\kappa \in \bar{B}} \left(z_{\lambda\kappa} - \sum_{e_\mu \in B} z_{\lambda\mu} N_\kappa^\mu \right) \xi^\kappa + \tilde{v}_\lambda \tag{4.3}$$

for $e_\lambda \in E$.

(iii) Substitute the expressions (4.3) of the voltages in terms of the currents ξ^κ in $\bar{B}$ into Kirchhoff's voltage law (2.2). Since the circuit matroid $M(G)$ and the cocircuit matroid $M^*(G)$ of a graph G are dual of each other, the matrix R in (2.2) has the form $[R] = [-N^T | I]$ if we choose $\bar{B}$ as the base of $M^*(G)$. Thus, we have

$$\eta_\lambda - \sum_{e_\nu \in B} N_\lambda^\nu \eta_\nu = 0 \quad (e_\lambda \in \bar{B}),$$

or

$$\sum_{e_\kappa \in \bar{B}} Z_{\lambda\kappa} \xi^\kappa = \tilde{\tilde{v}}_\lambda \quad (e_\lambda \in \bar{B}), \tag{4.4.1}$$

where

$$Z_{\lambda\kappa} = z_{\lambda\kappa} - \sum_{e_\mu \in B} z_{\lambda\mu} N_\kappa^\mu - \sum_{e_\nu \in B} N_\lambda^\nu z_{\nu\kappa} + \sum_{e_\mu, e_\nu \in B} N_\lambda^\nu z_{\nu\mu} N_\kappa^\mu \quad (e_\kappa, e_\lambda \in \bar{B}), \tag{4.4.2}$$

and

$$\tilde{\tilde{v}}_\lambda = \tilde{v}_\lambda - \sum_{e_\nu \in B} N_\lambda^\nu \tilde{v}_\nu \quad (e_\lambda \in \bar{B}). \tag{4.4.3}$$

In this manner, any fundamental system of equations of current-variable type has k unknowns and k equations (k being the rank of $M^*(G)$, i.e., the nullity of the graph G), and, therefore, the size of a fundamental system of equations of current-variable type does not depend on the choice of independent unknowns. The matrix Z in the equations (4.4) is called the "circuit (mesh, loop or tie-set) impedance matrix". It should also be noted that the expression (4.4.2) of Z is essentially the triple matrix product

$$R z R^T, \tag{4.5}$$

where R is the matrix in (2.2), R^T its transpose, and z the "branch impedance matrix" of (4.1).

Entirely in the dual manner, we can construct a fundamental system of voltage-variable type as follows.

(i′) Choose a base B of the circuit matroid $M(G)$ of G, and express the voltages across $\bar{B}(=E-B)$ in terms of the voltages across B using Kirchhoff's voltage law (2.2).

(ii′) Assuming $E_C = \emptyset$ and $E_V = E$, express all the currents in E in terms of the voltages across B using the expressions of (i′) and (4.1).

(iii′) Substitute the expressions of (ii′) in Kirchhoff's current law (2.1). The resulting equations are:

$$\sum_{e_\lambda \in B} Y^{\kappa\lambda} \eta_\lambda = \tilde{\tilde{c}}^\kappa \quad (e_\kappa \in B), \tag{4.6.1}$$

$$Y^{\kappa\lambda} = y^{\kappa\lambda} + \sum_{e_v \in \bar{B}} y^{\kappa v} N_v^\lambda + \sum_{e_\mu \in \bar{B}} N_\mu^\kappa y^{\mu\lambda} + \sum_{e_\mu, e_v \in \bar{B}} N_\mu^\kappa y^{\mu v} N_v^\lambda \quad (e_\kappa, e_\lambda \in \bar{B}), \tag{4.6.2}$$

$$\tilde{\tilde{c}}^\kappa = \tilde{c}^\kappa + \sum_{e_\mu \in \bar{B}} N_\mu^\kappa \tilde{c}^\mu \quad (e_\kappa \in B), \tag{4.6.3}$$

with the voltages across B as the independent unknowns.

The size of the equations of this type is again independent of the choice of unknowns, i.e., it is equal to the rank m of $M(G)$, or the rank of the network graph G. The matrix Y in (4.6) is called the "cocircuit (or cutset) admittance matrix", and is essentially the triple matrix product

$$D y D^T, \tag{4.7}$$

where D is the matrix in (2.1), D^T its transpose, and y the "branch admittance matrix" of (4.1).

Although the author does not know how far back we can trace the origin of the idea (cf. [Amari 1962], [Kron 1939], [Kron 1963]), there is a third type of fundamental equations which is a common generalization of the classical two types. It is the so-called "mixed-type equations", and is constructed as follows.

(i″) Dissect E into E_C and E_V ($E = E_C \cup E_V$, $E_C \cap E_V = \emptyset$), choose an arbitrary base B_V of $M(G)|E_V$ (the reduction of $M(G)$ to E_V) and an arbitrary base $\bar{B}_C$ of $M^*(G)|E_C$ ($=(M(G)\times E_C)^*$; $M(G)\times E_C$ being the contraction of $M(G)$ to E_C), and express the currents in B_C ($=E_C - \bar{B}_C =$ a cobase of $M^*(G)|E_C$) in terms of the currents in $\bar{B}_C$ using part of Kirchhoff's current law (2.1) and the voltages across $\bar{B}_V$ ($=E_V - B_V =$ a cobase of $M(G)|E_V$) in terms of the voltages across B_V using part of Kirchhoff's voltage law (2.2). (This is possible since $B = B_C \cup B_V$ (resp. $\bar{B} = E - B = \bar{B}_C \cup \bar{B}_V$) is a base of $M(G)$ (resp. $M^*(G)$) and since, if the equations (2.1) (resp. (2.2)) are written out with respect to the base B (resp. $\bar{B}$) of $M(G)$ (resp. $M^*(G)$), the currents in B_C (resp. the voltages across $\bar{B}_V$) are related only to the currents in $\bar{B}_C$ (resp. the voltages across B_V)).

(ii″) Assuming that $E_C^{(1)} \cup E_V^{(2)}$ is a cobase of the matroid represented by the matrix F of (2.8), express the voltages across E_C and the currents in E_V in terms of the currents in $\bar{B}_C$ and the voltages across B_V using (4.1) and the expressions of (i″).

(iii'') Substitute the expressions of (ii'') in the equations of Kirchhoff's laws which have not been used in (i''). (Note that all the currents and voltages have been expressed in terms of the currents in $\bar{B}_C$ and the voltages across B_V in (i'') and (ii'').)

The resulting equations are of the form:

$$\sum_{e_\mu \in \bar{B}_C} Z_{\lambda\mu} \xi^\mu + \sum_{e_\nu \in B_V} G_\lambda^\nu \eta_\nu = \tilde{v}_\lambda \quad (e_\lambda \in \bar{B}_C),$$
$$\sum_{e_\mu \in \bar{B}_C} H_\mu^\kappa \xi^\mu + \sum_{e_\nu \in B_V} Y^{\kappa\nu} \eta_\nu = \tilde{c}^\kappa \quad (e_\kappa \in B_V) \quad (4.8)$$

with the currents in $\bar{B}_C$ and the voltages across B_V as the independent unknowns, where the matrices Z, Y, G and H can be written as a triple matrix product like (4.5) and (4.7) in terms of z, y, g, h of the branch immittance matrix in (4.1) and of appropriate submatrices of D and R in (2.1) and (2.2). (The Z, Y, G and H are linear in z, y, g and h, respectively.)

What is most noteworthy is that the size of a mixed-type fundamental system of equations does depend on the choice of independent unknowns, or more exactly, on the dissection $E = E_C \cup E_V$. In fact, the size, i.e. the number of independent unknowns, of the equations is equal to the sum of the rank of $M^*(G)|E_C$ and that of $M(G)|E_V$.

Here arises a problem of fundamental importance in electric network theory: "What is the minimum size of a mixed-type fundamental system of equations?" (We call the minimum size the number of the *topological degrees of freedom* of an electric network.) (This problem was recognized in [Amari 1962], solved in [Iri 1968b], [Kishi-Kajitani 1968], [Ohtsuki-al. 1968] without reference to matroids. The relation with matroids was noted in [Iri-Tomizawa 1975a].)

Based upon the preliminaries so far prepared, the problem can readily be formulated in the language of matroids. Let $\rho: 2^E \to \mathbb{Z}_+$ be the rank function of the circuit matroid $M(G)$ of G. The rank function ρ^* of $M^*(G)$ is then

$$\rho^*(X) = |X| - \rho(E) + \rho(E - X) \quad (X \subseteq E). \quad (4.9)$$

Therefore, the size of the equations corresponding to a dissection $E = E_C \cup E_V$ is equal to $\rho(E_V) + \rho^*(E_C)$, and, noting that $E_C = E - E_V$, the problem is reduced to that of finding the following minimum value.

$$\min\{\rho(X) + \rho^*(E - X) | X \subseteq E\}$$
$$= \min\{\rho(X) + \rho(X) - |E - X| | X \subseteq E\} - \rho(E)$$
$$= \min\{2\rho(X) - |X| | X \subseteq E\} + [|E| - \rho(E)]. \quad (4.10)$$

The first member of (4.10), when combined with the min-max theorem for the matroid intersection problem:

$$\min\{\rho(X) + \rho^*(E - X) | X \subseteq E\}$$
$$= \max\{|X| | X \subseteq E, |X| = \rho(X) = \rho^*(X)\}, \quad (4.11)$$

tells us that the problem is to find a maximum common independent set $X (\subseteq E)$ in $M(G)$ and $M^*(G)$. Since

$$\mu(Y) = \min\left\{\sum_{i=1}^{r} \rho_i(X) + |Y - X| \,\Big|\, X \subseteq Y\right\} \quad (Y \subseteq E) \tag{4.12}$$

is the rank function of the union matroid composed of the matroids with rank functions $\rho_1, \ldots, \rho_r$, the second member of (4.10) tells us that the problem is equivalent to finding the rank (or, more concretely, a base) of the union matroid $M(G) \vee M(G)$. The last member of (4.10) is useful for the decomposition of the problem as will be discussed in section 9.

Finally, it should be remarked that the above arguments are applicable mutatis mutandis to elastic structures as well.

The problem of minimum fundamental equations is, as has been explicated, purely topological (or geometrical), and not physical at all. Historically, the problem was initially recognized and solved without reference to matroids, but, once it is formulated in terms of matroids, the theories and techniques developed in matroid theory are readily applied to the analysis and solution of the problem.

A simple example of the topological degrees of freedom is shown in Fig. 4.1. Fig. 4.1(a) is a sample network, where there is a voltage source (denoted by ⊣⊢) and a current source (denoted by ─⊕─). Its graph G is as shown in Fig. 4.2(b), where the current source has been deleted by opencircuiting and the voltage source by shortcircuiting since a voltage source has a prescribed voltage and is equivalent to a shortcircuit with respect to a perturbation and, similarly a current source is equivalent to an opencircuit. The size of a fundamental system of equations of voltage-variable type for this network is, therefore, equal to the rank of the network graph G, i.e. 10, and the size of a fundamental system of equations of current-variable type is equal to the nullity, 12. However, if we choose dissection of the edge set E of G into E_C and E_V as shown in the figure, the rank of $M(G)|E_V$ is equal to 5 and that of $M^*(G)|E_C$ is 4, so that the size of the mixed-type fundamental system of equations with respect to this dissection is $5 + 4 = 9$. This number, 9, can be proved to give the minimum of (4.10), so it is the number of the topological degrees of freedom of the network of Fig. 4.1(a).

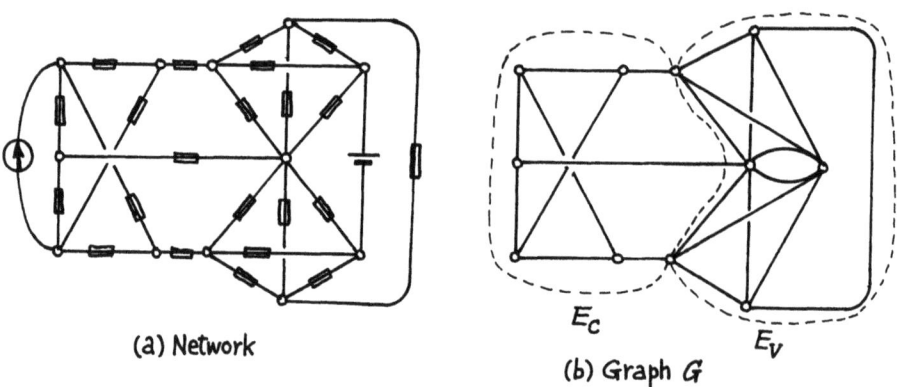

(a) Network

(b) Graph G

Fig. 4.1. An example of the topological degrees of freedom of a network (rank of $G = 10$, nullity of $G = 12$, rank of $M(G)|E_V = 5$, rank of $M^*(G)|E_C =$ nullity of $M(G) \times E_C = 4$)

5. Oono's Problem as a Pathology of Physical Matroids

In this section, we shall consider a problem which concerns the physical matroid only. The physical characteristics of branches of an electric network are

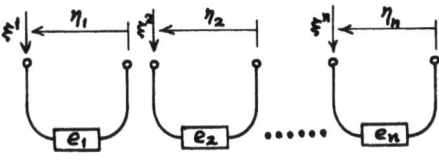

Fig. 5.1. $E=\{e_1, \ldots, e_n\}$

described by a system of equations like (2.7) with a coefficient matrix (2.8). Ordinarily, (2.7) admits an immittance-matrix expression, as was assumed in the beginning of the preceding section. That there is an immittance matrix corresponding to a dissection $E = E_C \cup E_V$ means physically that we can connect arbitrary current sources to branches of E_C and voltage sources to branches of E_V, separately, i.e. we can prescribe the values of currents in branches of E_C and those of voltages across branches of E_V, arbitrarily, and, if we connect sources to branches in that way, we shall have voltages across branches of E_C and currents in branches of E_V as response quantities. In other words, $E_C^{(1)} \cup E_V^{(2)}$ forms a cobase of the matroid on $E^{(1)} \cup E^{(2)}$ represented by the matrix F of (2.8), where $E_C^{(1)}$ and $E_V^{(2)}$ are the subsets of the current replica $E^{(1)}$ and the voltage replica $E^{(2)}$ of E, respectively, corresponding to E_C and E_V.

Although most of classical "passive elements" had the F-matrix of (2.8) for which any dissection $E = E_C \cup E_V$ corresponded to a cobase, a number of seemingly singular, or pathological, elements have recently come in use in electronics devices, as the progress of electronics technology enabled us to make various "active elements". For example, for the small-signal model of a "vacuum tube" (triode; see the upper half of Fig. 2.3), the possible cobases of the F-matrix are $\{e_2^{(1)}, e_1^{(2)}\}$, $\{e_2^{(1)}, e_2^{(2)}\}$ and $\{e_1^{(2)}, e_2^{(2)}\}$, where $E = \{e_1, e_2\}$ and e_1 is the grid-to-cathode branch and e_2 the anode-to-cathode branch, but, among them, only $\{e_2^{(1)}, e_1^{(2)}\}$ and $\{e_1^{(2)}, e_2^{(2)}\}$ have a corresponding dissection (i.e., $E_C = \{e_2\}$, $E_V = \{e_1\}$ and $E_C = \emptyset$, $E_V = E = \{e_1, e_2\}$), hence an immittance matrix:

$$\begin{pmatrix}\xi^1\\\eta_2\end{pmatrix}=\begin{pmatrix}0 & 0\\r_p & -\mu\end{pmatrix}\begin{pmatrix}\xi^2\\\eta_1\end{pmatrix}\quad\text{and}\quad\begin{pmatrix}\xi^1\\\xi^2\end{pmatrix}=\begin{pmatrix}0 & 0\\g_m & g_p\end{pmatrix}\begin{pmatrix}\eta_1\\\eta_2\end{pmatrix}, \tag{5.1}$$

where

$$g_p = 1/r_p, \quad g_m = \mu g_p. \tag{5.2}$$

Furthermore, in active network design, even elements having no immittance matrix are sometimes in use. For example, an "ideal operational amplifier" of infinite gain (Fig. 5.2(a)) has the F-matrix expression Fig. 5.2(c) of its characteristics. In fact, since the input impedance of an ideal amplifier is assumed to be infinite, we must have $\xi^1 = 0$; since the voltage gain, or the voltage amplification factor, is infinite, the input voltage must be zero, $\eta_1 = 0$, in order to have

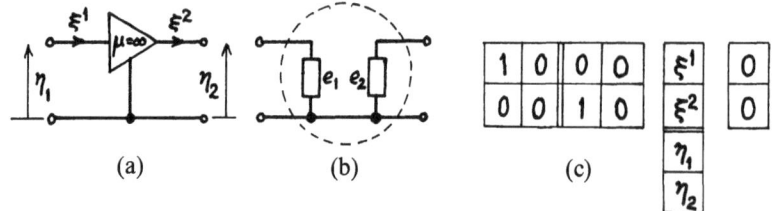

Fig. 5.2. (a) An ideal operational amplifier with infinite gain
(b) Its expression as a pair of nullator (e_1) and norator (e_2)
(c) F-matrix expression of its characteristics

a finite output voltage η_2: since the output impedance of an ideal amplifier is zero, the output current ξ^2 can take any value depending on the impedance of a load to be connected to the output port; and, finally, the output voltage η_2 is indeterminate because it is the product of $\eta_1 = 0$ and the gain ∞. If we regard an ideal operational amplifier of infinite gain as two partially connected branches as shown in Fig. 5.2(b) (as we did for a vacuum tube; see Fig. 2.3), there appear strange elements called a *nullator* (e_1 in Fig. 5.2(b)) and a *norator* (e_2 in Fig. 5.2(b)). A nullator is a branch whose current and voltage both are constrained to zero, and a norator is a branch whose current and voltage are subject to no constraint. However, so long as we adopt the assumption that the F-matrix has $2n$ columns and rank n, as we did in section 2 according to the degrees-of-freedom arguments, a nullator or a norator cannot appear separately, but they must appear in a pair with each other.

Y. Oono raised a question [Oono 1960]: "What conditions are needed to ensure the existence of an immittance matrix for an F-matrix?" He analysed the question and gave an answer based upon an algorithm. Although his algorithm was not effective, i.e., not of polynomial order, from the standpoint of the theory of computational complexity of today, he made use of an expression which corresponds to the matrix representation (over a sufficiently large extension field) of the union of two matroids represented by matrices, and, furthermore, proposed a kind of "probabilistic algorithm" (i.e., the technique of replacing transcendentals in the extension field which are algebraically independent over the ground field by "random numbers" from the ground field).

Oono's problem may now be formulated in the matroidal language as follows. For a matroid F on a set $E^{(1)} \cup E^{(2)}$ ($E^{(1)} \cap E^{(2)} = \emptyset$) where $E^{(i)}$ is a replica of a finite set E (i.e., there is a bijection $\psi^{(i)}: E^{(i)} \to E$) ($i = 1, 2$), we want to find as large an independent set $X (\subset E^{(1)} \cup E^{(2)})$ as possible such that $\psi^{(1)}(X \cap E^{(1)}) \cap \psi^{(2)}(X \cap E^{(2)}) = \emptyset$. In fact, if we take the matroid represented by the F-matrix of (2.8) for F, and if we have $|\hat{X}| = \max |X| = |E| = n$, then $\hat{X}$ is a base of F and $(E^{(1)} \cup E^{(2)}) - \hat{X}$ a cobase, so that $E_C = \psi^{(1)}(E^{(1)} - \hat{X})$ and $E_V = \psi^{(2)}(E^{(2)} - \hat{X})$ is a dissection of E for which we have an immittance matrix. Otherwise, if $|\hat{X}| = \max |X| < |E| = n$, then we may define $s \equiv n - \max |X|$ as the *degree of singularity* of the constitutive equations. (Since the rank of F is assumed to be n, there is an independent set of cardinality not less than $n/2$ either in $E^{(1)}$ or in $E^{(2)}$. Therefore, we should have $0 \leq s \leq [n/2]$.) The above example of nullator-norator pair is the simplest singular case for which we have

II. Applications of Matroid Theory

$n=2$ and $s=1$. (Note that any rank-2 matrix F with $2 \times 2 = 4$ columns for which $s=1$ can be brought to the form of Fig. 5.2(c) by suitable row operations.) This matroidal problem can further be reduced to one of the typical problems in matroid theory. The most direct way will be to regard it as the problem of finding a maximum-cardinality independent matching in the bipartite graph which has $E^{(1)} \cup E^{(2)}$ as the left vertex set, E as the right vertex set, and edges representing the bijections $\psi^{(1)}$ and $\psi^{(2)}$ and on whose left vertex set the matroid F is defined. (On the right vertex set no matroid is defined, i.e., the free matroid is defined.) It is evident that the left end vertices of a maximum independent matching give a solution X.

It is possible to consider a similar problem not only for separate branches but also for the "terminal characteristics" of a network (see section 6).

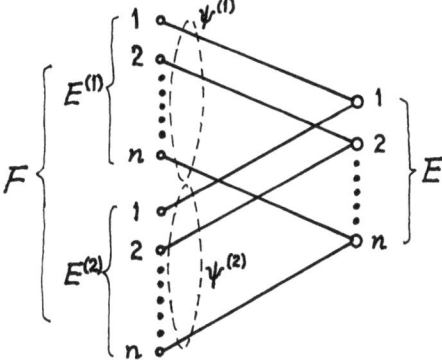

Fig. 5.3. The bipartite graph with matroid F for Oono's problem

6. Solvability of the Entire System

Once we describe the structural and the constitutive equations for an electric network (or an elastic structure, or any other system of a similar kind; see section 2), we have to go into the analysis of that system. Our final goal is to analyse it quantitatively, i.e., by numerical computation. However, numerical computation cannot be free from some kind of ambiguity such as rounding errors, and, what is more fatal, we cannot expect the constitutive equations to be the "exact" mathematical expression of the physical properties of elements, as was remarked in section 3. Therefore, it would be important to extract as much information about the performance of the system as possible before resorting to numerical manipulation of data.

Constructing a fundamental system of equations (which was explained in section 4) may be regarded as one such step in the sense that we use up all the structural equations (2.1), (2.2) to reduce the size of the system of equations by eliminating as many variables as possible. There, we have no serious computational problems, because the numerical computations performed on physical data are only additions and subtractions, as is evident from (4.4.2), (4.4.3),

(4.6.2), (4.6.3). (The same is true for the mixed-type equations although we did not write out the corresponding formulas in section 4.) Moreover, if we stop at an expression such as (4.5) and (4.7), there is no rounding-error problem at all up to that stage.

The condition that the determinant of the coefficient matrix of a system of equations shall not vanish is necessary and sufficient for that system of equations to have a unique solution, if the equations are linear. Even if some of the equations are nonlinear, the nonvanishing Jacobian ensures the local unique solvability of the linearized perturbation equations. So, we shall say that the system has a unique solution, or is *uniquely solvable,* if and only if the determinant does not vanish. The validity of this condition for the unique solvability could not be examined without numerical computation unless we made further assumptions. However, we can take advantage of the ambiguity of physical characteristics of elements, which we noted in section 3; i.e., we shall adopt a certain "generality assumption" on the physical characteristics of elements. Roughly speaking, we shall assume that, among the parameters which characterize the constitutive equations, some have "exact" values but the others have "general" values – general in the sense that they are not in special relation to one another.

To be specific, let us consider the triple matrix product expression (4.7) of the cocircuit admittance matrix of the fundamental system of equations of voltage-variable type, and adopt a typical generality assumption for the n^2 entries $y^{\kappa\lambda}$'s of the branch admittance matrix y:

Among $y^{\kappa\lambda}$'s, the zeroes are exact whereas the nonvanishing $y^{\kappa\lambda}$'s satisfy no algebraic equations with integer coefficients.

This is equivalent to the assumption that:

The nonvanishing entries are algebraically independent transcendentals over the field K to which the coefficients of structural equations belong, where $K = \mathbb{Q}$ (the field of rational numbers) in the present case. Then, let us look at the expansion of the determinant:

$$\det[DyD^T] = \sum_{\kappa_i}\sum_{\lambda_j} D^{[1\ldots m]}_{\kappa_1\ldots\kappa_m} y^{\kappa_1\lambda_1} \ldots y^{\kappa_m\lambda_m} D^{[1\ldots m]}_{\lambda_1\ldots\lambda_m}, \tag{6.1}$$

where $D^{[1\ldots m]}_{\kappa_1\ldots\kappa_m}$ is the subdeterminant of the matrix D^a_{κ} with all the m rows ($a = 1, \ldots, m$) and a subset of m columns $\kappa_1, \ldots, \kappa_m$, $D^{[1\ldots m]}_{\lambda_1\ldots\lambda_m}$ is a similiar subdeterminant, and the sum is taken over all the possible permutations of m distinct κ_i's chosen from $\{1, \ldots, n\}$ and all the possible combinations of m distinct λ_j's chosen from $[1, \ldots, n\}$ ($\lambda_1 > \lambda_2 > \ldots > \lambda_m$). Due to the generality assumption, different nonvanishing terms in the right-hand sum of (6.1) do not cancel one another, so that $\det[DyD^T]$ does not vanish *if and only if* there is a nonvanishing term on the right-hand side of (6.1). Furthermore, in each term of the right-hand side, the factor $D^{[1\ldots m]}_{\kappa_1\ldots\kappa_m}$ does not vanish if and only if $\{e_{\kappa_1}, \ldots, e_{\kappa_m}\}$ is a base of $M(G)$, and similarly for $D^{[1\ldots m]}_{\lambda_1\ldots\lambda_m}$ (since the matroid defined on the column set of D is the circuit matroid $M(G)$ of G of rank m). Thus, we are naturally led to independent matching problem on the bipartite graph of Fig. 6.1. The bipartite graph has the left vertex set $E^{(1)}$, and the right vertex set $E^{(2)}$, each of which is a replica of the edge set E of graph G of the electric network, and it

II. Applications of Matroid Theory

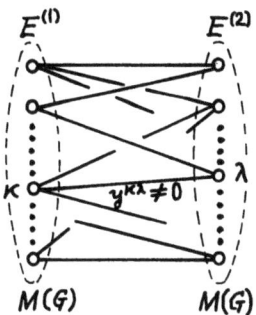

Fig. 6.1. The bipartite graph on which the maximum-cardinality independent matching problem discerns whether $\det[DyD^T]$ vanishes, or not, under the generality assumption on $y^{\kappa\lambda}$'s

has an edge connecting the κ-th element of $E^{(1)}$ to the λ-th element of $E^{(2)}$ if and only if $y^{\kappa\lambda} \neq 0$. On $E^{(1)}$ as well as on $E^{(2)}$, the matroid (isomorphic to) $M(G)$ is defined. An "independent matching" is a matching on the bipartite graph such that its left end vertices are independent in $M(G)$ on $E^{(1)}$ and its right end vertices are independent in $M(G)$ on $E^{(2)}$. Therefore, if $\det[DyD^T] \neq 0$, i.e., if there are bases $\{\kappa_1, \ldots, \kappa_m\}$ and $\{\lambda_1, \ldots, \lambda_m\}$ of $M(G)$ such that $y^{\kappa_1\lambda_1} \ldots y^{\kappa_m\lambda_m} \neq 0$, then there is the independent matching $\{(\kappa_1, \lambda_1), \ldots, (\kappa_m, \lambda_m)\}$ of cardinality m, which is obviously a maximum-cardinality independent matching on the bipartite graph. Conversely, if the maximum cardinality of independent matchings is equal to m, then there is a nonvanishing term on the right-hand side of (6.1), and hence $\det[DyD^T] \neq 0$ due to the generality assumption. Thus, the unique solvability problem for the voltage-variable type fundamental equations is completely reduced to the independent matching problem. (This formulation as well as the solution is due to [Tomizawa-Iri 1974]. This kind of problem was first solved in [Iri-Tomizawa 1974a] with matroids and in [Ozawa 1974] purely graphically.) Entirely the same method applies to the unique solvability problem for the current-variable type fundamental equations. For the mixed-type fundamental equations, the expansion of the determinant corresponding to (6.1) is more complicated (see, e. g., [Numata-Iri 1973]), but, based upon it, we can get a similar results [Iri-Tomizawa 1975a].

There is another approach which is methodologically interesting, too [Petersen 1978], [Petersen 1979a], [Recski 1976b], [Recski 1979a]. First, we write equations (2.1), (2.2) and (2.7) in combined form:

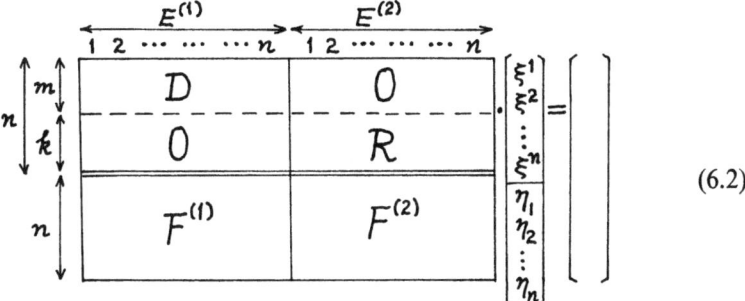

(6.2)

This expression reminds us of the following fact in matroid theory [Welsh 1976]. If two matroids M_1 and M_2 on the same set E are represented by matrices A and B (each with the column set E), respectively, on a field K, and if we choose transcendentals $\sigma_1, \ldots, \sigma_n$ ($n = |E|$) which are algebraically independent over K and denote by $\tilde{B}$ the matrix obtained by multiplying each column of B by σ_i, then the union $M_1 \vee M_2$ of the two matroids is represented by the matrix C over the extension field $K(\sigma_1, \ldots, \sigma_n)$ defined as follows:

$$C = \boxed{\begin{array}{c} A \\ \hline \tilde{B} \end{array}}. \tag{6.3}$$

Thus, if the entries in the F-part of the coefficient matrix of (6.2) are so "general" in value that the union of any two subsets of columns of the entire coefficient matrix, of which one is independent with respect to the upper half matrix and the other is independent with respect to the lower half matrix, is independent with respect also to the entire matrix, then the matroid defined on $E^{(1)} \cup E^{(2)}$ by the coefficient matrix of (6.2) is the union of the "topological matroid" (which is the direct sum of $M(G)$ and $M^*(G)$) and the "physical matroid" (i.e., the matroid represented by the F-matrix). There are several ways of making an assumption on generality of the F-matrix [Recski-Iri 1980], among which the simplest is to assume that

the nonvanishing entries of the F-matrix are algebraically independent transcendentals over the field of rational numbers.

(Evidently, this assumption is trivially a little too restrictive. In fact, we may multiply each row of the F-matrix by an arbitrary number, so that we can make at least one nonvanishing entry of each row rational, or even unity.) Under this simplest generality assumption, the physical matroid defined by the F-matrix may be regarded as a transversal matroid in a natural manner. Then, the union in question is the union of a transversal matroid and the direct sum of a graphic and a cographic matroid, so that relevant matroidal algorithms are easy to perform. Now, the condition for the unique solvability can be stated in a simple expression: "The rank of the union matroid be equal to $2n$," or "The union matroid be the free matroid on $E^{(1)} \cup E^{(2)}$."

This latter approach to the solvability can be extended to the "terminal solvability" problem [Recski 1979b]. (The terminal solvability problem had been studied by a Hungarian group led by A. Csurgay.) In electrical engineering, we deal not only with electrical networks which are closed systems but also with networks which are "open" for connection with other elements or networks through a certain number of "terminal pairs" or "ports". When we consider an open network, it is convenient to introduce the concept of *port-branches* to make the network closed. A port-branch is a branch with which a current and a voltage are associated; it connects the two terminals of a terminal pair through which the network is connected with the outside; but it has no physical characteristics. Therefore, if there are p port-branches and $n-p$ *internal branches,* i.e., non-port-branches, there are $2n$ currents and voltages, among which we have n equations of Kirchhoff's laws (m for currents and k for voltages, m and k being the rank and the nullity of the network graph, respectively) and $n-p$ constitu-

II. Applications of Matroid Theory

tive equations. Hence, by the degrees-of-freedom argument, we may expect that there will remain $2n-n-(n-p)=p$ degrees of freedom, and that, when viewed from outside, the p currents in and the p voltages across the port-branches, $2p$ variables in total, are constrained by a rank-p system of equations. (More rigorous arguments will follow below). In this manner, the set of port-branches will look like a set of p two-terminal elements whose constitutive equations are obtained from the $n+(n-p)$ equations on the $2n$ variables mentioned above by eliminating the $2(n-p)$ "internal" variables. If we denote the set of port-branches by E_P, that of internal branches by E_I ($E=E_P \cup E_I$, $E_P \cap E_I = \emptyset$), and their current replicas by $E_P^{(1)}$, $E_I^{(1)}$ and the voltage replicas by $E_P^{(2)}$, $E_I^{(2)}$, then the entire system of equations has a form like (6.4) (cf. (6.2)),

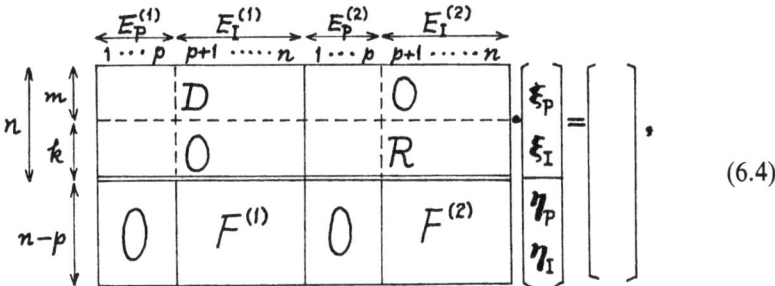

(6.4)

where ξ_P and η_P are the vectors with the currents in and the voltages across the port-branches as the components, respectively, and ξ_I and η_I are those for the internal branches. Usually, the right-hand side of (6.4) is put equal to zero, since electric sources are usually located outside of the network when we consider this kind of problem. Fig. 6.2. illustrates such a network, which is to be compared with the network of Fig. 2.1.

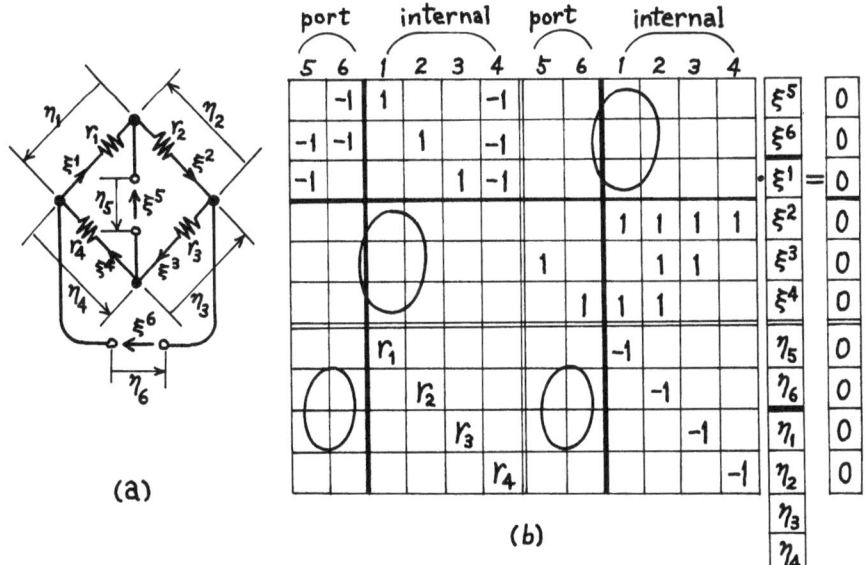

Fig. 6.2. A two-port network and the equations for it.

The coefficient matrix of (6.4) defines a matroid M on $E_P^{(1)} \cup E_I^{(1)} \cup E_P^{(2)} \cup E_I^{(2)}$. The unique solvability of this kind of port network is most adequately defined as follows. The network is solvable iff there are p variables (currents and/or voltages) of port-branches such that all the other variables are expressible as their functions. This means that the unique solvability is *nothing but the existence of a base X of the matroid M such that* $E_I^{(1)} \cup E_I^{(2)} \subseteq X$ and $|X - (E_I^{(1)} \cup E_I^{(2)})| = p$. If the network is uniquely solvable, the linear dependence relation defined among port-currents and port-voltages, i.e., the matroid defined on $E_P^{(1)} \cup E_P^{(2)}$, is the contraction of M to $E_P^{(1)} \cup E_P^{(2)}$, as is obvious. Even if the network is not uniquely solvable as a whole, we may consider the contraction $M_P \equiv M \times (E_P^{(1)} \cup E_P^{(2)})$ and regard it as the "terminal characteristics", saying that the network is *terminal-solvable* when M_P has rank p.

Even if the network is uniquely solvable or terminal-solvable, there still remains a problem of the kind discussed in section 5, i.e., we have to check whether the matroid M_P on $E_P^{(1)} \cup E_P^{(2)}$ admits an immittance-matrix representation or not, or to determine the degree of singularity.

All those concepts are effectively checkable either numerically (in case we concern quantitative data in the F-matrix) or combinatorially (under a generality assumption for the F-matrix). The basic ideas are the same as those in the preceding sections.

In a similar vein, the so-called "transfer function" is studied in [Petersen 1979b]. Similar techniques have been used for the problem of identifying the values of parameters in the constitutive equations when part of currents and voltages are observed for different values of port-voltages and/or port-currents [Ozawa-Kajitani 1979]. (The authors call this identification problem the problem of "diagnosis".)

7. Dynamical Degrees of Freedom

As is a commonsense in engineering mathematics, if we work on the field consisting of operators (of Heaviside, of Mikusiński, or the like) which are rational functions (over, say, the field of real numbers) of a symbol s standing for d/dt, the differentiation with respect to time, a fundamental system of equations for an electric network is nothing but an operator expression of the system of ordinary differential equations describing the dynamical performance of the network, and, when the determinant of the coefficient matrix is written as a fraction of two relatively prime polynomials in s, the zeros of its numerator polynomial give the complex frequencies of the eigen modes of the dynamics of the network. Therefore, the degree of the numerator polynomial of the determinant, which is equal to the number of zeros, gives the number of eigen modes, i.e., the number of dynamical degrees of freedom, of the network.

To be specific [Iri-Tomizawa 1974b], let us consider the coefficient matrix (4.7) of a fundamental system of equations of voltage-variable type. Ordinarily, the admittances $y^{\kappa\lambda}$ of elementary branches are of a form:

$$cs^0, \; cs \; \text{or} \; cs^{-1} \quad (c \in \mathbb{R}),$$

II. Applications of Matroid Theory

but we shall assume a more general case where

$$y^{\kappa\lambda} = c^{\kappa\lambda} s^{p^{\kappa\lambda}}, \tag{7.1}$$

the nonvanishing $c^{\kappa\lambda}$'s are algebraically independent transcendentals over $\mathbb{Q}$, and, for a (κ, λ) such that $c^{\kappa\lambda} \neq 0$, $p^{\kappa\lambda}$ is an integer (positive, zero or negative). ($p^{\kappa\lambda} = 0$ represents a resistive element, $p^{\kappa\lambda} = 1$ a capacitive element, and $p^{\kappa\lambda} = -1$ an inductive element.) Then, comparing the expansion (6.1) of the determinant and the bipartite graph of Fig. 6.1, we see that if $\det[DyD^T] \neq 0$, then, corresponding to each independent matching of cardinality m, there is a term

$$\pm c^{\kappa_1 \lambda_1} \ldots c^{\kappa_m \lambda_m} s^{p^{\kappa_1 \lambda_1} + \ldots + p^{\kappa_m \lambda_m}}, \tag{7.2}$$

and that the determinant is a polynomial in s and s^{-1}. Hence, it is natural to take $p^{\kappa\lambda}$ for the "weight" of the edge connecting the κ-th left vertex to the λ-th right vertex of the bipartite graph. Then, the exponent to s of the term (7.2) is equal to the weight of the corresponding independent matching, and the exponent of the highest or lowest power of s in the determinant polymoninal is equal, respectively, to the maximum or minimum weight of independent matchings of cardinality m. Let these values be p_{max} and p_{min}. Evidently, they can be obtained by solving the "weighted independent matching problem", or the "independent assignment problem", on the bipartite graph of Fig. 6.1, where edges are assigned weights $p^{\kappa\lambda}$ [Fujishige 1977a], [Iri-Tomizawa 1975b], [Lawler 1975], [Lawler 1976]. The number of dynamical degrees of freedom of the network is given by $p_{max} - \min(0, p_{min})$. Sometimes, we neglect the eigen modes of zero frequency. In such a case, we may take $p_{max} - p_{min}$ instead of $p_{max} - \min(0, p_{min})$. This number is called the "order of complexity" of the network.

It should be noted that the problem of dynamical degrees of freedom can in principle be treated by a weighted version of any technique for the solvability problem. See [Petersen 1979a], [Recski 1979a] for a method corresponding to the second approach to the solvability problem in section 6.

8. Signal-Flow Graph or a System of Equations

Since the signal-flow graph model of a system is not a main object of the present essay, we shall not concern ourselves with it but make a brief comment, referring to [Iri-Tsunekawa-Murota 1982], [Murota 1982], [Murota-Iri 1982] for further details.

When we describe the structure of a network or other kind of system in the form of a signal-flow graph, we should pick up as "elementary" variables and as "elementary" relations among them as possible, because it will make a generality assumption on the parameters contained in the elementary relations more likely to hold valid.

Suppose the system of equations corresponding to a signal-flow graph is written in a matrix form:

$$Ax = b. \tag{8.1}$$

(As has been remarked in section 2, the matrix A is the Jacobian matrix if some of the relations are nonlinear.) The unique solvability of the system may then be defined as the condition $\det A \neq 0$. If the nonvanishing entries of A are algebraically independent transcendentals, the rank of A is equal to its term-rank, so that the solvability problem is reduced to the maximum matching problem on a bipartite graph (with the left vertex set corresponding to the row set of A, the right vertex set corresponding to the column set of A, and the edge set corresponding to the nonvanishing entries of A) in an obvious way.

However, in most cases, many elementary relations among variables are simple additions, subtractions or the like, which will certainly make us hesitate to adopt a strong generality assumption such as the above. A more realistic assumption would be to consider that some part of the nonvanishing entries of A belong to $\mathbb{Z}$ (the ring of integers) or $\mathbb{Q}$ (the field of rational numbers) and the remaining part of them are algebraically independent transcendentals over $\mathbb{Q}$. Let the matrix consisting of the former entries be Q and that consisting of the latter be T:

$$A = Q + T. \tag{8.2}$$

Fortunately, under this assumption, it is possible to reduce the problem of determining the rank of A to the problem of finding the rank of a union matroid. In fact, we have

$$\operatorname{rank} A = \max\{r(X) + \tau(E-X) \mid X \subseteq E\} - |R|, \tag{8.3}$$

where $E = R \cup C$, R is the row set of A, C the column set of A, r is the rank function of the matroid defined on the column set E of the matrix $[I|Q]$ (the columns of the unit matrix I are identified with the rows of Q), and τ is the rank function of the matroid defined on the column set E of $[I|T]$. If we can perform numerical computations for the linear dependence-independence of columns of Q without serious error (as is the case when Q is totally unimodular), the minimum of (8.3) can be determined exactly by the help of an effective combinatorial algorithm since the matroid associated with τ is a transversal matroid and can be manipulated by means of matchings on the bipartite graph corresponding to the matrix T.

This approach is in close connection with, or generalizes, some previous results. In fact, as was noted in [Iri 1968b], [Iri 1969b], the problem of the minimum fundamental equations (section 4) is equivalent to the problem of determining the "2-block rank" of a matrix (a concept proposed in [Iri 1969b]; see also [Bruno-Weinberg 1971], [Maurer 1975]), and the 2-block rank of a matrix A is "formally" the rank which we should have if we could put $T = Q$ in (8.2) and (8.3). Furthermore, the matrix (6.2) appearing in the solvability argument in section 6 admits a very special additive decomposition (8.2), where Q has the upper half of the matrix of (6.2) in the upper half and zeros in the lower half, and T has the F-part of (6.2) in the lower half and zeros in the upper half. The solvability-testing algorithm based upon this additive decomposition proves to be essentially the same as that based upon the second formulation of the solvability problem mentioned in section 6.

9. Other Applications

Possible applications of mathematical techniques related to matroids, polymatroids, submodular functions are by no means limited to electric networks and elastic structures. The following are examples of other possible fields to which ideas similar to those explained so far as well as the decomposition technique to be explained in section 10 have been applied to get significant results.

The controllability/observability of a linear dynamical system with combinatorial constraints imposed on the manner of controlling/observing the system [Iri-Tomizawa-Fujishige 1977].

The consistency analysis of line drawings of polyhedra [Sugihara 1979].

The structural analysis of a set of correlated information sources based on the polymatroidal properties of entropy functions [Fujishige 1978b].

The analysis of communication nets based on the structural properties of a directed graph with respect to arborescences [Nakamura-Iri 1980a].

The analysis of balance among supplies, demands and transportation capacities of a transportation network system [Nakamura-Iri 1980] (by means of the decomposition of polylinking systems [Schrijver 1978], or bipolymatroids [Kung 1978]).

The problem of evacuation from a big city [Fujishige-Ohkubo-Iri 1978] as an application of the concept of optimal base of a polymatroid [Fujishige 1980c].

It will also be noted that the sum of squares appearing in statistical data analysis, such as in the anlysis of variance, shares common properties with the entropy function in information theory, i.e., the sum of squares, when suitably scaled with respect to the degrees of freedom, is a submodular (or supermodular) function defined on the set of relevant factors.

In the reference list at the end of this essay are listed some of the papers which discuss or survey applications of matroid theory from a little different standpoints, as well as those which treat other interesting but less systematic applications of matroid theory and related mathematical fields.

10. Decomposition Techniques

In applications, it is important to recognize a special hierarchical structure of a system under consideration by means of any decomposition technique. It is useful sometimes for getting a better understanding of the specific structure, sometimes for checking inconsistencies in the design (or modelling) of the system or errors in input data, and sometimes for reducing computational work in numerical analysis.

A decomposition technique which is the most basic and has the widest applicability would be that based upon the decomposition of a directed graph into strongly connected components with a partial order among the components, as is described in any textbook on graph theory.

The next oldest but less obvious technique would be the decomposition of a bipartite graph into partially ordered blocks, which was exploited in [Dulmage-

Mendelsohn 1958, 1959, 1962, 1963] and which has a close connection with the applications of matroidal methods such as have been discussed in this essay. (Although there is a claim [Tomizawa-Fujishige 1982] that the basic idea of the Dulmage-Mendelsohn (or, DM-) decomposition can be traced back to O. Ore, finding formulas in mathematics is one thing and recognizing their methodological significance and establishing a technological formalism is another.)

Without noting the intrinsic relation to the DM-decomposition at the early stage, a series of researches for system decomposition have been carried on, mostly in Japan, under the name of "principal partition" [Bruno-Weinberg 1971], [Iri 1971], [Kishi-Kajitani 1968], [Ozawa 1976b], [Tomizawa 1976]. (There are some more or less independent works outside of Japan [Kel'mans-al. 1976], [Narayanan 1974], [Narayanan-Vartak 1981], [Poleskii 1976]. See also [Iri 1979b], [Tomizawa-Fujishige 1982] for more details.)

The essence of the technique may be summarized as follows (in the style of [Iri 1982], [Nakamura-Iri 1981], making full use of lattice-theoretical concepts [Aigner 1979], [Birkhoff 1935], [Birkhoff 1967], [Crapo-Rota 1971]).

Consider a finite distributive lattice K and several submodular functions $\mu_\alpha: K \to \mathbb{R}$ ($\alpha = 1, \ldots, m$) on it. For any sublattice L of K, the family of all segments of L (where a segment means a pair of elements of L such that one is an immediate upper element of the other) is classified into the equivalence classes $F_1, \ldots, F_n$ with respect to the projectivity. These classes can be identified with classes of projective intervals of K. Adding to those classes of intervals the interval E^+ between the maximum of K and the maximum of L (if nonempty) and that E^- between the minimum of L and the minimum of K (if nonempty), we have a "decomposition" $\{E^+, F_1, \ldots, F_n, E^-\}$ of K with respect to L, where a partial order is naturally defined among E^+, F_i's, E^-. (If $K = 2^E$ with a finite set E, projective intervals of K correspond to the same subset of E, and under that correspondence, we have $E = E^+ \cup F_1 \cup \ldots \cup F_n \cup E^-$, which is why we use the term of "decomposition".) If (and only if) all the submodular functions μ_α are modular when restricted to L (in which case we call L a $(\mu_1, \ldots, \mu_m)$-skeleton of K), the submodular function $\mu_{\alpha[a,b]}(x) = \mu_\alpha(x) - \mu_\alpha(b)$ ($x \in [a, b]$) induced on an interval $[a, b]$ ($= E^+$, $\in F_i$, or $= E^-$) of K is isomorphic with that induced on any interval projective to it. Thus, if L is a skeleton of K, we have a decomposition of K into classes of projective intervals on which submodular functions are consistently induced and among which a partial order is defined. Furthermore, any skeleton L can be extended to the maximum superlattice $L_{\max}$ of L (still, a skeleton of K) in such a way that the classes of projective intervals as well as the submodular functions induced on them remain unchanged (but that the partial order may be different, i.e. weaker). The decomposition with respect to the maximum skeleton $L_{\max}$ is proved to be usually more useful in application than that with respect to L itself.

A skeleton appears when we deal with a problem of minimizing one or more submodular functions on a lattice. Furthermore, by considering the parametrized problem

$$\min \left\{ \sum_{\alpha=1}^{m} c_\alpha \mu_\alpha(x) \,\middle|\, x \in K \right\} \tag{10.1}$$

II. Applications of Matroid Theory

with positive parameters c_α, we can refine the decomposition. In fact, the x's giving the minimum (10.1) for fixed parameter values constitute obviously a skeleton $L(c)$. It is shown that, for a certain subregion S of the parameter space satisfying some monotonicity conditions, $L = \bigcup_{c \in S} L(c)$ also makes a skeleton.

Therefore, all the problems concerning (poly-)matroids, union matroids, (poly-)matroid intersections, independent matchings, etc. give rise to a skeleton. For the problems discussed in the preceding sections, this kind of decomposition will certainly afford much information about important structural properties hidden in the problem. Some problems have been analysed fairly well from this viewpoint, but many still await detailed analysis. For example, this kind of viewpoint enabled us to enumerate all the possible ways of choosing independent unknowns (currents and voltages) for a minimum fundamental system of equations (section 4); to find what kind of inconsistencies lie, and where they do, with respect to the solvability and to decompose the entire system of equations into subsystems to be solved one by one (section 6); and so forth.

Let us show only one example here. We consider the solvability problem expressed in the form of a bipartite graph (Fig. 6.1) with matroids defined on the two vertex sets, and define a Boolean lattice $2^{E^{(1)}} \times 2^{E^{(2)}}*$ which is the direct product of the Boolean lattice $2^{E^{(1)}}$ of the power set of the left vertex set $E^{(1)}$ and the dual (i.e., the order relation being reversed) of the Boolean lattice $2^{E^{(2)}}$ of the power set of the right vertex set. Then, the collection of covers $(X, Y)(X \subseteq E^{(1)}, Y \subseteq E^{(2)})$ of the bipartite graph forms a distributive lattice K which is a sublattice of $2^{E^{(1)}} \times 2^{E^{(2)}}*$. The problem dual to the problem of finding a maximum-cardinality independent matching is, as is now well known, to solve the following minimization problem:

$$\min\{\rho(X) + \rho(Y) | (X, Y) \in K\}, \tag{10.2}$$

where $\rho: E \to \mathbb{Z}_+$ is the rank function of $M(G)$. Here it should be noted that $\rho(X)$ is a nondecreasing submodular function on K whereas $\rho(Y)$ is a nonincreasing submodular function on K since we defined the dual order on $E^{(2)}$. The parametrized form of (10.2):

$$\min\left\{\frac{1+\lambda}{2}\rho(X) + \frac{1-\lambda}{2}\rho(Y) \Big| (X, Y) \in K\right\} \quad (-1 < \lambda < 1) \tag{10.3}$$

will give more useful information. In fact, we can prove that

$$L_{\text{all}} = \bigcup_\lambda L(\lambda) \tag{10.4}$$

is a skeleton of K, where $L(\lambda)$ is a collection of (X, Y)'s giving the minimum of (10.3), so that we can consider $(L_{\text{all}})_{\max}$.

For a concrete example, let us take up an electric network of Fig. 10.1(a) (voltage sources and current sources having been suppressed for simplicity's sake), where the branch admittance matrix is as shown in Fig. 10.1(b). The network graph G is shown in Fig. 10.2(a), and two trees (i.e. bases of $M(G)$) B_1 and B_2 in Fig. 10.2(b). B_1 has been chosen in accordance with the above-men-

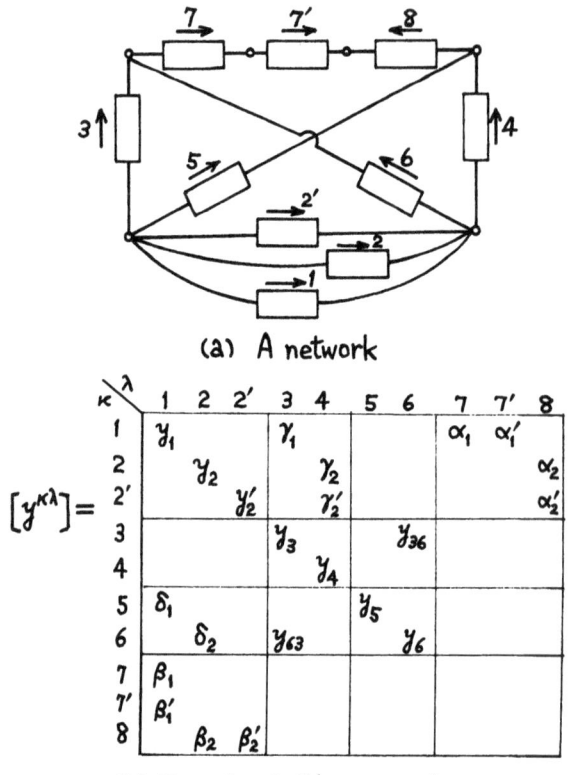

(a) A network

(b) Branch admittance matrix

Fig. 10.1. An electric network

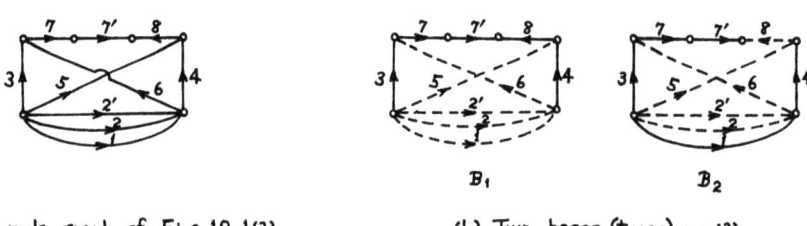

(a) Network graph of Fig. 10.1(a)   (b) Two bases (trees) on (a)

(c) Fundamental cocircuit matrices (cutset matrices) associated with the bases of (b)

Fig. 10.2. Different choices of bases for the network of Fig. 10.1 and the corresponding cocircuit matrices

II. Applications of Matroid Theory

$D_2 y D_2^T =$

a \ b	3	4	7	7'	8
3	$y_1+y_2+y_2'$ $+y_3+y_5$ $+\gamma_1+\delta_1$	$-y_1-y_2-y_2'$ $+y_{36}$ $+\gamma_2+\gamma_2'-\delta_1$	$y_1+y_2+y_2'$ $+y_5-y_{36}$ $+\alpha_1+\delta_1$	$y_1+y_2+y_2'$ $+y_5-y_{36}$ $+\alpha_1'+\delta_1$	$-y_1-y_2-y_2'$ $-y_5+y_{36}$ $+\alpha_2+\alpha_2'-\delta_1$
4	$-y_1-y_2-y_2'$ $+y_{63}$ $-\gamma_1+\delta_2$	$y_1+y_2+y_2'$ $+y_4+y_6$ $-\gamma_2-\gamma_2'-\delta_2$	$-y_1-y_2-y_2'$ $-y_6$ $-\alpha_2+\delta_1$	$-y_1-y_2-y_2'$ $-y_6$ $-\alpha_1'+\delta_2$	$y_1+y_2+y_2'$ $+y_6$ $-\alpha_2-\alpha_2'-\delta_2$
7	$y_1+y_2+y_2'$ $+y_5-y_{63}$ $+\beta_1+\gamma_1+\delta_1-\delta_2$	$-y_1-y_2-y_2'$ $-y_6-\beta_1+\gamma_2$ $+\gamma_2'-\delta_1+\delta_2$	$y_1+y_2+y_2'$ $+y_5+y_6$ $+\alpha_1+\beta_1+\delta_1-\delta_2$	$y_1+y_2+y_2'$ $+y_5+y_6$ $+\alpha_1'+\beta_1+\delta_1-\delta_2$	$-y_1-y_2-y_2'$ $-y_5-y_6+\alpha_2$ $+\alpha_2'-\beta_1-\delta_1+\delta_2$
7'	$y_1+y_2+y_2'$ $+y_5-y_{63}$ $+\beta_1'+\gamma_1+\delta_1-\delta_2$	$-y_1-y_2-y_2'$ $-y_6-\beta_1'+\gamma_2$ $+\gamma_2'-\delta_1+\delta_2$	$y_1+y_2+y_2'$ $+y_5+y_6$ $+\alpha_1+\beta_1'+\delta_1-\delta_2$	$y_1+y_2+y_2'$ $+y_5+y_6$ $+\alpha_1'+\beta_1'+\delta_1-\delta_2$	$-y_1-y_2-y_2'$ $-y_5-y_6+\alpha_2$ $+\alpha_2'-\beta_1'-\delta_1+\delta_2$
8	$-y_1-y_2-y_2'$ $-y_5+y_{63}+\beta_2$ $+\beta_2'-\gamma_1-\delta_1+\delta_2$	$+y_1+y_2+y_2'$ $+y_6-\beta_2-\beta_2'$ $-\gamma_2-\gamma_2'+\delta_1-\delta_2$	$-y_1-y_2-y_2'$ $-y_5-y_6-\alpha_1$ $+\beta_2+\beta_2'-\delta_1+\delta_2$	$-y_1-y_2-y_2'$ $-y_5-y_6-\alpha_1'$ $+\beta_2+\beta_2'-\delta_1+\delta_2$	$y_1+y_2+y_2'$ $+y_5+y_6-\delta_2-\delta_2'$ $-\beta_2-\beta_2'+\delta_1-\delta_2$

Fig. 10.3. Cocircuit admittance matrix with a bad base

$D_1 y D_1^T =$

a \ b	1	3	4	7	7'
1	$y_1+y_2+y_2'+y_5$ $+y_6-\alpha_2-\alpha_2'$ $-\beta_2-\beta_2'+\delta_1-\delta_2$	$-y_6-y_{63}$ $+\alpha_2+\alpha_2'$ $+\gamma_1$	y_5 $-\alpha_2-\alpha_2'$ $+\gamma_2+\gamma_2'$	$\alpha_1+\alpha_2+\alpha_2'$	$\alpha_1'+\alpha_2+\alpha_2'$
3	$-y_6-y_{36}$ $+\beta_2+\beta_2'$ $+\delta_2$	y_3+y_6 $+y_{36}+y_{63}$	0	0	0
4	y_5 $-\beta_2-\beta_2'$ $+\delta_1$	0	y_4+y_5	0	0
7	$\beta_1+\beta_2+\beta_2'$	0	0	0	0
7'	$\beta_1'+\beta_2+\beta_2'$	0	0	0	0

Fig. 10.4. Cocircuit admittance matrix with a good base

tioned decomposition of the bipartite graph (Fig. 6.1) constructed from the data of Fig. 10.1, whereas B_2 is a base chosen arbitrarily. The fundamental cocircuit matrices associated with these bases are shown in Fig. 10.2(c). The cocircuit (cutset) admittance matrix $Y = DyD^T$ appears as shown in Fig. 10.3, if we choose the base B_2. We can see nothing useful in this expression, although the fact is that this matrix is singular, having rank 4 (even under the generality assumption on branch admittances).

In contrast, if we choose the base B_1, we get the cocircuit admittance matrix of Fig. 10.4, where it is quite evident not only that the matrix is singular but also that the singularity comes from the cocircuits associated with branches 7 and 7' (and possibly branch 1). Once one see wherefrom the inconsistencies come, one can modify the design (or modelling) of the network. Fig. 10.5 shows such an improved network as well as the branch admittances.

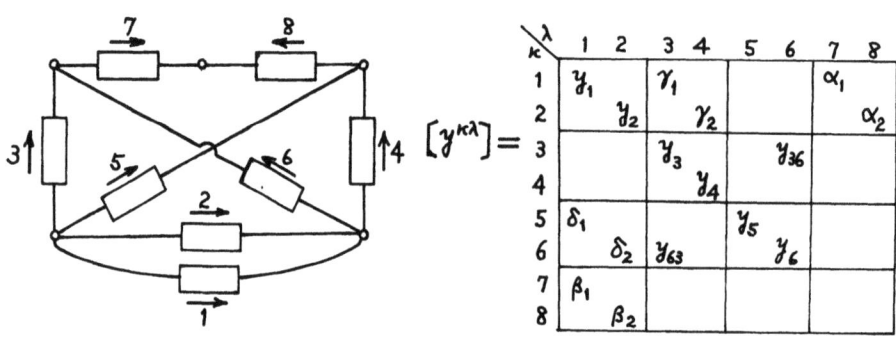

Fig. 10.5. An improved model

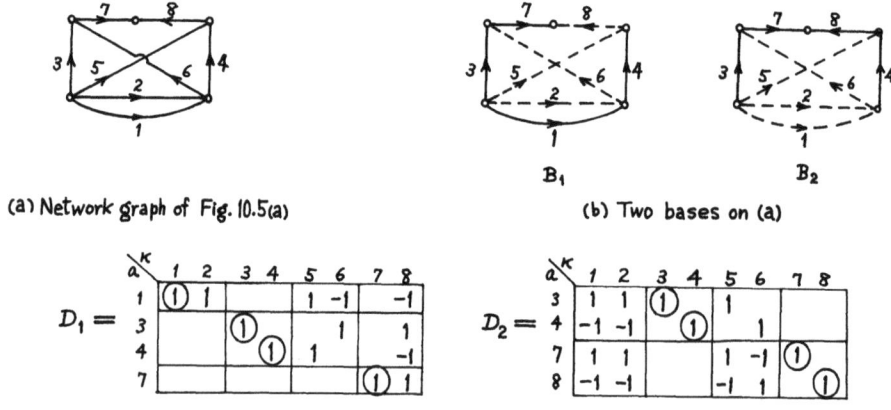

Fig. 10.6. Bases and cocircuit matrices for the improved model

II. Applications of Matroid Theory

Fig. 10.6. shows (a) the network graph, (b) two bases of $M(G)$ and (c) the corresponding cocircuit matrices for the improved network. The new bases B_1 and B_2 correspond to the old B_2 and B_1, respectively, in Fig. 10.2. The cocircuit admittance matrices associated with these two bases are, respectively, as shown in Fig. 10.7(a) and (b).

$D_1 y D_1^T =$

a\b	3	4	7	8
3	$y_1+y_2+y_3$ $+y_5$ $+\gamma_1+\delta_1$	$-y_1-y_2$ $+y_{36}$ $+\gamma_2-\delta_1$	$y_1+y_2+y_5$ $-y_{36}$ $+\alpha_1+\delta_1$	$-y_1-y_2-y_5$ $+y_{36}$ $+\alpha_2-\delta_1$
4	$-y_1-y_2$ $+y_{63}$ $-\gamma_1+\delta_2$	$y_1+y_2+y_4$ $+y_6$ $-\gamma_2-\delta_2$	$-y_1-y_2-y_6$ $-\alpha_1+\delta_2$	$y_1+y_2+y_6$ $-\alpha_2-\delta_2$
7	$y_1+y_2+y_5$ $-y_{63}+\beta_2$ $+\gamma_1+\delta_1-\delta_2$	$-y_1-y_2-y_6$ $-\beta_2+\gamma_2$ $-\delta_1+\delta_2$	$y_1+y_2+y_5$ $+y_6+\alpha_1$ $+\beta_1+\delta_1-\delta_2$	$-y_1-y_2-y_5$ $-y_6+\alpha_2$ $-\beta_1-\delta_1+\delta_2$
8	$-y_1-y_2-y_5$ $+y_{63}+\beta_2$ $-\gamma_1-\delta_1+\delta_2$	$y_1+y_2+y_6$ $-\beta_2-\gamma_2$ $+\delta_1-\delta_2$	$-y_1-y_2-y_5$ $-y_6-\alpha_1$ $+\beta_2-\delta_1+\delta_2$	$y_1+y_2+y_5$ $+y_6-\alpha_2$ $-\beta_2+\delta_1-\delta_2$

(a) A bad base

$D_2 y D_2^T =$

a\b	1	3	4	7
1	$y_1+y_2+y_5$ $+y_6-\alpha_2$ $-\beta_2+\delta_1-\delta_2$	$-y_6$ $-y_{63}$ $+\alpha_2+\gamma_2$	y_5 $-\alpha_2$ $+\gamma_2$	$\alpha_1+\alpha_2$
3	$-y_6$ $-y_{36}$ $+\beta_2+\delta_2$	y_3+y_6 $+y_{36}+y_{63}$	0	0
4	y_5 $-\beta_2$ $+\delta_1$	0	y_4+y_5	0
7	$\beta_1+\beta_2$	0	0	0

(b) A good base

Fig. 10.7. Cocircuit admittance matrices for the improved network of Fig. 10.5.

The matrix with the good base clearly indicates that it is regular so long as the generality assumption holds valid and that we may get the solution of the entire system of equations by solving the four 1×1 problems in an appropriate order. In contrast with this, we can hardly get any useful information from the matrix with the bad base.

11. Comments on "Algorithms"

The matroidal problems to which various problems in practical fields were shown to be reducible in the preceding sections are known to be solvable by efficient, i.e. polynomial-order, algorithms [Aigner 1979], [Edmonds 1965], [Edmonds 1971], [Iri-Tomizawa 1975b], [Knuth 1973], [Lawler 1976], [Tomizawa 1976], [Tomizawa-Iri 1974], [Welsh 1976], etc. (Indeed, only those applications which are backed up by efficient algorithms have been taken up here.) They are (i) matroid intersection problem (to find a maximum-cardinality common independent set of two matroids on the same set), (ii) matroid union problem (to find a base of the union of two matroids on the same set), (iii) independent matching problem, and their weighted versions. (The extensions where polymatroids appear in place of matroids are to be treated when we concern the parametrized skeletal structure in the decomposition (section 10).) They are said to be equivalent to one another. However, if a (poly-)matroidal technique is to be applied to the automatic analysis/design of a class of very large systems, the algorithm for it should be polished to adapt to characteristic features of that class of systems. In other words, a "theoretical" algorithm has usually been designed for a larger class of problems – the larger the class is, the more abstract is the theory –, and an algorithm for a larger class of problems is likely to be less efficient for a narrower class.

Some ten years ago, the polynomiality of an algorithm was a new concept both theoretically nice and practically useful. It stimulated the development of many good combinatorial and graphical algorithms. However, the more experience on applications of "good" algorithms to large practical problems we have gained, the more concerned have we become with finer differences in performance among different algorithms. We have learned that the mere polynomiality does not always imply practical efficiency at all (the Khachian ado is a typical example), and that even a small exponent p in complexity function $O(n^p)$ can occur with practically nonsense algorithms (a series of works, initiated by V. Strassen, which have reduced the exponent p to about 2.5 for matrix multiplication). In the field of computational geometry where problems of size greater than hundreds or even thousands are of practical concern, it seems that discrepancy between theoretical complexity (especially, the worst-case complexity) and practical efficiency is becoming more and more apparent.

Even if two classes of problems are polynomially equivalent to each other, i.e., if one is reducible to the other in polynomial time, the size of a problem substantially increases by such a reduction, in general. So, we should prepare a special algorithm for a special class of problems [Iri 1978]. Furthermore, in comparison with combinatorial algorithms for graphical problems, matroidal algorithms are higher in abstractness so that general theoretical algorithms cannot but be constructed on the basis of some kind of oracle, or oracles. (From the theoretical standpoint, oracles themselves, as well as their interrelations, are of proper significance and interest [Hausmann-Korte 1978], [Jensen-Korte 1982], [Korte-Hausmann 1978].) So, we must be more careful in applying those theoretical algorithms to practical problems. One might be tempted to write only a subroutine for an oracle for a specific problem and to apply a general algorithm to it, but, in most cases, one would not get a practically good algorithm, because it is most crucial for the practical efficiency what kind of oracle to adopt, how to represent a matroid, and how to organize, or correlate with one another, the jobs done by the oracle at different stages of the solution process. (See how the greedy algorithm for a minimum-weight base of a matroid can be accelerated to a minimum spanning tree algorithm for a graph [Edmonds 1970], [Edmonds 1971], [Welsh 1968], [Yao 1975].)

It is dubious that many of the algorithms which have been "proposed" in theoretical papers might have ever been coded into a practical programme, say, for commercial use. Evaluating different theoretical algorithms for matroids, polymatroids, submodular functions, etc. from the practical standpoint, choosing a most appropriate one and polishing it up for a specific practical problem is by no means a trivial work, if we want "matroids" to be recognized as a powerful tool by a large circle of practitioners. Even in Japan, the situation is far from being ideal. There is a big commercial program for the simulation of static and/or dynamic behaviour of a chemical process which makes partial use of the techniques mentioned in section 8 [Iri-Tsunekawa-Murota 1982], [Iri-Tsunekawa-Yajima 1971], [Thambynayagam-al. 1981], [Yajima-al. 1981], but the techniques incorporated in the program is not so much matroidal as graphical. The generalization (i.e., refinement) of the DM-decomposition along the lines of thought summarized in section 10 has been programmed [Sugihara 1981], and has been used for a number of system-analytic problems, but it has still a lot of points to be improved from the point of view of computational efficiency. The practical problem-oriented study of more general decomposition algorithms (see section 10) has just started [Imai 1983].

12. On Dualities

This is a very short remark on the concepts usually called "duality", of which the abuse is confusing people in practical engineering fields. (See [Iri 1968d], [Iri 1981a], [Iri-Recski 1979, 1980a, 1980b] for more details and for different views on it.)

We assume the concept of "duality" and that of "dual matroid" to be known. The former concept concerns the "structure of the theory" of matroids that every concept or theorem in matroid theory has its dual counterpart in the theory, and, in that sense, the concept of duality belongs to the level of "metatheory", i.e., a "theory for the structure of a theory or theories". In contrast with this, the concept of "dual matroid", or the dual counterpart of any other concept defined in matroid theory, belongs to the level of the theory of matroids itself. These two concepts of different levels are in so good an accord with each other that we need never care about the difference of their levels, which is why the difference is rarely emphasized in the literature.

However, this neglect of the difference of levels of the two kinds of dualities makes some confusion in electric network theory. In electric network theory, one often speaks of "duality", which is the "current-voltage duality", in the metatheory level. Every concept or theorem in the theory has its dual counterpart which is obtained by exchanging "voltage" and "current" with each other and other relevant concepts accordingly. The fact that this "current-voltage" duality holds in electric network theory is easy to see from the expression (6.2). The structural equations as well as the constitutive equations are "form-invariant" through the exchange of "current" and "voltage", i.e., of $E^{(1)}$ and $E^{(2)}$. In the structural equations, D and R are exchanged accordingly, i.e., the circuit matroid $M(G)$ of the network graph G and its cocircuit matroid $M^*(G)$ are exchanged, and, in the constitutive equations, the $F^{(1)}$-part and the $F^{(2)}$-part of the F-matrix are exchanged. Since the current-voltage duality holds valid in the very basic equations, it prevails throughout the whole theory. It is worth noting, in particular, that this current-voltage duality gives rise to the correspondence between $M(G)$ and $M^*(G)$, the dual matroids. (If one confines oneself to graphs, here arises a question of whether there is a dual of a graph or not, but, if one considers things on matroids, such a question never occurs.) Here is the origin of a traditional confusion. One is apt to consider that the current-voltage duality exchanges a matroid with its dual, or, in the traditional language of vector spaces, exchanges a subspace of the underlying vector space with the subspace of the dual vector space which is orthogonal to (or, more precisely, which annihilates) the former subspace. As has been mentioned, this is true for the structural equations since they are the direct sum of dual (or, "contragredient" in matrix language) systems of equations, one for currents and the other for voltages. However, this is not necessarily true for the constitutive equations. There is no reason why the exchange of the $F^{(1)}$-part and the $F^{(2)}$-part makes the dual matroid. Fortunately or unfortunately, the current-voltage duality transforms a matroid connected with the constitutive equations into one which is "very close" to the dual. For the simplicity of exposition, let us consider the case where the constitutive equations admit an impedance-matrix expression, i.e., where (2.7) can be expressed as

$$\left[\begin{array}{c|c} Z & -I \end{array}\right] \cdot \begin{bmatrix} \xi \\ \eta \end{bmatrix} = \begin{bmatrix} c \end{bmatrix}. \qquad (12.1)$$

The current-voltage duality transforms this expression into

$$\left[\begin{array}{c|c} -I & Y \end{array}\right] \cdot \begin{Bmatrix} \xi \\ \eta \end{Bmatrix} = \{c\}. \tag{12.2}$$

(12.2) is an admittance-matrix expression of the constitutive equations, so that the admittance matrix and the impedance matrix correspond to each other under the current-voltage duality. On the other hand, the dual of the matroid represented by $[Z|-I]$ is represented by $[I|Z^T]$ or $[-I|-Z^T]$ (or $[-I|Z^T]$), as is well known. This shows that the dual of the constitutive equations having the impedance matrix Z has the admittance matrix $Y = -Z^T$, or, if one deals with reciprocal elements alone, i.e., if $Z = Z^T$, the admittance matrix of the dual of the branch characteristics represented by the impedance matrix Z coincides with the Z itself up to a sign. Thus, the difference is very subtle, and could be found only in the case where the reciprocity fails. Once one recognizes this difference, one can clarify the interrelations among various network concepts from the matroidal point of view [Iri-Recski 1979, 1980a, 1980b]. For example, the matroidal dual of an electric network is not the electrical dual (i.e., one under the current-voltage duality), but is the "adjoint" of the electrical dual. (The concept of an adjoint network had already been introduced in network theory with respect to the sensitivity analysis, independently of matroids.)

The above arguments will apply conceptually to the theory of elastic structures described in section 2. However, since nonreciprocal elements are hardly realistic in elastic structure, the difference of the two dualities might be theoretically less manifest and physically less significant there.

13. Concluding Remarks and Acknowledgements

In this essay, the author laid emphasis upon the relation between the physical or engineering viewpoints and those mathematical in recognizing a problem, formulating it and solving it, rather than upon details in mathematical techniques. He intended to show by examples that the matroidal approach to engineering system-theoretic problems is not a mere sophistication but is inevitable to understand well what structure a system has and what happens there. However, he does not consider the present mathematical theory on matroids (or polymatroids or submodular functions) to be satisfactory for applications. It deals mainly with single-dimensional problems, although most system problems are intrinsically related to phenomena in various dimensions [Iri 1968c]. For example, an electric network is now formulated as a closed system where one-dimensional elements, branches, play the principal rôle, but, if we want to take a little account of its ambience, we are necessarily led to face zero-dimensional phenomena taking place at "nodes" (or vertices) (in fact, nodes are important not only physically and technologically but also algorithmically), and two- or three-dimensional phenomena such as linking magnetic fluxes [Mizoo-al. 1958].

As for the duality in electric network theory, there seems to be a lot awaiting further investigation. What is the physically most proper definition of a "dual network" still arouses discussions (see, e.g., [Meetz-Engl 1980] for a field-theoretic approach which seemingly indicates that the conventional definition of a dual network requires some modification.)

Finally, the author would like to thank Dr. Kazuo Murota for his help in preparing the manuscript, not to mention individually those names (to be found in the reference list) with whom the author has had the nice opportunity of jointly working on applications of matroids.

References

Relevant books and papers not explicitly cited in the text are also included.

Aigner, M. (1979): *Combinatorial Theory*. Springer-Verlag.
Amari, S. (1962): Topological foundations of Kron's tearing of electric networks; Information-theoretical foundations of diakoptics and codiakoptics. *RAAG Memoirs*, Vol. 3, F-VI; F-VII, pp. 322-350; 351-371.
Asimov, L., and Roth, B. (1978): Rigidity of graphs. *Transactions of the American Mathematical Society*, Vol. 245, pp. 279-289.
Asimov, L., and Roth, B. (1979): Rigidity of graphs, II. *Journal of Mathematical Analysis and Applications*, Vol. 68, pp. 171-190.
Birkhoff, G. (1935): Sur les espaces discrets. *Computes Rendus de l'Academie des Sciences*, Paris, Tome 201, pp. 19-20.
Birkhoff, G. (1967): *Lattice Theory* (third edition). American Mathematical Society Colloquium Publications, Providence.
Bixby, R. E. (1981): Hidden structure in linear programs. In *Computer-Assisted Analysis and Model Simplification* (Greenburg, H. J., and Maybee, J. S., eds.), Academic Press, pp. 327-360.
Bolker, E. D., and Crapo, H. (1977): How to brace a one-story-building. *Environment and Planing B*, Vol. 4, pp. 125-152.
Bruno, J., and Weinberg, L. (1970): A constructive graph-theoretic solution of the Shannon switching game. *IEEE Transactions on Circuit Theory*, Vol. CT-17, pp. 74-81.
Bruno, J., and Weinberg, L. (1971): The principal minors of a matroid. *Linear Algebra and Its Applications*, Vol. 4, pp. 17-54.
Bruno, J., and Weinberg, L. (1976): Generalized networks: networks embedded on a matroid, I and II. *Networks*, Vol. 6, pp. 53-94, pp. 231-272.
Bruter, C. P. (1974): *Eléments de la Théorie des Matroïdes*. Lecture Notes in Mathematics 387, Springer-Verlag, Berlin.
Bryant, P. R. (1962): The explicit form of Bashkow's A-matrix. *IRE Transactions on Circuit Theory*, Vol. CT-9, pp. 303-306.
Camerini, P. M., Fratta, L., and Maffioli, F. (1979): A note on finding optimum branchings. *Networks*, Vol. 9, pp. 309-312.

II. Applications of Matroid Theory

Chu, Y.-J., and Liu, T.-H. (1965): On the shortest arborescence of a directed graph. *Scientia Sinica*, Vol. 14, pp. 1396–1400.

Crapo, H. (1979): Structural rigidity. *Structural Topology*, Vol. 1, pp. 26–45.

Crapo, H., and Rota, G.-C. (1971): *Combinatorial Geometries*. M.I.T. Press.

Csurgay, A., Kovács, Z., and Recski, A. (1974): Transient analysis of lumped-distributed nonlinear networks. *Proceedings of the 5th International Colloquium on Microwave Communication*, Budapest.

Dulmage, A. L., and Mendelsohn, N. S. (1958): Coverings of bipartite graphs. *Canadian Journal of Mathematics*, Vol. 10, pp. 517–534.

Dulmage, A. L., and Mendelsohn, N. S. (1959): A structure theory of bipartite graphs of finite exterior dimension. *Transactions of the Royal Society of Canada*, Third Series, Section III, Vol. 53, pp. 1–13.

Dulmage, A. L., and Mendelsohn, N. S. (1962): On the inversion of sparse matrices. *Mathematics of Computation*, Vol. 16, pp. 494–496.

Dulmage, A. L., and Mendelsohn, N. S. (1963): Two algorithms for bipartite graphs. *Journal of the Society of the Industrial and Applied Mathematics*, Vol. 11, pp. 183–194.

Edmonds, J. (1965): Minimum partition of a matroid into independent subsets. *Journal of Research of the National Bureau of Standards*, Vol. 69 B, pp. 73–77.

Edmonds, J. (1967): Optimum branchings. *Journal of Research of the National Bureau of Standards*, Vol 71 B, pp. 233–240.

Edmonds, J. (1970): Submodular functions, matroids and certain polyhedra. *Proceedings of the International Conference on Combinatorial Structures and Their Applications* (Guy, R., et al. eds.), Gordon and Breach, New York, pp. 69–87.

Edmonds, J. (1971): Matroids and the greedy algorithm. *Mathematical Programming*. Vol. 1, pp. 127–136.

Ford, L. R., Jr., and Fulkerson, D. R. (1962): *Flows in Networks*. Princeton University Press, Princeton, New Jersey.

Fujishige, S. (1977 a): A primal approach to the independent assignment problem. *Journal of the Operations Research Society of Japan*, Vol. 20, pp. 1–15.

Fujishige, S. (1977 b): An algorithm for finding an optimal independent linkage. *Journal of the Operations Research Society of Japan*, Vol. 20, pp. 59–75.

Fujishige, S. (1978 a): Algorithms for solving the independent-flow problems. *Journal of the Operations Research Society of Japan*, Vol. 21, pp. 189–203.

Fujishige, S. (1978 b): Polymatroidal dependence structure of a set of random variables. *Information and Control*, Vol. 39, pp. 55–72.

Fujishige, S. (1980 a): Principal structures of submodular systems. *Discrete Applied Mathematics*, Vol. 2, pp. 77–79.

Fujishige, S. (1980 b): An efficient PQ-graph algorithm for solving the graph-realization problem. *Journal of Computer and System Sciences*, Vol. 21, pp. 63–86.

Fujishige, S. (1980 c): Lexicographically optimal base of a polymatroid with respect to a weight vector. *Mathematics of Operations Research*, Vol. 5, pp. 186–196.

Fujushige, S., Ohkubo, K., and Iri, M. (1978): Optimal flows in a network with several sources and sinks (in Japanese). *Proceedings of the 1978 Fall Conference of the Operations Research Society of Japan*, A-2, pp. 22–23.

Fulkerson, D. R. (1971): Blocking and anti-blocking pairs of polyhedra. *Mathematical Programming*, Vol. 1, pp. 168–194.

Han, T.-S. (1979 a): The capacity region of general multi-access channel with certain correlated sources. *Information and Control*, Vol. 40, pp. 37–60.

Han, T.-S. (1979 b): Source coding with cross observation at the encoders. *IEEE Transactions on Information Theory*, Vol. IT-25, pp. 360–361.

Han, T.-S. (1980): Slepian-Wolf-Cover theorem for networks of channels. *Information and Control,* Vol. 47, pp. 67–83.

Harary, F., and Welsh, D. (1969): Matroids versus graphs. *The Many Facets of Graph Theory,* Springer Lecture Notes 110, p. 115.

Hausmann, D., and Korte, B. (1978): Lower bounds on the worst-case complexity of some oracle algorithms. *Discrete Mathematics,* Vol. 24, pp. 261–276.

Holzmann, C. A. (1977): Realization of netoids. *Proceedings of the 20th Midwest Symposium on Circuits and Systems,* Lubbock, Texas, pp. 394–398.

Holzmann, C. A. (1979): Binary netoids. *Proceedings of the International Symposium on Circuits and Systems,* Tokyo, pp. 1000–1003.

Imai, H. (1983): *Efficient Solutions for Combinatorial Optimization Problems by Means of Network-Flow Algorithms* (in Japanese). Master's Thesis, Department of Mathematical Engineering and Instrumentation Physics, University of Tokyo.

Iri, M. (1962): A necessary and sufficient condition for a matrix to be the loop or cut-set matrix of a graph and a practical method for the topological synthesis of networks. *RAAG Research Notes,* Third Series, No. 50.

Iri, M. (1966): A criterion for the reducibility of a linear programming problem to a linear network-flow problem. *RAAG Research Notes,* Third Series, No. 98.

Iri, M. (1968 a): On the synthesis of loop and cutset matrices and the related problems. *RAAG Memoirs,* Vol. 4, A-XIII, pp. 4–38.

Iri, M. (1968 b): A min-max theorem for the ranks and term-ranks of a class of matrices: An algebraic approach to the problem of the topological degrees of freedom of a network (in Japanese). *Transactions of the Institute of Electronics and Communication Engineers of Japan,* Vol. 51 A, pp. 180–187.

Iri, M. (1968 c): A critical review of the matroid-theoretical and the algebraic-topological theory of networks. *RAAG Memoirs,* Vol. IV, Division A, pp. 39–46.

Iri, M. (1968 d): Metatheoretical considerations on duality. *RAAG Research Notes,* Third Series, No. 124.

Iri, M. (1969 a): *Network Flow, Transportation and Scheduling – Theory and Algorithms.* Academic Press, New York.

Iri, M. (1969 b): The maximum-rank minimum-term-rank theorem for the pivotal transforms of a matrix. *Linear Algebra and Its Applications,* Vol. 2, pp. 427–446.

Iri, M. (1971): Combinatorial canonical form of a matrix with applications to the principal partition of a graph (in Japanese). *Transactions of the Institute of Electronics and Communication Engineers of Japan,* Vol. 54A, pp. 30–37.

Iri, M. (1978): A practical algorithm for the Menger-type generalization of the independent assignment problem. *Mathematical Programming Study,* Vol. 8, pp. 88–105.

Iri, M. (1979 a): Survey of recent trends in applications of matroids. *Proceedings of the 1979 International Symposium on Circuits and Systems,* Tokyo, p. 987.

Iri, M. (1979 b): A review of recent work in Japan on principal partitions of matroids and their applications. *Annals of the New York Academy of Sciences,* Vol. 319 (Proceedings of the Second International Conference on Combinatorial Mathematics, 1978), pp. 306–319.

Iri, M. (1981 a): "Dualities" in graph theory and in the related fields viewed from the metatheoretical standpoint. In *Graph Theory and Algorithms* (Saito, N., and Nishizeki, T., eds.). Lecture Note in Computer Science 108, Springer-Verlag, pp. 124–136.

Iri, M. (1981 b): Application of matroid theory to engineering systems problems. *Proceedings of the Sixth Conference on Probability Theory* (September, 1979; Bereanu, B., et al., eds.), Editura Academiei Republicii Socialiste Romania, pp. 107–127.

Iri, M. (1982): Structural theory for the combinatorial systems characterized by submodular functions. Silver-Jubilee Conference on Combinatorics, The University of Waterloo, June.

Iri, M., and Fujishige, S. (1981): Use of matroid theory in operations research, circuits and systems theory. *International Journal of Systems Science*, Vol. 12, pp. 27-54.

Iri, M., and Han, T.-S. (1977): *Linear Algebra - Standard Forms of Matrices* (in Japanese). Kyoiku-Shuppan Co., Tokyo.

Iri, M., and Recski, A. (1979): Reflection on the concepts of dual, inverse and adjoint networks (in Japanese). *Papers of the Technical Group on Circuits and Systems*, Institute of Electronics and Communication Engineers of Japan, CAS 79-78 (English translation available).

Iri, M., and Recski, A. (1980a): Reflection on the concepts of dual, inverse and adjoint networks, II - Towards a qualitative theory (in Japanese). *Papers of the Technical Group on Circuits and Systems*, Institute of Electronics and Communication Engineers of Japan, CAS 79-133 (English translation available).

Iri, M., and Recski, A. (1980b): What does duality really mean? *Circuit Theory and Applications*, Vol. 8, pp. 317-324.

Iri, M., and Tomizawa, N. (1974a): A practical criterion for the existence of the unique solution in a linear electric network with mutual couplings (in Japanese). *Transactions of the Institute of Electronics and Communication Engineers of Japan*, Vol. 57A, pp. 599-605.

Iri, M., and Tomizawa, N. (1974b): An algorithms for solving the 'independent assignment problem' with application to the problem of determining the order of complexity of a network (in Japanese). *Transactions of the Institute of Electronics and Communication Engineers of Japan*, Vol. 57A, pp. 627-629.

Iri, M., and Tomizawa, N. (1975a): A unifying approach to fundamental problems in network theory by means of matroids (in Japanese). *Transactions of the Institute of Electronics and Communication Engineers of Japan*, Vol. 58A, pp. 33-40.

Iri, M., and Tomizawa, N. (1975b): An algorithm for finding an optimal "independent assignment". *Journal of the Operations Research Society of Japan*, Vol. 19, pp. 32-57.

Iri, M., Tomizawa, N., and Fujishige, S. (1977): On the controllability and observability of a linear system with combinatorial constraints (in Japanese). *Transactions of the Society of Instrument and Control Engineers, Japan*, Vol. 13, pp. 225-242.

Iri, M., Tsunekawa, J., and Murota, K. (1982): Graph-theoretic approach to large-scale systems - Structural solvability and block-triangularization (in Japanese). *Transactions of Information Processing Society of Japan*, Vol. 23, pp. 88-95.

Iri, M., Tsunekawa, J., and Yajima, K. (1971): The graphical techniques used for a chemical process simulator "JUSE GIFS". *Information Processing 71* (Proceedings of the 1971 IFIP Congress), Vol. 2 (Applications), pp. 1150-1155.

Jensen, P. M., and Korte, B. (1982): Complexity of matroid property algorithms. *SIAM Journal on Computing*, Vol. 11, pp. 184-190.

Kel'mans, A. K., Lomonosov, N. V., and Polesskii, V. P. (1976): Minimum matroid coverings (in Russian). *Problemy Peredachi Informatsii*, Vol. 12, pp. 94-107.

Kishi, G., and Kajitani, Y. (1968): Maximally distinct trees in a linear graph (in Japanese). *Transactions of the Institute of Electronics and Communication Engineers of Japan*, Vol. 51A, pp. 196-203.

Knuth, D. E. (1973): Matroid partitioning. Report Stan-CS-73-342, Stanford University.

Korte, B., and Hausmann, D. (1978): An analysis of the greedy heuristic for independence systems. *Annals of Discrete Mathematics*, Vol. 2, pp. 65-74.

Kron, G. (1939): *Tensor Analysis of Networks*. John Wiley and Sons, New York.

Kron, G. (1963): *Diakoptics – The Piecewise Solution of Large-Scale Systems*. Macdonald, London.

Kuh, E. S., Layton, D., and Tow, J. (1968): Network analysis via state variables. In *Network and Switching Theory* (Biorci, G., ed.), Academic Press, New York.

Kuh, E. S., and Rohrer, R. A. (1965): The state variable approach to network analysis. *Proceedings of the IEEE*, Vol. 53, pp. 672–686.

Kung, J. P. S. (1978): Bimatroids and invariants. *Advances in Mathematics*, Vol. 30, pp. 238–249.

Laman, G. (1970): On graphs and rigidity of plane skeletal structures. *Journal of Engineering Mathematics*, Vol. 4, pp. 331–340.

Lawler, E. L. (1975): Matroid intersection algorithms. *Mathematical Programming*, Vol. 9, pp. 31–56.

Lawler, E. L. (1976): *Combinatorial Optimization: Networks and Matroids*. Holt, Rinehart and Winston, New York.

Lehman, A. (1964): A solution to the Shannon switching game. *Journal of the Society of the Industrial and Applied Mathematics*, Vol. 12, pp. 687–725.

Lovász, L. (1977): Flats in matroids and geometric graphs. In *Combinatorial Surveys* (Cameron, P., ed.), (Proceedings of the 6th British Combinatorial Conference), Academic Press, pp. 45–86.

Lovász, L. (1980): Matroid matching and some applications. *Journal of Combinatorial Theory*, Series B, Vol. 28, pp. 208–236.

Lovász, L. (1981): The matroid matching problem. In *Algebraic Methods in Graph Theory* (Lovász, L., and Sós, V. T., eds.), Colloquia Mathematica Societatis János Bolyai 25, North-Holland, pp. 495–517.

Lovász, L., and Yemini, Y. (1982): On generic rigidity in the plane. *SIAM Journal on Algebraic and Discrete Methods*, Vol. 3, pp. 91–98.

Manabe, R., and Kotani, S. (1973): On the minimal spanning arborescence of a direct graph (in Japanese). *Keiei-Kagaku* (Official Journal of the Operations Research Society of Japan), Vol. 17, pp. 269–278.

Maurer, S. B. (1975): A maximum-rank minimum-term-rank theorem for matroids. *Linear Algebra and Its Applications*, Vol. 10, pp. 129–237.

Meetz, K., and Engl, W. L. (1980): *Elektromagnetische Felder*. Springer-Verlag, Berlin.

Megiddo, N. (1974): Optimal flows in networks with multiple sources and sinks. *Mathematical Programming*, Vol. 7, pp. 97–107.

McDiarmid, C. J. H. (1975): Rado's theorem for polymatroids. *Mathematical Proceedings of the Cambridge Philosophical Society*, Vol. 78, pp. 263–281.

Milic, M. (1974): General passive networks: solvability, degeneracies and order of complexity. *IEEE Transactions on Circuits and Systems*, Vol. CAS-21, pp. 173–183.

Minty, G. J. (1966): On the axiomatic foundations of the theories of directed linear graphs, electrical networks and network programming. *Journal of Mathematics and Mechanics*, Vol. 15, pp. 485–520.

Mizoo, Y., Iri, M., and Kondo, K. (1958): On the torsion and linkage characteristics and the duality of electric, magnetic and dielectric networks. *RAAG Memoirs*, Vol. 2, A-VIII, pp. 84–117.

Murota, K. (1982): Menger-decomposition of a graph and its application to the structural analysis of a large-scale system of equations. *Kokyuroku 453*, Research Institute of Mathematical Sciences, Kyoto University, pp. 127–173.

Murota, K., and Iri, M. (1982): Matroidal approach to the structural solvability of a system of equations. XIth International Symposium on Mathematical Programming, Universität Bonn.

II. Applications of Matroid Theory

Nakamura, M. (1982a): Boolean sublattices connected with minimization problems on matroids. *Mathematical Programming*, Vol. 22, pp. 117-120.

Nakamura, M. (1982b): *Mathematical Analysis of Discrete Systems and Its Applications* (in Japanese). Doctor's dissertation, Faculty of Engineering, University of Tokyo.

Nakamura, M., and Iri, M. (1979): Fine structures of matroid intersections and their applications. *Proceedings of the International Symposium on Circuits and Systems*, Tokyo, pp. 996-999.

Nakamura, M., and Iri, M. (1980a): On the decomposition of a directed graph with respect to arborescences and related problems. *Kokyuroku* 396 (Symposium on Graphs and Combinatorics III), Research Institute of Mathematical Sciences, Kyoto University, pp. 104-118.

Nakamura, M., and Iri, M. (1980b): Polylinking systems and their principal partition (in Japanese). *Proceedings of the 1980 Spring Conference of the Operations Research Society of Japan*, B-6, pp. 56-57.

Nakamura, M., and Iri, M. (1981): A structural theory for submodular functions, polymatroids and polymatroid intersections. *Research Memorandum RMI 81-06*, Department of Mathematical Engineering and Instrumentation Physics, University of Tokyo.

Narayanan, H. (1974): *Theory of Matroids and Network Analysis*. Ph. D. Thesis, Department of Electrical Engineering, Indian Institute of Technology, 117, Bombay.

Narayanan, H., and Vartak, M. N. (1981): An elementary approach to the principal partition of a matroid. *Transactions of the Institute of Electronics and Communication Engineers of Japan*, Vol. E 64, pp. 227-234.

Numata, J., and Iri, M. (1973): Mixed-type topological formulas for general linear networks. *IEEE Transactions on Circuit Theory*, Vol. CT-20, pp. 488-494.

Ohtsuki, T., Ishizaki, and Watanabe, H. (1968): Network analysis and topological degrees of freedom (in Japanese). *Transactions of the Institute of Electronics and Communication Engineers of Japan*, Vol. 51 A, pp. 238-245.

Oono, Y. (1960): Formal realizability of linear networks. *Proceedings of the Symposium on Active Networks and Feedback Systems*, Polytechnic Institute of Brooklyn, New York, pp. 475-586.

Ozawa, T. (1972): Order of complexity of linear active networks and common tree in the 2-graph method, *Electronics Letters*, Vol. 8, pp. 542-543.

Ozawa, T. (1974): Common trees and partition of two-graphs (in Japanese). *Transactions of the Institute of Electronics and Communication Engineers of Japan*, Vol. 57 A, pp. 383-390.

Ozawa, T. (1976a): Topological conditions for the solvability of linear active networks. *Circuit Theory and Applications*, Vol. 4, pp. 125-136.

Ozawa, T. (1976b): Structure of 2-graphs (in Japanese). *Transactions of the Institute of Electronics and Communication Engineers of Japan*, Vol. J 59 A, pp. 262-263.

Ozawa, T., and Kajitani, Y. (1979): Diagnosability of linear active networks. *Proceedings of the International Symposium on Circuits and Systems*, Tokyo, pp. 866-869.

Petersen, B. (1978): Investigating solvability and complexity of linear active networks. *Proceedings of the 1978 European Conference on Circuit Theory and Design*, Lausanne.

Petersen, B. (1979a): Investigating solvability and complexity of linear active networks by means of matroids. *IEEE Transactions on Circuits and Systems*. Vol. CAS-26, No. 5, pp. 330-342.

Petersen, B. (1979b): The qualitative appearance of linear active network transfer functions by means of matroids. *Proceedings of the International Symposium on Circuits and Systems*, Tokyo, pp. 992-995.

Picard, J.-C., and Queyranne, M. (1980): On the structure of all minimum cuts in a network and applications. *Mathematical Programming Study,* Vol. 13, pp. 8–16.

Polesskii, V. P. (1976): Isthmus structure in a summary matroid (in Russian). *Problemy Peredachi Informatsii,* Vol. 12, pp. 95–104.

Prim, R. C. (1957): Shortest connection networks and some generalizations. *Bell System Technical Journal,* Vol. 36, pp. 1389–1401.

Purslow, E. J. (1970): Solvability and analysis of linear active networks by use of the state equations. *IEEE Transactions on Circuit Theory,* Vol. CT-17, pp. 469–475.

Recski, A. (1976a): Matroids and independent state variables. *Proceedings of the 2nd European Conference on Circuit Theory and Design,* Genova.

Recski, A. (1976b): Contributions to the n-port interconnection problem by means of matroids. *Proceedings of the 5th Hungarian Combinatorial Conference* (Colloquia Mathematica Societatis János Bolyai 18), Hungary, pp. 877–892.

Recski, A. (1979a): Unique solvability and order of complexity of linear active networks containing memoryless n-ports. *Circuit Theory and Applications,* Vol. 7, pp. 31–42.

Recski, A. (1979b): Terminal solvability and the n-port interconnection problem. *Proceedings of the 1979 International Symposium on Circuits and Systems,* Tokyo, pp. 988–989.

Recski, A. (1980a): Sufficient conditions for the unique solvability of networks containing linear memoryless 2-ports. *Circuit Theory and Applications,* Vol. 8, pp. 95–103.

Recski, A. (1980b): Engineering applications of matroids – A survey. To appear in the *Proceedings of the Conference on Matroid Theory and Its Applications,* Varenna, Italy.

Recski, A., and Iri, M. (1980): Network theory and transversal matroids. *Discrete Applied Mathematics,* Vol. 2, pp. 311–326.

Schrijver, A. (1978): *Matroids and Linking Systems.* Mathematical Centre Tracts 88, Amsterdam.

Steward, D. V. (1962): On an approach to techniques for the analysis of the structure of large systems of equations. *SIAM Review,* Vol. 4, pp. 321–342.

Sugihara, K. (1978): A step toward man-machine communication by means of line drawings of polyhedra. *Bulletin of the Electrotechnical Laboratory of Japan,* Vol. 42, pp. 20–43.

Sugihara, K. (1979): Studies on mathematical structures of line drawings of polyhedra and their applications to scene analysis (in Japanese). *Researches of the Electrotechnical Laboratory,* No. 800.

Sugihara, K. (1980): On redundant bracing in plane skeletal structure. *Bulletin of Electrotechnical Laboratory,* Vol. 44, No. 5/6, pp. 78–89.

Sugihara, K. (1981): Program system for the principal partition of a bipartite graph – Reference manual. *Research Memorandum RMI* 81-03, Department of Mathematical Engineering and Instrumentation Physics, University of Tokyo.

Sugihara, K., and Iri, M. (1978): A mathematical approach to the determination of the structure of concepts. *Proceedings of the International Conference on Cybernetics and Society,* Tokyo-Kyoto, pp. 421–426.

Sugihara, K., and Iri, M. (1980): A mathematical approach to the determination of the structure of concepts. *Matrix and Tensor Quarterly,* Vol. 30, pp. 62–75.

Thambynayagam, R. K. M., Wood, R. K., and Winter, P. (1981): DPS – An engineers' tool for dynamic process analysis. *The Chemical Engineer,* No. 365, pp. 58–65.

Tomizawa, N. (1975): Irreducible matroids and classes of r-complete bases (in Japanese). *Transactions of the Institute of Electronics and Communication Engineers of Japan,* Vol. 58 A, pp. 793–794.

Tomizawa, N. (1976): Strongly irreducible matroids and principal partitions of a matroid into strongly irreducible minors (in Japanese). *Transactions of the Institute of Electronics and Communication Engineers of Japan,* Vol. J59A, pp. 83–91.

Tomizawa, N., and Fujishige, S. (1982): Historical survey of extensions of the concept of principal partition and their unifying generalization to hypermatroids. *Systems Science Research Report,* No. 5, Department of Systems Science, Graduate School of Science and Engineering, Tokyo Institute of Technology.

Tomizawa, N., and Iri, M. (1974): An algorithm for determining the rank of a triple matrix product AXB with application to the problems of discerning the existence of the unique solution in a network (in Japanese). *Transactions of the Institute of Electronics and Communication Engineers of Japan,* Vol. 57A, pp. 834–841.

Tomizawa, N., and Iri, M. (1976): Matroids (in Japanese). *Journal of the Institute of Electronics and Communication Engineers of Japan,* Vol. 59, pp. 1350–1352.

Tomizawa, N., and Iri, M. (1977): On matroids (in Japanese). *Journal of the Society of Instrument and Control Engineers, Japan,* Vol. 16, pp. 455–468.

Weinberg, L. (1962): *Network Analysis and Synthesis.* McGraw-Hill.

Weinberg, L. (1977): Matroids, generalized networks, and electric network synthesis. *Journal of Combinatorial Theory,* Series B, Vol. 23, pp. 106–126.

Welsh, D. J. A. (1968): Kruskal's theorem for matroids. *Mathematical Proceedings of the Cambridge Philosophical Society,* Vol. 64, pp. 3–4.

Welsh, D. J. A. (1970): On matroid theorems of Edmonds and Rado. *Journal of the London Mathematical Society,* Vol. 2, pp. 251–256.

Welsh, D. J. A. (1976): *Matroid Theory.* Academic Press, London.

Whitney, H. (1935): On the abstract properties of linear dependence. *American Journal of Mathematics,* Vol. 57, pp. 509–533.

Wilson, R. J. (1973): An introduction to matroid theory. *American Mathematical Monthly,* Vol. 80, pp. 500–525.

Yajima, K., Tsunekawa, J., and Kobayashi, S. (1981): On equation-based dynamic simulation. *Proceedings of the 2nd World Congress of Chemical Engineering,* Vol. V, pp. 469–480.

Yao, A. C.-C. (1975): An $O(|E|\log\log|V|)$ algorithm for finding minimum spanning trees. *Information Processing Letters,* Vol. 4, pp. 21–25.

Zimmermann, U. (1981a): Minimization of some nonlinear functions over polymatroidal flows. Report 81-5, Mathematisches Institut, Universität zu Köln.

Zimmermann, U. (1981b): Minimization on submodular flows. *Discrete Applied Mathematics,* Vol. 4, pp. 303–323.

Recent Results in the Theory of Machine Scheduling

E. L. Lawler
University of California, College of Engineering, Computer Science Division, 591 Evans Hall, Berkeley, CA 94720, USA

Abstract. The state of the art of deterministic machine scheduling is reviewed. Emphasis is placed on efficient, i.e. polynomial-bounded, optimization algorithms. A few of the more significant NP-hardness results are highlighted, and some open problems are mentioned.

I. Introduction

The theory of sequencing and scheduling encompasses a bewilderingly large variety of problem types. In the present paper, I shall not attempt to survey more than a small part of this active and rapidly growing area of research. In particular, I shall confine my attention to *machine* scheduling problems, excluding from consideration such worthy topics as project scheduling, timetabling, and cyclic scheduling of manpower. I shall concentrate on strictly *deterministic* models, with emphasis on *efficient,* i.e. polynomial-bounded, algorithms for *optimization,* as opposed to approximation. I shall mention certain significant NP-completeness results and point out several open problems.

This selection of topics reflects my own interests in the field, which I have pursued since the early 1960's but most vigorously since 1974 when I began working with Jan Karel Lenstra and Alexander Rinnooy Kan. One of the objectives of our collaboration has been to delineate, as closely as possible, the boundary between those machine scheduling problems which are *easy* (solvable in polynomial time), and those which are *NP-hard*. We have produced two surveys (Graham, et al, 1977), (Lawler, Lenstra & Rinnooy Kan, 1982 A), and a detailed tabulation of the status of problem types (Lageweg, et al, 1981). The reader is referred to (Lawler, Lenstra & Rinnooy Kan, 1982 B), for an anecdotal account of our collaboration, which will eventually result in a book.

The present paper differs from the surveys mentioned above in that I have chosen to emphasize only those algorithms which I believe are most interesting, elegant or important. Moreover, I have tried to provide something of a historical perspective, in that for each type of problem considered, I first try to state a "classical" algorithm (usually meaning one obtained prior to 1960) and then show how this algorithm has been generalized or improved (or how generalization has been blocked by NP-hardness). There are four principal parts to the paper dealing respectively with single machines, identical and uniform parallel machines, open shops and unrelated parallel machines, and flow shops and job shops.

It greatly facilitates any discussion of machine scheduling to have an appropriate notation for problem types. Such a notation is detailed in the references mentioned above, and is of the form $\alpha|\beta|\gamma$, where α indicates *machine environment* (single machine, parallel machine, open shop, flow shop, job shop), β indicates *job characteristics* (independent vs. precedence constrained, etc.), and γ indicates the *optimality criterion* (makespan, flow time, maximum lateness, total tardiness, etc.). Instead of defining this notation at the outset, I shall introduce it a bit at a time, and hope that the reader will find this natural and not distracting.

II. Single-Machine Scheduling

1. Minimizing Maximum Cost

Consider the following simple sequencing problem. There are n jobs to be processed by a *single machine* which can execute at most one job at a time. Each job j requires a *processing time* p_j and has a specified *due date* d_j. If the jobs are executed without interruption and without idle time between them, with the first job beginning at time $t=0$, then any given sequence induces a well defined *completion time* C_j and *lateness* $L_j = C_j - d_j$ for each job j. What sequence will minimize maximum lateness? That is, minimize

$$L_{\max} = \max_j \{L_j\}.$$

The "earliest due date" or EDD rule of (Jackson, 1955) provides a simple and elegant solution to this problem: Any sequence is optimal that puts the jobs in order of nondecreasing due dates. This result can be proved by a simple interchange argument. Let π be any sequence and π^* be an EDD sequence. If $\pi \neq \pi^*$ then there exist two jobs j and k such that j immediately precedes k in π, but k precedes j in π^*. Since $d_k \leq d_j$, interchanging the positions of j and k in π cannot increase the value of $L_{\max}$. A finite number of such transpositions transforms π to π^*, showing that π^* is optimal.

One generalization of this problem is to allow a monotone nondecreasing *cost function* f_j to be specified for each job j, and to attempt to minimize $f_{\max}$, where

$$f_{\max} = \max_j \{f_j(C_j)\}.$$

Another generalization is to allow *precedence constraints* to be specified in the form of a partial order $\rightarrow$; if $j \rightarrow k$ then job j has to be completed before job k can start. (In the original problem, jobs were *independent*, i.e. the relation $\rightarrow$ was assumed to be empty.)

We denote the problem solved by Jackson by $1||L_{\max}$ and the generalized problem by $1|prec|f_{\max}$ ("1" for single machine; "prec" for arbitrary precendence constraints; $f_{\max}$ for the objective function). The problem $1|prec|f_{\max}$ is solved by the following simpe rule (Lawler, 1973): From among all jobs that are eligible to be sequenced last, i.e. that have no successors under $\rightarrow$, put that job last which will incur the smallest cost in that position. Then repeat on the set of $n-1$ jobs remaining, and so on. (For $1\|L_{\max}$, this rule says, "Put that job last which has largest due date. Then repeat.")

The correctness of this rule is proved as follows. Let $N=\{1,\ldots,n\}$ be the set of all jobs, let $L\subseteq N$ be the set of jobs without successors, and for any $S\subseteq N$ let $f^*(S)$ be the maximum job completion cost in an optimal schedule for S. We know that the last job of any sequence is completed at time $P=p_1+p_2+\ldots+p_n$. If job $l\in L$ is chosen such that

$$f_l(P) = \min_{j\in L}\{f_j(P)\},$$

then the optimal value of a sequence subject to the condition that job l is processed last is given by

$$\max\{f^*(N-\{l\}), f_l(P)\}.$$

Since both $f^*(N-\{l\})\leqslant f^*(N)$ and $f_l(P)\leqslant f^*(N)$, the rule is proved.

The algorithm can be implemented to run in $O(n^2)$ time, under the assumption that each f_j can be evaluated in unit time for any value of the argument. This contrasts with $O(n\log n)$ time for sorting due dates under Jackson's rule.

A further natural generalization is obtained by specifying a *release time* r_j for each job j, prior to which the job cannot be performed. (Prior to this we have assumed each job is available at time zero, i.e. $r_j=0$ for all j.) If it is required that jobs be executed without interruption, then the introduction of release dates makes things quite difficult. It is an NP-complete problem even to determine if a set of independent jobs can be completed by specified due dates. However, if *preemption* is permitted, i.e. the processing of any job may arbitrarily often be interrupted and resumed at a later time without penalty, then the problem is much easier, and the procedure of (Lawler, 1973) has been further generalized in (Baker, et al, 1982) to apply to the problem $1|pmtn, prec, r_j|f_{\max}$ ("pmtn" for preemption, "r_j" for release dates). For brevity we do not describe this algorithm here, but mention that it can be implemented to run in $O(n^2)$ time.

We should mention that although the nonpreemptive problem $1|r_j|L_{\max}$ is NP-hard, the special case of unit-time jobs, $1|r_j, p_j=1|L_{\max}$ is easy. Moreover, it remains easy even if release dates and due dates are not integers (Simons, 1978).

2. Smith's Rule and Interchange Arguments

Now suppose that we have a single-machine scheduling problem in which each job has a specified *processing time* p_j and *weight* w_j, and the objective is to find a sequence minimizing the *weighted sum of job completion times*, $\Sigma w_j C_j$. This problem, $1\|\Sigma w_j C_j$ in our notation, is solved by another classical result of scheduling theory, the "ratio rule" of (Smith, 1956): Any sequence is optimal that puts the jobs in order of nondecreasing ratios $\rho_j=p_j/w_j$. As a corollary, if all jobs have equal weight, any sequence is optimal which places the jobs in nondecreasing order of processing times. (This is known as the "shortest processing time" or SPT rule.)

Let us pose a very general type of sequencing problem that includes $1\|L_{max}$ and $1\|\Sigma w_j C_j$ as special cases. Given a set of n jobs and a real-valued function f which assigns a value $f(\pi)$ to each permutation π of the jobs, find a permutation π^* such that

$$f(\pi^*) = \min_\pi \{f(\pi)\}.$$

If we know nothing of the structure of the function f, there is clearly nothing to be done except to evaluate $f(\pi)$ for each of the $n!$ permutations π. However, we may be able to find a transitive and complete relation $\leqslant$ (i.e. a quasi-total order) on the jobs with the property that for any two jobs b, c and any permutation of the form $\alpha b c \delta$ we have

$$b \leqslant c \Rightarrow f(\alpha b c \delta) \leqslant f(\alpha c b \delta).$$

Such a relation is called a *job interchange relation*. It says that whenever b and c occur as adjacent jobs with c before b, we are at least as well off to interchange their order. Hence, this relation is sometimes referred to as the "adjacent pairwise interchange property." It is a simple matter to verify the following.

Theorem 1. If f admits of a job interchange relation $\leqslant$, then an optimal permutation π^* can be found by ordering the jobs according to $\leqslant$, with $O(n \log n)$ comparisons of jobs with respect to $\leqslant$.

Note that both Jackson's rule and Smith's rule are based on job interchange relations. Job interchange realtions have been found for a number of other sequencing problems. There is such a relation for the weighted sum of *discounted* job completion times (Rothkopf, 1966), for the "least cost testing sequence" problem (Garey, 1973), and for the two-machine flow shop problem (Johnson, 1954; Mitten, 1958). More will be said about such problems in the next section in the context of precedence constraints.

Let us now consider the generalization of $1\|\Sigma w_j C_j$ to allow release dates and deadlines. It turns out that $1|r_j|\Sigma C_j$ (and *a fortiori*, $1|r_j|\Sigma w_j C_j$) is NP-hard (Lenstra, 1977), as is $1|pmtn, r_j|\Sigma w_j C_j$ (Labetoulle, et al, 1979). However, $1|pmtn, r_j|\Sigma C_j$ admits of a very simple solution (Baker, 1974): Schedule over time, starting at the first release date. At each decision point (whenever a job is released or a job is completed) choose to process next from among the available jobs (those whose release dates have been met and for which processing is not yet complete) a job whose remaining processing time is minimal. Only $O(n \log n)$ time is required to construct an optimal schedule in this way.

A *deadline* $\bar{d}_j$ (as opposed to a due date, which is used for computing the cost of a schedule) imposes a constraint that $C_j \leqslant \bar{d}_j$. In (Smith, 1956), a solution is offered to the problem $1|\bar{d}_j|\Sigma C_j$: From among all jobs that are eligible to be sequenced last, i.e. are such that $\bar{d}_j \geqslant p_1 + \ldots + p_n$, put that job last which has the largest possible processing time. Then repeat on the set of $n-1$ jobs remaining, and so on. This rule yields an $O(n \log n)$ algorithm. (It is left to the reader to verify that if there exists any sequence in which all jobs meet their deadlines, then Smith's algorithm produces such a sequence.)

Although $1|\bar{d}_j|\Sigma C_j$ is easy, $1|\bar{d}_j|\Sigma w_j C_j$ is NP-hard (Lenstra, 1977). (An incorrect algorithm is proposed in (Smith, 1956).) It is easily shown that there is no advantage to preemption, in that any solution that is optimal for $1|\bar{d}_j|\Sigma w_j C_j$ is optimal for $1|pmtn, \bar{d}_j|\Sigma w_j C_j$. It follows that $1|pmtn, \bar{d}_j|\Sigma C_j$ is easy and $1|pmtn, \bar{d}_j|\Sigma w_j C_j$ is NP-hard.

The cited results leave us with one unresolved issue in this area.

Open Problem: What is the status of $1|pmtn, r_j, \bar{d}_j|\Sigma C_j$? Easy or NP-hard?

3. Series Parallel Scheduling

Over a period of years, investigators considered the effect of precedence constraints on the $1||\Sigma w_j C_j$ problem. In (Conway, et al, 1967) precedence constraints in the form of parallel chains were dealt with, but only subject to the condition that all $w_j = 1$. In (Horn, 1972), an $O(n^2)$ algorithm was proposed for precedence constraints in the form of rooted trees. In (Adolphson & Hu, 1973) an $O(n \log n)$ algorithm was proposed for the same case. In (Lawler, 1978 A), an $O(n \log n)$ algorithm was given for "series parallel" precedence constraints (with rooted trees as a special case), and the problem was shown to be NP-hard for arbitrary precedence constraints, even if all $w_j = 1$ or if all $p_j = 1$. Other contributions to the problem were made by (Sidney, 1975) and (Adolphson, 1977).

The concept of series parallelism, as introduced in (Lawler, 1978 A), may be unfamiliar to the reader. A digraph is said to be *series-parallel* if its transitive closure is *transitive series-parallel*, as given by the recursive definition below:
(1) A digraph $G = (\{j\}, \phi)$ with a single vertex j and no arcs is transitive series-parallel.
(2) Let $G_1 = (V_1, A_1)$, $G_2 = (V_2, A_2)$ be transitive series-parallel digraphs with disjoint vertex sets. Both the series composition $G_1 \to G_2 = (V_1 \cup V_2, A_1 \cup A_2 \cup (V_1 \times V_2))$ and the parallel composition $G_1 \| G_2 = (V_1 \cup V_2, A_1 \cup A_2)$ are transitive series-parallel digraphs.
(3) No digraph is transitive series-parallel unless it can be obtained by a finite number of applications of Rules (1) and (2).

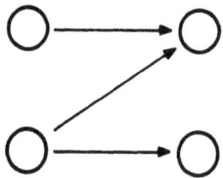

Fig. 1. Z-diagram

A variety of interesting and useful digraphs (and their corresponding partial orders) are series-parallel. In particular, rooted trees, forests of such trees and level digraphs are series-parallel. The smallest acyclic digraph which is not series-parallel is the *Z-digraph* shown in Figure 1. An acyclic digraph is series-parallel if and only if its transitive closure does not contain the Z-diagraph as an induced subgraph. It is possible to determine whether or not an arbitrary digraph $G=(V, A)$ is series-parallel in $O(|V|+|A|)$ time (Valdes, et al, 1982).

The structure of a series-parallel digraph is displayed by a *decomposition tree* which represents one way in which the transitive closure of the digraph

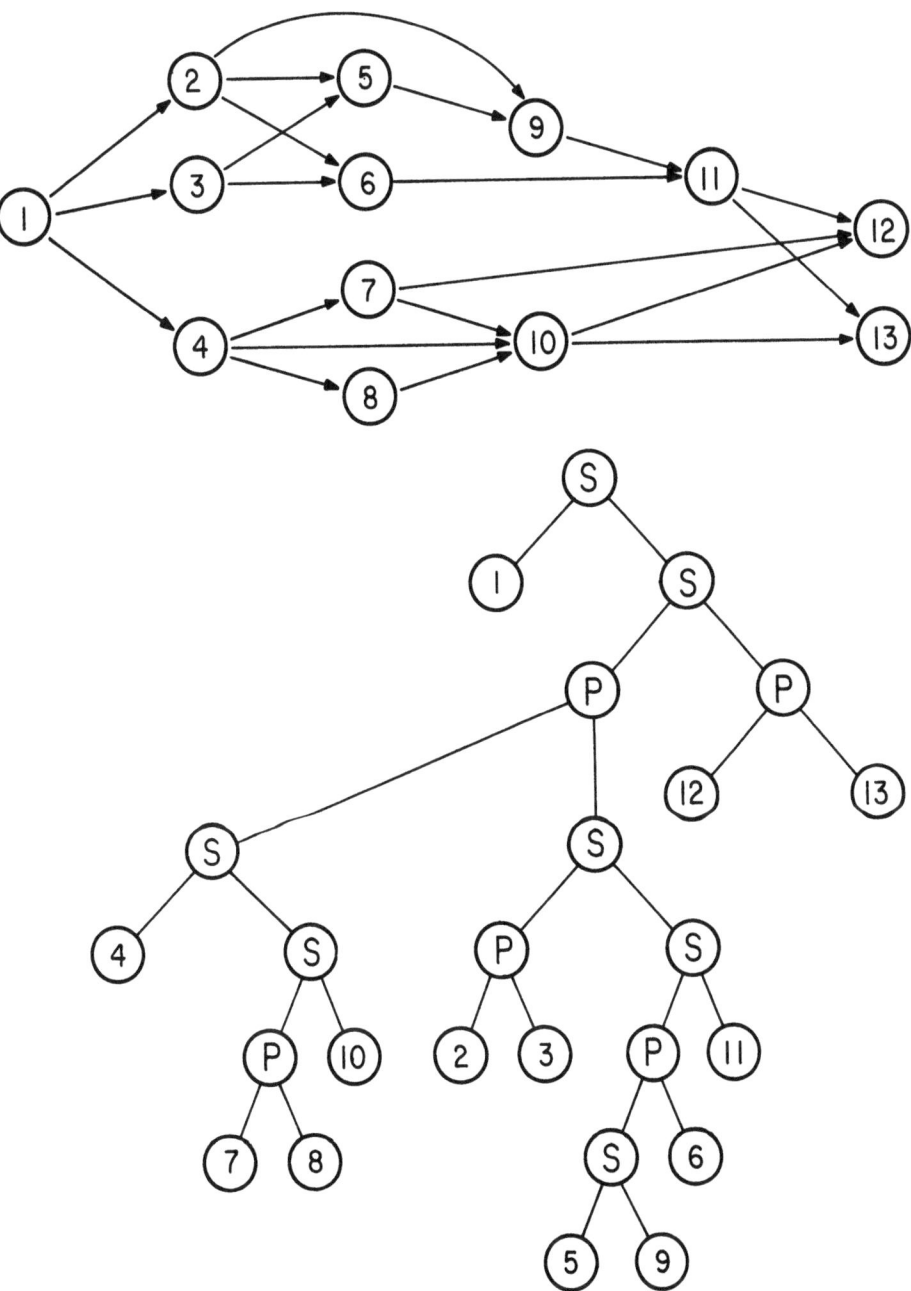

Fig. 2. Series-parallel diagraph and decomposition tree

can be obtained by successive applications of Rules (1) and (2). A series-parallel digraph and its decomposition tree are shown in Figure 2. Each leaf of the decomposition tree is identified with a vertex of the digraph. An S-node represents the application of series composition to the subdigraphs identified with its children; the ordering of these children is important: We adopt the convention that left precedes right. A P-node represents the application of parallel composition to the subdigraphs identified with its children; the ordering of these children is unimportant. The series or parallel relationship of any pair of vertices can be determined by finding their least common ancestor in the decomposition tree.

It was observed by (Monma and Sidney, 1979) that the series-parallel algorithm of (Lawler, 1978 A) has much broader application than just to the weighted completion time problem. We now indicate this by a very general problem formulation, as in (Lawler, 1978 B).

As in the previous section let f be a real-valued function of permutations. But now suppose that precedence constraints are specified in the form of a partial order $\to$. A permutation π is *feasible* if $j \to k$ implies that job j precedes job k under π. The objective is to find a feasible permutation π^* such that

$$f(\pi^*) = \min_{\pi \text{ feasible}} \{f(\pi)\}.$$

We need something stronger than a job interchange relation to solve this problem. A transitive and complete relation $\leq$ on subpermutations or *strings* of jobs with the property that for any two disjoint strings of jobs β, γ and any permutation of the form $\alpha\beta\gamma\delta$ we have

$$\beta \leq \gamma \Rightarrow f(\alpha\beta\gamma\delta) \leq f(\alpha\gamma\beta\delta)$$

is called a *string interchange relation*. Smith's rule generalizes to such a relation in a fairly obvious way: For any string α we define $\rho_\alpha = \sum_{j \in \alpha} p_j / \sum_{j \in \alpha} w_j$. However, it is not true that every function f which admits of a job interchange relation also has a string interchange relation.

The remainder of this section is devoted to an intuitive justification of the following result. Details can be found in (Lawler, 1978 B).

Theorem 2. If f admits of a string interchange relation $\leq$ and if the precedence constraints $\to$ are series-parallel, then an optimal permutation π^* can be found by an algorithm which requires $O(n \log n)$ comparisons of strings with respect to $\leq$.

Suppose we are to solve a sequencing problem for which a string interchange relation $\leq$ exists and a decomposition tree for the series-parallel constraints $\to$ is given. We can solve the problem by working from the bottom of the tree upward, computing a set of strings of jobs for each node of the tree from the sets of strings obtained for its children. Our objective is to obtain a set of strings at the root node such that sorting these strings according to $\leq$ yields an optimal feasible permutation.

We will accomplish our objective if the sets S of strings we obtain satisfy two conditions:

(i) Any ordering of the strings in a set S according to $\leqslant$ is feasible with respect to the precedence constraints $\rightarrow$.
(ii) At any point in the computation, let $S_1, \ldots, S_k$ be the sets of strings computed for nodes such that sets have not yet been computed for their parents. Then some ordering of the strings in $S_1 \cup \ldots \cup S_k$ yields an optimal feasible permutation.

If we order the strings computed at the root according to $\leqslant$, then condition (i) ensures that the resulting permutation is feasible and condition (ii) ensures that it is optimal.

For each leaf of the tree, we let $S=\{j\}$, where j is the job identified with the leaf. Condition (i) is satisfied trivially and condition (ii) is clearly satisfied for the union of the leaf-sets.

Suppose S_1 and S_2 have been obtained for the children of a P-node in the tree. There are no precedence constraints between the strings in S_1 and the strings in S_2. Accordingly, conditions (i) and (ii) remain satisfied if for the P-node we let $S=S_1 \cup S_2$.

Suppose S_1 and S_2 have been obtained for the left child and the right child of an S-node, respectively. Let

$$\sigma_1 = \max_{\leqslant} S_1, \quad \sigma_2 = \min_{\leqslant} S_2.$$

If $\sigma_2 \not\leqslant \sigma_1$, then conditions (i) and (ii) are still satisfied if for the S-node we let $S=S_1 \cup S_2$. If $\sigma_2 \leqslant \sigma_1$, we assert that there exists an optimal feasible permutation in which σ_1 and σ_2 are replaced by their concatenation $\sigma_1 \sigma_2$. (The proof of this assertion involves simple interchange arguments; see (Lawler, 1978 B).) This suggests the following procedure:

begin
    $\sigma_1 := \max_{\leqslant} S_1; \sigma_2 := = \min_{\leqslant} S_2;$
    if $\sigma_2 \not\leqslant \sigma_1$
    then $S := S_1 \cup S_2$
    else $\sigma := \sigma_1 \sigma_2$
        $S_1 := S_1 - \{\sigma_1\}; \sigma_1 := \max_{\leqslant} S_1;$
        $S_2 := S_2 - \{\sigma_2\}; \sigma_2 := \min_{\leqslant} S_2;$
        while $\sigma \leqslant \sigma_1 \lor \sigma_2 \leqslant \sigma$
        do if $\sigma \leqslant \sigma_1$
            then $\sigma := \sigma_1 \sigma;$
                $S_1 := S_1 - \{\sigma_1\}; \sigma_1 := \max_{\leqslant} S_1$
            else $\sigma := \sigma \sigma_2;$
                $S_2 := S_2 - \{\sigma_2\}; \sigma_2 := \min_{\leqslant} S_2$
        fi
        od;
        $S := S_1 \cup \{\sigma\} \cup S_2$
    fi
end.

(We make here the customary assumption that $\max_{\leqslant} \phi$ and $\min_{\leqslant} \phi$ are very small and large elements, respectively.) It is not difficult to verify that conditions (i) and (ii) remain satisfied if for an S-node we compute a set of strings according to the above procedure.

The entire algorithm can be implemented so as to require $O(n \log n)$ time plus the time for the $O(n \log n)$ comparisons with respect to $\leq$.

In addition to the total weighted completion time problem, several other sequencing problems admit of a string interchange relation and hence can be solved efficiently for series-parallel precedence constraints. Among these are the problems of minimizing *total weighted discounted completion time* (Lawler & Sivazlian, 1978), *expected cost of fault detection* (Garey, 1973; Monma & Sidney, 1979), *minimum initial resource requirement* (Abdel-Wahab & Kameda, 1978; Monma & Sidney, 1979), and the *two-machine permutation flow shop problem with time lags* (Sidney, 1979).

4. The Total Tardiness Problem

The *tardiness* of job j with respect to due date d_j is $T_j = \max\{0, C_j - d_j\}$. It is a far more difficult task to minimize total tardiness than to minimize total completion time. (Note that $\Sigma C_j = \Sigma L_j + \Sigma d_j$, where Σd_j is a constant.) Over the years a large number of papers have been written about the problem $1\|\Sigma T_j$, among the most important of which is (Emmons, 1969), in which certain "dominance" conditions were established. These are conditions under which there exists an optimal sequence in which one job j precedes another job k. (A simple case is that in which $p_j \leq p_k$ and $d_j \leq d_k$.)

In (Lawler, 1977), a "pseudopolynomial" algorithm with running time $O(n^4 \Sigma p_j)$ was described, and a "strong" NP-hardness proof (due to M. R. Garey and D. S. Johnson) was presented for $1\|\Sigma w_j T_j$. This pseudopolynomial algorithm also serves as the basis for a fully polynomial approximation scheme for the problem $1\|\Sigma T_j$ (Lawler, 1982 B).

The existence of a pseudopolynomial algorithm rules out the possibility that the total tardiness problem is NP-hard in the "strong" sense (unless $P = NP$). However, it leaves unresolved the foolowing question.

Open Problem: What is the status of $1\|\Sigma T_j$? Easy or NP-hard (in the "ordinary" sense)?

5. Minimizing the Number of Late Jobs

Another possible objective of single-machine scheduling is to find a sequence minimizing the number of jobs j which are late with respect to specified due dates d_j. We let

$$U_j(C_j) = \begin{cases} 0 & \text{if } C_j \leq d_j \\ 1 & \text{if } C_j > d_j \end{cases}$$

and denote this problem $1\|\Sigma U_j$.

The problem $1\|\Sigma U_j$ is clearly equivalent to that of finding a subset $S \subseteq N = \{1, \ldots, n\}$ such that all the jobs in S can be completed on time and $|S|$ is maximal. An optimal sequence then consists of the jobs in S, in EDD order, followed by the jobs in $N-S$ in arbitrary order. (Note that once a job is late it does not matter how late it is.) An elegant algorithm for finding a maximum subset S is given in (Moore, 1968). We state this without proof.

begin
    order the jobs so that $d_1 \leq \ldots \leq d_n$.
    $S := \phi$
    $P := 0$
    for $j := 1$ **to** n
        do $S := S \cup \{j\}$
            $P := P + p_j$
            if $P > d_j$
            then let $p_l = \max\{p_i | i \in S\}$
                    $S := S - \{l\}$
                    $P := P - p_l$
            fi
        od
end.

This algorithm can be easily implemented to run in $O(n \log n)$ time.

The weighted version of this problem, $1 \| \Sigma w_j U_j$, is easily shown to be NP-hard, but can be solved in pseudopolynomial time $O(n \max\{d_j\})$ by a dynamic-programming computation similar to that used for the knapsack problem (Lawler & Moore, 1969). However, if job weights are *agreeable*, i.e. there is an ordering of the jobs so that

$$p_1 \leq p_2 \leq \ldots \leq p_n,$$
$$w_1 \geq w_2 \geq \ldots \geq w_n,$$

then the problem can be solved in $O(n \log n)$ time by a simple modification of Moore's algorithm (Lawler, 1976).

If $1 \| \Sigma U_j$ is generalized to allow arbitrary release dates, it immediately becomes NP-hard. If release dates and due dates are *compatible*, i.e. there is an ordering of the jobs so that

$$r_1 \leq r_2 \leq \ldots \leq r_n,$$
$$d_1 \leq d_2 \leq \ldots \leq d_n,$$

then the problem is easy. An $O(n^2)$ algorithm was proposed in (Kise, et al, 1978). An $O(n \log n)$ algorithm, generalizing Moore's procedure, has been provided for this case in (Lawler, 1982C).

If preemption is permitted, then introducing arbitrary release dates does not cause NP-hardness. In (Lawler, 1982C) a rather complicated dynamic programming procedure for $1 | pmtn, r_j | \Sigma w_j U_j$ is presented. This procedure has running time $O(n^3 W^3)$ where $W = \Sigma w_j$. (Job weights are assumed to be integers.) Thus for $1 | pmtn, r_j | \Sigma U_j$ the running time becomes $O(n^6)$.

We should also mention the case in which each job has both a due date d_j and a deadline $\bar{d}_j$ (where without loss of generality $d_j \leq \bar{d}_j$). The problem $1 | \bar{d}_j | \Sigma U_j$ is shown to be NP-hard in (Lawler, 1982C).

6. Minimizing the Sum of Cost Functions

Suppose, as in Section 1, for each job j there is specified a cost function f_j. Except that now we wish to find a sequence that minimizes the sum of the costs, $\sum_j f_j(C_j)$, instead of the maximum of the costs. A general dynamic programming procedure for this problem, which we denote by $1\|\Sigma f_j$, was suggested in (Held & Karp, 1962) and (Lawler, 1964):

For any subset $S \subseteq N$, let $p(S) = \sum_{j \in S} p_j$ and let $F(S)$ denote the minimum total cost for a sequence of the jobs in S. Then we have

$$F(\phi) = 0$$
$$F(S) = \min_{j \in S} \{F(S - \{j\}) + f_j(p(S))\}. \tag{1}$$

The optimum value for the complete set of jobs N is $F(N)$ and this value can be obtained in $O(n2^n)$ time.

Note that this solution method does not require that the functions f_j be nondecreasing. It is only necessary to assume that jobs are performed without preemption and without idle time between them.

Interestingly, precedence constraints can only make the dynamic programming computation easier, not harder. In solving $1|prec|\Sigma f_j$, it is necessary to solve equations (1) only for subsets S that are *initial* sets of the partial order $\rightarrow$, in the sense that $j \rightarrow k$, $k \in S$ imply that $j \in S$. Moreover, the minimization in (1) need be carried out only over jobs $j \in S$ that have no successors in S. If the precedence constraints $\rightarrow$ are at all significant, then the number of initial sets is likely to be *very much* smaller than 2^n. One can then hope to solve much larger problems than in the absence of such constraints.

Dominance relations, such as those given by (Emmons, 1969) for the total tardiness problem, can be viewed as precedence constraints. Thus there is real motivation for finding dominance relations for various types of problems. This approach has been taken by (Fisher, et al, 1981) and others for solving the total tardiness problem and other related problems.

It should be noted that it is a nontrivial problem to devise suitable data structures for solving $1|prec|\Sigma f_i$ by dynamic programming. In (Baker, & Schrage, 1978), various "labeling" schemes have been proposed for assigning addresses to initial sets. The difficulty with these schemes is that the address space is not necessarily fully utilized and that memory may be wasted. In (Lawler, 1979), a time- and memory-efficient procedure is proposed for generating the necessary data structures, and tests by (Kao & Queyranne, 1980) seem to confirm its advantages.

There have been many proposals for solving problems of the form $1\|\Sigma f_i$ and $1|prec|\Sigma f_j$, by other dynamic programming schemes (Fisher, 1976), and by branch-and-bound, using a variety of bounding methods. Some bounding techniques, are quite sophisticated, as in (Fisher, 1976), where Lagrangean techniques are applied to problem relaxations. The author believes that "hybrid" techniques, combining the best features of branch-and-bound and dynamic programming have potential.

III. Identical and Uniform Parallel Machines

1. Introduction

As before, there are n jobs to be scheduled and job j has a *processing requirement* p_j. But we now suppose there are m *parallel machines* available to do the processing. Each job can be worked on by at most one machine at a time and each machine can work on at most one job at a time. Machine i processes job j with *speed* s_{ij}. Hence if machine i does all the work on job j, it requires a total amount of time p_j/s_{ij} for its processing.

We distinguish three types of parallel machines:

(P) *Identical Machines:* All s_{ij} are equal. In this case we may assume, without loss of generality, that $s_{ij} = 1$, for all i, j.

(Q) *Uniform Machines:* $s_{ij} = s_{ik}$, for all i, j, k. In other words, each machine i performs all jobs at the same speed s_i. Without loss of generality, we assume $s_1 \geqslant s_2 \geqslant \ldots \geqslant s_m$.

(R) *Unrelated Machines:* There is no particular relationship between the s_{ij} values.

In the next several sections, we shall consider problems involving identical and uniform machines. Unrelated machines, which require rather different algorithmic techniques, are dealt with in Part IV of this paper.

The difficulty of parallel machine problems is profoundly affected by whether or not preemption is permitted. It is easy to see that even the problem of nonpreemptively scheduling two identical machines so as to minimize makespan, $P2\|C_{\max}$ in out notation, is NP-hard. However, in recent years there have been many new good algorithms developed for preemptive scheduling of parallel machines.

2. Nonpreemptive Scheduling to Minimize the Sum of Completion Times

The solution to the problem $P\|\Sigma C_j$ is quite easy (Conway, et al, 1967): Order the jobs so that $p_1 \leqslant p_2 \leqslant \ldots \leqslant p_n$. Having scheduled jobs $1, 2, \ldots, j$, find the earliest available machine and schedule job $j+1$ on that machine, thereby completing it as soon as possible. This greedy approach, clearly a variant of the SPT rule, yields a schedule like that indicated by the Gannt chart in Figure 3.

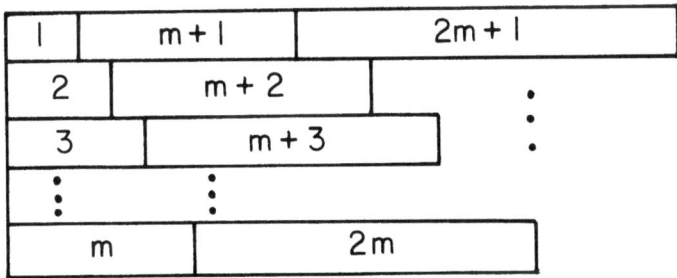

Fig. 3. Gannt chart for SPT schedule

As justification, consider the jobs to be performed on any single machine and for simplicity let these be $1, 2, \ldots, k$ in that order. For these jobs we have

$$\Sigma C_j = kp_1 + (k-1)p_2 + \ldots + 2p_{k-1} + p_k. \quad (2)$$

Considering all m machines, it is apparent that we have available as coefficients m 1's, m 2's, $\ldots, mn$'s, and the n smallest of these mn coefficients should be assigned to the n jobs, with the smallest coefficients being assigned to the largest jobs. This is precisely what the SPT rule accomplishes.

Now consider $Q\|\Sigma C_j$, and suppose jobs $1, 2, \ldots, k$ are performed on machine i. Then instead of (2) we have

$$\Sigma C_j = \frac{k}{s_i} p_1 + \frac{(k-1)}{s_i} p_2 + \ldots + \frac{2}{s_i} p_{k-1} + \frac{1}{s_i} p_k.$$

It is now apparent that what we have are mn coefficients of the form k/s_i, $k=1, 2, \ldots, n$; $i=1, 2, \ldots, m$ and our task is to assign the n smallest of these coefficients to the n jobs, with the smallest coefficients being assigned to the largest jobs. This can be done in $O(n\log n)$ time with the algorithm of (Howowitz & Sahni, 1976):

begin
    order the jobs so that $p_1 \leq p_2 \leq \ldots \leq p_n$
    initialize a priority queue Q with the values $\frac{1}{s_i}$, $i=1, 2, \ldots, m$.
    for $j=n$ **to** 1
        do let $\frac{k}{s_i}$ be a smallest value in Q
        assign $\frac{k}{s_i}$ as the coefficient of p_j
        replace $\frac{k}{s_i}$ in Q by $\frac{k+1}{s_i}$
    od
end

Unfortunately, the problem of nonpreemptively scheduling identical parallel machines so as to minimize *weighted* total completion time, i.e. $P\|\Sigma w_j C_j$, is NP-hard.

3. Preemptive Scheduling to Minimize the Sum of Completion Times

The following theorem is proved by rearrangement arguments.

Theorem 3. (McNaughton, 1959). *For any instance of $P|pmtn|\Sigma w_j C_j$ there exists a schedule with no preemptions for which the value of $\Sigma w_j C_j$ is as small as that for any schedule with a finite number of preemptions.*

(The finiteness restriction can be removed by results implicit in (Lawler & Labetoulle, 1978).)

From McNaughton's theorem it follows immediately that the procedure of the previous section solves $P|pmtn|\Sigma C_j$, since there is no advantage to preemption. Also, from the fact that $P||\Sigma w_j C_j$ is NP-hard, it follows that $P|pmtn|\Sigma w_j C_j$ (and *a fortiori* $Q|pmtn|\Sigma w_j C_j$) is NP-hard.

A simple example suffices to show that there *is* advantage to preemption in $Q|pmtn|\Sigma C_j$, i.e. the procedure of (Horowitz & Sahni, 1976) does not necessarily yield an optimal solution. We now state a lemma which is also proved by rearrangement arguments.

Lemma 4. (Lawler & Labetoulle, 1978). Let $p_1 \leq p_1 \leq \ldots \leq p_n$. Then there is an optimal schedule for $Q|pmtn|\Sigma C_j$ in which $C_1 \leq C_2 \leq \ldots \leq C_n$.

This lemma is essential to prove the validity of a solution to $Q|pmtn|\Sigma C_j$ presented in (Gonzalez, 1977). Let $p_1 \leq p_2 \leq \ldots \leq p_n$. Having scheduled jobs $1, 2, \ldots, j$, schedule job $j+1$ to be completed as easily as possible. This variant of the SPT rule yields a schedule as shown in Figure 4.

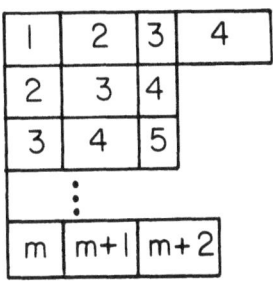

Fig. 4. SPT rule for preemptive scheduling of uniform machines

We comment that the same strategy is optimal for certain stochastic scheduling problems. Cf. (Weiss & Pinedo, 1980).

In (Gonzalez, 1977), a procedure is presented for minimizing total completion time, subject to a common deadline for all jobs. It seems quite feasible to extend the same ideas to the situation in which the deadlines $\bar{d}_j$ and the processing times p_j are *compatible*, i.e. there is a numbering of the jobs so that

$$p_1 \leq p_2 \leq \ldots \leq p_n$$
$$\bar{d}_1 \leq \bar{d}_2 \leq \ldots \leq \bar{d}_n.$$

However, the status of the problem for arbitrary deadlines is open.

Open Problem: What is the status of $Q|pmtn, \bar{d}_j|\Sigma C_j$? Easy or NP-hard?

Perhaps even more perplexing, almost nothing is known about the case of release dates.

Open Problem: What is the status of $Q|pmtn, r_j|\Sigma C_j$? (Or, for that matter, $P2|pmtn, r_j|\Sigma C_j$?) Easy or NP-hard?

4. Preemptive Scheduling to Minimize Makespan

The problem of preemptively scheduling m identical parallel machines so as to minimize makespan is quite easy. Clearly we must have

$$C_{\max} \geq \max\left\{\max_j \{p_j\}, \frac{1}{m}\Sigma p_j\right\}.$$

Yet one can achieve the lower bound given by the right hand side of the inequality by applying the "wraparound" rule of (McNaughton, 1959). Imagine the processing times of the n jobs are laid out end-to-end, in arbitrary order, in a strip. Now cut the strip at intervals whose length is given by the lower bound. Each of the shorter strips becomes a row of the Gannt chart for an optimal schedule, as shown in Figure 5.

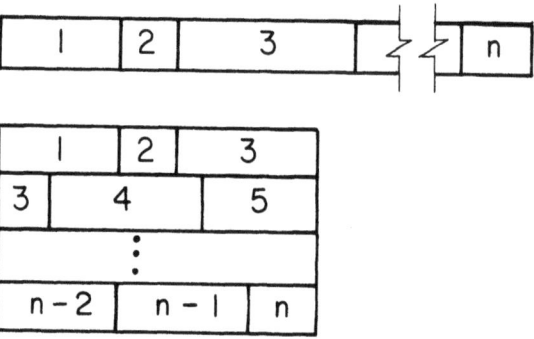

Fig. 5. McNaughton's "wraparound" rule

It is easy to see that this procedure requires $O(n)$ time and creates at most $m-1$ preemptions.

Now let us consider the case of uniform machines. Assuming $s_1 \geq s_2 \geq \ldots \geq s_m$, it is clear that we must have

$$C_{\max} \geq \max\left\{\max_{1 \leq k \leq m-1}\left\{\sum_{j=1}^{k} p_j \Big/ \sum_{i=1}^{k} s_i\right\}, \sum_{j=1}^{n} p_j \Big/ \sum_{i=1}^{m} s_i\right\} \quad (3)$$

It is possible to achieve the lower bound given by the right-hand size of this inequality, and to show this most easily, we now generalize the problem a bit.

Let us suppose that the speeds of the machines are time-varying and that the speed of machine i at time t is $s_i(t)$. Assume $s_1(t) \geq s_2(t) \geq \ldots \geq s_m(t)$, for all t. Define the *capacity* of machine i in the interval $[0, d]$ as

$$S_i = \int_0^d s_i(u)\,du.$$

Assume $p_1 \geq p_2 \geq \ldots \geq p_n$. Then, by the same reasoning behind (3), a necessary condition for there to exist a feasible schedule of length d is that all of the following inequalities be satisfied:

$$\begin{aligned} S_1 &\geq p_1 \\ S_1+S_2 &\geq p_1+p_2 \\ &\vdots \\ S_1+S_2+\ldots S_{m-1} &\geq p_1+p_2+\ldots+p_{m-1} \\ S_1+S_2+\ldots S_{m-1}+S_m &\geq p_1+p_2+\ldots+p_{m-1}+\ldots+p_n. \end{aligned} \qquad (4)$$

The inequalities simply assert that, for $k=1, 2, \ldots, m-1$, the sum of the capacities of the k fastest machines is at least as large as the sum of the processing requirements of the k largest jobs, and that the sum of the capacities of all m machines is at least as large as the sum of the requirements of all n jobs. Satisfaction of inequalities (4) is also a sufficient condition for the existence of a feasible schedule. Our proof is based on ideas of (Gonzalez & Sahni, 1978).

Let us choose an arbitrary job j to schedule in the interval $[0, d]$. Find the largest index k such that $S_k \geq p_j$. If $k=m$, then let machine $m+1$ be a dummy machine with zero speed in the discussion which follows. Let

$$f(t) = \int_0^t s_k(u)\,du, \quad g(t) = \int_t^d s_{k+1}(u)\,du.$$

From $f(0)=0$, $f(d) \geq p_j$, $g(0) < p_j$, $g(d)=0$, and the fact that f and g are continuous functions, it follows that there exists a t', $0 < t' \leq d$, such that $f(t')+g(t')=p_j$. Accordingly, schedule job j for processing by machine k in the interval $[0, t']$ and by machine $k+1$ in the interval $[t', d]$.

Next replace elementary machines k and $k+1$ by a *composite* machine formed from the remaining available time on those machines. This new machine has speed $s_{k+1}(t)$ in the interval $[0, t']$, speed $s_k(t)$ in $[t', d]$, and total processing capacity $S_k+S_{k+1}-p_j$ in the interval $[0, d]$. If $k<m$, create a dummy machine with zero speed. (We already created such a machine if $k=m$.) This gives us m machines with capacities S_i' as follows. If $k<m$, then

$$S_i' = \begin{cases} S_i & \text{for } i=1, \ldots, k-1 \\ S_k - S_{k+1} - p_j & \text{for } i=k, \\ S_{i+1} & \text{for } i=k+1, \ldots, m-1 \\ 0 & \text{for } i=m. \end{cases}$$

And if $k=m$, then

$$S_i' = \begin{cases} S_i & \text{for } i=1, \ldots, m-1 \\ S_m - p_j & \text{for } i=m. \end{cases}$$

The problem thus reduces to one involving m machines and $n-1$ jobs. The new machine capacities $S_i', i=1, 2, \ldots, m$, and processing requirements $p_1, p_2, \ldots, p_{j-1}, p_{j+1}, \ldots, p_n$ satisfy inequalities (4). (This can be verified by an analysis of the cases $j \leq k < m$, $k < j \leq m$, $k=m$, which we omit.) Induction on the number of jobs proves that repeated application of the scheduling rule yields a schedule in which all jobs are completed on time. (It can also be shown that such a schedule requires no more than $2(m-1)$ preemptions.) Satisfaction of inequalities (4) is thus a sufficient condition for the existence of a feasible schedule and we have proved the following result.

Theorem 5. Satisfaction of inequalities (4) is a necessary and sufficient condition for the existence of a schedule in which all n jobs are completed within the interval $[0, d]$.

5. Preemptive Scheduling with Due Dates

Now suppose the n given jobs have arbitrary due dates. Is it possible to schedule the jobs so that they are are all completed on time? The procedure we shall describe for solving this problem is adapted from (Sahni & Cho, 1980).

Let $d_0 = 0$, assume $d_0 < d_1 \leq d_2 \leq \ldots \leq d_n$ and let S_i denote the processing capacity of machine i in the interval $[0, d_1]$. If there exists a schedule in which all jobs are completed on time, then in any such schedule *all* of the processing requirement of job 1 and some amount $p'_j \leq p_j$ of the requirement of each job $j, j = 2, 3, \ldots, n$, is processed in the interval $[0, d_1]$. The unknown values $p'_j, j \neq 2$, together with the known value p_1 (appropriately reordered) satisfy inequalities (4). It follows that, if there exists a feasible schedule, there exists a feasible schedule in which job 1 is scheduled in accordance with the rule presented in the previous section.

So suppose we find the largest index k such that $S_k \geq p_1$, determine the value t', $0 < t' \leq d_1$, and schedule job 1. This leaves us with m machines with processing capacities S'_i, $i = 1, 2, \ldots, m$, in the interval $[0, d_1]$. We now also have m (composite) machines with well-defined speeds $s'_i(t)$, in the interval $[0, d_n]$, as follows. If $k < m$, then

$$s'_i(t) = \begin{cases} s_i(t) & (0 \leq t \leq d_n) \quad \text{for } i = 1, \ldots, k-1, \\ \left. \begin{array}{ll} s_{k+1}(t) & (0 \leq t \leq t') \\ s_k(t) & (t' < t \leq d_n) \end{array} \right\} & \text{for } i = k, \\ \left. \begin{array}{ll} s_{i+1}(t) & (0 \leq t \leq d_1) \\ s_i(t) & (d_1 < t \leq d_n) \end{array} \right\} & \text{for } i = k+1, \ldots, m-1 \\ \left. \begin{array}{ll} 0 & (0 \leq t \leq d_1) \\ s_m(t) & (d_1 < t \leq d_n) \end{array} \right\} & \text{for } i = m. \end{cases}$$

And if $k = m$, then

$$s'_i(t) \begin{cases} s_i(t) & (0 \leq t \leq d_n) \quad \text{for } i = 1, \ldots, m-1, \\ \left. \begin{array}{ll} 0 & (0 \leq t \leq t') \\ s_m(t) & (t' < t \leq d_n) \end{array} \right\} & \text{for } i = m. \end{cases}$$

The problem thus reduces to one involving m machines and $n-1$ jobs: There exists a feasible schedule for the n jobs $1, 2, \ldots, n$ on machines with speeds $s_i(t)$, $i = 1, 2, \ldots, m$, if and only if there exists a feasible schedule for the $n-1$ jobs $2, 3, \ldots, n$ on machines with speeds $s'_i(t)$, $i = 1, 2, \ldots, m$.

6. Preemptive Scheduling to Minimize the Number of Late Jobs

The algorithm described in the previous section provides a key to the solution of the problem of minimizing the weighted number of jobs that do not meet their deadlines, i.e. the problem $Q|pmtn|\Sigma w_j U_j$. The algorithm is rather complicated to explain and details can be found in (Lawler, 1979). We state only that the running time is $O(W^2 n^2)$ for $m=2$ and $O(W^2 n^{3m-5})$ for $m \leqslant 3$, where $W = \Sigma w_j$.

Note that for *fixed m* the algorithm is pseudopolynomial and that for fixed m and unweighted jobs, i.e. $W = n$, the algorithm is strictly polynomial This is probably the best that we can hope for, since the problem of minimizing the (unweighted) number of jobs is NP-hard, even for *identical* machines, if m is *variable*, i.e. specified as part of the problem instance.

Theorem 6. The problem $P|pmtn|\Sigma U_j$ is NP-hard.

Proof: An instance of the scheduling problem consists of n pairs of positive integers (p_j, d_j), $j = 1, 2, \ldots, n$, representing processing requirements and due dates of n jobs, a positive integer m, representing the number of identical machines, and a positive integer $k \leqslant n$. The question asked is: Does there exist a feasible preemptive schedule in which k (or more) jobs are on time?

We shall describe a polynomial transformation from the known NP-complete problem PARTITION. An instance of this problem consists of t positive integers a_i, $i = 1, 2, \ldots, t$, with $\Sigma a_i = 2b$. Is there a subset S such that

$$\sum_{i \in S} a_i = \sum_{i \notin S} a_i = b?$$

Given an instance of PARTITION, create an instance of the scheduling problem with $n = 5t - 2$ jobs as follows:
(1) t jobs with $p_j = b + a_j$, $d_j = b + 2a_j$ $(j = 1, \ldots, t)$;
(2) t jobs with $p_j = b$, $d_j = b$ $(j = t+1, \ldots, 2t)$;
(3) $3t - 2$ jobs with $p_j = b$, $d_j = 4b$ $(j = 2t+1, \ldots, 5t-2)$.
Let $m = t$ and $k = 4t - 1$.

If there exists a solution S to PARTITION, then there exists a solution to the scheduling problem. Let $|S| = s$ and construct a set of jobs consisting of the s jobs of type (1) whose indices are in S, any $t - s + 1$ jobs of type (2), and all $3t - 2$ jobs of type (3). We assert htat there is a feasible schedule for this set of $4t - 1$ jobs, by the following argument.

Each of the s jobs of type (1) and $t - s$ of the jobs of type (2) can be scheduled on a separate machine. Each type (2) job is processed continuously in the interval $[0, b]$. Each type (1) job j is processed continuously in the interval $[b, b + 2a_j]$, leaving a_j units of idle time on its machine in the interval $[0, b]$. The idle time remaining in $[0, b]$ can be distributed so that the $(t+1)$st type (2) job can be scheduled. For example, if $S = \{1, 2, 3\}$, with $a_1 + a_2 + a_3 = b$, the last type (2) job can be scheduled on one machine in the interval $[0, a_1]$, on a second machine in the interval $[a_1, a_1 + a_2]$, and on a third machine in the interval $[a_1 + a_2, b]$. After the $t+1$ jobs of type (1) and type (2) are scheduled, a total of

$(3t-2)b$ units of idle time remain on the t machines in the interval $[b, 4b]$, with at least $3b - 2\max_j\{a_j\} \geqslant b$ units of idle time on each machine. The $3t-2$ jobs of type (3) can be scheduled in "wrap-around" fashion in the idle time in the interval $[b, 4b]$.

Conversely, if there exists a solution to the scheduling problem, then there exists a solution to PARTITION, by the following argument. Suppose there exists a feasible set of size $4t-1$. Type (3) jobs have later due dates and no greater processing requirements than jobs of type (1) or (2). Hence by a simple substitution argument we may assume, without loss of generality, that the feasible set contains all $3t-2$ of the type (3) jobs. Let S denote the set of indices of the type (1) jobs. There are s, where $s = |S|$, type (1) jobs and $t-s+1$ type (2) jobs. The type (1) jobs require at least $sb - \sum_{j \in S} a_j$ units of processing in the interval $[0, b]$. It follows that the type (2) jobs can be scheduled only if $\sum_{j \in S} a_j \geqslant b$. The total amount of processing required by the type (1) and type (2) jobs is $(t+1)b + \sum_{j \in S} a_j$, which leaves $3(t-1)b - \sum_{j \in S} a_j$ units in the interval $[0, 4b]$ for the type (3) jobs. It follows that the type (3) jobs can be scheduled only if $\sum_{j \in S} a_j \leqslant b$ and we have $\sum_{j \in S} a_j = b$.

This NP-hardness proof differs significantly from previous proofs of NP-hardness for other preemptive scheduling problems. Virtually all other such proofs have proceeded by first showing that there is "no advantage" to preemption for the problem at hand, or for a restricted set of instances of the problem. That is, if there exists a feasible preemptive schedule there must also exist a feasible nonpreemptive schedule. The proof then proceeds by showing that the corresponding nonpreemptive scheduling problem (or an appropriately restricted set of instances of it) is NP-hard. We did not do this. And, in fact, there *is* advantage to preemption in our problem and the nonpreemptive version is well known to be NP-hard even if the number of identical machines is fixed at two.

7. A "Nearly On-Line" Algorithm for Preemptive Scheduling with Release Dates

The algorithm of Section 5 can be applied to solve the problem $Q|pmtn, r_j|C_{\max}$. (Simply turn the problem around and deadlines assume the role of release dates.) However, the algorithm is then "off line"; full knowledge of all problem parameters must be known in advance of computation. It can be shown that there is no fully "on-line" algorithm for the problem. However a "nearly" on-line algorithm is possible. By this we mean that at each release date r_j, the next release date r_{j+1} is known. However, there need be no knowledge of the processing times $p_{j+1}, \ldots, p_n$ nor the release dates $r_{j+2}, \ldots, r_n$.

What we want to accomplish in this. At time r_n we want the remaining processing requirements of jobs $1, 2, \ldots, n-1$, together with the processing requirement of job n, to be such that the value of C_{max} can be minimized. This will clearly be the case if the k longest of these requirements, for $k=1, 2, \ldots, m-1$, and for $k=n$, is as small as possible. The interesting fact is that this objective can be achieved at each release date $r_j, j=1, 2, \ldots, n$.

So the strategy is as follows. Having determined the schedule up to time r_j, we then determine the schedule for the interval $[r_j, r_{j+1}]$ so that the remaining processing requirements of jobs $1, 2, \ldots, j$ at time r_{j+1} are as "uniform" as possible. Without going into details, we simply indicate that at time r_j we have a "staircase" of remaining processing times, from longest to shortest, as indicated in Figure 6. We then choose to process an amount of each job in the interval $[r_j, r_{j+1}]$, as indicated by the shaded region in the figure. These amounts are chosen to satisfy the requirement that the sum of the k longest remaining processing requirements be as small as possible, for all k.

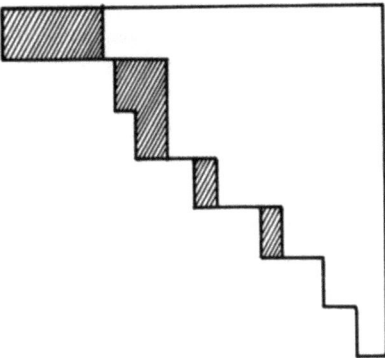

Fig. 6. "Staircase" of remaining processing times

Details may be found in (Sahni & Cho, 1979; and Labetoulle, et al, 1979). The nearly on-line algorithm requires $O(n \log n + mn)$ time. At most $O(mn)$ preemptions are introduced into the schedule.

8. Preemptive Scheduling with Release Dates and Due Dates

Consider the problem of preemptively scheduling jobs on identical parallel machines subject to release dates and deadlines. This was given a network flow formulation in (Horn, 1974).

Let $E = \{r_j\} \cup \{\bar{d}_j\}$ and order the numbers in E. This yields at most $2n-1$ time intervals $[e_h, e_{h+1}]$. Now create a flow network with a node for each job j, a node for each time interval $[e_h, e_{h+1}]$, a dummy source s and a dummy sink t. There is an arc (s, j) from the source to each job node j, and the capacity of this arc is p_j. There is an arc from job node j to the node for interval $[e_h, e_{h+1}]$ if and only if job j can be processed in the interval, i.e. $r_j \leq e_h$ and $e_{h+1} \leq \bar{d}_j$, and the capacity of this arc is $e_{h+1} - e_h$. There is an arc from each interval node $[e_h, e_{h+1}]$ to the sink, and the capacity of this arc is $m(e_{h+1} - e_h)$. We assert that there exists a feasible preemptive schedule if and only if there exists a flow of value $P = \Sigma p_j$ in this network.

In the case of uniform parallel machines, it is not possible to give such a simple formulation to the problem. What are needed are capacity constraints on *sets* of arcs into a given interval node $[e_h, e_{h+1}]$. In particular, it is necessary that the sum of the k largest flows not exceed $\left(\sum_{i=1}^{k} s_i\right)(e_{h+1} + e_h)$ for $k = 1, 2, \ldots, m-1$, and the arc to the sink is given capacity $\left(\sum_{i=1}^{m} s_i\right)(e_{h+1} - e_h)$.

What is interesting is that these capacity constraints are *submodular* and the network is an example of a so-called "polymatroidal" flow network. See (Martel, 1981) for a solution to the specific scheduling problem and (Lawler & Martel, 1982 A, B) for a discussion of polymatroidal network flows in general.

9. Nonpreemptive Scheduling of Unit-Time Jobs Subject to Precedence Constraints

Suppose that n unit-time jobs are to be processed on m identical parallel machines. Each job can be assigned to any machine, and the schedule has to respect given precedence constraints. The objective is to find a schedule which minimizes the maximum of job completion times.

This problem is NP-hard in general, but it can be solved in polynomial time if either the precedence constraints are in the form of a rooted tree or if there are only two machines.

First, let us assume that the precedence constraints are in the form of an *intree*, i.e. each job has exactly one immediate succesor, except for one job which has no successors and which is called the *root*. It is possible to minimize the maximum completion time in $O(n)$ time by applying an algorithm due to Hu (Hu, 1961). The *level* of a job is defined as the number of jobs in the unique path to the root. At the beginning of each time unit, as many available jobs as possible are scheduled on the m machines, where highest priority is granted to the jobs with the largest levels. Thus, Hu's algorithm is a *list scheduling* algorithm, whereby at each step the available job with the highest ranking on a priority list is assigned to the first machine that becomes available. It can also be viewed as a *critical path scheduling* algorithm: the next job chosen is the one which heads the longest current chain of unexecuted jobs.

To validate Hu's algorithm, we will show that, if it yields a schedule of length t^*, then no feasible schedule of length $t < t^*$ exists.

Choose any $t < t^*$ and define a label for each job by subtracting its level from t; note that the root has label t and that each other job has a label one less than its immediate sucessor. The algorithm gives priority to the jobs with the smallest labels. Since it yields a schedule of length larger than t, in some unit-time interval s a job is scheduled with a label smaller than s. Let s be the earliest such interval and let there be a job with label $l < s$ scheduled in it. We claim that there are m jobs scheduled in each earlier interval $s' < s$. Suppose there is an interval $s' < s$ with fewer than m jobs scheduled. If $s' = s - 1$, then the only reason that the job with label l was not scheduled in s' could have been that an immediate predecessor would have label $l-1 \leq s-1$; which contradicts the definition of s. If $s' < s-1$, then there are fewer jobs scheduled in s' than in $s'+1$, which is impossible from the structure of the intree. Hence, each interval $s' \leq s$ has m jobs scheduled. Since each of these jobs has a label smaller than s, at least one job with a label smaller than s must be scheduled in interval s, so that there is no feasible schedule of length $t < t^*$ possible. This completes the correctness proof of Hu's algorithm.

An alternative linear-time algorithm for this problem has been proposed by Davida and Linton (Davida & Linton, 1976). Assume that the precedence constraints are in the form of an *outtree*, i.e. each job has at most one immediate predecessor. The *weight* of a job is defined as the total number of its successors. The jobs are now scheduled according to decreasing weights.

If the problem is to minimize the *maximum lateness* with respect to given due dates rather than the maximum completion time, then the case that the precedence constraints are in the form of an *intree* can be solved by an adaptation of Hu's algorithm, but the case of an *outtree* turns out to be NP-hard (Brucker, et al, 1977). Polynomial-time algorithms and NP-hardness results for the maximum completion time problem with various other special types of precedence constraints are reported in (Dolev, 1981; Garey, et al, 1981; Warmuth, 1980).

Next, let us assume that there are *arbitrary precedence constraints* but only *two machines*. It is possible to minimize the maximum completion time by a variety of algorithms.

The earliest and simplest approach is due to Fujii, Kasami and Ninomiya (Fujii, et al, 1969, 1971). A graph is constructed with vertices corresponding to jobs and edges $\{j, k\}$ whenever jobs j and k can be executed simultaneously, i.e. $j \not\to k$ and $k \not\to j$. A *maximum cardinality matching* in this graph, i.e. a maximum number of disjoint edges, is then used to derive an optimal schedule; if the matching contains c pairs of jobs, the schedule has length $n-c$. Such a matching can be found in $O(n^3)$ time.

A completely different approach (Coffman & Graham, 1972) leads to a *list scheduling* algorithm. The jobs are labeled in the following way. Suppose labels $1, \ldots, l$ have been applied and S is the subset of unlabeled jobs all of whose successors have been labeled. Then a job in S is given the label $l+1$ if the labels of its immediate successors are *lexicographically minimal* with respect to all jobs in S. The priority list is formed by ordering the jobs according to decreasing labels. This method requires $O(n^2)$ time.

Recently, an even more efficient algorithm has been developed by Gabow (Gabow, 1980). His method uses labels, but with a number of rather sophisticated embellishments. The running time can be made strictly linear in $n+a$, where a is the number of arcs in the precedence graph (Gabow, 1982 B).

If the problem is to find a *feasible* two-machine schedule under arbitrary precedence constraints when each job becomes available at a given integer *release date* and has to meet a given integer *deadline*, polynomial-time algorithms exist (Garey & Johnson, 1976, 1977). These algorithms can be applied to minimize *maximum lateness* in polynomial time.

It was shown in (Ullman, 1975) that the problem of minimizing maximum completion time for n unit-time jobs on m identical parallel machines is NP-hard. (Cf. (Lenstra & Rinnooy Kan, 1978) or (Lawler & Lenstra, 1982) for a simpler proof.) However, this NP-completeness proof requires that the number of machines m be specified as part of the problem instance. The status of the problem for any *fixed* number of machines, in particular three, is open.

Open Problem: What is the status of the problem $P3|prec, p_j=1|C_{max}$? (Scheduling unit-time jobs on three identical parallel machines subject to arbitrary precedence constraints, so as to minimize maximum completion time.) Easy or NP-hard?

10. Preemptive Scheduling Subject to Precedence Constraints

In the previous section we discussed problems involving the *nonpreemptive* scheduling of *unit-time* jobs. There has been a parallel investigation of problems concerning *preemptive* scheduling of jobs of *arbitrary length*. Interesting, there is a preemptive counterparts for virtually all of the unit-time scheduling algorithms. In particular, in (Gonzalez & Johnson, 1980), a preemptive algorithm is given which is analogous to that of (Davida & Linton, 1976). In (Lawler, 1982A) algorithms are proposed that are the preemptive counterparts of those found in (Brucker, et al, 1977) and (Garey & Johnson, 1976, 1977). These preemptive scheduling algorithms employ essentially the same techniques for dealing with precedence constraints as the corresponding algorithms for unit-time jobs. However, they are considerably more complex, and we shall not deal with them here.

IV. Open Shops and Unrelated Parallel Machines

1. Introduction

Problems involving unrelated parallel machines have been solved primarily by linear programming techniques, the validity of which depends on results from the theory of open shops. Moreover, it turns out that it is possible to solve scheduling problems for models that generalize both open shops and unrelated parallel machine shops (Lawler, Luby & Vazirani, 1982). Accordingly, we shall consider these two types of shops together.

2. Open Shops

In an *open shop* there are n *jobs,* to be scheduled for processing by m *machines.* Each machine is to work on job j for a total *processing time* p_{ij}. A machine can work on only one job at a time and a job can be worked on by only one machine at a time. There is no restriction on the order in which a given job can be worked on by the different machines or on the order in which a machine can work on different jobs. (Hence the term "open shop".)

We let the letter "O" denote an open shop. It is easily established that $O\|C_{max}$ is NP-hard. However, the two-machine problem, $O2\|C_{max}$, can be solved in $O(n)$ time. Moreover, it can be established that there is no advantage to preemption in the two-machine case, so the linear time algorithm of (Gonzalez & Sahni, 1978) for $O2\|C_{max}$ solves $O2\|pmtn|C_{max}$ as well.

We shall now describe the algorithm of (Gonzalez & Sahni, 1978) for $O|pmtn|C_{max}$. The description is adapted from (Lawler & Labetoulle, 1978).

As noted, an instance of the $O|pmtn|C_{max}$ problem is defined by an $m \times n$ matrix $P = (p_{ij})$. It is evident that a lower bound on the length of a feasible schedule is given by

$$C_{max} = \max\left\{\max_i\left\{\sum_j p_{ij}\right\}, \max_j\left\{\sum_i p_{ij}\right\}\right\}. \tag{5}$$

We shall show how to construct a feasible schedule with this value of C_{max}. Since no schedule can be shorter, the constructed schedule will clearly be optimal.

Let us call row i (column j) of matrix P *tight* if $\sum_j p_{ij} = C_{max}$ ($\sum_i p_{ij} = C_{max}$), and *slack* otherwise. Suppose we are able to find a subset of strictly positive elements of P, with exactly one element of the subset in each tight row and in each tight column and no more than one element in any slack row or column. We shall call such a subset a *decrementing set*, and use it to construct a *partial schedule* of length δ, for some suitably chosen $\delta > 0$. In this partial schedule processor i works on job j for $\min\{p_{ij}, \delta\}$ units of time, for each element p_{ij} in the decrementing set. We then replace p_{ij} by $\max\{0, p_{ij} - \delta\}$, for each element in the decrementing set, thereby obtaining a new matrix P', for which $C'_{max} = C_{max} - \delta$ satisfies condition (5).

For example, suppose $C_{max} = 11$ and

$$P = \begin{pmatrix} 3 & ④ & 0 & 4 \\ ④ & 0 & 6 & 0 \\ 4 & 0 & 0 & ⑥ \end{pmatrix} \begin{matrix} 11 \\ 10 \\ 10 \end{matrix},$$

$$\begin{matrix} 11 & 4 & 6 & 10 \end{matrix}$$

with row and column sums as indicated on the margins of the matrix. One possible decrementing set is indicated by the encircled elements. Choosing $\delta = 4$, we obtain $C'_{max} = 7$ and P' as shown below, with the partial schedule indicated to the right:

$$P' = \begin{pmatrix} ③ & 0 & 0 & 4 \\ 0 & 0 & ⑥ & 0 \\ 4 & 0 & 0 & ② \end{pmatrix} \begin{matrix} 7 \\ 6 \\ 6 \end{matrix} \quad \begin{matrix} \boxed{2} \\ \boxed{1} \\ \boxed{4} \end{matrix}$$

$$\begin{matrix} 7 & 0 & 6 & 6 \end{matrix} \qquad\qquad\quad 4$$

A decrementing set of P' is indicated by the encircled elements.

There are various constraints that must be satisfied by δ, in order for $C'_{max} = C_{max} - \delta$ to satisfy condition (5) with respect to P'. First, if p_{ij} is an element of the decrementing set in a tight row or column, then clearly it is necessary that $\delta \leq p_{ij}$, else there will be a row or column sum of P which is strictly greater than $C_{max} - \delta$. Similarly, if p_{ij} is an element of the decrementing set in a slack row (slack column), then it is necessary that

$$\delta \leq p_{ij} + C_{max} - \sum_k p_{ik} \quad \left(\delta \leq p_{ij} + C_{max} - \sum_k p_{kj}\right).$$

And if row i (column j) contains no element of the decrementing set (and is therefore necessarily slack), it is necessary that

$$\delta \leq C_{max} - \sum_j p_{ij} \quad \left(\delta \leq C_{max} - \sum_i p_{ij}\right).$$

Thus for the example above we have

$$\delta \leq p_{12} = 4, \quad \delta \leq p_{21} = 4,$$
$$\delta \leq p_{34} + C_{max} - \sum_k p_{3k} = 7, \quad \delta \leq t_{34} + C_{max} - \sum_k p_{k4} = 7$$
$$\delta \leq C_{max} - \sum_k p_{k3} = 5.$$

Suppose δ is chosen to be maximum, subject to conditions indicated above. Then either P' will contain at least one less strictly positive element than P or else P' will contain at least one more tight column or tight row (with respect to C'_{max}) than P. It is thus apparent that no more than $r + m + n$ iterations, where r is the number of strictly positive elements in P, are necessary to construct a feasible schedule of length C_{max}.

To illustrate this point, we continue with the example. Choosing $\delta = 3$, we obtain from P' the matrix P', with the augmented partial schedule shown to the right:

$$P'' = \begin{pmatrix} 0 & 0 & 0 & \textcircled{4} & 4 \\ 0 & 0 & \textcircled{3} & 0 & 3 \\ \textcircled{4} & 0 & 0 & 0 & 4 \\ 4 & 0 & 3 & 4 \end{pmatrix} \quad \begin{array}{|c|c|} \hline 2 & 1 \\ \hline 1 & 3 \\ \hline 4 & 4\,\emptyset \\ \hline \end{array}$$
$$\qquad\qquad\qquad\qquad\qquad 4\;6\;7$$

(The symbol "$\emptyset$" indicates idle time.) The final decrementing set yields the following complete schedule:

2	1	4	
1	3	3	$\emptyset$
4	4 $\emptyset$	1	

$$\qquad 4\;\;6\;\;7\;\;10\;\;11$$

To complete our proof, we need the following lemma.

Lemma 7. *For any nonnegative matrix P and C_{max} satisfying condition (5), there exists a decrementing set.*

Proof: From the $m \times n$ matrix P construct an $(m+n) \times (m+n)$ matrix U, as indicated below:

$$U = \left(\begin{array}{c|c} P & D_m \\ \hline D_n & P^t \end{array} \right)$$

Here P^t denote the transpose of P, D_m and D_n are $m \times m$ and $n \times n$ diagonal matrices of nonnegative "slacks", determined in such a way that each row sum and column sum of U is equal to $C_{\max}$. It follows that $(1/C_{\max}) U$ is a doubly stochastic matrix. The well-known Birkhoff-von Neumann theorem states that a doubly stochastic matrix is a convex combination of permutation matrices. It is easily verified that any one of the permutation matrices in such a convex combination is identified with a decrementing set of P. □

There are several possible ways to construct a decrementing set. For our purposes, it is sufficient to note that one can construct the matrix U from P and then solve an assignment problem over U, which can be done in polynomial time. This observation, together with the observation that no more than a polynomial number of such assignment problems need be solved, is sufficient to establish a polynomial bound for the schedule construction procedure. In (Gonzalez & Sahni, 1976) time bounds of $O(r^2)$ and $O(r(\min\{r, m^2\} + m \log n))$, are obtained, where r is the number of strictly positive elements in P.

3. Preemptive Scheduling of Unrelated Parallel Machines

Let us now consider $R|pmtn|C_{\max}$, the problem of finding a minimum-length preemptive schedule for unrelated parallel machines. Let p_{ij} denote the total amount of time that machine i is to work on machine j. We assume that the processing requirements have been normalized so that for each job j we must have $\sum_i s_{ij} p_{ij} = 1$. It is evident that the values of $C_{\max}$ and p_{ij} for any feasible schedule must constitute a feasible solution to the following linear programming problem (Lawler & Labetoulle, 1978):

$$\text{minimize } C_{\max}$$
$$\text{subject to}$$
$$\sum_i s_{ij} p_{ij} = 1 \quad j = 1, \ldots, n$$
$$\sum_i p_{ij} \leq C_{\max} \quad j = 1, \ldots, n$$
$$\sum_j p_{ij} \leq C_{\max} \quad i = 1, \ldots, m$$
$$p_{ij} \geq 0.$$

It follows from the results of the previous section that the converse is also true: For any feasible solution to this linear programming problem there is a feasible preemptive schedule with the same values of p_{ij} and $C_{\max}$. Moreover, such a schedule can be constructed in polynomial time.

Because of the existence of the ellipsoid method for linear programming, the linear programming formulation of $R|pmtn|C_{\max}$ can be regarded yielding a polynomial-bounded algorithm for the problem. In the case of two machines, $R2|pmtn|C_{max}$, the linear programming problem can be solved in $O(n)$ time by special techniques (Gonzalez, et al, 1982).

It is a fairly straightforward matter to formulate a linear programming problem for $R|pmtn|L_{max}$ and $R|r_j, pmtn|C_{max}$. Moreover, $R|r_j, pmtn|L_{max}$ can be solved as a series of linear programming problems.

Open Problem: An upper bound of $4m^2 - 5m + 2$ on the number of preemptions required for an optimal schedule for $R|pmtn|C_{max}$ is established in (Lawler & Labetoulle, 1978). However, this upper bound is certainly not tight. It is known that no more than two preemptions are required for the case $m = 2$ (Gonzalez, et al, 1981), and no example is known to require more than $2(m-1)$ preemptions. What is a tight upper bound?

4. Minimizing the Sum of Completion Times

The problem $R||\Sigma C_j$ is solved by the technique of (Bruno & Coffman, 1976). Note that if jobs $1, \ldots, k$ are performed by machine i in that order, we have for the sum of the completion times of the jobs performed on that machine,

$$\Sigma C_j = \frac{k}{s_{i1}} + \frac{(k-1)}{s_{i2}} + \ldots + \frac{2}{s_{i,k-1}} + \frac{1}{s_{ik}}.$$

In other words, if job j is the last job performed on machine i, its contribution to the objective function is $\frac{1}{s_{ij}}$, if j is the second-to-last job on machine i, its contribution $\frac{2}{s_{ij}}$, and so forth. A little reflection shows that we can formulate the problem $R||\Sigma C_j$ as follows.

Let $S' = \left(\frac{1}{s_{ij}}\right)$ and let S'' be the matrix obtained by stacking multiples of S', as shown in Figure 7. A solution to the problem is obtained by finding a subset of elements of S'' of minimum total value, subject to the conditions that at least one element is chosen from each column of S'' (each job is performed by some machine) and at most one element is chosen from each row (no two jobs are

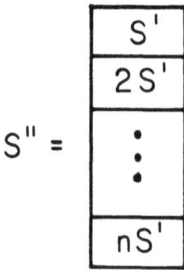

Fig. 7. Matrix S''

performed in the same position on a given machine). In other words, the problem reduces to a simple transportation problem.

Curiously, no good algorithm is known for $R|pmtn|\Sigma C_j$. If a "birdie" were to tell us the order in which jobs are to be completed in an optimal schedule, we could easily formulate and solve a linear programming problem to find such an optimal schedule. But this observation is not much help.

Open Question: What is the status of $R|pmtn|\Sigma C_j$? Easy of NP-hard? (If this problem is NP-hard, it will be the first such problem known to the author in which the nonpreemptive version is easy.)

V. Flow Shops and Job Shops

1. Introduction

Flow shops and *job shops* are much like open shops, except that for each job there is a prescribed order in which the job must be processed by the m machines. In a flow shop this order is the same for all jobs, i.e. each job must first be processed by machine 1, then machine 2, ... and finally by machine m. In a job shop, the prescribed order may differ from job to job. (Quite commonly, job shops are defined in terms of "operations" which must be performed on a given job in a prescribed order, and two or more operations may be performed by the same machine. However, we are not concerned with this distinction here.)

We shall not have much to say about flow shops or job shops in this survey paper, because nearly all flow shops and job shop problems are NP-hard. Essentially the only known polynomial algorithms are for special cases of the two-machine flow shop problem. Moreover, empirical work has shown that flow shop and job shop problems, other than these special cases, are extremely difficult to solve.

2. The Two-Machine Flow Shop

One of the most significant pioneering works of scheduling theory is (Johnson, 1954), in which an $O(n\log n)$ procedure was given for minimizing makespan in the two-machine flow shop. Johnson's results have been extended to encompass "time-lags" (Mitten, 1958), and series parallel precedence constraints (Sidney, 1979).

3. The "No-Wait" Two-Machine Flow Shop

The "no wait" flow shop is one in which each job must start its processing on machine $i+1$ the instant its processing is completed on machine i. Interestingly, the two-machine "no wait" problem is actually a special case of the traveling salesman problem that can be solved efficiently (Reddi & Ramamoorthy, 1972; Gilmore & Gomory, 1964).

4. The General Job Shop Problem

The general job shop problem is one of the most computationally intractable combinatorial problems existing. One indication of this is given by the fact that a certain 10-job, 10-machine problem formulated in 1963 (Muth & Thompson, 1963) still has not been solved, despite the efforts of many investigators. This challenge for future research perhaps provides us with an appropriate note on which to end this survey.

References

D. L. Adolphson, "Single Machine Job Scheduling With Precedence Constraints", *SIAM J. Comput.*, 6 (1977) 40–54.

D. L. Adolphson, T. C. Hu, "Optimal Linear Ordering", *SIAM J. Appl. Math.*, 25 (1973) 403–423.

H. M. Abdel-Wahab, T. Kameda, "Scheduling to Minimize Maximum Cumulative Cost Subject to Series Parallel Precedence Constraints", *Operations Res.*, 26 (1978) 141–158.

K. R. Baker, *Introduction to Sequencing and Scheduling*, Wiley, New York, 1974.

K. R. Baker, L. E. Schrage, "Finding an Optimal Sequence by Dynamic Programming: An Extension to Precedence-Constrained Tasks", *Operations Res.*, 26 (1978) 111–120.

K. R. Baker, E. L. Lawler, J. K. Lenstra, A. H. G. Rinnooy Kan, "Preemptive Scheduling of a Single Machine to Minimize Maximum Cost Subject to Release Dates and Precedence Constraints", (1982) to appear in *Operations Res.*

P. Brucker, M. R. Garey, D. S. Johnson, "Scheduling Equal-Length Tasks under Tree-Like Precedence Constraints to Minimize Maximum Lateness", *Math. Operations Res.*, 2 (1977) 275–284.

E. G. Coffman, Jr., R. L. Graham, "Optimal Scheduling for Two-Processor Systems", *Acta Informat.*, 1 (1972) 200–213.

R. W. Conway, W. L. Maxwell, L. W. Miller, *Theory of Scheduling*, Addison-Wesley, Reading, Mass. 1967.

G. I. Davida, D. J. Linton, "A New Algorithm for the Scheduling of Tree Structured Tasks", *Proc. Conf. Inform. Sci. and Systems.*, Baltimore, MD., 1976.

D. Dolev, "Scheduling Wide Graphs", (1981), unpublished manuscript.

H. Emmons, "One-Machine Sequencing to Minimize Certain Functions of Job Tardiness", *Operations Res.*, 17 (1969) 701–715.

M. L. Fisher, "A Dual Algorithm for the One-Machine Scheduling Problem", *Math. Programming*, 11 (1976) 229–251.

M. L. Fisher, B. J. Lageweg, J. K. Lenstra, A. H. G. Rinnooy Kan, "Surrogate Duality Relaxation for Job Shop Scheduling", Report, *Mathematisch Centrum*, Amsterdam, 1981.

M. Fujii, T. Kasami, K. Ninomiya, "Optimal Sequencing of Two Equivalent Processors", *SIAM J. Appl. Math.*, 17 (1969) 784-789; *Erratum*, 20 (1971) 141.

H. N. Gabow, "An Almost-Linear Algorithm for Two-Processor Scheduling", *J. Assoc. Comput. Mach.*, 29 (1982A) 766-780.

H. N. Gabow, private communication, 1982B.

M. R. Garey, "Optimal Task Sequencing with Precedence Constraints", *Discrete Math.*, 4 (1973) 37-56.

M. R. Garey, D. S. Johnson, "Scheduling Tasks with Nonuniform Deadlines on Two Processors", *J. Assoc. Comput. Mach.*, 23 (1976) 461-467.

M. R. Garey, D. S. Johnson, "Two-Processor Scheduling with Start-Times and Deadlines", *SIAM J. Comput.*, 6 (1977) 416-426.

M. R. Garey, D. S. Johnson, R. E. Tarjan, M. Yannakakis, "Scheduling Opposing Forests", unpublished manuscript, (1981).

P. C. Gilmore, R. E. Gomory, "Sequencing a One-State Variable Machine: A Solvable Case of the Traveling Salesman Problem", *Oper. Res.*, 12 (1964) 655-679.

T. Gonzalez, "Optimal Mean Finish Time Preemptive Schedules", Technical Report 220, (1977) Computer Science Department, Pennsylvania State University.

T. Gonzalez, "A Note on Open Shop Preemptive Schedules", *IEEE Trans. Computers*, C-28 (1979) 782-786.

T. Gonzalez, D. B. Johnson, "A New Algorithm for Preemptive Scheduling of Trees", *J. Assoc. Comput. Mach.*, 27 (1980) 287-312.

T. Gonzalez, E. L. Lawler, S Sahni, "Optimal Preemptive Scheduling of Two Unrelated Processors in Linear Time", (1981), to appear.

T. Gonzalez, S. Sahni, "Open Shop Scheduling to Minimize Finish Time", *J. Assoc. Comput. Mach.*, 23 (1976) 665-679.

T. Gonzalez, S. Sahni, "Preemptive Scheduling of Uniform Processor Systems", *J. Assoc. Comput. Mach.*, 25 (1978) 92-101.

R. L. Graham, E. L. Lawler, J. K. Lenstra, A. H. G. Rinnooy Kan, "Optimization and Approximation in Deterministic Sequencing and Scheduling : A Survey", *Ann. Discrete Math.*, 5 (1979) 287-326.

M. Held, R. M. Karp, "A Dynamic Programming Approach to Sequencing Problems", *SIAM J. Appl. Math.*, 10 (1972) 196-210.

W. A. Horn, "Single-Machine Job Sequencing with Treelike Precedence Ordering and Linear Dealy Penalties", *SIAM J. Appl. Math.*, 23 (1972) 189-202.

W. A. Horn, "Minimizing Average Flow Time with Parallel Machines", *Oper. Res.*, 21 (1973) 846-847.

W. A. Horn, "Some Simple Scheduling Algorithms", *Naval Res. Logist. Quart.*, 21 (1974) 177-185.

E. Horowitz, S. Sahni, "Exact and Approximate Algorithms for Scheduling Nonidentical Processors", *J. Assoc. Comput. Mach.*, 23 (1976) 317-327.

T. C. Hu, "Parallel Sequencing and Assembly Line Problems", *Oper. Res.*, 9 (1961) 841-848.

J. R. Jackson, "Scheduling a Production Line to Minimize Maximum Tardiness", Research Report 43, (1955) Management Science Research Project, University of California, Los Angeles.

J. R. Jackson, "An Extension of Johnson's Results on Job Lot Scheduling", *Naval Res. Logist. Quart.*, 3 (1956) 201-203.

S. M. Johnson, "Optimal Two- and Three-Stage Production Schedules with Setup Times Included", *Naval Res. Logist. Quart.*, 1 (1954) 61-68.

S. M. Johnson, "Discussion: Sequencing n Jobs on Two Machines with Arbitrary Time Lags", *Management Sci.*, 5 (1958) 299-303.

E. P. C. Kao, M. Queyranne, "On Dynamic Programming Methods for Assembly Line Balancing", working paper, Dept Quantitative Management Sci., University of Houston, Texas, (1980).

H. Kise, T. Ibaraki, H. Mine, "A Solvable Case of the One-Machine Scheduling Problem with Ready and Due Times", *Oper. Res.*, 26 (1978) 121–126.

J. Labetoulle, E. L. Lawler, J. K. Lenstra, A. H. G. Rinnooy Kan, "Preemptive Scheduling of Uniform Machines Subject to Release Dates", Report BW 99, (1979) Mathematisch Centrum, Amsterdam.

B. J. Lageweg, E. L. Lawler, J. K. Lenstra, A. H. G. Rinnooy Kan, "Computer Aided Complexity Classification of Deterministic Scheduling Problems", Report BW 138, (1981) Mathematisch Centrum, Amsterdam.

E. L. Lawler, "On Scheduling Problems with Deferral Costs", *Management Science*, 11 (1964) 270–288.

E. L. Lawler, "Optimal Sequencing of a Single Machine Subject to Precedence Constraints", *Management Sci.*, 19 (1973) 544–546.

E. L. Lawler, "Sequencing to Minimize the Weighted Number of Tardy Jobs", *RAIRO Rech. Oper.*, 10 suppl. (1976) 27–33.

E. L. Lawler, "A 'Pseudopolynomial' Algorithm for Sequencing Jobs to Minimize Total Tardiness", *Ann. Discrete Math.*, 1 (1977) 331–342.

E. L. Lawler, "Sequencing Jobs to Minimize Total Weighted Completion Time Subject to Precedence Constraints", *Ann. Discrete Math.*, 2 (1978 A) 75–90.

E. L. Lawler, "Sequencing Problems With Series Parallel Precedence Constraints", to appear in *Proc. Confer. on Combinatorial Optimization*, (N. Christofides, ed.), Urbino, Italy, 1978 B.

E. L. Lawler, "Efficient Implementation of Dynamic Programming Algorithms for Sequencing Problems", Report BW 106/79, (1979 A) Mathematisch Centrum, Amsterdam.

E. L. Lawler, "Preemptive Scheduling of Uniform Parallel Machines to Minimize the number of Late Jobs", Report BW 105, (1979 B) Mathematisch Centrum, Amsterdam.

E. L. Lawler, "Preemptive Scheduling of Precedence-Constrained Jobs on Parallel Machines", in *Deterministic and Stochastic Scheduling*, (M. A. H. Dempster, et al., eds.), D. Reidel, Dordrecht, Holland, 1982 A, pp. 101–124.

E. L. Lawler, "A Fully Polynomial Approximation Scheme for the Total Tardiness Problems", (1982 B), submitted for publication.

E. L. Lawler, "On Scheduling a Single Machine to Minimize the Number of Late Jobs", (1982 C), submitted for publication.

E. L. Lawler, J. Labetoulle, "On Preemptive Scheduling of Unrelated Parallel Processors by Linear Programming", *J. Assoc. Comput. Mach.*, 25 (1978) 612–619.

E. L. Lawler, J. K. Lenstra, A. H. G. Rinnooy Kan, "Minimizing Maximum Lateness in a Two-Machine Open Shop, *Math. Oper. Res.*, 6 (1981) 153–158.

E. L. Lawler, C. U. Martel, "Scheduling Periodically Occurring Tasks on Multiple Professors", *Info. Proc. Letters*, 12 (1981) 9–12.

E. L. Lawler, C. U. Martel, "Computing 'Polymatroidal' Network Flows", *Math. of Operations Res.*, 7 (1982 A) 334–347.

E. L. Lawler, C. U. Martel, "Flow Network Formulations of Polymatroid Optimization Problems", *Annals Discrete Math.*, 16 (1982 B) 189–200.

E. L. Lawler, J. K. Lenstra, "Machine Scheduling with Precedence Constraints", in *Ordered Sets*, (I. Rival, ed.), D. Reidel, Dordrecht, Holland, 1982, pp. 655–675.

E. L. Lawler, J. K. Lenstra, A. H. G. Rinnooy Kan, "Recent Developments in Deterministic Sequencing and Scheduling: A Survey", in *Deterministic Sequencing and Sched-

uling, (M. A. H. Dempster, et al., eds.), D. Reidel Co., Dordrecht, Holland, 1982 A, pp. 35-74.

E. L. Lawler, J. K. Lenstra, A. H. G. Rinnooy Kan, "A Gift for Alexander!: At Play in the Fields of Scheduling Theory", *OPTIMA,* Mathematical Programming Society Newsletter, No. 7, (1982 B).

E. L. Lawler, M. G. Luby, V. V. Vazirani, "Scheduling Open Shops with Parallel Machines", (1982), to appear in *Operations Res. Letters.*

E. L. Lawler, J. M. Moore, "A Functional Equation and Its Application to Resource Allocation and Sequencing Problems", *Management Sci.,* 16 (1969) 77-84.

E. L. Lawler, B. D. Sivazlian, "Minimization of Time Varying Costs in Single Machine Scheduling", *Operations Res.,* 26 (1978) 563-569.

J. K. Lenstra, *Sequencing by Enumerative Methods,* Mathematical Centre Tracts 69, Mathematisch Centrum, Amsterdam, 1977.

J. K. Lenstra, A. H. G. Rinnooy Kan, "Complexity of Scheduling under Precedence Constraints", *Operations Res.,* 26 (1978) 22-35.

C. Martel, "Scheduling Uniform Machines With Release Times, Deadlines and Due Times", *J. Assoc. Comput. Mach.,* (1981), to appear.

R. McNaughton, "Scheduling With Deadlines and Loss Functions", *Management Sci.,* 6 (1959) 1-12.

L. G. Mitten, "Sequencing n Jobs On Two Machines With Arbitrary Time Lags", *Management Sci.,* 5 (1958) 293-298.

C. L. Monma, A. H. G. Rinnooy Kan, "Efficiently Solvable Special Cases of the Permutation Flow-Shop Problem", Report 8105, (1981) Erasmus University, Rotterdam.

C. L. Monma, J. B. Sidney, "Sequencing With Series-Parallel Precedence Constraints", *Math. Oper. Res.,* 4 (1979) 215-224.

J. M. Moore, "An n Job, One Machine Sequencing Algorithm for Minimizing the Number of Late Jobs", *Management Sci.,* 15 (1968) 102-109.

J. F. Muth, G. L. Thompson, eds., *Industrial Scheduling,* Prentice Hall, Englewood Cliffs, N. J., 1963, p. 236.

S. S. Reddi, C. V. Ramamoorthy, "On the Flow-Shop Sequencing Problem With No Wait in Process", *Oper. Res. Quart.,* 23 (1972) 323-331.

M. H. Rothkopf, "Scheduling Independent Tasks on Parallel Processors", *Management Sci.,* 12 (1966) 437-447.

S. Sahni, Y. Cho, "Nearly On Line Scheduling of a Uniform Processor System With Release Times", *SIAM J. Comput.,* 8 (1979) 275-285.

S. Sahni, Y. Cho, "Scheduling Independent Tasks With Due Times on a Uniform Processor System", *J. Assoc. Comput. Mach.,* 27 (1980) 550-563.

J. B. Sidney, "An Extension of Moore's Due Date Algorithm", in *Symposium on the Theory of Scheduling and Its Applications,* Lecture Notes in Economics and Mathematical Systems 86, (S. E. Elmaghraby, ed.), Springer, Berlin, 1973, pp. 393-398.

J. B. Sidney, "Decomposition Algorithms for Single-Machine Sequencing With Precedence Relations and Deferral Costs", *Oper. Res.,* 23 (1975) 283-298.

J. B. Sidney, "The Two-Machine Maximum Flow Time Problem With Series Parallel Precedence Relations", *Oper. Res.,* 27 (1979) 782-791.

B. Simons, "A Fast Algorithm for Single Processor Scheduling", *Proc. 19th Annual IEEE Symp. Foundations of Computer Science,* (1978) 50-53.

W. E. Smith, "Various Optimizers for Single-Stage Production", *Naval Res. Logist. Quart.,* 3 (1956) 59-66.

J. D. Ullman, "NP-Complete Scheduling Problems", *J. Comput. System Sci.,* 10 (1975) 384-393.

J. Valdes, R. E. Tarjan, E. L. Lawler, "The Recognition of Series Parallel Digraphs", *SIAM J. Computing,* 11 (1982) 298–313.

M. K. Warmuth, "M Processor Unit-Execution-Time Scheduling Reduces to M-1 Weakly Connected Components", M. S. Thesis, (1980) Department of Computer Science, University of Colorado, Boulder.

G. Weiss, M. Pinedo, "Scheduling Tasks with Exponential Service Times on Non Identical Processors to Minimize Various Functions", *J. Appl. Probab.,* 17 (1980) 187–202.

Submodular functions and convexity

L. Lovász
Eötvös Loránd University, Department of Analysis I, Múzeum krt. 6-8, H-1088
Budapest, Hungary

0. Introduction

In "continuous" optimization convex functions play a central role. Besides elementary tools like differentiation, various methods for finding the minimum of a convex function constitute the main body of nonlinear optimization. But even linear programming may be viewed as the optimization of very special (linear) objective functions over very special convex domains (polyhedra). There are several reasons for this popularity of convex functions:

— Convex functions occur in many mathematical models in economy, engineering, and other sciencies. Convexity is a very natural property of various functions and domains occuring in such models; quite often the only non-trivial property which can be stated in general.

— Convexity is preserved under many natural operations and transformations, and thereby the effective range of results can be extended, elegant proof techniques can be developed as well as unforeseen applications of certain results can be given.

— Convex functions and domains exhibit sufficient structure so that a mathematically beautiful and practically useful theory can be developed.

— There are theoretically and practically (reasonably) efficient methods to find the minimum of a convex function.

It is less apperant, but we claim and hope to prove to a certain extent, that a similar role is played in discrete optimization by *submodular set-functions*. These functions do not enjoy the nice geometric image of convex functions, and accordingly their significance has been discovered only gradually. The first class of submodular functions which was studied thoroughly was the class of matroid rank functions. Generalizing certain basic properties of matroid polyhedra, Edmonds (1970) began the systematical study of submodularity. Let us remark, however, that approaching from quite a different angle, Choquet (1955) also introduced these set-functions. He proved that the newtonian "capacity" of a subset of $\mathbb{R}^3$ defines a submodular set-function. Quite a few proof techniques in graph theory, but also in probability, geometry and lattice theory have made implicit use of submodularity of certain set-functions, thus forecasting the importance of this property.

In this paper we survey some of the most important aspects of submodularity. In particular, we shall see that submodularity shares the above-listed four important properties of convexity. But besides this formal analogy, we shall develop a more fundamental connection between these two concepts. This connection will enable us to apply some basic facts concerning convex functions to obtain similarly basic results for submodular functions. In particular, a polynomial-time algorithm to find the minimum of a submodular set-function and a "sandwich theorem" for submodular functions can be obtained this way. (A refined integral version of this sandwich theorem, due to A. Frank (1982), needs a direct proof; but this powerful result is also motivated clearly by the convex-submodular analogy.)

Somewhat surprisingly, submodular functions are in some respects also similar to concave functions. This suggests that the maximization problem for submodular functions may also lead to interesting results. We shall see that this problem is substantially more difficult than the minimization problem, and in general no solution exists; but solutions for special cases, as well as reasonable heuristics, can be obtained.

Submodularity gives rise to polyhedra with very nice properties. Following Edmonds (1970), we treat various linear programming problems associated with submodular functions. For a single submodular function, these linear programming problems can be solved by appropriate versions of the *greedy algorithm*. For two submodular functions similar polyhedral considerations lead to very deep min-max results which can be examplified by the Matroid Intersection Theorem.

Several recent combinatorial studies involving submodularity fit into the following pattern. Take a classical graph-theoretical result (e. g. the Marriage Theorem, the Max-flow-min-cut Theorem etc.), and replace certain linear functions occuring in the problem (either in the objective function or in the constraints) by submodular functions. Often the generalizations of the original theorems obtained this way remain valid; sometimes even the proofs carry over. What is important here to realize is that these generalizations are by no means l'art pour l'art. In fact, the range of applicability of certain methods can be extended enormously by this trick. Choosing the submodular functions from among the many submodular functions arising from graphs and other combinatorial structures, various and often surprising results can be obtained this way. As an example, one can obtain, as a "submodular" generalization of the Marriage Theorem, a version of the famous Matroid Intersection Theorem, which in turn is known to imply Menger's Theorem, the Disjoint Spanning Tree Theorem and many other graph-theoretical results.

These are the main aspects of submodularity this paper will survey. We shall concentrate on some of the fundamental ideas and constructions; only a few proof will be described in detail, in cases when no appropriate reference can be given. This paper cannot undertake the task of a comprehensive survey of all aspects of submodularity, in particular if we consider matroid theory as a special case; for this we refer the interested reader to Welsh (1976), and shall assume that the reader is at least in part familiar with it. A forthcoming book of A. Schrijver is also strongly recommended for further reading on related subjects.

II. Submodular functions and convexity

1. Definitions and examples

Let S be a finite set and $f: 2^S \to \mathbb{R}$ a real valued function defined on the subsets of S. The set-function f is called *submodular* if the following inequality holds for any two subsets X and Y of S:

$$f(X \cap Y) + f(X \cup Y) \leq f(X) + f(Y). \tag{1}$$

The setfunction f is called *supermodular* if the reversed inequality holds true for every pair of subsets. Finally, f is called *modular* if it is both submodular and supermodular, i.e. the following equality holds identically for any two subsets X and Y of S:

$$f(X \cap Y) + f(X \cup Y) = f(X) + f(Y). \tag{2}$$

Let us remark immediately that modular functions are very simple. In fact, (2) implies easily by induction that

$$f(X) = f(\emptyset) + \sum_{x \in X} f(\{x\}). \tag{3}$$

Conversely, if $f(\emptyset)$ and $f(\{x\})$ $(x \in S)$ are prescribed arbitrarily, then the set-function defined by (3) is modular. If we identify every subset of S with its incidence vector, then modular functions will correspond to linear functions (restricted to 01-vectors).

Submodular (and supermodular) set-functions are much more interesting then modular ones, and there are many highly non-trivial examples of such set-functions. To verify submodularity of a set-function, the following characterization of submodularity often helps.

1.1. Proposition. Let f be a set-function defined on all subsets of S. Then f is submodular if and only if the derived set-functions

$$f_a(X) = f(X \cup \{a\}) - f(X) \quad (X \subseteq S - \{a\})$$

are monotone decreasing for all $a \in S$.

Obviously, an analogous characterization of supermodular set-functions can be formulated, by replacing "decreasing" by "increasing".

We continue with a number of examples of submodular and supermodular set-functions. As mentioned in the introduction, the main aim of these examples is to support the thesis that submodularity shares the first important property of convexity, namely that submodular functions occur in many mathematical models in a very natural way.

1.2. Example. Let S be the set of columns of a matrix A and let, for each $X \subseteq S$, $r(X)$ be the rank of the matrix formed by the columns in X. Then r is submodular. This fact follows by elementary linear algebra. For the reader familiar with matroid theory it is clear at this point that this example is just a special case of matroid rank functions, and so many more examples could be manufactured by taking other classes of matroids (graphic, transversal etc.). Some of these will occur below in different contexts.

1.3. Example. Let G be a (directed or undirected) graph and $X \subseteq V(G)$. Let $\delta(X)$ denote the number of lines of G connecting X to $V(G) - X$. Then a simple counting argument shows that $\delta(X)$ is a submodular setfunction on the subsets of $S = V(G)$. This fact has been the cornerstone of many studies in graph theory concerning connectivity, flows, and other problems. We can generalize this construction as follows. Let $w: E(G) \to \mathbb{R}_+$ be any weighting of the lines of G with non-negative real numbers, and define $\delta_w(X)$ as the sum of weights of the lines of G connecting X to $V(G) - X$. Instead of graphs, we could also consider hypergraphs.

1.4. Example. Let G be a bipartite graph, with bipartition $\{A, B\}$. For each $X \subseteq A$, let $\Gamma(X)$ denote the set of neighbors of X. Then $|\Gamma(X)|$ is a submodular set-function on the subsets of A. The submodularity of this set-function plays role in some treatments of the bipartite matching problem, see e.g. Ore (1955). From $|\Gamma(X)|$ we can construct another submodular set-function on the subsets of A by the formula

$$v(X) = |X| + \min\{|\Gamma(Y)| - |Y|: Y \subseteq X\}.$$

From the König-Hall Theorem we can see that $v(X)$ gives the maximum number of nodes in X which can be matched with nodes in B by disjoint lines of G. Thus v is in fact the rank function of the transversal matroid on A defined by the bipartite graph G.

1.5. Example. Let G be a graph. For each $X \subseteq E(G)$, let $c(X)$ denote the number of connected components of the subgraph $(V(G), X)$. Then $c(X)$ is a supermodular set-function, which can be easily verified e.g. by using Proposition 1.1. If one wishes to see a submodular function, one may consider $-c(X)$ or (if one wants it non-negative) one may consider $|V(G)| - c(X)$. This function is then the rank function of the circuit matroid of the graph G.

1.6. Example. We may modify the previous example as follows. For each $X \subseteq V(G)$, let $d(X)$ denote the number of connected components of the graph $G - X$. Then $d(X)$ is in general neither submodular nor supermodular. But if we restrict its domain to the subsets X of an independent set A of points of G, then $d(X)$ will be supermodular.

1.7. Example. Let B be a bar-and-joint structure, i.e. a graph whose nodes are points of the euclidean 3-space, and whose lines are considered rigid bars which are attached to the nodes by flexible joints. Let, for each $X \subseteq V(G)$, $f(X)$ denote the degree of freedom of the subset X, i.e. the dimension of the infinitesimal motions of the nodes in X which extend to an infinitesimal motion of all nodes compatible with the given bars. Thus, $f(\emptyset) = 0$, $f(\{x\}) = 3$ for every $x \in V(G)$, and $f(\{x, y\}) = 5$ or 6 depending on whether or not the whole structure forces x and y to stay at the same distance, etc. It follows from the ele-

II. Submodular functions and convexity

ments of the theory of rigid bar-and-joint structures that f is a submodular set-function of the subsets of $V(G)$. See Crapo (1979).

1.8. Example. Let $A_1, \ldots, A_n$ be random events, $S = \{A_1, \ldots, A_n\}$, and let, for $X = \{A_{i_1}, \ldots, A_{i_k}\} \subseteq S$,

$$f(X) = \text{Prob}(A_{i_1} \ldots A_{i_k})$$

denote the probability that all events in X occur. Then f is a supermodular set-function on the subsets of S. This fact is used implicity in some sieving techniques.

1.9. Example. Let G and H be two graphs, $X \subseteq (G)$ and $Y \subseteq E(H)$. Let $h(X)$ denote the number of homomorphisms of the graph $(V(G), X)$ into H, and let $g(Y)$ denote the number of homomorphisms of the graph G into the graph $(V(H), Y)$. Then g and h are supermodular set-functions. This fact (which can be verified easily using Proposition 1.1) can be generalized in many ways to categories etc., becoming more and more trivial. (This last remark shows that submodularity is not a deep property, but it is often difficult to recognize the circumstances under which it occurs.)

1.10. Example. This last example comes from geometry. Let $a_1, \ldots, a_n$ be linearly independent vectors in $\mathbb{R}^n$, and let, for $X = \{a_{i_1}, \ldots, a_{i_k}\}$,

$$f(X) = \log \text{vol}_k(a_{i_1}, \ldots, a_{i_k}),$$

where vol_k denotes the k-dimensional volume of the parallelepiped spanned by $a_{i_1}, \ldots, a_{i_k}$. Then

$$f_a(X) = \log \text{vol}_{k+1}(a_{i_1}, \ldots, a_{i_k}, a) - \log \text{vol}_k(a_{i_1}, \ldots, a_{i_k})$$

$$= \log \frac{\text{vol}_{k+1}(a_{i_1}, \ldots, a_{i_k}, a)}{\text{vol}_k(a_{i_1}, \ldots, a_{i_k})}$$

is the logarithm of the "height" of a above the plane spanned by $a_{i_1}, \ldots, a_{i_k}$. Hence Proposition 1.1 easily implies that f is submodular.

The first five examples above are well-understood and well-studied submodular functions. The last four examples show that submodularity arises in geometry, probability and other branches of mathematics and one could also mention sporadic use of these examples. Nevertheless, these examples and in particular the significance of submodularity are much less understood. One may hope that an in-depth study of submodularity will have some contribution to "classical" mathematical problems as well.

Often it is useful to extend the definition of submodularity and supermodularity to partial set-functions, i.e. to functions which are only defined on a family $\mathcal{H}$ of subsets of S. Then of course (1) is required to hold only if all four of $X, Y, X \cap Y$, and $X \cup Y$ belong to $\mathcal{H}$.

If we study submodularity of a set-functioned defined only on a family $\mathcal{H} \subseteq 2^S$, then it is natural to assume that there are "fairly many" pairs X, Y for which (1) is defined (and therefore is assumed to hold). There are three types of set-systems $\mathcal{H}$ defined by this kind of condition which have been studied extensively. The family $\mathcal{H} \subseteq 2^S$ is called a *lattice family* if X, Y implies that $X \cap Y \in \mathcal{H}$ and $X \cup Y \in \mathcal{H}$. $\mathcal{H}$ is called an *intersecting family* if $X, Y \in \mathcal{H}$, $X \cap Y \neq \emptyset$ implies that $X \cap Y \in \mathcal{H}$ and $X \cup Y \in \mathcal{H}$. Finally, $\mathcal{H}$ is called a *crossing family* if $X, Y \in \mathcal{H}$, $X \cap Y \neq \emptyset$, $X \cup Y \neq S$ implies that $X \cap Y \in \mathcal{H}$ and $X \cup Y \in \mathcal{H}$. For an intersecting family we shall assume that $\emptyset \notin \mathcal{H}$, and for a crossing family we shall assume that $\emptyset \notin \mathcal{H}$ and $S \notin \mathcal{H}$. Many results on submodular set-functions defined on all subsets of a set extend without any change to submodular set-functions defined on lattice families, and with natural modifications to submodular set-functions defined on intersecting or crossing families.

Let us mention two examples of submodular functions which are defined on subfamilies of 2^S only.

1.11. Example. Let G be a directed graph, $S = V(G)$, and let $\mathcal{H}$ be the family of those subsets of S which determine a directed cut, i.e. those subsets X for which no line of G has head in X and tail in $S - X$. Then the function $\delta(X)$ introduced in Example 1.3, when restricted to $\mathcal{H}$, will not only be submodular but even modular. Further, $\mathcal{H}$ is a lattice family.

1.12. Example. The most common intersecting and crossing families are the families of all non-empty subsets and of all non-empty proper subsets of S, respectively. It is easy to see that every submodular set-function defined on 2^S, when restricted to the non-empty subsets of S, gives a submodular set-function on this latter family and conversely, every submodular set-function defined on the non-empty subsets of S can be extended to a submodular set-function defined on all subsets (by letting $f(\emptyset)$ be a very small number). However, if we restrict our attention to, say, non-negative submodular functions then we obtain essentially more general problems if we consider set-functions defined on the non-empty subsets of S only. Similar remarks apply to the family of non-empty proper subsets.

We close this section with a few more definitions. A submodular function which is integral valued, monotone increasing, and has value 0 on $\emptyset$, is called a *polymatroid function*. The submodular function arising from a bar-and-joint structure (Example 1.7) is a polymatroid function. Further, every matroid rank function is a polymatroid function. Let us see one more example.

1.8. Example. Let S be a set of subspaces of a linear space and let, for each $X \subseteq S$, $f(X)$ denote the dimension of the linear subspace generated by the subspaces in X. Then f is a polymatroid function. More generally, let (E, r) be a matroid (r being its rank function), and let $S \subseteq 2^E$. For each $X \subseteq S$, let

$$f(X) = r(\cup X).$$

Then f is a polymatroid function. It will follow from the discussions in the next section that every polymatroid function, in fact, arises by this construction from some matroid (Helgason 1974, McDiarmid 1975).

A set-function f defined on the subsets of S will be called *t-smooth*, if

$$|f(X\cup\{a\})-f(X)|\leq t$$

for every $a\in S$ and $X\subseteq S-\{a\}$. This is equivalent to saying that

$$|f(X)-f(Y)|\leq t|X\triangle Y|$$

holds for all $X, Y\subseteq S$. Clearly, 1-smooth polymatroid functions are just matroid rank functions.

2. Operations on submodular functions

Following our line of establishing submodular set-functions as discrete analogs of convex functions, we turn to the second property of convexity stated in the introduction, namely that there are many natural operations which manufacture new submodular functions from old ones. Such operations are useful to extend the effective range of certain theorems (by applying them to modified functions), to prove submodularity (by recognizing that the function arises by a known construction from known submodular functions), to construct examples etc. But as it turns out, the study of various properties of some of these constructions yields sufficient insight into submodularity so that the study of these constructions may also be considered as the structure theory of submodular functions. So the results to be discussed in this section should also prove that submodular set-functions have sufficient structure so that a mathematically beautiful and practically useful theory can be developed.

Let us start with the trivialities. If f is submodular then the set-function $f(S-X)$ is also submodular, while $-f(X)$ is supermodular. If f and g are submodular then $f+g$ is also submodular.

Let f_i be a submodular set-function defined on the subsets of S_i ($i=1,\ldots,k$), where the S_i' are mutually disjoint. Then the *direct sum* of $f_1,\ldots,f_k$ is the set-function f defined on $S=S_1\cup\ldots\cup S_k$ by the formula

$$f(X)=f_1(S_1\cap X)+\ldots+f_k(S_k\cap X).$$

The following construction makes a monotone function from a non-monotone; cf. example 1.4. If f is any set-function defined on the subsets of S, we set

$$f_{\text{mon}}(X)=\min\{f(Y):Y\subseteq X\}.$$

Clearly, f_{mon} is monotone decreasing. The following is slightly less obvious:

2.1. Proposition. *If f is a submodular set-function then f_{mon} is also submodular.*

Neither the minimum nor the maximum of two submodular set-functions is in general submodular. But the following fact is useful in many constructions:

2.2. Proposition. Let f and g be submodular set-functions such that $f-g$ is either monotone increasing or monotone decreasing. Then $\min(f,g)$ is also submodular.

We now come to a very important and non-trivial construction. Let f be any set-function defined on the subsets of X. Define the *Dilworth truncation of f* as the set-function

$$f_*(\emptyset) = 0,$$
$$f_*(X) = \min\{f(X_1) + \ldots + f(X_k): X_1 \cup \ldots \cup X_k = X, X_i \neq \emptyset, X_i \cap X_j = \emptyset\}$$

This construction generalizes the notion of Dilworth truncation of matroids; hence the name (Dunstan 1976). For properties and geometric background, see also Mason (1977, 1981) and Lovász (1977).

2.3. Proposition. If f is a submodular set-function then its Dilworth truncation f_* is also submodular.

Note that if $f(\emptyset) \geq 0$ then $f_* = f$ on the non-empty subsets of S. In general, the relationship between f_* and f is somewhat more complicated:

2.4. Proposition. Let f be a submodular set-function defined on the subsets of S. Then f_* is the unique maximal set-function among all set-functions g with the following properties:
 a) g is submodular,
 b) $g(\emptyset) = 0$,
 c) $g(X) \leq f(X)$ for all non-empty $X \subseteq S$.

We can define the "upper Dilworth truncation" f^* for any set-function by replacing "min" by "max" in the definition of Dilworth truncation. Then we find that upper Dilworth truncation preserves supermodularity.

Let f and g be set-functions defined on the subsets of S. We define their *convolution* $h = f * g$ by

$$h(X) = \min\{f(Y) + g(X-Y): Y \subseteq X\}.$$

The reader familiar with the Matroid Intersection Theorem will notice that if f and g are two matroid rank functions then h is the rank function of their intersection. Since the intersection of two matroids is not a matroid in general, it follows that convolution does not always preserve submodularity. However, the following is true.

2.5. Theorem. The convolution of a submodular and a modular set-function is submodular.

Proof. Let f be a submodular, g a modular set-function, both defined on the subsets of S, and let $h = f * g$ be their convolution. Let $X, Y \subseteq S$, we claim that

$$h(X) + h(Y) \geq h(X \cap Y) + h(X \cup Y). \tag{4}$$

By definition, there exist subsets $U \subseteq X$ and $V \subseteq Y$ such that

$$h(X) = f(U) + g(X-U), \quad h(Y) = f(V) + g(Y-V).$$

II. Submodular functions and convexity

Thus
$$h(X)+h(Y)=f(U)+g(X-U)+f(V)+g(Y-V)$$
$$\geq f(U\cap V)+f(U\cup V)+g((X-U)\cap(Y-V))$$
$$+g((X-U)\cup(Y-V)).$$

However, $(X-U)\cap(Y-V)=((X\cap Y)-(U\cap V))\cap((X\cup Y)-(U\cup V))$ and $(X-Y)\cup(Y-V)=((X\cap Y)-(U\cap V))\cup((X\cup Y)-(U\cup V))$, and so using the modularity of g,
$$h(X)+h(Y)\geq f(U\cap V)+f(U\cup V)+g((X\cap Y)-(U\cap V))$$
$$+g((X\cup Y)-(U\cup V)).$$

But here we have, by the definition of h,
$$f(U\cap V)+g((X\cap Y)-(U\cap V))\geq h(X\cap Y)$$
and
$$f(U\cup V)+g((X\cup Y)-(U\cup V))\geq h(X\cup Y),$$
and so (4) follows.

Note that $f_{\text{mon}}=f*0$, and so proposition 2.1 is a special case of Theorem 2.5. We shall see more significant applications below.

Theorem 2.5 implies the following description of the convolution of a submodular and a modular function:

2.6. Corollary. Let f be a submodular set-function and g a modular set-function defined on the subsets of S. Assume that $f(\emptyset)=g(\emptyset)=0$. Then $f*g$ is the unique largest submodular set-function h satisfying $h(\emptyset)=0$, $h\leq f$ and $h\leq g$.

A special case of the convolution construction which we shall use is the following. Let the modular set-function g be defined by $g(\emptyset)=0$, $g(a)=\alpha$, and $g(x)=+\infty$ if $x\in S-a$ (where $a\in S$ is a fixed element and α any real number). Then it follows easily that
$$(f*g)(X)=\min\{f(X),f(X-a)+\alpha\},$$
in particular $(f*g)(X)=f(X)$ if $a\in X$. The set-function $f*g$ will be called the *local truncation of f at the point a with value α*. If $f(\emptyset)=0$, then this local truncation is the unique largest submodular set-function h such that $h(\emptyset)=0$, $h\leq f$ and $h(a)\leq \alpha$. In particular, $h=f$ if $f(a)\leq \alpha$.

If $(S,r_1),\ldots,(S,r_k)$ are matroids on the same set S, then the rank function of their sum can be written, by the results of Edmonds-Fulkerson (1965), as
$$r=(r_1+\ldots+r_k)*\text{card},$$
where $\text{card}(X)=|X|$ is the cardinality function. In fact it follows from the above results that r is the rank function of a matroid, and that this is the unique free-est matroid satisfying
$$r\leq r_1+\ldots+r_k.$$

We come to the problem of extending submodular set-functions – in two different senses. First we investigate how a submodular function defined on a lattice (intersecting, crossing) family of sets can be extended to all subsets. The case of a lattice family is easy. Let $\mathcal{H}$ be a lattice family of subsets of S and f a submodular function on $\mathcal{H}$. $\mathcal{H}$ contains a unique largest subset S'. For every $X \subseteq S'$, $\mathcal{H}$ contains a unique smallest subset including X; let this subset be denoted by $\bar{X}$. Then the following formula defines an extension of f to all subsets:

$$\bar{f}(X) = f(\overline{X \cap S'}).$$

The extension procedure is similar for crossing families. Let $S_1, \ldots, S_k$ be the maximal members of the crossing family $\mathcal{H}$. Clearly $S_1, \ldots, S_k$ are disjoint. For every $X \subseteq S_i$, $X \neq \emptyset$, there is a unique smallest set in $\mathcal{H}$ including X; let this be denoted by $\bar{X}$. For $1 \leq i \leq k$ and $X \subseteq S$, let

$$f_i(X) = \begin{cases} f(\overline{X \cap S_i}) & \text{if } X \cap S_i \neq \emptyset, \\ -c & \text{if } X \cap S_i = \emptyset. \end{cases}$$

(where c is a parameter to be chosen appropriately). Then

$$\bar{f}(X) = \sum_{i=1}^{k} f_i(X) + (k-1)c$$

is an extension of f and is submodular provided c is large enough.

In the case of a crossing family, we first extend the definition of f to those subsets $X \subseteq S$, $X \neq S$ which can be written as the union of sets in H, by the formula

$$f'(X) = \min \left\{ \sum_{i=1}^{k} f(X_i) + (k-1)c : \{X_1, \ldots, X_k\} \text{ is a partition of } X \right\}$$

(where again c is a parameter). In particular,

$$f'(\emptyset) = -c.$$

The complements of these subsets form an intersecting family, and so f' can be extended to a submodular set-function defined on all subsets of S by a procedure which is "dual" to the one used in the case of intersecting families.

The main properties of these extensions are given in the following proposition:

2.7. Proposition. *A submodular function defined on a crossing family of subsets can be extended to a submodular set-function defined on all subsets. If the original set-function is integral valued, so is its extension. If the original set-function is non-negative, then its extension is also non-negative provided the family is lattice, and is non-negative on the non-empty subsets if the family is intersecting.*

Let us consider now extension in a different sense, namely extending a submodular set-function to a larger underlying set. This is in general a very difficult problem and even in the case of matroids it cannot be considered well-solved. Crapo (1965) gave a description of single-element extensions of matroids, but his conditions are often difficult to apply. Some extension proce-

II. Submodular functions and convexity

dures, like principal extensions, are however well-understood and very useful tools in matroid theory.

We shall not attempt to develop a general theory of extensions of submodular functions, but shall confine ourselves with offering a few constructions.

A set-function f' defined on the subsets of S' is called an *extension* of the set-function f defined on the subsets of the set S if $S' \subseteq S$ and $f(X) = f'(X)$ for every $X \subseteq S$.

We have already defined the direct sum of submodular functions. This may also be viewed as a very simple extension procedure.

A very simple but important extension procedure in matroid theory is *parallel extension*. For submodular set-functions, this procedure can be carried out more generally – with some restrictions. Let f be a set-function defined on the subsets of S. Let $T \subseteq S$, and define a new set-function as follows. Take a new element $a_T \in S$, set $S' = S \cup \{a_T\}$, and define the *extension of f by an element a_T parallel to the subset T* by

$$f'(X) = f(X),$$
$$f'(X \cup \{a_T\}) = f(X \cup T) \quad \text{for } X \subseteq S.$$

Unfortunately, the parallel extension of a submodular set-function is not always submodular. Let an element $a \in S$ be called *increasing* with respect to the set-function f if $f(X \cup \{a\}) \geq f(X)$ for all $X \subseteq S$. If f is a monotone increasing set-function then of course every element of S is increasing with respect to f. Using this notion, the following result characterizing submodularity of parallel extensions can be proved easily.

2.7. Proposition. Let f be a set-function defined on the subsets of S, let $T \subseteq S$ and f' an extension of f by an element parallel to T. Then f' is submodular if and only if f is submodular and every element of T is increasing with respect to f.

An extension parallel to the subset T followed by a local truncation at the new element with value α is called a *principal extension on the subset T with value α*. It follows that if T consists of increasing elements and the original function is submodular, then so is this principal extension. This construction generalizes the principal extension of matroids.

We remark the following important property of principal extensions. Let f be a submodular set-function defined on the subsets of S, and let T_1 and T_2 be two subsets of S consisting of increasing elements. Let, further, α_1 and α_2 be real numbers. Construct the principal extension of f on T_1 with value α_1, and then construct the principal extension of the resulting set-function on T_2 with value α_2. This way we obtain a submodular set-function f'. Let us carry out the same construction but with the indices 1 and 2 interchanged, to get a submodular set-function f''. It is easy to verify that $f' = f''$. This fact can be expressed compactly as follows:

2.8. Proposition. Principal extensions commute.

A special case of the parallel extension has been used in example 1.8. In fact, given a matroid and a set of subsets, we may extend it by elements parallel to the given subsets, and then delete the original elements. This way we obtain a polymatroid function. Let us show now that conversely, every polymatroid function arises this way (Helgason 1974, McDiarmid 1975):

2.9. Thorem. Given any polymatroid function f defined on the subsets of a set S, there exists a matroid (E, r), and a mapping $\varphi: S \to 2^E$, such that

$$f(X) = r(\cup \varphi(X)) \tag{5}$$

holds for every $X \subseteq S$.

Proof (sketch). For every $a \in S$, construct $f(a)$ times the principal extension of f on the subset a with value 1. It follows from proposition 2.8 that the order in which this is done is irrelevant. Then delete the original elements. This way a matroid is obtained. Let, for each $a \in S$, $\varphi(a)$ denote the set of new elements added on a. Then it is easy to verify that (5) holds.

Principal extensions and Dilworth truncations have a very nice geometric meaning. Here we only sketch the background. Let S be a collection of subspaces of a linear space and let $f(X)$ denote the dimension of the subspace generated by the members of X (example 1.8).

Then a parallel extension of f by an element parallel to $T \subseteq S$ can be obtained by adding the linear span of T to S. A local truncation of f at $a \in S$ with value α (where $0 < \alpha \leq f(a)$) can be constructed by replacing a by an α-dimensional subspace of a "in general position". This latter phrase can be made precise by requiring that the new subspace is not contained in the span of any subset of S unless the whole of a is. It is not difficult to see that if the underlying field is large enough then such a "general" subspace always exists.

Combining these two observations we see that a principal extension of f on T with value α (where $0 < \alpha \leq f(a)$) can be obtained by adding to S a "general" α-dimensional subspace of the linear span of T.

The function $f - 1$ is also submodular (and monotone), but its value is negative on $\emptyset$. It is easy to see that $(f-1)_*$ is a polymatroid function. A representation of this can be constructed from a representation of f by linear subspaces as follows. Take a "general" hyperplane h in the whole space, and replace each $a \in S$ by $a \cap h$. Here, however, the notion of "general" is more involved. The easiest way out is to extend the underlying field with infinitely many algebraically independent transcendentals, and then take algebraically independent transcendentals as the coefficients of the equation of h. For details see Lovász (1977, 1982) and Mason (1977, 1981).

3. Polyhedra associated with submodular functions

Let f be a submodular set-function defined on the subsets of S and let $f(\emptyset) = 0$. Consider the following polyhedron:

$$P_f = \{x \in \mathbb{R}^S : x(T) \leq f(T) \quad \text{for all } T \subseteq S\}.$$

II. Submodular functions and convexity

(Here, as usual $x(T) = \sum_{t \in T} x(t)$.) We shall be concerned with finding the maximum of a linear objective function $c \cdot x$ over $x \in P_f$. To exclude trivial cases, assume that $c \geq 0$.

Note that this problem is a linear programming problem and that it has a finite optimum solution. Therefore we may consider its dual and this will have the same optimum value. We shall have a variable y_T for every $T \subseteq S$, and the constraints

$$y_T \geq 0,$$
$$\sum_{T \ni t} y_T = c(t) \quad \text{for all } t \in S.$$

The dual objective function is

$$\sum_T f(T) y_T.$$

The fact that this is a linear program, however, is of little help here since the number of constraints in the primal program (and hence the number of variables in the dual) is exponentially large. Both problems can be solved however by a very simple procedure called the *greedy algorithm*.

To formulate this, let us first label the elements of $S = \{a_1, \ldots, a_n\}$ so that $c(a_1) \geq c(a_2) \geq \ldots \geq c(a_n)$. Then an optimal primal solution is given by

$$x(a_k) = f(\{a_1, \ldots, a_k\}) - f(\{a_1, \ldots, a_{k-1}\}) \quad (1 \leq k \leq n). \tag{6}$$

Furthermore, an optimum dual solution is given by

$$y_T = \begin{cases} c(a_k) - c(a_{k+1}), & \text{if } T = \{a_1, \ldots, a_k\} \text{ for some } k, 1 \leq k \leq n, \\ 0, & \text{otherwise}. \end{cases} \tag{7}$$

Why are these solutions called greedy? An explanation is given by the following interpretations. We can arrive at the primal solution described above as follows. First, choose the value of the variable $x(a_1)$ as large as possible. There is a trivial upper bound on $x(a_1)$, namely $f(\{a_1\})$, so let

$$x(a_1) = f(\{a_1\}).$$

Next, choose the value of $x(a_2)$ as large as possible. There are now two trivial upper bounds: $f(\{a_2\})$ and $f(\{a_1, a_2\}) - f(\{a_1\})$. From the submodularity of f it follows that the second one of these is always the smaller, so let

$$x(a_2) = f(\{a_1, a_2\}) - f(\{a_1\})$$

etc. This argument also shows that the vector given by (6) is indeed a feasible point of P_f. (Note that optimality is not yet proved; the greedy solution is only optimal for very special linear programs.)

Second, we show how a greedy interpretation of the dual solution (7) can be given. Consider first the dual variable y_T with the largest T, namely $T = S$, and choose its value as large as possible. Clearly this largest possible value is $c(a_n)$, so fix the value

$$y_S = c(a_n).$$

Once this value is fixed, it follows immediately from the constraint corresponding to $t = a_n$ that $y_T = 0$ for every other subset T containing a_n. Consider the next largest subset T of S whose dual variable is not yet fixed; this is clearly $S - a_n = \{a_1, \ldots, a_{n-1}\}$; and chose y_{S-a_n} as large as possible, which clearly means

$$y_{S-a_n} = c(a_{n-1}) - c(a_n)$$

etc. Again, this argument proves that the greedy dual solution (7) is indeed a feasible dual solution, but does not prove optimality. However, the optimality of both the primal and the dual greedy solution follows now easily from the fact that they give the same objective value:

$$\sum_{k=1}^{n} c(a_k)(f(\{a_1, \ldots, a_k\}) - f(\{a_1, \ldots, a_{k-1}\}))$$

$$= \sum_{k=1}^{n} f(\{a_1, \ldots, a_k\})(c(a_k) - c(a_{k+1})).$$

The fact that (6) and (7) are optimal solutions has some important consequences. First note that if f is integral valued then (6) is integral and if c is integral then (7) is integral. Hence it follows:

3.1. Theorem. If f is integral valued then P_f has integral vertices. The system defining P_f is total dual integral.

A much stronger reult is due to Edmonds and Giles (1977).

3.2. Theorem. Let f and g be integral valued submodular set-functions defined on the subsets of the same set S, and assume that $f(\emptyset) = f(\emptyset) = 0$. Then $P_f \cap P_g$ has integral vertices and the system

$$x(T) \leq \min\{f(T), g(T)\}$$

(which describes $P_f \cap P_g$) is total dual integral.

Many graph-theoretical and combinatorial results follow from theorem 3.2, most notably the Matroid Intersection Theorem. We shall discuss further applications in section 6.

Let $\hat{f}(c)$ denote the maximum of $c \cdot x$ subject to $x \in P_f$. Let us consider the case when c is the incidence vector of a set $T \subseteq S$. Then it follows from either the primal or the dual greedy algorithm that $\hat{f}(c) = f(T)$. Thus $\hat{f}(c)$ may be considered as an extension of the function f, which originally is defined on 01-vectors, to all non-negative vectors.

Such an extension can be defined for an arbitrary set-function. For let f be any set-function defined on the subsets of a set S. For the incidence vector a of a set $T \subseteq S$, set

$$\hat{f}(a) = f(T).$$

So $\hat{f}$ is defined for 01-vectors. Next let c be any non-negative vector. Then c can be written uniquely in the following form:

$$c = \sum_{i=1}^{k} \lambda_i a_i,$$

II. Submodular functions and convexity

where $\lambda_i > 0$ and $a_1 \geq a_2 \geq \ldots \geq a_k$ are distinct 01-vectors. Define

$$\hat{f}(C) = \sum_{i=1}^{k} \lambda_i \hat{f}(a_i).$$

It is clear that this way a well-defined function is obtained.

Using this construction, we can re-formulate Theorem 3.2 in a form closely resembling the Matroid Intersection Theorem:

3.3. Theorem. Let f and g be integral valued submodular set-functions defined on the subsets of the same set S, and assume that $f(\emptyset) = g(\emptyset) = 0$. Further let $c \geq 0$ be an integral vector. Then

$$\max\{c \cdot x : x \text{ integral}, x(T) \leq f(T) \text{ for all } T \subseteq S\}$$
$$= \min\{\hat{f}(x) + \hat{g}(c-x) : x \text{ integral}, 0 \leq x \leq c\}.$$

In the special case when $c = 1$, which is most important from the combinatorial point of view, the right hand side is even simpler:

3.4. Corollary. Let f and g be integral valued submodular set-functions defined on the subsets of the same set S, and assume that $f(\emptyset) = g(\emptyset) = 0$. Then

$$\max\{1 \cdot x : x \text{ integral}, x(T) \leq f(T) \text{ for all } T \subseteq S\}$$
$$= \min\{f(X) + g(S - X) : X \subseteq S\}.$$

Note that the right hand side can be written as $(f * g)(S)$. In the special case when f and g are the rank functions of two matroids, the left hand side is just the cardinality of a maximum common independent set of the two matroids, and so we obtain the Matroid Intersection Theorem.

4. Submodularity and convexity

Are submodular set-functions more like convex or like concave functions? In this section we discuss some properties of them which are analogous to properties of convex functions; in the next section we shall survey some aspects of submodularity which relate it to concavity. The reader may then decide how he or she would answer the question above.

A very important connection between convexity and submodularity concerns the extension of a submodular set-function discussed at the end of the last chapter.

4.1. Proposition. Let f be any set-function defined on the subsets of a set S and let $\hat{f}$ be its extension to non-negative vectors. Then $\hat{f}$ is convex (concave) if and only if f is submodular (supermodular).

A trivial consequence of this statement is that $\hat{f}$ is linear if and only if f is modular. Conversely, the restriction of a linear set-function to 01-vectors is always a modular set-function. The restriction of a convex function to 01-vectors, however, does not always yield a submodular set-function; consider e.g. $f(x_1, x_2, x_3) = \max\{x_1, x_2 + x_3\}$.

This proposition can be used very well in the study of submodularity. As a first application, we shall use it to prove the first half of the following "sandwich theorem" due to A. Frank (1982).

4.2. Theorem. Let f be a submodular and g a supermodular set-function, both defined on the subsets of the same set S, and assume that $g \leq f$.

(a) Then there exists a modular set-function h on the subsets of S such that $g \leq h \leq f$.

(b) If f and g are integral valued then this separating set-function h can also be chosen integral valued.

Proof. (a) By proposition 4.1, $\hat{f}$ is convex and $\hat{g}$ is concave, and it follows from the construction of $\hat{f}$ and $\hat{g}$ that $\hat{g} \leq \hat{f}$. By an elementary result in convexity, there exists a linear function $\hat{h}$ such that $\hat{g} \leq \hat{h} \leq \hat{f}$ for all non-negative vectors. The restriction of $\hat{h}$ to 01-vectors yields a modular set-function separating f and g.

(b) This statement is considerably deeper and more interesting from the combinatorial point of view. It is in fact equivalent to Corollary 3.4. It can also be proved by mimicking (mutatis mutandis) a standard proof of the Hahn-Banach Theorem.

Perhaps the most significant application of Proposition 4.1 is the minimization of a submodular set-function. Let f be a submodular set-function defined on the subsets of S. We want to find the set $X \subseteq S$ for which $f(X)$ is minimum. It turns out that this is a very general and important problem in combinatorial optimization and very many combinatorial problems can be reduced to it – we shall see some in section 6.

Speaking about algorithms concerning submodular set-functions, we have to specify first how the set-function is given. Most problems which arise would become trivial if the set-function involved were given by listing all of its 2^n values. Therefore, we shall assume that our set-function is given by an oracle, i.e. subroutine which evaluates the function at any subset. One call of this subroutine counts as one step only, and the subroutine cannot be "broken open", i.e. the algorithm must work for any subroutine whose output is a real number for every input (a subset of S), provided these outputs form a submodular set-function, regardless of the way of calculating this value.

The key to the minimization of a submodular set-function is the following lemma.

4.3. Lemma. Let f be a set-function defined on the subsets of S such that $f(\emptyset) = 0$. Then

$$\min\{f(X): X \subseteq S\} = \min\{\hat{f}(x): x \in [0, 1]^S\}.$$

The proof of this lemma is straightforward. So instead of minimizing f over the subsets of S, it suffices to minimize $\hat{f}$ over the cube. Now this is made possible by the following two nice properties of the function $\hat{f}$:

a) for every rational vector $x \in [0, 1]^S$, $\hat{f}(x)$ can be evaluated in polynomial time;

b) if f is submodular then $\hat{f}$ is convex.

Thus we can apply an algorithm due to Judin and Nemirovskii (1976) and find the minimum of f in polynomial time. Since the hypothesis that $f(\emptyset) = 0$ is clearly irrelevant (we can replace, if necessary, f by $f - f(\emptyset)$), we have proved the following result (Grötschel, Lovász and Schrijver 1981):

4.4. Theorem. Let f be a submodular set-function defined on the subsets of S. Then the subset of S minimizing f can be found in polynomial time.

This theorem implies the polynomial time solvability of many problems in combinatorial optimization. Let us mention some.

4.5. Corollary. Let f be a submodular set-function such that $f(\emptyset) = 0$. Then $x \in P_f$ can be checked in polynomial time for every rational vector x.

4.6. Corollary. Let f and g be submodular set-functions defined on the subsets of S. Then $(f * g)(X)$ can be evaluated for every $X \subseteq S$ in polynomial time. In particular, $f_{\text{mon}}(X)$ can be evaluated in polynomial time.

Finally, the following results concerning the Dilworth truncation was shown by Grötschel, Lovász and Schrijver (1981).

4.7. Theorem. Let f be a submodular set-function defined on the subsets of S. Then $f_*(X)$ can be evaluated for every $X \subseteq S$ in polynomial time.

Thus we have seen that the operations introduced in section 2, when applied to submodular set-functions given by an oracle, yield polynomially computable set-functions. In particular, these operations preserve the polynomial-time computability of the set-functions involved.

5. Submodularity and concavity

Let us hear now some arguments of the advocate for concavity. He will start with pointing out the analogy between propositions 1.1 and 2.2 and some properties of concave functions. A closer connection is represented by the following proposition:

5.1. Proposition. Let f be a real-valued function defined on non-negative integers. Then $f(|X|)$ is submodular on the subsets of an arbitrary set if and only if f is concave.

We have seen that the problem of minimizing a submodular set-function is an important one, whose solution implies the solution of many very general combinatorial optimization problems. Let us show now that the problem of maximizing a submodular set-function is also very important.

Let H be a hypergraph all whose edges are r-tuples of the set V, and let S be the set of its edges. It is a very important combinatorial problem to determine the maximum number of disjoint edges of H. This can be formulated as a submodular function maximization problem as follows. For every $X \subseteq S$, let $f(X)$ denote the number of points of V covered by the edges in X. As remarked in example 1.4, this set-function is submodular. Therefore so is the set-function $f(X)-(k-1)|X|$. It is easy to verify that the maximum of this set-function is equal to the maximum number of disjoint edges of H.

One could show by a similar construction that the problem of finding maximum common independent set of k matroids is also a special case of submodular function maximization.

But these remarks are bad news rather than good. Since several NP-complete problems can be reduced to the problem of maximizing a submodular set-function, there is no hope to find a polynomial-time algorithm solving this problem (the problem of maximizing a submodular set-function is not in NP, so we cannot conclude that it is NP-complete, but it is certainly NP-hard). Further, it follows from the results of Hausmann-Korte (1978) and Lovász (1981) (independently of the $P \neq NP$ hypothesis) that every algorithm finding the maximum of submodular functions (even of 1-smooth submodular functions) takes exponential time in the worst case.

But this last mentioned special case of 1-smooth submodular functions is not as bad as it seems. No special choice of the submodular function is known to lead to an NP-complete problem, and quite a few special choices are solvable in polynomial time. This problem is in fact equivalent to the matroid matching (matroid parity, or matchoid) problem, which can be stated as follows.

Let f be a polymatroid function such that $f(\{x\})=2$ for every $x \in S$. We shall call f a *2-polymatroid function*. It is clear that $f(X) \leq 2|X|$ holds for every $X \subseteq S$. The subset X is called a *matching* if $f(X)=2|X|$. The *matroid matching problem* asks for the maximum size of a matching for a given 2-polymatroid function f.

It is easy to see that the maximum size of a matching is equal to the maximum of the submodular set-function $f(X)-|X|$.

Now the good news is that a maximum matching can be found in polynomial time provided the 2-polymatroid is given

(a) as a set of subspaces of a linear space (Example 1.8; Lovász 1981), or

(b) as a set of pairs of points of a transversal matroid (Lawler and Po Tong 1982).

(One may wonder why to state (b) as transversal matroids are known to be linear and therefore the 2-polymatroids (b) can also be represented in the form of (a). No efficient way is known, however, to construct such a representation.)

The problem of maximizing a t-smooth submodular function is NP-hard for every $t \geq 2$. But submodularity and t-smoothness do help in finding reasonably good approximations of the optimum; see Fisher-Nemhauser-Wolsey (1978).

6. Submodular objective functions and constraints

The set-function which occurs by far the most often in combinatorics is the cardinality function, which is a modular and 1-smooth set-function. In weighted and polyhedral versions of combinatorial optimization problems we deal with more general modular functions. In many proof techniques, however, one can further generalize and use only sub- (or super-) modularity of the set-functions in question. It turns out that this generalization is not all all l'art pour l'art, but it yields beautiful results which contain many deep combinatorial min-max results as special cases, while in their statements and proofs, as well as in the algorithms that go with them, they preserve the simplicity and heuristic value of the original purely combinatorial results.

As an early example of this approach, let us consider a version of the Matroid Intersection Theorem, due to Aigner and Dowling (1971). Let G be a graph and let $(V(G), \mathcal{M})$ be a matroid defined on its nodes. Let us say that a matching of G is a *matroid matching* if the set of nodes of G it meets is independent in the matroid. Then the following generalization of the König Theorem can be proved.

6.1. Theorem. Let G be a bipartite graph and $(V(G), \mathcal{M})$ a matroid such that the color-classes of G are separators of $(V(G), \mathcal{M})$. Then the maximum size of a matroid matching is equal to the minimum rank of a point-cover of G.

Next we discuss some more recent results on graphs featuring submodular constraints. Edmonds and Giles (1977) discuss *submodular flows*. Let G be a directed graph, $\mathcal{H} \subseteq 2^{V(G)}$ a crossing family and f a submodular function defined on $\mathcal{H}$. Let, further, two functions $c, d: E(G) \to \mathbb{R}$ be given such that $c(e) \geq d(e)$ for all $e \in E(G)$. A *submodular flow* defined on G is an assignment of real numbers $x(e)$ to the lines e such that

$$d(e) \leq x(e) \leq c(e) \quad \text{for all } e \in E(G), \tag{8}$$

$$x(V^+(T)) - x(V^-(T)) \leq f(T) \quad \text{for all } T \in \mathcal{H}. \tag{9}$$

(Here $V^+(T)$ and $V^-(T)$ denote the sets of lines leaving and entering T, respectively.) The result of Edmonds and Giles states:

6.2. Theorem. *If f, c and d are integral valued functions then the system (8)-(9) is total dual integral. Consequently, for any weighting of the lines with integers there exists a maximum weight submodular flow wiht integral values.*

Perhaps the most important special case of this result is the Lucchesi-Younger Theorem on diconnecting sets. Let $\mathcal{H}$ consist of those non-empty proper subsets T of $V(G)$ which determine a directed cut, i.e. which have $V^-(T) = \emptyset$. It is easy to see that this is a crossing family. Further, the function $V^+(X) - 1$ is submodular on $\mathcal{H}$. Take $d = 0$ and $c = 1$ for every line. Then any submodular flow with integral values may be viewed as a subgraph, and (9) says that the complement of this subgraph meets every directed cut. So the integral submodular flows are precisely the complements of diconnecting sets. Taking all weights equal to 1, and applying linear programming duality together with the primal and dual integrality implied by the Edmonds-Giles theorem, we obtain the Lucchesi-Younger Theorem (1978):

6.3. Corollary. *The minimum size of a diconnecting set in a digraph is equal to the maximum number of mutually disjoint directed cuts.*

Lawler and Martell (1980) introduce a similar problem. While in the problem of Edmonds and Giles classical network flows are generalized by retaining capacity constraints but generalizing the "Kirchhoff" laws, Lawler and Martell retain the Kirchhoff laws but replace capacity constraints by submodular ones. To be more precise, let G be a directed graph, s and t two specified nodes of G, and let, for each node $v \in V(G)$, two submodular functions α_v, β_v be given: α_v on the subsets of $V^+(v)$ and β_v on the subsets of $V^-(v)$. The functions α_v yield a submodular set-function on the set $E(G)$ by taking their direct sum, and let β be similarly the direct sum of the functions β_v. Then a *polymatroidal network flow* is defined as an st-flow in the usual sense, which in addition satisfies the following conditions:

$$x(T) \leq \min\{\alpha(T), \beta(T)\} \quad \text{for all } T \subseteq E(G)$$

(it would suffice to reqire this for the subsets of in- and out-stars of nodes). Let, for each st-cut C, the capacity of C be defined as $(\alpha * \beta)(C)$.

With this definition, the following generalization of the Max-flow-min-cut Theorem can be proved by an appropriate refinement of augmenting path techniques:

6.4. Theorem. *The maximum value of a polymatroidal network flow is equal to the minimum capacity of an st-cut.*

A further closely related class of results is due to Frank (1979, 1980). We only state one of these. Let G be a graph and let f be a supermodular set-function defined on the subsets of $V(G)$. Let us say that an orientation $\vec{G}$ of G satiesfies demand f if

$$|V_{\vec{G}}^+(T)| \geq f(T)$$

holds true for every subset $T \subseteq V(G)$, $T \neq \emptyset$, $V(G)$. Thus an orientation which satiesfies demand 1 is just strongly connected.

6.5. Theorem. *The graph G has an orientation which satisfies demand f if and only if*

$$\sum_{i=1}^{k} f(T_i) \leq |E(G)| \quad \text{and} \quad \sum_{i=1}^{k} f(S - T_i) \leq |E(G)|$$

holds for every partition $\{T_1, \ldots, T_k\}$ of $V(G)$.

Note that the first condition says that $f^*(V(G)) \leq |E(G)|$, and the second is obtained from the first by complementation in the variable.

We conclude with a different version of the same idea. Let us return to Theorem 6.1. What happens if we have a non-bipartite graph G or a bipartite graph whose color-classes are not separators of the underlying matroid? Already for ordinary graphs, the problems of maximum matching and minimum point-cover are different in the non-bipartite case, so we have to discuss them separately.

The problem of finding a maximum matroid matching is equivalent to the matroid matching problem as formulated in section 5. Thus, this problem is well-solved e.g. if the matroid is given as the columns of a matrix.

The problem of finding the minimum rank of a point-cover is NP-complete already in the case when the underlying matroid is free. So no complete solution of this problem can be expected. It is, however, possible to generalize various results on the point-covering number to this matroidal point-covering number. Applying these matroidal results to situations where the underlying matroids are not free (e.g. they are transversal), further graph-theoretic results can be proved. In some cases this generalization to matroids is the key to the proof. This is the case with a finite basis theorem for τ-critical graphs (see Lovász 1977).

7. Concluding remarks

In this last section we mention some open problems concerning submodularity, without aiming at a complete list.

1. The algorithm presented in section 4 to minimize a submodular function (which is probably the key problem in the algorithmic theory of submodularity) is not at all satisfactory. It is polynomial in running time, but it involves the ellipsoid method which renders it more or less useless in practical applications. It is an outstanding open problem to find an algorithm to minimize a submodular set-function in polynomial time, which would operate by combinatorial means and which would be also practical.

2. The results discussed in section 6 indicate that a more thorough study of combinatorial optimization problems with submodular objective functions and constraints is in progress. Do further standard combinatorial optimization problems (matching, chromatic number, feedback etc.) have generalizations in this spirit which lead to interesting results?

3. The submodular functions defined on a given set form a convex cone. Some properties of this cone have been studied (Nguyen 1978), but its description is by far not satisfactory.

4. The functions δ_w of example 1.3 also form a convex cone. As opposed to the cone of submodular functions, for this cone the extreme rays are easy to find (they are the functions δ_w where w is 0 on all lines except one), so the question is to find a system of linear inequalities describing this cone. More modestly, we may ask for interesting valid inequalities. Submodularity is one of them, but there are others, e.g. the following family of "generalized submodular inequalities":

$$\sum_{i=1}^{k} f(X_i) \geq \sum f(X_1^{e_1} \cap \ldots \cap X_k^{e_k}),$$

where $X^1 = X$, $X^{-1} = S - X$, and the summation on the right hand side extends to all ± 1-sequences $(e_1, \ldots, e_k)$ with $e_1 \ldots e_k = 1$.

5. The Dilworth truncation of a submodular function may be viewed as the most economical way to make $f(\emptyset)$ non-negative (at the expense of decreasing the function value on other sets). Is there a similar construction to make $f(\emptyset)$ and $f(\{x\})$ non-negative for every $x \in S$? More generally, which systems of subsets of S have the property that there is a canonical way to increase the value of a submodular function on these sets at the expense of decreasing it on the others?

6. Let f be a polymatroid function defined on the subsets of a set S and let $x, y \in S$. When does f have an extension f' to the set $S \cup \{z\}$ such that $f'(z) = 1$, $f'(\{x, z\}) = f(\{x\})$ and $f'(\{y, z\}) = f(\{y\})$? Geometrically, this would mean to place a point on the intersection of the subspaces x and y. This is probably a very difficult problem in general, but as the Dilworth truncation shows, it can be solved if e.g. x is a "general hyperplane". Can less restrictive conditions be found under which this extension problem can be solved?

7. One may also raise the question whether any reasonable class of set-functions more general than submodular ones admits a similarly deep theory. There are some examples which direct in the direction of a positive answer. The Polymatroid Intersection Theorem shows that the minimum of two submodular set-functions, even though not submodular in general, gives rise to a polyhedron which still has nice properties (e.g. integral vertices). Results of Hoffman and Gröflin (1981) show that an analogue of the Matroid Partition Theorem is true for the rank function of the intersection of two matroids, which is not submodular but the convolution of two submodular functions. Does there exist a property similar to, but more general than, submodularity which plays role in these examples?

References

M. Aigner, T. A. Dowling (1971), Matching theory for combinatorial geometries, *Trans. Amer. Math. Soc.* 158, 231–245.

L. Choquet (1955), Theory of capacities, *Ann. Inst. Fournier Grenoble* 5, 131–295.

H. Crapo (1965), Single-element extensions of matroids, *J. Res. Nat. Bur. Stand.* 69 B, 55–65.

H. Crapo (1979), Structural rigidity, *Structural Topology* 1, 26–45.

R. P. Dilworth (1944), Dependence relations in a semimodular lattice, *Duke Math. J.* 11, 575–586.

F. D. J. Dunstan (1976), Matroids and submodular functions, *Quart. J. Math. Oxford* 27, 339–348.

J. Edmonds (1970), Submodular functions, matroids, and certain polyhedra, in: *Combinatorial Structures and their Applications* (eds. R. Guy, H. Hanani, N. Sauer, J. Schönheim) Gordon and Breach, 69–87.

J. Edmonds, D. R. Fulkerson (1965), Transversals and matroid partition, *J. Res. Nat. Bur. Stand.* 69 B, 147–153.

J. Edmonds, R. Giles (1977), A min-max relation on submodular functions on graphs, *Annals of Discrete Math.* 1, 185-204.

M. L. Fisher, G. L. Nemhauser, L. A. Wolsey (1978), Analysis of approximations for maximizing a submodular setfunction II, *Math. Prog. Study* 8, 73-87.

A. Frank (1982), An algorithm for submodular functions on graphs, *Annals of Discrete Math.* 16, 97-120.

A. Frank (1980), On the orientation of graphs, *J. Comb. Theory* B28, 251-261.

A. Frank (1979), Kernel systems of directed graphs, *Acta Sci. Math. Univ. Szeged* 41, 63-76.

H. Gröflin, A. J. Hoffman (1981), On matroid intersections, *Combinatorica* 1, 43-47.

M. Grötschel, L. Lovász, A. Schrijver (1981), The ellipsoid method and its consequences in combinatorial optimization, *Combinatorica* 1, 169-197.

P. M. Jensen, B. Korte (1978), Complexity of matroid property algorithms, Rept. No. 7124-OR, Inst. Ökon. Oper. Res. Univ. Bonn.

T. Helgason (1974), Aspects of the theory of hypermatroids, in: *Hypergraph Seminar* (eds. C. Berge, D. K. Ray-Chaudhuri), Lecture Notes in Math. 411, Springer, 191-214.

D. B. Iudin, A. S. Nemirovskii (1976), Informational complexity and effective methods of solution for convex extremal problems, *Ekon. i Mat. Met.* 12, 357-369; *Matekon* 13 (3), 24-45.

E. L. Lawler, C. U. Martell (1980), Computing maximal "polymatroidal" network flows, Res. Rep. UCB/ERL M80/52.

E. L. Lawler, Po Tong (1982), Lecture at the Conference on Combinatorics and Graph Theory, Univ. of Waterloo.

L. Lovász (1977), Flats in matroids and geometric graphs, in: *Combinatorial Surveys,* (ed. P. Cameron), Acad. Press, 45-86.

L. Lovász (1981), The matroid matching problem, in: *Algebraic Methods in Graph Theory* (eds. L. Lovász, V. T. Sós), North-Holland, 495-517.

L. Lovász, Y. Yemini (1982), On generic rigidity in the plane, SIAM J. on Alg. Discr. Meth. 3, 91-99.

C. J. H. McDiarmid (1975), Rado's Theorem for polymatroids, Math. Proc. Cambridge Phil. Soc. 78, 263-281.

J. H. Mason (1977), Matroids as the study of geometrical configurations, in: *Higher Combinatorics* (ed. M. Aigner) Reidel, 133-176.

J. H. Mason (1981), Glueing matroids together: a study of Dilworth truncations and matroid analogues of exterior and symmetric powers, in: *Algebraic Methods in Graph Theory* (eds. L. Lovász, V. T. Sós), North-Holland, 519-561.

H. Q. Nguyen (1978), Semimodular functions and combinatorial geometries, *Trans. AMS* 238, 355-383.

O. Ore (1955), Graphs and matching theorems, *Duke Math. J.* 22, 625-639.

D. Welsh (1976), *Matroid Theory,* Academic Press.

Recent Developments in Algorithms and Software for Trust Region Methods*

J. J. Moré
Argonne Nat. Lab., Mathematics and Computer Science Division, 9700 S Cass Avenue, Argonne, IL 60439, USA

Abstract. Trust region methods are an important class of iterative methods for the solution of nonlinear optimization problems. Algorithms in this class have been proposed for the solution of systems of nonlinear equations, nonlinear estimation problems, unconstrained and constrained optimization, nondifferentiable optimization, and large scale optimization. Interest in trust region methods derives, in part, from the availability of strong convergence results and from the development of software for these methods which is reliable, efficient, and amazingly free of ad-hoc decisions. In this paper we survey the theoretical and practical results available for trust region methods and discuss the relevance of these results to the implementation of trust region methods.

1. Introduction to Trust Region Methods

The development of trust region methods can be traced back to the work of Levenberg [1944] and Marquardt [1963] on nonlinear least squares problems. Given a continuously differentiable function $F: R^n \to R^m$, the nonlinear least squares problem is to find a minimizer of the function

$$\|F(x)\|_2 \equiv \left(\sum_{i=1}^{m} f_i(x)^2 \right)^{\frac{1}{2}}.$$

The Levenberg-Marquardt algorithm can be viewed as a method for generating a sequence $\{x_k\}$ of iterates where the step s_k between iterates is a solution to the problem

(1.1) $\qquad \min \{\|F(x_k) + F'(x_k) w\|_2 : \|D_k w\| \leq \Delta_k\}.$

for some bound Δ_k and scaling matrix D_k. The norm $\|\cdot\|$ is arbitrary but it is usually chosen as the l_2 norm because for this choice Marquardt [1963] proved that if the step s_k is determined by solving the linear system

$$(F'(x_k)^T F'(x_k) + \lambda_k D_k^T D_k) s_k = - F'(x_k)^T F(x_k)$$

with some parameter $\lambda_k \geq 0$, then s_k solves (1.1) with $\Delta_k = \|D_k s_k\|$.

---

* Work supported in part by the Applied Mathematical Sciences Research Program (KC-04-02) of the Office of Energy Research of the U.S. Department of Energy under Contract W-31-109-Eng-38.

The guiding principle of a trust region method is that the step s_k is a solution to a subproblem with a bound on the step. The subproblem is chosen so that the solution s_k is likely to yield an improvement in the current approximation to the optimization problem. This principle appears in many forms and thus it is worthwhile reviewing some of the variations that have been used.

In the Goldfeld, Quandt, and Trotter [1966] version of Newton's method for the unconstrained minimization of a function $f: R^n \to R$, the step s_k solves a subproblem of the form

(1.2) $$\min\{\nabla f(x_k)^T w + \tfrac{1}{2} w^T \nabla^2 f(x_k) w : \|D_k w\| \leq \Delta_k\}$$

for some bound Δ_k and scaling matrix D_k. Goldfeld, Quandt, and Trotter [1966] chose $\|\cdot\|$ to be the l_2 norm and proved that if s_k satisfies

$$(\nabla^2 f(x_k) + \lambda_k D_k^T D_k) s_k = -\nabla f(x_k)$$

for some parameter $\lambda_k \geq 0$ with $\nabla^2 f(x_k) + \lambda_k D_k^T D_k$ positive semidefinite, then s_k solves (1.2) with $\Delta_k = \|D_k s_k\|$.

The Goldfeld-Quandt-Trotter and Levenberg-Marquardt algorithms are similar. In both algorithms the subproblem is of the form

(1.3) $$\min\{\psi_k(w) : \|D_k w\| \leq \Delta_k\}$$

where ψ_k is a quadratic approximation to the actual reduction in the objective function; in the Levenberg-Marquardt algorithm $\psi_k(w) - \psi_k(0)$ is a first order approximation, while in the Goldfeld-Quandt-Trotter algorithm $\psi_k(w) - \psi_k(0)$ is a second order approximation. A difference is that for the Levenberg-Marquardt algorithm ψ_k is convex and thus it is relatively easy to obtain the global solution to (1.3).

Early versions of the Levenberg-Marquardt and Goldfeld-Quandt-Trotter algorithms controlled s_k indirectly by changing λ_k. This does not fit the philosophy of trust region methods which requires direct control of Δ_k and determines s_k as the solution of subproblem (1.3). Direct control of λ_k has a number of problems. One of the problems is that there does not seem to be a reasonable, automatic choice of λ_0. On the other hand, a reasonable value of Δ_0 is quite often a small fraction of the size $\|D_0 x_0\|$ of the starting point. Another problem occurs when $x_k + s_k$ leads to an increase in the objective function. In this case we can use the function and derivative information obtained at x_k and $x_k + s_k$ to estimate the required decrease in Δ_k, but it is not clear that this information can be used to estimate a reasonable value for $\lambda_{k+1} > \lambda_k$. For example, if we set

$$\varphi(\tau) = \|F(x_k + \tau s_k)\|_2$$

in the Levenberg-Marquardt algorithm then we know $\varphi(0)$, $\varphi(1)$, and $\varphi'(0) < 0$. If $\varphi(1) \geq \varphi(0)$ then the quadratic which interpolates this data has a minimizer $\gamma < 1$, and it is reasonable to set $\Delta_{k+1} = \gamma \|D_k s_k\|$ in most cases.

Hebden [1973], in connection with the Goldfeld-Quandt-Trotter algorithm, was the first to propose a reasonable algorithm for the approximate solution of (1.3). His algorithm is basically sound but encounters difficulties when ψ_k is not convex. Moré [1978] presented theoretical and numerical results which showed that Hebden's ideas could be used to produce a reliable and efficient implementation of the Levenberg-Marquardt algorithm. This implementation, in particular, did not have many of the ad-hoc decisions of previous implementations.

The algorithms of Hebden and Moré did not seek an accurate solution of (1.3) but instead were content with a nearly optimal solution of (1.3). Strong theoretical and numerical results can be obtained if the step s_k satisfies

(1.4) $\quad \psi_k(s_k) \leq \beta_1 \min\{\psi_k(w): \|D_k w\| \leq \Delta_k\}, \quad \|D_k s_k\| \leq \beta_2 \Delta_k,$

for some positive constants β_1 and β_2. For many problems the cost of obtaining a step s_k which satisfies these conditions is quite reasonable, but for some problems (for example, large scale problems) the cost may be prohibitive. An alternative is to determine the step s_k by solving a restricted subproblem of the form

(1.5) $\quad \min\{\psi_k(w): w \in S_k, \|D_k w\| \leq \Delta_k\}$

for some subspace S_k. An advantage of this subproblem is that the solution of (1.5) can be considerably less expensive than (1.3). Moreover, if S_k is chosen appropriately then good theoretical and numerical results can be obtained. Convergence results can be obtained if

(1.6) $\quad \psi_k(s_k) \leq \beta_1 \min\{\psi_k(w): w = v s_k^G, \|D_k w\| \leq \Delta_k\}, \quad \|D_k s_k\| \leq \beta_2 \Delta_k,$

for some positive constants β_1 and β_2, and where s_k^G is the steepest descent direction for ψ_k with respect to the norm $\|D_k(\cdot)\|$. For future reference, it is convenient to define a Cauchy point s_k^C by

$$\psi_k(s_k^C) = \min\{\psi_k(w): w = v s_k^G, \|D_k w\| \leq \Delta_k\}.$$

Thus, convergence results can be obtained if s_k solves (1.5) and if the subspace S_k contains a vector w_k such that $\|D_k w_k\| \leq \Delta_k$ and $\psi_k(w_k) \leq \beta_1 \psi_k(s_k^C)$.

Powell [1970a] was the first to propose a trust region method with a subproblem of the form (1.5). In Powell's algorithm for the solution of a system of nonlinear equations $F: R^n \to R^n$, the step s_k solves a subproblem of the form

(1.7) $\quad \min\{\|F(x_k) + J_k w\|_2: w = \tau s_k^C + (1-\tau) s_k^N, \|D_k w\| \leq \Delta_k\},$

where s_k^C is the Cauchy step, s_k^N is the Newton step, and the matrix J_k is either a difference approximation to the Jacobian matrix $F'(x_k)$ or a scaled version of the quasi-Newton approximation obtained with the update of Broyden [1965]. This algorithm was extended by Powell [1970b] to the unconstrained minimization problem by requiring that the step s_k solve a subproblem of the form

(1.8) $\quad \min\{\nabla f(x_k)^T w + \tfrac{1}{2} w^T B_k w: w = \tau s_k^C + (1-\tau) s_k^N, \|D_k w\| \leq \Delta_k\}.$

The matrix B_k is usually the quasi-Newton approximation obtained with the symmetrization of the update of Broyden, but other choices are possible. If $B_k = \nabla^2 f(x_k)$ in subproblem (1.8) we then obtain a version of Newton's method for unconstrained minimization. Similarly, setting $J_k = F'(x_k)$ in subproblem

(1.7) leads to a version of Newton's method for systems of nonlinear equations.

Powell chose $\|\cdot\|$ as the l_2 norm in subproblems (1.7) and (1.8), but it is clearly possible to consider other choices of norm. An advantage of Powell's choice is that the optimal τ can be obtained by solving a quadratic equation.

Trust region methods have also been developed for the solution of constrained optimization problems. Fletcher [1972] proposed an algorithm for the linearly constrained optimization problem

$$\min\{f(x): C^T x \leq d\}$$

in which the subproblem is of the form

(1.9) $\quad \min\{\nabla f(x_k)^T w + \tfrac{1}{2} w^T B_k w: C^T(x_k+w) \leq d, \|D_k w\| \leq \Delta_k\}$

for some symmetric matrix B_k. Fletcher suggested that $\|\cdot\|$ be chosen as the l_∞ norm because then subproblem (1.9) is a quadratic programming problem. This is also the case if $\|\cdot\|$ is the l_1 norm, but with the l_∞ norm the constraint on the step leads to simple bounds on the quadratic programming problem. Fletcher's algorithm is a generalization of the algorithm of Griffith and Stewart [1961] in which $B_k \equiv 0$ and $\|\cdot\|$ is the l_∞ norm. An advantage of choosing $B_k = 0$ is that subproblem (1.9) is then a linear programming problem. The choice of $B_k = 0$, however, may lead to an unacceptably low rate of convergence.

In all of the trust region methods that we have mentioned, ψ_k is a quadratic function, but this is not always the case. We illustrate this point with the algorithm of Madsen [1975] for the minimax solution of a system of nonlinear equations $F: R^n \to R^m$. The minimax problem requires a minimizer of the function

$$\|F(x)\|_\infty = \max\{|f_i(x)|: 1 \leq i \leq m\}.$$

In the algorithm of Madsen [1975], the step s_k solves the subproblem

(1.10) $\quad \min\{\|F(x_k)+F'(x_k)w\|_\infty: \|D_k w\| \leq \Delta_k\}.$

The norm $\|\cdot\|$ is arbitrary, but Madsen uses the l_∞ norm because then (1.10) can be formulated as a linear programming problem. Note that subproblem (1.10) is of the form (1.3) with ψ_k a polyhedral convex function.

Fletcher [1981] extended Madsen's algorithm to the nondifferentiable optimization problem

(1.11) $\quad \min\{f(x)+\varphi(F(x)): x \in R^n\},$

where $f: R^n \to R$ and $F: R^n \to R^m$ are smooth but $\varphi: R^m \to R$ is only assumed to be convex. The subproblems in Fletcher's algorithm are of the form (1.3) with

(1.12) $\quad \psi_k(w) = \nabla f(x_k)^T w + \tfrac{1}{2} w^T B_k w + \varphi(F(x_k)+F'(x_k)w)$

for some symmetric matrix B_k. In this subproblem ψ_k is the sum of a quadratic function and a general convex function.

The nondifferentiable optimization problem (1.11) includes smooth minimization problems when $\varphi \equiv 0$ and convex minimization problems when $f \equiv 0$.

In particular, many nonlinear estimation problems are of the form (1.11) with $f \equiv 0$ and φ a polyhedral norm. See, for example, Anderson and Osborne [1977]. More generally, exact penalty functions for the general constrained optimization problem are of the form (1.11) with φ a polyhedral convex function. See, for example, Fletcher [1981].

As can be seen from the above papers, trust region methods can be applied to a wide range of problems. It is not our intent to survey all of this work, but rather to mention some of the recent developments which have had an important impact on trust region methods. These developments fall in three areas: scaling and preconditioning techniques, algorithms for the computation of the trust region step, and convergence results.

Scaling and preconditioning techniques dictate the choice of the scaling matrix D_k in a trust region method. These techniques have always played an important role in the development of optimization algorithms but in recent years there has been renewed interest in the use of these techniques. Preconditioned conjugate gradient methods, in particular, are emerging as powerful methods for the solution of linear and nonlinear large scale optimization problems. Our intention in Section 2 is to review, in a general setting, some of the basic ideas used in connection with scaling and preconditioning techniques. These ideas are then used throughout the paper.

The algorithm used for the computation of the trust region step should determine, reliably and efficiently, a step s_k in a finite number of steps. In addition, the step s_k should produce a sufficient decrease in the model ψ_k. At present there are algorithms which satisfy these requirements for subproblems (1.3) and (1.5) when ψ_k is a quadratic and $\|\cdot\|$ is the l_2 norm. We review and compare these algorithms in Section 3; in particular, we show that they satisfy the sufficient decrease conditions (1.4) and (1.6). Sufficient decrease conditions for more general subproblems, such as (1.9), (1.10), and (1.12), have not been formulated. Moreover, the convergence theory associated with these subproblems requires that s_k be the global solution. Since these subproblems are usually solved by quadratic programming techniques which do not guarantee a global solution unless ψ_k is convex, there is clearly a wide gap between theory and practice.

Convergence and rate of convergence results for trust region methods are presented in Section 4. These results are unusually strong. If the sufficient decrease condition (1.6) is satisfied then we can show that

$$\lim_{k \to +\infty} \nabla f(x_k) = 0$$

holds under reasonable conditions; if the stronger decrease condition (1.4) is satisfied then we can show, for example, that if x^* is an isolated limit point of $\{x_k\}$ then $\nabla^2 f(x^*)$ is positive semidefinite. Although results of this type are available for other algorithms, for trust region methods these results follow naturally from the sufficient decrease conditions (1.4) and (1.6). As a consequence, software for trust region methods tends to be free of ad-hoc decisions.

We end the paper with a brief discussion of some possible directions for research in trust region methods.

2. Scaling and Preconditioning

The notion of a scale invariant optimization algorithm is basic to the study of scaling and preconditioning techniques. We first consider unconstrained problems.

Given a function $f: R^n \to R$ and a starting vector $x_0 \in R^n$, an unconstrained minimization algorithm can be defined as a method for generating a sequence of iterates $\{x_k\}$ which hopefully converges towards a minimizer of f. Now consider optimization problems related by the transformation

(2.1) $$\hat{f}(x) = \alpha f(D^{-1}x), \quad \hat{x}_0 = Dx_0,$$

where D is a nonsingular matrix and α is a positive scalar. If the iterates $\{x_k\}$ and $\{\hat{x}_k\}$ generated by the algorithm satisfy

(2.2) $$\hat{x}_k = Dx_k, \quad k > 0,$$

then the algorithm is *scale invariant*. The *invariance group* of the algorithm is the set of all matrices D and scalars α which satisfy (2.2) for all suitable functions f and starting points x_0.

The concept of scale invariance extends easily to the general constrained minimization problem

$$\min\{f(x): c_i(x) \leq 0, i \in I, c_i(x) = 0, i \in E\}.$$

Given a constrained minimization problem defined by the objective function f, the constraint functions c_i, and the starting point x_0, we can consider a transformed problem where $\hat{f}$ and $\hat{x}_0$ are defined by (2.1), while $\hat{c}_i$ is defined by setting

$$\hat{c}_i(x) = \beta_i c_i(D^{-1}x), \quad i \in I \cup E,$$

for some constants β_i with $\beta_i > 0$ for $i \in I$ and $\beta_i \neq 0$ for $i \in E$. If the iterates $\{x_k\}$ and $\{\hat{x}_k\}$ generated by the algorithm satisfy (2.2) then the constrained minimization algorithm is scale invariant.

Several authors have explicitly used scale invariance ideas in the design of optimization algorithms. The following papers are representative of work in three different areas of unconstrained optimization. Deuflhard [1974] proposed a globally convergent version of Newton's method for the solution of the system of nonlinear equations $H(x) = 0$ which is invariant under the transformation

$$\hat{H}(x) = D H(x)$$

where D is a nonsingular matrix. Fletcher [1971] proposed a version of the Levenberg-Marquardt method for nonlinear least squares which is scale invariant for diagonal matrices D. Davidon [1980] proposed an unconstrained minimization algorithm which is invariant under the transformation

$$\hat{f}(x) = \alpha f(S^{-1}(x)), \quad \hat{x}_0 = S(x_0),$$

where $S: R^n \to R^n$ is an invertible collinear mapping, that is, S is the ratio of affine mappings.

Scaling considerations have also played a role in the design of constrained optimization algorithms. For example, Hanson and Hiebert [1981], and Fletcher and Jackson [1974] note that Lagrange multipliers are affected by the scaling of the variables and have devised scale invariant active set strategies for, respectively, linear programming and quadratic programming problems. As a last example note that the choice of exact penalty function made by Powell [1978] is of the form

$$\Psi(x) = f(x) + \sum_{i \in E} \mu_i |c_i(x)| + \sum_{i \in I} \mu_i \max\{0, c_i(x)\}.$$

If the choice of μ_i is such that $\hat{\mu}_i = (\alpha/|\beta_i|)\mu_i$ then $\hat{\Psi}(x) = \alpha \Psi(D^{-1}x)$, and thus the use of this form of penalty function does not destroy the scale invariance of the underlying algorithm.

We claim that the better optimization algorithms tend to have a large invariance group. This is sometimes justified by pointing out that if an algorithm is scale invariant then the convergence properties are not affected by transformation (2.1). In further support of this claim, note that the unconstrained minimization version of Newton's method

$$x_{k+1} = x_k - \alpha_k \nabla^2 f(x_k)^{-1} \nabla f(x_k)$$

is scale invariant for any nonsingular matrix D. Similarly, variable metric methods for unconstrained minimization are scale invariant for any nonsingular matrix D provided the intial metric reflects the change of variables (2.1). The steepest descent algorithm

$$x_{k+1} = x_k - \alpha_k \nabla f(x_k)$$

has a smaller invariance group; it is scale invariant if $D^T D = \alpha I$. Standard conjugate gradient methods have the same invariance group.

We are assuming that the optimization algorithm is well-defined. For the version of Newton's method above, this means that $\nabla^2 f(x_k)$ is assumed to be positive definite so that only local convergence can be guaranteed. Globally convergent versions of Newton's method usually have a smaller invariance group.

In the above examples we have also assumed that the line search algorithm which generates the steplengths α_k is invariant under the transformation (2.1). This is not an unreasonable assumption. Given an iterate x_k and a direction p_k, a line search algorithm is a method for producing a sufficient decrease in the function $\varphi: R \to R$ defined by

$$\varphi(\tau) = f(x_k + \tau p_k).$$

The use of transformation (2.1) on Newton and variable metric methods produces directions p_k and $\hat{p}_k$ related by $\hat{p}_k = D p_k$. This also holds for the steepest descent and conjugate gradient methods if $D^T D = \alpha I$. We thus have that $\hat{\varphi}(\tau) \equiv \alpha \varphi(\tau)$, and since most line search algorithms are invariant under this transformation, this justifies the invariance assumption.

We have considered optimization algorithms at a fairly abstract level. Scale invariance is harder to obtain if, for example, we consider the termination criteria as part of the algorithm. To illustrate this point consider the gradient test

$$\|\nabla f(x_k)\|_2 \leq \varepsilon |f(x_k)|.$$

This test may destroy the scale invariance properties of the algorithm because it is not invariant under transformation (2.1) unless D is an orthogonal matrix, and $\hat{x}_k = Dx_k$. A better test is obtained if we require that

$$|\min\{\nabla f(x_k)^T w + \tfrac{1}{2} w^T \nabla^2 f(x_k) w : \|D_k w\| \leq \Delta_k\}| \leq \varepsilon |f(x_k)|$$

This is a natural test for the Goldfeld-Quandt-Trotter algorithm because it requires that the decrease predicted by the quadratic is a fraction of $|f(x_k)|$. A similar test is to require that

$$|\min\{\nabla f(x_k)^T w : \|D_k w\| \leq \Delta_k\}| \leq \varepsilon |f(x_k)|.$$

If $\hat{x}_k = Dx_k$ then both of these test are invariant under transformation (2.1) provided $\hat{D}_k = v D_k D^{-1}$ and $\hat{\Delta}_k = v \Delta_k$ for some $v > 0$. These tests, therefore, preserve the scale invariance properties of the underlying algorithm. In practice these tests need to be supplemented; for an interesting discussion of other termination criteria for trust region methods, see Gay [1982].

At the software level it is only reasonable to expect scale invariance if D is a diagonal matrix. This remark can be illustrated by considering the solution process for the Newton step. If we use the Cholesky decomposition in this process then we must compute upper triangular matrices R_k and $\hat{R}_k$ such that

$$\nabla^2 f(x_k) = R_k^T R_k, \qquad \nabla^2 \hat{f}(\hat{x}_k) = \hat{R}_k^T \hat{R}_k.$$

For a scale invariant algorithm we have $\hat{x}_k = Dx_k$ and thus

$$\nabla^2 \hat{f}(\hat{x}_k) = \alpha D^{-T} \nabla^2 f(x_k) D^{-1}.$$

It follows that $\hat{R}_k = \beta R_k D^{-1}$ if D is a diagonal matrix and $\alpha = \beta^2$. This is the expected transformation of the Cholesky decomposition. The above derivation is not valid in the presence of roundoff, but it is not difficult to verify that if β and the entries of D are powers of the base of the machine, then $\hat{R}_k = \beta R_k D^{-1}$ holds.

It is possible to make an algorithm scale invariant by *preconditioning* the algorithm. Consider, for example, the steepest descent algorithm. If we apply the steepest descent algorithm to f and if we insist on having $\hat{\alpha}_k = \alpha_k$ and $\hat{x}_k = Dx_k$, then we obtain

(2.3) $$x_{k+1} = x_k - \alpha_k C^{-1} \nabla f(x_k)$$

where $C = (1/\alpha) D^T D$. In this way we obtain the *preconditioned steepest descent* algorithm. An important point about this derivation is that although we have used transformation (2.1) to obtain (2.3), the preconditioned steepest descent algorithm is defined for any positive definite matrix C.

Preconditioning improves the convergence properties of the steepest descent algorithm. For example, it is now possible to choose the matrix C so as to improve the rate of convergence of the algorithm. Also note that the preconditioned steepest descent algorithm is scale invariant provided

(2.4) $$\hat{C} = \alpha D^{-T} C D^{-1}.$$

Of course, it is now necessary to choose a positive definite C so that (2.4) holds, but this can be done automatically in some cases. For instance, the choice $C = \nabla^2 f(x_0)$ satisfies (2.4) for any nonsingular matrix D, but may not result in a positive definite C. The choice

$$C = \text{diag}(|\partial_{ii} f(x_0)|)$$

satisfies (2.4) for all diagonal matrices D with positive diagonal entries and is usually positive definite.

The technique used to derive the preconditioned steepest descent algorithm is quite general and can be applied to any algorithm. As a further example of this technique consider the conjugate gradient algorithm

$$x_{k+1} = x_k + \alpha_k d_k,$$
$$\beta_k = \left(\frac{\|\nabla f(x_{k+1})\|_2}{\|\nabla f(x_k)\|_2} \right)^2,$$
$$d_{k+1} = -\nabla f(x_{k+1}) + \beta_k d_k,$$

with the Fletcher-Reeves choice of β_k. The *preconditioned conjugate gradient* algorithm is then defined by the equations

$$x_{k+1} = x_k + \alpha_k d_k,$$
$$\beta_k = \frac{\nabla f(x_{k+1})^T C^{-1} \nabla f(x_{k+1})}{\nabla f(x_k)^T C^{-1} \nabla f(x_k)},$$
$$d_{k+1} = -C^{-1} \nabla f(x_{k+1}) + \beta_k d_k.$$

The similarity between the standard conjugate gradient method and the preconditioned version is reinforced if β_k is written in terms of the elliptical norm

(2.5) $$\|x\|_A = (x^T A x)^{\frac{1}{2}},$$

where A is symmetric and positive definite. In this notation the choice of β_k is

$$\beta_k = \left(\frac{\|\nabla f(x_{k+1})\|_{C^{-1}}}{\|\nabla f(x_k)\|_{C^{-1}}} \right)^2.$$

Note that the standard conjugate gradient algorithm is scale invariant if $D^T D = \alpha I$, that is, if (2.4) holds with $\hat{C} = C = I$. The preconditioned conjugate gradient algorithm is scale invariant if (2.4) holds.

For nonlinear problems, the preconditioned conjugate gradient method given above is modified in several ways by exploiting the relationship between conjugate gradient methods and variable metric methods. Some of these modifications are explored by Gill and Murray [1979] and Nazareth and Nocedal

[1982]. Buckley and LeNir [1982] present numerical results for a particularly appealing modification.

Numerical results for several of the preconditioning techniques used in conjugate gradient and Lanczos algorithms for the solution of linear equations and least squares problems are given by Björck and Elfving [1979], Paige and Saunders [1978], Saunders [1979], and Munksgaard [1980]. These results are of special interest because conjugate gradient and Lanczos algorithms can be used for the solution of the subproblems associated with trust region methods.

There is no doubt that scaling and preconditioning techniques are useful in the solution of optimization problems. There are, however, very few results which indicate how to scale the problem or how to precondition the algorithm. It is perhaps for this reason that optimization software tends to assume that the user has scaled his problem properly. This practice is not universal. Some codes allow the user to specify the scaling matrix D and then implicitly or explicitly work with the scaled problem (2.1). Most of these codes also have a default option which depends on the problem. The codes of Fletcher [1971], and Hiebert and Hanson [1981], for example, have these options. Other codes have the additional option of adaptively changing the scaling. The codes of Moré, Garbow, and Hillstrom [1980], Gay [1980], and Dennis, Gay, and Welsch [1981] have this option. The numerical results of Moré [1978] indicate that adaptive scaling is superior to fixed scaling, but this conclusion is restricted to the Levenberg-Marquardt algorithm for nonlinear least squares problems. More research is needed on adaptive scaling techniques. Some results in this direction are presented in Section 4.

3. Solutions of Subproblems

The computation of the step in a trust region method requires an algorithm for the approximate solution of the problem

(3.1) $$\min\{\psi(w): \|Dw\| \leq \Delta\},$$

where

$$\psi(w) = g^T w + \tfrac{1}{2} w^T B w$$

is the local quadratic model of the function and $\|\cdot\|$ is a norm in R^n. We assume that $\|\cdot\|$ is the l_2 norm. It is possible to consider other norms, but an advantage of this choice is that the cost of obtaining a nearly global solution of (3.1) is quite reasonable.

An approximate solution s of (3.1) should produce a sufficient decrease in the quadratic ψ, that is, for some positive constants β_1 and β_2

(3.2) $\quad \psi(s) \leq \beta_1 \min\{\psi(w): D^T D w = vg, \|Dw\| \leq \Delta\}, \quad \|Ds\| \leq \beta_2 \Delta.$

All of the algorithms that we consider satisfy (3.2) with $\beta_1 = 1$ and $\beta_2 \in [1,2]$. It is certainly possible to allow other values of β_2, but in practice it is desirable to restrict $\beta_2 \leq 2$ so that the step does not go too far out of the trust region.

The simplest choice of a step s which satisfies condition (3.2) is obtained if we restrict s to be a multiple of the *preconditioned* gradient direction

(3.3) $$s_G \equiv -C^{-1}g, \quad C \equiv D^T D.$$

A geometrical interpretation of s_G is that of a steepest descent direction for ψ in the norm $\|D(\cdot)\|$, that is, νs_G solves the problem

$$\min\{g^T w : \|Dw\| \leq \Delta\}$$

for some $\nu > 0$. The step s that minimizes ψ along s_G and subject to the restriction $\|Ds\| \leq \Delta$ is the *Cauchy step*. It can be determined by setting

$$s_C = \left(\frac{-g^T s_G}{s_G^T B s_G}\right) s_G,$$

if $s_G^T B s_G > 0$ and if this results in a step with $\|Ds_C\| < \Delta$, and

$$s_C = \frac{\Delta}{\|Ds_G\|} s_G$$

otherwise. A straightforward computation verifies that the Cauchy step s_C satisfies (3.2) with $\beta_1 = \beta_2 = 1$.

Although the Cauchy step satisfies (3.2), the sole use of this step is not recommended because it can lead to slow convergence unless the preconditioning matrix C is a good approximation to the Hessian of the function. The *preconditioned dogleg* algorithm attempts to improve the convergence properties of the Cauchy step by computing the step s as a linear combination of the Newton step

(3.4) $$s_N \equiv -B^{-1}g,$$

and the Cauchy step. If we define $s(\tau) = s_N + \tau(s_C - s_N)$, then the preconditioned dogleg algorithm sets $s = s(\tau)$ where

$$\psi(s) = \min\{\psi(s(\tau)) : \|Ds(\tau)\| \leq \Delta\}.$$

The standard dogleg algorithm of Powell [1970a, 1970b] is obtained if we set $D = I$. The preconditioned dogleg algorithm is used by the nonlinear equation solvers in MINPACK-1 (Moré, Garbow, and Hillstrom [1980]).

The term dogleg is appropriate when B is positive definite. In this case the quadratic $\psi(s(\tau))$ is convex and nondecreasing for $\tau \geq 0$. Thus, if we consider the piecewise linear curve – the dogleg – that joints the origin, the Cauchy step, and the Newton step, then the step s minimizes ψ on the portion of the dogleg that lies inside the trust region. In particular, the step s is a convex combination of the Cauchy step and the Newton step. If B is indefinite then the quadratic $\psi(s(\tau))$ may be concave. If this is the case, then the optimal value of τ occurs outside of the interval $[0, 1]$, and thus s is not a convex combination of the Cauchy step and the Newton step. Note that, since $s(1) = s_C$, the preconditioned dogleg step satisfies (3.2) in all cases.

The preconditioned dogleg algorithm is reasonably efficient and cost effective. The main requirements are the computation of the Cauchy and Newton steps. Usually C is chosen so that the computation of the Cauchy step is not costly. The computation of the Newton step, however, can be costly. If B is a quasi-Newton approximation to the Hessian then the computation of s_N requires $O(n^2)$ operations, but if B is the Hessian matrix or a difference approximation to the Hessian matrix, then the computation of s_N usually requires $O(n^3)$ operations. For large-scale problems the matrix B is usually structured and then it is possible to reduce the cost of computing s_N. For example, if B is a banded matrix with bandwidth $2\beta+1$ then $O(\beta^2 n)$ operations are needed to compute s_N. For many large problems, however, the computation of the Newton step at each iteration cannot be justified.

A possibility for reducing the computational cost of the Newton step is to replace s_N by a truncated Newton step. A step s_T is a *truncated* Newton step if

$$\|g + B s_T\| \leq \xi \|g\|, \quad \xi \in (0, 1),$$

for some norm $\|\cdot\|$ in R^n. The term truncated is particularly appropriate if s_T is obtained by terminating some iterative process. Dembo, Eisenstat, and Steihaug [1982] prefer the term inexact Newton step if the process for determining s_T is unspecified. Dembo and Steihaug [1980] used the term truncated Newton step because in their algorithm s_T was obtained by a conjugate gradient algorithm.

The convergence analysis of trust region methods that we shall present in Section 4 shows that it is natural to consider truncated Newton methods in an ellipsoidal norm. In this section a step s_T is a truncated Newton step if

(3.5) $$\|g + B s_T\|_{C^{-1}} \leq \xi \|g\|_{C^{-1}}, \quad \xi \in (0, 1),$$

where C is defined in (3.3) and the ellipsoidal norm is defined via (2.5). The global convergence properties of the preconditioned dogleg algorithm are not affected because (3.2) still holds with $\beta_1 = \beta_2 = 1$. The only change involves the ultimate rate of convergence of the trust region method. If we use the Newton step (3.4) then we expect second order convergence; with a truncated Newton method we expect at least linear convergence, and usually superlinear convergence. The rate of convergence depends on ξ.

If B is positive definite then we can compute a truncated Newton step by applying a conjugate gradient algorithm to the system $Bw = -g$. If B is indefinite then we can use a variation of the conjugate gradient method on the normal equations, or a Lanczos type method on the system $Bw = -g$. These algorithms can be preconditioned by applying the standard algorithm to the linear system $\hat{B}\hat{w} = -\hat{g}$ and then transforming back to the original variables via

(3.6) $$\hat{w} = Dw, \quad \hat{g} = D^{-T}g, \quad \hat{B} = D^{-T}BD^{-1}.$$

Algorithms of this type are discussed by Paige and Saunders [1975, 1982] and Björck and Elfving [1979].

There are other dogleg algorithms available. For example, the *double-dogleg* algorithm of Dennis and Mei [1979]. A preconditioned version of the double-dogleg algorithm is used by some of the unconstrained minimization algorithms of Gay [1980]. The common link between these dogleg algorithms is that they can be viewed as attempting to compute a step which minimizes ψ on the 2-dimensional ellipse

$$\{w: \|Dw\| \leq \Delta, w \in \text{span}\{s_C, s_N\}\}.$$

Later on in this section we shall discuss an algorithm which is able to solve this problem, but it would seem that such algorithms are not likely to be a vast improvement over the preconditioned dogleg algorithm. A more desirable approach is to compute a step s such that

(3.7) $\quad \psi(s) \leq \beta_1 \min\{\psi(w): \|Dw\| \leq \Delta, w \in S\}, \quad \|Ds\| \leq \beta_2 \Delta,$

for some positive constants β_1 and β_2, and some subspace S which contains the Cauchy point s_C.

A natural approach to computing a step s which satisfies (3.7) is to use a preconditioned conjugate gradient method. This is the approach used and studied by Steihaug [1981].

Algorithm (3.8). *Preconditioned Conjugate Gradient Method.*
Let $w_0 \in R^n$ be given. Set $r_0 = -(g + Bw_0)$ and $d_0 = q_0$ where $Cq_0 = r_0$.
For $k = 0, 1, \ldots,$
  a) Compute $\alpha_k = (r_k^T q_k)/(d_k^T B d_k)$
  b) Update the iterate via $w_{k+1} = w_k + \alpha_k d_k$.
  c) Update the residual via $r_{k+1} = r_k - \alpha_k B d_k$.
  d) Solve $Cq_{k+1} = r_{k+1}$.
  e) Compute $\beta_k = (r_{k+1}^T q_{k+1})/(r_k^T q_k)$.
  f) Update the direction via $d_{k+1} = q_{k+1} + \beta_k d_k$.

The preconditioned conjugate gradient method can be obtained by applying the standard conjugate gradient method of Hestenes and Steifel [1952] to the linear system $\hat{B}\hat{w} = -\hat{g}$ and then transforming back to the original variables via (3.6). It can also be obtained by applying the preconditioned conjugate gradient algorithm of Section 2 to the quadratic ψ. Note that $r_k = -\nabla \psi(w_k)$ and that α_k is chosen so that it gives the least value of $\psi(w_k + \alpha d_k)$.

The preconditioned conjugate gradient method has many interesting properties. These properties can be obtained by using the properties of the standard conjugate gradient on the system $\hat{B}\hat{w} = -\hat{g}$, and then transforming back to the original system via (3.6).

One of the most important properties of Algorithm (3.8) is that if m is the largest integer such that $d_k^T B d_k > 0$ for $0 \leq k < m$ then $w_0, \ldots, w_m$ are defined by Algorithm (3.8) and

$$\psi(w_k) = \min\{\psi(w): w - w_0 \in \text{span}\{d_0, \ldots, d_{k-1}\}\}, \quad 0 \leq k \leq m.$$

If B is positive definite then $Bw_m = -g$, but if B is not positive definite then Algorithm (3.8) terminates when it computes a direction d_m with $d_m^T B d_m \leq 0$.

The easiest way to obtain (3.7) is to set $w_0 = 0$. Then d_0 is the preconditioned gradient direction s_G. Moreover, it is not difficult to show that in this case

$$\psi(w_l) = \min\{\psi(w): w \in S_l\}, \quad 0 \leq l \leq m,$$

where S_l is the l-dimensional Krylov space

(3.9) $$\text{span}\{s_G, (C^{-1}B)s_G, \ldots, (C^{-1}B)^{l-1}s_G\}.$$

It follows that if $0 < l \leq m$ then w_l satisfies (3.7) with S the l-dimensional Krylov space (3.9). Note, however, that if $\|Dw_l\| > \beta_2 \Delta$, or if $l = 0$ then w_l is not an acceptable step.

We now show that we can find a suitable step of the form $w_l + \tau d_l$ where $0 \leq l \leq m$ and $\tau \geq 0$. If for some $\xi \in (0, 1)$ and $0 \leq l \leq m$ we have

$$\|g + Bw_l\|_{C^{-1}} \leq \xi \|g\|_{C^{-1}}, \quad \|Dw_l\| \leq \Delta,$$

then we accept $s = w_l$ as the step. This termination criteria on Algorithm (3.8) guarantees that s satisfies the condition (3.5) of a truncated Newton step. Another termination criteria can be justified by first proving that the directions generated by Algorithm (3.8) satisfy

$$(D(w_{j+1} - w_j))^T (D(w_{i+1} - w_i)) = \alpha_j \alpha_i (Dd_j)^T (Dd_i) > 0.$$

Hence,

$$(D(w_k - w_j))^T (D(w_j - w_i)) > 0, \quad i < j < k,$$

and this implies that

$$\|D(w_k - w_i)\|^2 > \|D(w_k - w_j)\|^2 + \|D(w_j - w_i)\|^2, \quad i < j < k.$$

In particular, since we have chosen $w_0 = 0$, setting $i = 0$ and $j = k - 1$ in the above inequality yields

$$\|Dw_{k-1}\| < \|Dw_k\|, \quad 1 \leq k \leq m.$$

Now let $l \leq m$ be the largest integer such that $\|Dw_l\| < \Delta$, and let $\tau > 0$ be the unique solution to the quadratic equation

$$\|D(w_l + \tau d_l)\| = \Delta.$$

Note that either $d_l^T B d_l \leq 0$, or $d_l^T B d_l > 0$ and $\|Dw_l\| < \Delta \leq \|Dw_{l+1}\|$. In either case we have that

$$\psi(w_l + \tau d_l) \leq \min\{\psi(w_l), \psi(s_C)\},$$

and thus $s = w_l + \tau d_l$ satisfies (3.7) with $\beta_1 = \beta_2 = 1$, and S the l-dimensional Krylov space (3.9).

Steihaug [1981] suggested the trust region step described above. A possible difficulty with this approach is that there is no control over l. It may happen that $l = 0$, and in this case s is the Cauchy point s_C. It would be preferable if we could choose the space S and then compute a step s which satisfies (3.7).

We now describe an algorithm for computing a step s which satisfies (3.7). This algorithm is suitable for small and medium scale problems. An algorithm suitable for large-scale problems would be desirable, but at present a complete solution to this problem is not known.

There is no loss of generality in assuming that the subspace S in (3.7) is n-dimensional because if V is a basis for S then we can let $w = Vz$ and consider

(3.7) for arbitrary z. Thus we consider the problem of computing a step s which satisfies

(3.10) $\quad \psi(s) \leq \beta_1 \min\{\psi(w): \|Dw\| \leq \Delta\}, \quad \|Ds\| \leq \beta_2 \Delta,$

for some positive constants β_1 and β_2. Hebden [1973], Moré [1978], Fletcher [1980], Gay [1981], Sorensen [1982a], and Moré and Sorensen [1982] have all discussed this problem in connection with trust region methods. Gay [1981] was the first to show that the step produced by his algorithm usually satisfies (3.10). His algorithm, however, assumes that $g \neq 0$ and may require a large number of iterations in certain cases. These difficulties are avoided in the algorithm of Moré and Sorensen [1982]; in particular, they show that their algorithm always produces a step which satisfies (3.10).

One of the main ideas needed to produce an algorithm for the solution of (3.10) is contained in the following result.

Theorem (3.11). *A vector $w \in R^n$ is a global solution to problem (3.1) if and only if $\|Dw\| \leq \Delta$ and for some $\lambda^* \geq 0$,*

$$(B + \lambda^* D^T D) w = -g, \quad \lambda^*(\Delta - \|Dw\|) = 0,$$

with $B + \lambda^* D^T D$ positive semidefinite. If $B + \lambda^* D^T D$ is positive definite then (3.1) has a unique global solution.

Theorem (3.11) characterizes the global solutions of an optimization problem. This is unusual. In most optimization problems it is not even possible to characterize local minimizers. Theorem (3.11) was obtained, independently, by Gay [1981] and Sorensen [1982a], and it is based on earlier work by Goldfeld, Quandt, and Trotter [1966].

If $B + \lambda^* D^T D$ is positive definite it is relatively straightforward to obtain a nearly optimal solution to (3.1). Define the mapping

$$w(\lambda) = -(B + \lambda D^T D)^{-1} g,$$

and note that Theorem (3.11) shows that if $\lambda^* = 0$ then B is positive definite and $\|Ds^N\| \leq \Delta$, while if $\lambda^* > 0$ then λ^* is the unique solution to the 1-dimensional problem

$$\|Dw(\lambda)\| = \Delta, \quad \lambda \in (-\lambda_1, +\infty)$$

where λ_1 is the smallest eigenvalue of the matrix $\hat{B} = D^{-T} B D^{-1}$. A nearly optimal solution to (3.1) can be obtained by setting $\lambda = 0$ if B is positive definite and $\|Ds^N\| \leq (1+\sigma)\Delta$ for some $\sigma \in (0, 1)$, or by determining $\lambda > -\lambda_1$ such that

(3.12) $\quad |\|Dw(\lambda)\| - \Delta| \leq \sigma \Delta.$

Moré and Sorensen [1981] have shown that $s = w(\lambda)$ satisfies (3.10) with $\beta_1 = (1-\sigma)^2$ and $\beta_2 = 1+\sigma$.

It is very tempting to use Newton's method to determine a solution λ for (3.12). After all, if $g \neq 0$ then the function

$$\varphi_1(\lambda) = \|Dw(\lambda)\| - \Delta$$

is convex and strictly decreasing in $(-\lambda_1, +\infty)$, and this makes Newton's method a promising candidate for the solution of $\varphi_1(\lambda)=0$. Unfortunately φ_1 has a pole at $-\lambda_1$, and this tends to slow down Newton's method. Reinsch [1971] and Hebden [1973] independently observed that better results could be obtained by applying Newton's method to the function

$$\varphi_2(\lambda) = \frac{1}{\Delta} - \frac{1}{\|Dw(\lambda)\|}.$$

There are several reasons for the improved behavior. First of all, φ_2 has no poles and tends to be nearly linear on $(-\lambda_1, +\infty)$. In addition, Reinsch [1971] proved that if $g \neq 0$ then φ_2 is also convex and strictly decreasing on $(-\lambda_1, +\infty)$ so that global convergence of Newton's method is guaranteed. Numerical results obtained by applying Newton's method to φ_2 are quite satisfactory; in practice less than two iterations are needed to obtain a solution λ for (3.12).

It is also tempting to dismiss the case where $B + \lambda^* D^T D$ is singular as a low probability case. In practice, however, cases with $B + \lambda^* D^T D$ nearly singular are amazingly frequent and it is therefore important to take this case into account. Since this case causes numerical difficulties, we shall call it the *hard* case.

In the hard case $\lambda^* = -\lambda_1$ and therefore $w(\lambda)$ is defined for $\lambda > \lambda^* \geq 0$. Theorem (3.11) shows that global solutions to (3.1) satisfy a singular system and it may seem that $w(\lambda)$ may be a nearly optimal solution if $\lambda > \lambda^*$ is sufficiently close to λ^*. We can certainly determine $w(\lambda)$ for $\lambda > \lambda^*$, but $w(\lambda)$ may not be nearly optimal for any $\lambda > \lambda^*$. For example, if $g = 0$ then $w(\lambda) \equiv 0$, and if B is indefinite then $w(\lambda)$ is not nearly optimal. The following result shows, however, that $w(\lambda)$ can be used as a component of a nearly optimal solution.

Theorem (3.13). *Let $\sigma \in (0, 1)$ be given and suppose that $w \in R^n$ satisfies*

$$(B + \lambda D^T D)w = -g, \quad \lambda \geq 0.$$

Define $B_\lambda \equiv B + \lambda D^T D$. If $z \in R^n$ satisfies

$$\|D(w+z)\| = \Delta, \quad z^T B_\lambda z \leq \sigma(w^T B_\lambda w + \lambda \Delta^2),$$

then $s = w + z$ satisfies (3.10) with $\beta_1 = 1 - \sigma$ and $\beta_2 = 1$.

The main use of Theorem (3.13) is in connection with the hard case, but we emphasize that this result can be applied in other cases also. Theorem (3.13) is an improvement, due to Moré and Sorensen [1981], of a result of Gay [1981].

We now show how to satisfy the conditions of Theorem (3.13) in the hard case. Let $w = w(\lambda)$ for $\lambda > \lambda^*$. Then $\|Dw\| < \Delta$ and thus given any $\hat{z} \neq 0$ there is a scalar τ such that

$$\|D(w + \tau \hat{z})\| = \Delta, \quad \text{sign}(\tau) = \text{sign}\{(Dw)^T(D\hat{z})\}.$$

This implies, in particular, that $\|D(\tau \hat{z})\| \leq \Delta$. Now choose $\hat{z}$ so that

(3.14)
$$\frac{\hat{z}^T B_\lambda \hat{z}}{\|D\hat{z}\|^2}$$

is as small as possible. If $\lambda - \lambda^*$ is sufficiently small then B_λ is nearly singular and thus it is possible to choose $\hat{z}$ so that (3.14) is arbitrarily close to zero. It follows that $w = w(\lambda)$ and $z = \tau \hat{z}$ satisfy the conditions of Theorem (3.13) for $\lambda - \lambda^*$ small enough. If we compute the Cholesky factorization $B_\lambda = R_\lambda^T R_\lambda$, then the LINPACK technique (Cline, Moler, Stewart, and Wilkinson [1979]) for estimating the smallest singular value of a triangular matrix can be used to determine $\hat{z}$. This technique produces excellent numerical results. In addition, with this choice of $\hat{z}$ it can be shown that (3.14) is arbitrarily small provided $\lambda - \lambda^*$ is small enough.

The above description is not complete, but it should give an idea of the overall algorithm. A complete algorithm is described by Moré and Sorensen [1982]. They prove their algorithm produces a step which satisfies (3.10) in a finite number of steps. Their numerical results for a trust region version of Newton's method show that, on the average, 1.6 factorizations of B_λ are required to satisfy (3.10) with $\beta_1 = 0.8$ and $\beta_2 = 1.1$. Gay [1981] reported similar results for his algorithm.

These numerical results demonstrate that the cost of obtaining a step s which satisfies (3.7) is quite reasonable for small and medium scale problems. As mentioned in this section, at present there is no entirely satisfactory algorithm for large scale problems.

4. Convergence Results

Powell [1970a, 1970b, 1975] was the first to establish convergence of trust region methods for unconstrained optimization. Recent developments, however, have led to several modifications of the class of trust region methods studied by Powell, and thus it is necessary to re-analyze the convergence properties.

We first specify a general trust region method for an unconstrained minimization problem defined by a continuously differentiable function $f: R^n \to R$.

Algorithm (4.1). *Trust Region Method.*
Let $x_0 \in R^n$ and Δ_0 be given. Set $\mu \in (0,1)$.
For $k = 0, 1, \ldots,$
    a) Compute $f(x_k)$ and the model ψ_k.
    b) Determine an approximate solution s_k to subproblem (1.5).
    c) Compute $\rho_k = (f(x_k + s_k) - f(x_k))/\psi_k(s_k)$.
    d) If $\rho_k > \mu$ then $x_{k+1} = x_k + s_k$. Otherwise $x_{k+1} = x_k$.
    e) Update the model ψ_k, the scaling matrix D_k and Δ_k.

Given an optimization problem, it is necessary to define the model ψ_k, the solution s_k of (1.5), and the algorithms for updating ψ_k, D_k, and Δ_k. The general aim of the updating algorithms is to force $\rho_k > \mu$. An iteration with $\rho_k > \mu$ is *successful*, otherwise the iteration is *unsuccessful*.

A difference between Algorithm (4.1) and Powell's class of algorithms is that Powell sets $D_k \equiv I$. We prefer the scaled version of Algorithm (4.1) because in this form the trust region method can be made scale invariant under transformation (2.1). Another difference is that Powell chooses $\mu = 0$. With this choice it is only possible to show that some subsequence of $\{\nabla f(x_k)\}$ converges to zero. Thomas [1975] was the first to prove that if $\mu > 0$ then it is possible to show that the whole sequence $\{\nabla f(x_k)\}$ converges to zero.

Algorithm (4.1) also differs from Powell's class of algorithms in the assumptions on the model ψ_k. We assume that ψ_k is the quadratic

(4.2) $$\psi_k(w) = g_k^T w + \tfrac{1}{2} w^T B_k w$$

where g_k is an approximation to the gradient $\nabla f(x_k)$ and B_k is an approximation to the Hessian $\nabla^2 f(x_k)$. Powell assumes that $g_k = \nabla f(x_k)$, but we assume that if $\{x_k\}$ is a convergent sequence then

(4.3) $$\lim_{k \to +\infty} \|g_k - \nabla f(x_k)\| = 0.$$

This is a natural condition. Since $g_k = \nabla \psi_k(0)$, it implies that if the sequence is convergent then the first order information provided by the model ψ_k must be asymptotically correct.

The conditions imposed on the step s_k are fairly mild. We assume that there are positive constants β_1 and β_2 such that

(4.4) $$\psi_k(s_k) \leqslant \beta_1 \min\{\psi_k(w): D_k^T D_k w = v g_k, \|D_k w\| \leqslant \varDelta_k\}, \quad \|D_k s_k\| \leqslant \beta_2 \varDelta_k.$$

The sufficient decrease condition (4.4) requires that the decrease $\psi_k(s_k)$ must be at least a fraction of the optimal decrease in ψ_k along the steepest descent direction for ψ_k in the norm $\|D_k(\cdot)\|$. Since (4.4) implies that for $\beta_2 > 1$ we have

$$\psi_k(s_k) \leqslant \left(\frac{\beta_1}{\beta_2^2}\right) \min\{\psi_k(w): D_k^T D_k w = v g_k, \|D_k w\| \leqslant \beta_2 \varDelta_k\},$$

we could replace $\beta_2 \varDelta_k$ by $\varDelta_k$ and thus assume that $\beta_2 = 1$ in (4.4). We prefer (4.4) as it stands because it is closer to the algorithms of Section 3. Another interpretation of (4.4) can be obtained by defining $\hat{\psi}_k(v) \equiv \psi(D_k^{-1} v)$ and

(4.5) $$\hat{g}_k \equiv D_k^{-T} g_k, \quad \hat{B}_k \equiv D_k^{-T} B_k D_k^{-1}.$$

If $\hat{s}_k = D_k s_k$ then assumption (4.4) is equivalent to assuming that

$$\psi_k(s_k) = \hat{\psi}_k(\hat{s}_k) \leqslant \beta_1 \min\{\hat{\psi}_k(v): v = v \hat{g}_k, \|v\| \leqslant \varDelta_k\}, \quad \|\hat{s}_k\| \leqslant \beta_2 \varDelta_k.$$

Thus (4.4) requires that the decrease $\hat{\psi}_k(\hat{s}_k)$ must be at least a fraction of the optimal decrease in $\hat{\psi}_k$ along the steepest descent direction for $\hat{\psi}_k$.

Powell [1975] assumes that $\beta_2 = 1$ and that if $\|D_k s_k\| < \varDelta_k$ then B_k is positive definite and s_k is the Newton step

$$s_k^N = -B_k^{-1} \nabla f(x_k).$$

The assumption that s_k is the Newton step if $\|D_k s_k\| < \varDelta_k$ is unnecessary. This assumption is also undesirable for large scale optimization problems because in this case the computation of the Newton step can be quite costly.

From a theoretical point of view, the scaling matrices are a nuisance. It is much easier to set $D_k \equiv I$ since this simplifies the exposition of the results. From a practical point of view, however, they are necessary and important. For badly scaled problems a scaling procedure is often the only means to solve the problem. We thus explicitly retain the scaling matrices and attempt to obtain convergence results with minimal assumptions on the scaling matrices. We assume that there are constants σ_1 and σ_2 such that

(4.6) $$\|\hat{B}_k\| \leq \sigma_1, \quad \|D_k^{-1}\| \leq \sigma_2.$$

where $\hat{B}_k$ is defined in (4.5). Instead of assuming that $\|\hat{B}_k\| \leq \sigma_1$, Powell [1975] makes the weaker assumption that

$$\|\hat{B}_k\| \leq \sigma_3 + \sigma_4 \sum_{j=0}^{k-1} \|D_j s_j\|$$

for some positive constants σ_3 and σ_4. This assumption is necessary to deal with certain quasi-Newton methods, but is not necessary for many of the practical trust region methods. Moreover, with assumption (4.6) the analysis is considerably simplified and the results can be improved. Note that we do not assume that the sequence $\{D_k\}$ has uniformly bounded condition numbers. Also note that if $\{B_k\}$ and $\{D_k^{-1}\}$ are bounded then (4.6) holds.

Algorithm (4.7). *Updating Δ_k.*
Let $0 < \mu < \eta < 1$ and $\gamma_1 < 1 < \gamma_2$ be given.
  a) If $\rho_k \leq \mu$ then $\Delta_{k+1} \in (0, \gamma_1 \Delta_k]$.
  b) If $\rho_k \in (\mu, \eta)$ then $\Delta_{k+1} \in [\gamma_1 \Delta_k, \Delta_k]$.
  c) If $\rho_k \geq \eta$ then $\Delta_{k+1} \in [\Delta_k, \gamma_2 \Delta_k]$.

The algorithm used by Powell [1975] is similar to Algorithm (4.7) but with Δ_k replaced by $\|D_k s_k\|$. Since, in Powell's class of algorithms $\|D_k s_k\| = \Delta_k$ unless $s_k = s_k^N$, the two updating rules usually coincide. A disadvantage of Powell's choice is that it depends on the Newton step, and as mentioned above, this is not desirable. Also note that if the only restriction on s_k is (4.4), then it is not desirable to let Δ_{k+1} be proportional to $\|D_k s_k\|$ on the successful iterations because this would destroy the convergence properties of the trust region method. On the other hand, Algorithm (4.7) allows the choice

$$\Delta_{k+1} \in [\gamma_0 \|D_k s_k\|, \gamma_1 \Delta_k]$$

on the unsuccessful iterations. This is a reasonable choice. For example, we noted in the introduction that a choice of the form $\Delta_{k+1} = \gamma_0 \|D_k s_k\|$ is often used in the Levenberg-Marquardt algorithm. A choice of the form $\Delta_{k+1} = \gamma_1 \Delta_k$ is used in Powell's [1970a] algorithm because the first order information in the model ψ_k is changed during the unsuccessful iterations (this is not the case for the Levenberg-Marquardt algorithm), and setting $\Delta_{k+1} = \gamma_1 \Delta_k$ allows the model to incorporate information from a larger trust region.

We have now specified completely the trust region method. Our convergence results apply to Algorithm (4.1) under assumptions (4.2–4.6), and with Δ_k updated by Algorithm (4.7). This algorithm is scale invariant under transformation (2.1) provided $D_k = \nu D_k D^{-1}$ and $\hat{\Delta}_k = \nu \Delta_k$ for some $\nu > 0$. Also note that the assumptions on this algorithm are satisfied if $\|\cdot\|$ is not the l_2 norm; if $\|\cdot\|$ is any other norm then the equivalence of norms guarantees that assumption (4.4) holds with Δ_k replaced by $\nu \Delta_k$ for some constant ν.

One of the results needed to prove convergence is a technical lemma which expresses assumption (4.4) in a workable form.

Lemma (4.8). *If s_k satisfies (4.4) and if $\|\cdot\|$ is the l_2 norm then*

$$-\psi_k(s_k) \geq \tfrac{1}{2} \beta_1 \|\hat{g}_k\| \min\left\{\Delta_k, \frac{\|\hat{g}_k\|}{\|\hat{B}_k\|}\right\}.$$

Proof: Define $\varphi: R \to R$ by setting

$$\varphi(\tau) = \psi_k \left[\tau D_k^{-1}\left(\frac{\hat{g}_k}{\|\hat{g}_k\|}\right)\right],$$

and let τ_k^* be the minimum of φ on $[0, \Delta_k]$. We now estimate $\varphi(\tau_k^*)$. By definition,

$$\varphi(\tau) = -\tau \|\hat{g}_k\| + \tfrac{1}{2} \tau^2 \mu_k, \quad \mu_k \equiv \frac{\hat{g}_k^T \hat{B}_k \hat{g}_k}{\|\hat{g}_k\|^2}.$$

If $\tau_k^* \in (0, \Delta_k]$ then $\tau_k^* = \|\hat{g}_k\|/\mu_k$ and thus

$$\varphi(\tau_k^*) = -\tfrac{1}{2} \frac{\|\hat{g}_k\|^2}{\mu_k} \leq -\tfrac{1}{2} \frac{\|\hat{g}_k\|^2}{\|\hat{B}_k\|}.$$

If $\tau_k^* = \Delta_k$ then $\mu_k \Delta_k \leq \|\hat{g}_k\|$, and hence

$$\varphi(\tau_k^*) = \varphi(\Delta_k) \leq -\tfrac{1}{2} \Delta_k \|\hat{g}_k\|.$$

Since $\psi_k(s_k) \leq \beta_1 \varphi(\tau_k^*)$, the result follows from the above estimates. ☐

This result is due to Powell [1975]. He proved it for $D_k = I$, but in view of the equivalence between ψ_k and $\hat{\psi}_k(v) = \psi_k(D_k^{-1} v)$, the result also holds for the scaled quantities. Also note that if we adjust β_1 then Lemma (4.8) holds for any norm $\|\cdot\|$.

The convergence results still hold if assumption (4.4) is replaced by the weaker assumption that s_k satisfies the inequality in Lemma (4.8). The advantage of (4.4) derives from its geometrical interpretation. An important consequence of Lemma (4.8) and assumption (4.6) is that if the iteration is successful then

(4.9) $$f(x_k) - f(x_{k+1}) \geq \tfrac{1}{2} \beta_1 \mu \|\hat{g}_k\| \min\left\{\Delta_k, \frac{\|\hat{g}_k\|}{\sigma_1}\right\}.$$

This inequality is crucial to the convergence results.

Theorem (4.10). *If $f: R^n \to R$ is continuously differentiable and bounded below on R^n then*

$$\liminf_{k \to +\infty} \|g_k\|_{C_k^{-1}} = 0$$

where the norm is defined via (2.5) and $C_k = D_k^T D_k$.

Proof: We need to show that $\{\|\hat{g}_k\|_2\}$ in not bounded away from zero. Let $\|\cdot\|$ be the l_2 norm and assume, on the contrary, that there is an $\varepsilon > 0$ such that $\|\hat{g}_k\| \geq \varepsilon$ for all k sufficiently large. We now show that

$$\sum_{k=1}^{\infty} \Delta_k < +\infty. \tag{4.11}$$

If there are a finite number of successful iterations then $\Delta_{k+1} \leq \gamma_1 \Delta_k$ for all k sufficiently large and then (4.11) clearly holds. If there is an infinite sequence $\{k_i\}$ of successful iterations then inequality (4.9) shows that

$$\sum_{i=1}^{\infty} \Delta_{k_i} < +\infty.$$

Now, the updating rules of Algorithm (4.7) imply that

$$\sum_{k=1}^{\infty} \Delta_k \leq \left(1 + \frac{\gamma_2}{1-\gamma_1}\right) \sum_{i=1}^{\infty} \Delta_{k_i},$$

and thus (4.11) holds in this case also.

We now show that (4.11) implies that $\{|\rho_k - 1|\}$ converges to zero. As a first step in the proof note that

$$\|x_{k+1} - x_k\| \leq \|s_k\| \leq \sigma_2 \beta_2 \Delta_k,$$

and hence (4.11) shows that $\{x_k\}$ converges. Hence, (4.3) holds. We now estimate $\rho_k - 1$ by first noting that (4.6) implies that

$$|\psi_k(s_k) - \nabla f(x_k)^T s_k| \leq \sigma_2 \|g_k - \nabla f(x_k)\| \|D_k s_k\| + \tfrac{1}{2} \sigma_1 \|D_k s_k\|^2.$$

This inequality, assumptions (4.3) and (4.6), and the continuity of ∇f now show that there is a sequence $\{\varepsilon_k\}$ converging to zero such that

$$|f(x_k + s_k) - f(x_k) - \psi_k(s_k)| \leq \varepsilon_k \Delta_k.$$

Since Lemma (4.8) implies that

$$-\psi_k(s_k) \geq \tfrac{1}{2} \beta_1 \varepsilon \Delta_k,$$

we readily obtain that $\{|\rho_k - 1|\}$ converges to zero. However, the updating rules for Δ_k show that Δ_k is not decreased if $\rho_k \geq \eta$. Thus $\{\Delta_k\}$ cannot converge to zero. This contradicts (4.11) and establishes the result. □

Theorem (4.10) is due to Powell [1975], although as mentioned in this section, there are a number of differences between the class of algorithms analyzed by Powell and the trust region methods specified by Algorithm (4.1) under assumptions (4.2–4.6). One of the most important differences is that the convergence properties of trust region methods do not depend on the Newton step; it is only necessary to assume that s_k satisfies a sufficient decrease condition of the form (4.4). This observation was common knowledge soon after the appearance of Powell's result and became a folklore result. Another difference is that Powell sets $D_k \equiv I$ and $g_k = \nabla f(x_k)$. In this case Theorem (4.10) guarantees that

$$\liminf_{k \to +\infty} \|\nabla f(x_k)\| = 0.$$

This result, of course, still holds if $\{D_k\}$ is uniformly bounded.

The assumptions of Theorem (4.10) are particularly easy to satisfy if the matrices B_k are positive semidefinite. We illustrate this remark with the Levenberg-Marquardt algorithm. Let $F: R^n \to R^m$ be continuously differentiable and set

$$g_k = F'(x_k)^T F(x_k), \quad B_k = F'(x_k)^T F'(x_k).$$

If we follow Moré [1978] and choose

$$D_k = \text{diag}(d_{k,i}), \quad d_{k,i} = \max\{d_{k-1,i}, \|F'(x_k)e_i\|_2\}, \quad k > 0,$$

then $\|D_k^{-1}\|_2 \leq \|D_0^{-1}\|_2$ and $\|F'(x_k)D_k^{-1}\|_2 \leq n^{\frac{1}{2}}$. Thus assumption (4.6) holds. Theorem (4.14) now shows that

$$\liminf_{k \to +\infty} \|(F'(x_k)D_k^{-1})^T F(x_k)\| = 0.$$

Of course, if F' is bounded then we also have that

$$\liminf_{k \to +\infty} \|F'(x_k)^T F(x_k)\| = 0.$$

These results are due to Moré [1978]. Note that in these results we can replace $F'(x_k)$ by any matrix J_k such that

$$\lim_{k \to +\infty} \|J_k - F'(x_k)\| = 0$$

whenever $\{x_k\}$ is a convergent sequence. In this form these results apply to Powell's [1970a] algorithm for the solution of systems of nonlinear equations.

It would be desirable to replace lim inf by lim in Theorem (4.10). This improvement can be obtained if we set $g_k \equiv \nabla f(x_k)$, and if we assume a boundedness condition on the scaling matrices. We assume that there are positive constants δ_1 and δ_2 such that

(4.12) $\qquad \|x_k - x_m\| \leq \delta_1, \quad k = m, \ldots, l \Rightarrow \|D_l - D_m\| \leq \delta_2.$

It is clear that (4.12) is satisfied if $\{D_k\}$ is uniformly bounded or if D_k is a uniformly continuous function of x_k. For example, the choice

$$D_k = \text{diag}(|\partial_{ii} f(x_k)|^{\frac{1}{2}})$$

satisfies (4.12) if $\nabla^2 f$ is uniformly bounded or if $\nabla^2 f$ is uniformly continuous. This choice, however, may fail to satisfy $\|D_k^{-1}\| \leq \sigma_2$. If D_0 is a nonsingular diagonal matrix then the choice

(4.13) $\qquad D_k = \text{diag}(d_{k,i}), \quad d_{k,i} = \max\{d_{k-1,i}, |\partial_{ii} f(x_k)|^{\frac{1}{2}}\}, \quad k > 0,$

satisfies (4.12) if $\nabla^2 f$ is uniformly bounded or if $\nabla^2 f$ is uniformly continuous, and moreover, $\|D_k^{-1}\|_2 \leq \|D_0^{-1}\|_2 \equiv \sigma_2$. This choice is the unconstrained minimization analogue of the choice made by Moré [1978] for nonlinear least squares problems. An advantage of this choice is that it is scale invariant under transformation (2.1) if $\hat{D}_0 = \beta D_0 D^{-1}$ and $\alpha = \beta^2$.

The proof that the choice made in (4.13) satisfies (4.12) if $\nabla^2 f$ is uniformly continuous is not difficult. Note that if $m \leq l$ are given and if $d_{l,i} > d_{m,i}$ then there is an index k with $m \leq k \leq l$ such that

$$d_{l,i} = d_{k,i} = |\partial_{ii} f(x_k)|^{\frac{1}{2}}.$$

Hence,

$$d_{l,i}^2 - d_{m,i}^2 \leq |\partial_{ii} f(x_k)| - |\partial_{ii} f(x_m)| \leq |\partial_{ii} f(x_k) - \partial_{ii} f(x_m)|,$$

and since

$$d_{l,i} - d_{m,i} = \frac{d_{l,i}^2 - d_{m,i}^2}{d_{l,i} + d_{m,i}} \leq \frac{d_{l,i}^2 - d_{m,i}^2}{2 d_{0,i}},$$

it follows that if $\nabla^2 f$ is uniformly continuous then the choice made in (4.13) satisfies (4.12).

Theorem (4.14). *Assume that $g_k = \nabla f(x_k)$ and that (4.12) holds. If $f: R^n \to R$ is bounded below on R^n and ∇f is uniformly continuous then*

$$\lim_{k \to +\infty} \|\nabla f(x_k)\|_{C_k^{-1}} = 0.$$

Proof: The proof is by contradiction. Let ε_1 in $(0, 1)$ be given and assume that there is a sequence $\{m_i\}$ such that $\|\hat{g}_{m_i}\| \geq \varepsilon_1$ where $\|\cdot\|$ is the l_2 norm. Theorem (4.12) guarantees that for any ε_2 in $(0, \varepsilon_1)$ there is a subsequence of $\{m_i\}$ (without loss of generality assume that it is the full sequence) and a sequence $\{l_i\}$ such that

$$\|\hat{g}_k\| \geq \varepsilon_2, \quad m_i \leq k < l_i, \quad \|\hat{g}_{l_i}\| < \varepsilon_2.$$

Inequality (4.9) then implies that if $m_i \leq k < l_i$, and if the k-th iteration is successful then

$$f(x_k) - f(x_{k+1}) \geq \tfrac{1}{2} \beta_1 \mu \varepsilon_2 \min\left\{\Delta_k, \frac{\varepsilon_2}{\sigma_1}\right\}.$$

Since $\{f(x_k)\}$ converges, and since $\|x_{k+1} - x_k\| \leq \beta_2 \sigma_2 \Delta_k$, this shows that

$$f(x_k) - f(x_{k+1}) \geq \varepsilon_3 \|x_{k+1} - x_k\|, \quad m_i \leq k < l_i,$$

where $\varepsilon_3 = (\tfrac{1}{2} \beta_1 \mu \varepsilon_2)/(\beta_2 \sigma_2)$. Hence,

$$f(x_{m_i}) - f(x_{k_i}) \geq \varepsilon_3 \|x_{k_i} - x_{m_i}\|, \quad m_i \leq k_i \leq l_i.$$

Assumption (4.12), the uniform continuity of ∇f, and the convergence of $\{f(x_k)\}$ can now be used to deduce that

$$\|g_{m_i} - g_{l_i}\| \leq \varepsilon_2, \quad \|D_{m_i} - D_{l_i}\| \leq \delta_2,$$

for i sufficiently large. We use this result to obtain a contradiction. Note that a computation with the triangle inequality shows that for any indices m and l

$$\|\hat{g}_m\| \leq \|D_m^{-1}\| \|g_m - g_l\| + \|D_m^{-1}\| \|D_m - D_l\| \|\hat{g}_l\| + \|\hat{g}_l\|.$$

If we use this estimate in an obvious way we obtain that

$$\varepsilon_1 \leq (\sigma_2(1+\delta_2)+1)\varepsilon_2.$$

Since ε_2 can be any number in $(0, \varepsilon_1)$, this is a contradiction. □

This result is due to Thomas [1975] although once again there are important differences between the algorithms analyzed by Thomas and the trust region methods being considered here.

The assumptions of Theorem (4.14) are easy to satisfy if we assume that $\{B_k\}$ is bounded. In this case assumptions (4.2) and (4.12) are satisfied if, for example, we choose

$$D_k = \text{diag}(d_{k,i}), \quad d_{k,i} = \max\{d_{k-1,i}, |b_{i,i}|^{\frac{1}{2}}\}, \quad k > 0.$$

An advantage of this choice is that it can be made scale invariant. Moreover, now $\{D_k\}$ is bounded so that Theorem (4.14) guarantees that

$$\lim_{k \to +\infty} \|\nabla f(x_k)\| = 0.$$

This is a version of the result obtained by Thomas [1975].

One of the attractive aspects of trust region methods is that under suitable conditions we can prove that

$$\limsup_{k \to +\infty} \lambda_1 [\nabla^2 f(x_k)] \geq 0,$$

where $\lambda_1[A]$ is the smallest eigenvalue of the symmetric matrix A. This result can be obtained if we assume that $f: R^n \to R$ is twice continuously differentiable, that the model $\psi_k: R^n \to R$ is defined by

(4.15) $$\psi_k(w) = \nabla f(x_k)^T w + \tfrac{1}{2} w^T \nabla^2 f(x_k) w,$$

and that the step s_k satisfies

(4.16) $$\psi_k(s_k) \leq \beta_1 \min\{\psi_k(w): \|D_k w\| \leq \Delta_k\}, \quad \|D_k s_k\| \leq \beta_2 \Delta_k,$$

for some positive constants β_1 and β_2. The scaling matrices $\{D_k\}$ can be chosen by (4.13) or by any method which guarantees that if $\{\nabla^2 f(x_k)\}$ is bounded then $\{D_k\}$ and $\{D_k^{-1}\}$ are uniformly bounded.

Theorem (4.17). *Let $f: R^n \to R$ be twice continuously differentiable and bounded below on R^n, and assume that $\nabla^2 f$ is bounded on the level set*

$$\{x \in R^n : f(x) \leq f(x_0)\}.$$

Let $\{x_k\}$ be the sequence generated by Algorithm (4.1) under assumption (4.15) on the model ψ_k and assumption (4.16) on the step s_k. Assume that $\{D_k\}$ and $\{D_k^{-1}\}$ are uniformly bounded. Then
a) The sequence $\{\nabla f(x_k)\}$ converges to zero.

b) If $\{x_k\}$ is bounded then there is a limit point x^* with $\nabla^2 f(x^*)$ positive semidefinite.
c) If x^* is an isolated limit point of $\{x_k\}$ then $\nabla^2 f(x^*)$ is positive semidefinite.
d) If $\nabla^2 f(x^*)$ is nonsingular for some limit point x^* of $\{x_k\}$ then $\nabla^2 f(x^*)$ is positive definite, $\{x_k\}$ converges to x^*, all iterations are eventually successful, and $\{\Delta_k\}$ is bounded away from zero.

We have already noted that Theorem (4.14) shows that $\{\nabla f(x_k)\}$ converges to zero. The result that there is a limit point x^* with $\nabla^2 f(x^*)$ positive semidefinite was obtained independently by Fletcher [1980] and Sorensen [1982a]. The other two claims of Theorem (4.17) were established by Moré and Sorensen [1981] as extensions of earlier results of Sorensen [1982a].

The proof that there is a limit point x^* with $\nabla^2 f(x^*)$ positive semidefinite is not difficult. Assume that there is an $\varepsilon_1 > 0$ such that

$$-\lambda_1[\nabla^2 f(x_k)] \geqslant \varepsilon_1.$$

If q_k is an eigenvector associated with the smallest eigenvalue of $\nabla^2 f(x_k)$ and scaled so that $\nabla f(x_k)^T q_k \leqslant 0$ and $\|D_k q_k\| = \Delta_k$, then (4.16) shows that

$$\psi_k(s_k) \leqslant \beta_1 \psi_k(q_k) \leqslant \beta_1 \lambda_1[\nabla^2 f(x_k)](\|q_k\|_2)^2 \leqslant -\varepsilon_1 \beta_1 (\|q_k\|_2)^2.$$

Since $\{D_k\}$ is bounded, this shows that there is an $\varepsilon_2 > 0$ such that

$$-\psi_k(s_k) \geqslant \varepsilon_2 \Delta_k^2.$$

Hence, $\{\Delta_k\}$ converges to zero, and this forces $\{s_k\}$ to converge to zero. We now use these results to obtain a contradiction. A standard estimate shows that

$$|f(x_k + s_k) - f(x_k) - \psi_k(s_k)| \leqslant \tfrac{1}{2} \|s_k\|^2 \max_{0 \leqslant \tau \leqslant 1} \|\nabla^2 f(x_k + \tau s_k) - \nabla^2 f(x_k)\|,$$

and since $\|s_k\| \leqslant \sigma_2 \beta_2 \Delta_k$, the last two inequalities yield that

(4.18) $$|\rho_k - 1| \leqslant \tfrac{1}{2} \left(\frac{(\sigma_2 \beta_2)^2}{\varepsilon_2} \right) \max_{0 \leqslant \tau \leqslant 1} \|\nabla^2 f(x_k + \tau s_k) - \nabla^2 f(x_k)\|.$$

Thus, the continuity of $\nabla^2 f$ and the boundedness of $\{x_k\}$ imply that $\{|\rho_k - 1|\}$ converges to zero, and then the updating rules of Δ_k show that $\{\Delta_k\}$ is bounded away from zero. This contradiction establishes the result.

Schultz, Schnabel, and Byrd [1982], and Gay [1982] have obtained several interesting variations on Theorem (4.17). Schultz, Schnabel, and Byrd [1982], in particular, have shown that if $\{x_k\}$ converges to x^* then $\nabla^2 f(x^*)$ is positive semidefinite if s_k satisfies (4.4) and

$$\psi_k(s_k) \leqslant \beta_1 \min\{\psi_k(w): w = v q_k, \|D_k w\| \leqslant \Delta_k\},$$

where q_k is an eigenvector associated with the smallest eigenvalue of $\nabla^2 f(x_k)$.

The following variation on Theorem (4.17) is of interest because it does not require the full power of (4.16); it suffices to assume that (4.4) holds.

Theorem (4.19). *Let $f: R^n \to R$ satisfy the assumptions of Theorem (4.17), and let $\{x_k\}$ be the sequence generated by Algorithm (4.1) under assumption (4.15) on the* model ψ_k and assumption (4.4) on the step s_k. Assume that $\{D_k\}$ and $\{D_k^{-1}\}$ are uniformly bounded. If x^* is a limit point of $\{x_k\}$ with $\nabla^2 f(x^*)$ positive definite then $\{x_k\}$ converges to x^*, all iterations are eventually successful, and $\{\Delta_k\}$ is bounded away from zero.

Proof: We first prove that $\{x_k\}$ converges to x^*. Choose $\delta > 0$ so that $\nabla^2 f(x)$ is positive definite for $\|x - x^*\| \leq \delta$. Since $\psi_k(s_k) \leq 0$, there is an $\varepsilon_1 > 0$ such that $\|x_k - x^*\| \leq \delta$ implies that

(4.20) $$\varepsilon_1 \|s_k\| \leq \|\nabla f(x_k)\|.$$

Theorem (4.14) guarantees that $\{\nabla f(x_k)\}$ converges to zero, and thus there is an index $k_1 \geq 0$ for which

$$\|\nabla f(x_k)\| \leq \tfrac{1}{2} \varepsilon_1 \delta, \quad k \geq k_1.$$

Hence, (4.20) shows that if $\|x_k - x^*\| \leq \tfrac{1}{2} \delta$ for $k \geq k_1$, then $\|x_{k+1} - x^*\| \leq \delta$. Now, since $\nabla^2 f(x^*)$ is positive definite, since $\nabla f(x^*) = 0$, and since x^* is a limit point of $\{x_k\}$, there is an index $k_2 \geq k_1$ with $\|x_{k_2} - x^*\| \leq \tfrac{1}{2} \delta$ and such that

$$f(x) \leq f(x_{k_2}), \quad \|x - x^*\| \leq \delta \Rightarrow \|x - x^*\| < \tfrac{1}{2} \delta.$$

Hence, $\|x_k - x^*\| \leq \tfrac{1}{2} \delta$ for $k \geq k_2$. This proves that $\{x_k\}$ converges to x^*.

We now prove that all iterations are successful. Since $\|s_k\| \leq \sigma_2 \beta_2 \Delta_k$, Lemma (4.8) and (4.20) yield that there is an $\varepsilon_2 > 0$ with

$$-\psi_k(s_k) \geq \varepsilon_2 \|s_k\|^2.$$

We can now use the argument that led to (4.18) and show that $\{|\rho_k - 1|\}$ converges to zero, Hence, all iterations are eventually successful. □

Rate of convergence results can be obtained if we assume that $\{x_k\}$ converges to a point x^* with $\nabla^2 f(x^*)$ positive definite, and if there is a constant $\beta_3 > 0$ such that $\|D_k s_k\| < \beta_3 \Delta_k$ implies that

(4.21) $$\|\nabla f(x_k) + \nabla^2 f(x_k) s_k\| \leq \xi_k \|\nabla f(x_k)\|, \quad \xi_k \in (0, 1).$$

Thus s_k is a truncated Newton step if $\|D_k s_k\| < \beta_3 \Delta_k$. Conditions of this form have been used by Dembo, Eisenstat, and Steihaug [1982] in a local analysis of Newton's method for systems of nonlinear equations, and by Dembo and Steihaug [1980] in a global analysis of Newton's method for unconstrained minimization.

For each of the algorithms mentioned in Section 3 there is a constant $\beta_3 > 0$ such that s_k is a truncated Newton step if $\|D_k s_k\| < \beta_3 \Delta_k$. In the preconditioned dogleg algorithm $\beta_3 = 1$ because s_k is the Newton step if $\|D_k s_k\| < \Delta_k$. Similarly, $\beta_3 = 1$ in the algorithm of Steihaug, and $\beta_3 = 1 - \sigma$ in the algorithm of Moré and Sorensen.

Rate of convergence results are an immediate consequence of (4.21). Let $\{x_k\}$ be the sequence defined in Theorem (4.19) and assume that $\{x_k\}$ converges to a point x^* with $\nabla^2 f(x^*)$ positive definite. Then (4.21) implies that

(4.22) $$\limsup_{k \to +\infty} \frac{\|\nabla f(x_{k+1})\|}{\|\nabla f(x_k)\|} \leq \limsup_{k \to +\infty} \xi_k \equiv \xi^*.$$

The following result spells out the consequences of (4.22).

Theorem (4.23). *Let $f: R^n \to R$ satisfy the assumptions of Theorem (4.17) and let $\{x_k\}$ be the sequence defined in Theorem (4.19). Assume, in addition, that (4.21) is satisfied whenever $\|D_k s_k\| < \beta_3 \Delta_k$. If $\{x_k\}$ converges to some x^* with $\nabla^2 f(x^*)$ positive definite then (4.22) holds. In particular, if $\xi^* < 1$ then $\{x_k\}$ converges Q-linearly to x^*, and if $\xi^* = 0$ then $\{x_k\}$ converges Q-superlinearly to x^*.*

Conditions (4.16) and (4.21) are closely related. We claim that under the conditions of Theorem (4.19) and for all k sufficiently large, s_k is a truncated Newton step if and only if (4.16) holds for some $\beta_1 > 0$. To establish this claim define the ellipsoidal norm

$$v_k[z] = (z^T \nabla^2 f(x_k)^{-1} z)^{\frac{1}{2}}.$$

Since $\{x_k\}$ converges to x^* and $\nabla^2 f(x^*)$ is positive definite, this norm is well-defined for all k sufficiently large. Moreover, since $\nabla f(x^*) = 0$ and since $\{\Delta_k\}$ is bounded away from zero, we have that eventually $\|D_k s_k^N\| \leq \Delta_k$. Thus (4.16) holds if and only if

$$\psi_k(s_k) - \psi_k(s_k^N) \leq (1 - \beta_1)(-\psi_k(s_k^N)).$$

A calculation now shows that

$$\psi_k(s_k) - \psi_k(s_k^N) = \tfrac{1}{2} v_k[\nabla f(x_k) + \nabla^2 f(x_k) s_k],$$

and that

$$-\psi_k(s_k^N) = \tfrac{1}{2} v_k[\nabla f(x_k)].$$

Hence, (4.16) holds if and only if

$$v_k[\nabla f(x_k) + \nabla^2 f(x_k) s_k] \leq (1 - \beta_1) v_k[\nabla f(x_k)].$$

This is the desired result. It shows that if the proper norm is used, then conditions of the form (4.21) are quite natural.

5. Future Developments

We have reviewed some of the recent developments in algorithms and software for trust region methods. The choice of topics was necessarily limited, so in the remainder of this section I would like to mention three other topics which deserve attention.

We have already pointed out, in the introduction, that trust region methods for non-smooth and constrained optimization problems deserve further study. We need to formulate sufficient decrease conditions and to develop algorithms for the relevant subproblems.

Large scale optimization received some attention in this paper, but we only scratched the surface. We did not mention the tradeoffs between storage and computational efficiency that are necessary in the development of software for large scale problems, nor any of the techniques used to estimate sparse Jacobian and Hessian matrices.

Optimization of noisy functions is an important problem. It is necessary to understand the behavior of optimization algorithms when the computation of the functions is subject to noise, and to develop algorithms and software for noisy problems.

The interested reader will find additional information on these topics in the proceedings of the NATO Advanced Research Institute on nonlinear optimization; the papers of Dembo [1982], Fletcher [1982], Moré [1982], and Sorensen [1982b] are especially relevant.

References

Anderson, D. H. and Osborne, M. R. [1977]. Discrete, nonlinear approximation problems in polyhedral norms, Numer. Math. 28, 157–170.

Björck, A. and Elfving, T. [1979]. Accelerated projection methods for computing pseudoinverse solutions of systems of linear equations, BIT 19, 145–163.

Broyden, C. G. [1965]. A class of methods for solving nonlinear simultaneous equations, Math. Comp. 19, 577–593.

Buckley, A. and LeNir, A. [1982]. QN-like variable storage conjugate gradients, University of Arizona, Management and Information Systems Department, Report 2, Tucson, Arizona.

Cline, A. K., Moler, C. B., Stewart, G. W. and Wilkinson, J. H. [1979]. An estimate for the condition number of a matrix, SIAM J. Numer. Anal. 16, 368–375.

Davidon, W. C. [1980]. Conic approximations and collinear scalings for optimizers, SIAM J. Numer. Anal. 17, 268–281.

Dembo, R. S. [1982]. Large scale nonlinear optimization, Nonlinear Optimization 1981, M. J. D. Powell, ed., Academic Press.

Dembo, R. S., Eisenstat, S. C. and Steihaug, T. [1982]. Inexact Newton methods, SIAM J. Numer. Anal. 19, 400–408.

Dembo, R. S. and Steihaug, T. [1980]. Truncated Newton algorithms for large scale unconstrained optimization, Yale University, School of Organization and Management, Report 48, New Haven, Connecticut.

Dennis, J. E., Gay, D. M. and Welsch, R. E. [1981]. An adaptive nonlinear least squares algorithm, ACM Trans. Math. Software 7, 348–368.

Dennis, J. E. and Mei, H. H. W. [1979]. Two new unconstrained optimization algorithms which use function and gradient values, J. Optim. Theory Appl. 28, 453–482.

Deuflhard, P. [1974]. A modified Newton method for the solution of illconditioned systems of nonlinear equations with application to multiple shooting, Numer. Math. 22, 289–315.

Fletcher, R. [1971]. A modified Marquardt subroutine for nonlinear least squares, Atomic Energy Research Establishment, Report R6799, Harwell, England.

Fletcher, R. [1972]. An algorithm for solving linearly constrained optimization problems, Math. Programming 2, 133–165.

Fletcher, R. [1980]. Practical Methods of Optimization, Volume 1: Unconstrained Optimization, John Wiley & Sons.

Fletcher, R. [1981]. Practical Methods of Optimization, Volume 2: Constrained Optimization, John Wiley & Sons.

Fletcher, R. [1982]. Methods for nonlinear constraints, Nonlinear Optimization 1981, M. J. D. Powell, ed., Academic Press.

Fletcher, R. and Jackson, M. [1974]. Minimization of a quadratic function of many variables subject to only lower and upper bounds, J. Inst. Math. Appl. 14, 159–174.

Gay, D. M. [1980]. Subroutines for unconstrained minimization using a model/trust region approach, Massachusetts Institute of Technology, Center for Computational Research in Economics and Management Report 18, Cambridge, Massachusetts.

Gay, D. M. [1981]. Computing optimal locally constrained steps, SIAM J. Sci. Stat. Comput. 2, 186–197.

Gay, D. M. [1982]. On convergence testing in model/trust region algorithms for unconstrained optimization, manuscript in preparation.

Gill, P. E. and Murray, W. [1979]. Conjugate-gradient methods for large-scale nonlinear optimization, Stanford University, Systems Optimization Laboratory, Report SOL 79-15, Stanford, California.

Goldfeld, S., Quandt, R. and Trotter, H. [1966]. Maximization by quadratic hill climbing, Econometrica 34, 541–551.

Griffith, R. E. and Stewart, R. A. [1961]. A nonlinear programming technique for the optimization of continuous processing systems, Management Sci. 7, 379–392.

Hanson, R. J. and Hiebert, K. L. [1981]. A sparse linear programming subprogram, Sandia Laboratories, Report Sand 81-0297, Albuquerque, New Mexico.

Hebden, M. D. [1973]. An algorithm for minimization using exact second derivatives, Atomic Energy Research Establishment, Report T. P. 515, Harwell, England.

Hestenes, M. R. and Steifel, E. [1952]. Methods of conjugate gradients for solving linear systems, J. Res. Bur. Standards 49, 409–436.

Levenberg, K. [1944]. A method for the solution of certain nonlinear problems in least squares, Quart. Appl. Math. 2, 164–168.

Madsen, K. [1975]. An algorithm for minimax solution of overdetermined systems of nonlinear equations, J. Inst. Math. Appl. 16, 321–328.

Marquardt, D. W. [1963]. An algorithm for least squares estimation of nonlinear parameters, SIAM J. Appl. Math. 11, 431–441.

Moré, J. J. [1978]. The Levenberg-Marquardt algorithm: Implementation and theory, Proceedings of the Dundee Conference on Numerical Analysis, G. A. Watson, ed., Springer-Verlag.

Moré, J. J. [1982]. Notes on optimization software, Nonlinear Optimization 1981, M. J. D. Powell, ed., Academic Press.

Moré, J. J., Garbow, B. S. and Hillstrom, K. E. [1980]. User guide for MINPACK-1, Argonne National Laboratory, Report ANL-80-74, Argonne, Illinois.

Moré, J. J. and Sorensen, D. C. [1981]. Computing a trust region step, Argonne National Laboratory, Report ANL-81-83, Argonne, Illinois.

Munksgaard, N. [1980]. Solving sparse symmetric sets of linear equations by preconditioned conjugate gradients, ACM Trans. Math. Software 6, 206–219.

Nazareth, L. and Nocedal, J. [1982]. Conjugate direction methods with variable storage, Math. Programming 23, 326–340.

Paige, C. C. and Saunders, M. A. [1975]. Solution of sparse indefinite systems of linear equations, SIAM J. Numer. Anal. 12, 617–629.

Paige, C. C. and Saunders, M. A. [1978]. A bidiagonalization algorithm for sparse linear equations and least squares problems, Stanford University, Department of Operations Research, Report 19, Stanford, California.

Paige, C. C. and Saunders, M. A. [1982]. LSQR: An algorithm for sparse linear equations and linear squares, ACM Trans. Math. Software 8, 43–71.

Powell, M. J. D. [1970a]. A hybrid method for nonlinear equations, Numerical Methods for Nonlinear Algebraic Equations, P. Rabinowitz, ed., Gordon and Breach.

Powell, M. J. D. [1970b]. A new algorithm for unconstrained optimization, Nonlinear Programming, J. B. Rosen, O. L. Mangasarian and K. Ritter, eds., Academic Press.

Powell, M. J. D. [1975]. Convergence properties of a class of minimization algorithms, Nonlinear Programming 2, O. L. Mangasarian, R. R. Meyer and S. M. Robinson, eds., Academic Press.

Powell, M. J. D. [1978]. A fast algorithm for nonlinearly constrained optimization calculations, Proceedings of the Dundee Conference on Numerical Analysis, G. A. Watson, ed., Springer-Verlag.

Reinsch, C. H. [1971]. Smoothing by spline functions II, Numer. Math. 16, 451-454.

Saunders, M. A. [1979]. Sparse least squares by conjugate gradients: A comparison of preconditioning methods, Stanford University, Department of Operations Research, Report 5, Stanford, California.

Schultz, G. A., Schnabel, R. S. and Byrd, R. H. [1982]. A family of trust region based algorithms for unconstrained minimization with strong global convergence properties, University of Colorado, Department of Computer Science, Report 216, Boulder, Colorado.

Sorensen, D. C. [1982a]. Newton's method with a model trust region modification, SIAM J. Numer. Anal. 19, 409-426.

Sorensen, D. C. [1982b]. Trust region methods for unconstrained optimization, Nonlinear Optimization 1981, M. J. D. Powell, ed., Academic Press.

Steihaug, T. [1981]. The conjugate gradient method and trust regions in large scale optimization, Rice University, Department of Mathematical Sciences, Report 1, Houston, Texas.

Thomas, S. W. [1975]. Sequential estimation techniques for quasi-Newton algorithms, Ph. D. dissertation, Cornell University, Ithaca, New York.

Variable Metric Methods for Constrained Optimization

M. J. D. Powell
D.A.M.T.P., University of Cambridge, Silver Street, Cambridge, CB3 9EW, England

Abstract. Variable metric methods solve nonlinearly constrained optimization problems, using calculated first derivatives and a single positive definite matrix, which holds second derivative information that is obtained automatically. The theory of these methods is shown by analysing the global and local convergence properties of a basic algorithm, and we find that superlinear convergence requires less second derivative information than in the unconstrained case. Moreover, in order to avoid the difficulties of inconsistent linear approximations to constraints, careful consideration is given to the calculation of search directions by unconstrained minimization subproblems. The Maratos effect and relations to reduced gradient algorithms are studied briefly.

1. Introduction

The methods to be considered are intended to calculate the least value of a function $F(x)$, subject to equality and inequality constraints

$$c_i(x) = 0, \quad i = 1, 2, \ldots, m' \atop c_i(x) \geq 0, \quad i = m'+1, \ldots, m \qquad (1.1)$$

on the vector of variables $x \in \mathbb{R}^n$, where the functions $F(x)$ and $\{c_i(x); i = 1, 2, \ldots, m\}$ are real and differentiable. Either or both of the integers m' and $(m-m')$ may be zero. Each method generates a sequence of points $\{x_k; k = 1, 2, 3, \ldots\}$ in $\mathbb{R}^n$, where x_1 is provided by the user, and where the sequence should converge to a solution, x^* say, of the given nonlinear programming problem. It is assumed that the functions and their first derivatives can be calculated for any x.

In the unconstrained case, $m = m' = 0$, the calculation of x_{k+1} from x_k by a variable metric method depends on the quadratic approximation

$$\bar{F}_k(x) = F(x_k) + (x - x_k)^T \nabla F(x_k) + \tfrac{1}{2}(x - x_k)^T B_k (x - x_k) \qquad (1.2)$$

to $F(x)$, where B_k is an $n \times n$ positive definite symmetric matrix that is chosen automatically. Often x_{k+1} is the vector $[x_k - B_k^{-1} \nabla F(x_k)]$ in order to minimize $\bar{F}_k(x)$, provided that this value is acceptable to a "line search" or "trust region" technique, that is present to force convergence from poor starting approximations (see Fletcher [11] or Gill, Murray and Wright [16], for instance). The matrix B_k serves to provide information about second derivatives of $F(x)$ that was obtained on previous iterations, but no second derivatives are calculated expli-

II. Variable Metric Methods for Constrained Optimization

citly. The use of a positive definite matrix for this purpose in the unconstrained case characterizes a variable metric method.

When constraints are present, we define an algorithm to be a variable metric method if second derivative information from previous iterations, used in the calculation of x_{k+1} from x_k, is also held in a symmetric matrix B_k, which now may be positive definite or positive semi-definite, and whose dimensions are at most $n \times n$. Allowing the rank of B_k to be less than n is necessary if one is to enjoy the savings that can be obtained from "reduced second derivative matrices" (see Fletcher [12], for instance). It is important to note that only one matrix is used for curvature information from both the objective and constraint functions. Often it is suitable to regard B_k as an approximation to the Hessian of the Lagrangian function of the given calculation at the solution x^*.

The following basic method is used by some variable metric algorithms to calculate x_{k+1} from x_k in the usual case when the dimensions of B_k are $n \times n$. The search direction d_k is defined to be the vector d in $\mathbb{R}^n$ that minimizes the function

$$\tilde{F}_k(x_k + d) = F(x_k) + d^T \nabla F(x_k) + \tfrac{1}{2} d^T B_k d \tag{1.3}$$

subject to the linear constraints

$$\left. \begin{array}{ll} c_i(x_k) + d^T \nabla c_i(x_k) = 0, & i = 1, 2, \ldots, m' \\ c_i(x_k) + d^T \nabla c_i(x_k) \geq 0, & i = m'+1, \ldots, m \end{array} \right\} \tag{1.4}$$

Therefore the calculation of d_k is a convex quadratic programming problem. Then x_{k+1} is given the value

$$x_{k+1} = x_k + \alpha_k d_k, \tag{1.5}$$

where α_k is a positive step length that is chosen by a line search procedure to give a reduction

$$W(x_{k+1}, \mu) < W(x_k, \mu) \tag{1.6}$$

in the line search objective function

$$W(x, \mu) = F(x) + \sum_{i=1}^{m'} \mu_i |c_i(x)| + \sum_{i=m'+1}^{m} \mu_i \max[0, -c_i(x)]. \tag{1.7}$$

Here $\mu \in \mathbb{R}^m$ is a vector of positive parameters that are held constant, except perhaps for some automatic adjustments on the early iterations to provide suitable values. We let α_k be the first number in a monotonically decreasing sequence $\{\alpha_{k1}, \alpha_{k2}, \alpha_{k3}, \ldots\}$ that is allowed by a condition that is a little stronger than inequality (1.6), where $\alpha_{k1} = 1$, and where the ratios $\{\alpha_{kj+1}/\alpha_{kj}; j = 1, 2, 3, \ldots\}$ are all in a closed subinterval of $(0, 1)$, for example $[0.1, 0.5]$. In practice one might use the freedom in each ratio to minimize the predicted value of $W(x_k + \alpha_{kj+1} d_k, \mu)$.

It is proved in Section 2 that, under certain conditions, this basic method generates a sequence $\{x_k; k = 1, 2, 3, \ldots\}$ whose limit points are all Kuhn-Tucker points of the given calculation. These conditions are less severe than those that occur in Han's [17] original analysis of global convergence, but even weaker conditions have been found by Chamberlain [5]. The restrictions on B_k are very mild indeed. Because they allow $B_k = I$ on each iteration, variable metric methods can be related to projected gradient algorithms.

It is well known, however, that if $B_k = I$ on each iteration, and if there are no constraints on the variables, then a version of the steepest descent method is obtained, that is often useless because it converges too slowly. Therefore the more successful variable metric algorithms for unconstrained optimization generate the matrices $\{B_k; k = 1, 2, 3, \ldots\}$ in such a way that the ratio

$$\|[B_k - \nabla^2 F(x^*)]d_k\|/\|d_k\| \tag{1.8}$$

tends to zero as $k \to \infty$, if the sequence $\{x_k; k = 1, 2, 3, \ldots\}$ converges to x^*. In this case, if $\nabla^2 F(x^*)$ is positive definite, then Q-superlinear convergence,

$$\lim_{k \to \infty} \|x_k + d_k - x^*\|/\|x_k - x^*\| = 0, \tag{1.9}$$

is obtained (Dennis and Moré [9]). Section 3 gives analogous conditions on B_k for superlinear convergence in constrained calculations, much of the theory being taken from Powell [22]. Of course we find that, if the solution x^* is at the intersection of n constraint boundaries whose normals are linearly independent, and if the strict complementarity condition holds, then the final rate of convergence is independent of B_k, which suggests correctly that constrained calculations often converge more rapidly than unconstrained ones.

Unfortunately, even though the limit (1.9) may be obtained, condition (1.6) may prevent the choice $x_{k+1} = x_k + d_k$ for every value of k. This phenomenon, which is known as the "Maratos effect", does not occur often in practice. However, Section 4 gives an example to show that the effect can cause a variable metric algorithm to be highly inefficient, and it describes briefly some remedies that have been proposed recently.

A much more common difficulty is that the linearizations (1.4) of the given nonlinear constraints (1.1) may introduce inconsistencies. For example, let $n = 1$ and let the constraints $x \leq 1$ and $x^2 \geq 0$ be linearized at $x_k = 3$. We obtain the inequalities $3 + d \leq 1$ and $9 + 6d \geq 0$, which have no solution. Therefore it is sometimes important to choose a search direction that does not satisfy the conditions (1.4). Suitable methods are suggested by Bartholomew-Biggs [2], Fletcher [13] and Powell [21]. Each of the recommended search directions can be defined as the vector d that solves an unconstrained minimization problem, whose objective function is $\bar{F}_k(x_k + d)$ plus penalty terms that depend on the left hand sides of the linear constraints (1.4). This work is the subject of Section 5. Because it allows one to introduce further constraints on the search direction, it makes possible the use of trust regions in variable metric algorithms for constrained optimization.

In the final section some topics are mentioned briefly that are important to variable metric methods, but that cannot be given more consideration because of time and space restrictions. Particular attention is given to the relations between reduced gradient and variable metric algorithms.

2. Global Convergence

In this section we consider the convergence of the sequence $\{x_k; k=1, 2, 3, \ldots\}$ when each iteration calculates x_{k+1} from x_k by the basic method that is described in Section 1. Therefore we assume that the constraints (1.4) are consistent, and that each quadratic programming problem that determines a search direction has a bounded solution.

We make use of the Kuhn-Tucker conditions that hold at the solution of the quadratic programming problem that determines d_k. These conditions state the existence of Lagrange parameters $\{\lambda_i^{(k)}; i=1, 2, \ldots, m\}$ that satisfy the bounds

$$\lambda_i^{(k)} \geq 0, \quad i = m'+1, \ldots, m, \tag{2.1}$$

and the equations

$$\nabla F(x_k) + B_k d_k = \sum_{i=1}^{m} \lambda_i^{(k)} \nabla c_i(x_k), \tag{2.2}$$

$$\lambda_i^{(k)} \{c_i(x_k) + d_k^T \nabla c_i(x_k)\} = 0, \quad i = 1, 2, \ldots, m. \tag{2.3}$$

Further, the conditions (1.4) are obtained when $d = d_k$. It follows that, if d_k is zero, then x_k is a Kuhn-Tucker point of the main calculation. Because one usually terminates a variable metric algorithm in this case, we assume from now on that none of the calculated vectors of variables are Kuhn-Tucker points of the given nonlinear programming problem.

The parameters $\{\mu_i; i=1, 2, \ldots, m\}$ of the line search objective function (1.7) have to be such that the reduction (1.6) can be obtained for a positive value of the step length α_k. In order to identify suitable conditions on μ, we note that the Lagrangian function

$$\bar{L}_k(x_k + d) = \bar{F}_k(x_k + d) - \sum_{i=1}^{m} \lambda_i^{(k)} \{c_i(x_k) + d^T \nabla c_i(x_k)\} \tag{2.4}$$

is a quadratic function of d, that is convex because its second derivative matrix is B_k, and that takes its least value when $d = d_k$ because of equation (2.2). Thus we deduce the relation

$$\bar{L}_k(x_k) = \bar{L}_k(x_k + d_k) + \tfrac{1}{2} d_k^T B_k d_k$$
$$\geq \bar{L}_k(x_k + d_k). \tag{2.5}$$

Moreover the function

$$\bar{W}_k(x_k + d) = \bar{F}_k(x_k + d) + \sum_{i=1}^{m'} \mu_i |c_i(x_k) + d^T \nabla c_i(x_k)|$$
$$+ \sum_{i=m'+1}^{m} \mu_i \max[0, -c_i(x_k) - d^T \nabla c_i(x_k)] \tag{2.6}$$

is an approximation to $W(x_k + d, \mu)$ that satisfies the condition

$$\bar{W}_k(x_k + d) = W(x_k + d, \mu) + o(\|d\|), \tag{2.7}$$

provided that the objective and constraint functions have continuous first derivatives. Therefore we compare $\bar{W}_k(x_k + d)$ with $\bar{L}_k(x_k + d)$.

The difference between these functions is bounded below by the expression

$$\bar{W}_k(x_k+d) - \bar{L}_k(x_k+d) \geq \sum_{i=1}^{m'} \{\mu_i - |\lambda_i^{(k)}|\}|c_i(x_k) + d^T \nabla c_i(x_k)|$$
$$+ \sum_{i=m'+1}^{m} \{\mu_i - \lambda_i^{(k)}\} \max[0, -c_i(x_k) - d^T \nabla c_i(x_k)], \tag{2.8}$$

which depends on inequality (2.1). We require the components of μ to satisfy the bounds

$$\mu_i > |\lambda_i^{(k)}|, \quad i = 1, 2, \ldots, m, \tag{2.9}$$

in order that the relation

$$\bar{W}_k(x_k+d) > \bar{L}_k(x_k+d) \tag{2.10}$$

is obtained if at least one of the conditions (1.4) does not hold.

In this case, if x_k is infeasible with respect to the given nonlinear constraints (1.1), we have the bound

$$\bar{W}_k(x_k) > \bar{L}_k(x_k)$$
$$\geq \bar{L}_k(x_k + d_k) = \bar{W}_k(x_k + d_k), \tag{2.11}$$

while, if x_k is a feasible point, we have the condition

$$\bar{W}_k(x_k) = \bar{L}_k(x_k)$$
$$> \bar{L}_k(x_k + d_k) = \bar{W}_k(x_k + d_k), \tag{2.12}$$

where the strict inequality depends on the observation that the term $d_k^T B_k d_k$ of expression (2.5) is positive, because otherwise $d_k = 0$ would be a solution of the quadratic programming calculation. Hence the number

$$r_k = \bar{W}_k(x_k) - \bar{W}_k(x_k + d_k) \tag{2.13}$$

is positive. This remark is important because the convexity of the function (2.6) and equation (2.7) give the relation

$$W(x_k, \mu) - W(x_k + \alpha d_k, \mu) = \bar{W}_k(x_k) - \bar{W}_k(x_k + \alpha d_k) + o(\alpha)$$
$$\geq \alpha [\bar{W}_k(x_k) - \bar{W}_k(x_k + d_k)] + o(\alpha)$$
$$= \alpha r_k + o(\alpha), \quad 0 \leq \alpha \leq 1. \tag{2.14}$$

Therefore the required reduction (1.6) can be achieved by choosing α_k to be sufficiently small and positive.

Further, we may replace the condition (1.6) on the step length by the inequality

$$W(x_{k+1}, \mu) \leq W(x_k, \mu) - \sigma \alpha_k r_k, \tag{2.15}$$

where σ is any constant from the open interval $(0, 1)$, for example $\sigma = 0.1$ is usually suitable in practice. This is the stronger condition that is mentioned in Section 1, and it is important to the following convergence theorem.

II. Variable Metric Methods for Constrained Optimization

Theorem 1. If the sequence $\{x_k; k=1, 2, 3, \ldots\}$ is calculated in the way that has been described, if the points of this sequence and the points $\{x_k + d_k; k=1, 2, 3, \ldots\}$ remain in a closed, bounded and convex region of $\mathbb{R}^n$ in which the objective and constraint functions have continuous first derivatives, if the matrices $\{B_k; k=1, 2, 3, \ldots\}$ are uniformly bounded, and if the components of $\underline{\mu}$ satisfy the condition

$$\mu_i \geq |\lambda_i^{(k)}| + \rho \tag{2.16}$$

for all i and k, where ρ is a positive constant, then all limit points of the sequence $\{x_k; k=1, 2, 3, \ldots\}$ are Kuhn-Tucker points of the given nonlinear programming problem.

Proof. Let η be a small positive constant, and consider the iterations on which the number r_k, defined by equation (2.13), exceeds η. Because continuity of first derivatives in a compact domain is equivalent to uniform continuity, and because the vector d_k in expression (2.14) is bounded, it follows that there is a positive constant $\beta(\eta)$ such that inequality (2.15) holds for any α_k in the interval $[0, \beta(\eta)]$. Thus, remembering the way in which the step length is chosen (see Section 1), we deduce that $[W(x_k, \underline{\mu}) - W(x_{k+1}, \underline{\mu})]$ is bounded away from zero if $r_k > \eta$. However, the reductions in the line search objective function tend to zero, because $\{W(x_k, \underline{\mu}); k=1, 2, 3, \ldots\}$ is a monotonically decreasing sequence that is bounded below. Therefore r_k also tends to zero, which is the condition

$$\lim_{k \to \infty} [\bar{W}_k(x_k) - \bar{W}_k(x_k + d_k)] = 0. \tag{2.17}$$

Because at least one of the expressions (2.11) and (2.12) is satisfied for each k, equation (2.17) gives the limits

$$\lim_{k \to \infty} [\bar{W}_k(x_k) - \bar{L}_k(x_k)] = 0 \tag{2.18}$$

and

$$\lim_{k \to \infty} [\bar{L}_k(x_k) - \bar{L}_k(x_k + d_k)] = 0. \tag{2.19}$$

By letting $\underline{d} = 0$ in inequality (2.8), and by using condition (2.16), we deduce from expression (2.18) that every limit point of the sequence $\{x_k; k=1, 2, 3, \ldots\}$ satisfies the nonlinear constraints (1.1).

Our method of proof allows us to simplify the notation by assuming without loss of generality that $\{x_k; k=1, 2, 3, \ldots\}$ has only one limit point, x^* say. We define I^* to be the set of indices of the inequality constraints that are satisfied as equations at x^*, and for each k we let δ_k be the distance

$$\delta_k = \min_{y \in C} \|\nabla F(x_k) - y\|_2, \tag{2.20}$$

where $C \subset \mathbb{R}^n$ is the convex cone of points that can be expressed in the form

$$y = \sum_{i=1}^{m'} v_i \nabla c_i(x^*) + \sum_{i \in I^*} v_i \nabla c_i(x^*), \tag{2.21}$$

the coefficients $\{v_i; i=1, 2, \ldots, m'\}$ and $\{v_i; i \in I^*\}$ being real numbers that are unconstrained and non-negative respectively. It remains to prove that $\delta_k \to 0$, so it is sufficient to establish the condition

$$\lim_{k \to \infty} \|\nabla F(x_k) - \sum_{i=1}^{m'} \lambda_i^{(k)} \nabla c_i(x^*) - \sum_{i \in I^*} \lambda_i^{(k)} \nabla c_i(x^*)\| = 0, \qquad (2.22)$$

where the Lagrange parameters $\{\lambda_i^{(k)}; i=1, 2, \ldots, m\}$ have been defined already.

Of course we make use of the fact that equation (2.2) gives the expression

$$\lim_{k \to \infty} \|\nabla F(x_k) - \sum_{i=1}^{m} \lambda_i^{(k)} \nabla c_i(x_k) + B_k d_k\| = 0. \qquad (2.23)$$

Because the theorem states that the Lagrange multipliers are bounded, this limit is preserved if $\nabla c_i(x_k)$ is replaced by $\nabla c_i(x^*)$ for all i. Moreover, because equations (2.5) and (2.19) imply that $B_k^{1/2} d_k \to 0$, we deduce from the boundedness of the matrices $\{B_k; k=1, 2, 3, \ldots\}$ that expression (2.23) remains valid if the term $B_k d_k$ is deleted. It follows that equation (2.22) is true if $\lambda_i^{(k)}$ tends to zero as $k \to \infty$, where i is the index of any inequality constraint that is not in I^*.

Let i be such an index, and let k be so large that $c_i(x_k)$ is positive. By giving further attention to the *derivation* of inequality (2.8), we deduce the condition

$$\bar{W}_k(x_k) - \bar{L}_k(x_k) \geq \lambda_i^{(k)} c_i(x_k). \qquad (2.24)$$

Since $\lambda_i^{(k)} \geq 0$, and since $c_i(x_k) \to c_i(x^*) > 0$, it follows from expressions (2.18) and (2.24) that $\lambda_i^{(k)}$ tends to zero as $k \to \infty$, which completes the proof of the theorem. □

This theorem compares favourably with other global convergence results that have been published for variable metric algorithms, because it allows the matrices $\{B_k; k=1, 2, 3, \ldots\}$ to be positive semi-definite, and because the gradient vectors of the active constraints at x^* do not have to be linearly independent. We recall from Section 1, however, that the algorithm that is analysed has the disadvantage that useful search directions may exist when there is no solution to the quadratic programming problem that normally defines d_k. We return to this question in Section 5.

We consider next the choice of the parameters $\{\mu_i; i=1, 2, \ldots, m\}$ of the line search objective function. Large constant values are not recommended, because variable metric methods for constrained optimization are faster than reduced gradient algorithms only if the calculated points $\{x_k; k=1, 2, 3, \ldots\}$ are allowed to move away from the boundaries of curved active constraints. Therefore an ideal algorithm would adjust μ automatically, but the techniques that have been proposed already are rather crude. Chamberlain [4] shows that a method that I suggested can lead to cycling instead of convergence, so now I [23] prefer to set the components of μ to very small positive values initially, which are increased if necessary during the calculation so that a positive step length can give the reduction (1.6) in the line search objective function. It is

II. Variable Metric Methods for Constrained Optimization

easy to obtain the property that, if $\underline{\mu}$ remains bounded, then the number of iterations that change its value is finite. Thus convergence theorems that depend on constant $\underline{\mu}$ are valid. It seems however that, given any technique for choosing $\underline{\mu}$, it is possible to find pathological examples to show that the technique is inefficient, but serious difficulties are unusual in practice.

3. Superlinear Convergence

We assume in this section that a variable metric algorithm for constrained optimization calculates each search direction $\underline{d}_k$ by the basic method of Section 1, and that the sequence $\{\underline{x}_k; k=1, 2, 3, \ldots\}$ converges to a Kuhn-Tucker point $\underline{x}^*$ of the given nonlinear programming problem. We study the important question of choosing the matrices $\{B_k; k=1, 2, 3, \ldots\}$ so that, if the step length of the line search is set to one for all sufficiently large k, which is the condition

$$\underline{x}_{k+1} = \underline{x}_k + \underline{d}_k, \quad k \geq k_0, \tag{3.1}$$

where k_0 is a constant, then the rate of convergence of the sequence $\{\underline{x}_k; k=1, 2, 3, \ldots\}$ is superlinear. Our theory depends on several conditions that are usual in this kind of analysis.

Because superlinear convergence is obtained only if $\|\underline{d}_k\|$ tends to zero as $k \to \infty$, we assume without loss of generality that any inequality constraints of the given nonlinear programming problem are satisfied as equations at $\underline{x}^*$. We also make the usual assumption that any inequality constraints can be treated as equations for sufficiently large k, but it does lose generality unless the constraint gradients are linearly independent at $\underline{x}^*$ and the strict complementarity condition holds. Therefore in this section $m' = m$ and the constraints on the variables are the equations

$$c_i(\underline{x}) = 0, \quad i = 1, 2, \ldots, m. \tag{3.2}$$

We require three more conditions. They are that all functions are twice continuously differentiable, that the first derivatives of the constraints $\{\nabla c_i(\underline{x}^*); i=1, 2, \ldots, m\}$ are linearly independent, and that second order sufficiency holds. To state the details of this last condition, we define the Lagrange parameters $\{\lambda_i^*; i=1, 2, \ldots, m\}$ by the equation

$$\nabla F(\underline{x}^*) = \sum_{i=1}^{m} \lambda_i^* \nabla c_i(\underline{x}^*), \tag{3.3}$$

and we let G^* be the second derivative matrix

$$G^* = \nabla^2 F(\underline{x}^*) - \sum_{i=1}^{m} \lambda_i^* \nabla^2 c_i(\underline{x}^*). \tag{3.4}$$

Second order sufficiency states that, if $\underline{d}$ is any non-zero vector that is orthogonal to the gradients $\{\nabla c_i(\underline{x}^*); i=1, 2, \ldots, m\}$, then $\underline{d}^T G^* \underline{d}$ is positive. Two consequences of these conditions are that the matrix

$$J(\underline{x}) = \left(\begin{array}{c|c} G^* & -N(\underline{x}) \\ \hline -N(\underline{x})^T & 0 \end{array} \right) \tag{3.5}$$

is nonsingular at $\underline{x}=\underline{x}^*$, where $N(\underline{x})$ is the $n\times m$ matrix that has the columns $\{\nabla c_i(\underline{x}); i=1, 2, \ldots, m\}$, and that there exists a positive number τ such that the matrix

$$G^* + \tau \sum_{i=1}^{m} \nabla c_i(\underline{x}^*) \nabla c_i(\underline{x}^*)^T \tag{3.6}$$

is positive definite (see Fletcher [12], for instance).

We will find it useful to take the point of view that the equations

$$\left. \begin{array}{l} \nabla F(\underline{x}) - \sum\limits_{i=1}^{m} \lambda_i \nabla c_i(\underline{x}) = 0 \\ -c_i(\underline{x}) = 0, \quad i = 1, 2, \ldots, m \end{array} \right\} \tag{3.7}$$

are a square system in $(m+n)$ unknowns, namely the components of $\underline{x}$ and $\underline{\lambda}$. The point $(\underline{x}^*, \underline{\lambda}^*)$ is a solution of the system, and here the Jacobian is the nonsingular matrix $J(\underline{x}^*)$. Therefore, if Newton's method for solving nonlinear equations is applied inductively to the system, and if the calculated points in $\mathbb{R}^{m+n}$ converge to $(\underline{x}^*, \underline{\lambda}^*)$, then the rate of convergence is superlinear. Further, this statement remains true if the true Jacobian matrix is replaced by $J(\underline{x})$ for each $(\underline{x}, \underline{\lambda})$. Thus we deduce the limit

$$\lim_{k\to\infty} \frac{\|\underline{x}_k + \underline{\delta}_k - \underline{x}^*\| + \|\underline{\lambda}_k + \underline{\eta}_k - \underline{\lambda}^*\|}{\|\underline{x}_k - \underline{x}^*\| + \|\underline{\lambda}_k - \underline{\lambda}^*\|} = 0, \tag{3.8}$$

provided that $\underline{\lambda}_k$ is sufficiently close to $\underline{\lambda}^*$, where $\underline{\delta}_k \in \mathbb{R}^n$ and $\underline{\eta}_k \in \mathbb{R}^m$ are defined by the system

$$J(\underline{x}_k) \begin{pmatrix} \underline{\delta}_k \\ \underline{\eta}_k \end{pmatrix} = \begin{pmatrix} -\nabla F(\underline{x}_k) + N(\underline{x}_k)\underline{\lambda}_k \\ \underline{c}(\underline{x}_k) \end{pmatrix}. \tag{3.9}$$

However, $\underline{\delta}_k$ is independent of $\underline{\lambda}_k$ because expression (3.9) is equivalent to the equations

$$J(\underline{x}_k) \begin{pmatrix} \underline{\delta}_k \\ \underline{\eta}_k + \underline{\lambda}_k \end{pmatrix} = \begin{pmatrix} -\nabla F(\underline{x}_k) \\ \underline{c}(\underline{x}_k) \end{pmatrix}. \tag{3.10}$$

Therefore we may let $\underline{\lambda}_k = \underline{\lambda}^*$ in expression (3.8), in order to obtain the condition

$$\lim_{k\to\infty} \|\underline{x}_k + \underline{\delta}_k - \underline{x}^*\| / \|\underline{x}_k - \underline{x}^*\| = 0. \tag{3.11}$$

This limit is used in the second half of the proof of the following theorem, which extends to the constrained case the Q-superlinear convergence result of Dennis and Moré [9]. A similar theorem is proved by Boggs, Tolle and Wang [3], but their conditions on $\{B_k; k=1, 2, 3, \ldots\}$ and $\{\underline{x}_k; k=1, 2, 3, \ldots\}$ are stronger than ours.

II. Variable Metric Methods for Constrained Optimization

Theorem 2. Let the conditions of the first three paragraphs of this section be satisfied, and for each k let σ_k be the number

$$\sigma_k = \min_{\phi \in \mathbb{R}^m} \|(B_k - G^*)d_k - \sum_{i=1}^{m} \phi_i \nabla c_i(x_k)\|_2, \quad (3.12)$$

where G^* is the matrix (3.4). Then the Q-superlinear convergence condition

$$\lim_{k \to \infty} \|x_k + d_k - x^*\| / \|x_k - x^*\| = 0 \quad (3.13)$$

is equivalent to the limit

$$\lim_{k \to \infty} \sigma_k / \|d_k\| = 0. \quad (3.14)$$

Proof. First we assume that condition (3.13) holds. Thus, because the Lagrangian function

$$L(x) = F(x) - \sum_{i=1}^{m} \lambda_i^* c_i(x) \quad (3.15)$$

is stationary at x^*, we deduce the relation

$$\lim_{k \to \infty} \|\nabla L(x_k + d_k)\| / \|d_k\| = 0. \quad (3.16)$$

Equation (3.4), the continuity of second derivatives, equation (2.2) and the definition (3.12) imply the inequality

$$\|\nabla L(x_k + d_k)\| = \|\nabla L(x_k) + G^* d_k\| + o(\|d_k\|)$$

$$= \|\nabla F(x_k) - \sum_{i=1}^{m} \lambda_i^* \nabla c_i(x_k) + G^* d_k\| + o(\|d_k\|)$$

$$= \|(G^* - B_k)d_k + \sum_{i=1}^{m} (\lambda_i^{(k)} - \lambda_i^*) \nabla c_i(x_k)\| + o(\|d_k\|)$$

$$\geq \sigma_k + o(\|d_k\|). \quad (3.17)$$

Therefore the limit (3.14) is a consequence of expression (3.16).

To prove the converse result, we compare d_k with the vector δ_k that is defined by equation (3.10). Therefore we write equation (2.2) and the linear constraints on d_k in the form

$$\begin{pmatrix} B_k & -N(x_k) \\ -N(x_k)^T & 0 \end{pmatrix} \begin{pmatrix} d_k \\ \lambda^{(k)} \end{pmatrix} = \begin{pmatrix} -\nabla F(x_k) \\ c(x_k) \end{pmatrix}, \quad (3.18)$$

which we subtract from expression (3.10) to obtain the identity

$$J(x_k) \begin{pmatrix} \delta_k - d_k \\ \eta_k + \lambda_k - \lambda^{(k)} \end{pmatrix} = \begin{pmatrix} (B_k - G^*)d_k \\ 0 \end{pmatrix}. \quad (3.19)$$

Further, the argument that gives equation (3.10) from (3.9) also provides the identity

$$J(x_k) \begin{pmatrix} \delta_k - d_k \\ \eta_k + \lambda_k - \lambda^{(k)} + \phi_k \end{pmatrix} = \begin{pmatrix} (B_k - G^*)d_k - N(x_k)\phi_k \\ 0 \end{pmatrix}, \quad (3.20)$$

where we let ϕ_k be the value of ϕ that minimizes expression (3.12). Since $J(x_k)$ tends to a nonsingular matrix, it follows that the bound

$$\|\underline{\delta}_k - \underline{d}_k\| = O(\sigma_k) \tag{3.21}$$

is obtained as $k \to \infty$.

Therefore, if condition (3.14) holds, we have $\|\underline{\delta}_k - \underline{d}_k\| = o(\|\underline{d}_k\|)$. Because equation (3.11) can be expressed in the form $\|\underline{\delta}_k - (\underline{x}^* - \underline{x}_k)\| = o(\|\underline{x}_k - \underline{x}^*\|)$, it follows that the limit

$$\|\underline{d}_k - (\underline{x}^* - \underline{x}_k)\| = o(\|\underline{x}_k - \underline{x}^*\|) \tag{3.22}$$

is obtained. This limit is the same as the required condition (3.13), which completes the proof of the theorem. □

A corollary of this theorem is that, if the conditions of this section hold, then there exist positive definite matrices $\{B_k; k=1, 2, 3, \ldots\}$ that give superlinear convergence, even though G^* may have some negative eigenvalues. To prove this statement we recall that the matrix (3.6) is positive definite. Therefore there exists an integer k_1 such that we may make the choice

$$B_k = G^* + \tau \sum_{i=1}^{m} \nabla c_i(\underline{x}_k) \nabla c_i(\underline{x}_k)^T, \quad k \geq k_1. \tag{3.23}$$

In this case the numbers $\{\sigma_k; k \geq k_1\}$ are all zero, so superlinear convergence is a consequence of Theorem 2.

We note that the superlinear convergence condition (3.14) reduces to expression (1.8) in the unconstrained case, and that superlinear convergence is independent of B_k when $m=n$ and the constraint gradients are linearly independent.

We now take the point of view that $\{\nabla c_i(\underline{x}_k); i=1, 2, \ldots m\}$ are known, and that we wish to obtain superlinear convergence by choosing B_k in a way that is independent of $\underline{d}_k$. Condition (3.14) holds if B_k has the form

$$B_k = G^* + \sum_{i=1}^{m} \sum_{j=1}^{m} \nabla c_i(\underline{x}_k) S_{ij}^{(k)} \nabla c_j(\underline{x}_k)^T + o(\|\underline{d}_k\|), \tag{3.24}$$

where $S^{(k)}$ is a symmetric $m \times m$ matrix. Hence, ignoring the $o(\|\underline{d}_k\|)$ term, there are $\frac{1}{2}[n(n+1) - m(m+1)]$ degrees of freedom in B_k to be determined. However, the following theorem gives a superlinear convergence result that requires the elements of B_k to satisfy only $\frac{1}{2}(n-m)(n-m+1)$ conditions. Thus the reduction in the number of conditions is substantial when n is large and m is close to n. We note in Section 6 that it is related to reduced gradient methods.

This theorem is a little stronger than Theorem 1 of Powell [22], because it does not require $\underline{d}^T B_k \underline{d}$ to be bounded below by a constant positive multiple of $\|\underline{d}\|^2$, when $\underline{d}$ is any vector that is orthogonal to the gradients $\{\nabla c_i(\underline{x}^*); i=1, 2, \ldots, m\}$.

Theorem 3. Let the conditions of the first three paragraphs of this section be satisfied, let the matrices $\{B_k; k=1, 2, 3, \ldots\}$ be bounded, and for each k let P_k be the symmetric projection matrix such that, for any $\underline{y} \in \mathbb{R}^n$, $P_k \underline{y}$ is the vector

II. Variable Metric Methods for Constrained Optimization

of least Euclidean length of the form

$$P_k y = y - \sum_{i=1}^{m} \phi_i \nabla c_i(x_k). \tag{3.25}$$

If the condition

$$\lim_{k \to \infty} \|P_k(B_k - G^*)P_k d_k\|/\|d_k\| = 0 \tag{3.26}$$

holds, then the variable metric algorithm gives the two-step superlinear rate of convergence

$$\lim_{k \to \infty} \|x_{k+1} - x^*\|/\|x_{k-1} - x^*\| = 0. \tag{3.27}$$

Proof. First we show that $\|d_k\|$ is not much larger than $\|x_k - x^*\|$. Because the right hand side of equation (3.12) is just $\|P_k(B_k - G^*) d_k\|$, and because P_k is the operator

$$P_k = I - N(x_k)[N(x_k)^T N(x_k)]^{-1} N(x_k)^T, \tag{3.28}$$

the limit (3.26) implies the bound

$$\sigma_k = \|P_k(B_k - G^*) N(x_k)[N(x_k)^T N(x_k)]^{-1} N(x_k)^T d_k\| + o(\|d_k\|)$$
$$\leq M_1 \|x_k - x^*\| + o(\|d_k\|), \tag{3.29}$$

where M_1 is a constant, and where the last line depends on the constraint

$$c(x_k) + N(x_k)^T d_k = 0 \tag{3.30}$$

in the quadratic programming problem that determines d_k. Since equation (3.21) is still valid, expression (3.29) gives the relation

$$\|d_k\| - \|\delta_k\| \leq M_2 \|x_k - x^*\| + o(\|d_k\|) \tag{3.31}$$

for some constant M_2. It follows from the limit (3.11) that $\|d_k\|$ is bounded by the inequality

$$\|d_k\| \leq M_3 \|x_k - x^*\|, \tag{3.32}$$

where M_3 is another positive constant.

As in Theorem 2, our proof of superlinear convergence depends on the closeness of d_k to δ_k. In order to use equation (3.21) again, we note that the first part of expression (3.29) and the constraint (3.30) give the limit

$$\sigma_k = O(\|c(x_k)\|) + o(\|d_k\|)$$
$$= O(\|d_{k-1}\|^2) + o(\|d_k\|), \tag{3.33}$$

where the last line depends on the remark that $c(x_k)$ is the error of the approximation

$$c(x_{k-1} + d_{k-1}) \approx c(x_{k-1}) + N(x_{k-1})^T d_{k-1}. \tag{3.34}$$

It follows from expressions (3.21), (3.32) and (3.33) that the limit

$$\|\delta_k - d_k\| = o(\|x_{k-1} - x^*\| + \|x_k - x^*\|) \tag{3.35}$$

is obtained. Therefore the triangle inequality and equation (3.11) imply the relation
$$\|x_k + d_k - x^*\| \leq \|x_k + \delta_k - x^*\| + \|\delta_k - d_k\|$$
$$= o(\|x_{k-1} - x^*\| + \|x_k - x^*\|). \qquad (3.36)$$

The required result (3.27) now follows from the fact that $\|x_k - x^*\|$ is bounded above by the sum
$$\|x_{k-1} - x^*\| + \|d_{k-1}\| \leq (1 + M_3)\|x_{k-1} - x^*\|, \qquad (3.37)$$
so the proof of Theorem 3 is complete. $\square$

It is interesting to compare Theorems 2 and 3 in the case when the constraints (3.2) depend on only the first m components of x. We consider the partitioned matrices
$$B_k = \left(\begin{array}{c|c} B_{11}^{(k)} & B_{12}^{(k)} \\ \hline B_{21}^{(k)} & B_{22}^{(k)} \end{array}\right) \qquad (3.38)$$
and
$$G^* = \left(\begin{array}{c|c} G_{11}^* & G_{12}^* \\ \hline G_{21}^* & G_{22}^* \end{array}\right), \qquad (3.39)$$
where the dimensions of $B_{11}^{(k)}$ and G_{11}^* are $m \times m$. The superlinear convergence result of Theorem 2 is independent of the submatrix $B_{11}^{(k)}$, while Theorem 3 is independent of the submatrices $B_{11}^{(k)}$, $B_{12}^{(k)}$, and $B_{21}^{(k)}$, provided that B_k is uniformly bounded. Thus superlinear convergence can be obtained when the rank of B_k is only $(n-m)$.

A useful technique for generating the matrices $\{B_k; k=1, 2, 3, \ldots\}$ so that two-step superlinear convergence is obtained is described in [21] and analysed in [22]. It applies the well-known BFGS formula from unconstrained optimization, but the change in gradient of the objective function $[\nabla F(x_{k+1}) - \nabla F(x_k)]$ is replaced by the change in gradient of an estimate of the Lagrangian function. Specifically the difference
$$\nabla F(x_{k+1}) - \nabla F(x_k) - \sum_{i=1}^{m} \lambda_i^{(k)} \{\nabla c_i(x_{k+1}) - \nabla c_i(x_k)\} \qquad (3.40)$$
is used, where the multipliers $\{\lambda_i^{(k)}; i=1, 2, \ldots, m\}$ are still the Lagrange parameters of the quadratic programming problem that determines d_k, expect that a further modification is sometimes made to preserve positive definiteness. Thus superlinear convergence is achieved without the explicit calculation of any second derivatives.

4. The Maratos Effect

We consider the following calculation:
$$\left.\begin{array}{ll} \text{minimize} & F(x) = -x_1 + 10(x_1^2 + x_2^2 - 1), \\ \text{subject to} & x_1^2 + x_2^2 - 1 = 0, \\ \text{starting at} & (x_1, x_2) = (0.8, 0.6). \end{array}\right\} \qquad (4.1)$$

II. Variable Metric Methods for Constrained Optimization

Powell [23] reports that, if it is solved by a variable metric method of the type that is analysed in Section 2, then after 35 iterations the estimate of the solution (1,0) is (0.9887, 0.1613) and convergence occurs at a very slow linear rate. However, if one gives up line searches and uses the formula

$$x_{k+1} = x_k + d_k \tag{4.2}$$

instead, then six decimals accuracy are obtained in only five iterations. The reason is that the iteration (4.2), which gives superlinear convergence, is not allowed by the line search condition (1.6). This unfortunate phenomenon is called the "Maratos effect", because it is observed and considered in his Ph.D. dissertation [18].

It is easy to analyse the effect in the example (4.1) when x_k is feasible and $B_k = G^* = I$. If the components of x_k are $(\cos\theta, \sin\theta)$, then the components of $x_k + d_k$ are $(\cos\theta + \sin^2\theta, \sin\theta[1 - \cos\theta])$. Thus the iteration (4.2) would converge quadratically, but the line search objective function (1.7) has the value

$$W(x_k + d_k, \mu) = -\cos\theta + (9 + \mu)\sin^2\theta, \tag{4.3}$$

which exceeds $W(x_k, \mu) = -\cos\theta$ for every $\mu \geq 0$. Therefore the condition (1.6) demands a step length that prevents a superlinear rate of convergence.

The effect would not be serious if it occurred on only a few iterations of a constrained optimization calculation, but the example (4.1) shows that it can persist. However, if the starting point of the example is moved away from the constraint, then usually the line search objective function allows step lengths of one. Because I disagree with the writers who suggest that the effect is so rare that it can be ignored, the remainder of this section mentions some useful remedies.

The remedy that is proposed by Mayne and Polak [19] can be derived from the following remarks. Suppose that x_k is feasible and is close to the Kuhn-Tucker point x^*, and that B_k gives the superlinear convergence condition

$$\|x_k + d_k - x^*\| = o(\|d_k\|), \tag{4.4}$$

where d_k is calculated by the basic method of Section 1. As before, let $\{\lambda_i^*; i = 1, 2, \ldots, m\}$ be the Lagrange parameters at x^*, and let $L(x)$ be the Lagrangian function (3.15). Then, assuming that $\{\mu_i \geq |\lambda_i^*|; i = 1, 2, \ldots, m\}$ we have $W(x_k, \mu) \geq L(x_k)$, and, assuming the second order sufficiency condition, we have the bounds

$$\left. \begin{array}{l} L(x_k) \geq L(x^*) + \eta \|d_k\|^2 \\ L(x_k + d_k) = L(x^*) + o(\|d_k\|^2) \end{array} \right\}, \tag{4.5}$$

where η is a positive constant. It follows that the inequality

$$W(x_k + d_k, \mu) \leq W(x_k, \mu) - \eta \|d_k\|^2 + o(\|d_k\|^2) + V(x_k + d_k) \tag{4.6}$$

holds, where $V(x)$ is the function

$$V(x) = \sum_{i=1}^{m} \lambda_i^* c_i(x) + \sum_{i=1}^{m'} \mu_i |c_i(x)| + \sum_{i=m'+1}^{m} \mu_i \max[0, -c_i(x)]. \tag{4.7}$$

Thus the Maratos effect does not occur if $V(x_k+d_k)$ is $o(\|d_k\|^2)$, but usually it is $O(\|d_k\|^2)$, because the linear conditions (1.4) imply that any violations of the given nonlinear constraints (1.1) at (x_k+d_k) are of this magnitude.

Therefore the technique of Mayne and Polak [19] gives x_{k+1} the form

$$x_{k+1} = x_k + d_k + \tilde{d}_k, \qquad (4.8)$$

where $\tilde{d}_k$ is a small correction to d_k such that $V(x_k+d_k+\tilde{d}_k)$ is normally $O(\|d_k\|^3)$. Specifically, they define I_k to be the set of constraint indices

$$I_k = \{1, 2, \ldots, m'\} \cup \{i : \lambda_i^{(k)} > 0\}, \qquad (4.9)$$

and they let $\tilde{d}_k$ be the vector of least Euclidean length that satisfies the equations

$$c_i(x_k+d_k) + \tilde{d}_k^T \nabla c_i(x_k) = 0, \qquad i \in I_k. \qquad (4.10)$$

Thus, in non-degenerate cases, $\|\tilde{d}_k\|$ is of magnitude $\|d_k\|^2$, and, provided that the indices of the constraints that make positive contributions to $V(x_k+d_k+\tilde{d}_k)$ are all in I_k, we have $V(x_k+d_k+\tilde{d}_k) = O(\|d_k\|^3)$ as required. However, if x_k is not sufficiently close to x^*, the line search objective function may not allow the value (4.8). In this case the step length α_k is calculated as described in Section 1, but x_{k+1} has the form

$$x_{k+1} = x_k + \alpha_k d_k + \alpha_k^2 \tilde{d}_k, \qquad (4.11)$$

in order that condition (2.15) can still be satisfied by choosing sufficiently small positive step lengths.

When there are no inactive inequality constraints in the quadratic programming calculation that determines d_k, the following modification to the method of the previous paragraph is sometimes useful. We replace the conditions (1.4) by the equations

$$d^T \nabla c_i(x_k) = 0, \qquad i = 1, 2, \ldots, m, \qquad (4.12)$$

before calculating d_k, so the function values $\{c_i(x_k); i=1, 2, \ldots, m\}$ are not required. Then $\tilde{d}_k$ is determined as before to satisfy expression (4.10). Coleman and Conn [7] study the asymptotic convergence properties of this technique when x_{k+1} is given the value (4.8) on every iteration, and they show that the Maratos effect does not occur.

A different remedy is proposed by Chamberlain, Lemaréchal, Pedersen and Powell [6]. It is based on the observation that, if x_{k+1} has the value (4.2) on every iteration, and if the superlinear convergence of Section 3 is obtained, then, for sufficiently large k, the line search objective function satisfies the inequality

$$W(x_{k+2}, \mu) < W(x_k, \mu). \qquad (4.13)$$

Thus the disadvantages of the Maratos effect can be avoided by allowing an iteration to increase the line search objective function if, at the beginning of the iteration, the value of $W(x, \mu)$ is the least that has been calculated. The "watchdog technique" of Chamberlain et al. [6] applies this idea. If a new least value of $W(x, \mu)$ does not occur during a prescribed number of iterations, then the vector of variables is reset to the value that gave the least value of $W(x,\mu)$, and then one insists on a line search that will make a further reduction in $\bar{W}(x,\mu)$. In this way one usually obtains the property that a limit point of the calculated sequence $\{x_k; k=1, 2, 3, \ldots\}$ is a Kuhn-Tucker point of the given nonlinear programming problem. If an iteration increases $W(x,\mu)$, then a reduction is required in the estimate of the Lagrangian function

$$L_k(x) = F(x) - \sum_{i=1}^{m} \lambda_i^{(k)} c_i(x), \qquad (4.14)$$

in order to try to prevent excessive changes to the variables. It is proved that, even with this refinement, the Maratos effect does not prevent a superlinear rate of convergence. Moreover, the technique can avoid some of the inefficiencies, discussed in the last paragraph of Section 2, that may be caused by a poor choice of μ.

5. Unconstrained Calculations of the Search Direction

Instead of using the basic quadratic programming method of Section 1, the search direction of an iteration of a variable metric algorithm may be defined by an unconstrained minimization calculation. Usually the unconstrained problem depends on parameters in such a way that, if the parameters are sufficiently large, and if the constraints (1.4) are consistent, then d_k is obtained, where in this section we reserve the notation d_k for the solution of the quadratic programming problem of Section 1. However, two advantages of the unconstrained approach are that it can provide useful directions when the quadratic programming problem has no solution, and that it allows the use of "trust regions". These ideas are explained well by Fletcher [12], [13] and [14], and we consider them briefly.

The usual unconstrained approach is to let the search direction be the value of d that minimizes the function

$$\bar{W}_k(x_k+d) = \bar{F}_k(x_k+d) + \sum_{i=1}^{m'} \mu_i |c_i(x_k) + d^T \nabla c_i(x_k)|$$
$$+ \sum_{i=m'+1}^{m} \mu_i \max[0, -c_i(x_k) - d^T \nabla c_i(x_k)], \qquad (5.1)$$

where we are deliberately using the same notation as equation (2.6), and where the parameters $\{\mu_i; i=1, 2, \ldots, m\}$ are non-negative. This calculation has a unique solution, $\bar{d}_k$ say, whenever the function (5.1) is strictly convex. We note that $\bar{d}_k = d_k$ if and only if the terms under the summation signs of expression (5.1) are zero when $d = \bar{d}_k$. This condition is necessary because d_k satisfies the constraints (1.4), and it is also sufficient because otherwise the replacement of $\bar{d}_k$ by d_k would reduce the function (5.1).

It is plausible that, if the constraints (1.4) are consistent, then large values of $\{\mu_i; i=1, 2, \ldots, m\}$ force the terms under the summation signs of expression

(5.1) to be zero at $\underline{d} = \bar{d}_k$. In fact $\bar{d}_k = d_k$ if the bounds

$$\mu_i \geq |\lambda_i^{(k)}|, \quad i = 1, 2, \ldots, m, \tag{5.2}$$

are obtained, where $\{\lambda_i^{(k)}; i = 1, 2, \ldots, m\}$ are still the Lagrange parameters at the solution of the quadratic programming problem that determines d_k. To prove this assertion we note that the right hand side of expression (2.8) is non-negative, and we recall that the function $\{\bar{L}_k(x_k + \underline{d}); \underline{d} \in \mathbb{R}^n\}$ is least when $\underline{d} = d_k$. Thus we deduce the condition

$$\bar{W}_k(x_k + \underline{d}) \geq \bar{L}_k(x_k + d_k), \quad \underline{d} \in \mathbb{R}^n. \tag{5.3}$$

Since $\bar{W}_k(x_k + d_k) = \bar{F}_k(x_k + d_k) = \bar{L}_k(x_k + d_k)$, it follows that $\underline{d} = d_k$ minimizes $\bar{W}_k(x_k + \underline{d})$, which completes the proof.

The search direction $\bar{d}_k$ has the advantage over d_k that, if it is non-zero, it automatically satisfies the descent condition

$$W(x_k + \alpha \bar{d}_k, \underline{\mu}) < W(x_k, \underline{\mu}) \tag{5.4}$$

when α is small and positive, which is due to the bound (2.7) and the convexity of $\bar{W}_k(x_k + \underline{d})$. Therefore, one may be able to keep the parameters $\{\mu_i; i = 1, 2, \ldots, m\}$ of the line search objective function smaller than the values that occur in the theory of Section 2, which sometimes reduces the number of iterations. However, small values of μ are not always suitable because, if x^* is a Kuhn-Tucker point at which the gradients of the active constraints are linearly independent, and if λ^* is the vector of Lagrange multipliers at x^*, then the search directions $\{\bar{d}_k; k = 1, 2, 3, \ldots\}$ provide convergence to x^* only if $\mu_i \geq |\lambda_i^*|$ for $i = 1, 2, \ldots, m$. Clearly μ should be increased if $d_k = 0$ and x_k is not feasible, unless it is believed that the given constraints (1.1) are inconsistent. Also it is usually worthwhile to increase μ if the last two terms of expression (5.1) make a non-zero contribution to $\bar{W}_k(x_k + \bar{d}_k)$ on several consecutive iterations.

Another technique for generating search directions when the constraints (1.4) cannot be satisfied is suggested by Powell [21]. In order to describe it we let V_k be the set of indices of the inequality constraints that are violated at x_k, and we let the set S_k contain the indices of the remaining inequality constraints. The largest value of ξ in [0, 1], $\tilde{\xi}$ say, is found such that the conditions

$$\left. \begin{array}{ll} \xi c_i(x_k) + \underline{d}^T \nabla c_i(x_k) = 0, & i = 1, 2, \ldots, m' \\ \xi c_i(x_k) + \underline{d}^T \nabla c_i(x_k) \geq 0, & i \in V_k \\ c_i(x_k) + \underline{d}^T \nabla c_i(x_k) \geq 0, & i \in S_k \end{array} \right\} \tag{5.5}$$

hold for some $\underline{d} \in \mathbb{R}^n$. If $\tilde{\xi} = 0$, which is always allowed by $\underline{d} = 0$, or if $\tilde{\xi}$ is less than a small prescribed tolerance, then the calculation finishes because it is assumed that the constraints (1.1) have no feasible point. Otherwise the search direction is defined to be the value of $\underline{d}$ that minimizes the quadratic function (1.3) subject to the conditions (5.5), where ξ is a constant that is chosen from the interval $(0, \tilde{\xi}]$. I used to set $\xi = \tilde{\xi}$ on all iterations, but now I prefer the value $\xi = 0.9 \tilde{\xi}$ when $\tilde{\xi} < 1$ [24], in order that the conditions (5.5) allow some freedom in $\underline{d}$ to reduce the objective function (1.3).

II. Variable Metric Methods for Constrained Optimization

If the first line of expression (5.5) is replaced by the inequalities

$$|c_i(x_k) + d^T \nabla c_i(x_k)| \leq (1-\xi)|c_i(x_k)|, \quad i = 1, 2, \ldots, m', \tag{5.6}$$

which is a suitable change because it allows more freedom in d without increasing the predicted constraint violations $\{|c_i(x_k) + d^T \nabla c_i(x_k)|; i = 1, 2, \ldots, m'\}$, then the method that has just been described can be expressed as an unconstrained minimization calculation. The objective function of this calculation depends on the ratios

$$\left.\begin{array}{l} r_i^{(k)}(d) = |c_i(x_k) + d^T \nabla c_i(x_k)|/|c_i(x_k)|, \quad i = 1, 2, \ldots, m' \\ r_i^{(k)}(d) = \dfrac{\max[0, -c_i(x_k) - d^T \nabla c_i(x_k)]}{\max[0, -c_i(x_k)]}, \quad i = m'+1, \ldots, m \end{array}\right\}, \tag{5.7}$$

where $r_i^{(k)}(d)$ is zero if its numerator is zero, but in all other cases a zero denominator makes $r_i^{(k)}(d)$ unbounded. We note that, if $\xi \in [0, 1]$, then d satisfies condition (5.6) and the last two lines of expression (5.5) if the function

$$R_k(d) = \max_{1 \leq i \leq m} r_i^{(k)}(d) \tag{5.8}$$

is at most $(1-\xi)$. Conversely, if $R_k(d) \leq 1$, then the constraints on d allow $\xi = 1 - R_k(d)$. Hence $(1-\tilde{\xi})$ is the least possible value of $R_k(d)$, where $\tilde{\xi}$ is still the greatest ξ in $[0, 1]$ such that the conditions on d hold for some $d \in \mathbb{R}^n$. Therefore, if $\tilde{d}_k$ is the vector d that minimizes the convex objective function

$$\tilde{W}_k(x_k + d) = \bar{F}_k(x_k + d) + \tilde{\zeta} R_k(d), \tag{5.9}$$

where $\tilde{\zeta}$ is a large positive constant, then $R_k(\tilde{d}_k)$ is close to or equal to $(1-\tilde{\xi})$, and $\tilde{d}_k$ minimizes $\bar{F}_k(x_k + d)$ subject to the constraint $R_k(d) \leq R_k(\tilde{d}_k)$. Thus, except for the change to the first line of expression (5.5), $\tilde{d}_k$ is the search direction that would be given by the method of the previous paragraph if ξ were equal to $1 - R_k(\tilde{d}_k)$.

However, there are advantages in preferring $\bar{W}(x_k + d)$ to $\tilde{W}(x_k + d)$ for calculating the search directions of a variable metric algorithm. One is that, unless x_k is a stationary point of the line search objective function, then it is possible to reduce $W(x, \mu)$ by moving from x_k along d_k. Further, non-zero terms in the sums of expression (5.1) indicate when consideration should be given to increasing the components of μ. On the other hand, if one takes the point of view that a search direction is not acceptable if it makes a constraint violation larger when the step length is close to zero, then, for large ζ, expression (5.9) provides an acceptable direction if any exist, but expression (5.1) may fail to do so.

Because the line search objective function $W(x, \mu)$ is an L_1 penalty function for the given nonlinear programming problem, and because expression (5.9) is closely related to L_∞ penalty functions, we ask whether least squares penalty functions make a useful contribution to the calculation of search directions for variable metric algorithms. We consider this question when all the constraints are equations ($m' = m$), so we let $\hat{d}_k(\hat{\zeta})$ be the value of d that minimizes the convex quadratic function

$$\hat{W}_k(x_k + d) = \bar{F}_k(x_k + d) + \hat{\zeta} \sum_{i=1}^{m} \{c_i(x_k) + d^T \nabla c_i(x_k)\}^2, \tag{5.10}$$

where $\hat{\zeta}$ is a positive parameter. If the basic method of Section 1 has a solution d_k, then $\hat{d}_k(\hat{\zeta})$ tends to d_k if $\hat{\zeta}$ is made very large, but in general, unlike the other two unconstrained minimization procedures that have been mentioned for calculating search directions, $\hat{d}_k(\hat{\zeta})$ is not equal to d_k for finite parameter values. However, $\hat{d}_k(\hat{\zeta})$ is not a new choice of search direction, because it occurs in the very successful REQP algorithm of Bartholomew-Biggs [1], [2].

In order to prove this statement we recall that equations (11) and (12) of [2] and equations (6) and (7) of [1] state that the REQP search direction is the vector d that minimizes the function

$$\tilde{F}_k(x_k+d) = F(x_k) + d^T \nabla F(x_k) + \tfrac{1}{2} d^T B_k d, \tag{5.11}$$

subject to the constraints

$$c_i(x_k) + d^T \nabla c_i(x_k) = -\tfrac{1}{2} r u_i, \quad i=1, 2, \ldots, m, \tag{5.12}$$

where $r > 0$, and where $u \in \mathbb{R}^m$ is defined by the nonsingular system of equations

$$[\tfrac{1}{2} r I + N_k^T B_k^{-1} N_k] u = N_k^T B_k^{-1} \nabla F(x_k) - c(x_k). \tag{5.13}$$

Here N_k is the matrix $N(x_k)$ of Section 3, whose columns are the gradients $\{\nabla c_i(x_k); i=1, 2, \ldots, m\}$. We show that the REQP search direction is the value of d that gives the least value of the function (5.10) when $\hat{\zeta} = 1/r$.

It follows from expressions (5.10), (5.11) and (5.12) that $\hat{d}_k(\hat{\zeta}) = \hat{d}_k$, say, is the REQP search direction if and only if it satisfies the conditions

$$c_i(x_k) + \hat{d}_k^T \nabla c_i(x_k) = -(1/2\hat{\zeta}) u_i, \quad i=1, 2, \ldots, m, \tag{5.14}$$

which in vector form are the equation

$$c(x_k) + N_k^T \hat{d}_k = -(1/2\hat{\zeta}) u. \tag{5.15}$$

Therefore, since u is defined by the system (5.13), we have only to verify that, if the left hand side of expression (5.15) is multiplied by the matrix $[-I - 2\hat{\zeta} N_k^T B_k^{-1} N_k]$, we obtain the vector $[N_k^T B_k^{-1} \nabla F(x_k) - c(x_k)]$. Because the gradient of the function (5.10) is zero at $d = \hat{d}_k$, we have the identity

$$\nabla F(x_k) + B_k \hat{d}_k + 2\hat{\zeta} N_k [c(x_k) + N_k^T \hat{d}_k] = 0, \tag{5.16}$$

which implies the relation

$$-2\hat{\zeta} N_k^T B_k^{-1} N_k [c(x_k) + N_k^T \hat{d}_k] = N_k^T B_k^{-1} \nabla F(x_k) + N_k^T \hat{d}_k. \tag{5.17}$$

Thus we deduce the equation

$$-[I + 2\hat{\zeta} N_k^T B_k^{-1} N_k][c(x_k) + N_k^T \hat{d}_k] = -c(x_k) + N_k^T B_k^{-1} \nabla F(x_k), \tag{5.18}$$

which is the required result.

The minimization of the function (5.10) gives a definition of the REQP search direction that is easier to understand than the usual definition that depends on expressions (5.11), (5.12) and (5.13). In particular, it is clear that the search direction is well defined if the constraint gradients $\{\nabla c_i(x_k); i=1, 2, \ldots, m\}$ are linearly dependent, which, as Bartholomew-Biggs [1] points out, is "not so obvious" when his definition is employed. The main advantage of calculating the search direction from the function (5.10), instead of from expression (5.1) or (5.9), is that one only has to solve a single system of linear equations. There is the disadvantage, however, that the REQP search direction is not zero when x_k is a Kuhn-Tucker point and $\hat{\zeta}$ is finite, unless $\nabla F(x_k)$ happens to be zero.

As well as providing search directions when linear approximations to constraints are inconsistent, the unconstrained minimization techniques of this section allow bounds on search directions to be imposed. For example, it is straightforward to minimize any of the functions (5.1), (5.9) and (5.10) subject to the conditions

$$|d_i| \leq h_k, \quad i=1, 2, \ldots, n, \tag{5.19}$$

on the components of d, where h_k is a positive parameter. Algorithms that include such bounds are called "trust region methods", and they have several advantages. For example, the use of trust regions can improve greatly the global convergence properties of a variable metric method for constrained optimization. For detailed information, including the description of an algorithm that chooses the parameters $\{h_k; k=1, 2, 3, \ldots\}$ automatically, the paper [13] by Fletcher is recommended.

6. Further Considerations

There are many important questions on variable metric algorithms that have not been considered so far. Some of them are mentioned briefly in this section, but the lack of attention that they receive here does not imply that they are less important than the subjects of the earlier sections.

At present the number of variables that can occur in practice is restricted mainly by the work of calculating each search direction d_k. The usual defence to this disadvantage is to point out that in many important applications the calculation of functions and gradients is so laborious that the time to solve quadratic programming subproblems is insignificant, but faster ways of choosing search directions should be investigated. For small values of n, Schittkowski's [25] extensive comparison of computer programs for constrained optimization shows that, of the methods considered, the REQP method [2] of Bartholomew-Biggs requires least computing time, while Powell's algorithm [21] uses the smallest number of function and gradient evaluations. Thus variable metric algorithms perform very well, but the present state of development is such that the gains in computer time over other methods for constrained optimization are usual only when the number of variables is small.

The reported [25] speed of the REQP algorithm is due to the use of an "active set" method for inequality constraints. An active set method provides and revises automatically a list of inequality constraints that are to be treated as equations in the calculation of each search direction, the remaining inequalities being ignored. Thus the quadratic programming subproblem of Section 1 to determine d_k is reduced to the solution of a system of linear equations, which usually saves much computer time on each iteration. However, as indicated in the previous paragraph, the use of an active set method may increase the total number of iterations. An interesting discussion of this subject is given by Murray and Wright [20].

Active set methods have several other uses, because they are needed whenever one wishes to apply to inequality constraints a technique that is intended for the case when all constraints are equations. In particular they occur in reduced gradient algorithms for constrained optimization. If there are m independent active constraints, these algorithms allow $(n-m)$ of the components of $\underline{x}$ to be unconstrained, and the remaining components are calculated to satisfy the active constraints.

Theorem 3 is closely related to reduced gradient algorithms when its conditions hold, which include the assumption that all constraints are equations $(m'=m)$. In order to explain this remark, we compare the "reduced second derivative matrix" of the reduced gradient algorithm at $\underline{x}=\underline{x}^*$ with the matrix $P_k G^* P_k$ of Theorem 3 in the limit as $k\to\infty$. The reduced second derivative matrix, $\hat{G}^*$ say, has dimensions $(n-m)\times(n-m)$, and it depends on the choice of the $(n-m)$ unconstrained variables, which we suppose are the first $(n-m)$ components of $\underline{x}$. For any sufficiently small $\hat{\underline{h}}\in\mathbb{R}^{n-m}$, we let $\underline{h}\in\mathbb{R}^n$ be the vector whose first $(n-m)$ components are $\{\hat{h}_i; i=1, 2, \ldots, n-m\}$, and whose last m components are such that the constraints are satisfied at $(\underline{x}^*+\underline{h})$. Then $\hat{G}^*$ is defined by the relation

$$F(\underline{x}^*+\underline{h})=F(\underline{x}^*)+\tfrac{1}{2}\hat{\underline{h}}^T\hat{G}^*\hat{\underline{h}}+o(\|\hat{\underline{h}}\|^2). \tag{6.1}$$

Further, because $F(\underline{x})$ is equal to the Lagrangian function (3.15) when $\underline{x}$ is feasible, we may also define $\hat{G}^*$ by the equation

$$L(\underline{x}^*+\underline{h})=L(\underline{x}^*)+\tfrac{1}{2}\hat{\underline{h}}^T\hat{G}^*\hat{\underline{h}}+o(\|\hat{\underline{h}}\|^2). \tag{6.2}$$

Since the reduced gradient algorithm seeks the unconstrained minimum of the function $\Phi(\hat{\underline{h}})\equiv F(\underline{x}^*+\underline{h})$, it can achieve superlinear convergence from calculated first derivatives when a suitable approximation to $\hat{G}^*$ is available.

In order to relate $\hat{G}^*$ to G^*, we compare expression (6.2) with the equation

$$L(\underline{x}^*+\underline{h})=L(\underline{x}^*)+\tfrac{1}{2}\underline{h}^T G^*\underline{h}+o(\|\underline{h}\|^2), \tag{6.3}$$

and, for $i=1, 2, \ldots, n-m$, we let $\underline{h}_i$ be the vector

$$\underline{h}_i = \lim_{\varepsilon\to 0}\underline{h}_i(\varepsilon)/\varepsilon, \tag{6.4}$$

where $\underline{h}_i(\varepsilon)$ is the value of $\underline{h}$ that is defined by the method of the previous paragraph when $\hat{\underline{h}}$ is ε times the i-th coordinate vector in $\mathbb{R}^{n-m}$. It follows that the elements of $\hat{G}^*$ are given by the equation

$$\hat{G}^*_{ij}=\underline{h}_i^T G^*\underline{h}_j. \tag{6.5}$$

Now, since $\underline{x}^*$ and $[\underline{x}^*+\underline{h}_i(\varepsilon)]$ are both feasible, the limit (6.4) implies that $\underline{h}_i$ is orthogonal to the constraint gradients $\{\nabla c_i(\underline{x}^*); i=1, 2, \ldots, m\}$. In other words each $\underline{h}_i$ is in the column space of P^*, where P^* is the limit of the symmetric projection matrices $\{P_k; k=1, 2, 3, \ldots\}$ that occur in Theorem 3. Therefore equation (6.5) shows that $\hat{G}^*$ can be derived from the matrix $P^* G^* P^*$. Conversely, because the vectors $\{\underline{h}_i; i=1, 2, \ldots, n-m\}$ span the column space of P^*, $P^* G^* P^*$ can be obtained from $\hat{G}^*$.

These remarks suggest that reduced gradient and variable metric algorithms require equivalent second derivative information to achieve superlinear convergence, but the main difference between these classes of methods is that reduced gradient procedures include an inner iteration in order to satisfy nonlinear constraints sufficiently accurately. Because of the usefulness of reduced gradient procedures for solving large structured problems, it might be very valuable to merge these two approaches to constrained optimization calculations. The papers of Coleman and Conn [7], [8] give further attention to this subject.

Another question that deserves more attention is the choice of the multipliers $\{\lambda_i^{(k)}; i=1, 2, \ldots, m\}$ in the expression (3.40) that is used to calculate B_{k+1} from B_k. As Chamberlain [4] shows, the method that is suggested in Section 3 may make the sequence of matrices $\{B_k; k=1, 2, 3, \ldots\}$ unbounded. Several estimates of Lagrange parameters are dicussed by Gill and Murray [15]. Moreover, Dixon [10] and Schittkowski [26] have investigated recently the possibility of replacing the line search objective function (1.7) by a function that has continuous first derivatives.

For information on programming considerations, when implementing a variable metric method for constrained optimization, the book by Gill, Murray and Wright [16] is recommended. It gives careful attention to suitable procedures for the matrix calculations that occur. A Fortran listing of a variable metric algorithm is available from the author [24].

It has been shown in this paper that the ideas and theory of variable metric methods are an important part of the subject of mathematical programming, and that there is no clear dividing line between these methods and several other classes of algorithms for the solution of nonlinear optimization problems.

References

[1] M. C. Bartholomew-Biggs, "Recursive quadratic programming methods for nonlinear constraints", in *Nonlinear Optimization 1981*, ed. M. J. D. Powell (Academic Press, London, 1982) pp. 213-221.

[2] M. C. Biggs, "Constrained minimization using recursive quadratic programming: some alternative subproblem formulations", in *Towards Global Optimization*, eds. L. C. W. Dixon and G. P. Szegö (North Holland Publishing Co., Amsterdam, 1975) pp. 341-349.

[3] P. T. Boggs, J. W. Tolle and P. Wang, "On the local convergence of quasi-Newton methods for constrained optimization", *SIAM Journal on Control and Optimization*, Vol. 20 (1982) pp. 161-171.

[4] R. M. Chamberlain, "Some examples of cycling in variable metric algorithms for

constrained optimization", *Mathematical Programming*, Vol. 16 (1979) pp. 378–383.

[5] R. M. Chamberlain, *The Theory and Application of Variable Metric Methods to Constrained Optimization Problems*, Ph. D. Dissertation (University of Cambridge, 1980).

[6] R. M. Chamberlain, C. Lemaréchal, H. C. Pedersen and M. J. D. Powell, "The watchdog technique for forcing convergence in algorithms for constrained optimization", *Mathematical Programming Study*, Vol. 16 (1982) pp. 1–17.

[7] T. F. Coleman and A. R. Conn, "Nonlinear programming via an exact penalty function: asymptotic analysis", Report CS-80-30 (University of Waterloo, 1980).

[8] T. F. Coleman and A. R. Conn, "Nonlinear programming via an exact penalty function: global analysis", Report CS-80-31 (University of Waterloo, 1980).

[9] J. E. Dennis and J. J. Moré, "A characterization of superlinear convergence and its application to quasi-Newton methods", *Mathematics of Computation*, Vol. 28 (1974) pp. 549–560.

[10] L. C. W. Dixon, "On the convergence properties of variable metric recursive quadratic programming methods", Report NOC-110 (The Hatfield Polytechnic, 1980).

[11] R. Fletcher, *Practical Methods of Optimization, Vol. 1, Unconstrained Optimization* (John Wiley & Sons, Chichester, 1980).

[12] R. Fletcher, *Practical Methods of Optimization, Vol. 2, Constrained Optimization* (John Wiley & Sons, Chichester, 1981).

[13] R. Fletcher, "Numerical experiments with an exact L_1 penalty function method", in *Nonlinear Programming 4*, eds. O. L. Mangasarian, R. R. Meyer and S. M. Robinson (Academic Press, New York, 1981) pp. 99–129.

[14] R. Fletcher, "Methods for nonlinear constraints", in *Nonlinear Optimization 1981*, ed. M. J. D. Powell (Academic Press, London, 1982) pp. 185–211.

[15] P. E. Gill and W. Murray, "The computation of Lagrange multiplier estimates for constrained minimization", *Mathematical Programming*, Vol. 17 (1979) pp. 32–60.

[16] P. E. Gill, W. Murray and M. H. Wright, *Practical Optimization* (Academic Press, London, 1981).

[17] S. P. Han, "A globally convergent method for nonlinear programming", *Journal of Optimization Theory and Applications*, Vol. 22 (1977) pp. 297–309.

[18] N. Maratos, *Exact Penalty Function Algorithms for Finite Dimensional and Control Optimization Problems*, Ph. D. Dissertation (Imperial College, University of London, 1978).

[19] D. Q. Mayne and E. Polak, "A superlinearly convergent algorithm for constrained optimization problems", *Mathematical Programming Study*, Vol. 16 (1982) pp. 45–61.

[20] W. Murray and M. H. Wright, "Computation of the search direction in constrained optimization algorithms", *Mathematical Programming Study*, Vol. 16 (1982) pp. 62–83.

[21] M. J. D. Powell, "A fast algorithm for nonlinearly constrained optimization algorithms", in *Numerical Analysis, Dundee, 1977, Lecture Notes in Mathematics 630*, ed. G. A. Watson (Springer-Verlag, Berlin, 1978) pp. 144–157.

[22] M. J. D. Powell, "The convergence of variable metric methods for nonlinearly constrained optimization calculations", in *Nonlinear Programming 3*, eds. O. L. Mangasarian, R. R. Meyer and S. M. Robinson (Academic Press, New York, 1978) pp. 27–63.

[23] M. J. D. Powell, "Extensions to subroutine VFO2AD", in *System Modeling and Optimization, Lecture Notes in Control and Information Sciences 38*, eds. R. F. Drenick and F. Kozin (Springer-Verlag, New York, 1982) pp. 529–538.

[24] M. J. D. Powell, "VMCWD: a Fortran subroutine for constrained optimization", Report DAMTP-1982/NA4 (University of Cambridge, 1982).
[25] K. Schittkowski, *Nonlinear Programming Codes, Lecture Notes in Economics and Mathematical Systems 183* (Springer-Verlag, Berlin, 1980).
[26] K. Schittkowski, "On the convergence of a sequential quadratic programming method with an augmented Lagrangian line search function", Report SOL-82-4 (Stanford University, 1982).

Polyhedral Combinatorics

W. R. Pulleyblank
Department of Combinatorics and Optimization, University of Waterloo, Waterloo, Ontario N2L 3G1, Canada

Abstract. Polyhedral combinatorics deals with the application of various aspects of the theory of polyhedra and linear systems to combinatorics. Over the past thirty years a great many researchers have shown how a large number of polyhedral concepts and results have elegant combinatorial consequences. Here, however, we concentrate our attention on developments over the last ten years.

We survey the relationship between linear systems which define a polyhedron and combinatorial min-max theorems obtained via linear programming duality. We discuss two methods of improving these theorems, first by reducing the number of dual variables (facet characterizations) and second, by requiring that dual variables take on integer values (total unimodularity and total dual integrality).

One consequence of the ellipsoid algorithm for linear programming has been to show the equivalence of optimization and separation (determining whether or not a given point belongs to a polyhedron, and if not, finding a separating hyperplane). We discuss the theoretical importance of this as well as several instances where in this approach was used to solve successfully "real world" optimization problems (before the development of the ellipsoid method).

Finally, we briefly discuss the concepts of adjacency and dimension, both of which recently have been shown to have interesting combinatorial consequences.

1. Introduction

Polyhedral combinatorics is the application of the theory of linear systems and linear algebra to combinatorial problems. Even though many of its results are of a pure "combinatorial" flavour, its roots are in the development of algorithms for combinatorial problems. In fact, it is possible to divide the development of polyhedral combinatorics into three periods, demarcated by major developments in the theory of algorithms.

The earliest explicit use of the methods of polyhedral combinatorics appears to be in a paper of Richard Rado [Rado, 1943]. Here he made use of a "theorem which follows most naturally from the theory of convex sets of points" which was essentially Farkas' lemma. However the first period of development of polyhedral combinatorics really began in the 1950's, following the discovery of the simplex algorithm for linear programming. It was observed that many combinatorial problems, in particular various forms of network flow problems, could be formulated as integer linear programs and because of the total unimodularity of the constraint matrix (see Section 4) these "integer" programs were really linear programs. Consequently the simplex algorithm and, more importantly, linear programming duality could be applied.

The second period of development began in the middle 1960's when Jack Edmonds showed that it was possible to handle real integer programs, i.e. problems for which the "obvious" set of necessary constraints was not sufficient. By adding possibly exponentially many constraints to the system, he was able to obtain a complete linear description to which linear programming duality could be applied, even if the simplex algorithm could not. Then polynomially bounded algorithms were designed for these problems, making use of, and generally proving, the duality relationship. These successes prompted a number of (usually unsuccessful) efforts to obtain complete linear descriptions for many outstanding "hard" problems, which were later shown to be NP-complete. However, even without a complete linear description it was possible to show that certain classes of inequalities would have to be included in any complete system. Moreover, it was demonstrated that these incomplete linear systems were still sufficient to prove optimality in some "real world" problems.

The third period began in 1979 with the development of the ellipsoid method for linear programming. What was significant for polyhedral combinatorics was not just that linear programs could be solved in polynomial time, but more importantly, the form of the algorithm. Its success depended on being able to efficiently verify whether or not a given point belonged to the solution set of a linear system, and if not, being able to produce a violated inequality. Thus optimization problems could be solved efficiently if this separation problem could be solved efficiently. Moreover, the converse implication was also true: if we could optimize efficiently over a linear system, so too could we solve the separation problem. This prompted a great interest in so-called separation algorithms, especially because they seem in general to look very different from the associated optimization algorithms.

In the next section we provide a brief summary of polyhedral definitions and basic properties which will be used throughout the remainder of the paper. In Section 3 we show the relationship between algorithms, min-max theorems and polyhedral characterizations. We also introduce a main technique used for improving these theorems, that of facet characterizations of polyhedra. In Section 4 we study a second technique for improving min-max theorems: dual integrality results. In particular we discuss total unimodularity and total dual integrality and their importance in combinatorial optimization.

In Section 5 we discuss several examples of successful application of the theory presented in Sections 3 and 4. In particular, we discuss matroid intersections and matchings, two of the outstanding successes of polyhedral combinatorics. In Section 6 we discuss separation, both from a theoretical and practical viewpoint. We show how knowledge of several classes of facets for the travelling salesman polytope has enabled solving several large (>100 city) problems.

Finally, in Section 7 we briefly discuss adjacency and dimension, two topics that have recently attracted attention.

There currently exist several excellent surveys of various areas and applications of polyhedral combinatorics. Lovász (1979) is a good introduction to the applicability of linear programming to combinatorial problems. Schrijver (1983) (this volume) provides a comprehensive survey of many of the areas successfully treated by polyhedral combinatorics. Hoffman (1982) presents an excellent overview of the applicability of many of the methods discussed here to the theory of ordered sets. Grötschel and Padberg (1982, 1982a) give a detailed discussion of the various aspects of polyhedral combinatorics, both theoretical and applied, with respect to the travelling salesman problem. Finally, Edmonds and Giles (1982) present a thorough introduction to "total integrality" of linear systems, which is what much of polyhedral combinatorics is about.

2. Basic Polyhedral Theory

In this section we discuss the basic terminology and results of polyhedral theory which will be used in later sections. Good fundamental references are the books of Stoer and Witzgall [1970] and Rockafellar [1970]. A very up to date reference on polyhedra theory is Bachem and Grötschel [1982] which, in particular, gives a good overview of the significance of the ellipsoid algorithm for linear programming.

For any set S, we let $\mathbb{R}^S$ denote the set of all real vectors indexed by S. Thus if I and J are finite sets, then $A \in \mathbb{R}^{I \times J}$ is a real matrix having $|I|$ rows and $|J|$ columns and $b \in \mathbb{R}^I$ is a real vector having $|I|$ components. We let A_{ij} denote the element of $A \in \mathbb{R}^{I \times J}$ indexed by (i,j) and for any $H \subseteq I$ and $K \subseteq J$ we let $A[H; K]$ denote the submatrix $(A_{ij}; i \in H, j \in K)$ of A. We let $A[H;]$ and $A[; K]$ denote $A[H; J]$ and $A[I; K]$ respectively. Similarly, b_i is the element of $b \in \mathbb{R}^I$ indexed by i and $b[H] = (b_i : i \in H)$ for $H \subseteq I$. For $i \in I, j \in J$ we abbreviate $A[\{i\};]$ by $A[i;]$ and $A[; \{j\}]$ by $A[;j]$.

A *polyhedron* is the solution set of a finite system of linear inequalities and equations. Equivalently, it is the intersection of a finite number of halfspaces. (The set of points on the hyperplane $\{x \in \mathbb{R}^J : ax = \alpha\}$ is the intersection of the halfspaces $\{x \in \mathbb{R}^J : ax \leq \alpha\}$ and $\{x \in \mathbb{R}^J : ax \geq \alpha\}$). A *polytope* is a bounded polyhedron. That is, a polyhedron $P \subseteq \mathbb{R}^J$ is a polytope if and only if there exist $l, u \in \mathbb{R}^J$ such that $l \leq x \leq u$ for all $x \in P$.

A *face* of a polyhedron P is either the empty set, or else the polyhedron F obtained by replacing some of the inequalities that define P with equations. Thus, in particular, P is a face of itself; all other faces are called *proper faces*. It is easy, using linear programming duality, to prove the following:

Theorem (2.1). *Let $P \subseteq \mathbb{R}^J$ be a polyhedron and let F be a nonempty subset of P. Then F is a face of P if and only if there exist $c \in \mathbb{R}^J$ and $\alpha \in \mathbb{R}$ such that $c^T x = \alpha$ for $x \in F$ and $c^T x < \alpha$ for $x \in P \backslash F$.*

Thus the nonempty faces of P are those subsets for which some linear objective function attains its maximum value over P. This also shows that the faces of P are independent of the set of linear inequalities chosen to represent P. A maximal nonempty proper face of P is called a *facet* of P; if $\{v\}$ is a face of

II. Polyhedral Combinatorics   315

P for some $v \in P$ then v is called a *vertex* of P. Thus v is the unique member of a polyhedron P which maximizes some cx over P, if and only if v is a vertex of P.

Let X be a finite subset of $\mathbb{R}^J$. Then X is said to be *linearly independent* if whenever we have $\Sigma(\lambda_x x : x \in X) = 0$ for some $\lambda \in \mathbb{R}^X$ we have $\lambda = 0$. Similarly X is said to be *affinely independent* if whenever $\Sigma(\lambda_x x : x \in X) = 0$ and $\Sigma(\lambda_x : x \in X) = 0$ for $\lambda \in \mathbb{R}^X$ we have $\lambda = 0$. Thus a set of vectors is affinely independent if and only if the set of vectors obtained by adding a new component having value one to each is linearly independent. For polyhedral combinatorics, affine independence is more useful than linear independence for it is invariant under translations of the origin. Thus we have:

Proposition 2.2. *A finite subset X of $\mathbb{R}^J$ is affinely independent if and only if*
i) *for any $w \in \mathbb{R}^J$, $\{x - w : x \in X\}$ is affinely independent;*
ii) *for any $\hat{x} \in X$, $\{x - \hat{x} : x \in X \setminus \{\hat{x}\}\}$ is linearly independent.*

The *affine hull* of a set $X \subseteq \mathbb{R}^J$ is the set of all $\tilde{x} \in \mathbb{R}^J$ which can be expressed as $\Sigma(\lambda_x x; x \in \overline{X})$ for $\lambda \in \mathbb{R}^{\overline{X}}$ satisfying $\Sigma(\lambda_x : x \in \overline{X}) = 1$ for some finite $\overline{X} \subseteq X$. Such an $\tilde{x}$ is said to be an *affine combination* of X. For any $S \subseteq \mathbb{R}^J$ there exists a finite $X \subseteq S$ such that every $s \in S$ is an affine combination of X. The cardinality of a smallest such X is called the *affine rank* of S and denoted by $r_a(S)$. It is related to the standard linear rank of S, denoted by $r_l(S)$, by the following, where 0 denotes the zero vector of $\mathbb{R}^J$:

Proposition 2.3. *For any $S \subseteq \mathbb{R}^J$,*
i) *if 0 is in the affine hull of S, then*

$$r_a(S) = r_l(S) + 1;$$

ii) *if 0 is not in the affine hull of S then*

$$r_a(S) = r_l(S).$$

The *dimension* of $S \subseteq \mathbb{R}^J$, denoted by $\dim(S)$, is defined to be $r_a(S) - 1$. This corresponds to our intuitive understanding of dimension: in $\mathbb{R}^3$ a tetrahedron has dimension 3, a triangle has dimension 2, a line segment has dimension 1 and a point has dimension 0.

Let $A \in \mathbb{R}^{I \times J}$, $b \in \mathbb{R}^I$ and suppose P is the polyhedron $\{x \in \mathbb{R}^J : A[I^=;]x = b[I^=]; A[I^\leqslant;]x \leqslant b[I^\leqslant]\}$ for a partition $I^= \cup I^\leqslant$ of I. The *equality set* of the linear system that defines P is the set of all those $i \in I$ for which $A[i;]x = b_i$, for all $x \in P$. If I^* is this equality set then certainly $I^= \subseteq I^*$, but in general we may have some $i \in I^\leqslant$ belonging to I^*. (In particular, any polyhedron is the solution of a linear system having $I^= = \emptyset$.) For any $i \in I \setminus I^*$ there is some $x^i \in P$ such that $A[i;]x^i < b^i$. The following relationship between the dimension of a polyhedron and the equality set of a defining linear system is frequently used in polyhedral combinatorics.

Theorem 2.4. *Let $P = \{x \in \mathbb{R}^J : A[I^=;]x = b[I^=]; A[I^\leqslant;]x \leqslant b[I^\leqslant]\}$ and let I^* be the equality set. Then, where ρ is the linear rank of the rows of $A[I^*;]$, we have $\dim(P) = |J| - \rho$.*

Note that a consequence of Theorem 2.4 is that even though there are many different linear systems that define a given polyhedron, for all such systems the linear rank of the submatrix of the coefficients in the rows indexed by the equality set is determined only by the polyhedron.

A polyhedron $P \subseteq \mathbb{R}^J$ is said to be of *full dimension* if $\dim(P) = |J|$ or equivalently, in view of Theorem 2.4, if there is no linear equation $ax = \alpha$ satisfied by all $x \in P$. If $P = \emptyset$ then we have $\dim(P) = -1$ and if $\{v\}$ is a vertex of a nonempty polyhedron P then $\dim\{v\} = 0$. An *edge* of P is a proper face of dimension 1. From the point of view of linear systems, however, the most interesting faces are the facets, i.e., the maximal nonempty proper faces.

Theorem 2.5. *Let F be a nonempty proper face of $P = \{x \in \mathbb{R}^J : A[I^=;]x = b[I^=]; A[I^\leq;]x \leq b[I^\leq]\}$ and let I^* be the equality set of P. Then the following statements are equivalent:*

i) *F is a facet of P.*
ii) *$\dim(F) = \dim(P) - 1$.*
iii) *If $a, \bar{a} \in \mathbb{R}^J$ and $\alpha, \bar{\alpha} \in \mathbb{R}$ satisfy $F = \{x \in P : a^T x = \alpha\} = \{x \in P : \bar{a}^T x = \bar{\alpha}\}$ and if $ax \leq \alpha$ and $\bar{a}x \leq \bar{\alpha}$ are valid for P then there exist $\lambda \in \mathbb{R}^{I^*}$ and positive $\gamma \in \mathbb{R}$ such that $\bar{a}^T = \gamma \cdot a^T + \lambda^T A[I^*;]$ and $\bar{\alpha} = \gamma \cdot \alpha + \lambda^T b[I^*]$.*

Note that what iii) asserts is that if two hyperplanes intersect P in the same facet, then the equation of one can be obtained from that of the other by first multiplying by a positive constant and then adding a linear combination of the rows indexed by the equality set.

An equality $ax \leq \alpha$ is *valid* for a polyhedron $P \subseteq \mathbb{R}^J$ if every $x \in P$ satisfies $ax \leq \alpha$. The face *induced* by the valid inequality $ax \leq \alpha$ is $\{x \in P : ax = \alpha\}$. The following gives the relationship between the equality set, facets and a minimal defining linear system for a polyhedron P.

Theorem 2.6. *Let P be a polyhedron and suppose $P = \{x \in \mathbb{R}^J : A[I^=;]x = b[I^=]; A[I^\leq;]x \leq b[I^\leq]\}$. Then this is a minimal linear system sufficient to define P if and only if*

i) *$I^=$ is the equality set and the rows of $A[I^=;]$ are linearly independent;*
and
ii) *for each $i \in I^\leq$, the inequality $A[i;]x \leq b[i]$ induces a distinct facet of P.*

Combining Theorems 2.5 and 2.6 we see that if P is of full dimension, then a defining linear system contains no equations and it is minimal if and only if there is a bijection between the inequalities in the system and the facets of P. Moreover, these facet inducing inequalities are unique up to positive multiples. Thus sometimes this determining of a minimal defining linear system is referred to as "determining the facets" of P. However when P is not of full dimension, although the bijection between the inequalities in a minimal system and the facets still exists, the uniqueness is lost.

A polyhedron C is called a *cone* if it is the solution set of a homogeneous linear system. In this case, for every $x \in C$ and for every positive $\gamma \in \mathbb{R}$ we have $\gamma x \in C$. For any finite $X \subseteq \mathbb{R}^N$ we let $\text{cone}(X) = \{w \in \mathbb{R}^J : \text{there exists } \lambda \in \mathbb{R}^X \text{ such that } \lambda \geq 0 \text{ and } w = \Sigma(\lambda_x : x \in X)\}$. Thus $\text{cone}(X)$ is the set of all nonnegative linear combinations of members of X. The following fundamental result is due to Weyl (1935).

II. Polyhedral Combinatorics

Theorem 2.7. *For every finite $X \subseteq \mathbb{R}^J$, $\operatorname{cone}(X)$ is a cone, i.e., the solution set of a homogeneous linear system.*

The converse of this theorem is due to Minkowski (see Bachem and Grötschel (1982)).

Theorem 2.8. *For every cone $C \subseteq \mathbb{R}^J$ there exists a finite set $X \subseteq \mathbb{R}^J$ such that $C = \operatorname{cone}(X)$.*

The *convex hull* of a finite set $X \subseteq \mathbb{R}^J$, denoted by $\operatorname{conv}(X)$, is the set of all $w \in \mathbb{R}^J$ for which there exists nonnegative $\lambda \in \mathbb{R}^X$ satisfying $w = \Sigma(\lambda_x x : x \in X)$ and $\Sigma(\lambda_x : x \in X) = 1$. In this case, we say that w is a *convex combination* of X. From Theorems 2.7 and 2.8 it is not difficult to deduce (using homogenization – see Bachem and Grötschel (1982)) the following fundamental result of polyhedra theory due to Goldman (1956). (The necessity, i.e. the fact that a polyhedron can be expressed as the sum of a polytope and a convex cone also follows from Motzkin (1936).)

Theorem 2.9. *$P \subseteq \mathbb{R}^J$ is a polyhedron if and only if there exist finite sets V and $R \subseteq \mathbb{R}^J$ such that $P = \{x \in \mathbb{R}^J : x = v + r \text{ for some } v \in \operatorname{conv}(V) \text{ and } r \in \operatorname{cone}(R)\}$.*

Moreover, P is a polytope if and only if the above set R is empty. Clearly if v is a vertex of P, and $P = \operatorname{conv}(V)$ then we must have $v \in V$, since a vertex cannot be a convex combination of other members of P. For a polytope P, the unique minimum set V such that $P = \operatorname{conv}(V)$ is the set of vertices of P.

Theorem 2.9 is probably the most basic result for polyhedral combinatorics. It implies that if we take the convex hull of a finite subset of $\mathbb{R}^J$, it will be a polyhedron, i.e., the solution set of a finite linear system. Much of polyhedral combinatorics consists of determining such linear systems.

Finally we discuss the notion of adjacency. Two distinct vertices v, w of a polyhedron P are said to be *adjacent* if $\operatorname{conv}(\{v, w\})$ is a face of P (which will necessarily be an edge). Thus we have the following:

Proposition 2.10. *Vertices v, w of P are adjacent if and only if for every $\lambda \in \mathbb{R}$ satisfying $0 < \lambda < 1$, we have $\lambda v + (1 - \lambda) w \notin \operatorname{conv}(V \setminus \{v, w\})$, where V is the set of vertices of P.*

Note that Proposition 2.10 provides in general a convenient means of showing that two vertices v, w are *not* adjacent: for some λ satisfying $0 < \lambda < 1$ express $\lambda v + (1 - \lambda) w$ as a convex combination of members of $V \setminus \{v, w\}$. What it does not provide is a practical means of showing that v and w *are* adjacent.

3. Linear Systems, Facets and Algorithms

One of the major problems facing some one interested in designing an efficient algorithm for a combinatorial optimization problem is that of determining an easily checkable optimality criterion. In many cases, even if one knows the optimum solution, but is forced to convince a nonbeliever of its optimality, the amount of work required is effectively as large as the amount of work required to find the optimum solution in the first place.

Polyhedral combinatorics provides a means of obtaining min-max theorems which can be used to prove optimality of a solution to a problem. Basically it involves a four step process:

i) Represent the feasible objects by vectors (usually their incidence vectors).
ii) Consider these vectors as points in $\mathbb{R}^J$ for suitable J and let P be their convex hull.
iii) Obtain a linear system sufficient to define this convex hull.
iv) Apply linear programming duality to this linear system in order to obtain a min-max theorem.

Generally the major difficulties encountered are with step iii). Frequently the necessary linear systems are very large and the complexity of the required inequalities may make the determination of the system very difficult. However, we shall see in Section 6 that even an incomplete linear system can be sufficient for solving "real world" problems.

As an illustration of this process, we consider the case of matroid polyhedra.

A *matroid* $M = (E, \mathscr{F})$ is an ordered pair consisting of a finite set E and a nonempty family $\mathscr{F}$ of so-called *independent* subsets of E which satisfy the following axioms:

(M1) A subset of an independent set is independent;
(M2) For any $A \subseteq E$, all maximal (with respect to inclusion) members of $\{I \in \mathscr{F} : I \subseteq A\}$ have the same cardinality, called the *rank* of A, and denoted by $r(A)$. Then $I \subseteq E$ belongs to $\mathscr{F}$ if and only if $|I| = r(I)$. A *basis* of $A \subseteq E$ is a set $I \subseteq A$ which satisfies $|I| = r(I) = (A)$.

Matroids were introduced by Whitney (1935) as an abstraction of the notion of linear independence of the sets of columns of a matrix. At the same time, van der Waerden (1937) discovered these structures when axiomatizing the notion of algebraic independence. For background information see Lawler (1976) or Welsh (1976). Probably the best known example of a matroid is the so-called *forest matroid* (or *cycle matroid*) of a graph $G = (V, E)$. In this case a subset of E is independent if and only if it contains no cycle of G, i.e. is a forest. This is a special case of the class of matroids which motivated Whitney, the so-called *matric* or *representable* matroids. Such a matroid is obtained by letting E be the set of columns of some matrix A over some field and then letting $\mathscr{F}$ be the family of all linearly independent subsets of the columns. If we consider the node-edge incidence matroid of an undirected graph over GF(2) we obtain the forest matroid of the graph.

If $(E, \mathscr{F})$ satisfies M1, but not necessarily M2, then it is called an *independence system*. A great many combinatorial optimization problems can be formulated as optimization problems over independence systems: let $c = (c_e : e \in E)$ be a vector of real (rational) costs; find $I^* \in \mathscr{F}$ which maximizes $\Sigma(c_e : e \in I)$ over all $I \in \mathscr{F}$. Perhaps the most obvious method of attack for such a problem is the so-called *greedy algorithm*. Sort the elements of E into the order $e_1, e_2, \ldots, e_n$ in such a way that $c_{e_1} \geqslant c_{e_2} \geqslant \ldots \geqslant c_{e_n}$. Then construct $\hat{I}$ as follows: Initially $\hat{I} := \emptyset$. For i going from 1 to n, if $\hat{I} \cup \{e_i\} \in \mathscr{F}$, and $c_{e_i} > 0$ then add i to $\hat{I}$.

II. Polyhedral Combinatorics

R. Rado (1957) and later independently J. Edmonds (1971), D. Gale (1968) and D. Welsh (1968) proved that if $(E, \mathscr{F})$ is a matroid, then for any c the set $\hat{I}$ produced by the greedy algorithm is indeed an optimal solution to the optimization problem. Moreover, if $(E, \mathscr{F})$ is not a matroid, then for some c, the greedy algorithm will obtain a suboptimal $\hat{I}$. (This latter fact follows easily from the axioms. If $(E, \mathscr{F})$ violates M2 then there are bases I_1, I_2 of $A \subseteq E$ such that $|I_1| < |I_2|$. For $\varepsilon > 0$ define

$$c_e := \begin{cases} 1 & \text{for } e \in I_1 \\ 1-\varepsilon & \text{for } e \in I_2 \setminus I_1 \\ -1 & \text{for } e \in E \setminus (I_1 \cup I_2). \end{cases}$$

The greedy algorithm will terminate with $\hat{I} = I_1$ but for $\varepsilon > 0$ sufficiently small the correct answer is I_2 which gives value $|I_2| - \varepsilon(|I_2 \setminus I_1|) > |I_1|$.)

Let $M = (E, \mathscr{F})$ be a matroid and let $\mathscr{X}$ be the set of all $0-1$ incidence vectors of independent subsets of E. We consider $\mathscr{X}$ as a set of vectors in $\mathbb{R}^E$ and define the matroid polytope $\mathscr{P}(M)$ to be the convex hull of the members of $\mathscr{X}$. Edmonds (1971) gave the following linear system sufficient to define $\mathscr{P}(M)$.

Theorem 3.1. $\mathscr{P}(M) = \{x \in \mathbb{R}^E :$

(3.1) $x_e \geq 0$ for all $e \in E$,

(3.2) $\Sigma(x_e : e \in S) \leq r(S)$ for all $S \subseteq E\}$.

It is easy to see that (3.1) and (3.2) are valid inequalities for $\mathscr{P}(M)$ and moreover that any *integer* solution to (3.1), (3.2) must be a member of $\mathscr{X}$. Thus proving Theorem 3.1 amounts to showing that every vertex of the polyhedron P defined by (3.1), (3.2) is integer valued. Edmonds proved this in the following fashion, which has become one of the fundamental methods of polyhedral combinatorics. It suffices to show that for any vector $c = (c_e : e \in E)$ of element costs, there exists an optimum solution x^* to the problem of maximizing $\{cx : x \in P\}$ which is integer valued. The greedy algorithm will produce an integer $\hat{x} \in P$ for which it proves optimality by also producing a feasible solution $\hat{y}$ to the dual linear program of maximizing cx subject to (3.1), (3.2) such that $\hat{x}$ and $\hat{y}$ satisfy the complementary slackness optimality conditions of linear programming.

The dual linear program to maximizing cx subject to (3.1), (3.2) is

$$\text{minimize} \sum_{S \subseteq E} y_s r(S)$$

subject to

$$y_s \geq 0 \quad \text{for all } S \subseteq E,$$

$$\Sigma(y_s : S \ni e) \geq c_e \quad \text{for all } e \in E.$$

Complementary slackness asserts that feasible $\hat{x}$ and $\hat{y}$ are optimal if and only if

(3.3) $\hat{x}_e > 0$ implies $\Sigma(\hat{y}_s : S \ni e) = c_e$

and

(3.4) $\qquad \hat{y}_s > 0$ implies $\Sigma(\hat{x}_e : e \in S) = r(S)$.

Recall that the greedy algorithm first sorted the elements of E into an order such that $c_{e_1} \geq c_{e_2} \geq \ldots \geq c_{e_n}$. Let k be the last i for which $c_{e_i} > 0$ and for $1 \leq i \leq k$, define $A^i = \{e_1, e_2, \ldots, e_i\}$. Then let $\hat{y}$ be defined by

$$\hat{y}_s = \begin{cases} c_{e_i} - c_{e_{i+1}} & \text{if } S = A^i \text{ for } 1 \leq i < k \\ c_{e_k} & \text{if } S = A^k \\ 0 & \text{otherwise.} \end{cases}$$

(If $c_{e_1} < 0$, i.e., no k as above exists, we simply take $y \equiv 0$.) Then it can be straightforwardly verified that $\hat{y}$ is dual feasible, and if $\hat{x}$ is the incidence vector of the solution produced by the greedy algorithm, then $\hat{x}$ and $\hat{y}$ satisfy (3.3) and (3.4).

We remark at this point that if c is integer valued, then the optimal dual solution $\hat{y}$ defined above will also be integer valued.

More recently, "pure" polyhedral proofs have been found for many polyhedral theorems such as Theorem 3.1. M. Grötschel (private communication) showed how an argument of Lovász (1979) could be adapted to provide the following proof of this result.

We can assume, with no loss of generality, that for every $e \in E$, $\{e\} \in \mathcal{F}$. Then the set of unit vectors of $\mathbb{R}^E$ together with the zero vector is a set of $|E| + 1$ affinely independent members of $\mathcal{P}(M)$ so $\mathcal{P}(M)$ is of full dimension. Therefore a valid inequality $ax \leq \alpha$ is facet inducing (i.e., necessary) for $\mathcal{P}(M)$ if and only if, for any valid inequality $bx \leq \beta$ which is not a positive multiple of $ax \leq \alpha$, there is some $\hat{x} \in \mathcal{P}(M)$ satisfying $a\hat{x} = \alpha$ and $b\hat{x} < \beta$. So suppose that $ax \leq \alpha$ is facet inducing for $\mathcal{P}(M)$ and let $\mathcal{W} = \{S \subseteq E:$ the incidence vector x of S satisfies $ax = \alpha\}$. Suppose, for some $e \in E$, that $a_e < 0$. If there existed $S \in \mathcal{W}$ such that $e \in S$ then by (M1), $S \setminus \{e\} \in \mathcal{F}$ but the incidence vector x' of $S \setminus \{e\}$ satisfies $ax' > \alpha$, contradicting the validity of $ax \leq \alpha$. Therefore the incidence vector $\bar{x}$ of any $S \in \mathcal{W}$ satisfies $\bar{x}_e = 0$ and so $ax \leq \alpha$ must be a positive multiple of the constraint $-x_e \leq 0$ which is of the form (3.1).

Otherwise, $a_j \geq 0$ for all $j \in E$. Let $A := \{e \in E : a_e > 0\}$. Then every $S \in \mathcal{W}$ must satisfy $|S \cap A| = r(A)$. For suppose $|S \cap A| < r(A)$. Then by (M2) there exists $j \in A \setminus S$ such that $S \cup \{j\} \in \mathcal{F}$. But then the incidence vector x' of $S \cup \{j\}$ satisfies $ax' > \alpha$, a contradiction. So the incidence vector $\bar{x}$ of every $S \in \mathcal{W}$ satisfies $\sum_{e \in A} \bar{x}_e = r(A)$ and so $ax \leq \alpha$ is a positive multiple of the inequality $\Sigma(x_e : e \in A) \leq r(A)$ which is of the form (3.2). □

Now we interpret the min-max theorem obtained by combining linear programming duality and Theorem 3.1 to the forest matroid of a graph $G = (V, E)$. It states that for any vector $(c_e : e \in E)$ of edge weights, the maximum sum of the weights on the edges of a forest of G equals the minimum weight of a cover of the edges defined in the following fashion. A nonnegative vector $y = (y_s : S \subseteq E)$ is said to be a *cover* if $\Sigma(y_s : S \ni e) \geq c_e$ for every $e \in E$. For any $J \subseteq E$ the rank of J, denoted by $\rho(J)$, is equal to the number of nodes incident with members of J

II. Polyhedral Combinatorics

minus the number of components of the subgraph induced by J. The *weight* of a cover y is then $\Sigma(\rho(J)y_J : J \subseteq E)$.

Now let us consider one way in which the previous result can be sharpened, namely by reducing the number of variables used in order to obtain the equality. That is, we want to reduce the number of dual variables used, or equivalently, reduce the number of primal constraints. Consequently, we would like to replace our linear system (3.1), (3.2) with the smallest possible system, sufficient to define $\mathcal{P}(M)$.

For any matroid $M = (E, \mathcal{F})$, a set $A \subseteq E$ is said to be *closed* if $r(A \cup \{e\}) > r(A)$ for every $e \in E \setminus A$. That is, it is maximal for its rank. A set $A \subseteq E$ is said to be *nonseparable* if whenever $S \subseteq A$ satisfies $r(S) + r(A \setminus S) = r(A)$ we have $S = \emptyset$ or $S = A$. Any inequality (3.2) for which S is not both closed and nonseparable can be dropped, for such an inequality is implied by others. However the converse is also true, namely that these are the only inequalities (3.2) which can be discarded.

Theorem 3.2 (Edmonds, see Giles (1975)). *For a matroid $M = (E, \mathcal{F})$, an inequality (3.2) induces a facet of $\mathcal{P}(M)$ if and only if S is closed and nonseparable.*

Proof. Again we assume that $\{e\} \in \mathcal{F}$ for every $e \in E$. Then $\mathcal{P}(M)$ is of full dimension and so the essential (facet inducing) inequalities are unique, up to positive multiples. As we remarked, if S is not both closed and nonseparable, it is easy to show that the inequality (3.2) is implied by other inequalities and so is not facet inducing. Conversely, suppose that S is closed and nonseparable and let $\mathcal{W} := \{I \in \mathcal{F} : |I \cap S| = r(S)\}$. We will show that the unique (up to positive multiples) equation satisfied by all incidence vectors of members of $\mathcal{W}$ is $\Sigma(x_e : e \in S) = r(S)$. Therefore this inequality induces a face of $\mathcal{P}(M)$ of dimension $|E| - 1$, i.e. a facet, by Theorem 2.5 iii).

Let $ax = \alpha$ be an equation satisfied by all incidence vectors of members of $\mathcal{W}$. We will show
i) $a_j = a_k$ for all $j, k \in S$,
ii) $a_j = 0$ for all $j \in E \setminus S$
which will imply the result.

Suppose that i) does not hold; let $S_1 := \{e \in S : a_e \text{ takes on the maximum value over } S\}$ and let $S_2 := S \setminus S_1$. Let I_2 be a basis of S_2 and let I be a basis of S obtained by extending I_2. Let $I_1 := I \setminus I_2$. Since S is nonseparable, $|I_1| < r(S_1)$ so there exists $e \in S_1 \setminus I_1$ such that $I_1 \cup \{e\} \in \mathcal{F}$. Therefore there exists $e' \in I_2$ such that $I' := I \cup \{e\} \setminus \{e'\}$ is a basis of S. But then $I, I' \in W$ and $\Sigma(a_j : j \in I') > \Sigma(a_j : j \in I)$ a contradiction, so i) is established.

Now let $j \in E \setminus S$ and let I be a basis of S. Then, since $I \in \mathcal{W}$, $\Sigma(a_e : e \in I) = \alpha$. Since S is closed, $I \cup \{j\} \in \mathcal{F}$ and so $\Sigma(a_e : e \in I \cup \{j\}) = \alpha$ implying $a_j = 0$ and establishing ii). □

Now let us apply Theorem 3.2 to the case of the forest matroid of a graph $G = (V, E)$. Then $J \subseteq E$ is closed if $J = G[S]$ for some $S \subseteq V$, i.e., J is the edge set of a node-induced subgraph of G. Moreover J is nonseparable provided that the graph $G[J]$ induced by J is itself nonseparable in the graph theoretic sense.

(That is, $G[J]$ is connected and contains no cutnode.) Thus in view of Theorem 3.2, we can improve our forest min-max equality by restricting our covering to y_s for $S \subseteq E$ such that $S = G[V']$ for some $V' \subseteq V$ and $G[V']$ is nonseparable. *Moreover if this set of variables were decreased any further, there would be some cost vectors for which equality would no longer hold.*

4. Dual Integrality

In this section we consider a second way in which polyhedrally derived min-max theorems can be improved, namely by restricting the set of possible values for dual variables. More precisely, we are interested in those cases when an integer optimum dual solution can be obtained. We discuss two problems of this type:
(4.1) Given an integer matrix A, when is it true that for every integral b and c the dual linear program to maximizing cx subject to $Ax \leq b$ has an integer optimum solution provided that an optimum exists?
(4.2) Given an integer matrix A and a fixed vector b (not necessarily integer), when is it true that for every integral c such that $\max\{cx: Ax \leq b\}$ exists, there exists an integer optimum solution to the dual linear program?

First we consider question (4.1). A matrix A is said to be *totally unimodular* if and only if every square submatrix has determinant 0, 1 or -1. (In particular, every entry must have value 0, 1 or -1.) The answer to (4.1) is provided by the following important result of Hoffman and Kruskal (1956).

Theorem 4.1. *Let A be an integral matrix. Then A is totally unimodular if and only if for every integral vector b the vertices of the polyhedron $\{x: Ax \leq b, x \geq 0\}$ are integral.*

From this it follows that the answer to (4.1) is "when A is totally unimodular". For the dual linear program to $\max\{c^T x: Ax \leq b\}$ is $\min\{y^T b: y^T A = c^T, y \geq 0\}$. This is equivalent to $\min\{y^T b: y^T A \leq c^T, y^T(-A) \leq -c^T, y \geq 0\}$. It is then easily verified that A is totally unimodular if and only if $\begin{bmatrix} A \\ -A \end{bmatrix}$ is totally unimodular which gives the result.

Totally unimodular matrices arise frequently in polyhedral combinatorics. In fact, the "first period" of polyhedral combinatorics mentioned in the introduction was largely a period of study of polyhedra defined by totally unimodular matrices. If A is the node-arc incidence matrix of a directed graph or the node-edge incidence matrix of a bipartite graph then A is totally unimodular. (This follows easily by induction on the size of a square submatrix S. If some column of S contains a single nonzero, i.e. ± 1, then we apply induction to the smaller submatrix obtained by deleting the row and column containing this entry. If not, then if A comes from a directed graph, the sum of the rows of S is zero. If A comes from a bipartite graph, then the sum of the rows of S corresponding to nodes in one part minus the sum of the rows corresponding to nodes in the other part is zero. Hence, in either case, A has determinant zero.)

II. Polyhedral Combinatorics

The total unimodularity of this matrix is used extensively in polyhedral combinatorics. In the next section we shall see how it implies the König theorem for bipartite matching. Here we note that it implies the famous max-flow min-cut theorem of Ford and Fulkerson (1956) and Elias, Feinstein and Shannon (1956). See Schrijver (1983) (in this volume).

Various characterizations of totally unimodular matrices have been obtained (see for example Padberg (1975) or Schrijver (1981a)). However until recently they all had the common feature that they provided an easy means of showing that a matrix was *not* totally unimodular but did not prove an efficient means of showing that a matrix was totally unimodular or of determining which was the case for a given matrix A. This defect was remedied by Seymour (1980). Earlier, Tutte (1965) had shown that a matrix A was totally unimodular if and only if the matric matroid defined by A over the reals was *regular*. (See Tutte (1965) or Welsh (1976)) for the definition of regular matroids.) Seymour showed that a matroid was regular if and only if it could be constructed via three simple types of composition starting from forest matroids, duals of forest matroids and one 10-element exception. (The dual matroid $M^* = (E, \mathscr{F}^*)$ of $M = (E, \mathscr{F})$ is obtained by letting $\mathscr{F}^* := \{S \subseteq E : E \backslash S \text{ contains a basis of } M\}$). Edmonds has described how this decomposition can be used to obtain a polynomial algorithm which will, given a $(0, \pm 1)$-matrix A, either find a method of constructing it via the Seymour decomposition or else show that it is not totally unimodular.

Now we turn to the question (4.2). We say that the linear system $Ax \leq b$ is *totally dual integral* or *TDI* if for every integer valued c such that $\max\{c^T x : Ax \leq b\}$ exists, the corresponding dual linear program has an integer optimum solution. Giles and Pulleyblank (1979) showed that for any rational linear system $Ax \leq b$, there exists a positive rational α such that $(\alpha A)x \leq \alpha b$ is a TDI system. However in general α is very small, and so αb is not integer valued. When it is integer valued, we obtain a useful consequence.

Theorem 4.2. (Edmonds and Giles (1977)). *If a polyhedron P is the solution set of a TDI system $Ax \leq b$ with b integer valued, then every nonempty face of P contains an integer valued point, in particular, every vertex of P is integer valued.*

This was proved by Hoffman (1974) for linear systems of the form $Ax \leq b$, $x \geq 0$ where A is a $0-1$ matrix. See also Hoffman (1982). This idea of using dual integrality to establish primal integrality was also used by Lehman (1979) and by Fulkerson (1971). In fact this theorem follows from linear programming duality and the following "primal" result.

Proposition 4.3. *Every nonempty face of a rational polyhedron $P \subseteq \mathbb{R}^n$ contains an integer point if and only if, for every integer $c \in \mathbb{R}^n$ such that $\max\{cx : x \in P\}$ exists, this maximum is integer valued.*

Notice the two ways in which this is stronger than Theorem 2.1, which stated that for every nonempty face of a polyhedron, there is an objective function maximized over the polyhedron by precisely the members of that face. First we restrict ourselves to considering integer c. But this is no real restriction, since the rational are dense in the reals and the elements of P that maximize cx

also maximize $(\alpha c)x$ for any positive α. Second, we do not consider the values of the individual components of an optimum solution x^* but rather only consider the objective value cx^*. If the polyhedron has vertices, then it is not difficult to see that if there is a fractional vertex x^* then there is some integer objective function whose optimum value is fractional.

For let $c^* = (c_j : j \in E)$ be an integer objective function maximized over P only by x^*. Then for integer M sufficiently large, the vector c^j defined by

$$c_i^j = \begin{cases} M \cdot c_i & \text{if } i \neq j \\ M \cdot c_j + 1 & \text{if } i = j \end{cases}$$

is also maximized by x^*, and the difference in objective value between c^j and $M \cdot c^*$ is exactly x_j^*. Thus if x_j^* is fractional, so too is at least one of $c^j x^*$ and $M \cdot c^* x^*$. (This is the idea of the proof of Hoffman (1974), see also Giles (1975).)

However, in the general case a more refined proof method is required. Edmonds and Giles (1977) proved Proposition 4.3 by making use of an "integer" Farkas' lemma: For a rational linear system $Ax = b$, either there exists an integer solution or else there exists λ such that $\lambda^T A$ is integer and $\lambda^T b$ is fractional. (Bachem and von Randow (1979) attribute this result to Kronecker (1899); it is also an easy consequence of the fact that an integer matrix can be put into Smith Normal form (see Bachem and Kannan (1979)).

For our purposes, however, the converse direction of Theorem 4.2 is at least as interesting.

Theorem 4.4 (Giles and Pulleyblank (1979)). *If $P = \{x : Ax \leq b\}$ is an integer polyhedron then there exists a TDI linear system $A'x \leq b'$ with b' integer such that $P = \{x : A'x \leq b'\}$.*

In fact, Schrijver showed that this could be strengthened to the following:

Theorem 4.5 (Schrijver (1981b)). *Let P be a full dimensional polyhedron defined by a rational linear system. Then there is a unique minimal TDI system $Ax \leq b$ with A integral such that $P = \{x : Ax \leq b\}$. Moreover, every nonempty face of P will contain an integral valued point if and only if b is integer.*

A minimal TDI defining linear system as in Theorem 4.5 will of course contain a facet inducing inequality for each facet of P. However in general a complete set of facet inducing inequalities will not be TDI, and extra inequalities must be added in order to obtain this stronger property. (See the discussion of b-matchings in the following section.) Consequently our objectives of improving polyhedrally derived min-max theorems by decreasing the number of dual variables and requiring them to take on only integer values may, in some cases, be contradictory. However what is perhaps surprising is that very often a minimal defining linear system is TDI, when each inequality is scaled so that all entries are integer and the greatest common divisor of the entries is 1. In particular, as we remarked following our description of the "extended" greedy algorithm in the previous section, this is true for matroid polyhedra. Thus we have the following best possible (in a certain sense) form of Theorem 3.1.

II. Polyhedral Combinatorics

Theorem 4.6. *Let $M=(E, \mathscr{F})$ be a matroid such that $\{e\}\in\mathscr{F}$ for all $e\in E$. Then the following is the unique minimal integer TDI system sufficient to define $\mathscr{P}(M)$. It is also a minimal defining system.*

$$x_e \geq 0 \quad \text{for all } e \in E,$$

$$\Sigma(x_e : e \in S) \leq r(S) \quad \text{for all } S \subseteq E \text{ such that } S \text{ is closed and nonseparable}.$$

Finally, we obtain the "best possible" min-max theorem for spanning forests in a graph as follows. Let $G=(V, E)$ be a graph and let $c=(c_j : j \in E)$ be a vector of integer edge costs. An *induced subgraph cover* $\mathscr{H}$ of G is a finite family $\mathscr{H}=(K_i \subseteq V : i \in I)$ of subsets of V which induce 2-connected subgraphs of G such that for each edge $[uv] \in E$, $|\{i \in I : u, v \in K_i\}| \geq c_{[uv]}$. In other words, each edge is covered at least as many times as its cost. The weight of $\mathscr{H}$ is $\sum_{i \in I}(|K_i|-1)$. Our min-max theorem then becomes: The maximum of the sum of the weights of the edges of a forest of G equals the minimum weight of an induced subgraph cover of G.

In the following section we give examples of this type of theorem for other problems of combinatorial optimization.

A property for a linear system $Ax \leq b$ which is even stronger than TDI is that for any vectors l and u of lower and upper bounds respectively, the linear system $Ax \leq b$, $l \leq x \leq u$ should be TDI. Edmonds and Giles (1977) call such a linear system $Ax \leq b$ *box TDI*, and say that a polyhedron is box TDI if it is defined by a box TDI linear system. They show that although every polyhedron is defined by a TDI system it is not true that every polyhedron is defined by a box TDI system. In particular, we have the following:

Theorem (Edmonds, Giles (1977)). *If P is a box TDI polyhedron then there exists a $(0, \pm 1)$-matrix A and a vector b such that $P=\{x : Ax \leq b\}$.*

If a matrix A is totally unimodular, then concatenating on an identity matrix or its negative preserves total unimodularity. Thus it follows that if A is totally unimodular, then for any rational b, the system $Ax \leq b$ is box TDI. However there are interesting examples of box TDI systems which do not come from totally unimodular systems. The reader is referred to Edmonds and Giles (1977, 1982).

Finally, we note that various algorithmic questions concerned with determining whether or not a given linear system is TDI have recently received study. See Chandresekaran (1981) or the survey article Edmonds and Giles (1982). Cook (1982) has shown that many operations which preserve integrality of the vertices of a polyhedron defined by a linear system also preserve total dual integrality.

5. Some Examples

In this section we discuss several examples of the theory discussed in the previous two sections and their relationship to some well known combinatorial theorems. For further examples, see Schrijver (1983) (in this volume).

Let $M_1 = (E, \mathscr{F}_1)$ and $M_2 = (E, \mathscr{F}_2)$ be matroids defined on the same set E. The *intersection* of M_1 and M_2, denoted by $M_1 \cap M_2$, is the independence system $(E, \mathscr{F}_1 \cap \mathscr{F}_2)$. Many problems of combinatorial optimization are matroid intersection problems. For example, a *matching* in a graph $G = (V, E)$ is a set $M \subseteq E$ meeting each node at most once. If we are given $c \in \mathbb{R}^E$, the maximum matching problem is to find a matching M^* of G for which $\Sigma(c_e : e \in M)$ is maximized over all matchings. Now let $G = (V_1 \cup V_2, E)$ be a bipartite graph and define matroids $M_1 = (E, \mathscr{F}_1)$ and $M_2 = (E, \mathscr{F}_2)$ as follows: For $i = 1, 2$, let $\mathscr{F}_i := \{J \subseteq E : J \text{ meets each node of } V_i \text{ at most once}\}$. Then for any $J \subseteq E$, the rank of J in M_i, denoted by $r_i(J)$, is the number of nodes of V_i incident with edges of J. The matchings of G are precisely the members of $\mathscr{F}_1 \cap \mathscr{F}_2$.

One of the most beautiful theorems of polyhedral combinatorics is Edmonds' matroid intersection theorem. It states that for matroids $M_1 = (E, \mathscr{F}_1)$ and $M_2 = (E, \mathscr{F}_2)$, the vertices of $\mathscr{P}(M_1) \cap \mathscr{P}(M_2)$ are the common vertices of $\mathscr{P}(M_1)$ and $\mathscr{P}(M_2)$. In terms of linear systems, the result can be stated as follows:

Theorem 5.1 (see Edmonds (1979)). *Let $M_1 = (E, \mathscr{F}_1)$ and $M_2 = (E, \mathscr{F}_2)$ be matroids having rank functions r_1 and r_2 respectively. Then the convex hull of the incidence vectors of the members of $\mathscr{F}_1 \cap \mathscr{F}_2$, denoted by $\mathscr{P}(M_1 \cap M_2)$, is given by*

(5.1) $x_e \geq 0$ *for all* $e \in E$,

(5.2) $\Sigma(x_e : e \in S) \leq \min\{r_1(S), r_2(S)\}$ *for all* $S \subseteq E$.

Theorem 5.1 is proved (see Edmonds (1979) or Lawler (1976)) by the same algorithmic/linear programming based method described for proving Theorem 3.1. (However, the algorithm is considerably more complicated.) It would be interesting to see if a proof analogous to Grötschel's proof of Theorem 3.1 could be constructed.

Let r be the rank function fo $M_1 \cap M_2$, i.e., for $S \subseteq E$, $r(S) = \max\{|I| : I \subseteq S, I \in \mathscr{F}_1 \cap \mathscr{F}_2\}$. In general it is not true that $r(S) = \min\{r_1(S), r_2(S)\}$. However, the strongest valid set of $(0-1)$-inequalities that we can write down for $\mathscr{P}(M_1 \cap M_2)$ (or in fact for any combinatorial polyhedron, having rank suitably defined) is

(5.3) $\qquad \Sigma(x_e : e \in S) \leq r(S)$ *for all* $S \subseteq E$.

Hence Theorem 5.1 remains true if we replace (5.2) with (5.3). However this system is, in general, far from minimal. Consider some inequality (5.2) and suppose $r_1(S) \leq r_2(S)$. Then we can drop this inequality unless S is closed and nonseparable in M_1. So let us return to bipartite matching. For any graph $G = (V, E)$, we let

$\delta(S) := \{j \in E : \text{one end of } j \text{ is in } S\}$,

$\gamma(S) := \{j \in E : \text{both ends of } j \text{ are in } S\}$ for all $S \subseteq V$,

$\delta(v) := \delta(\{v\})$ for $v \in V$.

Then $J \subseteq E$ will be closed and nonseparable in M_1 if and only if $J = \delta(v)$ for some $v \in V_1$, and similarly for M_2. Therefore, by Theorems 3.1 and 3.2, $\mathscr{P}(M_i) = \{x \in \mathbb{R}^E : x \geq 0, \Sigma(x_j : j \in \delta(v)) \leq 1 \text{ for all } v \in V_i\}$ for $i = 1, 2$.

II. Polyhedral Combinatorics

We let $\mathcal{M}(G)$ denote the *matching polytope of* G, i.e., the convex hull of the incidence vectors of the matchings of G. Then the matroid intersection theorem gives the following:

Theorem 5.2. *For any bipartite graph* $G = (V_1 \cup V_2, E)$, $\mathcal{M}(G) = \{x \in \mathbb{R}^E : x \geq 0,$ $\Sigma(x_j : j \in \delta(v)) \leq 1$ *for all* $v \in V_1 \cup V_2\}$.

This theorem, commonly called the Birkhoff-von Neumann theorem, (see Birkhoff (1946), von Neumann (1953)) is one of the best known results of polyhedral combinatorics and will be discussed later in this section. Also, there are many proofs of this theorem which are considerably more elementary than using matroid intersection. For example, it follows from the total unimodularity of the node-edge incidence matrix of a bipartite graph (see Section 4) or can easily be proved directly.

Recall that r is the rank function of $M_1 \cap M_2$. Then, as for matroids, $S \subseteq E$ is said to be *r-closed* if $r(S) < r(S \cup \{e\})$ for all $e \in E \setminus S$ and *r-nonseparable* if S cannot be partitioned into nonempty T and W such that $r(S) = r(T) + r(W)$. It is interesting that $\mathcal{P}(M)$ and $\mathcal{P}(M_1 \cap M_2)$ have the same facet characterizations.

Theorem 5.3 (Giles (1975)). *Let* M_1 *and* M_2 *be matroids and assume that* $\mathcal{P}(M_1 \cap M_2)$ *is of full dimension. Then an inequality* (5.3) *is facet inducing if and only if* S *is r-closed and r-nonseparable (where r is the rank function of* $M_1 \cap M_2$*).*

Note that the "full-dimensionality" requirement is very mild. It simply requires that each singleton be independent in both M_1 and M_2.

Edmonds (1979) observed that the linear system (5.1) (5.3) is TDI, from which it is easy to deduce that even when restricted to facet inducing inequalities, the system is TDI. Thus matroid intersection is another example of a minimal defining linear system being TDI.

Let $G = (V, A)$ be a directed graph. A *branching* is a subset J of the arcs such that J contains no (directed or undirected) cycles and, in addition, each node has at most one arc directed into it. It is easily checked that if we define $\mathcal{F}_1 := \{J \subseteq A : J$ has at most one arc directed into each node$\}$ then $M_1 = (A, \mathcal{F}_1)$ is a matroid. Therefore, if $M_2 = (A, \mathcal{F}_2)$ is the forest matroid of G (ignoring arc directions) then the branchings of G are the members of $\mathcal{F}_1 \cap \mathcal{F}_2$. Therefore the convex hull of the incidence vectors of the branchings, which we denote by $\mathcal{B}(G)$, can be obtained from the matroid intersection theorem. For any $v \in V$ we let $\delta^-(v)$ denote the set of arcs whose head is v.

Theorem 5.4 (Edmonds (1968)). *For any directed graph* $G = (V, A)$, $\mathcal{B}(G) = \{x \in \mathbb{R}^A :$

(5.4) $\quad x_j \geq 0 \quad$ *for all* $j \in A$,

(5.5) $\quad \Sigma(x_j : j \in \delta^-(v)) \leq 1 \quad$ *for all* $v \in V$,

(5.6) $\quad \Sigma(x_j : j \in \gamma(S)) \leq |S| - 1 \quad$ *for all* $S \subseteq V$ *with* $|S| \geq 2\}$.

Giles (1975, 1976) characterized the facet inducing inequalities in graph theoretic terms. For any $v \in V$, let $B(v) := \{w \in V : (w, v) \in A\}$. He showed that an inequality (5.5) is essential if either $|B(v)| \geq 2$, or if $B(v) = \{w\}$ for some $w \in V$, then $(v, w) \notin A$. An inequality (5.6) is essential provided that $G[S]$ is a strongly connected block of G. The matroid intersection result implies that the resulting minimal essential subset of (5.4)–(5.6) is TDI.

The general theory of matchings developed by Edmonds (1965, 1965a), Johnson (1965) and Edmonds and Johnson (1970) is one of the major successes of polyhedral combinatorics. Thm. 5.2 gave a linear system sufficient to define $\mathcal{M}(G)$ for a bipartite graph G. However if $G=(V, E)$ is nonbipartite this is not sufficient. Assigning $1/2$ to the edges of an odd cycle and 0 to all other edges gives a fractional vertex of the polytope $\{x \in \mathbb{R}^E : x \geq 0, \Sigma(x_j : j \in \delta(v)) \leq 1$ for all $v \in V\}$. In general, if $S \subseteq V$ is such that $|S|$ is odd then clearly the incidence vector x of every matching satisfies $\Sigma(x_j : j \in \gamma(S)) \leq 1/2(|S|-1)$. (In other words, not all nodes of S can be saturated using only edges of $\gamma(S)$.) Edmonds (1965a) showed that if such an inequality is added for each $S \subseteq V$ such that $|S| \geq 3$ and $|S|$ is odd, then this linear system is sufficient to define $\mathcal{M}(G)$. Again, his proof consisted of a (polynomially bounded) algorithm which, for any objective function $c \in \mathbb{R}^E$, constructed a matching and a feasible solution to the dual linear program which satisfied the complementary slackness conditions for linear programming optimality. (See also Lawler (1976) for an exposition of this algorithm.) Nonalgorithmic proofs of Edmonds' characterization have been obtained by Balinksi (1972) and more recently Lovász (1979), Seymour (1979) and Schrijver (1981).

A matching M of $G=(V, E)$ is said to be *perfect* if every node is incident with some edge. We say that $G=(V, E)$ is *1-critical (hypomatchable, factor critical)* if G does not have a perfect matching, but for every $v \in V$, $G \backslash v$ does have a perfect matching. Necessarily a 1-critical graph will have an odd number of nodes, be connected, and unless it consists of a single isolated node, be nonbipartite. For $i \in V$ we let $N(i)$ denote the subset of $V \backslash \{i\}$ adjacent to i.

Theorem 5.5 (Pulleyblank and Edmonds (1974)). *For any graph $G=(V, E)$, the following is the unique (up to positive multiples) minimal linear system sufficient to define $\mathcal{M}(G)$:*

(5.10) $x_j \geq 0$ *for all* $j \in E$;

(5.11) $\Sigma(x_j : j \in \delta(i)) \leq 1$ *for all* $i \in V$ *such that*
$|N(i)| \geq 3$
or
$|N(i)| = 2$ *and* $\gamma(N(i)) = \emptyset$
or
$|N(i)| = 1$ *and i is a node of a two node component of G;*

(5.12) $\Sigma(x_j : j \in \gamma(S)) \leq (|S|-1)/2$ *for all $S \subseteq V$ such that $|S| \geq 3$, $G[S]$ is 1-critical and nonseparable.*

Note that (5.12) prescribes for which of the exponentially many odd subsets of V we require an inequality. If G is bipartite, then none are required. If G is complete, then all are required.

Cunningham and Marsh (1978) showed that the linear system (5.10)–(5.12) is TDI. From this then we can obtain a number of well known combinatorial

II. Polyhedral Combinatorics

theorems. Let $c \in \mathbb{R}^E$. A *1-critical cover* of c is a family $(C_i : i \in I)$ of subsets of V such that for all $i \in I$, $G[C_i]$ is 1-critical and, for all $j \in E$, $c_j \leq |\{i \in I : (|C_i| = 1$ and $j \in \delta(C_i))$ or $(|C_i| \geq 3$ and $j \in \gamma(C_i)\}|$. The *weight* of the cover is $1/2 \Sigma(|C_i| - 1 : i \in I, |C_i| \geq 3) + |\{i \in I : |C_i| = 1\}|$.

Corollary 5.6. *For any $G = (V, E)$, for any integer $c \in \mathbb{R}^E$, the maximum of $\Sigma(c_j : j \in M)$ over all matchings of G equals the minimum weight of a 1-critical cover of c.*

If G is bipartite, then a 1-critical cover becomes a *node cover*, i.e., an assignment of a nonnegative integer y_v to each node v of G such that for each edge $[uv] \in E$, $y_u + y_v \geq c_{[uv]}$. If, in addition, $c_j = 1$ for all $j \in E$ then y_v will always be 0 or 1 so we obtain König's Theorem.

Corollary 5.7 (König (1931)). *The maximum cardinality of a matching of a bipartite graph equals the minimum cardinality of a set of nodes that meets every edge.*

From this it is easy to deduce Hall's theorem (Hall (1935)) which asserts that a bipartite graph $G = (V_1 \cup V_2, E)$ has a matching which saturates all nodes of V_1 if and only if, for every $S \subseteq V_1$, $|S| \leq |\{v \in V_2 : v$ is adjacent to some node of $S\}|$.

When G is nonbipartite, we obtain the following analogue of Corollary 5.7.

Corollary 5.8 (Berge (1958)). *The maximum cardinality of a matching in a graph $G = (V, E)$ equals the minimum weight of a cover of the edges with nodes and odd cardinality subsets of V.*

(In this case an edge j is covered if there is some node v in the cover with $j \in \delta(v)$ or if there is some odd set S in the cover with $j \in \gamma(S)$. The weight of the cover is the sum of the number of nodes in the cover and $\Sigma(|S| - 1)/2$, taken over all odd sets S of cardinality at least three in the cover). Note that this corollary does not require the facet version of the matching polytope characterization given in Theorem 5.5, Edmonds' original version is sufficient. However using Theorem 5.5 it is possible to derive a stronger version.

Let $\mathcal{M}(G)$ denote the *perfect matching polytope* of G, i.e., the convex hull of the incidence vectors of the perfect matchings. Then $\hat{\mathcal{M}}(G)$ is the face of $\mathcal{M}(G)$ obtained by turning the inequalities (5.11) into equations. W. Cook (1982) has shown that for any TDI linear system $Ax \leq b$, turning a subset of the inequalities into equations leaves a TDI system. Therefore (5.10), (5.11) as equations, (5.12) is a TDI defining system for $\hat{\mathcal{M}}(G)$, however it is in general far from minimal. Edmonds, Lovász, Pulleyblank (1982) describe a procedure for obtaining a minimal defining system for $\hat{\mathcal{M}}(G)$ for any graph G, but the dual integrality question has not been treated. However it is easy to deduce Tutte's theorem for the existence of a perfect matching in a graph from the not necessarily minimal TDI system described above.

Corollary 5.9 (Tutte (1952)). *$G = (V, E)$ has a perfect matching if and only if for every $X \subseteq V$, $G \setminus X$ has at most $|X|$ components with an odd number of nodes.*

Giles (1982, 1982a, 1982b) considered the following unification of branchings and matchings. A *mixed graph* $G=(V, E\cup A)$ is a graph having a set E of undirected edges and a set A of directed arcs. A *matching forest* is a subset F of $E\cup A$ which is a forest, if we replace all arcs with undirected edges, and such that each node is incident with at most one edge of F or the head of one arc of F but not both. If $E=\emptyset$, then matching forests are branchings, if $A=\emptyset$ then they are 1-matchings. Giles solved the problem of finding a maximum weight matching forest and proved analogues to Theorems 5.4 and 5.5.

We conclude this section by describing one of the few cases where we know both a minimal defining system and the minimal TDI system and they are different. Let $G=(V, E)$ be a graph and let $b=(b_i: i\in V)$ be a vector of positive integers. A *b-matching* is a nonnegative integral valued $x \in \mathbb{R}^E$ satisfying $\Sigma(x_j : j \in \delta(v)) \leq b_v$ for all $v \in V$. If we have $\Sigma(x_j : j \in \delta(v)) = b_v$ for all $v \in V$ then we call x a *perfect b-matching*. Edmonds (1965a) (see also Edmonds and Johnson (1970), Green-Krótki (1980), Pulleyblank (1973)) showed that $\mathcal{M}(G, b)$, the convex hull of the b-matchings of G, is given by the following linear system:

(5.13) $\quad\quad\quad\quad\quad\quad x_j \geq 0 \quad \text{for all } j \in E,$

(5.14) $\quad\quad\quad\quad\quad\quad \Sigma(x_j : j \in \delta(v)) \leq b_v \quad \text{for all } v \in V,$

(5.15) $\Sigma(x_j : j \in \gamma(S)) \leq 1/2(\Sigma(b_v : v \in S) - 1) \quad \text{for all } S \subseteq V \text{ such that } \Sigma(b_v : v \in S) \text{ is odd}.$

Again, this system is in general far from minimal. Suppose we are given $G=(V, E)$ and $b=(b_v : v \in V)$. For each $v \in V$, let the vector b^v be obtained from b by decreasing the value of b_v by one. We say that G is *b-critical* if G has no perfect b-matching, but for every $v \in V$, G does have a perfect b^v-matching. Pulleyblank (1973) showed that these b-critical graphs play a role for b-matchings analogous to the role played by 1-critical graphs for 1-matchings.

Theorem 5.10. *The following is the unique (up to positive multiples) linear system sufficient to define $\mathcal{M}(G, b)$:*

(5.16) $x_j \geq 0 \quad \text{for all } j \in E$;

(5.17) $\Sigma(x_j : j \in \delta(v)) \leq b_v \quad \text{for all } v \in V \text{ such that}$
$\quad\quad\quad \Sigma(b_i : i \in N(v)) \geq b_v + 2$
or
$\quad\quad\quad \Sigma(b_i : i \in N(v)) = b_v + 1 \quad \text{and} \quad \gamma(N(v)) = \emptyset$
or
$\quad\quad\quad v$ *belongs to a two node component of G, and $b_v = b_w$, where w is the other node*;

(5.18) $\Sigma(x_j : j \in \gamma(S)) \leq 1/2(\Sigma(b_v : v \in S) - 1)$ *for all $S \subseteq V$ such that $|S| \geq 3$, $G[S]$ is b-critical and contains no cutnode i having $b_i = 1$.*

(Note that if $b_v = 1$ for all $v \in V$, then a b-matching is just a 1-matching and so this result is a direct generalization of Theorem 5.5). However, this linear system is not TDI. For suppose that $G=(V, E)$ consists of a triangle having $b_v = 2$ for all $v \in V$. Then no inequalities (5.18) are present. If we take $c_j = 1$ for all

$j \in E$, then the maximum value of cx for a b-matching x is 3, which cannot be obtained with an integer dual solution.

A graph is said to be *b-bicritical* if $b_v \geq 2$ for all $v \in V$ and if whenever any node i has its coefficient b_i reduced by two, there exists a perfect b-matching. This implies that G itself has a perfect b-matching (Pulleyblank (1980)). These b-bicritical graphs are those for which new inequalities must be added to (5.16)–(5.18) in order to obtain a TDI system.

Theorem 5.11. (Pulleyblank (1981), Cook (1981)). *The unique minimal integer TDI system for $\mathcal{M}(G, b)$ is given by (5.16)–(5.18) and*

(5.19) $\quad \Sigma(x_j : j \in \gamma(S)) \leq 1/2 (b_v : v \in S)$ *for all $S \subseteq V$ such that $G[S]$ is b-bicritical, connected and every node $v \in V \setminus S$ which is adjacent to a node of S has $b_v \geq 2$.*

Informally, therefore, the situation for b-matchings is that if you want an integer min-max theorem, you must use considerably more dual variables than if you want an equality using the smallest possible set of variables. For a comprehensive survey of many other polyhedral theorems and min-max theorems the reader is referred to Schrijver (1983) (in this volume).

6. Separation

Probably one of the most interesting recent developments in polyhedral combinatorics has been the study of "separation" techniques. Simply stated, the problem is: given a polyhedron $P \subseteq \mathbb{R}^n$ and a point $\hat{x} \in \mathbb{R}^n$ determine whether or not $\hat{x} \in P$, and if not, find an inequality $ax \leq \alpha$ satisfied for every $x \in P$, but for which $a\hat{x} > \alpha$. There are both practical and theoretical reasons for interest in this problem. First, the idea underlying the cutting plane approach to integer programming is to start with a "simple" polyhedron P^0 which contains the (probably complex) polyhedron P^* of interest. The desired objective function c is maximized over P^0; let x^0 be the optimum point obtained. If $x^0 \in P^*$ then it must maximize cx over P^*. If not find an inequality $a^0 x \leq \alpha^0$ which is valid for P^* but which is violated by x^0. Let P^1 be the polyhedron thereby obtained and repeat the process. In this way we get a sequence of polyhedra $P^0 \supset P^1 \supset P^2 \supset \ldots \supset P^*$. Hopefully, for some n (not too large) the optimum solution x^* to maximizing cx over P^n will be in P^*, and hence solve the original problem. Note that if we are able to ensure that each inequality added induces a distinct facet of P^*, then we are guaranteed finite termination of this process. However if we do not have available a complete set of facet inducing inequalities, we cannot be sure that any x^n will belong to P^*.

Second, the polynomially bounded ellipsoid algorithm for linear programming can be applied to obtain a polynomial algorithm for optimizing over any polyhedron provided that the separation problem can be solved in polynomial time. Moreover, the converse is also true; if we can optimize polynomially over a class of polyhedra, then so too can we solve the separation problem polynomially. (These relationships were established independently by Grötschel, Lovász, Schrijver (1981), Karp and Papadimitriou (1980), Padberg and Rao (1981).) It should also be noted that the above statement hides certain technicalities required to make the optimization-separation equivalence precise for certain classes of polyhedra. If we restrict our attention to polyhedra with integer (or half integer) valued vertices, as is usually done in polyhedral combinatorics, then no problems arise. (See Grötschel, Lovász, Schrijver (1981, 1982).)

All the classes of facets of combinatorial polyhedra that we have described so far have had good (or "NP") descriptions. That is, given an inequality $ax \leq \alpha$, we are able to determine polynomially whether or not it belongs to a given class. For example, if we wish to see if it is of the form (5.12) we first verify that all nonzero coefficients are identical and positive, then we verify that the subgraph induced by the edges with positive coefficients is 1-critical and nonseparable. In fact, for what is termed an *NP-description* of a class of inequalities we actually demand even less. All we require is that given the inequality $ax \leq \alpha$ and some additional information whose length is polynomially bounded in the length required to encode $ax \leq \alpha$, we can determine whether or not this inequality is in our class. For example, given a graph $G = (V, E)$, let $\mathscr{F}_1 := \{J \subseteq E$: for each $j \in E \setminus J$, $J \cup \{j\}$ contains a hamilton cycle containing $j\}$ and let $\mathscr{F}_2 := \{J \subseteq E$: J is the edge set of a maximal nonhamiltonian subgraph$\}$. Then $\mathscr{F}_1$ is given by an NP-description, since all we need provide is for each $j \in E \setminus J$, a hamilton cycle whose edge set contains j and which is contained in $J \cup \{j\}$. However, as of 1982, $\mathscr{F}_2$ is not given by an NP-description for at present there is no general characterization known of nonhamiltonian graphs.

If we try to obtain an NP-description of a class of inequalities that induce every facet of the polytope of an NP-complete problem, then we are undertaking a provably difficult task:

Theorem 6.1. (Karp and Padpadimitriou (1980). *If there exists an NP-description of a class of valid inequalities that induce every facet of the polytopes of a class of NP-complete problems, then NP = co-NP.*

(Most researchers consider it to be almost as unlikely that NP = co-NP as that P = NP. See Garey and Johnson (1979) for an excellent introduction to NP-completeness, and Papadimitriou (1982) for a good exposition of the importance of complexity results for the polyhedral approach to combinatorics, and in particular, the travelling salesman problem.)

Consequently, perhaps the most that we can realistically hope for when dealing with an NP-complete problem is to obtain several classes of facets. What is interesting is that these admittedly incomplete linear characterizations have frequently proved sufficient to prove optimality in large "real world" instances of hard problems. We discuss this in connection with the travelling salesman problem.

The (symmetric) *travelling salesman problem* is the following: given a graph $G = (V, E)$ and a vector $c \in \mathbb{R}^E$ of edge costs, find a hamilton cycle, or *tour* in G such that its edge incidence vector $\hat{x}$ minimizes cx over all such incidence vectors x of hamilton cycles. Unlike the problems discussed in the previous sections, no polynomial algorithm is known for this problem in general graphs. It

II. Polyhedral Combinatorics

was one of the problems on Karp's (1972) original list of NP-complete problems. Consequently if the majority opinion that $P \neq NP$ is correct, then no polynomial algorithm exists.

We can add to G any nonexistent edges, giving them very large costs, without affecting the optimum solution. Consequently attention is often restricted to complete graphs. Thus we let Q^n denote the convex hull of the incidence vectors of the tours of K_n, the complete graph on n nodes. Unlike most of the other polyhedra previously described, Q^n is not of full dimension.

Proposition 6.2. (Grötschel, Padberg (1979)). *The dimension of Q^n equals $1/2 n(n-3)$ for $n \geq 3$.*

(This value $1/2n(n-3)$ is the number of edges of K_n minus the number of nodes. Since every $x \in Q^n$ must satisfy

(6.1) $$\Sigma(x_j : j \in \delta(i)) = 2 \quad \text{for all nodes of } K_n,$$

and these equations are easily seen to be linearly independent, it follows from Theorem 2.4 that $\dim(Q^n) \leq 1/2n(n-3)$. Showing that equality holds requires constructing $1/2n(n-3)+1$ tours with affinely independent incidence vectors.)

One result of Proposition 6.2 is that we can have two inequalities which appear very different inducing the same facet of Q^n. Thus we will also be concerned with characterizing which facet inducing inequalities induces the same facet of Q^n.

Throughout the remainder of this section we let V and E represent the node set and edge set respectively of K_n.

Theorem 6.3 (Grötschel, Padberg (1979)). *For all $n \geq 5$, the inequalities $x_j \geq 0$ induce distinct facets of Q^n.*

These are called the *trivial* facets of Q^n.

For any $S \subset V$ satisfying $2 \leq |S| \leq n-2$, we will have, for all $x \in Q^n$,

(6.2) $$\Sigma(x_j : j \in \gamma(S)) \leq |S| - 1.$$

This simply prohibits a "small" cycle being formed with nodeset S. This inequality is called a *subtour elimination constraint*.

Theorem 6.4 (Grötschel, Padberg (1979a)). *For all $n \geq 4$ and $S \subseteq V$ satisfying $2 \leq |S| \leq \lfloor n/2 \rfloor$ the subtour elimination constraints (6.2) induce distinct facets of Q^n.*

The constraints (6.2) for S and $V \setminus S$ induce the same facet of Q^n. Thus we only include these inequalities for $|S| \leq \lfloor n/2 \rfloor$. Also, none of these inequalities induce trivial facets. Taking $|S| = 2$, we obtain the edge upper bounds $x_j \leq 1$ for all $j \in E$.

There is another equivalent version of the inequality (6.2):

(6.3) $$\Sigma(x_j : j \in \delta(S)) \geq 2$$

where $\delta(S)$ is the set of edges with exactly one end in S. (To see this equivalence, multiply each of the equations (6.1) for $i \in S$ by $-1/2$ and add to (6.2).) This is called the *cut form* of the inequality.

A third class of facet inducing inequalities is given by the so called *combs*. A comb is a set of sets $H, T_1, T_2, \ldots, T_k \subseteq V$ such that

$$|H \cap T_i| \geq 1 \quad \text{for } i = 1, 2, \ldots, k;$$
$$|T_i \setminus H| \geq 1 \quad \text{for } i = 1, 2, \ldots, k;$$
$$T_i \cap T_j = \emptyset \quad 1 \leq i < j \leq k;$$
$$k \geq 3 \text{ and odd}.$$

We call H the *handle* of the comb and $\{T_i : i = 1, 2, \ldots, k\}$ the *teeth* (See Fig. (6.1)). The comb *inequality* is

$$(6.4) \quad \Sigma(x_j : j \in \gamma(H)) + \sum_{i=1}^{k} \Sigma(x_j : j \in \gamma(T_i)) \leq |H| + \sum_{i=1}^{k} (|T_i| - 1) - \frac{k+1}{2}.$$

Combs were introduced by Chvátal (1973) who also required $|H \cap T_i| = 1$ for $i = 1, 2, \ldots, k$. We call such combs *simple*. The above generalization is due to Grötschel and Padberg (1979). Note that if, for some i, $|H \cap T_i| \geq 2$, then all edges in $\gamma(H \cap T_i)$ will have coefficient 2 in (6.4). Otherwise, and in particular for simple combs, these are $0-1$ inequalities.

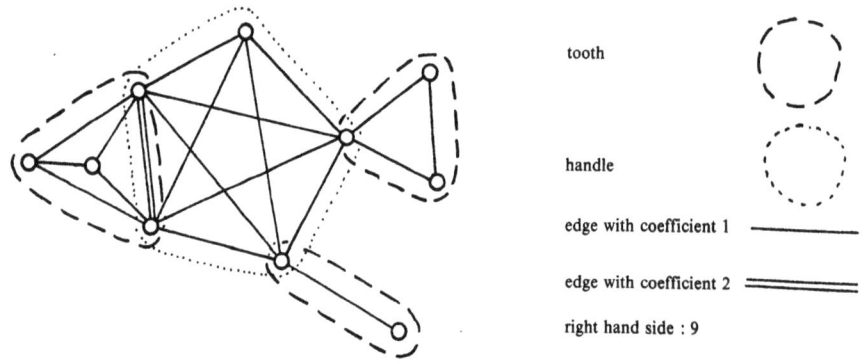

Figure 6.1. A comb

Theorem 6.5 (Grötschel, Padberg (1979a)). *Every comb inequality induces a facet of Q^n.*

If $\{H, T_1, T_2, \ldots, T_k\}$ is a comb, then so too is $\{V \setminus H, T_1, T_2, \ldots, T_k\}$ and these induce the same facet of Q^n. Grötschel (1977) showed that in every other case, distinct comb inequalities induce distinct facets of Q^n, and in no case are these facets induced by subtour elimination constraints or trivial inequalities. (See also Grötschel, Pulleyblank (1981).)

The final class of facet inducing equalities we describe properly contains the previous two classes. A *clique tree* is a set $\{H_1, H_2, \ldots, H_s, T_1, T_2, \ldots, T_t\}$ of subsets of V satisfying:
i) $T_i \cap T_j = \emptyset$ for $1 \leq i < j \leq t$;
ii) $H_i \cap H_j = \emptyset$ for $1 \leq i < j \leq s$;
iii) for each $i \in \{1, 2, \ldots, t\}$, $2 \leq |T_i| \leq n - 2$ and some $v \in T_i$ belongs to no H_j, for $1 \leq j \leq s$;

II. Polyhedral Combinatorics

iv) for each $j \in \{1, 2, \ldots, s\}$, the number of T_i having nonempty intersection with H_j is odd and at least three;

v) for $i \in \{1, 2, \ldots, s\}$ and $j \in \{1, 2, \ldots, t\}$, if $H_i \cap T_j \neq \emptyset$ then $H_i \cap T_j$ is an articulation set of the subgraph C of K_n with nodeset
$$\bigcup_{i=1}^{s} H_i \cup \bigcup_{i=1}^{t} T_i \text{ and edge set } \bigcup_{i=1}^{s} \gamma(H_i) \cup \bigcup_{i=1}^{t} \gamma(T_i);$$
moreover, C is connected.

(See Figure 6.2).

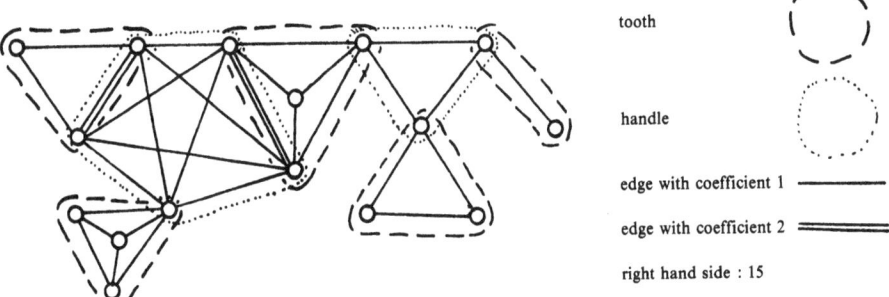

Figure 6.2. A clique tree

A comb is a clique tree with $s = 1$, a single clique W of K_n satisfying $2 \leq |W| \leq n - 2$ is a clique tree with $s = 0$ and $t = 1$.

It can be verified (Grötschel, Pulleyblank (1981)) that the *clique tree inequality*

$$(6.5) \quad \sum_{i=1}^{s} (x_j : j \in \gamma(H_i)) + \sum_{i=1}^{t} (x_j : j \in \gamma(T_i)) \leq \sum_{i=1}^{s} |H_i| + \sum_{i=1}^{t} (|T_i| - t_i) - \frac{t+1}{2}$$

is valid for Q^n, where for each $i \in \{1, 2, \ldots, t\}$, t_i is the number of handles which intersect T_i.

Theorem 6.6 (Grötschel and Pulleyblank (1981)). *Each clique tree inequality is facet inducing for Q^n. Moreover, distinct clique trees induce distinct facets, with the exception of the comb and subtour elimination cases already described.*

There are further classes of facets for Q^n which are related to hypohamiltonian or hypotraceable subgraphs of K_n. (See Maurras (1976), Grötschel (1980), Cornuéjols and Pulleyblank (1982).)

Consequently we know that the set of clique tree inequalities does not induce all facets of Q^n. Nonetheless, inequalities from this class have proved to be sufficient for solving real world problems. The general approach has been to use a commerical linear programming code and start with the constraints

(6.6)
$$0 \leq x_j \leq 1 \text{ for all } j \in E,$$
$$\Sigma(x_j : j \in \delta(v)) = 2 \text{ for all } v \in V.$$

The linear program of minimizing cx subject to (6.6) is solved and if an optimum solution x^* which is not the incidence vectors of a tour is obtained, then we try to find a constraint (6.2) or (6.4) violated by x^*, add it to the linear system (6.6), and repeat.

The first successful application of this approach was by Dantzig, Fulkerson and Johnson (1954). They constructed a 49 city problem by taking road distances between all pairs of the 48 U.S. state capitals (at that time) and Washington D.C. They used ad hoc arguments to reduce the number of nodes of the problem to 42 then solved it to optimality. In fact they were able to prove optimality of their solution using dual variables corresponding to only seven subtour elimination constraints (6.2), one simple comb inequality (6.4) and one so-called chain constraint (see Padberg and Hong (1980)).

This approach seems to have been then largely ignored until the 1970's when S. Hong (1972) automated some parts of the Dantzig, Fulkerson, Johnson procedure. Grötschel (1980a) constructed a 120 city problem by considering inter-city distances and solved it to optimality using this approach. He required thirteen LP runs and added a total of 36 subtour elimination constraints (6.2) and 60 comb inequalities (6.4). Of these 60 combs, 21 were non-simple.

The difficulty in applying this procedure lies in solving the separation problem for the polyhedron defined by (6.6), (6.2) and (6.4). Grötschel used entirely ad hoc methods in his solution of the 120-city problem. Padberg and Hong (1980) attempted to automate this procedure, and described remarkable success on a large number of problems. Even when the optimum LP solution at hand x^* was not the incidence vector of a tour and no separating hyperplane could be found, the resulting lower bounds on the tour were so close to the optimum that only a small number of branch and bound iterations were required.

At present, the largest problem solved to optimality using this approach was a 318 city problem (Crowder and Padberg (1980)). This was a problem introduced by Lin and Kernighan (1973) which arose in the routing of a numerically controlled laser drilling machine.

Another technique which can be used for introducing cuts to a problem is by means of Lagrange multipliers. Balas and Christofides (1981) reported successful application of an approach based on a combination of this method with branch and bound for asymmetric travelling salesman problems of up to 325 cities. (The asymmetric problem is normally posed for a complete directed graph, in which arcs (u, v) and (v, u) can have different costs.)

The separation problem for subtour elimination constraints can be solved using a polynomially bounded algorithm of Gomory and Hu (1961) (see also Hu (1969)) for finding the minimum cost cut in a graph. The idea is to assign the values x_j^* of the current solution to the edges as costs, then find a minimum cost cut. If this cut has cost at least two, then every constraint (6.3), and hence (6.2) is satisfied and if not we have found a most violated such constraint. Padberg and Rao (1982) showed that it was possible to modify the Gomory-Hu procedure to polynomially solve the separation problem for b-matching problems and for capacitated b-matching problems in which each edge j has a capacity α_j on the maximum allowable value x_j. Edmonds and Johnson (1970) had described a sufficient linear system for this problem. What is of particular interest in this latter case is that if all b_i's are taken to be two and all α_j's are equal to one, then the separation problem is equivalent to the separation problem for the travelling salesman problem over the class of all those comb inequalities for which all teeth are of cardinality two.

Certainly one of the outstanding open problems concerning the polyhedral approach to the travelling salesman problem is to find other classes of combs and/or clique trees for which the separation problem can be polynomially solved.

A question that arises naturally for any problem for which a polynomially bounded optimization algorithm is known is how to solve the associated separation problem. Of course this can be solved using the ellipsoid algorithm, as mentioned at the beginning of this section, however this approach is only of theoretical interest at present. The time required by this method, given the present state of the art for implementing ellipsoid algorithms, is polynomially bounded but so long as to be "practically" useless. However the fact that polynomial separation algorithms exist can provide a strong inducement for finding efficient, direct algorithms. For example, Cunningham (1981) gave a direct polynomially bounded separation algorithm for matroid polyhedra. (Note that separation for matroid intersection simply involves at most two applications of the matroid separation algorithm). The matroid separation algorithm, although polynomially bounded, is certainly more complex than the greedy algorithm for matroid optimization described in Section 3, but is considerably more efficient than combining the ellipsoid and greedy algorithms.

Finally, we describe one application of the equivalence between separation and optimization given in Grötschel, Lovász, Schrijver (1981). It states that whenever we can solve a combinatorial optimization problem, we "implicitly" know a complete linear description of the associated polytope.

Theorem 6.7 (Grötschel, Lovász, Schrijver (1981)). *Suppose we can polynomially optimize a linear objective function over a class $\mathcal{H}$ of rational polytopes. Then there exists a polynomially bounded algorithm which, given a polytope $P \in \mathcal{H}$, some $a_0 \in P$ and an integer objective function $c \in \mathbb{R}^n$, provides facet inducing inequalities $a_i^T x \leq b_i$ ($i = 1, 2, \ldots, n$) and rationals $\lambda_i \geq 0$ ($i = 1, 2, \ldots, n$) such that $\Sigma(\lambda_i a_i : i = 1, \ldots, n) = c$ and $\Sigma(\lambda_i b_i : i = 1, \ldots, n) = \max\{c^T x : x \in P\}$.*

For almost all classes of polyhedra for which we can solve the associated optimization problem we already explicitly know a defining linear system. Probably the outstanding case for which this is not true is that of stable sets in claw-free graphs. A *stable set* in $G = (V, E)$ is a set $S \subseteq V$ such that $\gamma(S) = \emptyset$. A graph is said to be *claw-free* if there is no stable set of size three (or more) all of whose members are adjacent to some single node. Minty (1980) (see also Sbihi (1978), (1980)) described an algorithm for finding a maximum weight stable set in a claw-free graph. However no sufficient linear system for the associated polytope is known. Giles and Trotter (1981) have shown that such a system will contain some inequalities having a rather complicated form.

7. Adjacency and Dimension

In this last section we discuss two other polyhedral concepts which do have combinatorial uses. The first of these is *adjacency*. The *graph*, or skeleton, $G(P)$ of a polyhedron P has a node for each vertex of P, and nodes u, v of $G(P)$ are adjacent if and only if the corresponding vertices of P belong to an edge. Recently there has been some interest in giving characterizations of adjacent pairs of vertices of a combinatorial polyhedron in terms of the properties of the combinatorial objects themselves. For example, we have the following characterization of adjacency on the 1-matching polytope $\mathcal{M}(G)$. The *symmetric difference* of sets $S, T \subseteq E$, denoted by $S \triangle T$, is defined to be $(S \setminus T) \cup (T \setminus S)$.

Theorem 7.1 (Chvátal (1975)). *Let v_1 and v_2 be vertices of $\mathcal{M}(G)$ and let M_1 and M_2 be the corresponding matchings. Then v_1 and v_2 are adjacent on $\mathcal{M}(G)$ if and only if $M_1 \triangle M_2$ is connected, i.e., either a single even cycle or single simple path.*

Adjacency on matroid polyhedra also has a well known combinatorial interpretation.

Theorem 7.2 (Edmonds, Hausmann and Korte (1978)). *Let I_1 and I_2 be independent sets of a matroid $M = (E, \mathcal{F})$. Then the vertices of $\mathcal{P}(M)$ corresponding to I_1 and I_2 are adjacent if and only if $|I_1 \triangle I_2| = 1$ or $|I_1 \triangle I_2| = 2$ and $I_1 \cup I_2 \notin \mathcal{F}$.*

A comprehensive survey of adjacency of combinatorial polyhedra is given by Hausmann (1980). See also Hausmann and Korte (1978).

It is much more difficult to prove that vertices v_1 and v_2 of a polyhedron P are adjacent than it is to prove that they are not. For we can prove non-adjacency by expressing some point on the line joining v_1 and v_2 as a convex combination of other vertices of P. In order to prove adjacency, we must show that this is not possible. Hausmann and Korte (1978) introduced a so-called "colouring criterion" which can be applied to any polyhedron $P \subseteq \mathbb{R}^J$ with $0-1$ vertices. In order to determine whether or not vertices v_1, v_2 are adjacent, we first colour each index $j \in J$ a different colour, then we systematically combine colours, following rules which depend on v_1 and v_2 as well as the structure of the polyhedron. If, at the end of this process, all $j \in J$ have the same colour then v_1 and v_2 are adjacent. If not, then in general v_1 and v_2 may or may not be adjacent. However they also showed that for many classes of combinatorial polyhedra, this colouring criterion is also sufficient, i.e., if it terminates with more than one colour then it can be concluded that v_1 and v_2 are not adjacent. See Hausmann and Korte (1978) or Hausmann (1980).

In many cases one can deduce certain properties of the graph $G(P)$ of the polyhedron P from a knowledge of the adjacency criterion. For example, in Naddef and Pulleyblank (1981) we defined a $0-1$ polyhedron to be *combinatorial* if vertices u and v are adjacent if and only if there exist no other pair of vertices x and y such that $u + v = x + y$. (Hausmann (1980) refers to this as the intersection-union property.) Matching polyhedra, matroid polyhedra and a great many other classes of polyhedra arising from combinatorial problems satisfy this property.

II. Polyhedral Combinatorics

Theorem 7.3 (Naddef and Pulleyblank (1981)). *If P is a combinatorial polyhedron, then either $G(P)$ is isomorphic to a hypercube or else it is hamilton connected (each pair of nodes is joined by a hamilton path).*

This implies, for example, that in a graph G which does not just consist of disjoint even cycles, it is possible to choose any two perfect matchings M and M' and then arrange all the perfect matchings of G into a sequence $M = M_1, M_2, M_3, \ldots, M_r = M'$ such that, for each $i = 1, 2, \ldots, r-1$, $M_i \triangle M_{i+1}$ is a single even cycle.

In some cases, it is not easy to decide whether or not two vertices of a polyhedron are adjacent.

Theorem 7.4 (Papadimitriou (1978)). *It is NP-complete to decide whether two given vertices v_1, v_2 of the travelling salesman polytope Q^n are adjacent.*

(This is also shown to hold for the asymmetric travelling salesman polytope.)

Note that it is not true that adjacency testing is necessarily hard for NP-complete problems. Chvátal (1975) showed that two vertices of the convex hull of the incidence vectors of the stable sets of a graph are adjacent if and only if the symmetric difference of the corresponding stable sets induces a connected subgraph.

It has been suggested that an adjacency criterion for a class of polytopes could provide a basis for an efficient algorithm which used some sort of local search technique. This is probably motivated in part by the operation of the simplex algorithm for linear programming, which passes from a vertex to an adjacent vertex on every *nondegenerate* pivot. In fact Trubin (1969) proved that two vertices of the convex hull of the incidence vectors of the stable sets of a graph were adjacent if and only if the corresponding vertices were adjacent on the polyhedron defined by a certain linear relaxation. He then suggested using the simplex algorithm on the relaxation, only restricting oneself to pivots that yield new integer solutions. Balas and Padberg (1972) observed that the problem with the approach is the high degree of degeneracy of the problem, a very common situation for combinatorial problems. That is, there exists a large set of feasible bases for each basic feasible solution (i.e. vertex) and it is not true that arbitrary bases corresponding to adjacent vertices will differ by a single simplex pivot. Thus the problem is how to find a "short" path through the set of feasible bases corresponding to a vertex, to enable one to make a nondegenerate pivot. They also showed that there did exist a "short" sequence of pivots to an optimum solution which satisfied certain properties; however it was not possible in general to produce these pivots without knowing the optimum in advance. (See also Ikura and Nemhauser (1981).) Cunningham (1979) showed that when the simplex algorithm is applied to network flow problems, exponentially long sequences of degenerate pivots (called "stalling") can be avoided if a certain pivot rule is chosen.

The *diameter* of a polyhedron P is the maximum length of a shortest path between any pair of nodes of $G(P)$. Padberg and Rao (1974) showed that the diameter of the asymmetric travelling salesman polytope is two. Thus any tour is either adjacent to the optimum tour or else there is some third tour adjacent to both. This makes a local search technique look rather attractive, but in view of Theorem 7.4 (for the asymmetric case), just seeing whether or not two given tours corresponding to adjacent vertices is as hard as the original problem. Moreover, once again the high degree of degeneracy would make it very difficult to carry out some sort of simplex pivot in a reasonable fashion, even if we did know a complete linear characterization.

Many of the polyhedra considered in polyhedral combinatorics are of full dimension, however for others the calculation of the dimension, or more precisely, a basis for the equality set (see Theorem 2.6) is an essential first step in obtaining a (minimal) defining linear system. Edmonds, Lovász, Pulleyblank (1982) (see also Padberg and Rao (1981)) show that for any $S \subseteq \mathbb{R}^n$, we can compute the dimension of S by solving $n+1$ linear optimization problems over S. In fact we also produce a set of $\dim(S)+1$ affinely independent members of S and a set of $n-\dim(S)$ linearly independent equations satisfied by all members of S.

Naddef (1982) considered the problem of computing the dimension of $\mathcal{M}(G)$, the perfect matching polytope, of an arbitrary graph G and proved a good characterization of this value. This was further developed in Edmonds, Lovász, Pulleyblank (1982) where it was shown that this dimension is directly related to the number of components of a certain decomposition of the graph. An interesting consequence of this result is that the formula for the dimension settled a conjecture of Lovász and Plummer (1975) on a lower bound on the number of perfect matchings in a bicritical graph. (A graph $G=(V,E)$ is bicritical if for every $x, y \in V$, $G\setminus\{x,y\}$ has a perfect 1-matching.) Thus at present the best known lower bound on the number of perfect matchings of such graphs is in fact a lower bound on the number of perfect matchings whose incidence vectors are affinely independent.

Acknowledgements

This paper has benefited from comments and suggestions of many persons, in particular, Egon Balas, Alan Hoffman, Lex Schrijver and Les Trotter. Moreover, I had the opportunity to participate in a series of four lectures on integrality of linear programs given by Jack Edmonds and Rick Giles in Waterloo in 1982, which influenced Section 4. I am grateful to Bill Cook, Gerd Reinelt and Mike Junger for careful reading of a preliminary version.

References

(1982) A. Bachem and M. Grötschel, "New aspects of polyhedral theory", in B. Korte (ed.) *Modern Applied Mathematics - Optimization and Operations Research*, North Holland (1982) 51-106.

(1979) A. Bachem and R. Kannan, "Applications of polynomial Smith normal form calculations" in L. Collatz et al. (eds.) *Numerische Methoden bei graphentheoretischen und kombinatorischen Problemen*, Band 2. Birkhäuser Verlag Basel (1979) 195-215.

II. Polyhedral Combinatorics

(1979) A. Bachem and R. von Randow, "Integer theorems of Farkas lemma type", *Operations Verfahren Vol. 32* (1979) 19-28.

(1981) E. Balas and N. Christofides, "A restricted Lagrangian approach to the travelling salesman problem", *Mathematical Programming* 21 (1981) 19-46.

(1972) E. Balas and M. Padberg, "On the set covering problem", *Operations Research* 20 (1972) 1152-1161.

(1972) M. Balinski, "Establishing the matching polytope", *Journal of Combinatorial Theory* B 13 (1972) 1-13.

(1958) C. Berge, "Sur le couplage maximum d'un graphe", *C. R. Académie des Sciences, Paris* 247 (1958) 258-259.

(1946) G. Birkhoff, "Observaciones sobre el algebra lineal", *Rev. Univ. Nac. Turcuman* (Ser. A) 5 (1946) 147-148.

(1981) R. Chandresekaran, "Polynomial algorithms for totally dual integral systems and extensions", in P. Hansen (ed.) *Studies on Graphs and Discrete Programming, Annals of Discrete Math.* 11 (1981) 39-51.

(1973) V. Chvátal, "Edmonds polytopes and weakly hamiltonian graphs", *Mathematical Programming* 5 (1973) 29-40.

(1975) V. Chvátal, "On certain polytopes associated with graphs", *Journal of Combinatorial Theory* B 18 (1975) 138-154.

(1981) W. Cook, "A minimal totally dual integral defining system for the b-matching polyhedron", Research report CORR 81-15, University of Waterloo (1981), to appear in *SIAM Journal on Algebraic and Discrete Methods*.

(1982) W. Cook, "Operations that preserve total dual integrality", Research report CORR 82-34, University of Waterloo (1982), to appear in *Operations Research Letters*.

(1982) G. Cornuéjols and W. R. Pulleyblank, "The travelling salesman polytope and {0, 2}-matching", *Annals of Discrete Mathematics* 16 (1982) 27-55.

(1980) H. P. Crowder and M. W. Padberg, "Solving large scale symmetric travelling salesman problems to optimality", *Management Science* 26 (1980) 495-509.

(1979) W. H. Cunningham, "Theoretical properties of the network simplex method", *Mathematics of Operations Research* 4 (1979) 196-208.

(1982) W. H. Cunningham, "Testing membership in matroid polyhedra", Research Report WP81207-OR, Inst. für Operations Research, Universität Bonn.

(1978) W. H. Cunningham and A. B. Marsh III, "A primal algorithm for optimum matching", *Mathematical Programming Study* 8 (1978) 50-72.

(1954) G. B. Dantzig, D. R. Fulkerson and S. M. Johnson, "Solution of a large-scale travelling salesman problem", *Operations Research* 2 (1954) 393-410.

(1965) J. Edmonds, "Paths, trees and flowers", *Canadian Journal of Mathematics* 17 (1965) 449-469.

(1965 a) J. Edmonds, "Maximum matching and a polyhedron with 0-1 vertices", *Journal of Research of the National Bureau of Standards* 69 B (1965) 125-130.

(1968) J. Edmonds, "Optimum branchings", in G. B. Dantzig and P. Veinott (eds.) *Mathematics of the Decision Sciences*, AMS (1968) 346-361, reprinted from *Journal of Research of the National Bureau of Standards* 71 B (1967) 233-240.

(1971) J. Edmonds, "Matroids and the greedy algorithm", *Mathematical Programming* 1 (1971) 127-136.

(1979) J. Edmonds, "Matroid intersection", *Annals of Discrete Mathematics* 4 (1979) 39-49.

(1977) J. Edmonds and R. Giles, "A min-max relation for submodular functions on graphs", *Annals of Discrete Mathematics* 1 (1977) 185-204.

(1982) J. Edmonds and R. Gilles, "Total integrality of linear inequality systems", *Proceedings of Silver Jubilee Conference on Combinatorics*, University of Waterloo (1982).

(1970) J. Edmonds and E. L. Johnson, "Matching: a well-solved class of integer linear programs" in R. K. Guy et al. (eds.) *Combinatorial Structures and their Applications*, Gordon and Breach, New York (1970), 89-92.

(1982) J. Edmonds, L. Lovász and W. R. Pulleyblank, "Brick decompositions and the matching rank of graphs", Research Report CORR 82-17, University of Waterloo (1982) to appear in *Combinatorica*.

(1956) P. Elias, A. Feinstein and C. E. Shannon, "A note on the maximum flow through a network", *IRE Transactions on Information Theory* IT2 (1956) 117-119.

(1956) L. R. Ford and D. R. Fulkerson, "Maximum flow through a network", *Canadian Journal of Mathematics* 8 (1956) 399-404.

(1971) D. R. Fulkerson, "Blocking and anti-blocking pairs of polyhedra", *Mathematical Programming* 1 (1971) 168-194.

(1968) D. Gale, "Optimal assignments in an ordered set: an application of matroid theory", *Journal of Combinatorial Theory* 4 (1968) 176-180.

(1979) M. R. Garey and D. S. Johnson, *Computers and Intractability, a Guide to the Theory of NP-Completeness*, W. H. Freeman & Co., San Francisco (1979).

(1975) R. Giles, *Submodular functions, graphs and integer polyhedra*, Ph.D. Thesis, University of Waterloo (1975).

(1976) R. Giles, "Facets and other faces of branching polyhedra", *Colloq. Math. Soc. János Bolyai* 18 Combinatorics (1976) 401-418.

(1982) R. Giles, "Optimum matching forests I: Special weights", *Mathematical Programming* 22 (1982) 1-11.

(1982a) R. Giles, "Optimum matching forests II: General weights", *Mathematical Programming* 22 (1982) 12-38.

(1982b) R. Giles, "Optimum matching forests III: Facets of matching forest polyhedra", *Mathematical Programming* 22 (1982) 39-51.

(1979) R. Giles and W. R. Pulleyblank, "Total dual integrality and integer polyhedra", *Linear Algebra and Its Applications* 25 (1979) 191-196.

(1981) R. Giles and L. Trotter, "On stable set polyhedron for $K_{1,3}$-free graphs", *Journal of Combinatorial Theory* B31 (1981) 313-326.

(1956) A. J. Goldman, "Resolution and separation theorems for polyhedral convex sets", in H. W. Kuhn and A. W. Tucker eds., *Linear Inequalities and Related Systems*, Annals of Mathematics Study 38 (1956) 41-51.

(1961) R. E. Gomory and T. C. Hu, "Multiterminal network flows", *Journal of S.I.A.M.* 9 (1961) 551-570.

(1980) J. J. Green-Krótki, *Matching polyhedra*, M.Sc. Thesis, Carleton University (1980).

(1977) M. Grötschel, *"Polyedrische Charakterisierungen kombinatorischer Optimierungsprobleme"*, Dissertation, Universität Bonn (1977), Verlag A. Hain, Meisenheim (1977).

(1980) M. Grötschel, "On the monotone symmetric travelling salesman problem: hypohamiltonian/hypotraceable graphs and facets", *Mathematics of Operations Research* 5 (1980) 285-292.

(1980a) M. Grötschel, "On the symmetric travelling salesman problem: solution of a 120 city problem, *Mathematical Programming Study* 12 (1980) 61-77.

(1981) M. Grötschel, L. Lovász, A. Schrijver, "The ellipsoid method and its consequences in combinatorial optimization", *Combinatorica* 1 (1981) 70-89.

II. Polyhedral Combinatorics

(1979)   M. Grötschel and M. Padberg, "On the symmetric travelling salesman problem I: inequalities", *Mathematical Programming* 16 (1979) 265-280.

(1979a)   M. Grötschel and M. Padberg, "On the symmetric travelling salesman problem II: lifting theorems and facets", *Mathematical Programming* 16 (1979) 281-302.

(1982)   M. Grötschel and M. Padberg, "Polyhedral aspects of the travelling salesman problem I: Theory", Research Report No. 82212, Institut für Operations Research, Universität Bonn (1982) to appear in E. Lawler, J. Lenstra, A. Rinnooy Kan (eds.) *The Travelling Salesman Problem,* Wiley (1983).

(1982a)   M. Grötschel and M. Padberg, "Polyhedral aspects of the travelling salesman problem II: Computation", Research Report No. 82214, Institut für Operations Research, Universität Bonn (1982) to appear in E. Lawler, J. Lenstra, A. Rinnooy Kan (eds.) *The Travelling Salesman Problem,* Wiley (1983).

(1981)   M. Grötschel and W. R. Pulleyblank, "Clique tree inequalities and the travelling salesman problem", Research Report 81196-OR, Institut für Operations Research, Universität Bonn (1981).

(1935)   P. Hall, "On representations of subsets", *Journal of the London Mathematical Society* 10 (1935) 26-30.

(1980)   D. Hausmann, *Adjacency on Polytopes in Combinatorial Optimization,* Anton Hain, Meisenheim (1980).

(1978)   D. Hausmann and B. Korte, "Colouring criteria for adjacency on 0-1 polyhedra", *Mathematical Programming Study* 8 (1978) 106-127.

(1974)   A. Hoffman, "A generalization of max flow-min cut", *Mathematical Programming* 6 (1974) 352-359.

(1982)   A. Hoffman, "Ordered sets and linear programming", in I. Rival (ed.), *Ordered Sets,* D. Reidel, Dordrecht, Holland (1982) 619-654.

(1956)   A. Hoffman and J. Kruskal, "Integral boundary points of convex polyhedra", in H. Kuhn and A. Tucker (eds.), *Linear Inequalities and Related Systems, Annals of Mathematics Studies* 38 (1956) 233-246.

(1972)   S. Hong, *A linear programming approach for the travelling salesman problem,* Ph. D. Thesis, Johns Hopkins University (1972).

(1969)   T. C. Hu, *Integer Programming and Network Flows,* Addison Wesley, Reading (1969).

(1981)   Y. Ikura and G. Nemhauser, "Simplex pivots on the set packing polytope", Tech. Report 513, S.O.R.I.E., Cornell University (1981).

(1965)   E. Johnson, *Programming in networks and graphs,* Ph. D. thesis, Operations Research Center, University of California Berkley (1965).

(1972)   R. Karp, "Reducibility among combinatorial problems", in R. Miller and J. Thatcher (eds.) *Complexity of Computer Computations,* Plenum Press, New York (1972) 85-103.

(1980)   R. Karp and C. Papadimitriou, "On linear characterizations of combinatorial optimization problems", *Proceedings of the Twenty-first Annual Symposium on the Foundations of Computer Science,* IEEE (1980), 1-9.

(1931)   D. König, "Graphs and matrices" (Hungarian) *Mat. Fiz. Lapok* 38 (1931) 116-119.

(1899)   L. Kronecker, "Näherungsweise ganzzahlige Auflösung linearer Gleichungen" in K. Hensel (ed.) *Leopold Kroneckers Werke* Band III, Teubner Leipzig (1899) 47-110.

(1976)   E. Lawler, *Combinatorial Optimization: Networks and Matroids,* Holt Rinehart and Winston, New York (1976).

(1979) A. Lehman, "On the width-length inequality", *Mathematical Programming* 17 (1979) 403-417.

(1973) S. Lin and B. Kernighan, "An effective heuristic algorithm for the travelling-salesman problem", *Operations Research* 21 (1973) 498-516.

(1979) L. Lovász, "Graph theory and integer programming", *Annals of Discrete Mathematics* 4 (1979) 141-158.

(1975) L. Lovász and M. Plummer, "On bicritical graphs", in A. Hajnal et al. (eds.) *Infinite and Finite Sets (Colloqu. Math. Soc. J. Bolyai* 10 (1975) 1051-1079.

(1976) J. Maurras, *Polytopes à sommets dans* $\{0,1\}^n$, Dissertation, Univ. Paris VII (1976).

(1980) G. Minty, "On maximal independent sets of vertices in claw-free graphs", *Journal of Combinatorial Theory* B (1980) 284-304.

(1936) T. S. Motzkin, "Beiträge zur Theorie der linearen Ungleichungen", Dissertation, Basel (1933) Jerusalem (1936).

(1982) D. Naddef, "Rank of maximum matchings in a graph", *Mathematical Programming* 22 (1982) 52-70.

(1981) D. Naddef and W. R. Pulleyblank, "Hamiltonicity and combinatorial polyhedra", *Journal of Combinatorial Theory* B (1981) 297-312.

(1953) J. von Neumann, "A certain zero-sum two person game equivalent to the optimum assignment problem", in A. Tucker and H. Kuhn (eds.), *Contributions to the Theory of Games II, Annals of Mathematics Study,* 38, Princeton University Press (1953) 5-12.

(1975) M. Padberg, "Characterizations of totally unimodular, balanced and perfect matrices", in B. Roy (ed.) *Combinatorial Programming: Methods and Applications,* Reidel, Boston (1975) 275-284.

(1980) M. Padberg and S. Hong, "On the symmetric travelling salesman problem: a computational study", *Mathematical Programming Study* 12 (1980) 78-107.

(1974) M. Padberg and M. R. Rao, "The travelling salesman problem and a class of polyhedra of diameter two", *Mathematical Programming* 7 (1974) 32-45.

(1981) M. Padberg and M. R. Rao, "On the Russian method for linear inequalities III: Bounded integer programs", Research Report, Inria, Paris (1981) to appear in *Mathematical Programming Studies.*

(1982) M. Padberg and M. R. Rao, "Odd minimum cuts and b-matching", *Mathematics of Operations Research* 7 (1982) 67-80.

(1978) C. Papadimitriou, "The adjacency relation on the travelling salesman polytope is NP-complete", *Mathematical Programming* 14 (1978) 312-324.

(1982) C. Papadimitiou, "Polytopes and complexity", *Proceedings of the Silver Jubilee Conference on Combinatorics,* University of Waterloo (1982).

(1973) W. R. Pulleyblank, *Faces of Matching Polyhedra,* Ph.D. Thesis, University of Waterloo (1973).

(1974) W. R. Pulleyblank and J. Edmonds, "Facets of 1-matching polyhedra" in C. Berge and D. Ray-Chaudhuri (eds.), *Hypergraph Seminar,* Springer-Verlag, Berlin (1974) 214-242.

(1980) W. R. Pulleyblank, "Dual integrality in b-matching problems", *Mathematical Programming Study* 12 (1980) 176-196.

(1981) W. R. Pulleyblank, "Total dual integrality and b-matchings", *Operations Research Letters* 1 (1981) 28-30.

(1943) R. Rado, "Theorems on linear combinatorial topology and general measure", *Annals of Mathematics.* 44 (1943) 228-270.

(1957) R. Rado, "Note on independence functions", *Proceedings of the London Mathematical Society* 7 (1957) 300-320.

II. Polyhedral Combinatorics

(1970) T. Rockafellar, *Convex Analysis*, Princeton University Press (1970).
(1978) N. Sbihi, *Etude des Stables dans les Graphes sans Etoile*, Doctoral disseration, University of Grenoble (1978).
(1980) N. Sbihi, "Algorithme de recherche d'un stable de cardinalité maximum dans un graphe sans étoile", *Discrete Mathematics* 29 (1980) 53–76.
(1981) A. Schrijver, "Short proofs on the matching polytope", Rapport AE17/81, Inst. Act. & Econ, Univ. van Amsterdam (1981), to appear in *Journal of Combinatorial Theory* B.
(1981a) A. Schrijver, "Theory of Integer Linear Programming", preprint.
(1981b) A. Schrijver, "On total dual integrality", *Linear Algebra and its Applications* 38 (1981) 27–32.
(1983) A. Schrijver, "Min-max Results in Combinatorial Optimization", this volume.
(1979) P. Seymour, "On multi-colourings of cubic graphs and conjectures of Fulkerson and Tutte", *Proceedings of the London Mathematical Society* 38 (1979) 423–460.
(1980) P. D. Seymour, "Decomposition of regular matroids", *Journal of Combinatorial Theory* B (1980) 305–359.
(1970) J. Stoer and C. Witzgall, *Convexity and Optimization in Finite Dimensions*, Springer, Berlin (1970).
(1969) V. Trubin, "On a method of solution of integer linear programming problems of a special kind", *Soviet Mathematics Doklody* 10 (1969) 1544–1546.
(1952) W. T. Tutte, "The factors of graphs", *Canadian Journal of Mathematics* 4 (1952) 314–328.
(1965) W. T. Tutte, "Lectures on matroids", *Journal of Research of the National Bureau of Standards* 69 B, 1–47.
(1937) B. L. van der Waerden, *Moderne Algebra* (2nd ed.) Springer, Berlin (1937).
(1968) D. Welsh, "Kruskal's theorem for matroids", *Proceedings of the Cambridge Philosophical Society* 64 (1968) 3–4.
(1976) D. Welsh, *Matroid Theory*, Academic Press, London (1976).
(1935) H. Weyl, "Elementare Theorie der konvexen Polyeder", *Commentarii Mathematici Helvetici* 7 (1935) 290–306.
(1935) H. Whitney, "On the abstract properties of linear dependence", *American Journal of Mathematics* 57 (1935) 509–533.

Generalized Equations

Stephen M. Robinson*
University of Wisconsin-Madison, Dept. of Industrial Engineering, 1513 University Avenue, Madison, WI 53706, USA

Abstract. The term "generalized equation" has recently been used to describe certain kinds of inclusions that involve multivalued functions, particularly normal-cone operators. Such problems include static generalized equations, which extend ordinary nonlinear equations, as well as generalized differential equations, which extend ordinary differential equations to the situation in which the defining relation contains multivalued functions (again, particularly normal-cone operators).

This survey gives an overview of what is known about these problems as of late 1982. Because of space limitations, it has not been practical to present an encyclopedic description of all recent work. Rather, this paper tries to exhibit samples of recent results in several different areas, and to lead the reader to works in the literature that explain those results in more detail and that contain references to other work not mentioned here. The choice of samples to be presented was based on the author's particular knowledge and interests.

1. What are Generalized Equations?

The term "generalized equation" has recently come to be used to describe certain types of relations that are similar to equations except that one side of the relation is multivalued; more specifically, the multivalued expression often involves a normal-cone operator. In this section we will define the terms needed to describe generalized equations and we will give some examples.

First, suppose that C is a closed convex set in $\mathbb{R}^n$. The *normal cone* to C at a point $x \in \mathbb{R}^n$ is defined to be:

$$\partial \psi_C(x) := \begin{cases} \emptyset, & \text{if } x \notin C \\ \{y | \langle y, c-x \rangle \leq 0 \text{ for each } c \in C\}, & \text{if } x \in C. \end{cases}$$

It is easy to see that $\partial \psi_C(x)$ is a closed convex cone; geometrically it is the cone of all outward normals to C at x_0.

The type of generalized equation with which we shall deal in the first part of this paper is of the form

$$0 \in F(x) + \partial \psi_C(x), \tag{1.1}$$

---

* Sponsored by the U.S. National Science Foundation under Grant No. MCS 8200632.

II. Generalized Equations

where F is a function from an open set $\Omega \subset \mathbb{R}^n$ to $\mathbb{R}^n$. In cases where the generalized equation formalism seems particularly useful, F is often a fairly smooth function, and the expression (1.1) is useful in separating the "smooth" part of the problem at hand from the part involving "corners". We shall see some examples of this later on.

To see what (1.1) means in terms of the set C, note that if (1.1) holds then the sum on the right is nonempty (it contains 0), so $\partial \psi_C(x)$ is nonempty, and this means that $x \in C$. Also, $-F(x)$ must belong to $\partial \psi_C(x)$, so for each $c \in C$,

$$\langle -F(x), c-x \rangle \leq 0.$$

Thus we can see that (1.1) will hold if and only if x satisfies the so-called *variational inequality*:

$$x \in C, \text{ and for each } c \in C \langle F(x), c-x \rangle \geq 0, \tag{1.2}$$

and this means geometrically that $F(x)$ is an inward normal to C at x.

One might then ask why we are interested in (1.1), since (1.2) is equivalent to it. The answer is that the form of (1.1) acts as an aid to analyzing problems. It recalls immediately the analogy with ordinary equations (which can be regarded as the special cases of (1.1) in which $C = \mathbb{R}^n$, since then $\partial \psi_C(x) \equiv \{0\}$, and (1.1) holds if and only if $F(x) = 0$). This analogy turns out to be quite helpful in developing results for generalized equations that are extensions of those already known for ordinary equations, such as implicit-function theorems and computational algorithms (e.g. Newton-type methods). We shall see a number of examples of this kind of extension later in the paper.

In the second part of the paper we extend our consideration from static problems to dynamic ones, in which we have some kind of evolution which, if the variable $x(t)$ were unrestricted, would be described by a relation like

$$0 = \dot{x}(t) + F[x(t)] \tag{1.3}$$

(i.e., $-\dot{x}(t) = F[x(t)]$, where $\dot{x}(t)$ denotes $\frac{d}{dt}x(t)$). However, if we suppose that $x(t)$ is to be confined to some closed convex set C, then (1.3) might not be appropriate, since if $x(t)$ were on the boundary of C it might be impossible to satisfy (1.3) while ensuring that $x(t)$ remained in C. In such a situation we can modify (1.3) by requiring that $\dot{x}(t)$ be the projection of $-F[x(t)]$ on the tangent cone to C at $x(t)$. Under appropriate conditions this can be stated equivalently as the requirement that $\dot{x}(t)$ be the smallest element of $-F[x(t)] - \partial \psi_C[x(t)]$; that is,

$$-\dot{x}(t) = \{F[x(t)] + \partial \psi_C[x(t)]\}^+, \tag{1.4}$$

where A^+ denotes the smallest element of the closed convex set A. In fact, by making appropriate extensions of the ideas of tangent and normal cones, we can extend this reasoning to sets C which may not be convex.

Note that if we ask for an "equilibrium" situation in 1.4 (i.e., one in which $\dot{x}(t) = 0$), then we are led to the problem of finding points x which satisfy (1.1). Thus it is appropriate to regard (1.4) as the dynamic extension of (1.1).

We shall return to the dynamic case in Section 6 of this paper. In the meantime, we consider in more detail the static problem (1.1), beginning in the next section with some examples of how (1.1) can be used to express familiar problems in optimization and in mathematical economics.

2. How are Generalized Equations Useful?

We will exhibit in this section some ways in which the "static" generalized equation (1.1) can be used as a unifying device to model relationships found in a number of applications in the areas of optimization, complementarity, and mathematical economics. In keeping with the interpretation of (1.1) as the equilibrium case of (1.4), we shall see that these relationships are typically of the "static equilibrium" type: they express the conditions for optimality in a mathematical programming problem or for some type of equilibrium in a problem from complementarity or from economics.

Let us first consider the mathematical programming problem

$$\begin{aligned} & \text{minimize } f(y) \\ & \text{subject to } g(y) \in K^\circ \\ & \quad y \in L, \end{aligned} \qquad (2.1)$$

where f and g are functions from an open subset U of $\mathbb{R}^p$ to $\mathbb{R}$ and $\mathbb{R}^m$ respectively, L is a closed convex set in $\mathbb{R}^p$, K is a closed convex cone in $\mathbb{R}^m$, and K° denotes the *polar cone* of K, defined by

$$K^\circ := \{y \in \mathbb{R}^m | \langle y, k \rangle \leq 0 \text{ for each } k \in K\}.$$

The formulation in (2.1) is general enough to express a great many of the specific optimization problems found in practice. For example, if the constraints of the problem one is dealing with are of the form

$$g_1(y) \leq 0, \ldots, g_k(y) \leq 0; \quad g_{k+1}(y) = 0, \ldots, g_{k+l}(y) = 0,$$

then one has only to set $K = \mathbb{R}_+^k \times \mathbb{R}^l$ (so that $K^\circ = \mathbb{R}_-^k \times \{0\}^l$), and $L = \mathbb{R}^p$.

If we define the *standard Lagrangian* associated with (2.1) to be $\mathscr{L}(y, u) := f(y) + \langle u, g(y) \rangle$ for $y \in U$ and $u \in \mathbb{R}^m$, then if y is a local optimizer of (2.1) at which an appropriate constraint qualification holds, there will exist $u \in \mathbb{R}^m$ which, with y, satisfies the *necessary optimality conditions*:

$$\begin{aligned} 0 &\in \frac{\partial}{\partial y} \mathscr{L}(y, u) + \partial \psi_L(y) \\ 0 &\in -\frac{\partial}{\partial u} \mathscr{L}(y, u) + \partial \psi_K(u). \end{aligned} \qquad (2.2)$$

See Robinson (1976a) for details of the derivation and of the kinds of constraint qualifications under which (2.2) can be expected to hold.

II. Generalized Equations

To express (2.2) as a generalized equation we need only set $x = (y, u)$,

$$F(x) = \left[\frac{\partial}{\partial y} \mathscr{L}(y, u), -\frac{\partial}{\partial u} \mathscr{L}(y, u)\right], \quad \text{and} \quad C = L \times K. \quad \text{Noting that}$$

$\partial \psi_{L \times K} = \partial \psi_L \times \partial \psi_K$, we then see that (2.2) is equivalent to

$$0 \in F(x) + \partial \psi_C(x). \tag{2.3}$$

It may help in understanding the structure of (2.2) to see what the generalized equation (2.3) looks like in the particular case of (generalized) quadratic programming, in which f and g in (2.1) have the special form

$$f(y) = \tfrac{1}{2} \langle y, Qy \rangle + \langle p, y \rangle$$
$$g(y) = Ay - a,$$

where Q and A are linear transformations from $\mathbb{R}^p$ to $\mathbb{R}^p$ and $\mathbb{R}^m$ respectively, $p \in \mathbb{R}^p$ and $a \in \mathbb{R}^m$. Note that here the variables y may be subject to implicit constraints, such as upper and lower bounds, which are enforced through the set L. In this case (2.3) takes the special form

$$0 \in \begin{bmatrix} Q & A^T \\ -A & 0 \end{bmatrix} \begin{bmatrix} y \\ u \end{bmatrix} + \begin{bmatrix} p \\ a \end{bmatrix} + \partial \psi_{L \times K} \begin{bmatrix} y \\ u \end{bmatrix}, \tag{2.4}$$

and except for the presence of the normal-cone operator, (2.4) looks like a system of linear equations (to which, indeed, it would reduce if $L = \mathbb{R}^p$ and $K = \mathbb{R}^m$: that is, if the variables y were unconstrained and the explicit constraints were linear equations). To specialize (2.4) to the case of linear programming, we just set $Q = 0$, and then the matrix in (2.4) becomes a skew matrix.

Given a linear transformation M from $\mathbb{R}^n$ to itself and a point $m \in \mathbb{R}^n$, if we set $F(x) = Mx + m$ then (1.1) becomes

$$0 \in Mx + m + \partial \psi_C(x), \tag{2.5}$$

and we shall call this a *linear generalized equation*. As we have just seen, linear generalized equations arise naturally from the optimality conditions for quadratic programming. They also occur in a number of other contexts; in fact, as we shall see in Section 4 below, linear generalized equations play a rôle with respect to nonlinear generalized equations that is analogous to the rôle played by linear equations with respect to nonlinear equations.

Another common problem that gives rise to generalized equations is that of *complementarity*. Given a function F from an open subset of $\mathbb{R}^n$ to itself, and a closed convex cone $K \subset \mathbb{R}^n$, the *generalized complementarity problem* for F and K is that of finding a point x such that

$$x \in K, \quad F(x) \in K^*, \quad \langle x, F(x) \rangle = 0, \tag{2.6}$$

where $K^* := -K^\circ$ is the *dual cone* of K. However, if we recall that because K is a cone,

$$\partial \psi_K(x) = \begin{cases} \phi & \text{if } x \notin K \\ \{y \in K^\circ | \langle y, x \rangle = 0\} & \text{if } x \in K, \end{cases}$$

then we can see at once that (2.6) is equivalent to

$$0 \in F(x) + \partial \psi_K(x). \tag{2.7}$$

The usual *nonlinear complementarity problem* found in the literature is the special case of (2.6) in which $K = \mathbb{R}^n_+$, and the *linear complementarity problem* is the special case of the nonlinear problem in which $F(x) = Mx + m$. Thus, linear complementarity problems give rise to linear generalized equations of a special type; namely, those in which the set whose normal-cone operator appears in (2.5) is actually a cone. For more information about linear and nonlinear complementarity problems, see the papers by Lemke (1970), Eaves (1971), Cottle (1976), Cottle and Dantzig (1968), and Karamardian (1969a, 1969b, 1972), among many others.

A number of models from mathematical economics can be expressed as generalized equations, and we shall examine two of these here. The first is the model proposed by Hansen and Koopmans (1972) of a capital stock invariant under optimization. In this model, Hansen and Koopmans consider an economic growth problem in which the technology is linear (i.e., has constant returns to scale), involving goods of three types: capital goods, resources, and consumption goods. The problem is to find a (technologically) feasible operating path over time which maximizes a sum of discounted utilities involving the consumption goods. More particularly, the authors ask whether such a path can be found in which the stock of capital goods is invariant over time, and they prove that the problem of finding such an invariant capital stock is substantially equivalent to a certain single-period problem. It is this single-period problem with which we shall be concerned here. By incorporating the consumption goods in the utility function, one can reduce the variables of the problem to the following classes, denoted by the letters shown opposite each class:

Capital goods: $z \in \mathbb{R}^L$

Resources: $w \in \mathbb{R}^M$

Activity levels: $x \in \mathbb{R}^I$.

These goods are related by the following inequalities, in which A, B, and C denote linear transformations on the appropriate spaces:

$$Ax \leq z \leq Bx$$
$$Cx \leq w \tag{2.8}$$
$$x \geq 0.$$

If one now poses the problem of maximizing, for a fixed $z \geq 0$, a concave differentiable function $v(x)$ subject to (2.8), the necessary optimality conditions will associate dual variables with the inequality constraints and will prescribe relations that must be satisfied by these dual variables. If we denote by q_A the dual variable associated with the inequality $Ax \leq z$, by q_B that associated with

II. Generalized Equations

$z \leq Bx$, and by r that associated with $Cx \leq w$, then the optimality conditions prescribe, in addition to (2.8), the following relationships:

$$v'(x) - q_A A + q_B B - rC \leq 0$$
$$\langle v'(x), x \rangle - \langle q_A, z \rangle + \langle q_B, z \rangle - \langle r, w \rangle = 0 \qquad (2.9)$$
$$q_A, q_B, r \geq 0.$$

The problem considered by Hansen and Koopmans is, given w, to solve (2.8) and (2.9) for x, z, q_A, q_B and r in such a way that $z \geq 0$ and $q_B = \alpha q_A$, where α is a prescribed discount factor in the interval $(0, 1)$. It is this particular requirement on the Lagrange multipliers that forces the one-period optimization problem to yield an invariant capital stock z.

In order to formulate this problem as a generalized equation, let us first consider the problem of maximizing $v(x)$ over all x and z satisfying (2.8) and the additional requirement that $z \geq 0$. Keeping the same notation for the Lagrange multipliers, we obtain the optimality conditions

$$v'(x) - q_A A + q_B B - rC \leq 0$$
$$q_A - q_B \leq 0$$
$$\langle v'(x) - q_A A + q_B B - rC, x \rangle = 0$$
$$\langle q_A - q_B, z \rangle = 0$$
$$\langle q_A, Ax - z \rangle = 0$$
$$\langle q_B, z - Bx \rangle = 0$$
$$\langle r, Cx - w \rangle = 0$$
$$q_A, q_B, r \geq 0,$$

together with (2.8) and the condition that $z \geq 0$. We can organize all of this rather extensive set of conditions into a single generalized equation by writing

$$0 \in \begin{bmatrix} 0 & 0 & A^T & -B^T & C^T \\ 0 & 0 & -I & I & 0 \\ -A & I & 0 & 0 & 0 \\ B & -I & 0 & 0 & 0 \\ -C & 0 & 0 & 0 & 0 \end{bmatrix} \begin{bmatrix} x \\ z \\ q_A \\ q_B \\ r \end{bmatrix} + \begin{bmatrix} -v'(x) \\ 0 \\ 0 \\ 0 \\ w \end{bmatrix} + \partial \psi_C \begin{bmatrix} x \\ z \\ q_A \\ q_B \\ r \end{bmatrix}, \qquad (2.11)$$

where C is the product $\mathbb{R}_+^l \times \mathbb{R}_+^l \times \mathbb{R}_+^l \times \mathbb{R}_+^l \times \mathbb{R}_+^M$. Note that the matrix in (2.11) is skew; indeed, this skewness serves as a guide in translating (2.8) and (2.10) into (2.11).

However, (2.11) does not express exactly the conditions required by Hansen and Koopmans, since (2.11) contains the complementary system

$$-q_A + q_B \geq 0, \quad z \geq 0, \quad \langle -q_A + q_B, z \rangle = 0 \qquad (2.12)$$

(i.e., $0 \in -q_A + q_B + \partial \psi_{\mathbb{R}_+^l}(z)$), instead of the relations $-\alpha q_A + q_B = 0$, $z \geq 0$ demanded by Hansen and Koopmans. At this point, we note that in the Hansen-Koopmans formulation the matrix A is required to have non-negative elements, so that the inequality $-Ax+z\geq 0$, already present in the model, together with $x\geq 0$, will guarantee that $z\geq 0$ even if z is not explicitly constrained. Let us therefore replace (2.12) by the complementary system $-\alpha q_A + q_B = 0$, $z \in \mathbb{R}^L$. This system is equivalent to

$$0 \in -\alpha q_A + q_B + \partial \psi_{\mathbb{R}^L}(z),$$

and we can make this change in (2.11) simply by replacing $-I$ in the (2,3) position of the matrix by $-\alpha I$ and the first copy of $\mathbb{R}^L_+$ in C by $\mathbb{R}^L$. With these changes, and with $C' := \mathbb{R}^I_+ \times \mathbb{R}^L \times \mathbb{R}^L_+ \times \mathbb{R}^L_+ \times \mathbb{R}^M_+$, the new generalized equation is

$$0 \in \begin{bmatrix} 0 & 0 & A^T & -B^T & C^T \\ 0 & 0 & -\alpha I & I & 0 \\ -A & I & 0 & 0 & 0 \\ B & -I & 0 & 0 & 0 \\ -C & 0 & 0 & 0 & 0 \end{bmatrix} \begin{bmatrix} x \\ z \\ q_A \\ q_B \\ r \end{bmatrix} + \begin{bmatrix} -v'(x) \\ 0 \\ 0 \\ 0 \\ w \end{bmatrix} + \partial \psi_{C'} \begin{bmatrix} x \\ z \\ q_A \\ q_B \\ r \end{bmatrix}. \quad (2.13)$$

Although (2.13) now expresses precisely the conditions demanded by Hansen and Koopmans, we note that its matrix is no longer skew, reflecting the fact that it no longer corresponds exactly to the optimality conditions for a linearly-constrained optimization problem; indeed, Hansen and Koopmans had to use a fixed-point algorithm to solve their problem. This illustrates the fact that generalized equations can be used effectively to model equilibrium type relations even when these do not correspond to optimization problems. Our next example is also of this type.

For the second example of modeling a problem from mathematical economics, we shall examine the structure of a model of energy equilibrium. This is a simplified model that retains the conceptual structure of the Project Independence Evaluation System model discussed by Hogan (1975). The model consists of two sectors; a production (and transportation) sector using a linear technology to produce a prescribed (vector) quantity q of different forms of energy at minimum cost, and a consumption sector which demands varying amounts of each energy form depending upon the prices, p, of all available forms. The key requirement is to find a pair (p, q) such that (i) q is the vector of energy forms demanded by consumers when the (given) prices p are in effect, and (ii) p is the dual (price) variable associated, in the suppliers' linear programming problem, with the constraint that the (given) amounts q of energy forms must be produced. Thus the pair (p, q) has to appear in both sides of the production-consumption system, and this is what makes the problem one of equilibrium rather than of optimization.

To formulate the problem in more precise terms, let us suppose that the energy production system has been modeled as a linear programming problem

$$\begin{aligned} &\text{minimize } \langle c, x \rangle \\ &\text{subject to } Ax = q \\ &\phantom{\text{subject to }} Bx = b \\ &\phantom{\text{subject to }} x \geq 0, \end{aligned} \quad (2.14)$$

II. Generalized Equations

where x is a vector of n non-negative activity levels, c is a vector of costs associated with the activities, and the l constraints $Bx = b$ represent material balance constraints, upper and lower bounds, and other structural properties of the production system. Of course, these constraints may include inequalities as well as equations, but we assume that the inequalities have already been transformed to equations by the use of appropriate slack variables. The k constraints $Ax = q$ in (2.14) express the relation between the activity levels x and the final output q of k different forms of energy.

The consumption of energy, on the other hand, is assumed to be modeled by a demand function $q_D(p)$ giving consumer demands for energy as functions of the prices in effect. This function might, for example, be estimated by econometric methods. We can now express in a different way the requirements placed on the pair (p, q) by saying that we want to find a solution of the linear programming problem (2.14) with the element q of the right-hand side equal to $q_D(p)$, in which p is an optimal dual variable corresponding to the first constraint. Note that this is a similar situation to that developed earlier in the Hansen-Koopmans model; we begin with an optimization model, then alter it by placing an additional requirement on the Lagrange multipliers.

If we have an optimal solution x of (2.14) with the specified right-hand side, then we must have

$$Ax = q_D(p)$$
$$Bx = b \qquad (2.15)$$
$$x \geq 0,$$

and the dual variables p and r corresponding to the two constraints must satisfy

$$c + pA + rB \geq 0$$
$$\langle c + pA + rB, x \rangle = 0. \qquad (2.16)$$

We can model the relations (2.15) and (2.16) as a generalized equation by writing

$$0 \in \begin{bmatrix} 0 & A^T & B^T \\ -A & 0 & 0 \\ -B & 0 & 0 \end{bmatrix} \begin{bmatrix} x \\ p \\ r \end{bmatrix} + \begin{bmatrix} c \\ q_D(p) \\ b \end{bmatrix} + \partial \psi_{\mathbb{R}^n_+ \times \mathbb{R}^k \times \mathbb{R}^l} \begin{bmatrix} x \\ p \\ r \end{bmatrix}. \qquad (2.17)$$

Note that this is almost like the linear generalized equation that would result from writing the optimality conditions for the linear programming problem (2.14); the difference is that the "constant term" is now no longer constant since it contains the function $q_D(p)$.

More details about this way of formulating such equilibrium models can be found in Josephy (1979c, 1979d). There the formulation is carried out in terms of the inverse function, $p_D(q)$, corresponding to $q_D(p)$. However, for conceptual purposes these formulations are equivalent.

We have shown in this section how generalized equations can be used to formulate optimality conditions, complementarity problems, and economic equilibrium problems in a conceptually simple, economical and unified way.

In the next two sections we turn from the question of formulation to questions of analysis and numerical solution. We shall ask when solutions of generalized equations exist, whether they are stable when they exist, and how we can compute them. We begin in the next section with some results that hold when monotonicity is present.

3. Existence and Stability: The Monotone Case

This section treats some results that are available to us when the expression $F(x) + \partial \psi_C(x)$ in (1.1) is a monotone operator in the variable x. We begin by reviewing the definition, and some properties, of monotone operators. For a multifunction (multivalued function) A we shall write $(x, y) \in A$ to mean $y \in A(x)$ (i.e., (x, y) belongs to the graph of A).

Definition 3.1: A multifunction $A: \mathbb{R}^n \to \mathbb{R}^n$ is a *monotone operator* if for each (x_1, y_1) and (x_2, y_2) in A, $\langle x_1 - x_2, y_1 - y_2 \rangle \geq 0$. A is a *maximal monotone operator* if its graph is not properly contained in that of any other monotone operator.

Examples of maximal monotone operators that occur naturally in connection with generalized equations include:

a. The normal cone operator $\partial \psi_C$ associated with a closed convex set C. In fact, the subdifferential mapping associated with any closed proper convex function is maximal monotone. For proofs and more details see Brezis (1973).

b. The linear operator represented by

$$\begin{pmatrix} Q & A^T \\ -A & 0 \end{pmatrix}$$

(cf. (2.4)), whenever Q is positive semidefinite. This will be the case when Q is the Hessian of a convex function. More generally, any positive semidefinite matrix represents a monotone operator, and in particular any skew matrix does.

Since our earlier examples of generalized equations involved sums of operators (e.g., a linear or nonlinear operator plus a normal-cone operator) it is of interest to determine when such sums will be monotone if their components are monotone. The following theorem gives a convenient criterion for such monotonicity. It specializes to $\mathbb{R}^n$ a result of Rockafellar (1970). We write dom F for $\{x | F(x) \neq \phi\}$, and ri for relative interior (interior relative to the affine hull).

Theorem 3.2: *Let F and G be maximal monotone operators from $\mathbb{R}^n$ to $\mathbb{R}^n$. If* ri dom $F \cap$ ri dom $G \neq \phi$, *then $F + G$ is maximal monotone.*

This result shows that under rather mild assumptions, if the function F appearing in (1.1) is monotone then (1.1) itself will be a problem of finding a zero of a monotone operator. Therefore, it will be of interest to us to review some known facts about existence and stability of such zeros. One of the simplest

II. Generalized Equations

such facts applies in case the operator involved is strongly monotone, and so we turn next to the definition of strong monotonicity.

Definition 3.3: A multifunction F from $\mathbb{R}^n$ to itself is *strongly monotone* if there exists a constant $\gamma > 0$ such that for each (x_1, y_1) and (x_2, y_2) in F,

$$\langle x_1 - x_2, y_1 - y_2 \rangle \geq \gamma \|x_1 - x_2\|^2.$$

Now suppose F is a maximal monotone operator which is strongly monotone with modulus γ. A fundamental result about monotone operators says that F is maximal monotone if and only if for any $\lambda > 0$, $(I + \lambda F)^{-1}$ is a contraction that is nonempty on the entire space. That is, for any z_1 and z_2 there exist (x_1, y_1) and (x_2, y_2) in F with $x_1 + \lambda y_1 = z_1$, $x_2 + \lambda y_2 = z_2$, and $\|x_1 - x_2\| \leq \|z_1 - z_2\|$. This result is discussed in Brezis (1973). It is clear that $(I + \lambda F)^{-1}$ is then single-valued (take $z_1 = z_2$). However, if F happens to be strongly monotone, we can deduce even more; suppose we form inner products as follows:

$$\begin{aligned}\langle x_1 - x_2, z_1 - z_2 \rangle &= \langle x_1 - x_2, (x_1 + \lambda y_1) - (x_2 + \lambda y_2) \rangle \\ &= \|x_1 - x_2\|^2 + \lambda \langle x_1 - x_2, y_1 - y_2 \rangle \geq (1 + \lambda \gamma) \|x_1 - x_2\|^2,\end{aligned} \quad (3.1)$$

where the inequality comes from strong monotonicity. Recalling that $\langle x_1 - x_2, z_1 - z_2 \rangle \leq \|x_1 - x_2\| \|z_1 - z_2\|$, we can rearrange (3.1) to yield

$$\|x_1 - x_2\| \leq (1 + \lambda \gamma)^{-1} \|z_1 - z_2\|,$$

which shows that $(I + \lambda F)^{-1}$ is actually a strong contraction since $(1 + \lambda \gamma)^{-1} < 1$. Such an operator has a unique fixed point by the contraction mapping theorem, and it is easy to see that the fixed points of $(I + \lambda F)^{-1}$ are exactly the zeros of F. Hence if we are dealing with a strongly monotone operator we will have a unique zero.

However, it is very often the case that the operator with which we have to deal is not strongly monotone. In such a case, another existence result can often be helpful: this result is local, rather than global, in nature. To state it we need another definition.

Definition 3.4: A multifunction A is said to be *locally bounded* at a point x_0 if there exists a neighborhood U of x_0 such that $A(U)(=\{y | y \in Ax \text{ for some } x \in U\})$ is a bounded set.

The following theorem translates to $\mathbb{R}^n$ a result of Rockafellar (1969).

Theorem 3.5: *Let G be a maximal monotone operator from $\mathbb{R}^n$ to itself. Then G is locally bounded at y_0 if and only if y_0 is not a boundary point of dom G.*

Now consider a case in which the maximal monotone operator G is known to be upper semicontinuous at $y_0 \in \text{dom } G$. If $G(y_0)$ is bounded, then so will be $G(U)$ for some neighborhood U of y_0 (by upper semicontinuity). But then G is locally bounded at y_0, and by the theorem we conclude that y_0 is not a boundary point of dom G. But as $y_0 \in \text{dom } G$ by assumption, we must have $y_0 \in \text{int dom } G$.

This reasoning can be particularly helpful if we take G to be the inverse of a maximal monotone operator F (and hence itself maximal monotone), with $y_0 = 0$. Then we conclude that if:

a. F^{-1} is upper semicontinuous at 0,

and

b. $F^{-1}(0)$ is bounded and nonempty

then the inclusion $y \in F(x)$ is solvable for all y in some neighborhood of the origin; moreover, if Q is any open set containing $F^{-1}(0)$, then for some neighborhood V of 0 we have $F^{-1}(V) \subset Q$.

This observation can be applied to yield, for example, a rather complete stability theory for linear generalized equations involving positive semidefinite matrices and polyhedral convex sets. These include problems of linear programming and of convex quadratic programming (for which the matrix has the special form shown in (2.4) above), as well as linear complementarity problems whose matrices are positive semidefinite (though not necessarily symmetric). This is discussed by Robinson (1979), where the following theorem is proved. In the statement of the theorem, B denotes the Euclidean unit ball.

Theorem 3.6: *Let A be a positive semidefinite $n \times n$ matrix, C be a nonempty polyhedral convex set in $\mathbb{R}^n$, and a be a point of $\mathbb{R}^n$. Then the following are equivalent:*

a. *The solution set of the linear generalized equation*

$$0 \in Ax + a + \partial \psi_C(x) \tag{3.2}$$

is nonempty and bounded.

b. *There exists $\varepsilon > 0$ such that for each $n \times n$ matrix A' and each $a' \in \mathbb{R}^n$ with*

$$\varepsilon' := \max\{\|A' - A\|, \|a' - a\|\} < \varepsilon, \tag{3.3}$$

the set

$$S(A', a') := \{x \mid 0 \in Ax + a + \partial \psi_C(x)\}$$

is nonempty.

Further, if these conditions hold, then for each open bounded set $Q \supset S(A, a)$ there are positive numbers η and μ such that for each (A', a') with ε' (defined by (3.3)) $< \eta$, one has

$$\emptyset \neq S(A', a') \cap Q \subset S(A, a) + \mu \varepsilon' B. \tag{3.4}$$

Finally, if (A', a') are restricted to values for which $S(A', a')$ is known to be connected (for example, if A' is restricted to be positive semidefinite), then Q can be replaced by $\mathbb{R}^n$.

The inclusion (3.4) means that for each solution, say x_1, of

$$0 \in A'x + a' + \partial \psi_C(x)$$

in Q, there is a solution of (3.2), say x_0, with $\|x_1 - x_0\| \leq \mu \max\{\|A' - A\|, \|a' - a\|\}$. Thus the solution sets obey a set-valued analogue of Lipschitz continuity, called "upper Lipschitz continuity" [see Robinson (1979, 1981)].

II. Generalized Equations

We may remark that the condition that A' be positive semidefinite (under which the "isolating" set Q is not required in Theorem 3.6) will hold automatically in a large class of practical applications of the theorem. Of course this will be true if A is actually positive definite, but it may well be true even if A is only semidefinite, because of the structure of A. For example, consider the linear programming problem

$$\text{minimize } \langle c, x \rangle$$
$$\text{subject to } Dx = d \qquad (3.5)$$
$$x \geq 0,$$

where $D: \mathbb{R}^n \to \mathbb{R}^m$, $d \in \mathbb{R}^m$, and $c \in \mathbb{R}^n$. We can formulate this problem as the generalized equation

$$0 \in \begin{bmatrix} 0 & D^T \\ -D & 0 \end{bmatrix} \begin{bmatrix} x \\ u \end{bmatrix} + \begin{bmatrix} -c \\ d \end{bmatrix} + \partial \psi_{\mathbb{R}^n_+ \times \mathbb{R}^m} \begin{bmatrix} x \\ u \end{bmatrix},$$

and we now observe that if (D', d', c') represent perturbed data close to (D, d, c) then the perturbed generalized equation is

$$0 \in \begin{bmatrix} 0 & (D')^T \\ -D' & 0 \end{bmatrix} \begin{bmatrix} x \\ u \end{bmatrix} + \begin{bmatrix} -c' \\ d' \end{bmatrix} + \partial \psi_{\mathbb{R}^n_+ \times \mathbb{R}^m} \begin{bmatrix} x \\ u \end{bmatrix},$$

whose matrix is positive semidefinite regardless of what D' is. Thus, in this problem we will always have a positive semidefinite matrix because of the particular structure imposed on the generalized equation by the optimality conditions of (3.5). Another such example, involving quadratic programming, is given in Robinson (1979).

4. Existence and Stability: The Case of Continuity or Differentiability

We now turn to the question of existence and stability of solutions for generalized equations in which we do not have monotonicity. Here we shall use a topological tool (the Brouwer fixed-point theorem) to establish some general existence results; then we shall investigate stability questions by employing local analytical methods analogous to the implicit-function theorem. In fact, we shall establish an implicit-function theorem for generalized equations and derive several related results.

Our first existence theorem is a simple but useful fact that has been noted by several authors [e.g., Hartman and Stampacchia (1966), Eaves (1971), Karamardian (1972)].

Theorem 4.1: *Let C be a compact convex set in $\mathbb{R}^n$ and let $F: C \to \mathbb{R}^n$ be a continuous function. Then the generalized equation*

$$0 \in F(x) + \partial \psi_C(x) \qquad (1.1)$$

has a solution (in C).

For a very easy proof of this theorem (Eaves (1971)), define a continuous self-map Φ of C by letting $\Phi(c)$ be the projection of $c - F(c)$ on C. As projections are nonexpansive, Φ is continuous, so by the Brouwer theorem it must have at least one fixed point. But its fixed points are precisely the solutions of (1.1).

We can extend this result to one involving unbounded sets C in a number of ways. Typically, one assumes some kind of condition on F at large elements of C, then uses Theorem 4.1. Several such results are discussed in Moré (1974a, 1974b) and elsewhere; we exhibit one as a sample of the sorts of conditions that may be imposed on F (see Moré (1974a), Theorem 2.4):

Theorem 4.2: *Let F be a continuous function from the closed convex set $C \subset \mathbb{R}^n$ to $\mathbb{R}^n$. Suppose that there is a positive number μ such that for each $x \in C$ with $\|x\| = \mu$, there is some $u \in C$ with $\|u\| < \mu$ and $\langle x - u, F(x) \rangle \geq 0$. Then (1.1) has a solution x with $\|x\| \leq \mu$.*

Of course, the hypothesis of Theorem 4.2 will be satisfied if one can show that the inequality $\langle x - x_0, F(x) \rangle \geq 0$ holds for some $x_0 \in C$ and all x with sufficiently large norm. As an illustration of how this may be applied, consider a linear generalized equation

$$0 \in Mx + m + \partial \psi_K(x) \qquad (4.1)$$

where K is a cone and the matrix M is strictly K-copositive: i.e., if $x \in K \setminus \{0\}$ then $\langle x, Mx \rangle > 0$. Then for some $\sigma > 0$, and all $x \in K$, $\langle x, Mx \rangle \geq \sigma \|x\|^2$. Given any $m \in \mathbb{R}^n$, take $\mu = \sigma^{-1} \|m\|$; then if $x \in K$ with $\|x\| = \mu$ we have, with $u = 0$,

$$\langle x - 0, Mx + m \rangle = \langle x, Mx \rangle + \langle x, m \rangle$$
$$\geq \sigma \|x\|^2 - \|x\| \|m\| = 0.$$

Thus, by Theorem 4.2 there is a solution x of (4.1) with $\|x\| \leq \sigma^{-1} \|m\|$. For example, if $K = \mathbb{R}^n_+$ and M is a non-negative matrix with positive diagonal, we can take σ to be the minimum diagonal element.

The above results are based on the Brouwer fixed-point theorem, but they can also be established by degree arguments (since Brouwer's theorem can be proved by such methods). We remark that Reinoza (1979) has developed a definition of degree for multivalued functions of the type appearing in (1.1), and Kojima (1980) made extensive use of degree arguments in his study of "strongly stable" solutions of nonlinear programming problems. These latter results are closely related to a class of stability results for generalized equations, which we shall discuss next.

We now shift our attention from results promising existence of a solution to results describing the stability of an existing solution when the problem is slightly perturbed. Here we will need to use differentiability properties of F, whereas in the first two theorems of this section we needed only continuity.

The generalized equation problem with which we shall deal is that of finding x so that

$$0 \in F(p, x) + \partial \psi_C(x), \qquad (4.2)$$

II. Generalized Equations

which differs from (1.1) in that the parameter p has been added. Its function is to introduce perturbations into F (but not C) so that we may study the dependence of the solution(s) on such perturbations. We shall ask whether, if x_0 solves (4.2) for a given value $p = p_0$, and if we then allow p to vary near p_0, there is some function $x(p)$ yielding a solution x of (4.2) for each p near p_0. If such a function exists, we should like to gain information about its behavior. In other words, we are seeking the type of information about (4.2) that could be provided by an implicit-function theorem.

We shall now show that an implicit-function theorem can indeed be established for (4.2). In order to do this, we have to introduce some definitions. The first idea is that of the linearization of (1.1): if we suppose that Ω is an open subset of $\mathbb{R}^n$ and that $F: \Omega \to \mathbb{R}^n$ is Fréchet differentiable at x_0, then the *linearization of $F(x) + \partial \psi_C(x)$ about x_0* is defined to be $F(x_0) + F'(x_0)(x - x_0) + \partial \psi_C(x)$: in other words, we just linearize the function appearing in (1.1) but leave the normal-cone operator alone.

At this point we might return briefly to the questions, considered earlier in the paper, of why the generalized-equation formalism can be helpful in dealing with problems. We shall see that the linearization defined here works, in the sense that good properties of this linearization guarantee (locally) good properties of the original nonlinear generalized equation. Thus, it seems to be an appropriate tool to use, and indeed it seems obvious when we look at (1.1) that this is the way in which we should linearize it. However, if C is a cone and if we then write (1.1) in the equivalent form of a complementary system

$$F(x) \in C^*, \quad x \in C, \quad \langle x, F(x) \rangle = 0, \tag{4.3}$$

then in looking at (4.3) one might be tempted to linearize not only the first inclusion, but also the complementarity equation on the right. In fact, the linearization obtained in this way does not work well, and so at least in this case the use of the generalized-equation symbolism (4.2) leads one naturally to the correct method of analysis.

Having introduced the idea of linearization, we next define a *regular solution* of (1.1).

Definition 4.3: Suppose x_0 is a point at which F is Fréchet differentiable, and that x_0 solves (1.1): i.e., $0 \in F(x_0) + \partial \psi_C(x_0)$. Define an operator T by $T(x) := F(x_0) + F'(x_0)(x - x_0) + \partial \psi_C(x)$. Then x_0 is a *regular solution* of (1.1) if there exist neighborhoods U of 0 and V of x_0 such that $(T^{-1} \cap V)|U$ is single-valued and Lipschitzian: in other words, the function that associates to each $u \in U$ the set of $v \in V$ such that $u \in T(v)$ is Lipschitzian on U.

This property was originally called "strong regularity" in Robinson (1980), because a weaker property had been analyzed in Robinson (1976b) under the name of "regularity." However, the present property has proven to be much more useful in a variety of situations, so well shall use the term "regularity" to refer to it.

One of the consequences of regularity is the following implicit-function theorem for (4.2). It is taken from Robinson (1980).

Theorem 4.4: *Let P be an open subset of a normed linear space and Ω be an open subset of $\mathbb{R}^n$, with $F: P \times \Omega \to \mathbb{R}^n$. Write $F_x(p, x)$ for $\dfrac{\partial}{\partial x} F(p, x)$, and suppose that:*

a. *F and F_x are continuous on $P \times \Omega$.*
b. *For each $x \in \Omega$, $F(\cdot, x)$ is Lipschitzian on P with Lipschitz modulus ν (independent of $x \in \Omega$).*
c. *x_0 is a regular solution of (4.2) (for $p = p_0$) with associated Lipschitz modulus λ.*

Then for any $\varepsilon > 0$ there exist neighborhoods N_ε of p_0 and W_ε of x_0, and a single-valued function $x: N_\varepsilon \to W_\varepsilon$ with Lipschitz modulus $\nu(\lambda + \varepsilon)$, such that for any $p \in N_\varepsilon$, $x(p)$ is the unique solution of (4.2) in W_ε.

This theorem says in effect that if the linearization has, locally, a Lipschitzian inverse then so does the original generalized equation. Moreover, as we shall see in the next theorem, the linearization can be used to approximate solutions of the nonlinear problem.

Theorem 4.5: *Assume the notation and hypotheses of Theorem 4.4. For each $\varepsilon > 0$ and for each $p \in N_\varepsilon$ let $\xi(p)$ be the (unique) solution in W_ε of the linear generalized equation*

$$0 \in F(x_0, p) + F_x(x_0, p_0)(\xi - x_0) + \partial \psi_C(\xi).$$

Then there exists a function $\alpha_\varepsilon: N_\varepsilon \to \mathbb{R}$ such that

$$\lim_{p \to p_0} \alpha_\varepsilon(p) = 0$$

and for any $p \in N_\varepsilon$,

$$\|x(p) - \xi(p)\| \leq \alpha_\varepsilon(p) \|p - p_0\|.$$

As Theorem 4.4 shows that a generalized equation with a regular solution remains solvable if slightly perturbed, we might wonder whether the regularity property is preserved for these perturbed solutions. The next theorem shows that the answer to this question is yes.

Theorem 4.6: *Let A be a linear transformation from $\mathbb{R}^n$ to itself, let $a \in \mathbb{R}^n$ and let C be a closed convex set in $\mathbb{R}^n$. Suppose that x_0 is a regular solution of*

$$0 \in Ax + a + \partial \psi_C(x)$$

with associated neighborhoods U of 0 and V of x_0 and Lipschitz modulus λ. Then there exist neighborhoods M of 0 and N of x_0, and a positive number ε, such that for any A' and a' with

$$\max\{\|A' - A\|, \|a' - a\|\} < \varepsilon,$$

if $T'(x) := A'x + a' + \partial \psi_C(x)$, then $[(T')^{-1} \cap N] | M$ is a single-valued function with Lipschitz modulus $\lambda' := \lambda(1 - \lambda\|A' - A\|)^{-1}$.

We remark that, by Theorem 4.4, for each pair (A', a') near (A, a), the generalized equation

$$0 \in A'x + a' + \partial \psi_C(x) \tag{4.4}$$

II. Generalized Equations

has a unique solution near x'. What Theorem 4.6 says is that this solution is in fact a regular solution of (4.4), and that its associated Lipschitz modulus is not much greater than λ. Moreover, the neighborhoods involved in the definition of regularity can be taken to be the same for all nearby versions of (4.4). It therefore shows that regularity is an "open" property, and it provides an analogue for generalized equations of the Banach lemma for linear operators (see, e.g., Kantorovich and Akilov (1964), Theorem 3 (2.V)).

For the particular case in which $C = \mathbb{R}^r \times \mathbb{R}^s_+$, which is very frequently seen in applications, a characterization of regularity for a linear generalized equation is given by Robinson (1980) (see Theorem 3.1 of that paper). Given a solution x_0, one removes the "inactive" variables (i.e., elements of x_0 that are non-negatively constrained but are in fact strictly positive), as well as those non-negatively constrained variables that must remain equal to zero because the corresponding function values are strictly positive. Regularity then holds if and only if the square matrix corresponding to the remaining "reduced" problem satisfies the property that its principal submatrix corresponding to the unconstrained variables is nonsingular and the Schur complement of that submatrix has positive principal minors. Here the Schur complement of the nonsingular principal submatrix A_{11} in the square matrix

$$A = \begin{pmatrix} A_{11} & A_{12} \\ A_{21} & A_{22} \end{pmatrix}$$

is defined to be

$$(A/A_{11}) := A_{22} - A_{21} A_{11}^{-1} A_{12}.$$

See Cottle (1974) for further information on Schur complements.

Among other results dealing with local existence and stability of solutions to generalized equations, we mention the "strong positivity conditions" of Reinoza (1981) and the work of Kummer (1982) on solvability of very general multivalued inclusions, as well as work of Spingarn (1977) on "cyrtohedra" and on perturbed optimization problems. In the next section, we shall see how the idea of linearization and some of the above stability results may be applied to develop efficient computational methods.

5. Computing Solutions: The Newton Method and some Variants

Given a generalized equation such as (1.1), we often want to compute a solution to it in order to solve some practical problem. In this section we describe methods of Newton type for computing such solutions. Although Newton methods were proposed by Robinson (1976b) and by Eaves (1978), the results that we give here were obtained by Josephy (1979a, 1979b, 1979c, 1979d). They extend to generalized equations the well known theorem of Kantorovich (see Kantorovich and Akilov (1964), Theorem 6 (1.XVIII)) for conventional equations.

Let us consider solving the generalized equation

$$0 \in F(x) + \partial \psi_C(x) \qquad (1.1)$$

by repeated linearization. That is, starting at some given point x_0, for each $k \geq 0$ we construct x_{k+1} from x_k by solving the linear generalized equation

$$0 \in F(x_k) + F'(x_k)(x - x_k) + \partial \psi_C(x). \qquad (5.1)_k$$

Obvious questions arise: can we be sure that the problems $(5.1)_k$ will be solvable? If they are solvable, will they have (at least locally) unique solutions? If so, will the sequence $\{x_k\}$ converge to a solution of (1.1)?

In general, one clearly could not expect positive answers to such questions. However, what Josephy showed was that if the first linearized problem $(5.1)_0$ had a regular solution x_1, and if certain inequalities held, then all of the problems $(5.1)_k$ would have (regular) solutions, and the sequence of solutions would converge R-quadratically to a solution of (1.1). The main result along these lines is Theorem 2 of Josephy (1979a), which we restate here as Theorem 5.1. In that theorem, we denote by $B(x, \rho)$ the closed ball of radius ρ about x.

Theorem 5.1 [Josephy (1979a)]: *Let F, C and Ω be as previously defined, and suppose further that Ω is convex and that F has a Fréchet derivative that is Lipschitz continuous on Ω with Lipschitz modulus L. Let $x_0 \in \Omega$, and suppose that the generalized equation*

$$0 \in F(x_0) + F'(x_0)(x - x_0) + \partial \psi_C(x) \qquad (5.1)_0$$

has a regular solution x_1 with associated Lipschitz modulus λ. Choose $r > 0$, $R > 0$ and $\rho > 0$ so that [Theorem 4.6] for any $x \in B(x_0, \rho)$ the operator $[F(x) + F'(x)[(\cdot) - x] + \partial \psi_C(\cdot)]^{-1} \cap B(x_1, r)$, restricted to $B(0, R)$, is single valued and Lipschitzian, with the Lipschitz modulus $\lambda[1 - \lambda \|F'(x) - F'(x_1)\|]^{-1}$. Define $\eta := \|x_1 - x_0\|$, and let $h := \lambda L \eta$. Assume that

a. $0 < h \leq \frac{1}{2}$,

and

b. $L\eta^2 \leq 2R$.

Define

$$t^* := \left[\frac{1 - (1 - 2h)^{\frac{1}{2}}}{h}\right] \eta,$$

and assume that

c. $B(x_0, t^*) \subset \Omega \cap B(x_0, \rho)$.

Then the sequence $\{x_k\}$ defined by letting x_{k+1} be the solution of $(5.1)_k$ in $B(x_1, r)$ is well defined and converges to $x^* \in B(x_0, t^*)$ with $0 \in F(x^*) + \partial \psi_C(x^*)$. Further, for each $k \geq 1$ one has

$$\|x^* - x_k\| \leq (2^n \lambda L)^{-1}(2h)^{(2^k)}.$$

II. Generalized Equations

Thus, Josephy's result infers the existence of a solution from the regularity condition and from the bounds expressed by assumptions (a), (b), and (c) of Theorem 5.1; it also establishes R-quadratic convergence in the sense of Ortega and Rheinboldt (1970) provided that $h < \frac{1}{2}$. Of course, if we assume the existence of a regular solution x^* of (1.1), we can obtain from Theorem 5.1 a "point of attraction" result to the effect that for any starting point x_0 close enough to x^*, the sequence $\{x_k\}$ defined by $(5.1)_k$ will be well defined and will converge quadratically to x^*. Theorem 1 of Josephy (1979a) establishes this result with the additional conclusion that $\{x_k\}$ converges Q-quadratically, as well as R-quadratically, to x^*.

Josephy tested his Newton method on a number of problems, among them a version of the Hansen-Koopmans capital stock model; the results are reported in Josephy (1979a). In all cases Lemke's method (see Cottle and Dantzig (1968)) was used to solve the subproblems, which in these cases were linear complementarity problems. He also tested the Newton algorithm on an example of an energy-equilibrium problem of PIES type, given by Hogan (1975). The tests are reported, and some properties of the model are analyzed in Josephy (1979c, 1979d).

It should be pointed out that if one applied Josephy's method to the nonlinear generalized equation resulting from the optimality conditions for a nonlinear programming problem (see Section 2), then the linearized problems will be the linear generalized equations arising from certain quadratic programming problems. These quadratic programming problems are precisely the approximating problems proposed by Wilson (1963) in his algorithm for solving nonlinear programming problems. Thus, in the case of nonlinear programming Josephy's work provides a proof of the implementability and convergence of Wilson's method under less stringent hypotheses than those previously known. In particular, it was previously shown by Robinson (1974) that Wilson's method would converge locally to a solution of a nonlinear programming problem satisfying the second-order sufficient condition, linear independence of the gradients of the binding constraints, and strict complementary slackness. However, in Robinson (1980) it is shown that the generalized equation associated with the optimality conditions of a nonlinear programming problem will have a regular solution if the corresponding solution of the nonlinear programming problem satisfies a strengthened second-order sufficient condition and linear independence of the gradients of the binding constraints (without strict complementary slackness). Therefore, Josephy's result shows that Wilson's method will converge quadratically to such solutions too.

In addition to his work on Newton's method, Josephy considered quasi-Newton methods in which $(5.1)_k$ is replaced by

$$0 \in F(x_k) + A_k(x - x_k) + \partial \psi_C(x), \qquad (5.2)_k$$

in which A_k is an approximation of some kind to $F'(x_k)$, chosen to reduce the computational labor of setting up $(5.1)_k$. He showed in Josephy (1979b) that two standard convergence theorems for quasi-Newton methods could be extended to generalized equations, again using the machinery of regularity. These theorems assert (1) local linear convergence to a regular solution, and (2) Q-superlinear convergence when the updates satisfy an appropriate limit condition. The results for ordinary equations are given by Dennis and Moré (1977), Theorems 3.1 and 5.1. Thus, Josephy's work makes available for the solution of generalized equations the use of approximations to $F'(x_k)$ via updates, such as are commonly used for ordinary equations. Again, Josephy tested some of these methods; some results are reported in Josephy (1979b).

In this section we have dealt with the properties of local, Newton-like, methods for solving generalized equations. We have not treated the global methods which go by the names of "simplicial", or "path-following", algorithms, simply because there is already an enormous literature on these methods. Although algorithms of this type can be, and have been, used to solve complementarity problems and related multivalued problems, there is no point in our duplicating here the excellent descriptions that have appeared elsewhere. In particular, for a very complete survey of this field the reader may consult the paper of Allgower and Georg (1980) and the more than 200 references contained therein.

6. A Brief Look at Generalized Differential Equations

In this concluding section we will survey very briefly some new results in the theory of generalized differential equations. Recall that in Section 1 we discussed problems such as

$$-\dot{x}(t) = \{F[x(t)] + \partial \psi_C[x(t)]\}^+, \qquad (1.4)$$

where A^+ denotes the smallest element of A. The results we discuss here apply to problems even more general than (1.4). They appear in papers of Cornet (1981a, 1981b), which contain many references to previous work in this area. Because of space limitations, we shall confine ourselves here to describing the main existence result of Cornet (1981b). We note, however, that particular cases of generalized differential equations, such as that in which the operators involved are monotone, have been studied for some time; see the references in Brézis (1973), for example.

Cornet's theorem deals with sets and generalized equations somewhat more general than those encountered earlier in this paper. For example, the sets involved need not be convex, and the functions involved may be multivalued. We therefore first quote a theorem of Cornet that characterizes the sets to which the existence theorem applies. We shall denote by $T_X(x_0)$ the Bouligand tangent cone to a subset X of $\mathbb{R}^n$ at a point $x_0 \in \operatorname{cl} X$, defined by

$$T_X(x_0) := \{y \in \mathbb{R}^n \mid \text{there exist } \{x_n\} \subset X, \{\lambda_n\} \subset (0, +\infty),$$
$$\text{with } x_n \to x_0 \text{ and } \lambda_n(x_n - x_0) \to y\}.$$

Having $T_X(x_0)$ we define the normal cone by

$$N_X(x_0) := T_X(x_0)^\circ.$$

II. Generalized Equations

If X happens to be convex then $N_X(x_0) = \partial \psi_X(x_0)$ as defined earlier. Finally, we need the Clarke tangent cone, defined by

$$TC_X(x_0) := \{y \in \mathbb{R}^n \mid \lim_{\substack{w \to x_0 \\ w \in X \\ \theta \to 0^+}} \theta^{-1} d[w + \theta y, X] = 0\},$$

where $d[\cdot, X]$ denotes the distance to X. Cornet first proves the following.

Theorem 6.1 [Cornet (1981a), Th. I.3.1]: *Let $x_0 \in X \subset \mathbb{R}^n$; suppose for some $\alpha > 0$, $X \cap B(x_0, \alpha)$ is compact.*
  a. *The following are equivalent:*
    i) $N_X(\cdot)$ *is closed at* x_0.
    ii) $T_X(\cdot)$ *is lower semicontinuous at* x_0.
    iii) $T_X(x_0) = TC_X(x_0)$.
  b. *If the equivalent properties in (a) hold, then $T_X(x_0)$ is convex.*

Having this characterization, Cornet next defines a set $X \subset \mathbb{R}^n$ to be *tangentially regular* if X is locally compact and the equivalent conditions in (a) of Theorem 6.1 hold at each $x_0 \in X$. This definition establishes a wide class of "nice" sets in $\mathbb{R}^n$, including in particular all locally compact convex sets. Cornet then extends to tangentially regular sets an existence theorem of Henry (1973) for generalized differential equations over convex sets [see Cornet (1981a), Th. II.4.1], and he proves the following key theorem:

Theorem 6.2 [Cornet (1981b), Th. 3.1]: *Let X be a nonempty, tangentially regular subset of $\mathbb{R}^n$, and let Φ be a continuous multifunction from X to $\mathbb{R}^n$ with nonempty compact convex values.*

Then for each $x_0 \in X$ there exist $T > 0$ and a Lipschitzian function $x: [0, T] \to X$ such that $x(0) = x_0$ and, for almost every $t \in [0, T]$,

$$-\dot{x}(t) = \{\Phi[x(t)] + N_X[x(t)]\}^+.$$

Theorem 6.2 thus establishes an existence result for a class of generalized differential equations even broader than that represented by (1.4). The application of these generalized differential equations to the modeling of dynamic systems, for example in economics, is the subject of current research efforts.

References

E. Allgower and K. Georg (1980), Simplicial and continuation methods for approximating fixed points and solutions to systems of equations. SIAM Review 22, pp. 28–85.

H. Brézis (1973), Opérateurs Maximaux Monotones (North-Holland Mathematics Studies No. 5). North-Holland, Amsterdam and London.

B. Cornet (1981a), Contributions à la théorie mathématique des mécanismes dynamiques d'allocation des ressources. Thèse d'état, Université Paris IX Dauphine.

B. Cornet (1981b), Existence of slow solutions for a class of differential inclusions. Cahiers de Mathématiques de la Décision No. 8131, Centre de Recherche de Mathématiques de la Décision, Université Paris IX Dauphine.

R. W. Cottle (1974), Manifestations of the Schur complement. Linear Algebra and its Applications 8, pp. 189–211.

R. W. Cottle (1976), Complementarity and variational problems. Symposia Mathematica XIX, pp. 177–208.

R. W. Cottle and G. B. Dantzig (1968), Complementary pivot theory of mathematical programming. Linear Algebra and its Applications 1, pp. 103–125. Reprinted in: Studies in Optimization (MAA Studies in Mathematics, Vol. 10); Mathematical Association of America, 1974.

J. E. Dennis and J. J. Moré (1977), Quasi-Newton methods: motivation and theory. SIAM Review 19, pp. 46–89.

B. C. Eaves (1971), On the basic theorem of complementarity. Math. Programming 1, pp. 68–75.

B. C. Eaves (1978), A locally quadratically convergent algorithm for computing stationary points. Technical Report SOL 78-13, Systems Optimization Laboratory, Stanford University.

T. Hansen and T. C. Koopmans (1972), On the definition and computation of a capital stock invariant under optimization. J. Econ. Theory 5, pp. 487–523.

P. Hartman and G. Stampacchia (1966), On some nonlinear differential-functional equations. Acta Mathematica 115, pp. 271–310.

C. Henry (1973), An existence theorem for a class of differential equations with multivalued right-hand sides. J. Math. Anal. Appl. 41, pp. 179–186.

W. W. Hogan (1975), Energy policy models for Project Independence. Computers and Operations Research 2, pp. 251–271.

N. H. Josephy (1979a), Newton's method for generalized equations. Technical Summary Report No. 1965, Mathematics Research Center, University of Wisconsin-Madison; available from National Technical Information Service under accession number AD A077 096.

N. H. Josephy (1979b), Quasi-Newton methods for generalized equations. Technical Summary Report No. 1966, Mathematics Research Center, University of Wisconsin-Madison; available from National Technical Information Service under accession number AD A077 097.

N. H. Josephy (1979c), A Newton method for the PIES energy model. Technical Summary Report No. 1971, Mathematics Research Center, University of Wisconsin-Madison; available from National Technical Information Service under accession number AD A077 102.

N. H. Josephy (1979d), Hogan's PIES example and Lemke's algorithm. Technical Summary Report No. 1972, Mathematics Research Center, University of Wisconsin-Madison; available from National Technical Information Service under accession number AD A077 103.

L. V. Kantorovich and G. P. Akilov (1964), Functional Analysis in Normed Spaces. Macmillan, New York [original in Russian: Fizmatgiz, Moscow, 1959].

S. Karamardian (1969a), The nonlinear complementarity problem with applications, Part I. J. Optimization Theory and Appl. 4, pp. 87–98.

S. Karamardian (1969b), The nonlinear complementarity problem with applications, Part II. J. Optimization Theory and Appl. 4, pp. 167–181.

S. Karamardian (1972), The complementarity problem. Math. Programming 2, pp. 107–129.

M. Kojima (1980), Strongly stable stationary solutions in nonlinear programs. Analysis and Computation of Fixed Points, ed. S. M. Robinson, Academic Press, New York, pp. 93–138.

II. Generalized Equations

B. Kummer (1982), Generalized equations: solvability and regularity. Preprint Nr. 30 (Neue Folge), Sektion Mathematik, Humboldt-Universität zu Berlin, DDR.

C. E. Lemke (1970), Recent results on complementarity problems. Nonlinear Programming, eds. J. B. Rosen, O. L. Mangasarian, K. Ritter, Academic Press, New York, pp. 349-384.

J. J. Moré (1974a), Coercivity conditions in nonlinear complementarity problems. SIAM Review 16, pp. 1-16.

J. J. Moré (1974b), Classes of functions and feasibility conditions in nonlinear complementarity problems. Math. Programming 6, pp. 327-338.

J. M. Ortega and W. C. Rheinboldt (1970), Iterative Solution of Nonlinear Equations in Several Variables. Academic Press, New York.

J. A. Reinoza (1979), A degree for generalized equations. Dissertation, Computer Sciences Department, University of Wisconsin-Madison.

J. A. Reinoza (1981), The strong positivity conditions. Preprint, Departamento de Matemáticas y Ciencias de la Computación, Universidad Simón Bolívar, Caracas, Venezuela.

S. M. Robinson (1974), Perturbed Kuhn-Tucker points and rates of convergence for a class of nonlinear-programming algorithms. Math. Programming 7, pp. 1-16.

S. M. Robinson (1976a), First order conditions for general nonlinear optimization. SIAM J. Appl. Math. 30, pp. 597-607.

S. M. Robinson (1976b), An implicit-function theorem for generalized variational inequalities. Technical Summary Report No. 1672, Mathematics Research Center, University of Wisconsin-Madison.

S. M. Robinson (1979), Generalized equations and their solutions, Part I: Basic Theory. Math. Programming Studies 10, pp. 128-141.

S. M. Robinson (1980), Strongly regular generalized equations. Math. Operations Res. 5, pp. 43-62.

S. M. Robinson (1981), Some continuity properties of polyhedral multifunctions. Math. Programming Studies 14, pp. 206-214.

R. T. Rockafellar (1969), Local boundedness of nonlinear, monotone operators. Michigan Math. Journal 16, pp. 397-407.

R. T. Rockafellar (1970), Maximality of sums of nonlinear monotone operators. Trans. Amer. Math. Soc. 149, pp. 75-88.

J. E. Spingarn (1977), Generic conditions for optimality in constrained minimization problems. Dissertation, Department of Mathematics, University of Washington, Seattle.

R. B. Wilson (1963), A simplicial algorithm for concave programming. Dissertation, Graduate School of Business Administration, Harvard University, Boston.

Generalized Subgradients in Mathematical Programming

R. T. Rockafellar*
University of Washington, Department of Mathematics, C138 Padelford Hall, GN-50, Seattle, WA 98195, USA

Abstract. Mathematical programming problems, and the techniques used in solving them, naturally involve functions that may well fail to be differentiable. Such functions often have "subdifferential" properties of a sort not covered in classical analysis, but which provide much information about local behavior. This paper outlines the fundamentals of a recently developed theory of generalized directional derivatives and subgradients.

Introduction

The theory of subgradients of convex functions is by now widely known and has found numerous applications in mathematical programming. It serves in the characterization of optimality conditions, sensitivity analysis, and the design and validation of algorithms. What is not so widely known is the surprisingly powerful extension of this theory to nonconvex functions. This has been achieved in full measure only in the last few years, following breakthroughs made by F. H. Clarke [13].

Generalized subgradients have already been applied to optimal control problems by Clarke [16], [17], [18], [19], [20], [21], and Thibault [57], to Lagrange multiplier theory and sensitivity analysis by Aubin [2], Aubin and Clarke [4], [5], Auslender [6], [7], Chaney [9], Clarke [22], Gauvin [26], Gauvin and Dubeau [27], Hiriart-Urruty [30], [31], [32], [34], Pomerol [42], Rockafellar [47], [48], [50], [52], [53], Mifflin, Nguyen and Strodiot [40], to nonlinear programming algorithms by Chaney and Goldstein [10], Goldstein [28], and Mifflin [38], [39], to stochastic programming by Hiriart-Urruty [33], and to game theory by Aubin [3]. Many more applications are surely possible and will be made when more researchers have gained familiarity with the concepts and results.

The purpose of the present article is to help matters along by reviewing the important role of nonsmooth, not necessarily convex functions in mathematical programming and describing briefly the central facts about such functions. Of course, much has to be left out. The reader can find further details in the lecture notes [49] and the many articles which are cited. Only the finite-dimensional case will be discussed, but the references often contain infinite-dimensional generalizations.

---

* Research supported in part by the Air Force Office of Scientific Research, United States Air Force, under grant no. F4960-82-K-0012.

1. The Role of Nonsmooth Functions

A real-valued function is said to be *smooth* if it is of class C^1, i.e. continuously differentiable. The functions which appear in classical problems in physics and engineering typically are smooth, but not so in economics and other areas where the operation of maximization and minimization are basic. Such operations give rise to quantities whose dependence on certain variables or parameters may not even be continuous, much less differentiable. Often these quantities nevertheless do exhibit some kind of generalized differentiability behavior, and this can be important in being able to work with them.

Let us look at a standard mathematical programming problem of the form

(P)
$$\text{minimize } f_0(x) \text{ over all } x \in C \text{ satisfying}$$
$$f_i(x) \begin{cases} \leq 0 & \text{for } i = 1, \ldots, s \\ = 0 & \text{for } i = s+1, \ldots, m, \end{cases}$$

where C is a subset of $\mathbb{R}^n$ and each f_i is a real-valued function on C, $i = 0, 1, \ldots, m$. Nonsmoothness can occur in connection with (P) not only because of the way the constraint and objective functions may be expressed, but also as an inescapable feature of auxiliary problems that may be set up as an aid to solving (P). For example, duality, penalty methods and decomposition techniques often require consideration of nonsmooth functions.

It is common in operations research to see problems in which the objective function f_0 is "piecewise linear":

(1.1)
$$f_0(x) = \max_{t=1, \ldots, r} \varphi_t(x),$$

where each φ_t is affine (i.e. linear + constant). Then the graph of f_0 can have "creases" and "corners" where differentiability fails. Sometimes f_0 is given as the supremum in an optimization problem in which x is a parameter vector:

(1.2)
$$f_0(x) = \sup_{t \in T} \varphi_t(x).$$

Here T could be a subset of some space $\mathbb{R}^d$, defined by a further system of constraints, or it could be any abstract set. With $T = \{1, \ldots, r\}$, we revert from (1.2) to (1.1). Such an f_0 may fail to be smooth, but if the functions φ_t are all affine, f_0 is at least *convex*. Indeed, this is almost a characterization of convexity: a real-valued function f_0 on $\mathbb{R}^n$ is convex if and only if it can be expressed in the form (1.2) for some collection of affine functions h_t. (Convexity relative only to a convex subset C or $\mathbb{R}^n$ can be characterized similarly when semicontinuity properties are present; see [43, §12].)

Thus, problems where merely the convexity of f_0 is a natural assumption, as in many economic applications, can well involve nonsmoothness. On the other hand, smooth convex functions do exist, so a formula of type (1.2) does not *preclude* smoothness of f_0.

How can one tell in a particular instance of (1.2) with smooth functions φ_t whether f_0 is smooth or not? More generally, what partial differentiability properties of f_0 can be deduced from (1.2)? Such questions are not covered by traditional mathematics, but they have attracted attention in optimization theory. Some answers will be provided below.

Much of what has just been said about f_0 also applies to the constraint functions in (P), at least for the inequality constraints. Such a function f_i could be expressed as in (1.1) or (1.2); in particular, f_i might be convex without being smooth. Of course, if

(1.3) $$f_i(x) = \sup_{t \in T_i} \varphi_{it}(x),$$

with smooth φ_{it}'s, the single constraint $f_i(x) \leq 0$ is equivalent to a system of smooth constraints:

(1.4) $$\varphi_{it}(x) \leq 0 \quad \text{for every } t \in T_i.$$

In this sense it may seem artificial to be worried about f_i being nonsmooth. Why not just write (1.4) rather than $f_i(x) \leq 0$?

Actually, the thinking can go just as well in the other direction. The set T_i may be infinite, or if finite, very large. Thus it may not be practical to treat (1.4) in full. Lumping (1.4) together as a single condition $f_i(x) \leq 0$ is a form of constraint *aggregation*. If enough is known about the behavior of formulas of type (1.3), it may be possible to treat f_i directly, generating only the particular φ_{it}'s needed locally at any time.

Note that a function of form (1.2) might have $+\infty$ as a value, unless restrictions are imposed on T and the way that $\varphi_t(x)$ depends on t. Minimizing such an extended-real-valued function is definitely a matter of interest too, for instance in connection with duality-based methods.

Duality is a major source of nonsmooth optimization problems. The ordinary dual associated with (P) is

(D) $\qquad\qquad$ maximize $g(y)$ over all $y \in Y$

where

(1.5) $$Y = \{y = (y_1, \ldots, y_m) \in R^m \mid y_i \geq 0 \text{ for } i = 1, \ldots, s\},$$

(1.6) $$g(y) = \inf_{x \in C} \varphi_x(y) \quad \text{for } \varphi_x(y) = f_0(x) + \sum_{i=1}^{m} y_i f_i(x).$$

Except for the reversal of maximization and minimization, we can identify this as an instance of a problem of type (P) where the objective function is represented as in (1.2) but might not be finite everywhere. Since $\varphi_x(y)$ is affine in y, g is a concave function, quite possibly nonsmooth.

Several important methods for solving (P) proceed by way of (D), despite nonsmoothness. In Dantzig-Wolfe decomposition, a cutting plane algorithm is used to maximize g. If C happens to be a convex polyhedron and the functions f_i are affine, g is piecewise linear and the maximization of g can be formulated simply as a problem in linear programming. The number of "pieces" of g can be so enormous, however, that this approach is impractical. More hope lies in treating g as a nonsmooth function which nevertheless has useful "subdifferential" properties, as will be discussed later. This is also the case in integer programming methods which solve (D) partially in order to obtain a lower bound for the minimum in (P) (cf. Held, Wolfe and Crowder [29]).

The dual of a "geometric" programming problem in the sense of Duffin, Peterson and Zener [25] provides another example. This consists of maximizing, subject to linear constraints, a certain finite, concave function of nonnegative variables which has a formula in terms of logarithms that looks harmless enough. The function fails, however, to be differentiable at boundary points of its domain.

Exact penalty methods, which replace (P) by a single minimization problem with no constraints, or at least simpler constraints, also lead to nonsmoothness. Under mild assumptions on (P), it will be true that for $r>0$ sufficiently large the optimal solutions to (P) are the same as those to the problem

$$\text{minimize } f(x) \text{ over all } x \in C, \text{ with}$$

(1.7) $$f(x) = f_0(x) + r\left(\sum_{i=1}^{s} [f_i(x)]_+ + \sum_{i=1}^{s+1} |f_i(x)|\right),$$

where

(1.8) $$[\alpha]_+ = \max\{\alpha, 0\},$$

and of course $|\alpha| = \max\{\alpha, -\alpha\}$. This type of penalty function has been considered by Zangwill [62], Petrzykowski [41], Howe [36], Conn [12] and Chung [11], for instance. If the functions f_i are smooth, the f in (1.7) can be represented as in (1.1) with smooth φ_l's. Thus f is piecewise smooth, not everywhere differentiable. Another exact penalty approach, based on the (quadratic-type) augmented Lagrangian for (P) (see [55]), preserves first-order smoothness of the f_i's but can create discontinuities in second derivatives.

So far we have been discussing nonsmoothness of the kind which arises from representations (1.2), (1.3). This is relatively easy to deal with, but a trickier kind of nonsmoothness is encountered in parametric programming and the study of Lagrange multipliers. Suppose that in place of (P) we have a problem which depends on a parameter vector $v = (v_1, \ldots, v_d) \in R^d$:

$$\text{minimize } f_0(v, x) \text{ over all } x \in C(v) \subset R^n$$

(P_v) $$\text{such that } f_i(v, x) \begin{cases} \geq 0 & \text{for } i=1, \ldots, s, \\ =0 & \text{for } i=s+1, \ldots, m. \end{cases}$$

Let $p(v)$ denote the optimal value in (P_v). Then p is a function on R^d whose values can in general be not just real numbers but $-\infty$ and $+\infty$ (the latter for v such that (P_v) is infeasible). What can be said about generalized differentiability properties of p?

Actually, p can be represented by

(1.9) $$p(v) = \inf_{x \in R^n} \varphi_x(v),$$

where

(1.10) $$\varphi_x(v) = \begin{cases} f_0(v, x) & \text{if } x \text{ is a feasible solution to } (P_v), \\ \infty & \text{otherwise}. \end{cases}$$

Thus p is the pointwise infimum of a collection of functions on R^d. These functions φ_x are not smooth, though, even if the f_i's are smooth, because of the jump to $+\infty$ when the parameter vector v shifts into a range where the feasibility of a given x is lost. For this reason the nonsmoothness of p is harder to analyze, yet strong results have been obtained (cf. [48] and its references). Generally speaking, directional derivative properties of p at v are closely related to the possible Lagrange multiplier vectors associated with optimality conditions for (P_v).

The simplest case is the one where $f_i(v, x)$ is convex jointly in (v, x), and the set of (x, v) satisfying $x \in C(v)$ is convex. Then p is a convex function. Even so, p may have nonfinite values.

The importance of p in sensitivity analysis is clear: we may need to know the rate of change of the optimal value $p(v)$, in some sense, as v varies. The components of v may be economic quantities subject to fluctuation or control. In Benders decomposition, (P_v) is just a subproblem; the real task consists of minimizing in x and v jointly, subject to the given conditions. The residual problem, or master problem, connected with this formulation, is that of minimizing $p(v)$ over all $v \in R^n$. Obviously, generalized differential properties of p can have much to do with success in this task.

Finally, we wish to point out that in the broad picture of optimization theory there are other nonsmooth functions worthy of consideration. Prime examples are the *indicators* ψ_C of sets $C \subset R^n$:

$$(1.11) \qquad \psi_C(x) = \begin{cases} 0 & \text{if } x \in C, \\ \infty & \text{if } x \notin C. \end{cases}$$

When C is a convex set, ψ_C is a convex function. "Tangential" properties of C correspond to "differential" properties of ψ_C. The study of these provides a bridge between geometry and analysis that leads to a deeper understanding of many topics, such as the characterization of optimal solutions.

2. Useful Classes of Functions

As we have observed, it is not enough just to consider real-valued functions: the values $+\infty$ and $-\infty$ sometimes need to be admitted. We shall speak of a function $f: R^n \to [-\infty, \infty]$ as *proper* if $f(x) > -\infty$ for all x, and $f(x) < \infty$ for at least one x. The set

$$(2.1) \qquad \text{dom} f = \{x \in R^n | f(x) < \infty\}$$

is called the *effective domain* of f, and

$$(2.2) \qquad \text{epi} f = \{(x, \alpha) \in R^{n+1} | f(x) \leq \alpha\}$$

the *epigraph* of f. For most purposes, there is little generality lost in concentrating on the case where f is *lower semicontinuous*:

$$(2.3) \qquad f(x) = \liminf_{x' \to x} f(x') \quad \text{for all } x \in R^n.$$

II. Generalized Subgradients in Mathematical Programming

This property holds if and only if $\mathrm{epi}\, f$ is a closed set.

Convex analysis [43] serves as a guide in developing a theory of differentiability in such a general context. The function f is convex if and only if $\mathrm{epi}\, f$ is a convex set. We shall see below that amazingly much of the theory of one-sided directional derivatives and subgradients of convex functions has a natural extension to the class of all lower semicontinuous proper functions on R^n. Incidentally, it is easy to give conditions on the parameterized problem (P_v) in § 1 which ensure that the corresponding optimal value function p is lower semicontinuous and proper; see [48].

Although a function f may be extended-real-valued in the large, we are often interested in situations where f is finite in a neighborhood of a certain point x and has stronger properties in such a neighborhood as well. Among the most important properties to be considered in such a context is Lipschitz continuity: f is *Lipschitzian* on an open set U if f is real-valued on U and there exists a number $\lambda \geq 0$ such that

(2.4) $\qquad |f(x') - f(x)| \leq \lambda |x' - x| \quad \text{when} \quad x \in U, x' \in U.$

(Here $|\cdot|$ stands for the Euclidean norm on R^n.) This condition can also be put in the form

(2.5) $\qquad [f(x+th) - f(x)]/t \leq \lambda |h| \quad \text{when} \quad x \in U, x+th \in U, t > 0.$

Thus it expresses a bound on difference quotients.

We shall say f is *Lipschitzian around* $\bar{x}$ if it is Lipschitz continuous in some neighborhood of $\bar{x}$, and that f is *locally Lipschitzian* on an open set U if f is Lipschitzian around each $\bar{x} \in U$. The distinction between this and simply being Lipschitzian on U is that the modulus λ in (2.4) need not be valid for all of U, but may change from one neighborhood to another. Two classical results on local Lipschitz continuity may be cited.

Theorem 2.1 [43, § 10]. *A convex function is locally Lipschitzian on any open subset of R^n where its values are finite.*

Theorem 2.2 (Rademacher; cf. [56]). *A locally Lipschitzian function f on an open set $U \subset R^n$ is differentiable except at a negligible set of points in U.*

A *negligible* set is a set of measure zero in the Lebesgue sense: for any $\varepsilon > 0$, it can be covered by a sequence of balls whose total n-dimensional volume does not exceed ε.

A more subtle form of Lipschitz continuity that will be important to us below is the following. Suppose f is a lower semicontinuous, proper function on R^n, and let $\bar{x} \in \mathrm{dom}\, f$. We say that f is *directionally Lipschitzian at $\bar{x}$ with respect to the vector $\bar{h}$* if

there exist $\varepsilon > 0$ and $\lambda \geq 0$ such that

(2.6) $\qquad [f(x+th) - f(x)]/t \leq \lambda |h| \quad \text{when} \quad f(x) \leq f(\bar{x}) + \varepsilon,$
$\qquad |x - \bar{x}| \leq \varepsilon, \; |h - \bar{h}| \leq \varepsilon, \; 0 < t < \varepsilon.$

If $\bar{h} = 0$, this reduces to f being Lipschitzian around $\bar{x}$.

We say simply that f is *directionally Lipschitzian at* $\bar{x}$ if (2.6) holds for at least one $\bar{h}$. Note that f need not be finite on a neighborhood of $\bar{x}$ for this condition to hold. For example, if f is *convex*, then f is directionally Lipschitzian at $\bar{x}$ with respect to any $\bar{h}$ such that $\bar{x}+t\bar{h}\in \text{int}(\text{dom} f)$ for $t>0$ sufficiently small [46]. Thus if $\text{int}[\text{dom} f]\neq\emptyset$, f is directionally Lipschitzian at every $\bar{x}\in\text{dom} f$.

Another example: if f is a *nondecreasing* function on R^n in the sense that

(2.7) $$f(x)\leq f(x') \quad \text{when } x'-x\in R^n_+,$$

then f is directionally Lipschitzian at every $\bar{x}\in\text{dom} f$. (Consider $\bar{h}\in\text{int} R^n_+$.)

From a geometrical point of view, f is directionally Lipschitzian at $\bar{x}$ if and only if epi f has "Lipschitzian boundary" in a neighborhood of $(\bar{x}, f(\bar{x}))$; see [44].

Moving to properties stronger than Lipschitz continuity but still short of actual smoothness, we come upon two highly significant classes of functions already suggested by the discussion in §1. Let us say that f is *subsmooth* (or *lower-$\mathscr{C}^1$*) *around* $\bar{x}$ if there is an open neighborhood X of $\bar{x}$ and a representation

(2.8) $$f(x)=\max_{t\in T}\varphi_t(x) \quad \text{for all } x\in X,$$

where T is a compact topological space, each φ_t is of class $\mathscr{C}^1$, and the values of φ_t and its first partial derivatives are continuous not only with respect to $x\in X$ but $(t,x)\in T\times X$. (In particular, T could be any finite index set in the discrete topology. Then the continuity requirements in t are trivial; f is just expressible locally as the pointwise maximum of a finite collection of smooth functions as in (1.1).) We shall say f is *subsmooth on* U, an open set in R^n, if f is subsmooth around every $\bar{x}\in U$. (The representation (2.8) may be different for different regions of U.) Obviously every smooth function is also subsmooth (take T to be a singleton).

In a similar vein, we shall call f *strongly subsmooth (of order r) on* U (or *lower-$\mathscr{C}^r$*), if in the local representations (2.8) around points of U the functions φ_t are actually of class $\mathscr{C}^r$ with $2\leq r\leq\infty$, and their partial derivatives up through order r depend continuously on (t,x).

Theorem 2.3 [13]. *If f is subsmooth (or strongly subsmooth) on an open set $U\subset R^n$, then f is locally Lipschitzian on U.*

Theorem 2.4 [51]. *The classes of strongly subsmooth functions of order r on U, for $2\leq r\leq\infty$, all coincide, so that one can speak simply of a single class of strongly subsmooth functions on U without reference to any particular r. There do exist subsmooth functions which are not strongly subsmooth, however.*

Theorem 2.5 [51]. *A real-valued function f on an open set $U\subset R^n$ is strongly subsmooth on U if and only if in some open convex neighborhood X of each $\bar{x}\in U$ there is a representation:*

(2.9) $$f(x)=g(x)+h(x) \quad \text{for all } x\in X,$$

with g convex, h of class $\mathscr{C}^2$.

Then there exist such representations with h actually of class $\mathscr{C}^\infty$, in fact with $h(x) = -\rho|x|^2, \rho > 0$.

Corollary 2.6. *A convex function is strongly subsmooth on any open subset of R^n where its values are finite.*

One other class of functions of great importance in optimization deserves mention: the *saddle functions*. Suppose $f(y, z)$ is convex in $y \in Y$ and concave in $z \in Z$, where $Y \times Z \subset R^n$ is convex. Then f is locally Lipschitzian on the interior of $Y \times Z$ [43, Theorem 35.1] and has many other properties, such as the existence of one-sided directional derivatives [43, Theorem 35.6], but f is not subsmooth. More generally, one could consider the class of all functions expressible locally as linear combinations of such saddle functions along with convex and concave functions. No abstract characterization of this class is known.

3. Sublinear Functions Representing Convex Sets

In the classical approach to differentiability, one seeks to approximate a function f around a point x by a *linear* function l:

(3.1) $$f(x+h) - f(x) = l(h) + o(|h|).$$

This l expresses directional derivatives of f at x with respect to various vectors h. Next, one uses the duality between linear functions l on R^n and vectors $y \in R^n$ to define the *gradient* of f at x: there is a unique y such that

(3.2) $$l(h) = y \cdot h \quad \text{for all } h \in R^n.$$

This y is the gradient $\nabla f(x)$.

We cannot limit ourselves merely to linear functions l as approximations, in trying to capture the generalized differentiability properties of functions f belonging to the various classes mentioned in § 2. A broader duality correspondence than the one between vectors and linear functions is therefore required. The correspondence about to be described replaces vectors $y \in R^n$ by closed convex sets $Y \subset R^n$.

A function $l: R^n \to [-\infty, +\infty]$ is said to be *sublinear* if l is convex, positively homogeneous ($l(\lambda h) = \lambda l(h)$ for $\lambda > 0$), and $l(0) < \infty$. These conditions mean that epi l is a closed convex cone in R^{n+1} containing the origin. Every linear function is in particular sublinear.

Theorem 3.1 [43, § 13]. *Let l be a sublinear function on R^n which is lower semicontinuous. Then either l is proper, with $l(0) = 0$, or l has no values other than $+\infty$ and $-\infty$. In the latter case, the set of points h where $l(h) = -\infty$ is a closed convex cone containing 0.*

Theorem 3.2 [43, § 13]. *There is a one-to-one correspondence between proper, lower semicontinuous, sublinear functions l on R^n and the nonempty, closed, convex subsets Y of R^n, given by*

(3.3) $$l(h) = \sup_{y \in Y} y \cdot h \quad \text{for all } h \in R^n$$

(3.4) $$Y = \{y \in R^n \mid y \cdot h \leq l(h) \text{ for all } h \in R^n\}.$$

The special case of singleton sets Y yields the classical correspondence between linear functions and vectors.

The Euclidean norm $l(h) = |h|$ is an example of a sublinear function which is not linear. It corresponds to the ball $Y = \{y \mid |y| \leq 1\}$. The sublinear function

(3.5) $$l(h) = \max\{a_1 \cdot h, \ldots, a_m \cdot h\}$$

corresponds to the polytope

(3.6) $$Y = \mathrm{co}\{a_1, \ldots, a_m\}.$$

The function

(3.7) $$l(h) = \begin{cases} 0 & \text{if } h \in K \\ \infty & \text{if } h \notin K, \end{cases}$$

where K is a closed convex cone containing 0, corresponds to the polar cone $Y = K^\circ$.

A finite sublinear function on R^n, being convex, is continuous (cf. Theorem 2.1), hence certainly lower semicontinuous. This gives the following special case of Theorem 3.2.

Corollary 3.3 [43]. *Formulas (3.3), (3.4), give a one-to-one correspondence between the finite sublinear functions l on R^n and the nonempty compact convex subsets Y of R^n.*

The duality between finiteness of l and boundedness of Y extends to a more detailed relationship. Here we recall that since Y is a closed convex set, if there is no halfspace having a nonempty bounded intersection with Y, then there is a vector $z \neq 0$ such that the line $\{y + tz \mid t \in R\}$ is included in Y for every $y \in Y$ [43, §13]. In the latter case, Y is just a bundle of parallel lines.

Theorem 3.4 [43, §13]. *Under the correspondence in Theorem 3.2, one has $h \in \mathrm{int}(\mathrm{dom}\, l)$ if and only if for some $\beta \in R$ the set $\{y \in Y \mid y \cdot h \geq \beta\}$ is bounded and nonempty. Thus the convex cone $\mathrm{dom}\, l$ has nonempty interior if and only if Y cannot be expressed as a bundle of parallel lines.*

The pattern we ultimately wish to follow is that of defining for a given function f and point x a kind of generalized directional derivative which is a *lower semicontinuous sublinear function l* of the direction vector h. The elements y of the corresponding closed convex set Y will be the "subgradients" of f at x.

4. Contingent Derivatives

For a function $f: R^n \to [-\infty, \infty]$ and a point x where f is finite, the ordinary one-sided *directional derivative of f at x with respect to h* is

(4.1) $$f'(x; h) = \lim_{t \downarrow 0} \frac{f(x + th) - f(x)}{t},$$

II. Generalized Subgradients in Mathematical Programming

if it exists (possible as $+\infty$ or $-\infty$). This concept of a derivative has its uses, but it is inadequate for dealing with functions that are not necessarily continuous at x. One of the drawbacks is that difference quotients are only considered along rays emanating from x; no allowance is made for "curvilinear behavior".

As an illustration, consider the indicator function of the curve $C = \{x = (x_1, x_2) \in R^2 | x_2 = x_1^3\}$:

(4.2) $$f(x_1, x_2) = \begin{cases} 0 & \text{if } x_2 = x_1^3 \\ \infty & \text{if } x_2 \neq x_1^3. \end{cases}$$

Obviously $f'(0, 0; h_1, h_2) = \infty$ for all nonzero vectors $h = (h_1, h_2)$. There is nothing to distinguish the vectors tangent ot the curve C at $(0, 0)$, namely the ones with $h_2 = 0$, from the others. Although the function (4.2) may be seen as an extreme case, it reflects a lack of responsiveness of ordinary directional derivatives that is rather widespread.

Certainly in handling optimal value functions like the p introduced towards the end of § 1, it is essential to consider more subtle limits of difference quotients. Instead of just sequences of points $x^k = x + t_k h$ with $t_k \downarrow 0$, one at least needs to look at the possibility of $x^k = x + t_k h^k$ with $t_k \downarrow 0$ and $h^k \to h$. (A sequence $\{x^k\}$ which can be expressed in the latter form with $h \neq 0$ is said to *converge to x in the direction of h*.)

A useful concept in this regard is that of the *contingent derivative* (also called the *lower Hadamard derivative*) of f at x with respect to h:

(4.3) $$f^*(x; h) = \liminf_{\substack{h' \to h \\ t \downarrow 0}} \frac{f(x + th') - f(x)}{t}$$

$$= \lim_{\varepsilon \downarrow 0} \left[\inf_{|h' - h| \leq \varepsilon} \frac{f(x + th') - f(x)}{t} \right].$$

In contrast to the ordinary derivative $f'(x; h)$, this kind of limit always exists, of course. In example (4.2) one gets

$$f^*(0, 0; h_1, h_2) = \begin{cases} 0 & \text{if } h_2 = 0, \\ \infty & \text{if } h_2 \neq 0. \end{cases}$$

Strong geometric motivation for the contingent derivative is provided by its epigraphical interpretation. To explain this, we recall that the *contingent cone* to a closed set $C \subset R^n$ at a point $x \in C$ is the set

(4.4) $$K_C(x) = \{h | \exists t_k \downarrow 0, h^k \to h, \text{ with } x + t_k h^k \in C\}.$$

Thus $K_C(x)$ consists of the zero vector and all vectors $h \neq 0$ giving directions in which sequences in C can converge to x. It is elementary that $K_C(x)$ is closed, and that $K_C(x)$ truly is a *cone*, i.e. $\lambda h \in K_C(x)$ when $h \in K_C(x)$ and $\lambda > 0$.

The contingent cone was introduced by Bouligand in the 1930's (cf. [56]). In the mathematical programming literature, it is often called the "tangent cone", but we prefer to reserve that term for another concept to be described in § 6 and keep to the classical usage. The following fact is easily deduced from the definitions.

Theorem 4.1. *For the contingent derivative function $l(h)=f^*(x;h)$, epi l is the contingent cone to epi f at the point $(x, f(x))$.*

Corollary 4.2. *The contingent derivative $f^*(x;h)$ is a lower semicontinuous, positively homogeneous function of h with $f^*(x;0) < \infty$.*

Since contingent cones to convex sets are convex, we also have the following.

Corollary 4.3. *Suppose f is a proper convex function, and let $x \in \text{dom} f$. Then the contingent derivative function $l(h)=f^*(x;h)$ is not only lower semicontinuous but sublinear.*

The convex case also exhibits a close relationship between ordinary directional derivatives and contingent derivatives.

Theorem 4.4 [43]. *Suppose f is a proper convex function on R^n, and let $x \text{dom} f$. Then $f'(x;h)$ exists for every h (possibly as $+\infty$ or $-\infty$) and*

$$(4.5) \qquad f^*(x;h) = \liminf_{h' \to h} f'(x;h').$$

In fact $f^*(x;h) = f'(x;h)$ for every h such that $x+th \in \text{int}(\text{dom} f)$ for some $t \to 0$.

The sublinearity in Corollary 4.3 is the basis of the well known subgradient theory of convex analysis [43], elements of which will be reviewed in §7. Unfortunately, contingent cones to nonconvex sets generally are *not* convex. By the same token, the contingent derivative function $l(h)=f^*(x;h)$ is generally *not* convex when f is not convex, except for certain important cases which will be noted in §6. The contingent derivative does not, therefore, lead to a robust theory of subgradients in terms of the duality correspondence in §3.

What is needed is a concept of directional derivative possessed of an inherent convexity. Such a concept has been furnished by Clarke [13] for locally Lipschitzian functions. We shall present it in the next section in an extended form which to a certain extent is useful also in connection with functions which are not locally Lipschitzian. The full concept of derivative needed in treating such general functions, the so-called "subderivative", will not be discussed until §6, however.

5. Clarke Derivatives

For a lower semicontinuous function f and a point x where f is finite, we define the *extended Clarke derivative* of f at x with respect to the vector h to be

$$(5.1) \qquad f^\circ(x;h) = \limsup_{\substack{x' \to_f x \\ h' \to h \\ t \downarrow 0}} \frac{f(x'+th') - f(x')}{t}$$

II. Generalized Subgradients in Mathematical Programming 379

where the notation is used that

(5.2) $$x \to_f x' \Leftrightarrow \begin{cases} x' \to x \\ f(x') \to f(x). \end{cases}$$

If f happens to be continuous at x, one has $f(x') \to f(x)$ when $x' \to x$, so the extra notation is unnecessary. The case where f is not necessarily continuous at x is of definite interest, however, as the following fact (immediate from the definition) well indicates.

Theorem 5.1. *Let f be a lower semicontinuous, proper function on R^n and let $x \in \text{dom } f$. Then for a vector h one has $f^\circ(x; h) < \infty$ if and only if f is directionally Lipschitzian at x with respect to h.*

Clarke's original definition of $f^\circ(x; h)$ in [13] applied only to the case where f actually is Lipschitzian around x. The formula then takes on a simpler form.

Theorem 5.2. *Suppose f is Lipschitzian around x. Then*

(5.3) $$f^\circ(x; h) = \limsup_{\substack{x' \to x \\ t \downarrow 0}} \frac{f(x' + th) - f(x')}{t}$$

A striking feature of the extended Clarke derivative is its inherent convexity.

Theorem 5.3. *Let f be a lower semicontinuous, proper function on R^n, and let $x \in \text{dom } f$. Then the extended Clarke derivative function $l(h) = f^\circ(x; h)$ is convex and positively homogeneous, but not necessarily lower semicontinuous (in fact $\text{dom } l$ is an open convex cone).*

If f is Lipschitzian around x, a property equivalent to having $\infty > l(0) = f^\circ(x; 0)$, then l is not only sublinear but finite everywhere (hence continuous).

This result too is simply a consequence of the definitions. We take note now of a very important case where extended Clarke derivatives, contingent derivatives and ordinary directional derivatives all have the same value.

Theorem 5.4. *Suppose f is subsmooth around x. Then $f'(x; h)$ exists as a real number (finite) for every $h \in R^n$, and*

(5.4) $$f'(x; h) = f^*(x; h) = f^\circ(x; h).$$

In fact for any local representation of f as in (2.8) (with φ_t smooth, and $\varphi'_t(x; h)$ continuous jointly in t and x for each h) one has

(5.5) $$f'(x; h) = \max_{t \in T_x} \varphi'_t(x; h) \quad \text{for all } h \in R^n,$$

where

(5.6) $$T_x = \{t \in T \mid \varphi_t(x) = f(x)\} = \arg\max_{t \in T} \varphi_t(x).$$

Formula (5.5) was first proved by Danskin [24], and the equation $f'(x; h) = f^\circ(x; h)$ was established by Clarke [13]. The proof of (5.5) shows that

$f^*(x;h)$ is given by the same maximum, in fact directional derivatives exist in the Hadamard sense:

(5.7) $$\lim_{\substack{h'\to h \\ t\downarrow 0}} \frac{f(x+th')-f(x)}{t} = \max_{t\in T_x} \varphi_t'(x;h) \quad \text{for all } h\in R^n$$

A subgradient version of (5.5) will be given in Theorem 7.3 and Corollary 7.4. It is good to note, however, that (5.4) associates with each x a finite, sublinear function l that depends only on f and not on any particular max representation (2.8).

For functions which are not locally Lipschitzian or even directionally Lipschitzian, the extended Clarke derivative conveys little information. Thus for the function in (4.2), one has $f°(0,0;h_1,h_2) = \infty$ for all (h_1,h_2). Indeed, if f is not directionally Lipschitzian at x, one necessarily has $f°(x;h) = \infty$ for all h by Theorem 5.1.

6. Subderivatives and the Clarke Tangent Cone

A type of directional derivative will now be described which is able to serve as the foundation for a theory of subgradients of very general functions. It agrees with the contingent derivative and extended Clarke derivative in the important cases where those derivatives yield a lower semicontinuous, sublinear function l of the direction vector h. But, it yields such an l no matter what the circumstances with the other derivatives.

Let f be lower semicontinuous and proper, and let $x \in \text{dom } f$. The *subderivative* of f at x with respect to h is defined to be

(6.1) $$f^\uparrow(x;h) = \lim_{\varepsilon\downarrow 0}\left[\limsup_{\substack{x'\to_f x \\ t\downarrow 0}} \left[\inf_{|h'-h|\le\varepsilon} \frac{f(x'+th')-f(x')}{t}\right]\right],$$

where again the notation (5.2) is used as a shorthand. This rather complicated limit is a sort of amalgam of the ones used in defining $f^*(x;h)$ and $f°(x;h)$; cf. (4.3) and (5.1). It was first given by Rockafellar [45], but as we shall see below, it is closely related to a geometrical notion of Clarke [13]. The initial difficulties in appreciating the nature of $f^\uparrow(x;h)$ are far outweighed by its remarkable properties. A convincing case can be made for this derivative as the natural one to consider for general functions f. Clearly

(6.2) $$f^*(x;h) \le f^\uparrow(x;h) \le f°(x;h) \quad \text{for all } h.$$

Theorem 6.1 [45]. *Let f be a lower semicontinuous, proper function on R^n, and let $x \in \text{dom } f$. Then the subderivative function $l(h) = f^\uparrow(x;h)$ is lower semicontinuous and sublinear.*

Just as the contingent derivative corresponds to a contingent cone to the epigraph of f (cf. Theorem 4.1), the subderivative corresponds to another geometrical concept. For a closed set $C \subset R^n$, the *Clarke tangent cone* at a point $x \in C$ is defined to be

(6.3) $\quad T_C(x) = \{h | \forall x^k \to x \text{ in } C, t_k \downarrow 0, \exists h^k \to h \text{ with } x^k + t_k h^k \in C\}$.

This is always a *closed convex cone containing* 0, a surprising fact in view of the lack of any convexity assumptions whatsoever on C. For a direct proof, see [44]. The cone $T_C(x)$ was originally defined by Clarke [13] in a more circuitous manner, but formula (6.3) turned out to be implicit in one of his results (see [13, Prop. 3.7]). We shall say more about the properties of this cone in Theorem 6.8. Obviously the Clarke tangent cone is a subset of the contingent cone in §4:

(6.4) $\quad T_C(x) \subset K_C(x)$.

Theorem 6.2 [45]. *Let f be a lower semicontinuous function on R^n, and let $x \in \text{dom} f$. Let $l(h) = f^\uparrow(x; h)$. Then the epigraph of l is the Clarke tangent cone to the epigraph of f at $(x, f(x))$.*

The relationship between subderivatives and the Clarke derivatives of the preceding section is quite simple.

Theorem 6.3 [46]. *Let f be a lower semicontinuous, proper function on R^n, and let $x \in \text{dom} f$. Then*

(6.5) $\quad \text{int}\{h \in R^n | f^\uparrow(x; h) < \infty\} = \{h \in R^n | f^\circ(x; h) < \infty\}$,

and on this open convex cone one has $f^\uparrow(x; h) = f^\circ(x; h)$.

Corollary 6.4. *Let f be a lower semicontinuous, proper function on R^n, and let $x \in \text{dom} f$. Then one has the epigraphical relationship*

(6.6) $\quad \text{int}\{(h, \beta) \in R^{n+1} | \beta \geq f^\uparrow(x; h)\} = \{(h, \beta) \in R^{n+1} | \beta > f^\circ(x; h)\}$.

Corollary 6.5. *Let f be a lower semicontinuous, proper function on R^n, and let $x \in \text{dom} f$. Let $\bar{h}$ be an element of the cone (6.5). Then*

(6.7) $\quad f^\uparrow(x; h) = \lim_{\varepsilon \downarrow 0} f^\circ(x; (1-\varepsilon)h + \varepsilon \bar{h}) \quad \text{for all } h \in R^n$.

Corollaries 6.4 and 6.5 follow from Theorem 6.3 by way of basic facts about the closures and interiors of epigraphs of convex functions [43, Lemma 7.3 and Corollary 7.5.1]. Of course, the cones (6.5) and (6.6) are nonempty only when f is directionally Lipschitzian at x; cf. Theorem 5.1. Thus in the directionally Lipschitzian case, $f^\uparrow$ can be constructed from the simpler function f° by taking $f^\uparrow(x; h) = f^\circ(x; h)$ on the cone (6.5) and limits (6.7) at boundary points, but when f is not directionally Lipschitzian, no help can be obtained from f° at all.

The relationship between subderivatives and contingent derivatives is not as easy to describe, although it does turn out that subderivatives can be expressed as certain limits of contingent derivatives. First of all, let us introduce the terminology: f is *subdifferentially regular* at x if $f^\uparrow(x; h) = f^*(x; h)$ for all h.

Theorem 6.6 [13], [46]. *If f is a lower semicontinuous and convex on a neighborhood of x (a point where f is finite), or if f is subsmooth around x, then f is subdifferentially regular at x.*

Theorem 6.7 [46]. *If f is a lower semicontinuous, proper function on R^n and subdifferentially regular at the point $x \in \text{dom} f$, then for every h in the cone (6.5) the ordinary directional derivative $f'(x; h)$ exists, and $f'(x; h) = f^\uparrow(x; h)$.*

Theorem 6.7 provides a natural generalization of Theorems 4.4 and 5.3, which correspond to the cases in Theorem 6.6. An example of a subdifferentially regular function not covered by Theorem 6.6 is given by (4.2); this function is not covered by Theorem 6.7 either, since it is not directionally Lipschitzian; the set (6.5) is empty for every $x \in \text{dom} f$.

To gain deeper understanding of the cases where f is subdifferentially regular, we can appeal to a result of Cornet [23] about the relationship between the Clarke tangent cone and the contingent cone.

Theorem 6.8 [23]. *Let C be a closed set in R^n and let $x \in C$. Then*

(6.8) $$T_C(x) = \liminf_{\substack{x' \to x \\ x' \in C}} K_C(x)$$

$$:= \{h \mid \forall x^k \to x \text{ with } x^k \in C, \, \exists h^k \to h \text{ with } h^k \in K_C(x^k)\}.$$

Thus $T_C(x) = K_C(x)$ if and only if the multifunction $K_C: x \mapsto K_C(x)$ is lower semicontinuous at x relative to C.

This can be applied to the epigraphs of $f^\uparrow(x; \cdot)$ and $f^*(x; \cdot)$ due to Theorems 4.1 and 6.2. The limit in (6.8) can be expressed in function terms using a fact in [46, Proposition 1]. We then obtain the following formula for $f^\uparrow$ in terms of f^*.

Theorem 6.9. *Let f be lower semicontinuous, and let x be a point where f is finite. Then for all h,*

(6.9) $$f^\uparrow(x; h) = \lim_{\varepsilon \downarrow 0} \left[\limsup_{x' \to_f x} \left[\inf_{|h' - h| \leq \varepsilon} f^*(x'; h') \right] \right].$$

This has force whether or not f is directionally Lipschitzian at x.

Corollary 6.10. *Let f be lower semicontinuous, and let x be a point where f is finite. Then f is subdifferentially regular at x if and only if for every sequence $x^k \to x$ with $f(x^k) \to f(x)$ and every h, there is a sequence $h^k \to h$ with*

$$\limsup_{k \to \infty} f^*(x^k; h^k) \leq f^*(x; h).$$

Incidentally, in infinite-dimensional spaces Theorem 6.8 generally fails (cf. Treiman [61]).

7. Generalized Subgradients

The sublinearity and lower semicontinuity of the subderivative function in Theorem 6.1 make it possible to invoke the duality correspondence in § 3. For

II. Generalized Subgradients in Mathematical Programming

an arbitrary lower semicontinuous function f on R^n and point x where f is finite, we define

(7.1) $$\partial f(x) = \{y \in R^n \mid y \cdot h \leq f^\uparrow(x;h) \text{ for all } h \in R^n\}.$$

The vectors $y \in \partial f(x)$ are called *subgradients* (or *generalized gradients*) of f at x. This terminology is totally in harmony with other usage, e.g. in convex analysis, as will be seen in a moment. From Theorems 3.1, 3.2 and 6.1 we immediately see the exact connection between subderivatives and subgradients.

Theorem 7.1. *Let f be a lower semicontinuous, proper function on R^n, and let $x \in \text{dom} f$. Then $\partial f(x)$ is a closed convex set. One has $\partial f(x) \neq \emptyset$ if and only if $f^\uparrow(x;0) > -\infty$, in which case*

(7.2) $$f^\uparrow(x;h) = \sup_{y \in \partial f(x)} y \cdot h \quad \text{for all } h.$$

Before drawing some general facts from these formulas, we look at some particular classes of functions.

Theorem 7.2. *Suppose f is convex, finite at x, and lower semicontinuous on a neighborhood of x. Then*

(7.3) $$\begin{aligned}\partial f(x) &= \{y \mid y \cdot h \leq f'(x;h) \text{ for all } h\} \\ &= \{y \mid f(x+th) \geq f(x) + y \cdot th \text{ for all } t > 0 \text{ and } h\}.\end{aligned}$$

The first equation is valid by Theorem 4.4 and the subdifferential regularity asserted in Theorem 6.6. The second equation then holds, because

(7.4) $$f'(x;h) = \inf_{t > 0} \frac{f(x+th) - f(x)}{t}$$

in the convex case. The second expression in (7.3) (where one could just as well take $t = 1$) is the definition customarily used for the subgradients of a convex function (cf. [43, § 23]). Thus Theorem 7.2 lays to rest any doubts about the present terminology versus terminology already established in convex analysis.

Theorem 7.3. *Suppose f is subsmooth around x. Then for any local representation of f as in (2.8) (with the gradient $\nabla \varphi_t(x)$ depending continuously not just on x but on (t,x)) one has*

(7.5) $$\partial f(x) = \text{co}\{\nabla \varphi_t(x) \mid t \in T_x\},$$

where T_x is the set in (5.6) and "co" stands for convex hull.

Theorem 7.3 follows at once from the subdifferential regularity of subsmooth functions (Theorem 6.6) and the derivative equations in Theorem 5.4. The set Y on the right side of (7.5) is compact and convex, and the formula in (5.5) asserts that the sublinear function corresponding to this Y as in Theorem 3.2 is $l = f'(x; \cdot)$.

Corollary 7.4. *Suppose* $f(x) = \max\{\varphi_1(x), \ldots, \varphi_r(x)\}$, *where each* φ_t *is smooth. Then* $\partial f(x)$ *is the polytope generated by the gradients* $\nabla \varphi_t(x)$ *of the functions* φ_t *which are active at* x, *in the sense that* $\varphi_t(x) = f(x)$.

Now we look at Lipschitzian cases. The next result combines Theorem 6.3 with Theorem 3.4 for $l(h) = f^\uparrow(x; h)$.

Theorem 7.5. *Let f be a lower semicontinuous, proper function on R^n, and let $x \in \operatorname{dom} f$. Then f is directionally Lipschitzian at x if and only if $\partial f(x)$ is nonempty but not expressible simply as a bundle of parallel lines. In that case one has*

(7.6) $$\partial f(x) = \{y \mid y \cdot h \leq f^\circ(x; h) \text{ for all } h\}.$$

Theorem 7.6 [44]. *Let f be a lower semicontinuous, proper function on R^n, and let $x \in \operatorname{dom} f$. Then f is Lipschitzian around x if and only if $\partial f(x)$ is nonempty and bounded.*

Of course (7.5) holds too when f is Lipschitzian around x, since that corresponds to f being directionally Lipschitzian with respect to $h = 0$. The necessity of the condition in Theorem 7.6 was observed by Clarke in his original paper [13]. Clarke also furnished (7.6) as a characterization of subgradients of locally Lipschitzian functions (see [14]), but his definition of $\partial f(x)$ for general lower semicontinuous functions, although equivalent to the one presented here (stemming from Rockafellar [45]) was rather circuitous. Not having the concept of subderivative at his disposal, he started with a special formula for subgradients of locally Lipschitzian functions (see Theorem 8.5) and used it to develop his notion of tangent cone by a dual method. Then he defined subgradients in general by a geometric version of formula (7.1) which corresponds to the epigraph relationship in Theorem 6.2.

In the locally Lipschitzian case there is a generalized mean value theorem which serves as a further characterization of the sets $\partial f(x)$; see Lebourg [37].

8. Relationship with Differentiability

A convex function f is differentiable at x if and only if $\partial f(x)$ consists of a single vector y, namely $y = \nabla f(x)$; cf. [43, Theorem 25.1]. What is the situation for nonconvex functions? Something more than ordinary differentiability is involved.

Recall that f is *differentiable* at x in the classical sense if and only if f is finite on a neighborhood of x and there exists a vector y such that

(8.1) $$\lim_{\substack{h' \to h \\ t \downarrow 0}} \frac{f(x + th') - f(x)}{t} = y \cdot h \quad \text{for all } h.$$

Then y is called the *gradient* of f at x and denoted by $\nabla f(x)$. The concept of *strict* differentiability of f at x is less well known; it means that

(8.2) $$\lim_{\substack{x' \to x \\ h' \to h \\ t \downarrow 0}} \frac{f(x' + th') - f(x')}{t} = y \cdot h \quad \text{for all } h.$$

II. Generalized Subgradients in Mathematical Programming

This is a localization of continuous differentiability: f is smooth on U (an open set in R^n) if and only if f is strictly differentiable at every $x \in U$.

Theorem 8.1. *Let f be lower semicontinuous and let x be a point where f is finite. Then f is strictly differentiable at x if and only if $\partial f(x)$ consists of a single vector y, namely the gradient $\nabla f(x)$. In this event f must be Lipschitzian around x.*

Clarke [13] proved this fact under the assumption that f is locally Lipschitzian. The general case follows from Clarke's result and Theorem 7.6.

Subdifferentially regular functions have an especially strong property in this regard.

Theorem 8.2 [51]. *Suppose f is locally Lipschitzian and subdifferentially regular on the open set $U \subset R^n$, as is true in particular if f is subsmooth on U. Then f is strictly differentiable wherever it is differentiable. Thus except for a negligible set of points x in U, the convex set $\partial f(x)$ reduces to a single vector.*

The final assertion is based on Rademacher's theorem (Theorem 2.2). Finite convex functions are in particular subsmooth (Corollary 2.6), so Theorem 8.2 explains the fact mentioned at the beginning of this section.

Corollary 8.3. *Suppose f is subsmooth around x and has a local representation (2.8) such that there is only one $t \in T$ with φ_t equal to f at x. Then f is strictly differentiable at x with gradient $\nabla f(x) = \nabla \varphi_t(x)$.*

The case in the corollary is obtained from Theorem 7.3. Note that this answers the question raised in §1 about when a function expressed as a maximum of a collection of smooth functions φ_t as in (2.8) can actually be smooth. It is smooth if the representation satisfies the assumptions in the definition of subsmoothness in §2, and if the maximum is attained for each x by a unique t. The latter is true, for instance, if T is a convex set and $\varphi_t(x)$ is strictly concave in t for each x.

Strongly subsmooth functions, as defined in §2, have an even nicer property. Recall that f is *twice-differentiable* at x if it is finite in a neighborhood of x and there exist $y \in R^n$ and $H \in R^{n \times n}$ such that

$$(8.3) \quad \lim_{\substack{h' \to h \\ t \downarrow 0}} \frac{f(x+th') - f(x) - y \cdot th'}{t} = \tfrac{1}{2} h \cdot Hh \quad \text{for all } h.$$

(This form of the definition does not require f to be once differentiable on a neighborhood of x.) A classical theorem of Alexandroff says that finite convex functions are twice differentiable almost everywhere. This translates by way of Theorem 2.5 into the following.

Theorem 8.4 [51]. *A strongly subsmooth function f on an open set $U \subset R^n$ is twice differentiable except on a negligible subset of U.*

Since subsmooth functions are quite common in optimization the preceding results can be applied in many situations.

Locally Lipschitzian functions which are not subdifferentially regular do not necessarily have $\partial f(x)$ consisting just of $\nabla f(x)$ at points where $\nabla f(x)$ exists, since f may be differentiable but not strictly differentiable at such points. An example exists of a locally Lipschitzian function f which is *nowhere* strictly differentiable [51], although f must be differentiable almost everywhere by Theorem 2.2. Nevertheless $\partial f(x)$ can by constructed from knowledge of $\nabla f(x')$ at points near to x, as the next theorem shows.

Theorem 8.5 [13], [14]. *Suppose f is Lipschitzian around x. Then*

(8.4) $\qquad \partial f(x) = \text{co}\{y | \exists x^k \to x \text{ with } f \text{ differentiable at } x^k \text{ and } \nabla f(x^k) \to y\}.$

This is the formula originally used by Clarke [13] for subgradients of locally Lipschitzian functions.

A generalization of Theorem 8.5 to arbitrary lower semicontinuous functions f has been furnished by Rockafellar [52], [49]. In such a setting one must consider not sequences of gradients $\nabla f(x^k)$, but "lower semigradients" or "proximal subgradients", and the notion of convex hull must be broadened to include "points at infinity".

9. Subdifferential Calculus

Not much could be accomplished with generalized subgradients if there were no rules for calculating or estimating them, beyond the definition itself. Such rules do exist in the convex case [43], and many of them have now been extended. We can only mention a few here.

A formula for the subgradients of the pointwise maximum of a collection of smooth functions φ_t, $t \in T$, has already been given in Theorem 7.3 and Corollary 7.4. This has been generalized to certain collections of nonsmooth functions φ_t by Clarke [13], [14].

An especially important operation to consider is that of addition of functions.

Theorem 9.1 [46]. *Let f_1 and f_2 be lower semicontinuous functions on R^n, and let x be a point where both f_1 and f_2 are finite. Suppose there exists an $h \in R^n$ such that $f_1^\uparrow(x; h) < \infty$ and f_2 is directionally Lipschitzian at x with respect to h. Then*

(9.1) $\qquad\qquad\qquad \partial (f_1 + f_2)(x) \subset \partial f_1(x) + \partial f_2(x).$

Moreover equality holds in (9.1) if f_1 and f_2 are subdifferentially regular at x.

Corollary 9.2. *Suppose f_1 is finite at x and lower semicontinuous around x, and f_2 is Lipschitzian around x. Then the conclusions of the theorem are valid.*

The corollary is the case of the theorem where $h = 0$. A number of applications of Theorem 9.2 have an indicator function in place of either f_1 or f_2. The following fact then comes into play.

Theorem 9.3 [13]. *Let f be the indicator ψ_C of a closed set $C \subset R^n$, and let $x \in C$. Then*

(9.2) $\qquad\qquad \partial f(x) = N_C(x), \text{ where } N_C(x) = T_C(x)^\circ \text{ (polar)}.$

II. Generalized Subgradients in Mathematical Programming

The polar $N_C(x)$ of the Clarke tangent cone $T_C(x)$ is the *Clarke normal cone to C at x*, and its elements are called *normal vectors*. A nonzero normal vector exists at x if and only if x is a boundary point of C (cf. Rockafellar [44]). For direct expressions of the normal cone as the convex hull of limits of more special kinds of normals at special points, see Clarke [13] and Treiman [61].

Theorem 9.1 is only a sample of the kind of calculus that can be carried out. The chain rule too has its generalizations; see Clarke [14], Rockafellar [44], [54]. Further rules are listed in [49].

As far as mathematical programming is concerned, the question of how to estimate the subgradients of an optimal value function p, as described toward the end of §1, is highly significant, and much effort has been expended on it (see [48] and its references). We must content ourselves here with indicating what the answer is in a special case.

Let us consider the problem

$$(P_u) \qquad \begin{array}{l} \text{minimize } f_0(x) \text{ subject to} \\ f_i(x) + u_i \leq 0 \text{ for } i = 1, \ldots, m, \end{array}$$

where $u = (u_1, \ldots, u_m)$ and the functions $f_0, f_1, \ldots, f_m$ are locally Lipschitzian on R^n. Suppose that for every $u \in R^n$ and $\alpha \in R$ the set of feasible solutions to (P_u) with $f_0(x) \leq \alpha$ is bounded (maybe empty). Let $p(u)$ denote the optimal value in (P_u). Then p is a lower semicontinuous, proper function on R^m (convex actually if every f_i is convex, but that is not the situation we want to restrict ourselves to at the moment). Since p is nondecreasing in u, it is also directionally Lipschitzian throughout its effective domain (cf. end of §2).

Theorem 9.4. *Let $p(u)$ be the optimal value in problem (P_u) under the above assumptions. Fix any u such that $p(u)$ is finite, and let $X(u)$ denote the corresponding set of optimal solutions. For each $x \in X(u)$, let $K(u, x)$ denote the set of all Lagrange multiplier vectors $y \in R^m$ satisfying the generalized Kuhn-Tucker conditions*

$$(9.3) \qquad \begin{array}{l} y_i \geq 0 \text{ and } y_i f_i(x) = 0 \text{ for } i = 1, \ldots, m, \\ 0 \in \partial f_0(x) + y_1 \partial f_1(x) + \ldots + y_m \partial f_m(x). \end{array}$$

Similarly let $K_0(u, x)$ be the set of vectors y which would satisfy (9.3) if the term $\partial f_0(x)$ were omitted. If $K_0(u, x)$ consists of just the zero vector for every $x \in X(u)$, then p is Lipschitzian around u and

$$(9.4) \qquad \partial p(u) \subset \text{co} \left[\bigcup_{x \in X(u)} K(x, y) \right] \quad (compact),$$

$$(9.5) \qquad p^\uparrow(u; h) = p^\circ(u; h) \leq \max_{x \in X(u)} \max_{y \in K(u, x)} y \cdot h \quad \text{for all } h.$$

For a stability condition ensuring that $\partial p(u) = K(x, u)$ in (9.4) see Pomerol [42]. A more abstract analysis of $\partial p(u)$ in the case of functions p of the form

$$p(u) = \inf_x f(u, x),$$

with f lower semicontinuous on $R^m \times R^n$ and extended-real-valued, is carried out in [50].

References

1. J. P. Aubin, "Gradients généralisés de Clarke", *Annales des Sciences Mathématiques Québec* 2 (1978) 197–252.
2. J. P. Aubin, "Further properties of Lagrange multipliers in nonsmooth optimization", *Applied Mathematics and Optimization* 6 (1980) 79–80.
3. J. P. Aubin, "Locally Lipschitzian cooperative games", *Journal of Mathematical Economics* 8 (1981) 241–262.
4. J. P. Aubin and F. H. Clarke, "Multiplicateurs de Lagrange en optimisation non convexe et applications", *Comptes Rendus de l'Academie des Sciences Paris* 285 (1977) 451–454.
5. J. P. Aubin and F. H. Clarke, "Shadow prices and duality for a class of optimal control problems, *SIAM Journal on Control and Optimization* 17 (1979) 567–586.
6. A. Auslender, "Stabilité différentiable en programmation non convexe nondifférentiable", *Comptes Rendus de l'Academie des Sciences Paris* 286.
7. A. Auslender, "Differential stability in nonconvex and nondifferentiable programming", *Mathematical Programming Study* 10 (1978).
8. A. Auslender, "Minimisation de fonctions localement lipschitziennes: Applications a la programmation mi-convex, mi-differentiable", in: O. L. Mangasarian et al., ed., Nonlinear Programming 3 (Academic Press, New York, 1978) 429–460.
9. R. W. Chaney, "Second-order sufficiency conditions for nondifferentiable programming problems", *SIAM Journal on Control and Optimization* 20 (1982) 20–33.
10. R. W. Chaney and A. A. Goldstein, " An extension of the method of subgradients", in C. Lemarechal and R. Mifflin, eds., Nonsmooth Optimization (Pergamon Press, New York, 1978).
11. S. M. Chung, "Exact penalty algorithms for nonlinear programming", in: O. L. Mangasarian et al., ed., Nonlinear Programming 3 (Academic Press, New York, 1978) 197–224.
12. A. R. Conn, "Constrained optimization using a nondifferentiable penalty function", *SIAM Journal on Numerical Analysis* 10 (1973) 760–784.
13. F. H. Clarke, "Generalized gradients and applications", *Transactions of the American Mathematical Society* 205 (1975) 247–262.
14. F. H. Clarke, "Generalized gradients of Lipschitz functionals", *Advances in Mathematics* 40 (1981) 52–67.
15. F. H. Clarke, "On the inverse function theorem", *Pacific Journal of Mathematics* 67 (1976) 97–102.
16. F. H. Clarke, "The Euler-Lagrange differential inclusion", *Journal of Differential Equations* 19 (1975) 80–90.
17. F. H. Clarke, "La condition hamiltonienne d'optimalité", *Comptes Rendus de l'Academie des Sciences Paris* 280 (1975) 1205–1207.
18. F. H. Clarke, "Optimal solutions to differential inclusions", *Journal of Optimization Theory and Applications* 19 (1976) 469–478.
19. F. H. Clarke, "The generalized problem of Bolza", *SIAM Journal on Control and Optimization* 14 (1976) 682–699.
20. F. H. Clarke, "The maximum principle under minimal hypotheses", *SIAM Journal on Control and Optimization* 14 (1976) 1078–1091.
21. F. H. Clarke, "Inequality constraints in the calculus of variations", *Canadian Journal of Mathematics* 39 (1977) 528–540.
22. F. H. Clarke, "A new approach to Lagrange multipliers", *Mathematics of Operations Research* 1 (1976) 165–174.
23. B. Cornet, "Regularity properties of normal and tangent cones", forthcoming.

24. J. M. Danskin, The theory of max-min and its application to weapons allocations problems (Springer-Verlag, New York, 1967).
25. R. J. Duffin, E. L. Peterson and C. Zener, Geometric programming (Wiley, New York, 1967).
26. J. Gauvin, "The generalized gradient of a marginal function in mathematical programming", *Mathematics of Operations Research* 4 (1979) 458–463.
27. J. Gauvin and F. Dubeau, "Differential properties of the marginal value function in mathematical programming", *Mathematical Programming Studies*, forthcoming.
28. A. A. Goldstein, "Optimization of Lipschitz continuous functions", *Mathematical Programming* 13 (1977) 14–22.
29. M. Held, P. Wolfe and H. P. Crowder, "Validation of subgradient optimization", *Mathematical Programming* 6 (1974) 62–88.
30. J.-B. Hiriart-Urruty, "Tangent cones, generalized gradients and mathematical programming in Banach spaces", *Mathematics of Operations Research* 4 (1979) 79–97.
31. J.-B. Hiriart-Urruty, "On optimality conditions in nondifferentiable programming", *Mathematical Programming*, forthcoming.
32. J.-B. Hiriart-Urruty, "Gradients généralisés de fonctions marginales", *SIAM Journal on Control and Optimization* 16 (1978) 301–316.
33. J.-B. Hiriart-Urruty, "Conditions nécessaires d'optimalité pour un programme stochastique avec recours", *SIAM Journal on Control and Optimization* 16 (1978) 317–329.
34. J.-B. Hiriart-Urruty, "Refinements of necessary optimality conditions in nondifferentiable programming, I", *Applied Mathematics and Optimization* 5 (1979) 63–82.
35. J.-B. Hiriart-Urruty, "Extension of Lipschitz functions", *Journal of Mathematical Analysis and Applications* 77 (1980) 539–554.
36. S. Howe, "New conditions for the exactness of a simple penalty function", *SIAM Journal on Control and Optimization* 11 (1973) 378–381.
37. G. Lebourg, "Valeur moyenne pour gradient généralisé", *Comptes Rendus de l'Academie des Sciences Paris* 281 (1975) 795–797.
38. R. Mifflin, "Semismooth and semiconvex functions in constrained optimization", *SIAM Journal on Control and Optimization* 15 (1977) 959–972.
39. R. Mifflin, "An algorithm for constrained optimization with semismooth functions", *Mathematics of Operations Research* 2 (1977) 191–207.
40. R. Mifflin, V. H. Nguyen and J. J. Strodiot, "On conditions to have bounded multipliers in locally Lipschitz programming", *Mathematical Programming* 18 (1980) 100–106.
41. T. Petrzykowski, "An exact potential method for constrained maxima", *SIAM Journal on Numerical Analysis* 6 (1969) 299–304.
42. C. Pomerol, "The Lagrange multiplier set and the generalized gradient set of the marginal function of a differentiable program in a Banach space", *Journal of Optimization Theory and Applications*, forthcoming.
43. R. T. Rockafellar, Convex analysis (Princeton University Press, Princeton, NJ, 1970).
44. R. T. Rockafellar, "Clarke's tangent cones and the boundaries of closed sets in R^n", *Nonlinear Analysis: Theory, Methods and Applications* 3 (1979) 145–154.
45. R. T. Rockafellar, "Generalized directional derivatives and subgradients of nonconvex functions", *Canadian Journal of Mathematics* 32 (1980) 257–280.
46. R. T. Rockafellar, "Directionally Lipschitzian functions and subdifferential calculus", *Proceedings of the London Mathematical Society* 39 (1979) 331–355.
47. R. T. Rockafellar, "Proximal subgradients, marginal values and augmented Lagrangians in nonconvex optimization", *Mathematics of Operations Research* 6 (1981) 424–436.

48. R. T. Rockafellar, "Lagrange multipliers and subderivatives of optimal value functions in nonlinear programming", *Mathematical Programming Study* 17 (1982) 28–66.
49. R. T. Rockafellar, "The theory of subgradients and its applications to problems of optimization: convex and nonconvex functions" (Heldermann-Verlag, West-Berlin, 1981).
50. R. T. Rockafellar, "Augmented Lagrangians and marginal values in parameterized optimization problems", in: A. Wierzbicki, ed., Generalized Lagrangian methods in optimization (Pergamon Press, New York, 1982).
51. R. T. Rockafellar, "Favorable classes of Lipschitz continuous functions in subgradient optimization", in: E. Nurminski, ed., Nondifferentiable Optimization (Pergamon Press, New York, 1982).
52. R. T. Rockafellar, "Proximal subgradients, marginal values, and augmented Lagrangians in nonconvex optimization", *Mathematics of Operations Research* 6 (1981) 427–437.
53. R. T. Rockafellar, "Marginal values and second-order necessary conditions for optimality", *Mathematical Programming*, to appear.
54. R. T. Rockafellar, "Directional differentiability of the optimal value function in a nonlinear programming problem", *Mathematical Programming Studies*, to appear.
55. R. T. Rockafellar, "Augmented Lagrange multiplier functions and duality in nonconvex programming", *SIAM Journal on Control* 12 (1974) 268–285.
56. S. Saks, "Theory of the integral" (Hafner Publishing Co., New York, 1937).
57. L. Thibault, "Problème de Bolza dans un espace de Banach séparable", *Compes de l'Academie des Sciences Paris* 282 (1976) 1303–1306.
58. L. Thibault, "Subdifferentials of compactly Lipschitzian vector-valued functions", *Annali di Matematica Pura e Applicata* 125 (1980) 157–192.
59. L. Thibault, "On generalized differentials and subdifferentials of Lipschitz vector-valued functions", *Nonlinear Analysis: Theory, Methods and Applications*, forthcoming.
60. L. Thibault, "Subdifferentials of nonconvex vector-valued functions", *Journal of Mathematical Analysis and Applications*, forthcoming.
61. J. Treiman, "Characterization of Clarke's tangent and normal cones in finite and infinite dimensions", *Nonlinear Analysis: Theory, Methods and Applications*, forthcoming.
62. W. I. Zangwill, "Nonlinear programming via penalty functions", *Management Science* 13 (1961) 34–358.

Nondegeneracy Problems in Cooperative Game Theory

J. Rosenmüller
Universität Bielefeld, Institut für Mathematische Wirtschaftsforschung, Postfach 8640,
4800 Bielefeld 1

Chapter 0. Introduction

Game Theory originally is rooting to a certain extent in the various fields of optimization and programming. Most students of elementary courses in linear programming techniques are familiar with the fact that there is a close relation between the optimal solutions of an L.P. and the optimal strategies of an associated matrix game. Hence, if we introduce the mixed extension of a finite two-person zero-sum game then, according to von Neumann's minimax theorem, there exist always optimal mixed strategies for both players and, in addition, it is easy to define a certain associated linear program such if we obtain the optimal solutions of this L.P., say, via the simplex algorithm, then this optimal solutions simultaneously yield the optimal strategies of the two-person game under consideration.

Similarly saddle point problems which arise in the context of zero-sum games which do not have a finite strategy space for both players are obiously closely related to nonlinear (convex or dynamic) programming; in particular techniques like the Pontrjagin-Theorem or gradient methods may be employed in order to obtain optimal strategies for such games.

Another subject to be mentioned is the computation of equilibrium points of a bimatrix game by means of the Lemke-Howson algorithm. This algorithm can also be viewed as method which is applicable for a linear quadratic optimization problem (or a certain bilinear optimization problem). But on the other hand it readily yields the equilibrium points of a bimatrix game – which generically are finitely many (and thus non connected). The Lemke-Howson algorithm is a somewhat advanced relative of the simplex algorithm. Hence this example shows again the close intimacy of the methods and techniques applied in optimization and game theory.

It is important, however, to realize that all the examples mentioned above are taken from the field of *non-cooperative* game theory. In the terminology of von Neumann and Morgenstern all the examples are dealing with the so called normal form; that is a representation of a game in terms of payoff functions or payoff matrices. In this version the strategic aspect of the game is already very coarse. A "strategy" is just a possible choice for a player involved in the game; there being several alternatives for him. But the structure of these alternatives, or rather the structure of a strategy, is not to be seen explicitly.

Game theoretically speaking, the extensive form is a more explicit description of the rules of the game – in other words a tree or a dynamic process such that the players do have influence on the development of the state variable during various instances of time. Again there are close roots to dynamic programming and stochastic (Markovian) optimization processes such that the optimal strategies of a game in extensive form may be found by recursive computations similar to the ones which are familiar from dynamic programming. In fact the theorem of Zermelo-von Neumann-Kuhn is of course the same as what is called an optimality theorem of dynamic programming. Similarly the techniques in differential games as well as in control theory (Hamiltonian methods or Pontrjagin's principle) are of course close relatives. It should not be overlooked, however, that game theory has additional problems (and solutions) to offer as far as the information problem is concerned which arises when more than one person is involved in controlling a certain stochastic or deterministic process.

The present paper will be dealing with neither of the two topics mentioned above. Rather it is our aim to show that the third version of a game as introduced by von Neumann and Morgenstern, that is the representation by a "characteristic function" has also roots and connections to various aspects of optimization and programming.

Cooperative game theory deals with non additive set functions; intuitively the notion of a strategy does not appear (at least not obviously) in this context, it is rather the notion of coalition forming and the strength of coalitions which is of interest in this context. Therefore, cooperative game theory has been of minor interest to researchers in operations research and optimality theory. It will be pointed out, however, that there are striking similarities between elementary aspects of optimization and cooperative game theory as well.

Clearly the question of finding the extreme points of a convex compact polyhedron is of interest to cooperative game theory as well: It is sometimes a formidable task to describe the extreme games, that is the extreme set functions among a certain class of set functions specified by certain inequalities which represent the economic or game theoretically interesting properties of such games.

Theorems concerning dividing hyperplanes or linear functionals and their relations to convex sets are also well familiar and it shall be pointed out that there is again a similarity in the domain of non additive set functions: If we consider the additive set functions dominating a certain non additive set function than this can readily be interpreted to be a problem similar to finding linear functionals dominating certain subadditive functionals. Hence, the Hahn-Banach theorem and its relatives can be applied in order to proof that the "core", that is the system of additive set functions dominating a non additive set function and having the same total mass, is not empty.

The next analogy to be mentioned and to be treated in the present paper is the property of convex functions to be representable as an envelope of supporting linear functions. There is a well defined notion of a convex set function (convex game) and exactly as in the "point function case" it is possible to represent convex set functions by the "supporting" affine set function – an affine set function being an additive set function minus a constant.

Furthermore it is quite conceivable that any optimization problem can be rewritten in a form which is of interest for cooperative game theory. Roughly

II. Nondegeneracy Problems in Cooperative Game Theory

speaking, the restrictions or constraints of a certain optimization problem might be economically interpreted as restrictions on resources of factors to be put into a certain production process. Now, if we imagine that there is not one decision maker but several ones ("the players"), all of them commanding a certain amount of the factors or resources, then certain groups, coalitions, of these players might combine their resources, put them into the production process and, by solving the optimization problem, reach a certain payoff to the coalition which is just acting. Hence, for every coalition the constraints vary and therefore the optimal solutions as well as the value of the optimum, depend on coalitions. This obviously defines a function on coalitions which can be interpreted in the usual game theoretical framework. It is then the task of game theory to evaluate the influence of the techniques necessary to solve the optimization problem (e.g. the dual variables or certain other objects of optimization theory) on the solution concepts of game theory (e.g. the core, or certain notions of equilibrium).

In the framework of game theory and of general equilibrium theory some of these solutions concepts have been seen to converge or to coincide once the player set increases or is represented by a non atomic measure space. If the game stems from some optimization problem as mentioned above what are the properties connecting the statements of general equilibrium theory ("for large games the core and the competitive equilibrium coincide") and the linear or non linear programming procedures? As turns out, it is sometimes useful to handle "programming games" in a finite framework. In this finite framework there are certain "surrogates" for non atomicity which again turn out to be problems which have their roots in optimization – this time in integer programming. In fact we shall point out that the non atomicity of an additive or non additive set function has a close analogon in the finite area. This analogon being called either "non-degeneracy" or "homogeneity". Evaluating the properties of non-degeneracy and homogeneity leads to combinatorical and number theoretical problems which also can be seen as relatives of certain methods of integer programming. Hence, some statements from game theory and general equilibrium, that is, that for large sets of players certain solution concepts coincide, can be rephrased in the framework of integer programming and combinatorics. Thus the notion of a large player space can be precisely expressed in terms of indicating certain distributions over types of players such that, if sufficiently many players of each typ are available, then the player set can be considered large and the coincidence of certain solution concepts can be proved.

Therefore techniques from programming and combinatorics resulting from studying non-degeneracy properties seem to be appropriate to replace the term "non-atomic" which can be used to formulate economic or game theoretic principles which have been of interest to researchers in various fields for the last 15 or 20 years.

We shall start by introducing some examples, discussing the general definition of games and their solution concepts in the first chapter. The second chapter deals with extreme points of a certain convex compact polyhedron of games and introduces the notion of non-degeneracy which is shown to be a relative of non-atomicity and to induce solutions of the extreme point problems under consideration. Chapter 3 again is dealing with extreme point problems but in a different framework: We are interested in super-additive games. Another version of "non-atomicity in the finite range" is introduced. Its relation to the extremality problem is studied and solution concepts of extreme games are again considered. Chapter 4 finally deals with the introduction of games as resulting from optimization problems: The linear programming game is studied extensively where players may combine their resources hence influencing the constraints of the L.P. and defining a certain non-additive set function. The core of this set function is not empty and shrinks (actually in finitely many steps) to the competitive equilibrium.

Chapter I. Optimization Problems and Cooperative Games

This chapter has an introductory character. It is appropriate to introduce several elementary definitions and to supply a certain bulk of examples which support the idea that the general principles which are formulated in the framework of optimization and hence ment to be guidelines for a decision maker, can also be formulated in a framework where several decision makers are concerned. As usual we shall call the decision makers "players". The intuitive idea is that the players can cooperate by forming coalitions and hence play various "optimization problems", the conditions under which these problems are being solved depending on the coalition to be formed.

As an example let us consider the generalization of a simple linear programming model, the linear programming game or linear production game ("L.P.-game"). These games have been studied by Owen [12] and also by Billera and Raanan [2], see also [15].

Let $A = (a_{jk})_{j=1,\ldots,m; k=1,\ldots,l}$ be an $m \times l$ matrix (input-output matrix). Also let $b^0 \in \mathbb{R}^m$ and $c \in \mathbb{R}^l$ be vectors, then the tripel (A, b^0, c) represents a linear program. It is our task to find

(1) $$\max\{cx \mid x \in \mathbb{R}^l_+, Ax \leq b^0\}.$$

Let us assume that all quantities indicated are strictly positive so that we have no problems with the existence of the solutions to the problem indicated by (1). The usual interpretation of the linear program is the control of a production process (represented by the input output matrix A) by a decision maker (a player). The vector b^0 defines the constraints to be put on the resources of vectors which can be inserted into the production process while c (representing the objective function) is to be interpreted as market prices of the outputs which can be varied in various quantities, represented by the vector $x = (x_1, \ldots, x_l)$.

Imagine now that instead of one player there are n players involved in the decision process. Let $\Omega = \{1, \ldots, n\}$ denote the set of these players. For $i \in \Omega$, let $b_i \in \mathbb{R}^m_+$ be the vector of resources of factors which is controlled by player i.

Imagine now that a coalition S is capable of combining its forces, that is, pooling its resources and than operating the production process jointly. The factors available to this coalition are given by

II. Nondegeneracy Problems in Cooperative Game Theory

(2) $$\sum_{i \in S} b_i =: b(S).$$

Because of

$$b(S) = (b^1(S), \ldots, b^m(S))$$

b can be interpreted as an additive function defined on the power set of Ω and taking values in $\mathbb{R}^m$, that is, b may be regarded as a vector valued measure. This measure represents the distribution of factors over the various players and coalitions. If coalition $S \subseteq \Omega$ chooses to cooperate than it will be able to maximize the profit of the output resulting from the linear program which is defined accordingly by

(3) $$v(S) = \max\{cx \mid x \in \mathbb{R}^l_+, Ax \leq b(S)\}.$$

Formula (3) defines a function $v = v^{A,b,c} : \underline{P} \to \mathbb{R}_+$ where $\underline{P}$ is the power set of Ω, that is the system of coalitions.

It is now the task of game theory to analyze the set function v. Specifically it is of interest to analyze the consequences which may be drawn e.g. by the main theorem of linear programming with respect to game theoretical problems.

An important "solution concept" for cooperative games is the core. Note that the function $v = v^{A,b,c}$ is not an additive one. The core is the system of all additive set functions dominating a given non additive set function v and having the same total measure. Thus

(4) $$\mathscr{C}(v) = \{x \mid x : \underline{P} \to \mathbb{R}, x \text{ additive}, x(\Omega) = v(\Omega), x \geq v\}.$$

Now using the dual program for the problem which is specified for various coalitions S we may construct elements of the core of the L.P.-game as follows:

Because of

(5) $$v(S) = v^{Abc}(S) = \min\{yb(S) \mid y \in \mathbb{R}^m_+, yA \geq c\}$$

a core element of this game is easily obtained as follows. Pick an *optimal* solution $\bar{y}$ for the "dual Ω-program", i.e., $\bar{y} \in \mathbb{R}^m_+$ such that

(6) $$v(\Omega) = \bar{y} b(\Omega) = \min\{yb(\Omega) \mid y \in \mathbb{R}^m_+, yA \geq c\},$$

then $\bar{y}$ is *feasible* for (5) (as the contraints do not depend on S) and hence

(7) $$\bar{y} b(S) \geq v(S).$$

But (6) and (7) show that $\bar{y} b(\cdot) = \bar{y}_1 b^1(\cdot) + \ldots + \bar{y}_m b^m(\cdot)$ is an element of the core of v, we write

(8) $$\bar{y} b \in \mathscr{C}(v).$$

The fact that linear programming may be applied to finding elements of the set of dominating additive set functions for a given non additive one is not constrained to the above example. Intuitively an additive set function is the analogue of a linear point function and a non-additive set function is the analogue of a non-linear function (imagine a convex or concave function) and finding a dominating additive set function with the same total mass somehow amounts to constructing a supporting affine or linear function to a given non linear one. The application of the main theorem of linear programming as demonstrated above illustrates the fact that separation theorems also enter the scene when we consider problems with respect to set functions.

That this is also the case for a general version of the proof that the core is not empty has been already observed by Scarf [19].

Let 1_S denote the indicator function of $S \in \underline{P}$ also let $\mathbb{V}$ denote the system of all set functions (in general non additive) which are defined on the power set $\underline{P}$ of Ω.

Let $\underline{S} \subseteq \underline{P}$ be a system of coalitions such that $\emptyset \notin \underline{S}, \Omega \notin \underline{S}$. $\underline{S}$ is said to be balanced if there are coefficients $c_s > 0 \, (S \in \underline{S})$ such that

$$\sum_{S \in \underline{S}} c_S 1_S(\cdot) = 1_\Omega(\cdot).$$

Let $\mathbb{V} = \{v | v: \underline{P} \to \mathbb{R}, \, v(\emptyset) = 0\}$.

$v \in \mathbb{V}$ is called balanced if, for any balanced $\underline{S} \subseteq \underline{P}$ with coefficients $(c_S)_{S \in \underline{S}}$ it follows that

$$\sum_{S \in \underline{S}} c_S v(S) \leq v(\Omega).$$

As it turns out balancedness is the apropriate condition in order to ensure that the core is non empty. Scarf's proof of this runs as follows: Denote by $\mathbb{A}$ the set of all additive set functions defined on $\underline{P}$.

Let $v \in \mathbb{V}$. Suppose we are to find the quantity

$$\min\{x(\Omega) | x \in \mathbb{A}, \, x(S) \geq v(S), \, (S \in \underline{P} - \{\Omega\})\}.$$

This problem can be viewed as to be represented by the linear program (A, b, d), where

$$A = (a_{iS})_{i \in \Omega, S \in \underline{P} - \{\Omega, \emptyset\}}, \quad a_{iS} = 1_S(i) = \delta_i(S),$$
$$b = (b_S)_{S \in \underline{P} - \{\Omega, \emptyset\}}, \quad b_S = v(S)$$
$$d = (d_i)_{i \in \Omega}, \quad d_i = 1,$$

since obviously

$$\min\{dx | xA \geq b\}$$

is the same problem as above. Clearly, this program does enjoy optimal solutions. Consider the dual program

$$\max\{cb | Ac = d, \, c \geq 0\}.$$

Here $c = (c_S)_{S \in \underline{P} - \{\emptyset, \Omega\}}$ is to be taken from $\mathbb{R}^{2^n - 2}$. Clearly, $Ac = d$ means

$$\sum_S c(S) 1_S(i) = 1 \quad (i \in \Omega),$$

i.e.

$$\sum_S c(S) 1_S = 1.$$

II. Nondegeneracy Problems in Cooperative Game Theory

(Because the primal problem allows no restrictions $x_1 \geq 0$, the dual problem has equalities correspondingly.) Now, let $\bar{c}$ be an optimal solution of the dual problem and let

$$\underline{S} := \{S \neq \emptyset, \bar{c}_S \geq 0\}.$$

Then (11) implies that $\underline{S}$ is a balanced system with coefficients $(\bar{c}_S)_{S \in \underline{S}}$. If $\bar{x}$ is an optimal solution of the primal problem, then

$$d\bar{x} = \bar{c}b \quad \text{or} \quad \bar{x}(\Omega) = \sum_{S \in \underline{S}} \bar{c}_S v(S).$$

If, in addition, v is balanced, then we have

$$\bar{x}(\Omega) = \sum_{S \in \underline{S}} \bar{c}_S v(S) \leq v(\Omega).$$

Because $\bar{x}(S) \geq v(S)$ $(S \in \underline{P} - \{\Omega\})$, it turns out that

$$\bar{x} + (v(\Omega) - \bar{x}(\Omega))m \in \mathbb{C}(v).$$

It is partly the aim of this paper to show that the above mentioned analogues are not a mere coincidence. For instance the fact that the convex point function is the envelope of its supporting affine function has its direct analogue in the territory of set functions $v \in V$ if the term "convex" is given a precise and appropriate meaning. The core of the L.P.-game as mentioned above contains at least one element which is obtained by computing the dual optimal variables of the grand coalition Ω (shadow prices for grand coalition), then attaching these shadow prices as rates to the measures of factors, distributions. As it turns out this element is the only element of the core of the L.P.-game if the set of players is increased sufficiently.

These are only a few hints of the various coincidences and analogues we want to discuss in the following chapters. Let us, however, finish the introductory chapter by offering a few formal definitions.

Let $\Omega = \{1, \ldots, n\}$ ("the set of players") and $\underline{B} := \mathfrak{p}(\Omega)$ ("the coalitions"). An additive set function $m: \underline{B} \to \mathbb{R}$ is tantamount to a vector $m = (m_1, \ldots, m_n) \in \mathbb{R}^n$ via $m(S) = \sum_{i \in S} m_i$ $(S \in \underline{B})$. $\mathbb{A} = \mathbb{A}$ is the system of such functions and $\mathbb{A}^+$, $\mathbb{A}_1$ is used to indicate nonnegativity, normalization $(m(\Omega) = 1)$, etc.

Generally, a game is a tripel $\Delta = (\Omega, \underline{B}, v)$ where $v: \underline{B} \to \mathbb{R}$, $v(\emptyset) = 0$, is a (non-additive) set function. Examples are

(9) the weighted majority; $v = 1_{[\alpha, 1]} \circ m$ where $m \in \mathbb{A}_1^+$, $\alpha \in (0, 1)$ (representing a voting committee);

(10) production games ("one factor"); $v = f \circ m$ where $m \in \mathbb{A}$ and $f: \mathbb{R} \to \mathbb{R}$, $f(0) = 0$;

(11) L.P.-games ("linear program"); defined by $v(S) = \max\{cx \mid x \in \mathbb{R}_+^l, Ax \leq b(S)\}$, $(S \in \underline{B})$ where $c \in \mathbb{R}_+^l$, A an $l \times m$-matrix and $b \in (\mathbb{A}^+)^m$ (all strictly positive);

(12) market games, given by $v(S) = \max\left\{\sum_{i \in S} u^i(x^i) \mid (x^i)_{i \in \Omega} \in \mathbb{R}_+^{m \times n}, \sum_{i \in S} x^i = \sum_{i \in S} a^i\right\}$ $(S \in \underline{B})$ where $a^i \in \mathbb{R}_+^m$ ("initial allocation of player $i \in \Omega$")

and $u^i: \mathbb{R}_+^m \to \mathbb{R}$ (usually concave and continuous, "player i's utility function"); here players $i \in S$ maximize their joint utility by "exchange of commodities". (cf. e.g. [14], [23].)

$\mathbb{V}$ is the system of all functions $v: \underline{B} \to \mathbb{R}$, $v(\emptyset) = 0$.

Certain classes of set functions v (and of games Δ) are of particular interest. E.g. v is superadditive (" $v \in \mathscr{S}$ ") if

(13) $\quad v(S) + v(T) \leqslant v(S+T) \quad$ for $S, T \in \underline{B}$,

(e.g. example (9) for $\alpha > \frac{1}{2}$, $S + T = S \cup T$ iff $S \cap T = \emptyset$); v is convex (" $v \in \mathscr{C}$ ") if

(14) $\quad v(S) + v(T) \leqslant v(S \cap T) + v(S \cup T) \quad$ for $S, T \in \underline{B}$,

(e.g. example (10) for f convex). These functions are also called "alternating capacities of 2nd order" (Choquet [5]) or "supermodular functions" (Edmonds-Rota [6]). v is balanced (" $v \in \mathscr{B}$ "), if, for every "partition of the unit"

(15) $\quad 1_\Omega = \sum_{S \in \underline{S}} c_S 1_S (\underline{S} \subseteq \underline{B}, c_S \geq 0)$, it follows that

$$\sum_{S \in \underline{S}} c_S v(S) \leqslant v(\Omega).$$

Note that (11) and (12) both provide examples for balanced games and functions. The "positive parts" $\mathscr{S}^+, \mathscr{B}^+, \mathscr{C}^+$ are convex polyhedral cones (not trivial for $\mathscr{B}^+$, see Shapley [21].

Let us now turn to the concept of a solution. Generally speaking, a "solution concept" is a mapping defined on some subsets of functions v, say $\mathbb{V}^0$ (e.g. $= \mathbb{A}, \mathscr{S}, \mathscr{C}, \ldots$) and taking values in $\mathfrak{p}(\mathbb{A})$.

Let us mention the Core, defined by

$$\mathscr{C}(v) = \{x \in \mathbb{A} \mid x \geqslant v, x(\Omega) = v(\Omega)\}$$

which is nonempty if and only if v is balanced (Shapley [21], Bondareva [3]); thus $\mathscr{C}: \mathbb{B} \to \mathfrak{p}(\mathbb{A})$. Stable sets or von Neumann-Morgenstern-solutions are somewhat more tedious to be defined: let us say that $x \in \mathbb{A}$ dominates $y \in \mathbb{A}$ if there is $S \in \underline{B}$ such that $x_i > y_i (i \in S)$, $x(S) \leqslant v(S)$. Then a von Neumann-Morgenstern-solution (for some specified v) is a subset $\mathscr{S} \subseteq \{x \in \mathbb{A} \mid x_i \geqslant v(\{i\})\} =: \mathscr{I}(v)$ such that

1. For $x, y \in \mathscr{S}$, x does not dominante y,
2. For $y \notin \mathscr{S}, y \in \mathscr{I}(v)$, there is $x \in \mathscr{S}$ such that x dominates y.

The class of functions v admitting at least one von Neumann-Morgenstern-solution is unknown, see e.g. von Neumann-Morgenstern [10], Lucas [9], Schmitz [20], for details.

A further solution concept, to be considered with respect to market games (example (12)) is the competitive ("Walrasian") equilibrium; an advanced treatment is offered by Hildenbrand [7], [8].

As a formal definition and discussion of the competitive equilibrium is a concept which is rather attached to general equilibrium theory and not to game theory, we shall refer the reader to the literature, e.g. [7], [8], [14].

Chapter II. Extreme Point Problems in the Convex Range, the Concept of Nondegeneracy

This chapter is dealing with the concept of an extreme convex game or rather an extreme convex set function. The term convexity applies to a certain class of non additive set function v to be defined later on. The convex set functions, if restricted to non-negative values, form a cone with vertex 0; we are interested in characterizing the extreme raise of this cone. Of course it suffices if we restrict ourself to a basis of this cone, which is easily obtained by normalizing the set functions such that the total mass $v(\Omega)=1$.

As convex set functions, much as convex point functions, can be represented as envelopes of affine set functions (measures plus a constant), additive or affine set functions play an important rôle in this chapter. In particular we are dealing with the concept of nondegeneracy of an additive set function with respect to a system of subsets of coalitions (subsets of Ω). Recall that $\mathbb{A}^+$ denotes the additive set functions on our finite set $\Omega = \{1, \ldots, n\}$.

Call $m \in \mathbb{A}^+$ homogeneous w.r.t. $\alpha \in \mathbb{R}_+$ ("m hom α") if, for $T \in \underline{B}$, $m(T) > \alpha$, there is $S \subseteq T$ such that $m(S) = \alpha$ (v. Neumann-Morgenstern [10]); this is a direct translation of nonatomicity from the continuous case (where e.g. $\Omega = [0, 1]$, $\underline{B} = \{\text{Borelian sets}\}$, m σ-additive).

E.G. $m = (\frac{3}{7}, \frac{1}{7}, \frac{1}{7}, \frac{1}{7}, \frac{1}{7})$ is homogeneous w.r.t. $\alpha = \frac{4}{7}$ and uniform distribution $(m_i = \frac{1}{n}, i \in \Omega)$ is homogeneous w.r.t. $\alpha = \frac{k}{n}$, $k = 0, 1, \ldots, n$. Counterexamples are straightforward.

Next, $m \in \mathbb{A}^+$ is said to be nondegenerate (cf. [16], [17]) w.r.t. $\underline{Q} \subseteq \underline{B}$ ("m n.d. $\underline{Q}$") if the linear system of equations in variables $y_1, \ldots, y_n$, given by

(16) $$\sum_{i \in S} y_i = m(S) \quad (S \in \underline{Q})$$

has the unique solution $m = (m_1, \ldots, m_n)$. In particular, for $\alpha \in \mathbb{R}_+$, we may consider $\underline{Q} = \underline{Q}_\alpha = \{S \in \underline{B} \mid m(S) = \alpha\}$; in this case we say "$m$ hom α" if m hom $\underline{Q}_\alpha$. (16) translates to

(17) $$\sum_{i \in S} y_i = \alpha \quad (m(S) = \alpha)$$

and "n.d." means that m is uniquely defined by its values on its α-constancy sets. In the above mentioned examples, m is n.d. α as well. For $m = (\frac{1}{8}, \frac{1}{8}, \frac{1}{8}, \frac{2}{8}, \frac{3}{8})$ clearly m hom $\frac{7}{8}$ but m n.d. $\frac{7}{8}$ is wrong. The relations between "hom" and "n.d." are discussed in [13] Chapter III. Note that a nonatomic probability is clearly n.d. α for $0 < \alpha < \frac{1}{2}$ (i.e. coincides with any σ-additive measure having identical values α on all α-constancy sets of m).

Let us start with the representation of convex set functions by affine set functions. As usual $\mathcal{C}$ denotes the set of convex set functions $\mathcal{C}_+^1$ is the set of convex nonnegative and normalized ($v(\Omega) = 1$) set functions.

Theorem. Let $c \in \mathcal{C}^1$. Then there is $t \in \mathbb{N}$ and (up to possible reordering) a unique $(m, \alpha) = (m^1, \ldots, m^t; \alpha_1, \ldots, \alpha_t) \in \mathbb{A}_+^t \times \mathbb{R}_+^t$ such that

(18) $$v(\cdot) = \max_{\tau = 1, \ldots, t} (m^\tau(\cdot) - \alpha_\tau).$$

In fact it is possible to define a unique representation of v by means of (m, α) such that m and α satisfy certain regularity conditions. This representation will be referred to as the canonical one; the definition is explicitly stated and explained in [13], [16], [17].

Formula (18) clearly represents the fact that a convex v can be considered as to be the envelope of its supporting affine set functions. We would like to capitalize on this fact by stating conditions to m and α such that v is extreme in $\mathcal{C}^1_+$.

To this end, let us first consider the case that $t=2$. Assume in addition that the second term in (18) vanishes identically. Hence by rewriting formula (18) it is at once seen that v may be written as

$$(19) \qquad v = e^\alpha = \frac{1}{1-\alpha}(m-\alpha)^+,$$

where now m is a normalized nonnegative real valued measure $(m \in \mathbb{A}^1_+)$ and $\alpha \in (0,1)$. Hence, in this particular case v is the composition of the real valued function f^α, defined by

$$(20) \qquad f^\alpha(s) = \frac{1}{1-\alpha}(s-\alpha)^+ \qquad (s \in \mathbb{R}).$$

As it turns out we may in addition assume that for every $i \in \Omega$, $m_i \leq 1-\alpha$ („canonical representation"), see [16]. Now, in this simple case the characterization of the extreme ones among the e^α is as follows:

Theorem. Let $e^\alpha = f^\alpha \circ m \in \mathcal{C}^1_+$ be canonically represented by $m \in \mathbb{A}^1_+$ and $\alpha \in (0, 1)$. Then e^α is extreme in $\mathcal{C}^1_+$ if and only if m is non degenerate with respect to α.

As the theorem is the intuitively most important clue to the following exposition, we shall proof the "if"-part of our statement. The reader may want to compare this proof with the one given in [16], which contains the complete proof of the theorem as well.

In order to prove sufficiency assume that

$$e^\alpha = \tfrac{1}{2}(v^1 + v^2)$$

where $v^1, v^2 \in \mathcal{C}^1_+$. Assuming that m n.d. α, we have to verify that $v^1 = v^2 = e^\alpha$.

1. Step: Observe that, for $i, j \in \Omega$

$$(21) \qquad 1 - v^k(\Omega - i) - v^k(\Omega - j) \geq 1 - v^k(\Omega - i - j)$$

follows from $v^k \in \mathcal{C}$ (i and $\{i\}$ are always identified). ($k = 1, 2$).

2. Step: Because of $m_i \leq 1-\alpha$, $m(\Omega - i) \geq \alpha$ (m being normalized!), we have

$$(22) \qquad 1 - e^\alpha(\Omega - i) = \frac{m_i}{1-\alpha} \qquad (i \in \Omega)$$

3. Step: *Define*, for $i \in \Omega$,

$$\mu_i^k = 1 - v^k(\Omega - i)$$

II. Nondegeneracy Problems in Cooperative Game Theory

such that $\mu^k \in \mathbb{A}$. Note that

(23)
$$\mu_i^1 + \mu_j^1 \geqslant 1 - v^1(\Omega - i - j)$$
$$\vdots$$
$$\mu_i^1 + \ldots \mu_l^1 \geqslant 1 - v^1(\Omega - i - \ldots - l)$$
$$\mu^1(S) \geqslant 1 - v^1(\Omega - S) \quad (S \in \underline{P})$$

follows from (20) by induction.

Observe that

(24)
$$\frac{1}{1-\alpha} m = \tfrac{1}{2}(\mu^1 + \mu^2)$$

follows immediately from (22).

4. Step: Therefore, for any $S \in \underline{P}$ such that $m(S) \geqslant \alpha$, we have

(25)
$$\frac{1}{1-\alpha} m(S^c) = 1 - e^\alpha(S)$$
$$\phantom{\frac{1}{1-\alpha} m(S^c)} = \tfrac{1}{2}[1 - v^1(S) + 1 - v^2(S)] \quad (23)$$
$$\phantom{\frac{1}{1-\alpha} m(S^c)} \leqslant \tfrac{1}{2}[\mu^1(S^c) + \mu^2(S^c)] \quad (24)$$
$$\phantom{\frac{1}{1-\alpha} m(S^c)} = \frac{1}{1-\alpha} m(S^c)$$

Obviously, all inequalities engaged in (25) must be equations, hence

(26) $\quad \mu^k(S^c) = 1 - v^k(S^c) \quad (m(S) \geqslant \alpha).$

5. Step: Recall

$$\underline{Q}_\alpha := \{S \in \underline{P} \mid m(S) = \alpha\}.$$

For $S \in \underline{Q}_\alpha$, $m(S) = \alpha$ and $e^\alpha(S) = 0$. Thus, $v^k(S) = 0$ $(k = 1, 2)$ and by (26) $\mu^k(S^c) = 1$, $\mu^k(S) = \mu^k(\Omega) - 1$.

Hence

$$\frac{\alpha}{\mu^k(\Omega) - 1} \mu^k(S) = \alpha \quad (S \in \underline{Q}_\alpha)$$

and, since m n.d. α,

$$\frac{\alpha}{\mu^k(\Omega) - 1} \mu^k = m \quad (k = 1, 2)$$

6. Step: Concluding, we find for all S such that $m(S) \geqslant \alpha$:

(27)
$$v^k(S) = 1 - \mu^k(S^c)$$
$$= \mu^k(S) - (\mu^k(\Omega) - 1)$$
$$= \frac{\mu(\Omega) - 1}{\alpha}(m(S) - \alpha).$$

Inserting $S=\Omega$ in (27) yields

$$\frac{\mu^k(\Omega)-1}{\alpha} = \frac{1}{1-\alpha},$$

thus

(28) $$v^k(S) = \frac{1}{1-\alpha}(m(S)-\alpha) = e^\alpha(S)$$

But for $m(S) > \alpha$, (28) holds trivially true (as $e^\alpha(S) = v^k(S) = 0$), hence

$$v^1 = v^2 = e^\alpha,$$

q.e.d.

The previous theorem clears the situation or the case that the convex set function v is represented by means of two measures and two constants, that is two affine set functions, one of them being identically 0.

This, however, generalizes at once to the case in which v is the envelope of several affine set functions. The following may serve as the intuitive background:

Intuitively, the crucial point of a function $f^\alpha = \frac{1}{1-\alpha}(\cdot-\alpha)^+$ is α, and thus the crucial sets of $m \in \mathbb{A}^1$ w.r.t. f^α are those which are thrown into α by m, i.e., the sets in $m^{-1}(\{\alpha\}) = Q_\alpha$. Similarly, a function defined by

$$f(s) = \max(s_1-\alpha_1, \ldots, s_{t-1}-\alpha_{t-1}, 0)$$

has certain "crucial areas" $Q_H := \{s \in \mathbb{R}^{t-1} | 0 \leqslant s_\tau - \alpha_\tau = s_\sigma - \alpha_\sigma \geqslant s_\kappa - a_\kappa (\tau, \sigma \in H, \kappa \in H^c)\}$ ($H \subseteq \{1, \ldots, t-1\}$) and $Q_\tau = \{s \in \mathbb{R}^{t-1} | s_\tau - \alpha_\tau = 0 \geqslant s_\sigma - \alpha_\sigma \ (\sigma \in H^c)\}$; and thus the "crucial" sets are those which are thrown into $\bigcup_{H \subseteq \{1, \ldots, t\}} Q_H$ by m, i.e.

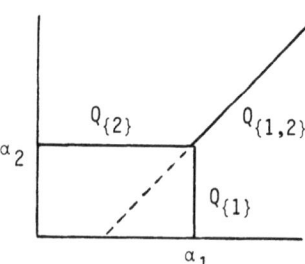

the sets from $\bigcup_{H \subseteq \{1, \ldots, t\}} Q_H$. If m is determined by its "crucial sets" we cannot vary m in order to generate partitions $v = \frac{1}{2}(v^1 + v^2)$ of v and hence v is extreme.

Our intuitive remarks can be given a precise meaning by means of the following theorem.

Theorem [17]. Let $v \in \mathcal{C}^1$ and assume

$$v(\cdot) = \max(m^1 - \alpha_1, \ldots, m^{t-1} - \alpha_{t-1}, 0)$$

II. Nondegeneracy Problems in Cooperative Game Theory 403

if m n.d. α then v is extreme in $\mathcal{C}^1$. If v is extreme in $\mathcal{C}^1$ then m n.d. α provided the representation is canonical.

As nondegeneracy is obviously the clue in order to render a convex set function v, represented in terms of (18) by affine set functions to be extreme we shall comment on this property as follows. In order to find the extrems of $\mathcal{C}_+^1$ it is obviously advisable to find sufficient conditions for a pair (m, α) to enjoy non degeneracy properties. Now, if m n.d. α $(m \in \mathbb{A}^1, \alpha \in (0,1))$, then m and α can easily be verified to be rational valued by solving the system (16). Thus, by rescaling, we find an integer valued measure $M \in \mathbb{A}$ and an integer $\lambda \in \mathbb{N}$ such that M n.d. λ. Let

$$0 < g_1 < \ldots < g_r$$

denote the different values taken by $M_i = M(\{i\})$ $(i \in \Omega)$ and define

$$K_\rho := \{i \mid M_i = g_\rho\} \quad \rho = 1, \ldots, r$$
$$k_\rho := |K_\rho|$$

Now, any equation $\sum_{i \in T} M_i = \lambda$ $(T \in \underline{\underline{Q}}_\lambda = \underline{\underline{Q}}_\lambda(M))$ rewrites

$$\lambda = \sum_{i \in T} M_i = \sum_{\rho=1}^r \sum_{i \in K_\rho \cap T} M_i$$

$$= \sum_{\rho=1}^r |K_\rho \cap T| g_\rho =: \sum_{\rho=1}^r a_\rho^T g_\rho$$

where $a_\rho^T := |K_\rho \cap T|$ as an integer satisfying

(29) $$0 \leq a_\rho^T \leq k_\rho.$$

Hence every $T \in \underline{\underline{Q}}_\lambda$ induces a representation

(30) $$\sum_{\rho=1}^r a_\rho^T g_\rho = \lambda$$

of the natural number λ as a weighted sum of the natural numbers g_ρ $(\rho = 1, \ldots, r)$ by means of natural coefficients a_ρ^T bounded by k_ρ $(\rho = 1, \ldots, r)$. As M n.d. α, the coefficient matrix of (16) has rank n and it is not hard to verify that equivalently the coefficient matrix of (30), i.e. $(a_\rho^T)_{T \in \underline{\underline{Q}}_\lambda,\ \rho=1,\ldots,r}$ has rank r.

Thus it seems that nondegeneracy of M w.r.t. λ is connected to finding r linear independent "representations" of λ by the g_ρ $(\rho = 1, \ldots, r)$, the coefficients being suitably bounded.

Indeed, suppose $(g_1, \ldots, g_r) \in \mathbb{N}^r$ and $(k_1, \ldots, k_r) \in \mathbb{N}^r$ are given (and thus M is specified up to permutations. Let us assume that we can solve the following problem:

"Find integers a_ρ^σ $(\sigma, \rho = 1, \ldots, r)$ such that

1. $0 \leq a_\rho^\sigma \leq k_\rho$ $(\sigma, \rho = 1, \ldots, r)$

(31) 2. $(a_\rho^\sigma)_{\sigma, \rho = 1, \ldots, r}$ is nonsingular

3. $\sum_{\rho=1}^r a_\rho^\sigma g_\rho = \lambda$ $(\sigma = 1, \ldots, r)$.

Then we may define $T^\sigma \in \underline{Q}_\lambda$ by choosing $T^\sigma_\rho \subseteq K_\rho$, $|T^\sigma_\rho| = a^\sigma_\rho$ ($\sigma, \rho = 1, \ldots, r$) and putting

$$T^\sigma := \sum_{\rho=1}^{r} T^\sigma_\rho.$$

Obviously, this definition would provide $a^{T^\sigma}_\rho = a^\sigma_\rho$ and hence $(a^T_\rho)_{\substack{T \in Q_\lambda \\ \rho = 1, \ldots, r}}$ has rank r, thus M rendering n.d. α. In other words, finding a solution to problem (31) is equivalent to proving M n.d. λ.

Now, let us (vaguely) quote a result from [17]. Given the vector $g = (g_1, \ldots, g_r)$ of natural numbers, there is a natural number R_r and "nice" "lower bounds" L_ρ ($\rho = 1, \ldots, r$) (all depending on g) such that problem (31) can be solved provided $k_\rho \geqslant L_\rho$ ($\rho = 1, \ldots, r$), $R_r \leqslant \lambda \leqslant M(\Omega) - R_r$, and $\lambda \equiv 0$ mod g.c.d. $(g_1, \ldots, g_r)$. In other words: We must require $|K_\rho| = \{i \mid M_i = g_\rho\}$ to be not too small, then there is a symmetric (nonempty) "interval" $[R_r, M(\Omega) - R_r]$ such that every λ within this interval, which is also an element of the ideal spanned by the g_ρ ($\rho = 1, \ldots, r$), admits of a solution of problem (31).

This type of result admits at once to list nice classes of pairs (M, λ) such that M n.d. λ. Needless to say that every such (M, λ) provides an extreme convex set function of the type $f^\alpha \circ m$ by a normalization procedure, i.e., by putting $v := \dfrac{(M - \lambda)^+}{M(\Omega) - \lambda}$. Let us remark that, in many cases, the "lower bounds" L_ρ "are equal to 1 for $\rho \geqslant 3$". As the "weights" g_ρ have been listed in increasing order, this means we need "sufficiently many small players" in order to ensure non-degeneracy and thus extremality of $f^\alpha \circ m$. This will nicely fit into later attempts to interpret "extreme games".

So far we have discussed nondegeneracy of an integer-valued measure. Switching to vector-valued measures $M = (M^1, \ldots, M^t)$ (M integer valued) and vectors $\lambda = (\lambda_1, \ldots, \lambda_t)$ of rational numbers, nondegeneracy of M w.r.t. λ becomes too complicated as to be discussed informally. As is suggested by the form of the systems $\underline{Q}_H$, we will have to study not just "representations"

$$\sum_{i \in T} M^\tau_i = \lambda_\tau$$

which can be rewritten $\sum_{\rho=1}^{r} a^T_\rho g^\tau_\rho = \lambda_\tau$ ($\tau = 1, \ldots, t$) but there are also inequalities

$$\sum_{i \in T} M^\kappa_i \leqslant \lambda_\kappa$$

which provide additional constraints of the type $\sum_{\rho=1}^{r} a^T_\rho g^\kappa_\rho \leqslant \lambda_\kappa$.

Thus, the t-dimensional version of problem (31) is somewhat more involved. As it turns out, integer programming techniques are closely related to the methods necessary to answer the n.d. question in this case. In any case there are "similar results" as compared with the case $t = 1$: Given the weights $(g^\tau_\rho)_{\substack{\rho = 1, \ldots, r \\ \tau = 1, \ldots, t}}$ there are lower bounds that should be respected by the number

$k_p^\tau = |\{i | M_i^\tau = g_p^\tau\}|$. Then there are nice regions in $\mathbb{N}^t$ such that the vectors λ within these regions admit of a solution of the t-dimensional version of problem (31) – thus yielding extreme convex set functions (see [17]).

It is nice to observe that the representation theory as developed above has also certain consequences for the shape of solution concepts to be delt with in the framework of cooperative games.

The appropriate solution concept for convex functions is the core as defined in Chapter I, that is the set of all additive set functions x dominating the v and having the same total measure, $x,(\Omega) = v(\Omega)$.

If v is convex then $\mathscr{C}(v)$ is always nonempty [22]. In particular the extremes of the compact convex polyhedron $\mathscr{C}(v)$ may be obtained by an intuitively appealing procedure which can be interpreted as a "band-waggon-process". For according to [22] an extreme point of $\mathscr{C}(v)$ is obtained as follows. Take an increasing family of coalitions

(32) $$\emptyset = S_0 \subseteq S_1 \subseteq \ldots \subseteq S_n = \Omega$$

such that

(33) $$|S_k - S_{k-1}| = 1 \quad (k = 1, \ldots, n)$$

and put

(34) $$\bar{x}_i := v(S_k) - v(S_{k-1})$$

if $S_k - S_{k-1} = i$; then $\bar{x} \in \mathscr{C}^e(v)$. On the other hand, every $\bar{x} \in \mathscr{C}^e(v)$ induces a sequence (32). Thus, extremes of $\mathscr{C}(v)$ correspond to a "coalition forming process" such that every player joining in the process receives his marginal value at this very instant.

Suppose now for the moment that $v = e^\alpha$ has been concidered, then the result of the "coalition forming process" is obtained at once. It turns out that the extreme points at the core of e^α are given by

(35) $$\mathscr{C}^e(e^\alpha) = \left\{ x^T := \frac{m_T}{1-\alpha} \mid T \in \underline{Q}_{1-\alpha}(m) \right\}$$
$$\cup \{ x^{S,i} := (m_S + ((1-\alpha) - m(S))\delta_i)/(1-\alpha)$$
$$0 < (1-\alpha) - m(S) < m_i, i \notin S \}.$$

It is clearly seen that if m is nondegenerate w.r.t. α then the core of e^α contains "more" extreme points of the type $x^T := m_T/(1-\alpha)$, where $T \in \underline{Q}_{1-\alpha}(m)$. These particular extreme points may be interpreted as to represent clear cut situations where the players are divided into two classes, those which obtain a utility according to their "resource distribution" m and those who obtain a utility which is 0. Hence it can be roughly argued that extreme games could induce an extreme core in the sense that the extremes of the core represent very unjust situations – at least as unjust as it is possible within the concept of the core. Concluding we may argue that the idea of representing convex set functions as an envelope of affine ones has been very fruitful. It turned out that extreme games can be characterized as well as extreme points of the core in a very straight forward manner. Moreover, it seems reasonable to expect that a canonically represented v is extreme if in its representation expressed by (18) there are many small players, which is precisely what nondegeneracy means. The situation is quite similar if $\underline{P}$ is assumed to be the system of Borel sets of the unit intervall and v is nonatomic. If we assume that v is also represented by nonatomic affine set functions $m^1 - \alpha_1, \ldots, m' - \alpha_l$, then v is verified to be extreme if in addition the range of the vector valued measure $m = (m^1, \ldots, m^l)$ is a full dimensional convex compact set in $\mathbb{R}^+$.

Hence, the fact that many players induce extremality of the convex set function represented canonically has its analogue in the nonatomic territory. This also supports the idea that nondegeneracy is a finite substitute for nonatomicity.

In the following we shall see that this idea is not confined to convex games and their solution concepts.

Chapter III. Extreme Functions in the Superadditive Range and Homogeneity

In this chapter we study a similar problem as in the previous one. Given the class $\mathscr{S}$ of superadditive set functions v, we want to characterize the extreme functions and, if possible study the solution concepts attached to them. Again "extreme" means that we are looking for extreme points of the compact convex polyhedron $\mathscr{S}^1_+$ (non negative and normalized functions). The procedure is rather similar: Find a suitable representation by means of "elementary" superadditive functions an prove that certain regularity conditions (like nondegeneracy or, in the present case, homogeneity) ensure the extreme point property of certain functions.

The rôle of the elementary functions is played by the weighted majority games $v = 1_{[\alpha, 1]} \circ m$ as defined by example (9) in chapter I.

By technical reasons we shall use the term "m is homogeneous from below w.r.t. α" if for any $T \in \underline{P}$, $m(T) \leq \alpha$, there is $S \supseteq T$ such that $m(S) = \alpha$. We have then the following lemma.

Lemma. Let $v = 1_{[\alpha, 1]} \circ m$ be a weighted majority. Suppose

1. m n.d. α
2. m hom $\uparrow \alpha$

Then, if $v = 1_{(\beta, 1]} \circ \mu$ $((\mu, \beta) \in \mathbb{A}^1 \times (0, 1))$ where μ n.d. β, it follows that $(m, \alpha) = (\mu, \beta)$. Thus, a "homogeneous n.d. representation" of v, if it exists, is the only n.d. representation.

Here the notation hom $\uparrow$ has been used for "homogeneous from below".

Next it turns out that the weighted majorities indeed play the rôle of the "tangencies" in the previous chapter. It is possible to represent superadditive games as envelopes of weighted majorities.

Call $v \in \mathscr{S}$ simple if it takes only the values 0 and 1. It is easily seen that there are simple games which are no weighted majorities (that is there exists no

II. Nondegeneracy Problems in Cooperative Game Theory

pair (m, α) such that $v = 1_{[\alpha, 1]} \circ m$ [10]. However, we have the following theorem.

Theorem [13]. Let $v \in \mathcal{S}$ be simple. Then there is a uniquely defined $t \in \mathbb{N}$ and (up to recording) a unique $(m, \alpha) \in \mathbb{A}^t \times (0, 1)^t$ such that

1. $v = \max_{\tau=1,\ldots,t} 1_{(\alpha_\tau, 1]} \circ m^\tau$
2. $1_{(\alpha_\tau, 1]} \circ m^\tau = 1_{(\alpha_\sigma, 1]} \circ m^\sigma \quad (\tau \neq \sigma)$
3. m^τ n.d. hom $\uparrow \alpha_\tau \quad (\tau = 1, \ldots, t)$
4. If, for some $(\mu, \beta) \in \mathbb{A} \times (0, 1)$, $1_{(\beta, 1]} \circ \mu \leq v$

where μ n.d. hom $\uparrow \beta$, then there is $\tau \{1, \ldots, t\}$ such that

$$1_{(\beta, 1]} \circ \mu \leq 1_{(\alpha_\tau, 1]} \circ m^\tau$$

holds true. If "=" prevails in the last inequality, then $(\mu, \beta) = (m^\tau, \alpha_\tau)$.

As a consequence it can be seen that every superadditive function v which is not simple but takes various values may be represented "canonically" as

(36) $$v = \max_{\tau=1,\ldots,t} f^\tau \circ m^\tau$$

where

$$f^\tau = \sum_{\kappa=1}^{K_\tau - 1} c_\kappa^\tau 1_{(\alpha_\kappa^\tau, \alpha_{\kappa+1}^\tau]}$$

and

$$m^\tau \text{ n.d. hom} \uparrow \alpha_\kappa^\tau \quad (\kappa = 1, \ldots, K_\tau)$$

We now want to employ the representation theory in order to obtain the extreme super-additve set functions.

To this end denote by F the system of all superadditive point-to-point functions $f: \left\{0, \frac{1}{N}, \frac{2}{N}, \ldots, 1\right\} \to [0, 1]$ such that $f(0) = 0$, $f(1) = 1$. Here N is some fixed natural number. We may then formulate the next theorem which tells us that indeed homogeneity is now the clue to the extreme point property. Indeed nondegeneracy has already been used in order to obtain the "canonical" representation and therefore it is the second property mentioned in chapter I which is now essential for extremality of set functions.

Theorem [13]. Let $(f, m) \in F \times \mathbb{A}^1$ satisfy the following conditions:

1. $\mathcal{R}(m) \subseteq \theta$, ($\mathcal{R}(m)$ is the range of m.)
2. $f = \sum_{\kappa=1}^{K-1} c_\kappa 1_{[\alpha_\kappa, \alpha_{\kappa+1})} + 1_{[\alpha_K, 1]}$
3. $m_i < \alpha_{\kappa+1} - \alpha_\kappa \ (i \in \Omega, \kappa = 1, \ldots, K-1)$
4. $m \text{ hom} \downarrow \alpha_\kappa \ (\kappa = 1, \ldots, K)$, $m \text{ hom} \downarrow \alpha_{\kappa+1} - \alpha_\kappa \ (\kappa = 1, \ldots, K-1)$.

If f is extreme in F, then $v = f \circ m$ is extreme in $\mathcal{S}^1$.

As the reader no doubt will be aware the previous theorem is the exact analogue of the case where $t = 2$ in the convex territory. Similarly we have a generalization to arbitrary t by the following theorem which specifies the extreme points in the super additive range.

Theorem [13]. Let $v = \max\limits_{\tau=1,\ldots,t} f^\tau \circ m^\tau \in \mathscr{S}$ where

1. $m^\tau \in \mathbb{A}^1$, $\mathscr{R}(m^\tau) = \theta$
2. $f^\tau = \sum\limits_{\kappa=1}^{K_t-1} c_\kappa^\tau 1_{[\alpha_\kappa^\tau, \alpha_\kappa^\tau)} + 1_{[\alpha_{K_\tau}^\tau, 1]}$
3. $m_i^\tau < \alpha_{\kappa+1}^\tau - \alpha_\kappa^\tau$ ($i \in \Omega$, $\tau\{1, \ldots, t\}$, $\kappa \in \{1, \ldots, K_\tau - 1\}$)
4. $m^\tau \text{ hom} \downarrow \alpha_\kappa^\tau$ ($\kappa = 1, \ldots, K_\tau$), $m^\tau \text{ hom} \downarrow \alpha_{\kappa+1}^\tau - \alpha_\kappa^\tau$ ($\kappa = 1, \ldots, K_\tau - 1$)

Also, let $\underline{U}^\tau := \{S \mid f^\tau \circ m^\tau \geq f^\sigma \circ m^\sigma \ (\sigma \in \{1, \ldots, \tau\})\}$ and assume

5. $S \in \underline{U}^\tau$ ($S \subseteq C(m^\tau)$).

If f^τ is extreme in F, then v is extreme in $\mathscr{S}^1$.

Both theorems tell us that we have to look for the extreme superadditive point functions in order to deal with extreme super-additive set functions, that is the extremes of $\mathscr{S}^1_+$. Note that extreme superadditive point functions are also to be specified – they are not as obvious as the kink functions in chapter II. There is a close similarity between the situations in $\mathscr{C}^1_+$ and $\mathscr{S}^1_+$: In both cases the generic function can be reduced to an envelop of "elementary" functions and the property of nondegeneracy or homogeneity respectively is essentially used in order to characterize the extreme functions.

Let us now shortly deal with the solution concept which is appropriate for $\mathscr{S}$. We are referring to the von Neumann-Morgenstern solution as defined in chapter I.

The following is already due to von Neumann and Morgenstern [10].

Theorem. Let $v = 1_{[\alpha, 1]} \circ m \in \mathscr{S}^1$ be a weighted majority. Assume in addition that v is a constant sum game, that is, $v(S) + v(S^c) = 1$ ($S \in \underline{P}$). Also, assume $m \text{ hom} \downarrow \alpha$. Then

(37)
$$\mathscr{S}^\alpha := \left\{ x^T := \frac{m_T}{\alpha} \mid T \in \underline{Q}_\alpha \right\}$$

is stable.

$\mathscr{S}^\alpha$ is called the main simple solution. One might argue that $\mathscr{S}^\alpha$ suggests a division into classes: for every $T \in \underline{Q}_\alpha$, $x^T \in \mathscr{S}^\alpha$ will allocate $\dfrac{m_i}{\alpha}$ to $i \in T$ and nothing to everybody else, that is, some people get money according to their voting power, and others (those not within the ruling coalition) receive zero.

For $N \in \mathbb{N}$ let

$$\theta = \theta_N = \left\{ 0, \frac{1}{N}, \frac{2}{N}, \ldots, 1 \right\}$$

Then the natural generalization of the theorem of von Neumann and Morgenstern concerning the main simple solution is as follows:

II. Nondegeneracy Problems in Cooperative Game Theory

Theorem. Let $K \in \mathbb{N}$, $\alpha \in \theta^K$, and suppose that there are natural numbers $A, D \in \mathbb{N}$ satisfying

$$K = AD, \quad N = D(K+1) - 1, \quad \alpha_\kappa = \kappa \frac{D}{N} \ (\kappa = 1, \ldots, K).$$

Consider the function

(38)
$$f_\alpha = \sum_{\kappa=1}^{K-1} \frac{\kappa}{K} 1_{[\alpha_\kappa, \alpha_{\kappa+1})} + 1_{[\alpha_K, 1]}$$

$$= \sum_{\kappa=1}^{K-1} \frac{\kappa}{K} 1_{\left[\alpha_\kappa, \alpha_{\kappa+1} - \frac{1}{N}\right]} + 1_{[\alpha_K, 1]} \in F.$$

If $m \in \mathbb{A}^1$ satisfies $\mathscr{R}(m) \subseteq \theta$ and $m \operatorname{hom} \downarrow \alpha_1 = \frac{D}{N}$, then

$$\mathscr{S}^{\alpha_\kappa} := \left\{ x^T := \frac{m_T}{\alpha_\kappa} \mid T \in \underline{Q}_{\alpha_\kappa} \right\}$$

is stable for $v = f^\alpha \circ m$.

It should be observed that $m \operatorname{hom} \downarrow \alpha_\kappa$ ($\kappa = 1, \ldots, K$) is implied by $m \operatorname{hom} \downarrow \alpha_1$.

The previous theorem is a very special one as far as the function f_α is concerned. If the function f_α is somewhat modified then various more or less complicated versions of the von Neumann-Morgenstern solution appear. To be more precise let $K \geq 2$ and $\alpha = (\alpha_1, \ldots, \alpha_K) \in \theta^K$ such that

$$0 < \alpha_1 < \ldots < \alpha_K < 1.$$

Given $m \in \mathbb{A}^1$, $\mathscr{R}(m) \subseteq \theta$, define

$$\mathscr{S}^{\alpha_\kappa} := \left\{ x^T := \frac{m_T}{\alpha_\kappa} \mid T \in \underline{Q}_{\alpha_\kappa} \right\}$$

$$\mathscr{S}_\kappa = \mathscr{S}^{\alpha_\kappa, \frac{K}{K-1} \alpha_\kappa} := \left\{ x^{U,V} := \frac{m_U}{\alpha_\kappa} + \frac{m_V}{K - \frac{K}{K-1}\alpha_\kappa} \middle| \begin{array}{l} U \in \underline{Q}_{\alpha_\kappa}, \\ V \in \underline{Q}_{\frac{K}{K-1}(\alpha_K - \alpha_\kappa)}, \ U \cap V = \emptyset \end{array} \right\} (\kappa = 1, \ldots, K-1) \quad \mathscr{S}^\alpha := \mathscr{S}^{\alpha_K} + \sum_{\kappa=1}^{K-1} \mathscr{S}_\kappa$$

Recall:
If $K \geq 2$, $\alpha \in \theta^K$, $0 < \alpha_1 < \ldots < \alpha_K < 1$, $f_\alpha := \sum_{\kappa=1}^{K-1} \frac{\kappa}{K} 1_{\left[\alpha_\kappa, \alpha_{\kappa+1} - \frac{1}{N}\right]} + 1_{[\alpha_K, 1]}$ and $m \in \mathbb{A}^1$, $\mathscr{R}(m) \in \Theta$, then $v = f_\alpha \circ m \in \mathscr{S}^1$ if $f_\alpha \in F$.

We have then the following theorem:

Theorem [13]. Let $K \geq 3$, $\alpha \in \theta^K$,

$$f_\alpha = \sum_{\kappa=1}^{K-1} \frac{\kappa}{K} 1_{\left[\alpha_\kappa, \alpha_{\kappa+1} - \frac{1}{N}\right]} + 1_{[\alpha_K, 1]} \in F$$

and suppose that there is $D \in \mathbb{N}$ satisfying

$$N = D(K^2+1) - 1$$

$$\alpha_\kappa = ((K-1)\kappa + 1)\frac{D}{N} = \frac{(K-1)\kappa + 1}{K^2+1}\frac{N+1}{N} \quad (\kappa = 1, \ldots, K).$$

If $m \operatorname{hom} \downarrow \frac{D}{N}$, then $\mathscr{S}^\alpha$ is stable for $v = f_\alpha \circ m$.

Clearly the theory of $\mathscr{S}^1$ is less developed than the one of $\mathscr{C}^1$. However, we have a pretty similar situation. We may represent any $v \in \mathscr{S}$ as an envelope of set functions, the extreme ones among the superadditive set functions being characterized by homogeneity. These extreme superadditive set functions admit of von Neumann-Morgenstern solutions. Close inspections of the results of chapters II and III reveales that indeed the solution concepts have a certain similarity: It is alsways a group of players which is discriminated while another group (the minimal winning coalition, the ruling coalition) is paid according to its voting measure.

It is obvious that homogeneity can also be seen as a surrogate for non-atomicity in the finite range. Therefore it would be nice to have characterizations of homogeneity. In other words what is e.g. a sufficient condition or a pair $(m, \alpha) \in \mathbb{A}^+ \times \mathbb{R}_+$ such that m is homogeneous w.r.t. α? Again in comparison to the nondegeneracy problem it would be nice to obtain results which lead to the conclusion that "many players imply homogeneity". Also we want to restrict ourselves to the case that m and α essentially are rationals. We may therefore at once assume that we are in fact dealing with a measure M and a natural number λ such that M takes natural numbered values and M is homogeneous w.r.t. λ. Similarly as to the previous chapter we may assume that there is a decomposition of Ω into non empty subsets K_i, $i = 1, \ldots, R$. Now, if

$$g = (g_1, \ldots, g_r) \in \mathbb{N}^r,$$

then an integer-valued measure $M > 0$ on Ω is specified via

(39) $$M(S) = \sum_{i=1}^{r} |S \cap K_i| g_i.$$

On the other hand, M specifies a decomposition of Ω and a vector $g \in \mathbb{N}^r$; intuitively, K_i represents the "players of type $i \in \{1, \ldots, r\}$" and g_i is the "weight" of players of type i.

Permutations of the "types" (i.e. of $\{1, \ldots, r\}$) and of all members of a type (i.e. of some K_i) do not alter any homogeneity M might enjoy w.r.t. some $\lambda \in \mathbb{N}$, thus, if we write

$$k_i := |K_i| \in \mathbb{N}, \quad k = (k_1, \ldots, k_r) \in \mathbb{N}^r,$$

then M and $(g, k) \in \mathbb{N}^{2r}$ determine each other up to permutations (provided of course $\sum_{i=1}^{r} k_i = n$).

Throughout the following we shall *always* assume that

$$0 < g_1 < g_2 < \ldots < g_r$$

holds true. Then $g \in \mathbb{N}^r$ induces a system of natural numbers

$$l_{ij} \quad (i,j=1,\ldots,r; i<j),$$

defined by

(40) $$l_{ij} g_i \leq g_j < (l_{ij}+1) g_i,$$

which will be used frequently without further reference.

The following lemma reduces the possibilities of "representing" a natural number λ by the weights ρ_i given the information that $M \hom \lambda$.

Lemma [18]. If $M \hom \lambda$, then there is $i_0 \in \{1, \ldots, r\}$ and $c \in \mathbb{N}$, $1 \leq c \leq k_{i_0}$, such that the following holds true:

1. $\lambda = c g_{i_0} + \sum_{i=i_0+1}^{r} k_i g_i$;

2. $k_{i_0} \leq c + l_{0i}$ for all $i \in \{i_0 + 1, \ldots, r\}$ satisfying $g_{i_0} \chi g_i$;

3. $k_i \leq l_{i i_0}$ for all $i \in \{1, \ldots, i_0 - 1\}$ satisfying $g_i \chi g_{i_0}$.

The following interpretation of our results as stated by the last theorem is offered. $k = (k_1, \ldots, k_r)$ represents a distribution of the number of players of the various types. A representation of λ (the majority level) as indicated by (I), corresponds to a specific distribution of the players over various types within a minimal winning coalition. Such a typical minimal winning coalition is composed by all "big players" $(i \geq i_0 + 1)$ and a few players of "medium size" $(i = i_0)$, while it contains none of the small players $(i < i_0)$.

From this it can be deduced that given the integer valued non negative additive set function M or rather the vector $(g, k) \in \mathbb{N}^{2r}$, then it is possible to define certain "intervalls" of integer numbers λ with the property that M is homogeneous w.r.t. λ. These intervals are of the form

$$I_\rho^r = \left\{ c g_\rho + \sum_{i=\rho+1}^{r} k_i g_i \,\middle|\, c_\rho^r \leq c \leq k_\rho, c \in \mathbb{N} \right\},$$

the numbers c_ρ^r ($\rho = 1, \ldots, r$) depending on (g, k). In other words, we have $m \hom \lambda$ if and only if $\lambda \in \bigcup_{\rho=1}^{r} I_\rho^r$ holds true.

The numbers $(c_\rho^r)_{\rho=1,\ldots,r}$ completely describe the intervalls of homogeneity w.r.t. a given (g, k).

A recursive formula for computing the numbers $(c_\rho^r)_{\rho=1,\ldots,r}$ is offered in [18]. Again it turns out that homogeneity is satisfied essentially if there are sufficiently many small players. This again supports the idea that homogeneity as well as nondegeneracy is a substitute for nonatomicity in the finite framework. Both concepts play an essential rôle the first one w.r.t. convex games, the second one with respect to the super-additive games. Both concepts imply that the canonically represented functions are extreme and admit of nice solution concepts.

Chapter IV. L.P.-Games and Nondegeneracy

The last chapter is devoted to a certain "generalized optimization problem". We would like to study a game which is obtained if the restrictions of the optimization problem depend on the various coalitions. To this end we shall shortly outline the theory of the L.P.-game. It should be mentioned however, that most likely some of the principles and methods do apply to a more general framework: The L.P.-game is a special case since, in view of its linearity properties, the convergence of core and competitive equilibrium is a finite one [12]. It can be expected that, therefore, the particular method to be explained in this chapter does not apply with respect to the core but presumably with respect to ε-cores. Also it should be stressed that the inherent sidepayment character of games as studied in this context is not a necessary requirement: It would seem that also in models with nontransferable utility nondegeneracy can be seen as a substitute for nonatomicity similar to what shall be explained in this chapter.

Recall the definition of the L.P.-game, the function $v = v^{A,b,c}$ is defined by (3) of chapter I.

If we take the dual program for coalition S, as specified by formula (5) then any optimal solution $\bar{y}$ for the "dual Ω-program" establishes an element of the core of v, namely the additive set function $\bar{m} := \bar{y} b(\cdot)$.

As the dual variable is roughly the gradient of function defined via the linear program when the constraints are varied, economists are used to interpret the dual variable as "shadow prices". Hence

(41) $$\bar{y} b(\cdot) = \bar{y}_1 b^1(\cdot) + \ldots + \bar{y}_m b^m(\cdot)$$

is at once interpreted as the "worth" or "value" of the various resources weighted by means of shadow prices ("expenditure"). Therefore it is not surprising that some simple computations reveal the fact that $\bar{y} b(\cdot)$ can be seen as the "competitive equilibrium" of an appropriate market which is defined by means of the L.P.-game.

In general it is not excluded that various different elements of the core exist. However, as it turns out for sufficiently many players the core and the competitive equilibrium coincide, that is we have

$$\mathscr{C}(v) = \{\bar{y} b(\cdot)\}.$$

For any $z \in \mathbb{R}^m_+$ consider the (A, z, c)-program (dual version) which is obtained by replacing within the dual program (5) the vector-valued measure b by the vector z. Thus the (A, z, c)-program is a simple linear program and does not define a game. Now define

(42) $Q_0 := \{z \in \mathbb{R}^m_+ \mid \bar{y} \text{ is an optimal dual solution for the } (A, z, c)\text{-program}\}$
$= \{z \in \mathbb{R}^m_+ \mid \bar{y} z = \min\{yz \mid y \in \mathbb{R}^m_+, yA \geqslant c\}\},$

(43) and let $\underline{\underline{Q}} = \{S \in \underline{\underline{B}} \mid b(S) \in Q_0, b(S^c) \in Q_0\}$.

Then it is not hard to see that $v = v^{A,b,c}$ is "additive on $\underline{\underline{Q}}$", i.e.,

(44) $$v(S) + v(S^c) = v(\Omega) \quad (S \in \underline{\underline{Q}}).$$

II. Nondegeneracy Problems in Cooperative Game Theory

In fact $\underline{Q}$ is the system of all those coalitions which generate the same shadow prices when producing their optimal output bundle. The system $\underline{Q}$ will play the rôle of the system Q_α which has been used in the context of chapter II in order to employ nondegeneracy as a tool for establishing extreme convex set functions. Indeed, let us assume that $\bar{y}$ is a unique solution of the dual Ω-program. Assume now that $\bar{m} = \bar{y}b(\cdot)$ is nondegenerate w.r.t. the system $\underline{Q}$ as defined above.

It is then established at once that nondegeneracy of $\bar{m}$ implies that the core collapses to $\bar{m}$. Indeed, if $\mu \in \mathscr{C}(v)$, it follows that for $S \in \underline{Q}$

(45) $$\mu(\Omega) = \mu(S) + \mu(S^c) \geq v(S) + v(S^c) = v(\Omega) = \mu(\Omega)$$

i.e.

(46) $$\mu(S) = v(S) = \bar{m}(S) \quad (S \in \underline{Q})$$

and

$$\mu = \bar{m} \text{ (by "n.d.")}, \quad \mathscr{C}(v) = \{\bar{m}\}.$$

Hence, nondegeneracy implies the coincidence of core and C.E. (see Owen [12] for the "replica" version and Billera-Raanan [2] for the "nonatomic" version).

We are used to attach the "equivalence" theorem to a concept of "large economies" or markets. It is, therefore, desirable to translate the surrogates for nonatomicity in a way such that "large games" can be identified with "games enjoying n.d.-properties".

To this end we introduce the notion of "types" similarly to what has been done in chapters II and III.

Suppose a decomposition $\Omega = \sum_{\rho=1}^{r} K_\rho$ of Ω has been specified such that K_ρ represents all players of type ρ having the same initial bundle of goods. Hence the vector valued measure b satisfies

(47) $$b(\cdot) = \sum_{\rho=1}^{r} |K_\rho \cap \cdot| g_\rho$$

where $g_\rho \in \mathbb{R}_+^m$ denotes the common initial allocation of all players of type ρ. Again let $k_\rho := |K_\rho|$.

The next step is to switch from (47) to an analogue of (31) (less directly in this case). To this end, introduce a suitable matrix Λ such that Q_0 (cf. (42)) is given by

$$Q_0 = \{z \in \mathbb{R}_+^m \mid \Lambda z \geq 0\}$$

In fact, Λ is the "tableau" associated by standard L.P. routine with the optimal solution $\bar{y}$ of the dual Ω-program (5). Then it turns out that

(48) $\bar{m}$ n.d. $\underline{Q}$ iff there are integers
(a_ρ^σ) $(\rho, \sigma = 1, \ldots, r)$ such that
1. $0 \leq a_\rho^\sigma \leq k_\rho \quad (\sigma, \rho = 1, \ldots, r)$
2. $(a_\rho^\sigma)_{\rho, \sigma = 1, \ldots, r}$ is nonsingular
3. $0 \leq \Lambda \sum_{\rho=1}^{r} a_\rho^\sigma g_\rho \leq \Lambda e$

"Lower bounds" for the numbers of players of each type may be (roughly) obtained as follows. By a transformation of bases assume that the g_ρ are unit vectors of $\mathbb{N}^r$. Then (48) reads:

Find r linearly independent $\mathbb{N}^r$ vectors a^σ ($\sigma = 1, \ldots, r$), bounded by $k \in \mathbb{N}^r$ and within a convex polyhedron determined by Λ.

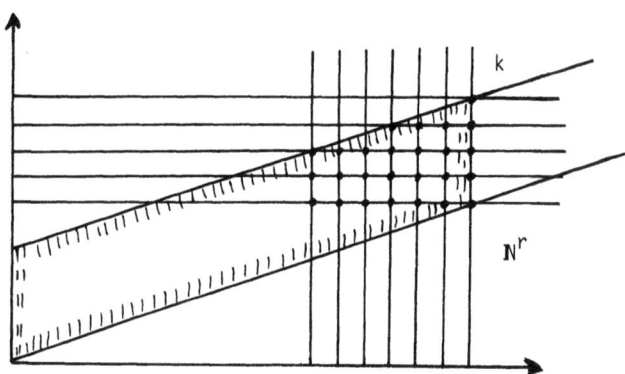

Given the data of the L.P.-game, (i.e., the tableau-matrix Λ) we may therefore specify regions in $\mathbb{N}^r$ such that if $k \in \mathbb{N}^r$ is an element of such a region then (48) can be solved, thus nondegeneracy takes place and the core and the C.E. coincide.

It is nice to observe that the tools for this task are provided by geometric number theory. Indeed the problem of finding r linearly independent $\mathbb{N}^r$ vectors within a certain convex polyhedron was already studied by Minkowski. Minkowski's "Second Theorem" is devoted to the evaluation of certain lattice constants: If a certain lattice (like $\mathbb{N}^r$) is given and we have a convex compact polyhedron, by how much do we have to blow up the polyhedron in order to obtain sufficiently many linearly independent lattice elements within the blown up version of the polyhedron? The factor of enlargement is a certain "lattice constant" depending on the polyhedron. Minkowski gave estimates connecting these lattice constants and the volume of the polyhedron provided the polyhedron has certain symmetry properties.

In our context we may use these estimates in order to specify regions of $k \in \mathbb{N}^r$ such that our convex polyhedron determined by Λ is sufficiently large to include r linearly independent $\mathbb{N}^r$ vectors, which implies nondegeneracy, which implies the coincidence of the core and the competitive equilibrium. Consequently, Minkowski's "Second Theorem" may be used for specifying regions of $k \in \mathbb{N}^r$ as mentioned above. As k resembles distributions of players over types, looking to our regions in $\mathbb{N}^r$ verifies again the basic result: "many" players induce the core and the C.E., to coincide – but how many can be said much more exactly compared to nonatomic theory if one is willing to apply Geometric Number Theory via nondegeneracy (see [15] for details).

Indeed it turns out that the estimates required by Minkowski's second theorem again explicitly involve the numbers of players of each type and if there are sufficiently many of each type then the core and the competitive equili-

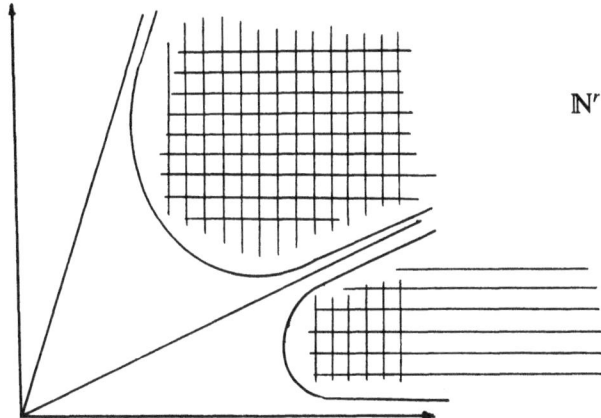

brium coincide. Specifying distributions over the types by integer valued vectors $k \in \mathbb{N}^r$, in other words specifying regions of $\mathbb{N}^r$, seems to be a much more precise method of stating the equivalence theorem compared to either the replica or the nonatomic case.

The fact that nondegeneracy plays the essential rôle similar to the extreme point problem in the convex range and similar to the rôle of homogeneity in the super-additive range again shows that nondegeneracy and homogeneity are appropriate substitutes in the finite range for nonatomicity.

References

[1] Aumann, J. and Shapley, L. S.: Values of nonatomic games, Princeton University Press, Princeton, N.J., (1974).
[2] Billera, L. J. and Raanan, J.: Cores of nonatomic linear production games, Math. of O.R., Vol. 6, 420–423, (1981).
[3] Bondareva, O. N.: Some applications of linear programming methods to the theory of cooperative games, Problemy Kibernet. 10, 119–139 (in Russion), (1963).
[4] Cassels, J. W. S.: An introduction to the geometry of numbers, Springer-Verlag, Berlin, Göttingen, Heidelberg, (1959).
[5] Choquet, G.: Theory of capacities, Ann. Inst. Fourier 5, 131–295, (1953–1954).
[6] Edmonds, J. and Rota, G.-C.: Submodular set functions (abstract, Waterloo Combinatorics Conference), University of Waterloo, Waterloo, Ontario, (1966).
[7] Hildebrand, W.: Core and equilibria of a large economy, Princeton University Press, Princeton, N.J., (1974).
[8] Hildebrand, W.: Core of an economy, Handbook of Mathematical Economics, Vol. II, edited by K. J. Arrow and M. D. Intriligator, North-Holland Publishing Co., (1982).
[9] Lucas, W. F.: The proof that a game may have no solution, Trans. Amer. Math. Soc. 137, 219–229, (1969).
[10] v. Neumann, J. and Morgenstern, O.: Theory of games and economic behavior, Princeton University Press, Princeton, N.J., (1944, 1953).

[11] Muto, S.: Symmetric solutions for $(n, n-2)$ Games with small values of $v(n-2)$, Internat. Journal of Game Theory, Vol. II, 43-52, (1982).
[12] Owen, G.: On the core of linear production games, Math. Programming, Vol. 9, 358-370, (1975).
[13] Rosenmüller, J.: Extreme games and their solutions, Lecture Notes in Economics and Math. Systems 145, Springer, Heidelberg, (1977).
[14] Rosenmüller, J.: The theory of games and markets, North-Holland Publishing Co., (1981).
[15] Rosenmüller, J.: L.P.-games with sufficiently many players, to appear: International Journal of Game Theory, (about 1982).
[16] Rosenmüller, J. and Weidner, H. G.: A class of extreme convex set functions with finite carrier, Advances in Mathematics, Academic Press, Vol. 10, No. 1, 1-38, (1973).
[17] Rosenmüller, J. and Weidner, H. G.: Extreme convex set functions with finite carrier: general theory, Discrete Math. 10, 343-382, (1974).
[18] Rosenmüller, J.: On homogeneous weights for simple games. Preprint series, Nr. 115; Inst. of Math. Economics, University of Bielefeld, 43 p., (1982).
[19] Scarf, H.: The core of an N-person game, Econometrica 35, p. 50-69, (1967).
[20] Schmitz, N.: Eine Klasse von Mehrpersonen-Spielen ohne von Neumann'sche Lösung, International Journal of Game Theory 5, (1976).
[21] Shapley, L. S.: On balanced sets and cores, Naval Rest. Logist. Quart. 14, 453-460, (1967).
[22] Shapley, L. S.: Cores of convex games, International Journal of Game Theory 1, 12-26, (1971).
[23] Shapley, L. S. and Shubik, M.: Quasi cores in a monetary economy with nonconvex preferences, Econometrica 34, 805-827, (1966).

Conic Methods for Unconstrained Minimization and Tensor Methods for Nonlinear Equations*

R. B. Schnabel
University of Colorado, Dept. of Computer Science, Boulder, CO 80309, USA

Abstract. Standard methods for nonlinear equations and unconstrained minimization base each iteration on a linear or quadratic model of the objective function, respectively. Recently, methods using two generalizations of the standard models have been proposed for these problems. Conic methods for unconstrained minimization use a model that is the ratio of a quadratic function divided by the square of a linear function. Tensor methods for nonlinear equations augment the standard linear model with a simple second order term. This paper surveys the research to date on methods for unconstrained minimization and nonlinear equations that use conic and tensor models. It begins with a brief summary of the standard methods, so that the paper is essentially selfcontained.

1. Introduction

The two major unconstrained nonlinear algebra problems are the nonlinear equations problem

$$\text{given } F: R^n \to R^n, \text{ find } x_* \in R^n \text{ such that } F(x_*) = 0 \tag{1.1}$$

where we assume $F \in C^1$, and the unconstrained minimization problem,

$$\underset{x \in R^n}{\text{minimize}} f: R^n \to R \tag{1.2}$$

where we assume $f \in C^2$. Computational methods exist that solve many such problems successfully and efficiently, but research aimed at improving these methods continues. In this paper, we discuss two recently introduced classes of algorithms for solving these problems, conic methods for unconstrained minimization and tensor models for nonlinear equations. Both classes contain interesting innovations and seem to offer advantages over the standard methods, although it is too early to access the ultimate importance of either one.

We assume the reader has at least some familiarity with computational methods for nonlinear equations and unconstrained minimization, although we briefly summarize the leading methods in Section 2. Some survey papers on these methods include Brodlie [1977], Dennis [1977], Schnabel [1982a] and Moré and Sorensen [1982]. The books by Fletcher [1980], Gill, Murray, and

---

* This research was supported by ARO contracts DAAG 29-79-C-0023 and DAAG 29-81-K-0108

Wright [1981], and Dennis and Schnabel [1983] contain a more detailed treatment.

We will denote the matrix of first partial derivatives of F at x, the Jacobian matrix, by $F'(x) \in R^{n \times n}$; here $F'(x)[i,j] = \partial f_i(x)/\partial x[j]$ where $f_i: R^n \to R$ is the ith component function of $F(x)$. We will denote the vector of first partial derivatives of f at x, the gradient vector, by $\nabla f(x) \in R^n$, and the symmetric matrix of second partial derivatives of f at x, the Hessian matrix, by $\nabla^2 f(x) \in R^{n \times n}$; $\nabla f(x)[i] = \partial f/\partial x[i]$ and $\nabla^2 f(x)[i,j] = \partial^2 f/\partial x[i] \partial x[j]$. Note that we are denoting the ith component of a vector x by $x[i]$ so that we can reserve the notation x_i for the ith iterate in a sequence of vectors $\{x_k \in R^n\}$.

The main difference between standard methods and conic and tensor methods is in the local model of the nonlinear function that the method uses in determining its iterates. Standard methods for nonlinear equations base the step from the current iterate x_c upon a linear model of $F(x)$ around x_c,

$$M(x_c + d) = F(x_c) + J_c d \tag{1.3}$$

where $d \in R^n$ and $J_c \in R^{n \times n}$ is $F'(x_c)$ or some approximation to it. Similarly, standard methods for unconstrained minimization base each iteration upon a quadratic model of $f(x)$ around x_c,

$$m(x_c + d) = f(x_c) + g_c^T d + \tfrac{1}{2} d^T H_c d \tag{1.4}$$

where $g_c \in R^n$ is $\nabla f(x_c)$ or a finite difference approximation to it, and $H_c \in R^{n \times n}$ is $\nabla^2 f(x_c)$ or some symmetric approximation to it. These two models are closely related because the minimizer of $f(x)$ must occur at a point x_* where $\nabla f(x_*) = 0$, and the gradient of the model (1.4),

$$\nabla m(x_c + d) = \nabla f(x_c) + H_c d \tag{1.5}$$

is a linear model of the system of nonlinear equations $\nabla f(x): R^n \to R^n$.

The two new classes of methods are based upon generalizations of (1.3) and (1.4). Conic methods for unconstrained minimization base each step on a model of the form

$$\hat{m}(x_c + d) = f(x_c) + \frac{g_c^T d}{1 + b_c^T d} + \frac{\tfrac{1}{2} d^T A_c d}{(1 + b_c^T d)^2} \tag{1.6}$$

where $A_c \in R^{n \times n}$ is symmetric and $b_c \in R^n$. Tensor methods base each iteration on a model of the form

$$\hat{M}(x_c + d) = F(x_c) + J_c d + \tfrac{1}{2} T_c d d \tag{1.7}$$

where $T_c \in R^{n \times n \times n}$ has a particularly simple form. Here we use the notation $T_c d d$ to denote the vector in R^n whose i^{th} component is

$$(T_c d d)[i] = \sum_{j=1}^{n} \sum_{k=1}^{n} T_c[i,j,k] \cdot d[j] \cdot d[k]. \tag{1.8}$$

Of course the justification for either of these models is not obvious and we explain it in this paper. Conic models were introduced by Davidon [1980] and also have been investigated by Bjorstad and Nocedal [1979], Sorensen [1980],

Stordahl [1980], Davidon [1982], Gourgeon and Nocedal [1982], and Schnabel [1982b]. Tensor models were introduced by Schnabel and Frank [1982] and also are discussed in Frank [1982]. The main goal of the developers of tensor methods is to improve the performance of existing methods on problems were $F'(x_*)$ is singular or ill-conditioned, while at least maintaining the performance of the existing methods on all other problems. The developers of conic methods do not seem to have a similarly limited objective.

The remainder of the paper is organized as follows. Section 2 provides a brief survey of the leading standard methods for nonlinear equations and unconstrained minimization, which are based on the models (1.3) and (1.4) respectively. These include both the derivative methods used when $F'(x)$ or $\nabla^2 f(x)$ are available analytically or from finite differences, and the secant methods that are used otherwise. We concentrate on the ideas and properties that are relevant to our discussion of conic and tensor methods. A reader familiar with these methods should skip or skim Section 2. In Section 3 we briefly discuss several extensions of the standard methods that help motivate conic and tensor methods. These are the methods of Barnes [1965] and Gay and Schnabel [1978] for nonlinear equations and of Davidon [1975] for unconstrained minimization. They all still use the standard models (1.3) and (1.4), but some of their objectives and techniques are similar to conic and tensor methods. We discuss conic methods in Section 4, and tensor methods in Section 5. We comment briefly on the application of these two classes of methods to other nonlinear problems in Section 6.

In our opinion, this paper covers most of the important methods for nonlinear equations and unconstrained minimization based on nonstandard models. There has been occasional other work along these lines, however. Perhaps most significant are the methods for homogeneous functions investigated by Jacobson and Oxsman [1972], Charalambous [1973], Kowalik and Ramakrishnan [1976], and others. These methods do not seem to have led to improved algorithms for general classes of problems.

2. Standard models and methods

The fundamental method for solving the nonlinear equations problem is Newton's method. It consists of choosing the new iterate, x_+, as the root of the linear model of $F(x)$ around x_c,

$$M(x_c+d)=F(x_c)+F'(x_c)d, \qquad (2.1)$$

the first two terms of the Taylor series. If $F'(x_c)$ is nonsingular, (2.1) has a unique root at

$$x_+ = x_c - F'(x_c)^{-1}F(x_c). \qquad (2.2)$$

If $F(x_*)=0$, $F'(x_*)$ is nonsingular, and $F'(x)$ is Lipschitz continuous in an open neighborhood containing x_*, then the sequence produced by iterating (2.2) is well-defined and converges q-quadratically to x_*, provided the starting point

$x_0 \in R^n$ is sufficiently close to x_*. A method that converges provided it is started sufficiently close to the solution is called *locally convergent*. (For our definitions of rates of convergence, see for example Ortega and Rheinboldt [1970] or Dennis and Schnabel [1983].)

There are four weaknesses of Newton's method as a computational procedure for solving systems of nonlinear equations that we wish to discuss. They are

(1) The sequence of iterates may not converge to any root if x_0 is not sufficiently close to a root.
(2) The iteration (2.2) is not well-defined computationally if $F'(x_c)$ is singular or ill-conditioned.
(3) Newton's method usually is slowly locally convergent or does not converge at all to a root where $F'(x_*)$ is singular.
(4) $F'(x)$ may not be available in practical applications.

The first difficulty is addressed by modifying (2.2) when necessary so that the method converges to a root from starting points outside the region of local convergence. This property is called *global convergence*. The most common modifications to achieve global convergence are the line search, where each x_+ is chosen by

$$x_+ = x_c - \lambda_c F'(x_c)^{-1} F(x_c) \tag{2.3}$$

for some $\lambda_c > 0$, and the trust region approach, where x_+, is chosen by

$$x_+ = x_c - (F'(x_c)^T F'(x_c) + \alpha_c I)^{-1} F'(x_c)^T F(x_c) \tag{2.4}$$

with $\alpha_c \geq 0$. In both cases, the real valued parameter λ_c or α_c is selected so that x_+ is a satisfactory next iterate, for example so that $\|F(x_+)\|_2 < \|F(x_c)\|_2$. In the line search, Newton's method corresponds to $\lambda_c = 1$, and it is guaranteed that $\|F(x_+)\|_2 < \|F(x_+)\|_2$ for sufficiently small positive λ_c. In the trust region formula (2.4), Newton's method is $\alpha_c = 0$, and $\|F(x_+)\|_2 < \|F(x_c)\|_2$ is guaranteed for sufficiently large positive α_c. Since the new algorithms use the same types of modifications to achieve global convergence, we do not discuss these strategies further. Many of the references listed in the second paragraph of Section 1 contain information on these strategies.

Various modification may be made to these methods when $F'(x_c)$ is singular or ill-conditioned. These include: i) replacing $F'(x_c)^{-1}$ in the line search formula (2.3) by the pseudo-inverse $F'(x_c)^+$, where the pseudo-inverse of $A \in R^{n \times n}$ may be defined by

$$A^+ = \lim_{\gamma \to 0+} (A^T A + \gamma I)^{-1} A^T; \tag{2.5}$$

ii) replacing $F'(x_c)^{-1}$ in the line search by $(F'(x_c)^T F'(x_c) + \gamma I)^{-1} F'(x_c)^T$ with an appropriate small positive value of γ; iii) using the trust region iteration (2.4) with α_c strictly positive. For further information, see Section 6.5 of Dennis and Schnabel [1983]. We mention this difficulty mainly because tensor models for nonlinear equations deal with it nicely.

If $F'(x_*)$ is singular, the convergence of the existing methods to x_* usually is linear at best, even with the above modifications. (See Decker and Kelley

[1980a, 1980b, 1982], Griewank [1980], Griewank and Osborne [1981], Reddien [1978, 1980], Rall [1966] for a discussion of the convergence of Newton's method on singular problems.) Some modifications have been proposed to speed convergence on singular problems (see many of the same references), but they mainly require apriori knowledge that $F'(x_c)$ is singular and do not seem suitable for general classes of problems.

If the Jacobian matrix $F'(x)$ is not available in analytic form, it may be approximated by finite differences, meaning that the jth column of $F'(x_c)$ is approximated by

$$(J_c)_{\text{column } j} = \frac{F(x_c + h e_j) - F(x_c)}{h} \tag{2.6}$$

for some small $h \in R$. (Here e_j denotes the jth unit vector.) If the expense of this approximation, n additional evaluations of $F(x)$ per iteration, is acceptable, this is done and the aforementioned methods are used with (2.6) in place of $F'(x_c)$. If the stepsizes h are chosen correctly, these is little or no deterioration in performance when changing from analytic to finite difference Jacobians.

If the additional cost of finite difference Jacobian approximation is unacceptable, then a class of methods referred to as *secant* (or quasi-Newton) methods is used instead. These methods replace $F'(x_c)$ in formulas (2.2), (2.3), or (2.4) by a less precise approximation J_c calculated as follows. At the first iteration, J_0 is the finite difference approximation to $F'(x_0)$. After the step from x_c to x_+ is determined, the approximation J_c to $F'(x_c)$ is updated into an approximation J_+ to $F'(x_+)$. The most commonly used updating rule is

$$J_+ = J_c + \frac{(y_c - J_c s_c) s_c^T}{s_c^T s_c} \tag{2.7}$$

where

$$s_c = x_+ - x_c, \quad y_c = F(x_+) - F(x_c). \tag{2.8}$$

This update was introduced by Broyden [1965]. It obeys the *secant equation*

$$J_+ s_c = y_c. \tag{2.9}$$

the multi-dimensional generalization of the standard one dimensional secant equation. For any J_+ that obeys (2.9), the new linear model of $F(x)$ around x_+,

$$\bar{M}(x_+ + d) = F(x_+) + J_+ d \tag{2.10}$$

obeys

$$\bar{M}(x_+) = F(x_+), \quad \bar{M}(x_c) = F(x_c). \tag{2.11}$$

Update (2.7) is selected because of all the matrices obeying (2.9), J_+ given by (2.7) is the closest to J_c in the Frobenius norm. (The Frobenius norm of a matrix or tensor is the square root of the sum of the squares of all the matrix's or tensor's components.)

The local method obtained by using (2.7) to calculate the Jacobian approximations with J_0 a finite difference approximation to $F'(x_0)$, and using

$$x_+ = x_c - J_c^{-1} F(x_c) \tag{2.12}$$

to calculate the steps is referred to as *Broyden's method*. It is locally q-superlinearly convergent to a root x_* under the same assumptions on $F(x)$ and x_* stated above for the q-quadratic convergence of Newton's method. Notice that a secant method for nonlinear equations requires the values of $F(x)$ at the iterates, and no other function or derivatives values. In general, secant methods for nonlinear equations or unconstrained minimization usually require more iterations to solve a particular problem than the corresponding analytic or finite difference derivative method, but they usually require fewer function evaluations than the finite difference method. Thus they usually are preferred for problems where function evaluation is expensive and analytic derivatives are unavailable.

The above discussion of secant methods, while cursory, contains the background required for our forthcoming consideration of conic and tensor models. In particular, we emphasize the interpolation property (2.11) that results from formulas (2.9) and (2.10). For further information on these methods, see Dennis and Moré [1977], or the references in paragraph 2 of Section 1.

Now let us turn to unconstrained minimization. Newton's method for unconstrained minimization is based on the quadratic model of $f(x)$ around x_c,

$$m(x_c+d)=f(x_c)+\nabla f(x_c)^T d+\tfrac{1}{2} d^T \nabla^2 f(x_c)d, \tag{2.13}$$

the first three terms of the Taylor series. If $\nabla^2 f(x_c)$ is positive definite, $m(x_c+d)$ has a unique minimizer at

$$x_+ = x_c - \nabla^2 f(x_c)^{-1} \nabla f(x_c). \tag{2.14}$$

Alternatively, the iteration (2.14) can be derived by considering the linear model of $\nabla f(x)$ around x_c,

$$\tilde{M}(x_c+d) = \nabla f(x_c) + \nabla^2 f(x_c)d \tag{2.15}$$

and selecting x_+ as the root of $\tilde{M}(x_c+d)$. Viewed in this way, (2.14) is just the application of Newton's method for nonlinear equations to the problem $\nabla f(x)=0$. Therefore it is locally q-quadratically convergent to any point x_* where $\nabla f(x_*)=0$, $\nabla^2 f(x_*)$ is nonsingular, and $\nabla^2 f(x)$ is Lipschitz continuous in an open neighborhood containing x_*. Such a point may be a minimizer, maximizer, or saddle point of $f(x)$.

The four weaknesses of Newton's method for nonlinear equations that we discussed carry over to Newton's method for unconstrained minimization, and the solutions are similar. Global convergence usually is achieved by modifying (2.14) to

$$x_+ = x_c - \lambda_c H_c^{-1} \nabla f(x_c) \tag{2.16}$$

or

$$x_+ = x_c - (\nabla^2 f(x_c) + \alpha_c I)^{-1} \nabla f(x_c). \tag{2.17}$$

In the first case, $H_c = \nabla^2 f(x_c)$ if $\nabla^2 f(x_c)$ is safely positive definite, otherwise H_c is some positive definite modification of $\nabla^2 f(x_c)$, for example $H_c = \nabla^2 f(x_c) + \gamma I$ with γ large enough to make H_c positive definite. Then it is guaranteed that $f(x_+) < f(x_c)$ for sufficiently small positive λ_c. In the second case, α_c is nonnegative if $\nabla^2 f(x_c)$ is positive definite, and larger than the magni-

II. Conic Methods for Unconstrained Minimization

tude of the most negative eigenvalue of $\nabla^2 f(x_c)$ otherwise. It is guaranteed that $f(x_+) < f(x_c)$ for sufficiently large positive α_c. The conic methods we discuss use the same strategies; no further understanding of these strategies is required for the purposes of this paper.

Modifications (2.16) or (2.17) sucessfully deal with the problem of defining a satisfactory step when $\nabla^2 f(x_c)$ is singular or ill-conditioned. However, standard methods still usually converge linearly at best to a point where $\nabla f(x_*) = 0$ and $\nabla^2 f(x_*)$ is singular.

Finally, Newton's method for unconstrained minimization requires both the gradient vector $\nabla f(x)$ and the Hessian matrix $\nabla^2 f(x)$. If the gradient is not available analytically, it must be approximated by finite differences since accurate gradient values are essential. If the Hessian matrix is not available, $\nabla^2 f(x)$ is replaced by a finite difference approximation if evaluation of $f(x)$ is inexpensive, by a secant approximation otherwise. Secant approximations for unconstrained minimization are derived similarly to Broyden's update for nonlinear equations. After a step from x_c to x_+, the approximation H_c to $\nabla^2 f(x_c)$ is updated into an approximation H_+ to $\nabla^2 f(x_+)$ obeying

$$H_+ s_c = y_c \tag{2.18}$$

where

$$s_c = x_+ - x_c, \quad y_c = \nabla f(x_+) - \nabla f(x_c). \tag{2.19}$$

Thus the quadratic model

$$\bar{m}(x_+ + d) = f(x_+) + \nabla f(x_+)^T d + \tfrac{1}{2} d^T H_+ d \tag{2.20}$$

satisfies the interpolation conditions

$$\bar{m}(x_+) = f(x_+), \quad \nabla \bar{m}(x_+) = \nabla f(x_+), \quad \nabla \bar{m}(x_c) = \nabla f(x_c). \tag{2.21}$$

In addition, H_+ is chosen to be symmetric since $\nabla^2 f(x)$ always is symmetric. Still, many symmetric H_+ satisfying (2.18) exist; the most used choice is the Broyden-Fletcher-Goldfarb-Shanno (BFGS) update

$$H_+ = H_c + \frac{s_c s_c^T}{y_c^T s_c} - \frac{H_c s_c s_c^T H_c}{s_c^T H_c s_c}. \tag{2.22}$$

If H_c is positive definite and

$$s_c^T y_c > 0, \tag{2.23}$$

H_+ is positive definite as well. In practice the initial approximant H_0 is chosen to be positive definite and the step selection strategy enforces (2.23), so all the BFGS approximations to the Hessian are positive definite. This simplifies the modifications required to achieve global convergence. The local method resulting from using (2.22) to define the Hessian approximations and

$$x_+ = x_c - H_c^{-1} \nabla f(x_c) \tag{2.24}$$

to define the steps is locally superlinearly convergent to a point x_* where $f(x)$ and x_* obey the conditions for the q-quadratic convergence of Newton's method, if in addition $\nabla^2 f(x_*)$ is positive definite and H_0 is sufficiently close to $\nabla^2 f(x_0)$.

In summary, we divide the methods considered in this paper into the four categories in Table 2.1 below. Table 2.1 also lists the information interpolated by the local models at x_c used by the standard methods for each category. We denote the iterate before x_c by x_{prev}.

Table 2.1. Categories of methods considered in this paper

---

(1) First derivative methods for nonlinear equations
    local model at x_c interpolates $F(x_c)$, $F'(x_c)$
(2) Secant methods for nonlinear equations
    local model at x_c interpolates $F(x_c)$, $F(x_{\text{prev}})$
(3) Second derivative methods for unconstrained minimization
    local model at x_c interpolates $f(x_c)$, $\nabla f(x_c)$, $\nabla^2 f(x_c)$
(4) Secant methods for unconstrained minimization
    local model at x_c interpolates $f(x_c)$, $\nabla f(x_c)$, $\nabla f(x_{\text{prev}})$

In Section 4 we also refer to another type of method for unconstrained minimization, *conjugate direction methods*. These methods are related to secant methods for unconstrained minimization in that they use function and gradient information only. They differ in that do not use any approximation to the Hessian, and require only $O(n)$ storage. Thus they are mainly intended for problems where n is large and the use of $O(n^2)$ storage locations is undesirable. Many of them have the property that if $f(x)$ is a positive definite quadratic, then the nth iterate of the method will be the minimizer x_*. Space does not permit us to describe conjugate direction methods further here; they are described thoroughly in Fletcher [1980], Hestenes [1980], and Gill, Murray, and Wright [1981].

The conic and tensor methods to be described are based on generalizations of the models discussed in this section. As motivation, Table 2.2 summarizes the properties of a good model.

Table 2.2. Properties of a good model for nonlinear equations or unconstrained minimization

---

(1) The model should interpolate useful information.
(2) The model should be a useful approximation to the problem.
(3) The model should be easy to form.
(4) The model should be easy to solve.

Conic and tensor models aim to improve properties 1 and 2 without seriously harming properties 3 and 4. The hope is that the additional costs incurred in items 3 and 4 will be offset by gains in the efficiency or succes rate of the algorithm. As a point of reference, Table 2.3 summarizes the costs that may be used to measure the efficiency of algorithms for nonlinear equations or unconstrained minimization, and where applicable, the costs incurred by the standard methods.

II. Conic Methods for Unconstrained Minimization

Table 2.3. Costs of solving nonlinear equations or unconstrained minimization problems by standard methods

---
(1) Algorithmic overhead (dominated by cost of solving the linear system to find the Newton or secant step; $O(n^3)$ for derivative methods, $O(n^2)$ for secant methods)
(2) Storage (between $n^2/2$ and $2n^2$ locations)
(3) Evaluations of $F(x)$ or $f(x)$, and possibly derivatives
---

In many practical applications, the evaluations of $F(x)$ or $f(x)$ are very expensive and are the dominant cost. Therefore when assessing the efficiency of new methods, it is desirable that they solve problems using fewer evaluations of the nonlinear function. It also is important, however, that they do not appreciably increase the algorithmic overhead or storage requirements of the standard algorithms.

3. Interpolating additional information using standard models

Conic and tensor models interpolate more function and derivative information than the standard models listed in Table 2.1, by using a more general model. In one of our four problem classes, secant methods for nonlinear equations, it is in fact possible to interpolate additional information using the standard model. For unconstrained minimization, this is only possible in general for quadratic objective functions. We briefly discuss these ideas to motivate further conic and tensor methods.

Secant methods for nonlinear equations use the model

$$M(x_+ + d) = F(x_+) + J_+ d \tag{3.1}$$

to model $F(x)$ around x_+. The secant equation

$$J_+ s_c = y_c \tag{3.2}$$

guarantees $M(x_c) = F(x_c)$. Suppose we also want the model to interpolate the function values $F(x_{-i})$ at some previous iterates x_{-i}, $i = 1, \ldots, p$. This requires

$$F(x_{-i}) = F(x_+) + J_+ (x_{-i} - x_+) \tag{3.3}$$

or

$$J_+ s_i = y_i \tag{3.4}$$

where

$$s_i = x_+ - x_{-i}, \quad y_i = F(x_+) - F(x_{-i}). \tag{3.5}$$

Since J_+ is an $n \times n$ matrix, we may satisfy (3.2), plus (3.4) for up to $n-1$ values of i, as long as $s_c, s_1, \ldots, s_{n-1}$ are linearly independent. This possiblity, and a generalization of Broyden's method that achieves it, was first proposed by Barnes [1965]. However the method was not very successful in practice; problems arose when the directions to the past points $s_c, s_1, \ldots, s_{n-1}$ were linearly dependent or close to being dependent. Gay and Schnabel [1978] revived and modified Barnes' idea. By limiting the set of additional past function values to be interpolated to $p<n$ points for which $s_c, s_1, \ldots, s_p$ are strongly linearly independent, they were able to construct a locally q-superlinearly convergent algorithm that appears quite competitive with a standard Broyden's method algorithm in practice. We do not discuss their method further here. We emphasize, however, the two ideas that will recurr in the forthcoming methods: i) using the model to interpolate function values at previous iterates, and ii) limiting these previous iterates, possibly in a fairly restrictive manner.

The obvious extension of the idea of Barnes and Gay and Schnabel to unconstrained minimization does not work in general. The extension would be to ask the secant model of $f(x)$ around x_+,

$$\nabla m(x_+ + d) = \nabla f(x_+) + H_+ d \tag{3.6}$$

to interpolate $\nabla f(x_{-i})$ at some previous iterates x_{-i} as well as interpolating $\nabla f(x_c)$ at x_c. The difficulty comes from the required symmetry of H_+. Suppose for the model (3.6), $\nabla m(x_c) = \nabla f(x_c)$ and $\nabla m(x_{-1}) = \nabla f(x_{-1})$. This would require

$$H_+ s_c = y_c \quad \text{and} \quad H_+ s_1 = y_1 \tag{3.7}$$

where s_c and s_1 are defined as above and

$$y_c = \nabla f(x_+) - \nabla f(x_c), \quad y_1 = \nabla f(x_+) - \nabla f(x_{-1}). \tag{3.8}$$

Since H_+ is symmetric, (3.7) would imply

$$s_1^T y_c = s_c^T y_1 \tag{3.9}$$

since both sides of (3.9) equal $s_1^T H_+ s_c$. While (3.9) is satisfied if $f(x)$ is a quadratic, it is not satisfied in general for nonquadratic functions.

Davidon [1975] proposed a secant method for unconstrained minimization that interpolates up to n past gradients when $f(x)$ is quadratic, and suggested an extension to general objective functions. Schnabel [1977] studied Davidon's method and proposed several other extensions. None of these have proven superior to the standard secant methods for unconstrained minimization in practice. The desire to interpolate additional previous function or gradient information partially motivates some of the conic methods for unconstrained minimization that we discuss next.

4. Conic methods for unconstrained minimization

The use of conic models in unconstrained minimization algorithms was first proposed by Davidon [1980] and much of the following background material is contained in his paper. A conic function has two equivalent algebraic forms. One is the ratio of a quadratic function divided by the square of a linear function,

$$c(d) = \frac{f + h^T d + \tfrac{1}{2} d^T B d}{(1 + b^T d)^2} \tag{4.1}$$

II. Conic Methods for Unconstrained Minimization

where $f \in R$, $b, d, h \in R^n$, $B \in R^{n \times n}$. Equation (4.1) is equivalent to

$$c(d) = f + \frac{g^T d}{1 + b^T d} + \frac{\tfrac{1}{2} d^T A d}{(1 + b^T d)^2} \tag{4.2}$$

where

$$g = h - fb, \quad A = B - gb^T - bg^T - 2fbb^T. \tag{4.3}$$

The form (4.2) is used for the remainder of this section. The function $c(d)$ is called a conic because its level sets are conic sections, i.e., circles, ellipses, parabolas, or hyperbolas. Figure 1 is an example of a conic function in one variable. Note the discontinuity in $c(d)$; in general, function (4.2) is discontinuous along the $n-1$ dimensional hyperplane $\{d \mid 1 + b^T d = 0\}$, called the *horizon* of $c(d)$. The vector b is called the *gauge vector*.

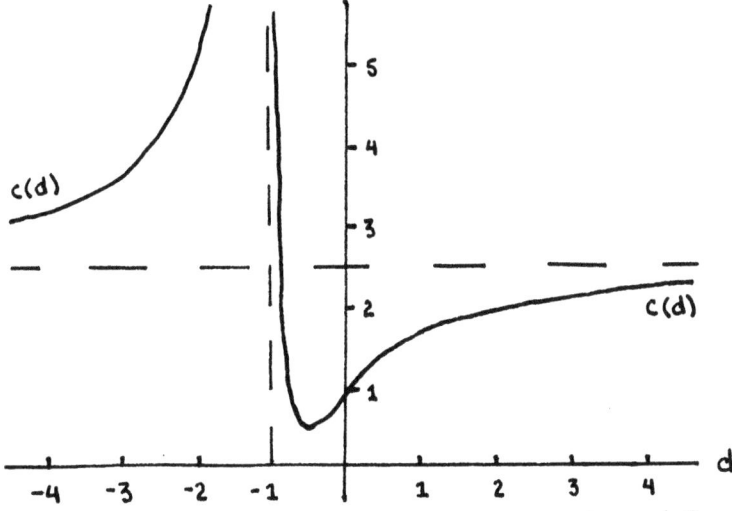

Fig. 1. The conic function of one variable $c(d) = 1 + \dfrac{d}{1+d} + \dfrac{\tfrac{1}{2} d^2}{(1+d)^2}$

The conic function (4.2) is related to the quadratic

$$c(s) = f + g^T s + \tfrac{1}{2} s^T A s \tag{4.4}$$

by

$$s = \frac{d}{1 + b^T d}. \tag{4.5}$$

Equation (4.5) is known as a *collinear scaling* of d; if d and s are related by (4.5) then

$$d = \frac{s}{1 - b^T s}. \tag{4.6}$$

Collinear scalings map straight lines to straight lines, affine subspaces to affine subspaces, and most generally, convex sets to convex sets. Furthermore, they are the most general transformation with these properties. (A collinear scaling may have a slightly more general form than (4.5); see Davidon [1980].) Of course, the mapping (4.5) has the discontinuity mentioned above. The relationship between conics and quadratics may be used to derive some of the conic algorithms described in this section, particularly the conjugate direction algorithms of Davidon [1982] and Gourgeon and Nocedal [1982].

To justify the use of the conic function (4.2) in unconstrained minimization algorithms, first we must ask whether a model of this form is appropriate for use in minimization algorithms, and then, to what use the extra degrees of freedom in the model may be put. The answer to the first question is obtained using the same technique that may be used to find the minimizer of a quadratic. Since for nonsingular A, (4.2) can be written

$$c(d) = \tfrac{1}{2} \left(\frac{d}{1+b^T d} + A^{-1} g \right)^T A \left(\frac{d}{1+b^T d} + A^{-1} g \right) + (f - \tfrac{1}{2} g^T A^{-1} g), \quad (4.7)$$

$c(d)$ has a unique minimer only if A is positive definite. In this case, the minimizer is any d satisfying

$$\frac{d}{1+b^T d} + A^{-1} g = 0 \quad (4.8)$$

if (4.8) has a solution. From (4.6), the solution to (4.8) is

$$d = \frac{-A^{-1} g}{1 + b^T A^{-1} g} \quad (4.9)$$

as long as $1 + b^T A^{-1} g \neq 0$. So roughly speaking, a conic model has a unique minimizer in almost the same cases as a quadratic does, that is, when the matrix in the model is positive definite.

To use a conic function of form (4.2) as a model in a minimization algorithm, presumably it should interpolate some function and derivative values of $f(x)$. The first two derivatives of $c(d)$ are

$$\nabla c(d) = \frac{1}{1+b^T d} \left[I - \frac{b d^T}{1+b^T d} \right] \left[g + \frac{A d}{1+b^T d} \right], \quad (4.10)$$

$$\nabla^2 c(d) = \frac{A - b g^T - g b^T}{(1+b^T d)^2} - \frac{2 A d b^T + 2 b d^T A}{(1+b^T d)^3} - \frac{b b^T}{(1+b^T d)^3} \left[2 g^T d + \frac{3 d^T A d}{1+b^T d} \right]. \quad (4.11)$$

Thus

$$\nabla c(0) = g, \quad \nabla^2 c(0) = A - b g^T - g b^T. \quad (4.12)$$

Therefore to model $f(x)$ around x_c with a conic model, one should use a model of the form

$$\hat{m}(x_c + d) = f(x_c) + \frac{\nabla f(x_c)^T d}{1 + b_c^T d} + \frac{\tfrac{1}{2} d^T A_c d}{(1 + b_c^T d)^2} \quad (4.13)$$

since it satisfies

$$\hat{m}(x_c) = f(x_c), \quad \nabla \hat{m}(x_c) = \nabla f(x_c), \quad (4.14)$$

for any $b_c \in R^n$ and $A_c \in R^{n \times n}$. From (4.12), the second derivative matrix of this model is

$$\nabla^2 \hat{m}(x_c) = A_c - b_c \nabla f(x_c)^T - \nabla f(x_c) b_c^T. \quad (4.15)$$

II. Conic Methods for Unconstrained Minimization

Most of the research on conic methods is based on a model of form (4.13). If A_c is positive definite and $1 + b_c^T A_c^{-1} \nabla f(x_c) \neq 0$, it has a unique minimizer at

$$x_c - \frac{A_c^{-1} \nabla f(x_c)}{1 + b_c^T A_c^{-1} \nabla f(x_c)}; \tag{4.16}$$

the minimizer is on the same side of the horizon as x_c if and only if $1 + b_c^T A_c^{-1} \nabla f(x_c) > 0$. In the remainder of this section, we describe how several authors have used conic models in unconstrained minimization algorithms. This includes the choice of the parameters A_c and b_c in (4.13). It is important to note that the matrix A_c will be different than the matrix used in standard methods for unconstrained minimization. Thus the direction as well as the magnitude of the step to the minimizer of the conic model will be different.

Davidon [1980] and Sorensen [1980] consider the use of a conic model in a secant method for unconstrained minimization. As in our discussion of secant methods in Section 2, assume we have taken a step from x_c to x_+, and are constructing a new conic model of $f(x)$ around x_+,

$$\hat{m}(x_+ + d) = f(x_+) + \frac{\nabla f(x_+)^T d}{1 + b_+^T d} + \frac{\tfrac{1}{2} d^T A_+ d}{(1 + b_+^T d)^2}. \tag{4.17}$$

Davidon and Sorensen show how to use this model to obtain

$$\hat{m}(x_c) = f(x_c), \quad \nabla \hat{m}(x_c) = \nabla f(x_c). \tag{4.18}$$

as well as $\hat{m}(x_+) = f(x_+)$ and $\nabla \hat{m}(x_+) = \nabla f(x_+)$ which are satisfied for all values of b_+ and A_+. Let $s_c = x_+ - x_c$. The first interpolation condition in (4.18) requires

$$f(x_c) = f(x_+) - \frac{\nabla f(x_+)^T s_c}{1 - b_+^T s_c} + \frac{\tfrac{1}{2} s_c^T A_+ s_c}{(1 - b_+^T s_c)^2}, \tag{4.19}$$

while the second is satisfied if

$$\nabla f(x_c) = \frac{1}{1 - b_+^T s_c} \left[I + \frac{b_+ s_c^T}{1 - b_+^T s_c} \right] \left[\nabla f(x_+) - \frac{A s_c}{1 - b_+^T s_c} \right]. \tag{4.20}$$

Let $\beta = 1 - b_+^T s_c$. Taking the inner product of (4.20) with s_c and then using (4.19) to eliminate the $s_c^T A_+ s_c$ term yields

$$\beta[\beta^2 s_c^T \nabla f(x_c) - 2\beta(f(x_+) - f(x_c)) + s_c^T \nabla f(x_+)] = 0. \tag{4.21}$$

If (4.21) has a nonzero real solution (this can be assured by the choice of λ_c in a line search algoritm), then any b_+ that satisfies

$$b_+^T s_c = 1 - \beta \tag{4.22}$$

together with any A_+ that satisfies

$$A_+ s_c = \beta \nabla f(x_+) - \beta^2 \nabla f(x_c) + \beta^2 b_+ s_c^T \nabla f(x_c) \tag{4.23}$$

causes the model (4.17) to satisfy (4.18).

Thus by using (4.21), (4.22), and (4.23), the conic model (4.17) satisfies the three interpolation conditions (2.21) satisfied by standard secant methods for unconstrained minimization, plus $\hat{m}(x_c) = f(x_c)$. This causes the step to the minimizer of the conic model to depend on function as well as gradient values, whereas in a standard secant method, the minimizer of the model is determined solely by gradient values. This fact is cited by several authors as a reason why algorithms based on conic models should be more efficient than the corresponding algorithms based on quadratic models.

Sorensen [1980] proves the local q-superlinear convergence of an algorithm of the type we have just described. His algorithm assures that all the matrices A_+ are positive definite through the choice of the line search parameter, the proper choice from among the two nonzero roots of (4.21), and by using BFGS updates. The parameter b_+ is chosen in the direction $\nabla f(x_c)$. Sorensen also presents some promising test results, but they are from a very limited set of tests, and considerably more testing would be required to establish the utility of this approach.

Davidon [1980] also shows how A_+ and b_+ may be chosen so that the model satisfies the additional interpolation conditions

$$\hat{m}(x_{-i}) = f(x_{-i}) \quad \text{and} \quad \nabla \hat{m}(x_{-i}) = \nabla f(x_{-i}), \quad i = 1, \ldots, n-1 \quad (4.24)$$

when $f(x)$ is a conic function, where as in Section 3 $\{x_{-i}\}$ are past iterates. For nonconic functions this is not usually possible, however, and to our knowledge, this idea has not yet been used to make a general purpose unconstrained minimization algorithm.

Schnabel [1982b] and Stordahl [1980] study the use of the model (4.13) when the Hessian matrix is available analytically or by finite differences. The model they use is

$$\hat{m}(x_c + d) =$$
$$f(x_c) + \frac{\nabla f(x_c)^T d}{1 + b_c^T d} + \frac{\frac{1}{2} d^T (\nabla^2 f(x_c) + b_c \nabla f(x_c)^T + \nabla f(x_c) b_c^T) d}{(1 + b_c^T d)^2}. \quad (4.25)$$

From (4.14) and (4.15), (4.25) satisfies

$$\hat{m}(x_c) = f(x_c), \quad \nabla \hat{m}(x_c) = \nabla f(x_c), \quad \nabla^2 \hat{m}(x_c) = \nabla^2 f(x_c) \quad (4.26)$$

for any $b_c \in R^n$. The vector b_c then may be chosen to allow the model to interpolate additional information. Schnabel and Stordahl impose the requirement

$$\hat{m}(x_{-i}) = f(x_{-i}), \quad i = 1, \ldots, p \quad (4.27)$$

at $p \leq n$ past iterates x_{-i}. Substitution into (4.25) shows that this model satisfies $\hat{m}(x_{-i}) = f(x_{-i})$ if

$$b_c^T s_i = \sigma_i = \quad (4.28)$$
$$1 + \frac{\nabla f(x_c)^T s_i - \sqrt{(\nabla f(x_c)^T s_i)^2 + (f(x_{-i}) - f(x_c))(\frac{1}{2} s_i^T \nabla f(x_c) s_i + \nabla f(x_c)^T s_i)}}{f(x_{-i}) - f(x_c)}$$

where

$$s_i = x_c - x_{-i}. \quad (4.29)$$

II. Conic Methods for Unconstrained Minimization

The term inside the square root in (4.28) may be negative but rarely is in practice. Since (4.28) only determines the projection of b_c in the direction s_i, the conic model (4.25) may interpolate up to n past function values $f(x_{-i})$ as long as the directions to the past points, s_i, are linearly independent. If $f(x)$ is quadratic, the right hand side of (4.28) is zero so the choice $b_c = 0$ is allowed.

Schnabel and Stordahl tested an algorithm based on the above model. They found it advantageous to limit the number of past function values interpolated at any iteration to $p \leq \sqrt{n}$, and to require the directions s_i to the past points to be strongly linearly independent of each other; roughly speaking, each direction s_i that is used must make an angle of at least 45 degrees with the linear subspace spanned by the other directions $\{s_k\}$ that are used. They chose b_c to be the minimum l_2 norm solution to the underdetermined system of equations

$$b_c^T s_i = \sigma_i, \quad i = 1, \ldots, p. \tag{4.30}$$

Schnabel and Stordahl compared their algorithm to an algorithm that used the standard second derivative quadratic model but was identical in all other respects, on the unconstrained minimization test problems in Moré, Garbow, and Hillstrom [1981]. Each algorithm used the strategy in Dennis and Schnabel [1983] to augment $\nabla^2 f(x_c)$ or A_c to be positive definite when necessary, and then chose the next iterate using a line search. Out of 32 test runs, the conic algorithm was more efficient (in iterations and function evaluations) on 19, less efficient on 11, and tied on 4; on the average, the conic algorithm required 18% fewer iterations and 21% fewer function evaluations than the quadratic algorithm. It is not clear whether these results justify the added complexity of the conic method.

Another possibility in using model (4.25) is to choose b_c so that $\nabla \hat{m}(x_{-1}) = \nabla f(x_{-1})$ where x_{-1} is the most recent past iterate. This is more in keeping with the philosophy of the algorithms of Davidon and Gourgeon and Nocedal discussed next, as it implies that $\hat{m}(x) = f(x)$ if $f(x)$ is a conic function. We currently are investigating this approach.

Davidon [1982] and Gourgeon and Nocedal [1982] have proposed a conjugate direction method based on conic models. It is a generalization of the standard conjugate gradient method for minimizing quadratics (Hestenes and Stiefel [1952]) and finds the minimizer of a conic function in at most n iterations. An extension of these methods to nonconic functions has not yet been developed, however, so no comparison with existing conjugate direction methods for unconstrained minimization is possible.

Since we do not include a thorough treatment of conjugate direction methods in this paper, we just give a brief description of Davidon's and Gourgeon and Nocedal's algorithms. The key feature is that they show that if $f(x)$ is a conic function, then given the values of $f(x)$ and $\nabla f(x)$ at any three collinear points, the value of the gauge vector b can be determined. The value of the gauge vector changes as the conic is expanded around different points, but only by scalar multiples that Davidon and Gourgeon and Nocedal also show how to calculate. Using this information, they are able to generalize the conjugate gradient algorithm to exactly minimize the conic function along a line d_k at the kth iteration, while also choosing d_k so that the next iterate x_{k+1} is the minimizer of the conic in the affine subspace spanned by the $k-1$ previous step directions $d_1, \ldots, d_{k-1}$ as well. Thus the algorithm minimizes a conic function in n or fewer iterations, without storing or using an $n \times n$ matrix. Gourgeon and Nocedal point out several possible numerical instabilities in implementing this algorithm, and seem to remedy them satisfactorily.

Davidon [1982] also proposes a related algorithm that again minimizes a conic function in n or fewer iterations, and also accumulates an approximation to the matrix A in (4.2) as it proceeds. If $f(x)$ is conic, after n iterations this approximation is $A = \nabla^2 f(x_*)$. Gourgeon and Nocedal state that the sequence of points generated by their algorithm and Davidon's two algorithms is identical for conic functions, if exact arithmetic is used. Davidon's secant algorithm probably could be extended into a new secant conic algorithm for unconstrained minimization. This extension would require the evaluation of $f(x)$ and $\nabla f(x)$ at three collinear points on some or all iterations, as would an extension of the conjugate direction algorithms of Davidon or Gourgeon and Nocedal to nonconics.

Finally, Bjorstad and Nocedal [1979] show that the secant conic algorithm originally proposed by Davidon converges quadratically under reasonable assumptions when applied to one dimensional problems. This rate is faster than the order $(1+\sqrt{5}/2) \cong 1.61$ convergence rate of the secant method on one dimensional problems. Similarly, the second derivative conic method of Schnabel and Stordahl converges with order $1+\sqrt{2} \cong 2.41$ on one dimensional problems, compared to the quadratic convergence of Newton's method. These results probably have practical importance only if they extend to multiple dimensions, which is very unlikely.

5. Tensor methods for nonlinear equations

Tensor models for nonlinear equations augment the standard linear model of $F(x)$ around x_c by a second order term. The most obvious tensor model is the first three terms of the Taylor series,

$$\bar{M}(x_c+d) = F(x_c) + F'(x_c)d + \tfrac{1}{2} F''(x_c)dd, \qquad (5.1)$$

where $F'' \in R^{n \times n \times n}$. However the use of model (5.1) in a nonlinear equations algorithm would violate many of the principles expressed at the end of Section 2. In particular, the model would require at least $O(n^3)$ additional operations to form and roughly $n^3/2$ additional locations to store, and finding a root of the model would require solving a system of n quadratic equations at n unknowns at each iteration. Any of these costs clearly is unacceptable.

Instead, Schnabel and Frank [1982] have proposed the use of a model of the form

$$\hat{M}(x_c+d) = F(x_c) + F'(x_c)d + \tfrac{1}{2} T_c dd \qquad (5.2)$$

where $T_c \in R^{n \times n \times n}$ has a particulary simple form. In particular, the additional costs of forming, storing, and finding a root of the tensor model all are small in

II. Conic Methods for Unconstrained Minimization

comparison to the corresponding costs already required by Newton's method. In this section, we summarize how Schnabel and Frank determine the term T_c in (5.2), how the resultant model is solved efficiently, and how a nonlinear equations method utilizes this method. Finally, we summarize some of their test results.

The main motivation for the work of Schnabel and Frank was to construct a general purpose method that would be more efficient than standard methods on problems where $F'(x_*)$ is singular or ill-conditioned, and at least as efficient as standard methods on all other problems. An early version of their work is reported in Frank [1982]. For practical purposes, it probably would be more desirable to have a secant tensor method that does not require values of $F'(x)$; we currently are working on such an extension.

To determine the tensor term T_c in (5.2), we require the model $\hat{M}(x_c+d)$ to interpolate the function values $F(x_{-i})$ at $p \leq \sqrt{n}$ previous iterates $x_{-1}, \ldots, x_{-p}$. Substituted into (5.2), this requirements is

$$F(x_{-i}) = F(x_c) - F'(x_c)s_i + \tfrac{1}{2} T_c s_i s_i, \quad i=1,\ldots,p \leq \sqrt{n} \qquad (5.3)$$

where
$$s_i = x_c - x_{-i}. \qquad (5.4)$$

The upper bound of $\sqrt{n}$ past points was suggested by the computational experience of Schnabel and Stordahl mentioned in Section 4 for conic methods that use past function values; it also is required to keep the costs of forming and solving the tensor model small. The past points are selected by the same method as is used in Schnabel and Stordahl's conic code. At each iteration, the most recent past point is selected. Then, for $i=2,\ldots,\sqrt{n}$, the ith past point is selected if it makes an angle of at least 45 degrees with the affine subspace spanned by the already selected subset of the 1st through $i-1^{st}$ past points.

The $p \leq \sqrt{n}$ interpolation conditions (5.3) result in np linear equations in the n^3 elements of T_c, meaning that T_c is underdetermined. Following successful precedent in determining secant updates for nonlinear equations (see e.g. Dennis and Schnabel [1979]), Schnabel and Frank choose the T_c that solves

$$\underset{T_c \in R^{n \times n \times n}}{\text{minimize}} \ \|T_c\|_F \qquad (5.5)$$

subject to $T_c s_i s_i = 2(F(x_{-i}) - F(x_c) + F'(x_c)s_i), \quad i=1,\ldots,p$

where $\|\cdot\|_F$ denotes the Frobenius norm. They show that the solution to (5.5) has the form

$$T_c = \sum_{i=1}^{p} a_i s_i s_i \qquad (5.6)$$

where $a_i \in R^n$, $i=1,\ldots,p$ and $a_i s_i s_i$ denotes the rank one tensor whose i,j,k element is $a_i[i] \cdot s_i[j] \cdot s_i[k]$. Thus T_c is a rank p tensor. The p vectors $a_i \in R^n$ are calculated by solving one symmetric and positive definite $p \times p$ system of linear equations with n different right hand sides, requiring a total of $np^2 \leq n^2$ each multiplications and additions.

Given the form (5.6) of T_c, the tensor model (5.2) becomes

$$\hat{M}(x_c+d) = F(x_c) + F'(x_c)d + \sum_{i=1}^{p} a_i (s_i^T d)^2. \qquad (5.7)$$

Therefore a maximum of $2n^{1.5}$ additional storage locations are required for the tensor term, to store the p a_i and s_i vectors. (The x_{-i} and s_i vectors can share storage.) The dominant arithmetic cost in forming the tensor term is the $n^2p \leqslant n^{2.5}$ multiplications and additions required to form the p right hand sides of (5.5). Neither of these costs is significant in comparison to the n^2 locations and $O(n^3)$ arithmetic operations required to store and solve the standard linear model.

To use $\hat{M}(x_c+d)$ given by (5.7) in an algorithm for nonlinear equations, presumably we need to be able to find a $d_* \in R^n$ for which $\hat{M}(x_c+d_*)=0$. However the model may not always have a real root, as is obvious by considering the one dimensional case, and in this case it seems reasonable to find a d_* that minimizes $\|\hat{M}(x_c+d)\|_2$. Thus in general we need an efficient procedure for solving

$$\text{minimize}_{d \in R^n} \|\hat{M}(x_c+d)\|_2 \tag{5.8}$$

for $\hat{M}(x_c+d)$ given by (5.7). Schnabel and Frank show how to reduce (5.8) to a much smaller problem of the form

$$\text{minimize}_{z \in R^p} \|Q(z)\|_2, \quad Q: R^p \to R^q \tag{5.9}$$

where the q equations in p unknowns $Q(z)$ are quadratic and $p \leqslant q \leqslant n$, with usually $q=p$. The reduction is accomplished using orthogonal transformations of the variable and function spaces. It requires the QR decomposition of $F'(x_c)$ which also usually is used in a Newton's method algorithm and takes $2n^3/3$ each multiplications and additions; the next leading term in the reduction is $2n^2p \leqslant 2n^{2.5}$ additional multiplications and additions which is insignificant in comparison.

Now to find a root or minimizer of the tensor model (5.7) we still must solve the nonlinear problem (5.9); however, this is inexpensive due to the reduced number of variables in (5.9). To solve (5.9) by a nonlinear least squares algorithm costs $O(p^2q)$ arithmetic operations per iteration, and in practice a maximum of $2p$ iterations suffices. Thus the total cost of solving (5.9) is $O(p^3q)$ operations, which is $\leqslant O(n^2)$ in the normal case when $p=q$, and $\leqslant O(n^{2.5})$ in all cases. Given the solution to (5.9), the solution to (5.8) is obtained by backsolving and requires $O(n^2)$ arithmetic operations. Therefore the total cost of finding a root or minimizer of the tensor model is not significantly more than calculating the root of the standard linear model using a QR factorization. It is twice as expensive as calculating the root of the standard linear model using Gaussian elimination; however many production codes do use the QR factorization because it facilitates the global portion of the algorithm and the modifications required when $F'(x_c)$ is ill-conditioned (see Dennis and Schnabel [1983]).

An important point is that the model (5.7) may have isolated solutions even when $F'(x_c)$ is singular and $F(x_c)$ is not contained in the subspace spanned by the columns of $F'(x_c)$; if so the method of Schnabel and Frank should find a solution. Similarly, (5.8) often is not an ill-conditioned problem when the solution of the linear model would be. It is hoped that these properties are beneficial on problems where $F'(x_*)$ is singular. Another interesting issue is that (5.7) usually has multiple roots or minimizers; we try to find the d_* closest to the Newton step, by appropriately choosing the starting point $z_0 \in R^n$ in the algorithm that solves (5.9).

Schnabel and Frank have tested an algorithm based on a preliminary version of the above techniques. At each iteration, after determining the solution d_* to (5.8), it takes $x_* = x_c + d_*$ if this is an acceptable next iterate. If not, it does a line search in the Newton direction if d_* is not a descent direction, or occasionally if it prefers the Newton direction for other reasons given in Frank [1982], otherwise it does a line search in the direction d_*. Schnabel and Frank compare their algorithm to an algorithm that uses the standard linear model exclusively but otherwise is identical to the tensor algorithm, using the test problems in Moré, Garbow, and Hillstrom [1981]. For all but one of these problems (Powell's singular function) $F'(x_*)$ is nonsingular, so they also construct singular versions of the problems in Moré, Garbow, and Hillstrom, as described in Schnabel and Frank [1982]. The dimensions of the problems range from 2 to 50, with most of the problems having dimension between 10 and 50.

Detailed results of these tests are given in Frank [1982] and in Schnabel and Frank [1982]. In summary, on 44 problems where $F'(x_*)$ is nonsingular, the tensor method requires fewer iterations and function and Jacobian evaluations than the standard method in 30 cases, the same number in 13 cases, and more in 1 case. The improvements rarely are by more than 20%, however most of these problems are quite easy when using analytic Jacobians, requiring 10 or fewer iterations. On a set of 37 problems where $F'(x_*)$ has rank $n-1$, the tensor method requires fewer iterations and function and Jacobian evaluations in 26 cases, the same number in 8 cases, and more in 3 cases. Here the improvement by the tensor method is far more dramatic; the standard method fails to solve 11 of the 37 problems in 150 iterations, while the tensor method solves all of these in at most 27 iterations. On a second differently constructed set of 25 problems where $F'(x_*)$ again has rank $n-1$, the tensor model requires fewer iterations and function and Jacobian evaluations in all cases. The standard method solves 24 of these; on these problems, the tensor algorithm requires an average of 53% of the iterations and 58% of the function evaluations used by the standard algorithm. Finally, on 37 problems where $F'(x_*)$ has rank $n-2$, the tensor model again does better in 26 cases, the same in 8, and worse in 3, again with substantial improvements in many cases. Clearly these results show promise for the tensor methods.

6. Future applications of conic and tensor models

Methods that use conic and tensor models are still in their infancy. Depending upon their success, conic and tensor models might eventually be used in many areas of optimization besides the ones described in Sections 4 and 5. In this section we comment briefly on some of these possibilities.

Conic models or collinear scalings do not seem suited to the nonlinear equations problem. The collinear transformation of the standard linear model for nonlinear equations,

$$\tilde{M}(x_c + d) = F(x_c) + \frac{F'(x_c)d}{1 + b_c^T d} \tag{6.1}$$

suffers from the fact that its root is in the Newton direction $-F'(x_c)^{-1} F(x_c)$ for all values of b_c; thus, only a steplength parameter is introduced. On the other hand, conic models certainly could be applied to nonlinear least squares or constrained optimization problems. They also may be useful for solving special classes of optimization problems, for example penalty functions where the horizon of the conic function might help reflect the shape of the penalty function.

Hopefully, tensor models can be applied to all four problem classes listed in Section 2, derivative and secant methods for both nonlinear equations and unconstrained minimization. In fact, our own main practical interest would be in a secant tensor method that quickly solves unconstrained minimization problems with $\nabla^2 f(x_*)$ singular or ill-conditioned, because we see many such problems in practice. An example is overparameterized data fitting problems, where in our experience the objective function usually is sufficiently expensive to evaluate that secant methods are preferred. The extension of the method of Section 5 to a secant method for nonlinear equations, and the extension to unconstrained minimization, both appear to present challenging problems. The tensor method of Section 5 generalizes virtually without change into a Gauss-Newton or Levenberg-Marquardt type method for solving the nonlinear least squares problem, and since the analytic or finite difference Jacobian always is used in this setting, such a method would be of practical interest. Tensor methods also may be useful in finding complex solutions to nonlinear equations or optimization problems.

Finally, there certainly are other nonstandard models besides conics and tensors that could be considered for nonlinear equations or optimization problems. The main contribution of conic and tensor models may be that they cause researchers to consider various nonstandard models, and that some of these prove to be useful in practice.

Acknowledgements

The impetus for much of the work on nonstandard models for unconstrained optimization has come from the work of William Davidon, especially his papers of 1975 and 1980. I am pleased to acknowledge my debt to Davidon and his ideas; Davidon's 1975 paper on projected updates for unconstrained minimization led to the paper of Gay and Schnabel, Davidon's 1980 paper on conics led to my work on conics and indirectly supplied the stimulus for the work on tensor models.

I also thank Richard Byrd, Paul Frank, Andreas Griewank, and Kari Stordahl for many helpful discussions on these subjects.

7. References

J. Barnes [1965], "An algorithm for solving nonlinear equations based on the secant method", *Computer Journal* 8, pp. 66-72.

P. Bjorstad and J. Nocedal [1979], "Analysis of a new algorithm for one dimensional minimization", *Computing* 22, pp. 93-100.

K. W. Brodlie [1977], "Unconstrained minimization", in *The State of the Art in Numerical Analysis*, D. Jacobs, ed., Academic Press, London, pp. 229-268.

C. G. Broyden [1965], "A class of methods for solving nonlinear simultaneous equations", *Mathematics of Computation* 19, pp. 577-593.

C. Charalambous [1973], "Unconstrained optimization based on homogeneous models", *Mathematical Programming* 5, pp. 189-198.

W. C. Davidon [1975], "Optimally conditioned optimization algorithms without line searches", *Mathematical Programming* 9, pp. 1-30.

W. C. Davidon [1980], "Conic approximations and collinear scalings for optimizers", *SIAM Journal on Numerical Analysis* 17, pp. 268-281.

W. C. Davidon [1982], "Conjugate directions for conic functions", in *Nonlinear Optimization 1981*, M. J. D. Powell, ed., Academic Press, London, pp. 23-28.

D. W. Decker and C. T. Kelley [1980a], "Newton's method at singular points I", *SIAM Journal on Numerical Analysis* 17, pp. 66-70.

D. W. Decker and C. T. Kelley [1980b], "Newton's method at singular points II", *SIAM Journal on Numerical Analysis* 17, pp. 465-471.

D. W. Decker and C. T. Kelley [1982], "Convergence acceleration for Newton's method at singular points", *SIAM Journal on Numerical Analysis* 19, pp. 219-229.

J. E. Dennis Jr. [1977], "Nonlinear least squares and equations", in *The State of the Art in Numerical Analysis*, D. Jacobs, ed., Academic Press, London, pp. 269-312.

J. E. Dennis Jr. and J. J. More' [1977], "Quasi-Newton methods, motivation and theory", *SIAM Review* 19, pp. 46-89.

J. E. Dennis Jr. and R. B. Schnabel [1979], "Least change secant updates for quasi-Newton methods", *SIAM Review* 21, pp. 443-459.

J. E. Dennis Jr. and R. B. Schnabel [1983], *Numerical Methods for Nonlinear Equations and Unconstrained Optimization*, Prentice-Hall, Englewood Cliffs, New Jersey.

R. Fletcher [1980], *Practical Method of Optimization, Vol. 1, Unconstrained Optimization*, John Wiley and Sons, New York.

P. Frank [1982], "A second-order local model for solution of systems of nonlinear equations", M. S. Thesis, Department of Computer Science, University of Colorado at Boulder.

D. M. Gay and R. B. Schnabel [1978], "Solving systems of nonlinear equations by Broyden's method with projected updates", in *Nonlinear Programming 3*, O. L. Mangasarian, R. R. Meyer, and S. M. Robinson, eds., Academic Press, New York, pp. 245-281.

P. E. Gill, W. Murray, and M. H. Wright [1981], *Practical Optimization*, Academic Press, London.

H. Gourgeon and J. Nocedal [1982], "A conic algorithm for optimization", preprint.

A. O. Griewank [1980], "Analysis and modification of Newton's method at singularities", Ph. D. thesis, Australian National University.

A. O. Griewank and M. R. Osborne [1981], "Newton's method for singular problems when the dimension of the null space is >1", *SIAM Journal on Numerical Analysis* 18, pp. 145-150.

M. R. Hestenes [1980], *Conjugate-Direction Methods in Optimization*, Springer-Verlag, New York.

M. R. Hestenes and E. Stiefel [1952], "Methods of conjugate gradients for solving linear systems", *Journal of Research of the National Bureau of Standards* 49, pp. 409-436.

D. Jacobsman and W. Oxman [1972], "An algorithm that minimizes homogeneous functions of n variables in $n+2$ iterations and rapidly minimizes general functions", *Journal of Mathematical Analysis and Applications* 38, pp. 533-552.

J. Kowalik and K. Ramakrishnan [1976], "A numerically stable optimization method based on a homogeneous function", *Mathematical Programming* 11, pp. 50-66.

J. J. Moré, B. S. Garbow, and K. E. Hillstrom [1981], "Testing unconstrained optimization software", *Transactions on Mathematical Software* 7, pp. 17-41.

J. J. Moré and D. C. Sorensen [1982], "Newton's method", Argonne National Laboratory report ANL-82-8.

J. M. Ortega and W. C. Rheinboldt [1970], *Iterative Solution of Nonlinear Equations in Several Variables,* Academic Press, New York.

L. B. Rall [1966], "Convergence of the Newton process to multiple solutions", *Numerische Mathematik* 9, pp. 23-37.

G. W. Reddien [1978], "On Newton's method for singular problems", *SIAM Journal on Numerical Analysis* 15, pp. 993-996.

G. W. Reddien [1980], "Newton's method and high order singularities", *Comput. Math. Appl.* 5, 79-86.

R. B. Schnabel [1977], "Analyzing and improving quasi-Newton methods for unconstrained optimization", Ph. D. thesis, Department of Computer Science, Cornell University.

R. B. Schnabel [1982a], "Unconstrained Optimization in 1981", in *Nonlinear Optimization 1981,* M. J. D. Powell, ed., Academic Press, London.

R. B. Schnabel [1982b], "Unconstrained minimization using conic models and second derivatives", to appear.

R. B. Schnabel and P. Frank [1982], "Tensor methods for solving systems of nonlinear euations", to appear.

D. C. Sorensen [1980], "The q-superlinear convergence of a collinear scaling algorithm for unconstrained optimization", *SIAM Journal on Numerical Analysis* 17, pp. 84-114.

K. A. Stordahl [1980], "Unconstrained minimization using conic models and exact derivatives", M. S. thesis, Department of Computer Science, University of Colorado at Boulder.

Min-Max Results in Combinatorial Optimization

A. Schrijver
Universiteit van Amsterdam, Instituut voor Actuariaat en Econometrie,
Jodenbreestraat 23, 1011 NH Amsterdam, The Netherlands

1. Introduction

Often the optimum of a combinatorial optimization problem is characterized by a *min-max relation,* asserting that the maximum value in one combinatorial optimization problem is equal to the minimum value in some other optimization problem. One of the best-known examples is the *max-flow min-cut theorem* of Ford and Fulkerson [1956] and Elias, Feinstein and Shannon [1956]: If "nodes" $1, \ldots, n$ and "capacities" $c_{ij} \geq 0$ $(i, j = 1, \ldots, n)$ are given, then the maximum value of a flow from "source node" 1 to "sink node" n, subject to c, is equal to the minimum capacity of a cut separating 1 from n, i.e., to

(1) $$\min \sum_{i \in T} \sum_{j \notin T} c_{ij},$$

where the minimum ranges over all subsets T of $\{1, \ldots, n\}$ with $1 \in T$ and $n \notin T$. Moreover, if the capacities are integer there exists an integer optimum flow. Here a *flow from 1 to n subject to c* is a vector $(x_{ij})_{i,j=1,\ldots,n}$ satisfying

(2) (i) $\sum_{i=1}^{n} x_{ij} = \sum_{i=1}^{n} x_{ji}$ $(j = 2, \ldots, n-1)$,

(ii) $0 \leq x_{ij} \leq c_{ij}$ $(i, j = 1, \ldots, n)$.

The *value* of the flow is the "net amount" of flow leaving node 1, which is

(3) $$\sum_{i=1}^{n} x_{1i} - \sum_{i=1}^{n} x_{i1}.$$

(This is clearly equal to the net amount of flow entering node n.)

Generally, a combinatorial min-max relation is closely related to the algorithmic solvability of the corresponding optimization problem. Often a min-max relation appears as a by-product of an algorithm, and it can serve as an optimality criterion, and as a "good characterization", for the optimization problem. In turn, with the ellipsoid method a min-max relation sometimes gives that the combinatorial optimization problem can be solved within time bounded by a polynomial in the problem size. Moreover, a min-max relation can be used in a sensitivity analysis, giving insight in the question in how much the optimum changes if we vary the initial constraints.

Besides, combinatorial min-max relations are of theoretical interest. Usually, they yield elegant combinatorial theorems, and they allow a geometrical representation of the problems in terms of polyhedra. Such theoretical considerations have turned out to be very useful in understanding the algorithmic practice.

Historically, the first combinatorial min-max relations were found by König [1916, 1932], Menger [1927] and Tutte [1947]. They deal with maximum matchings and path packings, and at first they were formulated and proved purely combinatorially. The algorithmic and polyhedral aspects of combinatorial min-max relations were revealed in the 1950s by the work of Ford, Fulkerson and Hoffman on bipartite matchings and network flows, and were founded and developed further in the 1960s by the pioneering results of Edmonds on matchings, matroids and branchings. Further significant min-max results were discovered in the 1970s by Lovász (perfect graphs, matroid matching), Seymour (binary hypergraphs), Lucchesi and Younger (directed cuts), Mader (S-paths).

We now first go further into the polyhedral and algorithmic aspects of combinatorial min-max relations. For a more comprehensive review of "polyhedral combinatorics", see W. R. Pulleyblank's survey.

Polyhedral aspects of min-max relations. Often a combinatorial min-max relation amounts to the fact that a certain linear program has an integer optimal (primal and/or dual) solution, without requiring integrality explicitly in advance. The duality theorem of linear programming then gives the min-max relation.

E.g., the max-flow min-cut theorem is equivalent to the fact that the linear program of maximizing (3) subject to (2) has an integer optimal dual solution. Moreover, if c is integer, the linear program has an integer optimal primal solution.

In this example the integrality, and hence the combinatorial min-max relation, follows from the total unimodularity of (2). Total unimodularity is one important tool in integer linear programming and combinatorial optimization, due to Hoffman and Kruskal. For most of the combinatorial min-max relations, however, the reduction to LP-duality requires more than just total unimodularity. Consider, e.g., the following problem.

Let nodes $1, \ldots, n$, together with (possibly asymmetric) "lengths" $l_{ij} \geq 0$ ($i, j = 1, \ldots, n$) be given. Suppose we wish to select certain directed trajects between the nodes such that each node $j \neq 1$ is reachable from node 1 by a directed path, and such that the total length of the selected trajects is as small as possible. That is, we wish to find a shortest 1-*arborescence*. It is easy to see that this amounts to the following integer linear programming problem:

(4)
$$\min \sum_{i,j=1}^{n} l_{ij} x_{ij}$$

subject to

$$\sum_{j \in T} \sum_{i \notin T} x_{ij} \geq 1 \quad (T \subseteq \{2, \ldots, n\}, T \neq \emptyset),$$

$$x_{ij} \geq 0 \quad (i, j = 1, \ldots, n),$$

$$x_{ij} \text{ integer} \quad (i, j = 1, \ldots, n).$$

Now it is a theorem of Edmonds [1967 a] that this minimum is not decreased if we delete the integrality condition. So the optimum value in (4) is equal to the common value in the linear programming duality equation:

(5) $\quad \min \sum_{i,j} l_{ij} x_{ij} \qquad\qquad = \max \sum_T y_T$

$\qquad$ subject to $\qquad\qquad\qquad$ subject to

$\qquad\qquad \sum_{j \in T} \sum_{i \notin T} x_{ij} \geq 1 \quad (\forall T), \qquad \sum_{T \ni i,\, T \ni j} y_T \leq l_{ij} \quad (\forall i,j),$

$\qquad\qquad x_{ij} \geq 0 \qquad\qquad (\forall i,j), \qquad y_T \geq 0 \qquad\qquad (\forall T),$

where i and j range over $1, \ldots, n$, and T ranges over all nonempty subsets of $\{2, \ldots, n\}$.

So we now have two equivalent formulations of Edmonds' result: as the assertion that the minimum in (5) has an integer optimum solution, or as a min-max relation equating the minimum in (4) to the maximum in (5). There is a third equivalent interpretation: the vertices of the polyhedron defined by the constraints for the minimum in (5) are integer (and hence the incidence vectors of 1-arborescences). This of course follows from the fact that the minimum in (5) has an integer optimum solution x_{ij} *for each* length function $l_{ij} \geq 0$. So Edmonds' min-max relation can be stated alternatively as a polyhedral result.

We emphasize here that this combinatorial min-max relation is not just reduced to LP-duality: the point is to prove the existence of integer optimum solutions. (Fulkerson [1974] extended Edmonds' result by showing that also the maximum in (5) has an integer optimum solution, if the lengths are integer.)

Algorithmic aspects of min-max relations. First, a min-max relation often follows as a by-product of an algorithm to solve the optimization problem. Typically the process is as follows. First one easily shows the inequality max ≤ min. Next an algorithm for, say, the maximization problem has as main step a procedure which either improves the current solution, or shows that this solution is optimum by constructing a feasible solution for the minimization problem with the same objective value as that of the current solution.

Thus the relation max ≤ min proves the optimality of the algorithm, while, in turn, the algorithm yields the inequality max ≥ min.

E.g., Ford and Fulkerson's maximum flow algorithm hangs on the subroutine which either finds an "augmenting path" improving the current flow, or finds a cut separating source and sink with capacity equal to the value of the current flow. So the max-flow min-cut theorem follows. Note that Ford and Fulkerson's algorithm is at the same time a minimum cut algorithm.

Similarly, Edmonds' min-max formula for minimum length 1-arborescences given above follows as a by-product of an algorithm finding a shortest 1-arborescence.

The above also shows the role of a min-max relation as an optimality criterion. E.g., in the case of max-flow min-cut, a given flow can be shown to be optimum by specifying a cut separating source and sink, with capacity equal to the value of the given flow.

Frequently (in fact, in all cases treated in this paper) a min-max relation yields a *good characterization*. This means that the problem has an answer, the correctness of which can be shown by a proof of length bounded by a polynomial in the size of the problem instance. (This concept of good characterization is due to Edmonds [1965a].)

Indeed, suppose we have a min-max relation for a combinatorial optimization problem. To give a good characterization it suffices to display a feasible solution for the *maximization* problem, together with a feasible solution for the *minimization* problem, with the same objective value. If such feasible solutions exist of polynomially bounded size (which usually is the case), we have a polynomial-length proof of the optimality of the solution.

In "NP-language", a problem has a good characterization if and only if it belongs to the complexity class NP∩co-NP. So if one believes that NP≠NP∩co-NP, one may not expect a satisfactory min-max relation for any NP-complete problem – satisfactory in the sense of supplying a good characterization.

Always, a polynomial algorithm yields a good characterization (i.e., P⊆NP∩co-NP): one can prove any answer to be correct just by writing down the steps of the algorithm (which takes polynomial space), together with a proof of the correctness of the algorithm (which takes only fixed space, independent of the special instance of the problem).

The reverse implication, that a good characterization yields a polynomial algorithm, is not known. However, the combinatorial optimization problems investigated so far have created the impression that if a problem has a good characterization, it also is polynomially solvable. Recently, this impression has got the added support of the ellipsoid method, which sometimes deduces polynomial solvability from a min-max relation, or from the associated polyhedral representation (cf. Grötschel, Lovász and Schrijver [1981]).

To explain this, suppose a certain combinatorial optimization problem is accompanied by a min-max relation which amounts to the fact that the minimum in the LP-duality equation

(6) $\min\{wx|Ax \geq b\} = \max\{yb|y \geq 0, yA = w\}$

has an integer optimum solution x. Now, roughly speaking, if for each vector x, the system $Ax \geq b$ can be checked in polynomial time, then with the ellipsoid method the minimum in (6) can be determined in polynomial time as well. So the polynomial solvability of the combinatorial optimization problem follows.

Here "checking" means testing whether $Ax \geq b$ holds, and, if not, finding a violated inequality. If the size of the system $Ax \geq b$ is bounded by a polynomial in the size of the original combinatorial optimization problem, we can check $Ax \geq b$ by testing the constraints one by one (and (6) can be solved with Khachiyan's method for linear programming). However, it is not necessary that the number of inequalities in $Ax \geq b$ is polynomially bounded. There may exist other ways of checking $Ax \geq b$ than testing the constraints one by one.

Consider, e.g., the example of 1-arborescences given above. Finding a shortest 1-arborescence means solving (4), which is, by Edmonds' theorem, the same as solving (5). Now in the minimum of (5) there are exponentially many constraints. Yet these constraints can be checked in polynomial time. Given a vector (x_{ij}), first test whether all components are nonnegative. This can be done easily in polynomial time. If one of the components is negative we have found a violated inequality. If they all are nonnegative, consider x_{ij} as a capacity function. Next find for each $j \neq 1$ (e.g. with Ford and Fulkerson's minimum cut algorithm) a cut C_j separating node 1 from node j, with minimum capacity. Taking the minimum of the capacities of $C_2, \ldots, C_n$ gives a nonempty subset T of $\{2, \ldots, n\}$ minimizing $\sum_{j \in T, i \notin T} x_{ij}$. If this minimum is less than 1, T determines a violated inequality. Otherwise we may conclude that the constraints for the minimum in (5) hold for (x_{ij}). So with ellipsoids also the minimum in (5), and hence a shortest 1-arborescence, can be found in polynomial time.

It must be admitted immediately that Edmonds' direct polynomial algorithm for the shortest 1-arborescence problem is much more efficient (and gives the polyhedral results as a by-product). With the ellipsoid method, however, the polynomial solvability can be shown quickly. Moreover, for some other combinatorial optimization problems the polynomial solvability could be derived, as yet, only by combining a min-max relation with the ellipsoid method.

Concluding, a min-max relation often follows from an algorithm, but can in turn be helpful in providing not only a good characterization, but sometimes even the polynomial solvability of the problem.

Finally we notice that the interpretation of a combinatorial min-max relation in terms of LP-duality also enables to extend the known sensitivity analysis for linear programming to a *sensitivity analysis* for the combinatorial optimization problem. Thus if we have solved both the maximization and the minimization problem, the optimum solutions give us some of the tight constraints.

E.g., in the max-flow min-cut problem each of the arcs in the minimum-capacitated cut has a tight capacity. If we tighten any of these capacities, the maximum flow value will decrease, whereas if we wish to increase the flow value we must relax at least one of these capacities.

Unfortunately, combinatorial optima considered as LP-optima often turn out to be highly degenerate, so that such a duality analysis provides not more than an upper bound for the new maximum, also marginally (lower bound for the new minimum, respectively). E.g., in the max-flow min-cut case there often exist several cuts of minimum capacity, and in each of these cuts we have to relax some capacity if we want to increase the flow value.

Below we give a survey of combinatorial min-max relations discovered so far, and we go into the methods for deriving them. Above we passed already on two types of methods: *algorithmic* methods, which consist of showing that the maximum value found by an algorithm is equal to the minimum value in some other problem, and *polyhedral* methods, which consist of showing that certain linear programs have integer optimum solutions. A third "type" of method could be named *combinatorial,* where purely combinatorial, mostly graph-theoretical techniques are used, generally without a direct algorithmic relevance.

We first give in Section 2 some preliminaries on terminology, notation and conventions. In Section 3 we briefly review the notions of total unimodularity and total dual integrality and the theory of blocking and anti-blocking polyhedra (for a more comprehensive survey, see Pulleyblank [1982]). In the Sections 4 and 5 we describe min-max relations for bipartite graphs and network flows, respectively, which turn out to be a basis, and a special case, of several of the min-max relations we discuss further. These are grouped around the subjects of (nonbipartite) matchings and coverings (Section 6), multicommodity flows (Section 7), arborescences and directed cuts (Section 8), perfect graphs (Section 9), clutters and blockers (Section 10), and matroids and submodular functions (Section 11).

Some references for background information are: for graph theory, Bondy and Murty [1976] or Wilson [1972]; for a survey of polynomial algorithms in combinatorial optimization, Lawler [1976]; for complexity theory (NP-completeness), Garey and Johnson [1979]; for both latter two subjects, Papadimitriou and Steiglitz [1982]; for polyhedral theory, Pulleyblank [1983] and Stoer and Witzgall [1970].

2. Some Terminology, Notation and Conventions

Throughout this paper we assume familiarity with the basic concepts from graph theory and linear programming.

An *(un)directed graph* is a pair (V, E) where V is a finite set (of *vertices* or *points*) and E is a collection of (un)ordered pairs (called the *edges* in the undirected case, and *arcs* in the directed case). The set E may contain a pair more than once, i.e., *parallel* edges or arcs are allowed.

Invited by the figurative character of graphs we shall often use loose language, which will shorten arguments, and which could be made as formal as we like. Thus we use expressions like: "add a new vertex", "replace an edge by two parallel edges", "replace an edge by a path of length three" (which means replace $\{u, v\}$ by $\{u, u'\}, \{u', v'\}, \{v', v\}$, where u' and v' are new vertices), etc.

A graph $G = (V, E)$ is *bipartite* if V can be split into classes V_1 and V_2 such that every edge or arc of G contains a vertex in V_1 and a vertex in V_2. We shall call V_1 and V_2 *the colour classes* of G, although this splitting need not be unique.

If $G = (V, E)$ is an undirected graph, and $V' \subseteq V$, the subgraph *induced by* V', denoted by $\langle V' \rangle$, is the graph $(V', \{e \in E \mid e \subseteq V'\})$.

If r and s are vertices of an (un)directed graph, an *r-s-path* is an (un)directed path from r to s. Sometimes, if no danger of confusion exists, we shall identify a subgraph or a path with the set of edges or arcs contained in it. Two subgraphs or paths are *edge-disjoint* (*arc-disjoint, vertex-disjoint*, respectively) if they have no edges (arcs, vertices, respectively) in common. Two paths are *internally vertex-disjoint* if they have no vertices in common, except possibly for the end points.

The complete undirected graph on n vertices is denoted by K_n, and the complete undirected bipartite graph with colour classes of sizes m and n, by $K_{m,n}$.

II. Min-Max Results in Combinatorial Optimization

If $G=(V, E)$ is an undirected graph and $V' \subseteq V$, then $\delta(V')$ or $\delta_E(V')$ denotes the set of edges of G with exactly one vertex in V'. If $D=(V, A)$ is a directed graph and $V' \subseteq V$, then $\delta^+(V')$ or $\delta_A^+(V')$ denotes the set of arcs leaving V'. Similarly, $\delta^-(V')$ or $\delta_A^-(V')$ denotes the set of arcs entering V'.

We often use just v instead of $\{v\}$.

If A is a matrix, and w and b are vectors, then three (equivalent) forms of the duality theorem of linear programming are:

(1) $\quad \max\{wx | x \geq 0, Ax \leq b\} = \min\{yb | y \geq 0, yA \geq w\},$

$\quad \max\{wx | Ax \leq b\} \quad\quad = \min\{yb | y \geq 0, yA = w\},$

$\quad \min\{wx | Ax \geq b\} \quad\quad = \max\{yb | y \geq 0, yA = w\}.$

Here any of these relations holds if at least one of the two optima is finite.

In (1), and in similar expressions throughout this paper, we use the following conventions. Concatenations wx and yb denote inner products. If v and w are vectors, $v \leq w$ denotes component-wise comparison. When using expressions like $wx, Ax \leq b, yA \geq w$ we implicitly assume compatibility of sizes. Moreover, 0 and 1 denote (also) the all-zero and the all-one vectors, of appropriate length.

For a vector w, $|w|$ denotes the sum of its components (so $|w|=w\mathbf{1}$).

$\mathbb{R}, \mathbb{Q}, \mathbb{Z}, \frac{1}{2}\mathbb{Z}$ denote the sets of reals, rationals, integers, half-integers, respectively, while the subscript $+$ restricts such sets to their nonnegative elements.

Let S be a finite set. We identify functions $x: S \to \mathbb{R}$ with vectors x in $\mathbb{R}^S$. Furthermore, we identify any subset S' of S with its incidence vector $\chi_{S'}$ in $\mathbb{R}^S$, defined by

(2) $\quad \chi_{S'}(s) = 1$, if $s \in S'$, and $\chi_{S'}(s) = 0$, if $s \notin S'$,

for s in S. Hence any collection $\mathscr{C}$ of subsets of S is also a collection of vectors in $\mathbb{R}^S$, and "the convex hull of (the elements of) $\mathscr{C}$" means the convex hull of the incidence vectors of the elements of $\mathscr{C}$. E.g., the convex hull of the matchings is the convex hull of the incidence vectors of the matchings.

If $c: S \to \mathbb{R}$, then we denote

(3) $\quad c(S') := \sum_{s \in S'} c(s),$

for $S' \subseteq S$. If c is called a "capacity" ("length", "weight", ...) function, then $c(S')$ is called the capacity (length, weight, ...) of S'.

Further nonstandard notation and terminology used below will be defined on the spot.

3. Total Unimodularity, Total Dual Integrality, and Blocking and Anti-Blocking Polyhedra

The concepts mentioned in the title of this section play an important role in the field of polyhedral methods for combinatorial min-max relations. Here we review these notions in brief. For a more extensive treatment, see the survey on "polyhedral combinatorics" by W. R. Pulleyblank.

Total unimodularity. A matrix A is called *totally unimodular* if each square submatrix of A has determinant 0, $+1$ or -1. In particular, each entry of A is 0, $+1$ or -1.

The interest of totally unimodular matrices for optimization was discovered by the following theorem of Hoffman and Kruskal [1956]: *if A is totally unimodular and b and w are integer vectors, then both sides of the LP-duality equation*

(1) $$\max\{wx|Ax \leq b\} = \min\{yb|y \geq 0, yA = w\}$$

have integer optimum solutions.

It follows that any linear program with totally unimodular constraint matrix and integer right hand sides has an integer optimum primal solution.

If A is totally unimodular, then also, e.g., the matrices $[-I\,A]$ and $[I\,-I\,A\,-A]$ are totally unimodular. Hence also each of the optima in

(2) $$\max\{wx|x \geq 0, Ax \leq b\} = \min\{yb|y \geq 0, yA \geq w\}$$

and in

(3) $$\max\{wx|c_1 \leq x \leq c_2, b_1 \leq Ax \leq b_2\} =$$
$$\min\{y_2 b_2 - y_1 b_1 + z_2 c_2 - z_1 c_1 | (y_2 - y_1)A + z_2 - z_1 = w\}$$

has an integer optimum solution, for integer b, w, b_1, b_2, c_1, c_2. In fact, Hoffman and Kruskal proved that a matrix A is totally unimodular if and only if it is integer and the maximum in (2) has an integer optimum solution x for all integer vectors b and w for which this maximum exists. That is, if and only if the polyhedron $\{x|x \geq 0, Ax \leq b\}$ has integer vertices, for each integer vector b.

Ghouila-Houri [1962] gave the following characterization of total unimodularity: *a matrix A is totally unimodular, if and only if each collection R of rows of A can be split into classes R_1 and R_2 such that the sum of the rows in class R_1 minus the sum of the rows in class R_2 is a vector with components 0, $+1$, -1 only.*

Total dual integrality. A second useful concept, introduced by Edmonds and Giles [1977], is defined as follows. A system $Ax \leq b$ of linear inequalities is called *totally dual integral* if the minimum in the LP-duality equation

(4) $$\max\{wx|Ax \leq b\} = \min\{yb|y \geq 0, yA = w\}$$

has an integer optimum solution y, for each integer objective function w for which the optima exist.

Motivation for this concept comes from the following result of Edmonds and Giles: *if $Ax \leq b$ is totally dual integral and the right hand side b is integer, then the maximum in (4) has an integer optimum solution for each objective function w for which the optima exist.* That is, each face of the polyhedron $\{x|Ax \leq b\}$ contains integer vectors.

So if $Ax \leq b$ is totally dual integral and if for some w the optima (4) exist, then the minimum in (4) has an integer optimum solution if w is integer, and the maximum in (4) has an integer optimum solution if b is integer.

Total dual integrality forms a useful proof technique, as showing the existence of integer optimum dual solutions in (4) suffices for showing integer optimum primal and dual solutions.

Several systems of linear inequalities have been shown to be totally dual integral, and below we shall meet a number of them.

A system $Ax \leq b$ is called *totally dual half-integral* if the minimum in (4) has a half-integer optimum solution y for each integer vector w for which the minimum exists.

Blocking and anti-blocking polyhedra. Fulkerson [1970, 1971, 1972] and Lehman [1979] developed a theory of "blocking" and "anti-blocking" polyhedra, which throws a new light on combinatorial min-max relations through the classical polarity of vertices and facets of polyhedra.

Let A and B be nonnegative matrices, both with n columns, and with rows $a_1, \ldots, a_t$ and $b_1, \ldots, b_m$, respectively. Suppose we have the following relation between A and B:

(5) $\quad \{x \in \mathbb{R}^n_+ \mid Bx \geq 1\} = \{x \in \mathbb{R}^n_+ \mid x \geq y$
for some convex combination y of $a_1, \ldots, a_t\}$.

Then the same holds if we interchange A and B:

(6) $\quad \{x \in \mathbb{R}^n_+ \mid Ax \geq 1\} = \{x \in \mathbb{R}^n_+ \mid x \geq y$
for some convex combination y of $b_1, \ldots, b_m\}$.

This duality principle (which can be shown easily, e.g., with Farkas' lemma or with the LP-duality theorem) can be formulated equivalently as follows. Suppose that

(7) $\quad \min\{wa_1, \ldots, wa_t\} = \max\{|y| \mid y \geq 0, \, yB \leq w\}$

for each w in $\mathbb{R}^n_+$. Then also

(8) $\quad \min\{wb_1, \ldots, wb_m\} = \max\{|y| \mid y \geq 0, \, yA \leq w\}$

for each w in $\mathbb{R}^n_+$. Indeed, (7) is equivalent to (5) (by applying LP-duality to the maximum in (7)), and similarly (8) is equivalent to (6).

So if we have proved one min-max relation, viz. (7), we get another min-max relation, viz. (8), as a present. And conversely.

Fulkerson gave several interesting combinatorial applications of this equivalence, especially if A and B are $\{0, 1\}$-matrices. He defined for any polyhedron $P \subseteq \mathbb{R}^n$ the *blocking polyhedron* $b(P)$ of P by

(9) $\quad b(P) := \{z \in \mathbb{R}^n_+ \mid zx \geq 1 \text{ for all } x \text{ in } P\}$,

and observed that the polyhedron (6) is the blocking polyhedron of (5), and vice versa.

Similarly a theory of *anti-blocking* polyhedra was obtained, where $\geq$ in (5), (6) and (9) is replaced by $\leq$.

4. Bipartite Graphs

We start with surveying min-max relations for bipartite graphs, as they form non-trivial special cases of many other combinatorial min-max relations, and as they exhibit several typical min-max phenomena.

Min-max relations for bipartite graphs are closely related, and for a part equivalent, to those for network flows to be treated in the next section.

First we give some standard notation and terminology. Let $G = (V, E)$ be an undirected graph. Then

(1) $\nu(G) =$ the *matching number* of $G =$ the maximum size of a matching in G [a *matching* is a set of pairwise disjoint edges];

$\tau(G) =$ the *vertex-cover number* of $G =$ the minimum size of a vertex-cover for G [a *vertex-cover* is a set of vertices intersecting every edge];

$\alpha(G) =$ the *coclique number* of $G =$ the maximum size of a coclique in G [a *coclique* is a set of pairwise non-adjacent vertices];

$\rho(G) =$ the *edge-cover number* of $G =$ the minimum size of an edge-cover for G [an *edge-cover* is a set of edges covering V].

It is easy to see that the inequalities $\nu(G) \leq \tau(G)$ and $\alpha(G) \leq \rho(G)$ always hold. The triangle K_3 shows that generally we do not have equality. In fact, equality in one of these relations implies equality in the other, as Gallai [1958, 1959] proved the following.

Theorem 1 (Gallai's theorem). *For any undirected graph* $G = (V, E)$ *one has*

(2) $$\alpha(G) + \tau(G) = |V| = \nu(G) + \rho(G)$$

(where the second equality holds if G has no isolated vertices).

Proof. (i) A set of vertices is a coclique, if and only if its complement is a vertex-cover. This proves $\alpha(G) + \tau(G) = |V|$.

(ii) We next show that $\nu(G) + \rho(G) \leq |V|$. Let E' be a collection of $\nu(G)$ pairwise disjoint edges. Hence E' covers $2\nu(G)$ vertices. Choose for any vertex not covered by E' an arbitrary edge containing that vertex, and let E'' be the set of these edges. So $|E''| = |V| - 2\nu(G)$, and $E' \cup E''$ covers all vertices. Therefore

(3) $$\rho(G) \leq |E' \cup E''| = \nu(G) + |V| - 2\nu(G) = |V| - \nu(G).$$

(iii) To show that $\nu(G) + \rho(G) \geq |V|$, let E' be a collection of $\rho(G)$ edges covering V. Then the graph (V, E') has no paths of lenght three (otherwise we could delete the middle edge to obtain a smaller edge-cover). Hence each component of (V, E') is a "star". It is easy to see that there are $|V| - \rho(G)$ such stars. Choosing from each component one edge gives us $|V| - \rho(G)$ pairwise disjoint edges. Hence $\nu(G) \geq |V| - \rho(G)$. □

Deming [1979], Sterboul [1979] and Lovász [1982] characterized the undirected graphs G with $v(G)=\tau(G)$ and $\alpha(G)=\rho(G)$. These graphs include the bipartite graphs, by the following theorem, which was presented in König [1931], but which finds its roots in earlier papers by Frobenius [1912, 1917] and König [1915, 1916].

Theorem 2 (König's matching theorem). *The maximum size of a matching in a bipartite graph is equal to the minimum number of vertices intersecting all edges, i.e., $v(G)=\tau(G)$ for bipartite graphs G.*

Proof. Let a matching M in the bipartite graph $G=(V, E)$ be given. We describe a procedure which either finds a larger matching, or finds a vertex-cover V' with $|V'|=|M|$. Obviously this proves the theorem.

Let G have colour classes V_1 and V_2. Orient the edges in M from V_2 to V_1, and orient all other edges of G from V_1 to V_2. Let V_0 be the set of vertices covered by M. Now there are two possibilities.

(i) There is a directed path from $V_1\backslash V_0$ to $V_2\backslash V_0$. Necessarily, the second, fourth, sixth, ... arc in this path belongs to M, and the other not. Since the set P of edges occurring in this path has odd cardinality, the symmetric difference $M \triangle P$ is a matching larger than M.

(ii) There is no directed path from $V_1\backslash V_0$ to $V_2\backslash V_0$. Let

(4) $\quad V':=\{v\in V_1 | \text{there is no directed path from } V_1\backslash V_0 \text{ to } v\} \cup$
$\qquad \{v\in V_2 | \text{there is a directed path from } V_1\backslash V_0 \text{ to } v\}.$

It is not difficult to see that V' is a vertex-cover, and that $|V'|=|M|$. □

This proof, due to Ford and Fulkerson [1956], is an example of an algorithmic proof: finding a maximum matching now amounts to finding repeatedly a directed path. The proof yields a polynomial algorithm.

In the sequel we shall meet several other min-max relations which generalize König's matching theorem, like Menger's theorem on disjoint paths in directed graphs, the Tutte-Berge formula for matchings in arbitrary graphs, the perfect graph theorem, the Lucchesi-Younger theorem on directed cuts, the matroid intersection theorem.

By Gallai's theorem, König's matching theorem is equivalent to another result of König [1932].

Corollary 2a (König's covering theorem). *The minimum number of edges covering al vertices of a bipartite graph is equal to the maximum size of a coclique, i.e., $\rho(G)=\alpha(G)$ for bipartite graphs G (without isolated vertices).*

Proof. Immediately by combining Theorems 1 and 2. □

It is characteristic for min-max relations that they produce, and often are equivalent to, a characterization of "necessary and sufficient" type. Thus, König's matching theorem yields that a bipartite graph $G=(V, E)$, with colour classes V_1 and V_2, has a matching covering V_1, if and only if $|\Delta(V')|\geq|V'|$ for each subset V' of V_1. Here $\Delta(V')$ denotes the set of vertices adjacent to at least one vertex in V'.

This corollary was found by Hall [1935] and is called Hall's *Marriage theorem*. In fact, it is not difficult to derive in turn König's theorem from Hall's. Hall's theorem is often formulated in terms of "transversals", "systems of distinct representatives", or "assignments" – see Mirsky [1971] for a comprehensive survey.

The theorems of König can be shown also with polyhedral methods. Let A be the incidence matrix of the bipartite graph $G=(V, E)$. That is, the rows and columns of A are indexed by V and E, respectively, where $A_{v,e} = 1$ or 0 according to whether or not v belongs to e.

Then Theorem 2 and Corollary 2a are equivalent to the LP-optima

(5) $$\max\{wx \,|\, x \geq 0, Ax \leq b\} = \min\{yb \,|\, y \geq 0, yA \geq w\}$$

(6) $$\min\{wx \,|\, x \geq 0, Ax \geq b\} = \max\{yb \,|\, y \geq 0, yA \leq w\}$$

having integer optimum solutions, if w and b are all-one vectors.

Now it is not difficult to prove that A is totally unimodular, and hence König's theorems follow also from Hoffman and Kruskal's result (cf. Section 3).

Of course, much more follows: for all integer vectors w and b, the optima in (5) and (6) have integer optimum solutions. This gives more general min-max relations. E.g., it yields that, given a bipartite graph $G=(V, E)$ and a weight function $w: E \to \mathbb{Z}$,

(7) the maximum weight of a matching is equal to the minimum of $\sum_{v \in V} y(v)$

where $y: V \to \mathbb{Z}_+$ such that $y(u) + y(v) \geq w(e)$ for all $e = \{u, v\}$ in E;

(8) the minimum weight of an edge-cover is equal to the maximum of $\sum_{v \in V} y(v)$

where $y: V \to \mathbb{Z}_+$ such that $y(u) + y(v) \leq w(e)$ for all $e = \{u, v\}$ in E.

These results, due to Egerváry [1931], can also be shown algorithmically by extending the proof method of Theorem 2 above. This amounts to Kuhn's *Hungarian method* [1955, 1956] for the *optimal assignment problem*: given a matrix $(w_{ij})_{i,j=1}^n$, find a permutation π of $\{1, \ldots, n\}$ such that $\sum_{i=1}^n w_{i\pi(i)}$ is as large (or small) as possible.

Similarly, this gives an algorithm and a min-max relation for the *(Hitchcock-Koopmans) transportation problem*: given a cost matrix $(c_{ij})_{i=1, j=1}^{m, n}$, a "supply" vector $b = (b_1, \ldots, b_m)$ and a "demand" vector $d = (d_1, \ldots, d_n)$, find nonnegative integers x_{ij} ($i=1, \ldots, m$; $j=1, \ldots, n$) such that $\sum_{j=1}^n x_{ij} \leq b_i$ ($i=1, \ldots, m$) and $\sum_{i=1}^m x_{ij} \geq d_j$ ($j=1, \ldots, n$), and such that $\sum_{i,j} c_{ij} x_{ij}$ is as small as possible.

In fact, one of the most general min-max relations in this direction follows from the fact that, for a bipartite graph $G=(V, E)$, with incidence matrix A, and functions $b_1, b_2: V \to \mathbb{Z}$, $c_1, c_2, w: E \to \mathbb{Z}$, the optima

II. Min-Max Results in Combinatorial Optimization

(9) $\max\{wx \mid b_1 \leq x \leq b_2, c_1 \leq Ax \leq c_2\} =$
$\min\{y_2 c_2 - y_1 c_1 + z_2 b_2 - z_1 b_1 \mid y_1, y_2, z_1, z_2 \geq 0, (y_1 - y_2)A + z_2 - z_1 = w\}$

have integer optimum solutions, again by the total unimodularity of A.

These integrality results can be formulated equivalently in terms of polyhedra. E.g., if A again is the incidence matrix of the bipartite graph $G = (V, E)$, the fact that $\max\{wx \mid x \geq 0, Ax \leq 1\}$ has an integer optimum solution for every w, is equivalent to the fact that the *matching polytope* of G (which is the convex hull of the matchings in G) is determined by the constraints

(10) $\qquad x(e) \geq 0 \qquad (e \in E),$
$\qquad x(\delta(v)) \leq 1 \qquad (v \in V).$

Similarly, the *perfect matching polytope*, being the convex hull of the perfect matchings, is determined by (10) after replacing $\leq$ in the second line by $=$. This is equivalent to a theorem of Birkhoff [1946] and Von Neumann [1953]: each doubly stochastic matrix is a convex combination of permutation matrices.

Edge-colourings. Moreover min-max relations have been found for edge-colourings in bipartite graphs, which seem not to follow from total unimodularity.

For any undirected graph $G = (V, E)$, let

(11) $\Delta(G) =$ the maximum degree of G,
$\chi(G) =$ the *edge-colouring number* of $G =$ the minimum number of colours needed to colour the edges of G such that no two intersecting edges have the same colour.

Equivalently, $\chi(G)$ is the minimum number of matchings needed to cover the edges of G.

Clearly always $\Delta(G) \leq \chi(G)$, and again the triangle K_3 shows that strict inequality can occur. It is a famous theorem of Vizing [1964] that $\chi(G) \leq \Delta(G) + 1$ if G has no parallel edges. König [1916] showed that bipartite graphs again have $\Delta(G) = \chi(G)$.

Theorem 3 (König's edge-colouring theorem). *The edge-colouring number of a bipartite graph G is equal to its maximum degree, i.e., $\chi(G) = \Delta(G)$.*

Proof. First notice that the theorem is easy if $\Delta(G) \leq 2$. In this case the graph consists of a number of vertex-disjoint paths and even circuits.

If $\Delta(G) \geq 3$, colour as many edges of G as possible with $\Delta(G)$ colours, without giving the same colour to two intersecting edges. Suppose edge $e = \{u, v\}$ is not coloured. At least one colour, say red, does not occur among the colours given to the edges containing v. Similarly, there is a colour, say blue, not occurring at w. Clearly, red $\neq$ blue, since otherwise we could colour the edge e red. Let G' be the subgraph of G consisting of the red and blue edges, together with the edge e. Now $\Delta(G') \leq 2$, and hence $\chi(G') \leq 2$. So the edges occurring in G'

may be recoloured red and blue, and in this way we have coloured more edges than before. □

A related theorem was proved by Gupta [1967, 1978] and by D. König (unpublished).

Corollary 3a (Gupta's theorem). *Let $G=(V, E)$ be a bipartite graph. Then the maximum number of pairwise disjoint edge-sets, each covering V, is equal to the minimum degree $\delta(G)$ of G.*

Proof. One may derive from G a bipartite graph H, with vertex-degrees $\delta(G)$ or 1, by repeated application of the following splitting procedure: choose a vertex v of degree more than $\delta(G)$, add a new vertex v', and replace one edge $\{v, w\}$ by the new edge $\{v', w\}$ (the other edges containing v remain unchanged). So the edges of the final graph H are in one-to-one correspondence with the edges of G.

Since H has maximum degree $\delta(G)$, by Theorem 3 the edges of H can be coloured with $\delta(G)$ colours such that no two edges of the same colour intersect. It follows that at any vertex of H of degree $\delta(G)$ all colours occur. In the original graph G this yields $\delta(G)$ pairwise disjoint edge-covers as required. □

If we understand the vertices of the two colour classes of a bipartite graph as teachers and as classes, respectively, and the edges as lessons to be given, we may interpret an edge-colouring as a time-table – each colour represents a period. For algorithms finding explicitly solutions, see De Werra [1970, 1972], Dempster [1971], McDiarmid [1972], Bondy and Murty [1976] pp. 96–100.

5. Network Flows

A second fundamental field of combinatorial min-max relations is that of flows in networks (directed graphs), with the max-flow min-cut theorem as landmark. Min-max relations for network flows are closely connected to those for bipartite graphs given in Section 4, and below we shall see that to a large extent they can be derived from each other by some simple constructions. For a survey on network flows, see Ford and Fulkerson [1962].

The classical min-max equality for directed graphs is Menger's theorem [1927].

Theorem 4 (Menger's theorem). *Let $D=(V, A)$ be a directed graph, and let r, $s \in V$. Then the maximum number of pairwise arc-disjoint r-s-paths is equal to the minimum size of an r-s-cut.*
[An *r-s-path* is a directed path from r to s. An *r-s-cut* is a set $\delta^-(V')$ with $r \notin V'$, $s \in V'$.]

Proof. It is easy to see that the maximum is not more than the minimum. Like in the proof of König's matching theorem (Theorem 2), to prove the reverse inequality we describe a procedure which, given k pairwise arc-disjoint r-s-paths, either finds $k+1$ pairwise arc-disjoint r-s-paths, or finds an r-s-cut of size k.

Therefore, let $A_1, \ldots, A_k$ be the arc sets of the given k pairwise arc-disjoint r-s-paths. Reverse the orientation of the arcs occurring in $A_1 \cup \ldots \cup A_k$, thus making the directed graph D'. Now there are two possibilities.

(i) There is an r-s-path in D'. Let A_0 be the set of arcs of D occurring in this path (reversed or not). Then it is easy to see that the symmetric difference $(A_1 \cup \ldots \cup A_k) \Delta A_0$ is a disjoint union of $k+1$ arc-disjoint r-s-paths.

(ii) There is no r-s-path in D'. Let V' be the set of points v for which there exists a v-s-path in D'. So $r \notin V'$, $s \in V'$. Moreover, $\delta_A^-(V')$ contains not more than k arcs. Indeed, no arc of D' enters V', so no arc in $A_1 \cup \ldots \cup A_k$ leaves V', and no arc in $A \setminus (A_1 \cup \ldots \cup A_k)$ enters V'. Hence exactly one arc in each of the A_i enters V', and no other arc of D enters V'. So $|\delta_A^-(V')| = k$. □

This is again an algorithmic proof, based on the famous augmenting path method of Ford and Fulkerson [1956], and yielding a polynomial algorithm. The proof extends that of König's matching theorem – see also Remark 1 below.

The following corollaries can be seen to be equivalent to Menger's theorem.

Corollary 4a (Menger's theorem – vertex form). *Let $D = (V, A)$ be a directed graph, and let $r, s \in V$. Then the maximum number of pairwise internally vertex-disjoint r-s-paths is equal to the minimum number of vertices in $V \setminus \{r, s\}$ intersecting all r-s-paths.*

Proof. By an easy construction from Theorem 4. □

The second corollary is due to Dantzig [1951] (the existence of integer optimum flows), Ford and Fulkerson [1956] and Elias, Feinstein and Shannon [1956].

Corollary 4b (Max-flow min-cut theorem). *Let $D = (V, A)$ be a directed graph, let $r, s \in V$ and let $c: A \to \mathbb{R}_+$ be a capacity function. Then the maximum value of an r-s-flow subject to the capacity c is equal to the minimum capacity of an r-s-cut. If all capacities are integer, there exists an integer optimum flow.*

[Here an r-s-flow is a vector $x: A \to \mathbb{R}$ such that

(1)   (i) $x(a) \geq 0$ $(a \in A)$,

            (ii) $x(\delta^-(v)) = x(\delta^+(v))$ $(v \in V, v \neq r, s)$.

The *value* of the flow is the net amount of flow leaving r, i.e., is

(2) $x(\delta^+(r)) - x(\delta^-(r))$

(which is equal to the net amount of flow entering s). The flow x is *subject to c* if $x(a) \leq c(a)$ for all a in A.]

Proof. If the capacities are integer, the corollary follows from Menger's theorem by splitting each arc a into $c(a)$ parallel arcs, and by observing that k

pairwise arc-disjoint r-s-paths in the new graph yield an integer r-s-flow of value k in the original graph, subject to c.

If c is rational, there is a natural number M such that Mc is integer. Since by multiplying c by M, also the maximum flow value and the minimum cut capacity are multiplied by M, the case of c rational follows from the case of c integer.

If c is real-valued, the corollary follows from the rational case by simple continuity and compactness arguments. □

Note that this proof of splitting arcs is non-polynomial: if c is integer, we obtain a graph of size proportional to $\sum_{a \in A} c(a)$, while the size of the original problem is proportional to $\sum_{a \in A} \log c(a)$. So the algorithm for Menger's theorem does not pass over in this way to a polynomial algorithm for max-flow min-cut. However, Ford and Fulkerson [1956] gave a direct maximum-flow algorithm, and Edmonds and Karp [1972] showed that this algorithm, with some modifications, is polynomial.

Above we saw that finding maximum packings of r-s-paths and finding r-s-flows of maximum value, essentially are two sides of the same problem. In the following remark we shall see that bipartite matching is a third way of looking at this same problem.

Remark 1. König's matching theorem, and its algorithmic proof, described in Section 4, are special cases of Menger's theorem and its algorithmic proof given above. Indeed, if $G = (V, E)$ is a bipartite graph, with colour classes V_1 and V_2, orient the edges from V_1 to V_2, and add new vertices r and s, and arcs (r, v) for v in V_1 and (v, s) for v in V_2. This makes the directed graph D. Then Menger's theorem for D, r, s is equivalent to König's theorem for G.

In fact, also the converse implication holds by a direct construction, given by Orden [1956] (where the "transshipment problem" is reduced to the "transportation problem") and Hoffman [1960] (cf. Ford and Fulkerson [1958] and Hoffman and Markovitz [1963]). We here describe the idea behind this construction.

We show how to derive from König's matching theorem the following equivalent form of Menger's theorem: if $D = (V, A)$ is a directed graph, and R and S are disjoint subsets of V, then the maximum number of pairwise vertex-disjoint directed R-S-paths (i.e., paths starting in R and ending in S), is equal to the minimum number of vertices intersecting all R-S-paths.

To this end, split each vertex v of D into two new vertices v' and v'', replace any arc (u, v) of D by the arc (u'', v'), and add arcs (v'', v') for v in V. Denote for any subset W of V, $W' := \{v' | v \in W\}$ and $W'' := \{v'' | v \in W\}$. Remove the vertices in R' and S''. This makes a (directed) bipartite graph G, with colour classes $V'' \setminus S''$ and $V' \setminus R'$.

We show that if k is the maximum number of pairwise vertex-disjoint R-S-paths in D, the matching number $v(G)$ of G is equal to $|V| - |R| - |S| + k$, and that maximum packings of R-S-paths in D and maximum matchings in G can be transformed easily to each other.

Let $A_1, \ldots, A_k$ be the arc sets of pairwise vertex-disjoint R-S-paths in D. Then the collection

(3) $\{(u'', v') \mid (u, v) \in A_1 \cup \ldots \cup A_k\} \cup \{(v'', v') \mid v \notin R \cup S,$
v is not covered by $A_1 \cup \ldots \cup A_k\}$

is a matching in G, of size $|V| - |R| - |S| + k$, and hence $\nu(G) \geq |V| - |R| - |S| + k$.

Conversely, let M be a matching in G of size $\nu(G)$. Consider the set A' of arcs (u, v) of D for which (u'', v') belongs to M, and the set V' of vertices of D for which (v'', v') belongs to M. Then A' forms a collection of pairwise vertex-disjoint directed paths and cycles, not intersecting V'. It is not difficult to see that at least $\nu(G) - |V| + |R| + |S|$ of these paths connect R and S, and hence $\nu(G) \leq |V| - |R| - |S| + k$.

So finding a maximum packing of R-S-paths is equivalent to finding a maximum matching in G, and it is not difficult to check that König's theorem yields the form of Menger's theorem given above.

This may illustrate the third angle of vision: packings of paths can be considered alternatively not only as flows but also as matchings in an appropriate bipartite graph. Keeping this in mind is often helpful in understanding network flow problems.

Remark 2. As is well-known, the incidence matrix of a directed graph is totally unimodular, which also implies the max-flow min-cut theorem. Here the *incidence matrix* of the directed graph $D = (V, A)$ is the matrix H with rows and columns indexed by V and A, respectively, and with

(4) $H_{v,a} = +1$ if $a = (v, u)$ for some u,
-1 if $a = (u, v)$ for some u,
0 otherwise,

for $v \in V, a \in A$.

Indeed, if $r, s \in V$, let $D_0 = (V, A_0)$ arise from D by adding the arc (s, r). Let H_0 be the incidence matrix of D_0. Then H_0 is totally unimodular, and hence for all $c, w: A_0 \to \mathbb{Z}_+$, both sides of the LP-duality equation

(5) $\max\{wx \mid 0 \leq x \leq c, H_0 x = 0\} = \min\{yc \mid y \geq 0, y + zH_0 = w \text{ for some } z \in \mathbb{R}^V\}$

have integer optimum solutions. Taking $w(s, r) = 1$, and $w(a) = 0$ for a in A, and $c(s, r) = \infty$ (or very large), gives the max-flow min-cut theorem. Below we shall see more applications.

There is an alternative way of viewing max-flow min-cut in terms of linear programming. Let $D = (V, A)$ be a directed graph, and let $r, s \in V$. Let K be the matrix with columns indexed by A, and with rows all incidence vectors of r-s-paths (considered as arc sets). Then the max-flow min-cut theorem says that for $c: A \to \mathbb{R}_+$, the minimum in

(6) $\min\{cx \mid x \geq 0, Kx \geq 1\} = \max\{\mathbf{1}y \mid y \geq 0, yK \leq c\}$

has an integer optimum solution. If moreover c is integer, also the maximum has an integer optimum solution.

Integer optimum solutions for the minimum in (6) are $\{0, 1\}$-vectors, which are exactly incidence vectors of r-s-cuts. Solutions for the maximum in (6) correspond to r-s-flows subject to c.

Here we could apply the ellipsoid method: the minimum in (6) can be found in polynomial time, as the inequality system $x \geq 0, Kx \geq 1$ can be checked in polynomial time. Indeed, checking $x \geq 0$ is easy. To check $Kx \geq 1$ we cannot enumerate all inequalities as there may exist exponentially many. However, for given $x \geq 0$ the system $Kx \geq 1$ can be checked as follows. Consider x as a length function, and find an r-s-path of minimum length (e.g., by Dijkstra's method). If this length is less than 1 this path gives a violated inequality in $Kx \geq 1$. Otherwise we conclude that $Kx \geq 1$ holds.

This application of the ellipsoid method is maybe not very convincing: the formulation (5) allows a direct application of Khachiyan's method to a concrete (polynomially bounded) LP-problem. In the next sections we shall see more satisfactory applications of the ellipsoid method, which extend the idea of (6).

There is an easier theorem, which is in a sense "polar" to the results above. Again this result can be formulated equivalently in combinatorial terms and in terms of "currents". The latter one we give first.

Theorem 5 (Min-potential max-work theorem). *Let $D = (V, A)$ be a directed graph, let $r, s \in V$, and let $l: A \to \mathbb{R}_+$ be a length function. Then the maximum potential difference of r and s is equal to the minimum length of an r-s-path. Moreover, if all length are integer, there is an integer optimum potential.*
Here a *potential* is a function $\phi: V \to \mathbb{R}$ such that $\phi(v) - \phi(u) \leq l(a)$ for $a = (u, v)$ in A. The *potential difference* of r and s is $\phi(s) - \phi(r)$.

Proof. The inequality max $\leq$ min is trivial. Defining $\phi(v) :=$ the minimum length of an r-v-path gives a potential for which $\phi(s) - \phi(r)$ equals the minimum length of an r-s-path. □

E.g. with Dijkstra's method one easily handles the optima of Theorem 5 algorithmically.

The equivalent combinatorial version is as follows.

Corollary 5a. *Let $D = (V, A)$ be a directed graph, and let $r, s \in V$. Then the minimum number of arcs in an r-s-path is equal to the maximum number of pairwise disjoint r-s-cuts.*

Proof. Apply Theorem 5 with $l = 1$. Let ϕ be a potential with $\phi(s) - \phi(r)$ equal to the minimum number of arcs in an r-s-path. If k is this minimum number, let $V_i := \{v \in V | \phi(v) > \phi(s) - i\}$ for $i = 1, \ldots, k$. Then $\delta^-(V_1), \ldots, \delta^-(V_k)$ are pairwise disjoint r-s-cuts. □

This theorem, observed by Fulkerson [1968], is "polar" to Menger's theorem: interchanging "r-s-paths" and "r-s-cuts" carries one to the other. Similarly as in Remark 2 above, the max-potential min-work theorem can be formulated in terms of linear programming in two ways, which is left to the reader. It will follow that, by the theory of blocking polyhedra, the max-flow min-cut theorem and the max-potential min-work theorem are equivalent, if we forget the integrality of optimum flows and potentials.

Remark 3. By the Bellman-Ford method one can solve also the shortest path problem if lengths are allowed to be negative, provided that all directed cycles have nonnegative length. A corresponding min-max relation and good characterization can be derived from Theorem 5 as follows.

Let $D = (V, A)$ be a directed graph, and let $l: A \to \mathbb{R}$ be a length function such that each directed cycle has a nonnegative length. Define, for each vertex v of D, $\psi(v)$ to be the shortest length of any directed path ending in v (starting whereever). Then for each arc (u, v) of D, $l(u, v) \geq \psi(v) - \psi(u)$, and hence $\tilde{l}(u, v) := l(u, v) + \psi(u) - \psi(v)$ is nonnegative. However, the shortest path problem for l is equivalent to that for $\tilde{l}$.

Note that the shortest path problem for undirected graphs is easily reduced to the directed case if all lenghts are nonnegative: replace each edge $\{u, v\}$ by two arcs (u, v) and (v, u) of the same length as $\{u, v\}$. However, if edges have a negative length, we would create a negative directed cycle in this way. Yet there is a min-max relation and a polynomial algorithm for the shortest path problem in undirected graphs, also if negative lengths occur, provided that there are no circuits of negative length – see the subsection on T-joins in Section 6.

Actually, the max-flow min-cut and max-potential min-work theorems can be combined to one min-max relation for minimum-cost flows. A combinatorial version of this theorem is as follows.

Theorem 6 (Min-cost flow theorem – combinatorial version). *Let $D = (V, A)$ be a directed graph, let $r, s \in V$, and let $k \in \mathbb{Z}_+$. Then the minimum number of arcs in k pairwise arc-disjoint r-s-paths is equal to*

$$(7) \qquad \max\left(kt + |A_1 \cup \ldots \cup A_t| - \sum_{i=1}^{t} |A_i|\right),$$

where the maximum ranges over all $t \geq 0$ and collections of r-s-cuts $A_1, \ldots, A_t$.

This theorem contains Corollary 5a on shortest paths by taking $k = 1$. It also contains Menger's theorem: the largest k for which there exist k pairwise arc-disjoint r-s-paths is equal to the largest k for which the maximum (7) is bounded, which can be seen to be the minimum size of an r-s-cut.

Theorem 6 in its flow-formulation follows from the total unimodularity of the incidence matrix H of D. This gives that both optima in:

$$(8) \qquad \min\{wx \mid c_1 \leq x \leq c_2, Hx = 0\} = \max\{y_1 c_1 - y_2 c_2 \mid y_1, y_2 \geq 0, y_1 - y_2 + zH = w\}$$

have integer optimum solutions, for $w, c_1, c_2 : A \to \mathbb{Z}$. By choosing w, c_1, c_2 appropriately, (8) gives a min-max relation for the minimum cost of an r-s-flow subject to a capacity constraint, and of a given value. It also follows that if the capacities are integer, there is an integer minimum-cost flow. This contains (and is actually equivalent to) Theorem 6 by taking all costs and capacities equal to one.

Fulkerson [1961] and Minty [1960] designed an algorithm for finding a minimum-cost flow. This *out-of-kilter method* also produces the min-max relation. Edmonds and Karp [1972] gave a polynomial-time version of this method.

Similarly the minimum-cost circulation problem can be treated, which consists of finding optimum solutions x and y for (8) (in its full generality). Here a *circulation* is a function $x: A \to \mathbb{R}$ satisfying

(9) $$x(\delta^-(v)) = x(\delta^+(v))$$

for all v in V. (Often one adds the condition $x \geq 0$, so that a circulation can be considered as an "r-r-flow", for arbitrary vertex r.)

The out-of-kilter method finds integer optimum circulations x if the "need" function c_1 and the capacity function c_2 are integer, and it finds an integer optimum solution y_1, y_2 if the cost function w is integer.

The min-max relation (8) for minimum-cost circulations also contains the following theorem of Hoffman [1960]: *given a directed graph $D = (V, A)$ and a need function $c_1: A \to \mathbb{R}$ and a capacity function $c_2: A \to \mathbb{R}$ with $c_1 \leq c_2$, there exists a circulation x satisfying $c_1 \leq x \leq c_2$ if and only if for each subset V' of V the capacity of $\delta^-(V')$ is not less than the need of $\delta^+(V')$.* [Note that this last condition is eqivalent to: if $y_1, y_2: A \to \mathbb{Z}_+$ and $z: V \to \mathbb{Z}$ such that $y_1 - y_2 + zH = 0$, then $y_1 c_1 \leq y_2 c_2$. So Hoffman's theorem follows from the fact that the minimum in (8) is feasible (for $w = 0$), if and only if the maximum in (8) is bounded (as it is obviously feasible).] If moreover c_1 and c_2 are integer, the polyhedron $\{x | c_1 \leq x \leq c_2, Hx = 0\}$ has integer vertices, and hence the existence of a circulation between c_1 and c_2 implies the existence of an integer circulation between c_1 and c_2.

We finally give one of the most general min-max relations in this direction, which again follows from total unimodularity. If H is the incidence matrix of the directed graph $D = (V, A)$, and if $b_1, b_2: V \to \mathbb{R}$ and $c_1, c_2, w: A \to \mathbb{R}$, then

(10) $\min\{wx | c_1 \leq x \leq c_2, b_1 \leq Hx \leq b_2\} =$
$\max\{y_1 c_1 - y_2 c_2 + z_1 b_1 - z_2 b_2 | y_1, y_2, z_1, z_2 \geq 0, y_1 - y_2 + (z_1 - z_2) H = w\}.$

If b_1, b_2, c_1, c_2 are integer the minimum has an integer optimum solution. If w is integer, the maximum has an integer optimum solution.

This min-max relation contains all min-max results described above. It may be derived in turn form the min-max relation (8) for minimum-cost circulations, by some direct constructions. These constructions also give that the optimum solutions in (10) can be found with Fulkerson and Minty's out-of-kilter method for the min-cost circulation problem.

A wide range of further min-max equalities are contained in (10) as special cases, e.g., for flows obeying lower bounds, and for maximum-"gain" flows. Also the following min-max relations for partially ordered sets can be derived from (10).

Theorem 7 (Dilworth's theorem). *The maximum size of an antichain in a partially ordered set $(P, \leq)$ is equal to the minimum number of chains needed to cover P.*
[Dilworth [1950], cf. Fulkerson [1956]. An *(anti)chain* is a set of pairwise (in)comparable elements. In fact, Dilworth's theorem is equivalent to König's matching theorem by a similar construction as the one described in Remark 1.]

Theorem 8. *The maximum size of a chain in a partially ordered set $(P, \leq)$ is equal to the minimum number of antichains needed to cover P.*
[This is an easy theorem.]

Theorem 9 (Greene-Kleitman theorem). *The maximum size of the union of k antichains in a partially ordered set $(P, \leq)$ is equal to $\min_{P' \subseteq P} |P \backslash P'| + k\gamma(P')$, where $\gamma(P')$ denotes the minimum number of chains needed to cover P'.*
[Greene and Kleitman [1976], cf. Hoffman and Schwartz [1977], Frank [1980], Cameron [1982].]

Theorem 10 (Greene's theorem). *The maximum size of the union of k chains in a partially ordered set $(P, \leq)$ is equal to $\min_{P' \subseteq P} |P \backslash P'| + k\alpha(P')$, where $\alpha(P')$ denotes the minimum number of antichains needed to cover P'.*
[Greene [1976], cf. Hoffman and Schwartz [1977], Frank [1980], Cameron [1982]. This theorem is the poset analogue of Theorem 6 on unions of paths.]

By applying a projection technique, Cameron [1982] derived the following result from the total unimodularity of the incidence matrix of a directed graph. Let $D = (V, A)$ be a directed graph, and let $c_1, c_2, d: A \to \mathbb{Z}$. Then the following system is totally dual integral:

(11) $\qquad c_1(a) \leq x(a) \leq c_2(a) \quad (a \in A),$

$\qquad\qquad x(C) \leq d(C) \qquad\quad (C \subseteq A, C \text{ directed cycle}).$

From this result again Dilworth's theorem and the theorems of Greene and Kleitman follow. Also the perfectness of certain graphs is included – see Section 9.

6. Matchings and Generalizations

In Section 4 we discussed matching and covering problems for bipartite graphs. We now consider these problems for the more difficult general case of not-necessarily bipartite graphs.

The fundamental result on matchings is Tutte's *perfect matching theorem* [1947], characterizing undirected graphs containing a perfect matching. Tutte [1952], 1954] and Edmonds and Johnson [1973] showed that this theorem is of a self-refining nature: by a series of elementary constructions it can be generalized to results on objects like (capacitated) b-matchings, Chinese postman routes, and "T-joins". Below we emphasize describing these constructions, rather than deriving the min-max relations in detail.

Matchings and coverings. Berge [1958] showed that Tutte's perfect matching theorem is equivalent to the following min-max relation for the maximum matching size.

Theorem 11 (Tutte-Berge formula). *Let $G=(V, E)$ be an undirected graph. Then the maximum size $v(G)$ of a matching in G is equal to*

(1) $$\min_{V' \subseteq V} \frac{|V|+|V'|-\mathcal{O}(V\setminus V')}{2},$$

where $\mathcal{O}(V\setminus V')$ denotes the number of components of odd size in the graph $\langle V\setminus V'\rangle$ induced by $V\setminus V'$.

Proof. The inequality max $\leq$ min is an easy exercise (observe that for $V'\subseteq V$, any matching leaves at least $\mathcal{O}(V\setminus V')-|V'|$ vertices uncovered). Equality is proved by induction on $|V|$, where $V=\emptyset$ is trivial. Consider the following two cases.

(i) There is a vertex v covered by every matching of size $v(G)$. Let G' arise from G by deleting v and the edges containing v. Then $v(G')=v(G)-1$, and moreover, by induction, $v(G')=\frac{1}{2}(|V\setminus v|+|V''|-\mathcal{O}((V\setminus v)\setminus V''))$ for some $V''\subseteq V\setminus v$. Taking $V':=V''\cup v$ gives $v(G)=v(G')+1=\frac{1}{2}(|V|+|V'|-\mathcal{O}(V\setminus V'))$, and we have the required equality.

(ii) No vertex is covered by every matching of size $v(G)$. We show that each matching of size $v(G)$ leaves at most one vertex in each of the components of G uncovered. This will prove equality in (1), since then $2v(G)=|V|-\mathcal{O}(V)$, and we can take $V'=\emptyset$ in (1).

Suppose to the contrary there are two vertices u and v in one component of G not covered by the maximum matching M, and assume furthermore that we have chosen M and u and v such that u and v have distance $d(u,v)$ as small as possible.

Obviously, $d(u,v)>1$, since otherwise we could augment M by the edge $\{u,v\}$. Let $w\neq u,v$ be a vertex on the shortest u-v-path, and let M' be a maximum matching not covering w. Then $M\cup M'$ forms a vertex-disjoint union of paths and circuits. Let P be the set of edges in the component of $M\cup M'$ containing w. Then P forms a path, starting in w, and not containing both u and v (as each of u,v and w is covered by at most one edge in $M\cup M'$). Say P does not cover u. Then the symmetric difference $M\Delta P$ again is a maximum matching (as $|M\Delta P|-|M|=|M'|-|M'\Delta P|\geq 0$), not covering u and w. However, $d(u,w)<d(u,v)$, contradicting our choice of M,u,v. □

It is not difficult to see that the Tutte-Berge formula generalizes König's matching theorem.

Combining the Tutte-Berge formula with Gallai's theorem (Theorem 1) directly gives the following.

II. Min-Max Results in Combinatorial Optimization

Corollary 11a. *Let $G=(V, E)$ be an undirected graph without isolated vertices. Then the minimum number $\rho(G)$ of edges covering V is equal to*

(2)
$$\max_{V' \subseteq V} \frac{|V'| + \mathcal{O}(V')}{2}.$$

Proof. Directly from Theorems 1 and 11. □

Another consequence is the original perfect matching theorem of Tutte [1947]: *an undirected graph $G=(V, E)$ has a perfect matching, if and only if for each subset V' of V, the size of V' is not less than the number of odd-sized components in the subgraph induced by $V \setminus V'$.*

The proof of the Tutte-Berge theorem given above, containing ideas described in Lovász [1979a], is an example of a strictly combinatorial proof, seemingly not implying directly a polynomial algorithm for finding a maximum matching. Edmonds [1965c] designed a polynomial algorithm finding a maximum matching, which yields as a by-product the Tutte-Berge theorem.

More generally, Edmonds [1965d] gave a polynomial algorithm finding a maximum *weighted* matching. This algorithm gives as a by-product Edmonds' *matching polytope theorem*: the *matching polytope* of $G=(V, E)$ (being the convex hull of the matchings in G) is determined by the inequalities

(3)
  (i) $x(e) \geq 0$ $(e \in E)$,
  (ii) $x(\delta(v)) \leq 1$ $(v \in V)$,
  (iii) $x(\langle V' \rangle) \leq \lfloor \frac{1}{2}|V'| \rfloor$ $(V' \subseteq V, |V'|$ odd$)$.

Here $\langle V' \rangle$ is the set of edges contained in V', and $\lfloor \, \rfloor$ denotes lower integer part.

The incidence vector of any matching clearly satisfies (3), and hence the matching polytope is contained in (3) – the content of Edmonds' theorem is the converse inclusion.

The matching polytope theorem is equivalent to the following. Denote the system (3) (ii) by $Ax \leq 1$, and the system (3) (iii) by $Bx \leq f$. So the columns of A and B are indexed by E, and their rows by V and by the odd subsets of V, respectively. Then the maximum in the LP-duality equation

(4) $\max\{wx \mid x \geq 0, Ax \leq 1, Bx \leq f\} = \min\{\mathbf{1}y + zf \mid y \geq 0, z \geq 0, yA + zB \geq w\}$

has an integer optimum solution (being a matching). The inequalities (3) (iii) can be considered as the Gomory cuts, the addition of which to (3) (i) (ii) transforms the integer linear program to a linear program.

So Edmonds' theorem is equivalent to a min-max relation for the maximum weight of a matching. Cunningham and Marsh [1979] showed that for integer w also the minimum in (4) has an integer optimum solution, i.e., that the system (3) is totally dual integral. This extends the Tutte-Berge theorem, which is the case $w = 1$.

The proof by Cunningham and Marsh is algorithmical. Schrijver and Seymour [1977] (cf. Schrijver [1981]) gave a polyhedral proof. We here give a combinatorial proof which extends the above proof of the Tutte-Berge theorem.

Theorem 12. *For integer w, both optima in (4) have integer optimum solutions.*

Proof. For $G=(V,E)$ and $w \in \mathbb{Z}_+^E$, let v_w denote the maximum weight of a matching. We have to show that there are integer vectors $y \geq 0$ and $z \geq 0$ such that $yA + zB \geq w$ and $v_w \geq |y| + zf$.

Suppose G and w contradict this, with $|V| + |E| + w(E)$ as small as possible. Then G is connected (otherwise one of the components of G will form a smaller counterexample), and $w(e) \geq 1$ for each edge e (otherwise we could delete e).

Now there are two cases.

(i) There is a vertex v covered by every maximum-weighted matching. In this case, let $w' \in \mathbb{Z}_+^E$ arise from w by decreasing the weights of the edges containing v by one. Then $v_{w'} = v_w - 1$. Since $w'(E) < w(E)$, there are integer vectors $y' \geq 0$, $z \geq 0$ such that $y'A + zB \geq w'$ and $|y'| + zf \leq v_{w'}$. Increasing y'_v by one gives $y \geq 0$, $z \geq 0$ such that $yA + zB \geq w$ and $|y| + zf \leq v_w$.

(ii) No vertex is covered by every maximum-weighted matching. Now let w' arise from w by decreasing all weights by one. We show that $v_{w'} \leq v_w - \lfloor \frac{1}{2}|V| \rfloor$. This will imply the theorem: since $w'(E) < w(E)$ there are integer vectors $y \geq 0$, $z' \geq 0$ such that $yA + z'B \geq w'$ and $|y| + z'f \leq v_{w'}$; increasing z'_V by one gives integer vectors $y \geq 0$, $z \geq 0$ such that $yA + zB \geq w$ and $|y| + zf \leq v_w$.

Assume $v_w < v_{w'} + \lfloor \frac{1}{2}|V| \rfloor$, and let M be a matching with $v_{w'} = w'(M)$, such that $w(M)$ is as large as possible. Then M leaves at least two vertices in V uncovered, since otherwise $v_w \geq w(M) \geq w'(M) + \lfloor \frac{1}{2}|V| \rfloor = v_{w'} + \lfloor \frac{1}{2}|V| \rfloor$.

Let u and v be not covered by M, and suppose we have chosen M, u and v such that the distance $d(u,v)$ is as small as possible. Then $d(u,v) > 1$, since otherwise we could augment M by $\{u,v\}$, thereby increasing $w(M)$. Let t be an internal vertex of the shortest path between u and v. Let M' be a matching with $w(M') = v_w$ not covering t.

Now $M \cup M'$ is a disjoint union of paths and circuits. Let P be the set of edges of the component of $M \cup M'$ containing t. Then P forms a path starting in t and not covering both u and v (as t, u and v have degree at most one in $M \cup M'$). Say P does not cover u. Now the symmetric difference $M \Delta P$ is a matching with $|M \Delta P| \leq |M|$, and therefore:

(5) $\quad w'(M \Delta P) - w'(M) = w(M \Delta P) - |M \Delta P| - w(M) + |M|$

$\quad \geq w(M \Delta P) - w(M) = w(M') - w(M' \Delta P) \geq 0$.

Hence $v_{w'} = w'(M \Delta P)$ and $w(M \Delta P) \geq w(M)$. However, $M \Delta P$ does not cover t and u, and $d(u,t) < d(u,v)$, contradicting our choice of M, u, v. $\square$

Edmonds' weighted matching algorithm can be easily modified to find minimum (or maximum) weighted *perfect* matchings in polynomial time. Moreover, the characterization (3) of the matching polytope directly gives that the *perfect matching polytope* (being the convex hull of the perfect matchings) is determined by:

(6) $\quad$ (i) $x(e) \geq 0 \quad (e \in E)$,

$\qquad$ (ii) $x(\delta(v)) = 1 \quad (v \in V)$,

$\qquad$ (iii) $x(\delta(V')) \geq 1 \quad (V' \subseteq V, |V'| \text{ odd})$.

Indeed, adding $x(E) = \frac{1}{2}|V|$ to (3) is equivalent to (6).

With LP-duality this gives a min-max relation for the minimum and for the maximum weight of a perfect matching. It follows from Theorem 12 that the system (6) is totally dual *half*-integral. The graph K_4 shows that (6) generally is not totally dual integral. Multiplying (6) (iii) by $\frac{1}{2}$ would make the system totally dual integral.

The polynomial solvability of the weighted (perfect) matching problem also follows with the ellipsoid method: Padberg and Rao [1982] showed that the inequalities (6) can be checked in polynomial time (by reduction to a series of minimum-cut problems), and hence the ellipsoid method gives polynomial solvability.

Similar results hold for weighted edge-covers.

b-**Matchings.** The results described above are of a certain self-refining nature. Given an undirected graph $G = (V, E)$ and a function $b: V \to \mathbb{Z}_+$, a *b-matching* is a vector $x: E \to \mathbb{Z}_+$ such that $x(\delta(v)) \leq b(v)$ for all v in V. If we have equality here for each v, the *b*-matching is called *perfect*. So if $b \equiv 1$, *b*-matchings reduce to ordinary matchings.

In case $b \equiv 2$, we obtain 2-matchings, which may be considered as a collection of vertex-disjoint paths and circuits (corresponding to edges e with $x(e) = 1$), and edges (corresponding to edges e with $x(e) = 2$). Define a *2-vertex-cover* as a function $y: V \to \mathbb{Z}_+$ such that $y(u) + y(v) \geq 2$ for each edge $\{u, v\}$ of G, and define the *size* of a vector to be the sum of its components. Now Tutte [1952, 1953] showed:

(7)   in any undirected graph, the maximum size of a 2-matching is equal to the minimum size of a 2-vertex-cover.

This can be derived from König's matching theorem for bipartite graphs (Theorem 2) by replacing each vertex v of G by two new vertices v' and v'', and each edge $\{u, v\}$ by two new edges $\{u', v''\}$ and $\{u'', v'\}$, and by applying König's theorem to the bipartite graph thus obtained.

One may derive similarly a min-max relation for "2-edge-covers" and "2-cocliques", and an associated Gallai-type theorem – see Lovász [1975].

The case of arbitrary b can be reduced to the Tutte-Berge theorem, which gives that the maximum size of a *b*-matching is equal to

(8) $$\min_{V' \subseteq V} \frac{b(V) + b(V') - \beta(V \setminus V')}{2},$$

where $\beta(V \setminus V')$ denotes the following: let I be the set of isolated vertices in the graph $\langle V \setminus V' \rangle$ induced by $V \setminus V'$, and let t be the number of components C of the graph $\langle V \setminus (V' \cup I) \rangle$ with $b(C)$ odd; then $\beta(V \setminus V') := b(I) + t$ (cf. Tutte [1952]).

To derive this formula from Theorem 11, split each vertex v of G into $b(v)$ new vertices, and replace any edge $\{u, v\}$ of G by $b(u)b(v)$ new edges connecting the new vertices corresponding to u and those corresponding to v. Next apply Theorem 11 to the new graph. This construction was given by Tutte [1954]. The min-max relation implies the original characterization of Tutte [1952] of graphs with a perfect *b*-matching.

Edmonds and Pulleyblank (cf. Pulleyblank [1973]) showed that the *b-matching polytope* (being the convex hull of the *b*-matchings) is determined by the inequalities

(9)   (i) $x(e) \geq 0$ $(e \in E)$,

             (ii) $x(\delta(v)) \leq b(v)$ $(v \in V)$,

             (iii) $x(\langle V' \rangle) \leq \lfloor \frac{1}{2} b(V') \rfloor$ $(V' \subseteq V)$.

This can be derived from characterization (3) of the matching polytope by the same method of splitting of vertices. Clearly, characterization (9) gives a min-max relation for the weighted *b*-matching problem. Pulleyblank [1980] strengthened this relation by showing that (9) is totally dual integral.

The method of splitting vertices described above does not carry over the polynomial solvability from the matching problem to the *b*-matching problem: the size of the *b*-matching problem is proportional to $\sum_{v \in V} \log b(v)$, while splittings create a graph of size proportional to $\sum_{v \in V} b(v)$. Such a splitting preserves polynomiality if $b(v)$ is bounded, e.g., in the case of 2-matchings.

Cunningham and Marsh (cf. Marsh [1979]) showed that also the general maximum weighted *b*-matching problem can be solved in polynomial time. Again, the polynomial solvability can be derived alternatively from the characterization (9) with the ellipsoid method – see Padberg and Rao [1982].

Perfect *b*-matchings can be treated similarly. It follows from (9) that the *perfect b-matching polytope* (being the convex hull of the perfect *b*-matchings) is determined by

(10)   (i) $x(e) \geq 0$ $(e \in E)$,

            (ii) $x(\delta(v)) = b(v)$ $(v \in V)$,

            (iii) $x(\delta(V')) \geq 1$ $(V' \subseteq V, b(V') \text{ odd})$.

So this gives a min-max relation for the weighted perfect *b*-matching problem. Generally (10) is not totally dual integral. However, replacing (10) (iii) by (9) (iii) makes (10) totally dual integral.

In particular, the perfect 2-matching polytope is determined by:

(11)   (i) $x(e) \geq 0$ $(e \in E)$,

               (ii) $x(\delta(v)) = 2$ $(v \in V)$,

and this system is totally dual integral, after adding the constraints $x(\langle V' \rangle) \leq |V'|$ for $V' \subseteq V$. A perfect 2-matching may be considered as a vertex-disjoint union of circuits and "double" edges, covering V.

The minimum weight perfect 2-matching problem is sometimes used as a relaxation for the traveling salesman problem (cf. Balas and Christofides [1981]). Here one would like to dispose of small circuits in the perfect 2-matching (in fact, of all circuits smaller than $|V|$). While removing "double" edges gives more problems (see below), Cornuéjols and Pulleyblank [1980] showed that just adding to (11) the constraints

(12) $x(\{u, v\}) + x(\{v, w\}) + x(\{u, w\}) \leq 2$

for all triples u, v, w gives the convex hull of the perfect 2-matchings without triangles. Also total dual integrality is maintained if we add again $x(\langle V'\rangle) \leq |V'| (V' \subseteq V)$ - cf. Cook and Pulleyblank [1982].

Capacitated b-matchings. Frequently when considering b-matching problems we wish that the b-matchings contain certain edges not too often, e.g. when using 2-matchings as relaxation for traveling salesman routes. It turns out that min-max results for such *capacitated b-matchings* can be derived from those for uncapacitated b-matchings given above, by an elementary construction given by Tutte [1954]. This shows again the self-refining nature of matching theory.

We first consider the 2-matching case. Let $G = (V, E)$ be an undirected graph. Call a 2-matching x *simple* if $x \leq 1$, i.e., if x is a $\{0, 1\}$-vector with $x(\delta(v)) \leq 2$ for each vertex v. A simple 2-matching can be identified with the set of edges e with $x(e) = 1$, being a union of pairwise vertex-disjoint circuits and simple paths.

Now Edmonds [1965d] showed the following min-max relation for simple 2-matchings.

Theorem 13. *Let $G = (V, E)$ be an undirected graph. The maximum size of a simple 2-matching is equal to*

(13) $$\min_{V' \subseteq V} |V| + |V'| - \tfrac{1}{2}(|K| + \kappa),$$

where K is the set of vertices having degree at most one in $\langle V \setminus V' \rangle$ and where κ denotes the number of components of $\langle V \setminus V' \rangle$ containing an odd number of vertices in K.

Proof. The theorem follows from formula (8) for the maximum size of an uncapacitated b-matching. Indeed, add for each edge $e = \{u, v\}$ of G two new vertices u_e and v_e, and replace e by the new edges $\{u, u_e\}, \{u_e, v_e\}, \{v_e, v\}$. Define $b(v) = 2$ if v is an "old" vertex, and $b(v) = 1$ if v is a "new" vertex. Then the maximum size of a b-matching in the new graph is exactly $|E|$ more than the maximum size of a simple 2-matching in the original graph. Applying (8) to the new graph gives the theorem. □

The construction described here occurs in Tutte [1954]. The theorem also presents a good characterization for the existence of a perfect simple 2-matching (cf. Belck [1950], Tutte [1952]). The construction given in this proof, together with the splitting technique of vertices for 2-matchings described before, also reduces algorithmically (in polynomial time) the maximum weighted simple 2-matching problem to the maximum weighted matching problem.

Similarly polyhedral results follow. E.g., the *perfect simple 2-matching polytope* (being the convex hull of the perfect simple 2-matchings) is determined by:

(14) $\quad 0 \leq x(e) \leq 1 \qquad (e \in E),$

$\qquad\quad x(\delta(v)) = 2 \qquad (v \in V),$

$\qquad\quad x(\delta(V')) + |E'| - 2x(E') \geq 1 \quad (V' \subseteq V, E' \subseteq \delta(V'), |E'| \text{ odd}).$

General capacitated b-matchings can be treated similarly. Given an undirected graph $G=(V,E)$, and functions $b\colon V\to \mathbb{Z}_+$ and $c\colon E\to \mathbb{Z}_+$, one derives a min-max relation for the maximum size or weight of a b-matching x *subject to c* (i.e., $x\leq c$) again with Tutte's construction. Replace any edge $e=\{u,v\}$ by a path $\{u,u_e\}$, $\{u_e,v_e\}$, $\{v_e,v\}$, and define $\tilde{b}(v)=b(v)$ for the old vertices v, and $\tilde{b}(u_e)=\tilde{b}(v_e)=c(e)$ for the new vertices. Then the maximum size of a $\tilde{b}$-matching in the new graph is exactly $c(E)$ more than the maximum size of a b-matching subject to c in the original graph. Applying (8) gives a min-max relation for capacitated b-matching.

Similarly, a good characterization for the existence of perfect capacitated b-matchings, the weighted case, and the polynomial solvability can be reduced to the uncapacitated case (see Belck [1950], Tutte [1952, 1954, 1981], Edmonds and Johnson [1973], Marsh [1979]).

In this way it follows that the convex hull of the b-matchings subject to c is given by:

(15) $\quad 0\leq x(e)\leq c(e) \hspace{3cm} (e\in E),$

$\quad x(\delta(v))\leq b(v) \hspace{3cm} (v\in V),$

$\quad x(\langle V'\rangle)+x(E')\leq \lfloor \tfrac{1}{2}(b(V')+c(E'))\rfloor \quad (V'\subseteq V, E'\subseteq \delta(V')),$

and moreover, that this system is totally dual integral.

Similarly, or alternatively from (15), one has that the convex hull of the *perfect b-matchings* subject to c is given by:

(16) $\quad 0\leq x(e)\leq c(e) \hspace{3cm} (e\in E),$

$\quad x(\delta(v))=b(v) \hspace{3cm} (v\in V),$

$\quad x(\langle V'\rangle)+c(E')-2x(E')\geq 1 \quad (V'\subseteq V, E'\subseteq \delta(V'), b(V')+c(E') \text{ odd}).$

Padberg and Rao [1982] showed that the constraints (16) can be checked in polynomial time. Hence the ellipsoid method also gives the polynomial solvability of weighted capacitated b-matching problems.

Lower and upper bounds. Also lower bounds can be fetched into the min-max framework. Let $G=(V,E)$ be an undirected graph, and let $b_1, b_2\colon V\to \mathbb{Z}_+$. Then the convex hull of the functions $x\colon E\to \mathbb{Z}_+$ satisfying

(17) $\hspace{3cm} b_1(v)\leq x(\delta(v))\leq b_2(v) \quad (v\in V)$

is given by the constraints

(18) (i) $x(e)\geq 0 \hspace{5cm} (e\in E),$

(ii) $b_1(v)\leq x(\delta(v))\leq b_2(v) \hspace{3cm} (v\in V),$

(iii) $x(\langle V'\rangle)-x(\langle V''\rangle)-x(\delta(V'')\setminus \delta(V'))\leq \lfloor\tfrac{1}{2}(b_2(V')-b_1(V''))\rfloor$

$\hspace{6cm} (V', V''\subseteq V, V'\cap V''=\emptyset)$

(cf. Schrijver and Seymour [1977]).

II. Min-Max Results in Combinatorial Optimization

This may be reduced to capacitated b-matchings as follows. Let for each vertex v of G, v' be a new vertex, and add to G the new edges $\{v, v'\}$ for $v \in V$, and $\{u', v'\}$ for all $u, v \in V$. Let $b(v) = b(v') = b_2(v)$ for $v \in V$, and let $c(\{v, v'\}) = b_2(v) - b_1(v)$ for $v \in V$, and $c(e) = \infty$ (or very large) for all other edges of the new graph. Then $x: E \to \mathbb{Z}_+$ satisfies (17), if and only if it is a projection of a perfect b-matching subject to c in the extended graph. In this way (18) follows from (15).

Similarly good characterizations, the polynomial solvability, and the total dual integrality of (18) follows.

We can add capacities on the edges: like in the case of capacitated b-matchings we can dispose of them by splitting edges to paths of length three. This construction gives a good characterization of Lovász [1970] for the existence of subgraphs with valencies obeying certain prescribed lower and upper bounds.

This also implies a min-max relation for the maximum "weight" $\sum_e w(e) x(e)$ of a (not-necessarily nonnegative) function $x: E \to \mathbb{Z}$ satisfying

(19)
$$c_1(e) \leq x(e) \leq c_2(e) \quad (e \in E),$$
$$b_1(v) \leq x(\delta(v)) \leq b_2(v) \quad (v \in V),$$

where $c_1, c_2: E \to \mathbb{Z}$ and $b_1, b_2: V \to \mathbb{Z}$. Indeed, this can be reduced to the case $c_1 = 0$, as replacing $c_1(e)$ by 0, $c_2(e)$ by $c_2(e) - c_1(e)$, $b_1(v)$ by $\max\{0, b_1(v) - c_1(\delta(v))\}$ and $b_2(v)$ by $\max\{0, b_2(v) - c_1(\delta(v))\}$, does not change the essence of the problem.

This last problem is the undirected analogue of the general transshipment problem (Section 5, (10)). In fact, Edmonds [1967] and Edmonds and Johnson [1973] mixed the directed and undirected case to one problem on "bidirected flows". Let A be a $\{0, \pm 1\}$-matrix, where each column has exactly two nonzero entries. Then for all vectors c_1, c_2, b_1, b_2, w the integer LP-problem:

(20)
$$\text{maximize } wx$$
$$\text{subject to } c_1 \leq x \leq c_2,$$
$$b_1 \leq Ax \leq b_2,$$
$$x \text{ integer,}$$

has a good characterization, and can in fact be solved in polynomial time.

Indeed, the matrix A can be considered as the incidence matrix of a "mixed" graph, where each edge has either two heads, or two tails, or a head and a tail. Clearly, if A has no entries -1, (20) reduces to (19), while if each column of A has both a $+1$ and a -1, we have (10) of Section 5.

Problem (20) can be reduced to (19) as follows (cf. Edmonds [1967]). Let $A = (a_{ij})$ have n rows and m columns. Make an undirected graph with vertices $v_1^-, \ldots, v_n^-$, $v_1^+, \ldots, v_n^+$, and edges $e_1, \ldots, e_m$ with $v_i^\pm \in e_j$ iff $a_{ij} = \pm 1$, (for $i = 1, \ldots, n$; $j = 1, \ldots, n$; $\pm = +, -$), and moreover edges $\{v_i^-, v_i^+\}$ (for $i = 1, \ldots, n$). Define

(21)
$$\tilde{b}_1(v_i^+) := (b_1)_i \text{ and } \tilde{b}_2(v_i^+) := (b_2)_i \qquad (i = 1, \ldots, n),$$
$$\tilde{b}_1(v_i^-) := \tilde{b}_2(v_i^-) := 0 \qquad (i = 1, \ldots, n),$$

$$\tilde{c}_1(e_j):=(c_1)_j \text{ and } \tilde{c}_2(e_j):=(c_2)_j \qquad (j=1,\ldots,m),$$
$$\tilde{c}_1(\{v_i^-,v_i^+\}):=-\infty \text{ and } \tilde{c}_2(\{v_i^-,v_i^+\}):=\infty \qquad (i=1,\ldots,n).$$

Then problem (20) is equivalent to finding a maximum weighted integer solution for (19) (after adding $\sim$). In this way a good characterization and the polynomial solvability follows. Note that this construction generalizes the one described in Remark 1 of Section 5, reducing network flow problems to bipartite matching.

T-joins and T-cuts. We are not yet at the end of our self-refining trip. It turns out that also certain parity constraints can be added. We first consider the case of T-joins and T-cuts.

Let $G=(V,E)$ be an undirected graph, and let $T\subseteq V$ be such that $|T|$ is even. A *T-join* is a set E' of edges such that T is exactly the set of vertices in the graph (V,E') of odd valency. A *T-cut* is a set of edges of the form $\delta(V')$ with $V'\subseteq V$ and $|V'\cap T|$ odd. It is immediate that each T-join intersects each T-cut. In fact, the minimal T-joins are exactly the minimal sets of edges intersecting all T-cuts, and vice versa (minimal under inclusion). Note that minimal T-joins do not contain circuits, and always form the edge-disjoint union of a collection of $\frac{1}{2}|T|$ paths, containing each vertex in T exactly once as an end point.

Edmonds and Johnson [1973] proved the following.

Theorem 14. *The minimum size of a T-join is equal to half of the maximum number t of T-cuts $C_1,\ldots,C_t$ (repetition allowed) such that no edge is contained in more than two of the C_i.*

Proof. This may be derived from the minimum weight perfect matching problem, and the total dual half-integrality of (6), as follows. Replace any edge $e=\{u,v\}$ of G by the new edge $e'=\{u_e,v_e\}$, where the new edges are pairwise disjoint. Next, for each v in V, add edges $\{v_e,v_f\}$ for all pairs of edges e,f containing v. Moreover, if $v\in T$ and has even degree, or if $v\notin T$ and has odd degree, add a new vertex v_0, and edges $\{v_0,v_e\}$ for all edges e containing v. So vertices v in T correspond to a clique of odd size in the new graph, and the other vertices to a clique of even size.

Define weights as follows. For e in E, $w(e'):=1$, and for the other edges the weight is 0. Then the minimum weight of a perfect matching in the new graph is equal to the minimum size of a T-join in G, and (6) applies. □

Similarly, the weighted case, and the algorithmic problem can be reduced to perfect matchings. Note that Theorem 14 itself is of a self-refining nature: the weighted case follows by replacing each edge e by a path of length equal to its weight (with the same end points as e).

It similarly follows that the convex hull of the subsets of E containing a T-join (i.e., intersecting all T-cuts) is determined by:

(22)
$$0 \leqslant x(e) \leqslant 1 \quad (e\in E),$$
$$x(E') \geqslant 1 \quad (E' \text{ T-cut}),$$

and that (22) is totally dual half-integral.

In general (22) is not totally dual integral, as is shown by taking $G=K_4$ and $T=V$. This is essentially the only counterexample, as Seymour [1977] showed that (22) is totally dual integral, if and only if V cannot be partitioned into subsets V_1, V_2, V_3, V_4, each intersecting T in an odd number of vertices, such that each V_i and each union of any two of the V_i induce a connected subgraph of G. Besides, Seymour [1979a] showed that if G is bipartite then the minimum size of a T-join is equal to the maximum number of pairwise disjoint T-cuts. This strengthens Theorem 14, and actually implies it as a special case.

Theorem 14 contains the following special cases. If $|T|=2$, minimal T-joins are exactly minimal paths. If $T=V$, and we add a large number to all weights, the minimum weight T-join problem is equivalent to the minimum weight perfect matching problem. If we take T to be the set of all vertices of odd degree in G, T-joins are exactly those sets of edges which make G eulerian by "doubling" them. That is, the minimum weight of a T-join is equal to the minimum weight of edges which should be traversed twice in order to make a tour through the graph traversing all edges at least once. Thus, it gives a min-max relation for the *Chinese postman problem*.

By the theory of blocking polyhedra (cf. Section 3), the characterization (22) implies that the convex hull of the sets of edges containing a T-cut (i.e., intersecting all T-joins) is determined by

(23)
$$0 \leqslant x(e) \leqslant 1 \quad (e \in E),$$
$$x(E') \geqslant 1 \quad (E' \text{ } T\text{-join}).$$

So this yields a min-max relation for the minimum capacity of a T-cut, which contains, e.g., for $|T|=2$ the fractional, undirected version of the max-flow min-cut theorem. Padberg and Rao [1982] showed that minimum-capacitated T-cuts can be found in polynomial time.

In general, (23) is not totally dual integral, even not totally dual half integral (cf. Seymour [1979a]). Seymour [1977] characterized pairs of graphs $G=(V, E)$ and subsets T of V for which (23) is totally dual integral.

Note that the construction given in the proof of Theorem 14 above also works in the weighted case, even if negative weights occur. Thus, also if negative weights occur, minimum weight T-joins are well characterized. This may be derived alternatively from the case of nonnegative weights as follows. Let $G=(V,E)$ be an undirected graph, and let $T \subseteq V$ with $|T|$ even. Let $w: E \to \mathbb{R}$ and let E_0 be the set of edges of negative weight. Let T_0 be the set of vertices of odd degree in the graph (V, E_0). Then a subset E' of E is a T-join, if and only if $E' \Delta E_0$ is a $T \Delta T_0$-join (Δ denoting symmetric difference). Moreover, let $\tilde{w}(e) := |w(e)|$ for e in E. Then $\tilde{w}(E') = w(E' \Delta E_0) - w(E_0)$, and hence finding a T-join E' with $w(E')$ minimal, is equivalent to finding a $T \Delta T_0$-join E'' with $\tilde{w}(E'')$ minimal.

So in case $|T|=2$ we obtain a min-max relation and a polynomial algorithm for shortest paths in undirected graphs, where lengths may be negative, provided that all circuits have nonnegative length (cf. Remark 3 in Section 5).

One similarly derives that the convex hull of the T-joins is given by:

(24) $\quad 0 \leqslant x(e) \leqslant 1 \quad (e \in E),$

$\quad x(\delta(V')) + |E'| - 2x(E') \geqslant 1 \quad (V' \subseteq V, E' \subseteq \delta(V'), |V' \cap T| + |E'| \text{ odd}).$

In case $T = \emptyset$ we obtain the convex hull of eulerian subgraphs.

Finally, by combining constructions like those describe in this section, one may derive a min-max relation, and a polynomial algorithm, for the following more general problem:

(25) $\qquad$ maximize $\sum_{e \in E} w(e) x(e),$

$\qquad$ such that $c_1(e) \leqslant x(e) \leqslant c_2(e) \quad (e \in E),$

$\qquad\qquad\qquad b_1(v) \leqslant x(\delta(v)) \leqslant b_2(v) \quad (v \in V),$

$\qquad\qquad\qquad x(\delta(v)) \equiv 0 \pmod{2} \quad (v \in S),$

$\qquad\qquad\qquad x(\delta(v)) \equiv 1 \pmod{2} \quad (v \in T),$

$\qquad\qquad\qquad x(e) \text{ integer} \quad (e \in E).$

where $G = (V, E)$ is an undirected graph, T and S are disjoint subsets of V, and $w, c_1, c_2 \colon E \to \mathbb{Z} \cup \{\pm \infty\}$ and $b_1, b_2 \colon V \to \mathbb{Z} \cup \{\pm \infty\}$.

More generally, one derives similarly from the weighted matching problem a min-max relation for the integer LP-problem

(26) $\qquad$ maximize wx

$\qquad$ subject to $c_1 \leqslant x \leqslant c_2,$

$\qquad\qquad\qquad b_1 \leqslant Ax \leqslant b_2,$

$\qquad\qquad\qquad x$ integer,

where A is a matrix with entries from $\{0, \pm 1, \pm 2\}$, such that the sum of the absolute values of the entries in any column is at most 2, and where w, c_1, c_2, b_1, b_2 are arbitrary vectors (of size compatible for (26)) – see Edmonds and Johnson [1970, 1973]. Similarly, the polynomial solvability of (26) follows. Problem (26) contains most of the problems treated before.

A min-max relation for (26) can be described as follows. Without loss of generality let c_1, c_2, b_1, b_2 be integer. Denote the inequality system given in (26), without the integrality constraint, by $Cx \leqslant d$. Then if we add to $Cx \leqslant d$ the inequalities

(27) $\qquad yC \leqslant \lfloor yd \rfloor \quad (y\{0, \tfrac{1}{2}\}\text{-vector with } yC \text{ integer}),$

we obtain the convex hull of the integer solutions of $Cx \leqslant d$, and we could leave out the integrality constraint. Adding (27) makes the system totally dual integral.

It is an open problem to characterize the matrices A enjoying this property (i.e., that for all c_1, c_2, b_1, b_2, the addition of (27) makes the system $c_1 \leqslant x \leqslant c_2$, $b_1 \leqslant Ax \leqslant b_2$ totally dual integral). Beside the matrices given above, also for any totally unimodular matrix B, the matrices $A = B$ and $A = 2B$ have this property. It is easy to see that any matrix with this property has entries only 0, ± 1, ± 2.

7. Multicommodity Flows

In Section 5 we studied network flows linking one "commodity", i.e., linking one pair of source and sink. Often, however, one is interested in *multicommodity flows*, linking several pairs of sources and sinks simultaneously (like in telephone networks). In general, such problems turned out to be hard (NP-complete), and no min-max relations could be derived. However, in some special cases polynomial algorithms and min-max relations have been found. These special cases all concern undirected graphs.

Here we represent a multicommodity flow as a packing of paths (which is for our purposes equivalent to the usual representation of multicommodity flows). Let $G = (V, E)$ be an undirected graph, and let $\{r_1, s_1\}, \ldots, \{r_k, s_k\}$ be different pairs of vertices of G. These pairs are called the *commodities*. Then a *multicommodity flow* is given by paths $P_1, \ldots, P_s$ and nonnegative numbers $\lambda_1, \ldots, \lambda_s$ such that, for $j = 1, \ldots, s$, P_j is an r_i-s_i-path for some $i = 1, \ldots, k$. Its *i-th value* is equal to the sum of those λ_j for which P_j is an r_i-s_i-path. The *total value* is the sum of these values, which equals $\lambda_1 + \ldots + \lambda_s$. If a capacity function $c: E \to \mathbb{R}_+$ is given, the multicommodity flow is said to be *subject to c* if for each edge e of G the sum of those λ_j for which P_j contains e is at most $c(e)$.

A *multicommodity cut* is a set of edges separating each of the pairs $\{r_1, s_1\}, \ldots, \{r_k, s_k\}$.

Now the following inequality always holds:

(1) given an undirected graph $G = (V, E)$, commodities $\{r_1, s_1\}, \ldots, \{r_k, s_k\}$, and a capacity function $c: E \to \mathbb{R}_+$, the maximum total value of a multicommodity flow subject to c, is *not more than* the minimum capacity of a multicommodity cut.

If $k = 1$ we have equality by the max-flow min-cut theorem. The graph $K_{1,3}$ (with commodities all pairs of vertices of valency one) shows that in general one does not have equality in (1). However, equality has been proved in the following special cases.

I. Two commodities. If there are only two commodities, (1) holds with equality. This is the content of Hu's two-commodity flow theorem [1963]. Hu also gave a polynomial algorithm finding an optimum multicommodity flow, thereby also showing that if the capacities are integer, there exists a half-integer optimum multicommodity flow (i.e., with the λ_j half-integer).

Rothschild and Whinston [1966a, 1966b] (cf. Lovász [1976b], Seymour [1979b]) showed that if $G = (V, E)$ is eulerian (i.e., all valencies are even), and $r_1, s_1, r_2, s_2 \in V$, then the maximum number of pairwise edge-disjoint paths, each connecting either r_1 and s_1 or r_2 and s_2, is equal to the minimum size of a cut separating both r_1 and s_1, and r_2 and s_2. It is not difficult to see that this generalizes Hu's result.

Seymour [1977] characterized undirected graphs $G=(V, E)$ and commodities $\{r_1, s_1\}, \{r_2, s_2\}$ for which there exists an integer optimum two-commodity flow for each integer capacity function on E. In general, finding an optimum integer two-commodity flow is NP-complete.

II. Commodities with a common source. If $r_1 = \ldots = r_k$, equality in (1) can be derived from the max-flow min-cut theorem: add a new vertex s, and add edges $\{s_1, s\}, \ldots, \{s_k, s\}$, with high capacity. It follows moreover that if all capacities are integer, there is an integer optimum multicommodity flow.

III. Kleitman, Martin-Löf, Rothschild and Whinston theorem. If the commodities form a union of complete bipartite graph, each two of which cover V, then (1) holds with equality. If all capacities are integer, there exists an integer optimum flow. This was proved by Kleitman, Martin-Löf, Rothschild and Whinston [1970].

In particular, in each of the cases I–III the convex hull of the multicommodity cuts is given by:

(2) $\qquad 0 \leq x(e) \leq 1 \quad (e \in E),$

$\qquad\qquad x(E') \geq 1 \quad (E'\ r_i\text{-}s_i\text{-path for some } i=1, \ldots, k).$

Actually, this system is totally dual half-integer for two commodities (while Seymour [1977] characterized when (2) is totally dual integral), and totally dual integral in cases II and III.

Remark 4. Beside the cases I–III, in some more general cases the following good characterization holds:

(3) given an undirected graph $G=(V, E)$, commodities $\{r_1, s_1\}, \ldots, \{r_k, s_k\}$, a capacity function $c: E \to \mathbb{R}_+$, and "requirements" $\rho_1, \ldots, \rho_k$, then there exists a multicommodity flow with i-th value at least ρ_i ($i=1, \ldots, k$), if and only if for all $V' \subseteq V$, the capacity of $\delta(V')$ is not less than the sum of the ρ_i for which r_i and s_i are separated by V'.

This statement is true in the following cases:
  (i) $|\{r_1, s_1, \ldots, r_k, s_k\}| \leq 4$ (Papernov [1976], Seymour [1980]);
  (ii) there is a pair $\{u, v\}$ of vertices intersecting each of $\{r_1, s_1\}, \ldots, \{r_k, s_k\}$ (Papernov [1976]);
  (iii) $\{r_1, s_1\}, \ldots, \{r_k, s_k\}$ form a circuit of length five (Lomonosov [1978]);
  (iv) G is planar, and $r_1, s_1, \ldots, r_k, s_k$ are on the boundary of the unbounded face (Okamura and Seymour [1981]);
  (v) the graph $(V, E \cup \{\{r_1, s_1\}, \ldots, \{r_k, s_k\}\})$ is planar (Seymour [1978b, 1981a]).

In each of these cases, if the capacities and requirements are integer, and if there exists a multicommodity flow satisfying the capacities and requirements, there exists a half-integer such flow.

Note that the graph $K_{1,3}$ (with commodities all pairs of end vertices) shows that none of the cases (i)–(v) in general has equality in (1). For an interesting characterization, see Lomonosov [1982].

If the commodities form a complete graph, there is a min-max relation for the maximum total value of an *integer* multicommodity flow, which relation was found by Mader [1978a], and which differs from the relation suggested by (1).

For an undirected graph $G=(V, E)$ and a subset S of V, a path is called an *S-path* if it has two distinct vertices in S as its end points.

Theorem 15 (Mader's theorem). *Let $S=\{s_1, \ldots, s_k\}$. Then the maximum number of pairwise edge-disjoint S-paths is equal to*

$$\text{(4)} \qquad \min \tfrac{1}{2}\left(\sum_{i=1}^{k} |\delta(S_i)| - \mathcal{O}'(V\setminus(S_1 \cup \ldots \cup S_k))\right),$$

where the minimum ranges over all collections of pairwise disjoint subsets $S_1, \ldots, S_k$ of V with $s_1 \in S_1, \ldots, s_k \in S_k$, and where $\mathcal{O}'(V\setminus(S_1 \cup \ldots \cup S_k))$ denotes the number of components C of the graph $\langle V\setminus(S_1 \cup \ldots \cup S_k)\rangle$ with $|\delta(C)|$ odd.

So if G is eulerian, the $\mathcal{O}'$-term in (4) disappears – see Lovász [1976b].

Mader [1978b] also gave a vertex-disjoint analogue: assuming (without loss of generality) that S is a coclique, *the maximum number of pairwise internally vertex-disjoint S-paths is equal to*

$$\text{(5)} \qquad \min |V_0| + \sum_{i=1}^{t} \lfloor \tfrac{1}{2}|V_i| \rfloor,$$

where the minimum ranges over all subsets $V_0, V_1, \ldots, V_t$ of V ($t \geq 0$) such that each S-path not intersecting V_0 contains at least two vertices in at least one of $V_1, \ldots, V_t$. This theorem contains the Tutte-Berge theorem (Theorem 11) as special case.

Finally, there is the "polar" problem of packing multicommodity *cuts*. By the theory of blocking polyhedra (cf. Section 3), from the characterization (2) it follows that, given an undirected graph $G=(V, E)$ and commodities $\{r_1, s_1\}, \ldots, \{r_k, s_k\}$, the convex hull of the subsets E' which connect at least one of the pairs $\{r_i, s_i\}$, is determined by:

$$\text{(6)} \qquad 0 \leq x(e) \leq 1 \quad (e \in E),$$
$$x(E') \geq 1 \quad (E' \text{ multicommodity cut}),$$

if we are in one of the cases I–III above.

Hu [1973] showed that this is not true in the general case. Moreover, Seymour [1978a] showed that in case there are only two commodities, (6) is totally dual half-integral. More generally, he showed: *if $G=(V, E)$ is a bipartite graph and r_1, s_1, r_2, s_2 are vertices of G, then the minimum length of a path connecting either r_1 and s_1, or r_2 and s_2, is equal to the maximum number of pairwise disjoint edge sets, each meeting every r_1-s_1-path and every r_2-s_2-path.*

Furthermore, Seymour [1977] characterized when (6) is totally dual integral for two commodities.

8. Arborescences and Directed Cuts

In this section we describe min-max relations which are centered around the concepts of arborescence, directed cut, and strong connector, and which again turn out to generalize those for bipartite graphs and network flows.

Arborescences and rooted cuts. Let $D = (V, A)$ be a directed graph, and let $r \in V$. An *r-arborescence* is a set A' of arcs of D forming a rooted directed spanning tree, with root r. That is, A' contains no circuit, and each vertex $s \neq r$ is entered by exactly one arc in A'. A *cut rooted at* r, or an *r-cut*, is a set of arcs of the form $\delta^-(V')$, with $\emptyset \neq V' \subseteq V \setminus r$.

It is immediate that each r-arborescence intersects each r-cut. Moreover, the minimal r-cuts are exactly the minimal sets intersecting all r-arborescences, and vice versa.

Fulkerson [1974] (cf. Edmonds [1967a]) proved the following min-max equation.

Theorem 16 (Fulkerson's optimum branching theorem). *Let $D = (V, A)$ be a directed graph, let $r \in V$ and let $l: A \to \mathbb{Z}_+$ be a length function. Then the minimum length of an r-arborescence is equal to the maximum number t of r-cuts $C_1, \ldots, C_t$ (repetition allowed) such that no arc a is in more than $l(a)$ of the C_i.*

Before we prove the theorem, observe the following. Let B be the matrix with columns indexed by A, and with rows the incidence vectors of all r-cuts. Then the theorem states that for any $l: A \to \mathbb{Z}_+$, the optima in

(1) $$\min\{lx \mid x \geq 0, Bx \geq 1\} = \max\{\mathbf{1}y \mid y \geq 0, yB \leq l\}$$

are attained by integer optimum solutions. By the theory of total dual integrality (cf. Section 3), it suffices to show that the maximum in (1) has an integer optimum solution.

Proof. Let y be an optimum solution for the maximum in (1), such that

(2) $$\sum_{\emptyset \neq V' \subseteq V \setminus r} y_{\delta^-(V')} \cdot |V'|^2$$

is as large as possible (such a y exists, by reason of compactness). Now let

(3) $$\mathscr{F} := \{V' \subseteq V \mid y_{\delta^-(V')} > 0\}.$$

Then $\mathscr{F}$ is *laminar*, i.e., if $U, W \in \mathscr{F}$ then $U \subseteq W$ or $W \subseteq U$ or $U \cap W = \emptyset$. For suppose to the contrary that $U \not\subseteq W \not\subseteq U$ and $U \cap W \neq \emptyset$. Let $0 < \varepsilon \leq \min\{y_{\delta^-(U)}, y_{\delta^-(W)}\}$. Let the vector y' be given by:

(4) $\quad y'_{\delta^-(U)} := y_{\delta^-(U)} - \varepsilon, \qquad y'_{\delta^-(W)} := y_{\delta^-(W)} - \varepsilon,$

$\quad y'_{\delta^-(U \cap W)} := y_{\delta^-(U \cap W)} + \varepsilon, \quad y'_{\delta^-(U \cup W)} := y_{\delta^-(U \cup W)} + \varepsilon,$

and let y' coincide with y in the other coordinates. Then $y' \geq 0$, $y'B \leq yB$, and $|y'| = |y|$, so y' again is an optimum solution in (1). However (2) is augmented, contradicting the maximality of (2).

Now let B_0 be the submatrix of B consisting of those rows of B corresponding to sets in $\mathscr{F}$. Then B_0 is totally unimodular (in fact, it is the incidence matrix of some directed graph). This can be seen with Ghouila-Houri's characterization of total unimodularity (cf. Section 3) as follows. The collection $\mathscr{F}$ can be partitioned into "levels" $\mathscr{F}_1, \mathscr{F}_2, \ldots$, by defining $\mathscr{F}_i$ inductively as the collection of maximal sets (under inclusion) in $\mathscr{F} \setminus (\mathscr{F}_1 \cup \ldots \cup \mathscr{F}_{i-1})$. By the laminarity of $\mathscr{F}$, each $\mathscr{F}_i$ consists of pairwise disjoint sets. Now let R_1 be the set of rows of B_0 corresponding to a set in $\mathscr{F}_1, \mathscr{F}_3, \mathscr{F}_5, \ldots$, and let R_2 be the set of the other rows of B_0. Then it is easy to see that the sum of the rows in R_1, minus the sum of the rows in R_2, is a $\{0, +1, -1\}$-vector. As such a splitting can be made similarly for each submatrix of B_0, by Ghouila-Houri's result B_0 is totally unimodular.

Now we have:

(5) $\qquad \max\{|y_0| \,|\, y_0 \geq 0, y_0 B_0 \leq l\} = \max\{|y| \,|\, y \geq 0, y B \leq l\}$.

Indeed, the inequality $\leq$ is trivial (by extending y_0 with zeroes), while the inequality $\geq$ follows from the fact that the second maximum in (5) is attained by the vector y above, which has zeroes outside of B_0.

Since B_0 is totally unimodular, the first maximum, and hence also the second maximum, has an integer optimum solution. □

So although the constraints $x \geq 0, Bx \leq 1$ generally are not totally unimodular, integer optimum solutions are shown by proving that in the optimum the active constraints can be chosen to be totally unimodular. This method of proof is an example of a general technique for deriving total dual integrality – see Edmonds and Giles [1977] (or Section 11 below). It consists of showing first that the active constraints in the optimum can be chosen to be "nice" (e.g., "laminar" or "cross-free"), and next that nice constraint sets are totally unimodular, so that integer optimum solutions follow from Hoffman and Kruskal's theorem.

Fulkerson's optimum branching theorem can be equivalently stated as the system

(6) $\qquad 0 \leq x(a) \leq 1 \quad (a \in A),$

$\qquad \qquad x(A') \geq 1 \quad (A' \text{ r-cut}),$

being totally dual integral. In particular, (6) determines the convex hull of the sets of edges containing an r-arborescence.

Edmonds [1967a] and Fulkerson [1974] designed a polynomial algorithm for finding a shortest arborescence. The polynomial solvability also follows with the ellipsoid method, as (6) can be checked in polynomial time, although there are exponentially many r-cuts: considering a given $x \in \mathbb{R}_+^A$ as a capacity function, we can find an r-cut of minimum capacity (e.g., with Ford and Fulkerson's algorithm); if this minimum is less than 1 the r-cut yields a violated inequality in (6), and otherwise (6) is satisfied.

By the theory of blocking polyhedra (cf. Section 3) it follows that if we replace "r-cut" in (6) by "r-arborescence", we obtain the convex hull of sets of edges containing an r-cut. Also total dual integrality is maintained, as follows from the following theorem of Edmonds [1973] (see Lovász [1976a] for a short proof).

Theorem 17 (Edmonds' disjoint branching theorem). *Let $D=(V,A)$ be a directed graph, and let $r\in V$. Then the maximum number of pairwise disjoint r-arborescences is equal to the minimum size of an r-cut.*

The weighted case follows by replacing arcs by parallel arcs.

Note that the (easy) max-potential min-work theorem, and the max-flow min-cut theorem (Theorems 5 and 4) follow from Theorems 16 and 17, respectively, by adding arrows (s,v) for each v in V, of length zero or high capacity.

Giles [1982a, 1982b, 1982c] considered *matching forests*, which generalize matchings and arborescences.

Directed cuts and their coverings. Let $D=(V,A)$ be a directed graph. A *directed cut* is a set of arcs of the form $\delta^-(V')$, where $\emptyset \neq V' \neq V$ and $\delta^+(V')=\emptyset$. A *(directed cut) covering* is a set of arcs intersecting all directed cuts. It follows that a set A' of arcs is a covering, if and only if the contraction of the arcs in A' makes D strongly connected.

By a method similar to that of proving Fulkerson's optimum branching theorem ("uncrossing" cuts), one can show that the system

(7) $\qquad 0 \leqslant x(a) \leqslant 1 \quad (a \in A),$

$\qquad\qquad x(A') \geqslant 1 \quad (A' \text{ directed cut}),$

is totally dual integral. Hence the polyhedron determined by (7) has integer vertices, and is the convex hull of the directed cut coverings.

The total dual integrality of (7) is equivalent to the following theorem of Lucchesi and Younger [1978] (cf. Lovász [1976a]).

Theorem 18 (Lucchesi-Younger theorem). *Let $D=(V,A)$ be a directed graph. Then the minimum size of a directed cut covering is equal to the maximum number of pairwise disjoint directed cuts.*

The weighted version follows by replacing arcs by directed path. If D is a complete bipartite graph, with all arcs directed from one colour class to the other, this weighted version is equivalent to the weighted version of König's covering theorem ((8) in Section 4).

Since the system (7) can be checked in polynomial time (as it amounts to finding a directed cut of minimum capacity (taking x as capacity), which can be done with Ford and Fulkerson's minimum cut algorithm), also minimum length directed cut coverings can be found in polynomial time, with ellipsoids. For direct algorithms, see Karzanov [1979], Lucchesi [1976], Frank [1981].

It is conjectured by Edmonds and Giles [1977] that also the "polar" minmax relation holds, i.e., that in any directed graph, the minimum size of a directed cut is equal to the maximum number of pairwise disjoint directed cut coverings. This conjecture is still unsettled. However, in Schrijver [1980] a counterexample to the weighted version is given. So the system

(8) $\qquad 0 \leqslant x(a) \leqslant 1 \quad (a \in A),$

$\qquad\qquad x(A') \geqslant 1 \quad (A' \text{ directed cut covering})$

is in general not totally dual integral, although, by characterization (7) and the theory of blocking polyhedra (Section 3), (8) determines the convex hull of the sets of arcs containing a directed cut.

Feofiloff and Younger, and Schrijver [1982a] showed that Edmonds and Giles' conjecture, and its weighted version, are true in the case formulated in the following theorem.

Theorem 19. *Let $D=(V,A)$ be an acyclic directed graph, such that each pair of source and sink is connected by a directed path. Let $c: A \to \mathbb{Z}_+$ be a capacity function. Then the minimum capacity of a directed cut is equal to the maximum number t of directed cut coverings $C_1, \ldots, C_t$ such that no arc a is in more than $c(a)$ of the C_i.*

The weighted version is also true if D comes from a directed tree by adding all arcs (u, v) for which there exists an u-v-path. Moreover, the cardinality version is true if the minimum size of a directed cut is two (but not the weighted version).

Strong connectors. Let $D_0 = (V, A_0)$ be a directed graph, and let $D = (V, A)$ be a second directed graph. A subset A' of A is called a *strong connector* (*for D_0*) if the directed graph $(V, A_0 \cup A')$ is strongly connected. A subset A' of A is called a *cut* (*induced by D_0*) if $A' = \delta_A^-(V')$ for some nonempty proper subset V' of V with $\delta_{A_0}^-(V') = \emptyset$.

It is easy to see that a set of arcs of D is a strong connector, if and only if it intersects each cut induced by D_0. The following two theorem were shown in Schrijver [1982a].

Theorem 20. *Suppose D_0 is acyclic, and for all (u, v) in A there exist u', v' in V and directed paths in D_0 from u to u', from v' to u', and from v' to v. Let $l: A \to \mathbb{Z}_A$ be a length function. Then the minimum length of a strong connector for D_0 is equal to the maximum number t of cuts induced by D_0 such that no arc a is in more than $l(a)$ of these cuts.*

Theorem 21. *Suppose D_0 is acyclic, and each pair of source and sink of D_0 is connected by a directed path in D_0. Then the maximum number of pairwise disjoint strong connectors for D_0 is equal to the minimum size of a cut induced by D_0.*

A capacitated version of Theorem 21 follows by replacing arcs by parallel arcs.

Theorems 20 and 21 contain the following special cases.

(i) If $r, s \in V$, and $A_0 = \{(u, v) | u = s \text{ or } v = r\}$, then Theorem 20 is equivalent to the max-potential min-work theorem (Theorem 5), and Theorem 21 to Menger's theorem (Theorem 4).

(ii) If V is the disjoint union of V_1 and V_2, $A_0 = \{(u, v) | u \in V_1, v \in V_2\}$, and $A \subseteq \{(v, u) | u \in V_1, v \in V_2\}$, then Theorem 20 is equivalent to (the weighted version of) König's covering theorem (Corollary 2a and (8) in Section 4), while Theorem 21 is equivalent to Gupta's edge-colouring theorem (Corollary 3a).

(iii) If $r \in V$ and $A_0 = \{(v, r) | v \in V\}$, then Theorem 20 is equivalent to Fulkerson's optimum branching theorem (Theorem 16), and Theorem 21 is equivalent to Edmonds' disjoint branching theorem (Theorem 17).
(iv) If $A \subseteq \{(u, v) | (v, u) \in A_0\}$, Theorem 20 is equivalent to the Lucchesi-Younger theorem (Theorem 18), while Theorem 21 is equivalent to Theorem 19.

Again, Theorem 20 can be shown by an uncrossing technique for cuts, like in the proof of Fulkerson's optimum branching theorem.

Note that the conditions for D_0 and D given in Theorem 21 are more restrictive than those given in Theorem 20.

We leave it to the reader to formulate these theorems in terms of polyhedra and total dual integrality. With the theory of blocking polyhedra it will follow from Theorem 20, that under the conditions of Theorem 20 a min-max relation for *fractional* packings of strong connectors holds.

The following result (Schrijver [1983a]) generalizes Theorem 21 and part of Theorem 20. Let $\mathscr{C}$ be a family of nonempty proper subsets of the finite set V such that $V' \cap V'' \in \mathscr{C}$ and $V' \cup V'' \in \mathscr{C}$ whenever $V', V'' \in \mathscr{C}$ and $V' \cap V'' \neq \emptyset$ and $V' \cup V'' \neq V$ (i.e., $\mathscr{C}$ is a *crossing family*). For any directed graph $D = (V, A)$, call a set A' of arcs a *cut induced by* $\mathscr{C}$ if $A' = \delta_A^-(V')$ for some V' in $\mathscr{C}$. Call A' a *covering for* $\mathscr{C}$ if it intersects all cuts induced by $\mathscr{C}$, i.e., if $\delta_{A'}^-(V') \neq \emptyset$ for all $V' \in \mathscr{C}$. Now the following are equivalent:

(9) (i) for each directed graph $D = (V, A)$ the minimum size of a cut induced by $\mathscr{C}$ is equal to the maximum number of pairwise disjoint coverings for $\mathscr{C}$;
(ii) for each directed graph $D = (V, A)$ and for each length function $l: A \to \mathbb{Z}_+$, the minimum length of a covering for $\mathscr{C}$ is equal to the maximum number t of cuts $C_1, \ldots, C_t$ induced by $\mathscr{C}$ such that no arc a is in more than $l(a)$ of these cuts;
(iii) there are no V_1, V_2, V_3, V_4, V_5 in $\mathscr{C}$ such that: $V_1 \subseteq V_3 \subseteq V_5$, $V_1 \subseteq V_2$, $V_2 \cup V_3 = V$, $V_3 \cap V_4 = \emptyset$, $V_4 \subseteq V_5$.

As directed graphs may have parallel arcs, property (i) can be seen to be equivalent to its capacitated version.

9. Perfect Graphs

We now consider a class of undirected graphs which is defined by a min-max-relation. This class of *perfect graphs* was introduced by Berge [1961, 1962]. What makes the class interesting is that it is closed under taking complementary graphs, so that one min-max relation implies another. This was conjectured by Berge and proved by Lovász. For surveys on perfect graphs, see Golumbic [1980], Berge and Chvátal [1982], Lovász [1981b].

II. Min-Max Results in Combinatorial Optimization

Let $G=(V, E)$ be an undirected graph, and consider the following numbers:

(1) $\omega(G)$ = the *clique number* of G = the maximum size of a clique [a *clique* is a set of pairwise adjacent vertices];

$\gamma(G)$ = the *colouring number* of G = the minimum number of colours needed to colour the vertices of G such that no two adjacent vertices have the same colour (i.e., the minimum number of cocliques needed to cover V);

$\alpha(G)$ = the *coclique number* of G = the maximum size of a coclique [a *coclique* is a set of pairwise nonadjacent vertices];

$\bar{\gamma}(G)$ = the *clique cover number* of G = the minimum number of cliques needed to cover V.

The following relations are trivial: $\omega(G) \leqslant \gamma(G)$, $\alpha(G) \leqslant \bar{\gamma}(G)$, $\alpha(G) = \omega(\bar{G})$, and $\bar{\gamma}(G) = \gamma(\bar{G})$, where $\bar{G}$ denotes the complementary graph of G (which has vertex set V, two vertices being adjacent in $\bar{G}$ iff they are nonadjacent in G). The circuit on five vertices shows that the inequalities can be strict.

Now G is called *perfect* if $\omega(G') = \gamma(G')$ for each induced subgraph G' of G.

Implicitly we have met several classes of perfect graphs: (i) bipartite graphs (trivially); (ii) complements of bipartite graphs (by König's covering theorem (Corollary 2a)); (iii) line graphs of bipartite graphs (by König's matching theorem (Theorem 2)); (iv) complements of line graphs of bipartite graphs (by König's edge-colouring theorem (Theorem 3)); (v) *comparability graphs* (which, by definition, arise from a partially ordered set $(V, \leqslant)$, two vertices being adjacent iff they are comparable; the perfectness is easy – cf. Theorem 8); (vi) complements of comparability graphs (by Dilworth's theorem (Theorem 7)).

It was conjectured by Berge [1961, 1962] and proved by Lovász [1972] that the complement of each perfect graph is perfect again. This result, the *perfect graph theorem*, implies König's covering theorem (as (ii) are the complements of (i)), hence also König's matching theorem (by Gallai's theorem), and therefore also König's edge-colouring theorem (as (iv) are the complements of (iii)). Similarly it implies Dilworth's theorem.

More classes of perfect graphs will be discussed after the proof of the perfect graph theorem. We give a polyhedral proof of this theorem, due to Fulkerson [1972], Lovász [1972] and Chvátal [1975]. To this end, define for any undirected graph $G = (V, E)$, the *clique polytope* as the convex hull of the cliques in G. Clearly, any vector x in the clique polytope satisfies:

(2)   (i) $x(v) \geqslant 0$ $(v \in V)$,

     (ii) $x(S) \leqslant 1$ $(S \subseteq V, S$ coclique$)$.

The circuit on five vertices shows that generally the polyhedron (2) can be larger than the clique polytope. Chvátal [1975] showed that the clique polytope coincides with (2) if and only if G is perfect. This can be seen to imply the perfect graph theorem.

First observe the following. Let $Ax \leqslant 1$ denote the inequality system (2) (ii). So the rows of A are the incidence vectors of cocliques. Then it follows from the definition of perfectness that G is perfect if and only if the optima in

(3) $\qquad \max\{wx \mid x \geqslant 0, Ax \leqslant 1\} = \min\{|y| \mid y \geqslant 0, yA \geqslant w\}$

have integer optimum solutions, for each $\{0, 1\}$-vector w.

Theorem 22. *A graph G is perfect if and only if its clique polytope is determined by* (2).

Proof. I. First let G be perfect. For $w: V \to \mathbb{Z}_+$, let c_w denote the maximum weight of a clique. To prove that the clique polytope is given by (2), it suffices to show that

(4) $\qquad c_w = \max\{wx \mid x \geqslant 0, Ax \leqslant 1\}$

for each $w: V \to \mathbb{Z}_+$. This will be done by induction on $w(V)$.

If w is a $\{0, 1\}$-vector (4) follows from the note made on (3). So we may assume that $w(u) \geqslant 2$ for a certain vertex u. Let $e(u) = 1$, and $e(v) = 0$ if $v \neq u$. Replacing w by $w - e$ in (3) and (4), gives by induction a vector $y \geqslant 0$ such that $yA \geqslant w - e$ and $|y| = c_{w-e}$. Since $(w-e)(u) \geqslant 1$, there is a coclique S with $y_S > 0$ and $u \in S$. Let a be the incidence vector of S (we may assume $a \leqslant w - e$).

Then $c_{w-a} < c_w$. For suppose $c_{w-a} = c_w$. Let C be any clique with $(w-a)C = c_{w-a}$. Since $c_{w-a} = c_w$, $a(C) = 0$. On the other hand, since $w - a \leqslant w - e \leqslant w$, we know that $(w-e)C = c_{w-e}$, and hence, by complementary slackness, $a(C) > 0$.

Therefore

(5) $\qquad c_w = 1 + c_{w-a} = 1 + \max\{(w-a)x \mid x \geqslant 0, Ax \leqslant 1\} \geqslant \max\{wx \mid x \geqslant 0, Ax \leqslant 1\}$,

implying (4).

II. Conversely, suppose that the clique polytope of G is determined by (2), i.e., that the maximum in (3) is attained by a clique for each w. To show that G is perfect, it suffices to show that also the minimum in (3) has an integer optimum solution for each w in $\{0, 1\}^V$. This is shown by induction on $w(V)$.

Let $w \in \{0, 1\}^V$, and let y' be a (not-necessarily integer) optimum solution for the minimum in (3). Let S be a coclique with $y'_S > 0$, and let a be its incidence vector (we may assume that $a \leqslant w$). Then the common value of

(6) $\qquad \max\{(w-a)x \mid x \geqslant 0, Ax \leqslant 1\} = \min\{|y| \mid y \geqslant 0, yA \geqslant w - a\}$

is less than the common value of (3), as decreasing component y'_S of y' a little bit, keeps it feasible in (6). However, the values in (3) and (6) are integer (as by assumption the maxima have integer optimum solutions). Hence they differ by exactly one. Moreover, by induction the minimum in (6) has an integer optimum solution y^*. Increasing component y^*_S of y^* by one, gives an integer optimum solution in (3). □

The theory of anti-blocking polyhedra now gives directly the perfect graph theorem as a corollary.

Corollary 22a (perfect graph theorem). *The complement of a perfect graph is perfect again.*

Proof. If $G=(V,E)$ is perfect, by Theorem 22 the clique polytope P is defined by (2). Hence, by the theory of anti-blocking polyhedra (cf. Section 3), the coclique polytope of G, i.e., the clique polytope of $\bar{G}$, is defined by (2) after replacing "coclique" by "clique", i.e., by "coclique for $\bar{G}$". Applying Theorem 22 again gives that $\bar{G}$ is perfect. □

Stated otherwise, if G is an undirected graph, then $\omega(G')=\gamma(G')$ for each induced subgraph G', if and only if $\alpha(G')=\bar{\gamma}(G')$ for each induced subgraph G'.

The following classes of perfect graphs have been found.
 (i) Bipartite graphs (König [1932]).
 (ii) Line graphs of bipartite graphs (König [1916, 1931]).
 (iii) Comparability graphs (Dilworth [1950]).
 (iv) Chordal graphs (where each circuit of length at least four has a chord; Dirac [1961], Hajnal and Surányi [1958], Berge [1960]).
 (v) Meyniel graphs (where each circuit of length odd and at least five has at least two chords; Meyniel [1976]).
 (vi) Chvátal graphs (the edges of which can be oriented to an acyclic directed graph with no induced subgraph isomorphic to $(\{u,v,w,x\}, \{(u,v),(v,w),(x,w)\})$; Chvátal [1981]).
 (vii) Unimodular graphs (where the incidence matrix of all maximal cliques form a totally unimodular matrix; Hoffman and Kruskal [1956]).
 (viii) Cameron graphs (which arise from a directed graph $D=(V,A)$, and subset T of V intersecting each directed cycle in exactly one vertex, as follows: the vertex set is $V\setminus T$, two vertices being adjacent iff they occur together in some directed cycle of D; Cameron [1982] – cf. (11) in Section 5).
 (ix) Edmonds-Giles graphs (which are graphs G arising as follows: let $D=(V,A)$ be a directed graph, and let $\mathscr{C}$ be a collection of subsets of V, such that $\delta^+(V')=\emptyset$ for $V'\in\mathscr{C}$, and such that if $V',V''\in\mathscr{C}$ with $V'\cap V''\neq\emptyset$ and $V'\cup V''\neq V$ then $V'\cap V''\in\mathscr{C}$ and $V'\cup V''\in\mathscr{C}$; let G have vertex set A, two distinct arcs a' and a' of D being adjacent in G if and only if $a',a''\in\delta^-(V')$ for some $V'\in\mathscr{C}$; Edmonds and Giles [1977] – cf. Section 11).
 (x) Complements of the graphs above.

We leave it to the reader to deduce the several inclusions between these classes.

It is conjectured by Berge [1969] that a graph is perfect if and only if it has no induced subgraph isomorphic to an odd circuit of odd length at least five or its complement. This *strong perfect graph conjecture* is still unsettled.

Suppose that we wish to derive, with the ellipsoid method, from the characterization (2) of the clique polytope, a polynomial algorithm determining a maximum clique in a perfect graph. Then we must be able to check the system (2) in polynomial time. However, this is equivalent to finding a maximum weighted coclique in G. This is the same as a maximum weighted clique in the complementary graph, which is perfect again. Hence, we cannot use (2) to obtain a polynomial algorithm for the clique problem for perfect graphs in general.

However, there is another, non-polyhedral min-max relation for $\omega(G)$. In Lovász [1979] it is shown that if G is perfect, say with vertices $1, \ldots, n$, then:

(7) $\omega(G) = \max \{ \sum_{i,j=1}^{n} b_{ij} | B = (b_{ij})$ is a symmetric positive semi-definite $n \times n$-matrix with trace 1, and with $b_{ij} = 0$ if $i \neq j$ and i and j are not adjacent in $G \} =$
$= \min \{ \Lambda(C) | C = (c_{ij})$ is a symmetric $n \times n$-matrix, with $c_{ij} = 1$ if $i = j$ or i and j are adjacent in $G \}$,

where $\Lambda(C)$ denotes the largest eigenvalue of C. Since the largest eigenvalue of a matrix can be approximated fast enough, (7) gives a good characterization for the clique problem for perfect graphs. Moreover, since the conditions in the maximum in (7) can be checked in polynomial time, the maximum itself, and hence $\omega(G)$ can be determined in polynomial time, again with ellipsoids. Similarly the weighted case can be handled – see Grötschel, Lovász and Schrijver [1981, 1981a]. In Lovász [1981b] it is shown that G is perfect if and only if (7) holds for all induced subgraphs of G.

Other classes of graphs with a good characterization for $\omega(G)$ or $\alpha(G)$. Beside the perfect graphs there are some other classes of graphs for which the coclique or clique problem is well-characterized, or even polynomial solvable.

We saw already in Section 6 that the coclique number of the line graph of an undirected graph can be found in polynomial time, as this is the matching number of the original graph. (Clearly, also the clique number of a line graph can be computed in polynomial time.)

Minty [1980] and Sbihi [1978] showed more generally that in any claw-free graph, a maximum weighted coclique can be found in polynomial time. Here a graph is *claw-free* if it has no $K_{1,3}$ as an induced subgraph (e.g., line-graphs are claw-free). So this implies a good characterization for the (weighted) coclique number. However, no min-max relation for this number is known. Key problem here is characterizing (in terms of inequalities) the convex hull of the cocliques in a claw-free graph (cf. Giles and Trotter [1981]).

Other classes for which the weighted coclique number can be found in polynomial time are: circular arc graphs, circle arc graphs, and their complements (a *circular arc graph (circle graph,* respectively) has a collection of connected intervals (chords, respectively) of a circle as vertex set, two of them being adjacent iff they intersect (Gavril [1973, 1974])); graphs without "long" odd circuits (Hsu, Ikura and Nemhauser [1979]).

Another class of, not-necessarily perfect, graphs G with $\omega(G) = \gamma(G)$ was found by Györi [1981]: let A be a $\{0, 1\}$-matrix, such that the ones in each row form a connected interval; then the graph G has vertex set $\{(i, j) | a_{ij} = 1\}$, where two pairs (i, j) and (i', j') are adjacent iff $a_{ij'} = 0$ or $a_{i'j} = 0$.

Finally, Sbihi and Uhry [1982] defined a graph $G = (V, E)$ to be *h-perfect* if its coclique polytope (the convex hull of the cocliques) is determined by:

(8) $\qquad x(v) \geq 0 \qquad (v \in V),$
$\qquad x(C) \leq 1 \qquad (C \subseteq V, C \text{ clique}),$
$\qquad x(C) \leq \lfloor \tfrac{1}{2}|C| \rfloor \qquad (C \subseteq V, C \text{ odd circuit}).$

By Theorem 22, each perfect graph is h-perfect. Sbihi and Uhry described a constructive class of h-perfect graphs, including the "series-parallel graphs". Clearly, if G is h-perfect, $\alpha(G)$ has a good characterization.

10. Clutters and Blockers

Many of the min-max relations dealt with above allow an interpretation in terms of "hypergraphs" or "clutters". Interesting theorems for these more general structures, extending these min-max results, were found by Fulkerson [1970, 1971, 1972], Lehman [1979], Lovász [1972, 1975, 1977], and Seymour [1977, 1979a].

Let S be a finite set. A collection $\mathscr{C}$ of subsets of S is called a *clutter* if no two sets in $\mathscr{C}$ are contained in each other. The *blocker* $b(\mathscr{C})$ of $\mathscr{C}$ is the collection of minimal sets (under inclusion) intersecting all sets in $\mathscr{C}$. So $b(\mathscr{C})$ is a clutter again.

Edmonds and Fulkerson [1970] observed that for any clutter $\mathscr{C}$ the following duality phenomenon holds:

(1) $\qquad\qquad b(b(\mathscr{C})) = \mathscr{C}.$

In the Sections 4–8 we have met several examples of a clutter $\mathscr{C}$ together with its blocker $b(\mathscr{C})$:

	given:	$\mathscr{C}$ contains the minimal:	$b(\mathscr{C})$ contains the minimal:
(2)	$G = (V, E)$	"stars"	edge coverings
(3)	$D = (V, A), r, s \in V$	r-s-cuts	r-s-paths
(4)	$G = (V, E), T \subseteq V$	T-cuts	T-joins
(5)	$G = (V, E), r_1, s_1, r_2, s_2 \in V$	two-commodity cuts	r_1-s_1-paths and r_2-s_2-paths
(6)	$D = (V, A), r \in V$	r-cuts	r-arborescences
(7)	$D = (V, A)$	directed cuts	directed cut coverings
(8)	$D_0 = (V, A_0), D = (V, A)$	cuts induced by D_0	strong connectors for D_0

where G is an undirected graph, and D and D_0 are directed graphs. A *star* is a collection of edges of the form $\delta(v)$ for some vertex v.

For any clutter $\mathscr{C}$ on a set S, let $M_\mathscr{C}$ denote the matrix with rows all incidence vectors of sets in $\mathscr{C}$. So the rows and columns of $M_\mathscr{C}$ are indexed by $\mathscr{C}$ and S, respectively. For $w: S \to \mathbb{Z}_+$ consider the linear programming duality equation:

(9) $\qquad \min\{wx \mid x \geq 0, M_\mathscr{C} x \geq 1\} = \max\{\mathbf{1}y \mid y \geq 0, y M_\mathscr{C} \leq w\}.$

One immediately sees that if the minimum has an integer optimum solution x, such a vector x is the incidence vector of a set in $b(\mathscr{C})$.

Now $\mathscr{C}$ is said to have the $\mathbb{Q}_+$-*max-flow min-cut property*, or the $\mathbb{Q}_+$-*MFMC-property*, if the minimum in (9) has an integer optimum solution (being the incidence vector of a set in $b(\mathscr{C})$), for each nonnegative vector w. Equivalently, if:

(10) $\{x \geq 0 \mid M_{\mathscr{C}} x \geq 1\} = \{x \mid x \geq z \text{ for some convex combination } z \text{ of the rows of } M_{b(\mathscr{C})}\}$.

By the theory of blocking polyhedra (cf. Section 3), (10) is equivalent to the same equation with $\mathscr{C}$ and $b(\mathscr{C})$ interchanged:

(11) $\{x \geq 0 \mid M_{b(\mathscr{C})} x \geq 1\} = \{x \mid x \geq z \text{ for some convex combination } z \text{ of the rows of } M_{\mathscr{C}}\}$.

Therefore, the following result of Lehman [1979] holds:

(12) $\mathscr{C}$ has the $\mathbb{Q}_+$-MFMC-property, if and only if $b(\mathscr{C})$ has the $\mathbb{Q}_+$-MFMC-property.

So if the minimum in (9) has an integer optimum solution for each nonnegative vector w, the same holds if we replace $\mathscr{C}$ by $b(\mathscr{C})$. Below we shall see that this gives several equivalences between combinatorial min-max relations.

First consider the following more general property. Let $\mathbb{K} \subseteq \mathbb{Q}_+$. The clutter $\mathscr{C}$ is said to have the $\mathbb{K}$-*MFMC-property*, if the minimum in (9) has an integer optimum solution x, and if the maximum has an optimum solution in $\mathbb{K}^S$, for each nonnegative integer vector w. In particular, $\mathscr{C}$ has the $\mathbb{Z}_+$-MFMC-property, if and only if the system $x \geq 0$, $M_{\mathscr{C}} x \geq 1$ is totally dual integral. Note that if $\mathbb{K}' \supseteq \mathbb{K}$, then the $\mathbb{K}$-MFMC-property implies the $\mathbb{K}'$-MFMC-property.

It is the content of König's (weighted) edge-covering theorem (Corollary 2a and (8) in Section 4), that for bipartite graphs the clutter $\mathscr{C}$ in (2) has the $\mathbb{Z}_+$-MFMC-property. The fact that for bipartite graphs also $b(\mathscr{C})$ in (2) has the $\mathbb{Z}_+$-MFMC-property is equivalent to Gupta's theorem (Corollary 3a).

The clutters $\mathscr{C}$ and $b(\mathscr{C})$ in (3) both have the $\mathbb{Z}_+$-MFMC-property, which is the content of the max-potential min-work and the max-flow min-cut theorem (Theorem 5 and Corollary 4b), respectively.

The clutter $\mathscr{C}$ in (4) has the $\frac{1}{2}\mathbb{Z}_+$-MFMC-property (Theorem 14), hence $b(\mathscr{C})$ has the $\mathbb{Q}_+$-MFMC-property (and maybe the $\frac{1}{4}\mathbb{Z}_+$-MFMC-property, but generally not the $\frac{1}{2}\mathbb{Z}_+$-MFMC-property – see Seymour [1979a]).

The clutters $\mathscr{C}$ and $b(\mathscr{C})$ in (5) both have the $\frac{1}{2}\mathbb{Z}_+$-MFMC-property, which was shown by Seymour [1978a] and Hu [1963], respectively – see Section 7.

The clutters $\mathscr{C}$ and $b(\mathscr{C})$ in (6) both have the $\mathbb{Z}_+$-MFMC-property, which is the content of Fulkerson's optimum branching theorem and Edmonds' disjoint branching theorem (Theorems 16 and 17), respectively.

The clutter $\mathscr{C}$ in (7) has the $\mathbb{Z}_+$-MFMC-property, by the Lucchesi-Younger theorem (Theorem 18), hence $b(\mathscr{C})$ has the $\mathbb{Q}_+$-MFMC-property. Generally, $b(\mathscr{C})$ does not have the $\mathbb{Z}_+$-MFMC-property (see Schrijver [1980]). It has this property under the conditions formulated in Theorem 19.

The clutters $\mathscr{C}$ and $b(\mathscr{C})$ in (8) have the $\mathbb{Z}_+$-MFMC-property under the conditions formulated in Theorems 20 and 21, respectively.

Note that, roughly speaking, the ellipsoid method gives the following:

(13) *if the clutter $\mathscr{C}$ on S has the $\mathbb{Q}_+$-MFMC-property, then: there exists a polynomial algorithm finding a minimum weight set in $\mathscr{C}$, for each nonnegative weight function, if and only if there exists a polynomial algorithm finding a minimum weight set in $b(\mathscr{C})$, for each nonnegative weight function.*

Indeed, if the minimum in (9) has an integer optimum solution, then a minimum weight set in $b(\mathscr{C})$ can be found in polynomial time, if the system $x \geq 0$, $M_{\mathscr{C}} x \geq 1$ can be checked in polynomial time. Considering x as a weight function, this last amounts to finding a minimum weight set in $\mathscr{C}$.

In fact (13) could be stated more generally (and more precisely) in terms of *classes* of clutters, for which there exists a *fixed* polynomial algorithm finding minimum weight sets in the clutters (like Ford and Fulkerson's algorithm for all clutters $\mathscr{C}$ of r-s-cuts) – see Grötschel, Lovász and Schrijver [1981]. It is still an open question whether there exists a fixed polynomial algorithm finding minimum weight sets in any given clutter with the $\mathbb{Q}_+$-MFMC-property (which would make (13) trivial).

The problem of characterizing clutters with the $\mathbb{Z}_+$-, $\frac{1}{2}\mathbb{Z}_+$-, or $\mathbb{Q}_+$-MFMC-property is still unsettled for the greater part. Seymour [1977] showed that if $\mathscr{C}$ has the $\mathbb{K}$-MFMC-property, then any minor of $\mathscr{C}$ has the same property. Here a *minor* $\mathscr{C}'$ arises from $\mathscr{C}$ by taking disjoint subsets S_1 and S_2 of S, and defining $\mathscr{C}'$ to be the collection of minimal sets in

(14) $$\{S'\backslash S_1 \mid S' \in \mathscr{C}, S' \subseteq S\backslash S_2\}.$$

So finding all minimal clutters (under taking minors) not having the $\mathbb{K}$-MFMC-property would characterize this property. As Seymour pointed out, this may be a difficult problem.

Seymour showed that the clutter Q_6, defined by

(15) $$Q_6 := \{\{1, 3, 5\}, \{1, 4, 6\}, \{2, 3, 6\}, \{2, 4, 5\}\}$$

(which could be identified with the triangles in the complete graph K_4), is a minimal clutter without the $\mathbb{Z}_+$-MFMC-property. However, Q_6 has the $\mathbb{Q}_+$-MFMC-property, and $b(Q_6)$ has the $\mathbb{Z}_+$-MFMC-property. It was conjectured that if $\mathscr{C}$ has no Q_6 minor, and if $b(\mathscr{C})$ has the $\mathbb{Z}_+$-MFMC-property, then $\mathscr{C}$ has the $\mathbb{Z}_+$-MFMC-property. This is, however, contradicted by the example in Schrijver [1980].

Note that Q_6, and the clutters in (7), show that the $\mathbb{Z}_+$-MFMC-property is not invariant under taking blockers. Seymour [1978a] gave an example showing that also the $\frac{1}{2}\mathbb{Z}_+$-property is not invariant under taking blockers.

Seymour [1977] was able to characterize the $\mathbb{Z}_+$-MFMC-property for an interesting class of clutters. A clutter $\mathscr{C}$ on S is called *binary* if for all $S_1, \ldots, S_k$ in $\mathscr{C}$ with k odd, the set $S_1 \Delta \ldots \Delta S_k$ includes a set in $\mathscr{C}$. It is not difficult to prove that the blocker of a binary clutter is binary again.

Examples of binary clutters are those of r-s-paths, r-s-cuts, T-joins, T-cuts, two-commodity paths and two-commodity cuts ((3), (4), (5) above).

Now Seymour proved:

Theorem 23. *A binary clutter has the $\mathbb{Z}_+$-MFMC-property, if and only if it has no Q_6 minor.*

Applied to the examples mentioned above, this theorem characterizes when the $\mathbb{Z}_+$-MFMC-property holds, i.e., when the system $x \geq 0$, $M_\mathscr{C} x \leq 1$ is totally dual integral. Another class of binary clutters is obtained as follows: let $G = (V, E)$ be an undirected graph, let V be split into classes R, S, R', S', and let the clutter $\mathscr{C}$ consist of all paths (considered as edge sets) joining either R and S, or R' and S'. Then Seymour's theorem implies a special case of Kleitman, Martin-Löf, Rothschild and Whinston's theorem (see Section 7).

The problem of characterizing similarly the binary clutters with the $\mathbb{Q}_+$-MFMC-property seems to be hard – see Seymour [1977, 1979a].

For some related min-max results, see Lovász [1974, 1975, 1976a, 1977], Seymour [1981b], Schrijver and Seymour [1979].

One may ask for similar results for the anti-blocking case, where we consider the *anti-blocker* $a(\mathscr{C})$ of $\mathscr{C}$, defined as the collection of all maximal subsets of S intersecting no set in $\mathscr{C}$ in more than one element. In fact the theory here is much more streamlined than in the blocking case, and turns out to reduce completely to perfect graph theory. E.g., if $\mathscr{C}$ is a clutter, and the polytope

(16) $\qquad 0 \leq x(s) \leq 1 \quad (s \in S),$

$\qquad\qquad x(S') \leq 1 \quad (S' \in \mathscr{C}),$

has integer vertices (being the incidence vectors of sets in $a(\mathscr{C})$), then (16) is totally dual integral, and $\mathscr{C}$ is the collection of all maximal cliques in a perfect graph, while the polytope (16) is just the convex hull of its cocliques – see Fulkerson [1971, 1972], Lovász [1972, 1974], Padberg [1975].

The blocking and the anti-blocking case come together in the so-called *balanced* matrices, which are $\{0, 1\}$-matrices having no square submatrix of odd order with exactly two ones in each row and in each column. Berge [1970, 1972], Berge and Las Vergnas [1970], and Fulkerson, Hoffman and Oppenheim [1974] showed that if M is a balanced matrix, and w is an integer vector, then each of the optima

(17) $\qquad \max\{wx \mid x \geq 0, Mx \leq 1\} = \min\{|y| \mid y \geq 0, yM \geq w\},$

$\qquad\qquad \min\{wx \mid x \geq 0, Mx \geq 1\} = \max\{|y| \mid y \geq 0, yM \leq w\}$

have integer optimum solutions. Moreover, the columns of M can be split into k classes, such that each of the classes contains at least one 1 in each of the rows, where k is the minimum number of ones in any row. A similar result holds if we replace "at least" by "at most", and "minimum" by "maximum".

These results can be formulated equivalently in terms of total dual integrality, clutters, blockers, $\mathbb{Z}_+$-MFMC-property, anti-blockers, and so on, which we leave to the reader.

As the incidence matrix of a bipartite graph is balanced, these results extend the results of König, Egerváry, Gupta on bipartite graphs given in Section 4. More generally, each nonnegative totally unimodular matrix is balanced. Another class of balanced matrices are the *totally balanced* matrices, containing no incidence matrix of a circuit of lenght at least three as a submatrix - see Anstee and Farber [1982], Brouwer and Kolen [1980], Hoffman, Kolen and Sakarovitch [1982], and Lubiw [1982].

11. Matroids and Submodular Functions

The concept of matroid, introduced by Whitney [1935] and Van der Waerden [1937] as a framework for graph-theoretic and algebraic studies, turned out to play also a unifying role in combinatorial optimization. This was revealed by the work of Rado [1957] and Edmonds [1970]. In the latter paper also optimization problems for the more general structure of submodular functions were studied.

Here we review briefly the min-max relations for matroids and submodular functions. For a more comprehensive discussion, see Lovász's survey on "Submodular functions and convexity". The standard text on matroid theory is the book of Welsh [1976].

A *matroid* M is a pair $(S, \mathscr{I})$, where S is a finite set and $\mathscr{I}$ is a collection of subsets of S such that

(1)   (i) $\emptyset \in \mathscr{I}$,
      (ii) if $S'' \subseteq S' \in \mathscr{I}$ then $S'' \in \mathscr{I}$,
      (iii) if $S', S'' \in \mathscr{I}$ and $|S'| < |S''|$ then $S' \cup s \in \mathscr{I}$ for some $s \in S'' \setminus S'$.

The sets in $\mathscr{I}$ are called the *independent* sets of the matroid. The *bases* are the maximal independent sets. It follows from (1) (iii) that all bases have the same size. Associated with M is its *rank function* $r: \mathscr{P}(S) \to \mathbb{Z}$, defined by:

(2) $r(S') := \max\{|S''| \mid S'' \in \mathscr{I}, S'' \subseteq S'\}$

for $S' \subseteq S$. The rank function uniquely determines M, as $S' \in \mathscr{I}$ iff $r(S') = |S'|$.

Examples of matroids are as follows.

I. *Graphic matroids* (Whitney [1935]). Let $G = (V, E)$ be an undirected graph, and let $\mathscr{I}$ consist of all subsets E' of E not containing any circuit. Then $M = (E, \mathscr{I})$ is a matroid, with rank function r given by: $r(E') = |V| - \kappa(V, E')$, where $\kappa(V, E')$ denotes the number of components of the graph (V, E'). If G is connected, the bases of M are exactly the spanning trees.

II. *Cographic matroids* (Whitney [1935]). Let $G = (V, E)$ be an undirected graph, and let $\mathscr{I}$ consist of all subsets E' of E such that the graph $(V, E \setminus E')$ has the same number of components as G has (i. e., E' does not contain a nonempty cut). Then $M = (E, \mathscr{I})$ is a matroid, with rank function $r(E')=|E'|+\kappa(V,E)-\kappa(V,E\setminus E')$. If G is connected, the bases of M are exactly the complements of spanning trees.

III. *Transversal matroids* (Edmonds and Fulkerson [1965]). Let $G=(V,E)$ be a bipartite graph, with colour classes S and T, and let $\mathscr{I}$ consist of all subsets S' of S for which there exists a matching in G covering S'. Then $(S,\mathscr{I})$ is a matroid, with rank function given by $r(S')=\min_{S''\subseteq S'}|S'\setminus S''|+|\Delta(S'')|$, where $\Delta(S'')$ denotes the set of vertices adjacent to at least one vertex in S'' (this min-max relation follows from König's matching theorem (Theorem 2)).

IV. *Gammoids* (Perfect [1968]). Let $D=(V,A)$ be a directed graph, and let S and T be subsets of V. Let $\mathscr{I}$ be the collection of subsets S' of S for which there exists $|S'|$ pairwise vertex-disjoint paths starting in S' and ending in T. Then $(S,\mathscr{I})$ is a matroid, and its rank function can be written as a min-max relation, using Menger's theorem. Clearly, each transversal matroid is a gammoid.

V. *(Linearly) representable matroids* (Whitney [1935], Van der Waerden [1937]). Let S be a collection of vectors in a vector space, and let $\mathscr{I}$ be the collection of linearly independent sets of vectors in S. Then $(S,\mathscr{I})$ is a matroid.

Moreover, Van der Waerden [1937] considered *algebraic matroids*.

Given a matroid $M=(S,\mathscr{I})$ and a weight function $w:S\to\mathbb{R}_+$, a maximum weighted independet set can be found with the following *greedy algorithm*: order $S=\{s_1,\ldots,s_n\}$ such that $w(s_1)\geq\ldots\geq w(s_n)$, and determine $S_0,S_1,\ldots,S_n$ inductively as follows:

(3)   (i) $S_0:=\emptyset$,

(ii) $S_{i+1}:=S_i\cup\{s_{i+1}\}$, if $S_i\cup\{s_{i+1}\}\in\mathscr{I}$,

$S_{i+1}:=S_i$,   otherwise,

for $i=0,\ldots,n-1$.

Then S_n is a maximum weighted independent set. Edmonds [1971] and Gale [1968] observed that this characterizes matroids: if $(S,\mathscr{I})$ satisfies (1) (i) (ii), then $(S,\mathscr{I})$ is a matroid if and only if the greedy algorithm finds a maximum weighted set in $\mathscr{I}$, for each weight function.

Edmonds [1970, 1971] derived from the greedy method the following min-max formula: given a matroid $M=(S,\mathscr{I})$, with rank function r, and weight function $w:S\to\mathbb{Z}_+$,

(4) the maximum weight of an independent set is equal to the minimum value of $r(S_1)+\ldots+r(S_k)$, where $S_1\subseteq\ldots\subseteq S_k\subseteq S$ (repetition allowed), such that each element s of S occurs in at least $w(s)$ of the S_i.

This is equivalent to the total dual integrality of:

(5) $x(s)\geq 0$ $(s\in S)$,

$x(S')\leq r(S')$ $(S'\in S)$.

In particular, (5) determines the *matroid polytope* of M, being the convex hull of the independent sets of M. E. g., (4) implies as a special case a good characterization for the maximum (or minimum) length of a spanning tree in a graph. It may be derived from (5) that the convex hull of the forests (i.e., edge sets containing no circuit) in an undirected graph $G=(V, E)$ is determined by:

(6) $$x(e) \geq 0 \quad (e \in E),$$
$$x(\langle V' \rangle) \leq |V'|-1 \quad (\emptyset \neq V' \subseteq V)$$

where $\langle V' \rangle$ denotes the set of edges contained in V'.

Matroid intersection. Even more interestingly, there is a min-max relation for the maximum weight of a common independent set in *two* matroids. First consider the cardinality case, for which Edmonds [1970] showed the following.

Theorem 24 (Edmonds' matroid intersection theorem). *Let $M_1 = (S, \mathcal{I}_1)$ and $M_2 = (S, \mathcal{I}_2)$ be matroids, with rank functions r_1 and r_2, respectively. Then the maximum size of a set in $\mathcal{I}_1 \cap \mathcal{I}_2$ is equal to*

(7) $$\min_{S' \subseteq S} (r_1(S') + r_2(S \setminus S')).$$

This generalizes König's matching theorem: if $G=(V, E)$ is a bipartite graph, with colour classes V_1 and V_2, let $\mathcal{I}_i = \{E' \subseteq E \mid \text{no two edges in } E' \text{ intersect in } V_i\}$, for $i = 1, 2$; then $(E, \mathcal{I}_1)$ and $(E, \mathcal{I}_2)$ are matroids, and here Theorem 24 reduces to König's theorem. Applying Theorem 24 to two transversal matroids gives a min-max formula for the maximum size of a common "system of distinct representatives".

Other consequences are the following min-max relations, due to Edmonds [1965 a, 1965 b] and Nash-Williams [1964].

Corollary 24a (Matroid partition theorems). *Let $M = (S, \mathcal{I})$ be a matroid, with rank function r. Then the minimum number of independent sets needed to cover S is equal to $\max_{S' \neq \emptyset} \lceil |S'|/r(S') \rceil$. Moreover, the maximum number of pairwise disjoint bases is equal to* $\min_{S' \subseteq S, r(S') \neq r(S)} \lfloor |S \setminus S'|/(r(S)-r(S')) \rfloor$.

We leave it to the reader to derive this corollary from Theorem 24 (cf. Welsh [1970]).

Specialized to graphic matroids, this corollary yields the following results of Nash-Willams [1961, 1964] and Tutte [1961]. Let $G = (V, E)$ be an undirected graph. Then the minimum number of forests needed to cover E is equal to

(8) $$\max \left\lceil \frac{|\langle V' \rangle|}{|V'|-1} \right\rceil,$$

where the maximum ranges over all subsets V' of V with at least two elements, and where $\langle V' \rangle$ denotes the set of edges contained in V'. If G is connected, the maximum number of pairwise disjoint spanning trees is equal to

(9) $$\min \left\lfloor \frac{|E \setminus E'|}{\kappa(V, E')-1} \right\rfloor,$$

where the minimum ranges over all subsets E' of E with $\kappa(V, E')$ (being the number of components of (V, E')) at least two.

The following weighted version of the matroid intersection theorem was shown also by Edmonds [1970]. Let $M_1=(S, \mathscr{I}_1)$ and $M_2=(S, \mathscr{I}_2)$ be matroids, with rank functions r_1 and r_2, respectively, and let $w: S\to\mathbb{Z}_+$ be a weight function. Then:

(10) the maximum weight of a set in $\mathscr{I}_1\cap\mathscr{I}_2$ is equal to the minimum value of $r_1(S_1)+\ldots+r_1(S_k)+r_2(T_1)+\ldots+r_2(T_l)$, where $S_1\subseteq\ldots\subseteq S_k\subseteq S$ and $T_1\subseteq\ldots\subseteq T_l\subseteq S$ such that each element s of S is contained in at least $w(s)$ of the $S_1,\ldots,S_k, T_1,\ldots, T_l$.

This implies the total dual integrality of the system

(11) $$x(s)\geq 0 \quad (s\in S),$$
$$x(S')\leq r_1(S') \quad (S'\subseteq S),$$
$$x(S')\leq r_2(S') \quad (S'\subseteq S).$$

So (11) determines the convex hull of the sets in $\mathscr{I}_1\cap\mathscr{I}_2$. Equivalently, the intersection of two matroid polytopes has integer vertices again. Polynomial algorithms for finding maximum weighted common independent sets were given by Edmonds [1979] and Lawler [1975].

Result (10) contains as special case the weighted version of König's matching theorem. Also Fulkerson's optimum branching theorem (Theorem 16) can be derived.

Submodular functions. In fact, Edmonds [1970] showed more generally the following. A function $f: \mathscr{P}(S)\to\mathbb{R}$ is called *submodular* if

(12) $$f(S'\cap S'')+f(S'\cup S'')\leq f(S')+f(S''),$$

for all $S', S''\subseteq S$. Now, for submodular $f_1, f_2: \mathscr{P}(S)\to\mathbb{R}$, the system

(13) $$x(S')\leq f_1(S') \quad (S'\subseteq S),$$
$$x(S')\leq f_2(S') \quad (S'\subseteq S),$$

is totally dual integral. Moreover, total dual integrality is maintained if we add $c_1\leq x\leq c_2$ to (13), for arbitrary c_1, c_2 (i.e., (13) is *box totally dual integral*). In particular, if f_1 and f_2 (and c_1 and c_2) are integer-valued, each face of the polyhedron (13) contains integer vectors.

This generalizes the results on matroids given above, as the rank function of any matroid is submodular. Other examples of submodular functions are: (i) let $G=(V, E)$ be a bipartite graph, with colour classes S and T, and let $w: T\to\mathbb{R}_+$; then $f(S'):=w(\Delta(S'))$ for $S'\subseteq S$ is submodular ($\Delta(S')$ denoting the set of vertices in T adjacent to at least one vertex in S'); (ii) let $D=(V, A)$ be a directed graph, and let $c: A\to\mathbb{R}_+$; then $f(V'):=c(\delta^+(V'))$ for $V'\subseteq V$ is submodular.

Edmonds and Giles [1977] gave a further generalization, containing also graph-theoretical results like the Lucchesi-Younger theorem as special cases.

Let $\mathscr{C}$ be a family of subsets of the finite set V, and let $f: \mathscr{C} \to \mathbb{R}$ be such that:

(14) if $V', V'' \in \mathscr{C}$ with $V' \cap V'' \neq \emptyset$ and $V' \cup V'' \neq V$, then $V' \cap V'' \in \mathscr{C}$ and $V' \cup V'' \in \mathscr{C}$, and $f(V' \cap V'') + f(V' \cup V'') \leq f(V') + f(V'')$.

Then for any directed graph $D = (V, A)$, and $c_1, c_2: A \to \mathbb{R} \cup \{\pm \infty\}$, the system

(15) $\qquad c_1(a) \leq x(a) \leq c_2(a) \qquad (a \in A),$
$\qquad\qquad x(\delta^-(V')) - x(\delta^+(V')) \leq f(V') \quad (V' \in \mathscr{C}),$

is totally dual integral. This contains König's matching theorem, the max-flow min-cut theorem, the Lucchesi-Younger theorem, the matroid intersection theorem, Dilworth's theorem, and the total dual integrality of (13) as special cases.

Moreover, if $c_1 \geq 0$, and if for all V_1, V_2, V_3 in $\mathscr{C}$ with $V_1 \subseteq V \setminus V_2 \subseteq V_3$ there is no arc of D entering both V_1 and V_3, then the system

(16) $\qquad c_1(a) \leq x(a) \leq c_2(a) \quad (a \in A),$
$\qquad\qquad x(\delta^-(V')) \geq -f(V') \quad (V' \in \mathscr{C}),$

is totally dual integral. This contains König's matching theorem, the max-potential min-work theorem, Fulkerson's optimum branching theorem, the Lucchesi-Younger theorem, the matroid intersection theorem, and the total dual integrality of (13) as special cases.

The total dual integrality of (15) and (16) can be shown by extending the method given in Section 8 to prove Fulkerson's optimum branching theorem. Related results, which can be proved similarly, are given in Frank [1979], Grishuhin [1981], Hassin [1978], Hoffman and Schwartz [1978], Lawler and Martel [1982], Schrijver [1982b]. For a survey, see Schrijver [1982c].

Matroid matching. It is not difficult to see that the following problem generalizes both the matching problem in undirected graphs and the matroid intersection problem: given a matroid $M = (V, \mathscr{I})$ and an undirected graph $G = (V, E)$, find an independent matching of maximum size. Here an *independent matching* is a set E' of pairwise disjoint edges with $\bigcup E'$ independent in M.

However, this *matroid matching problem* (Jenkyns [1974], Lawler [1971, 1976]) is in general NP-hard (B. Korte, and Lovász [1981a]), and a satisfactory min-max formula is likely not to exist.

On the other hand, if M is linearly representable, and given by an explicit representation of V as vectors $v_1, \ldots, v_n$ in a vector space, Lovász found a polynomial algorithm [1981a], and the following min-max formula [1980a].

Theorem 25. *The maximum size of an independent matching is equal to*

(17) $\qquad\qquad \min\left(\dim L_0 + \sum_{i=1}^{t} \left\lfloor \frac{\dim L_i - \dim L_0}{2} \right\rfloor \right)$

which minimum ranges over linear spaces $L_0, L_1, \ldots, L_t$ ($t \geq 0$) such that $L_0 \subseteq L_1 \cap \ldots \cap L_t$ and such that, for each edge e of G, the linear hull of e intersects L_0 or e is contained in L_i for some $i = 1, \ldots, t$.

In case $v_1, \ldots, v_n$ are linearly independent, Theorem 25 reduces to the Tutte-Berge theorem. Also the matroid intersection theorem for representable matroids is included. Moreover, Mader's theorem on vertex-disjoint S-paths (see Section 7) follows as a special case – see Lovász [1980b], where some more corollaries are described.

It is still an open problem to extend Theorem 25 to the weighted case. Also no polynomial algorithm finding a maximum weighted independent matching has been found as yet. In order to apply the ellipsoid method here, we need a description in terms of linear inequalities of the convex hull of the independent matchings, which is still unknown.

Colourings. Finally we mention the following min-max relations for "supermodular colourings". A collection $\mathscr{C}$ of subsets of the finite set S is called an *intersecting family*, and the function $g: \mathscr{C} \to \mathbb{R}$ is called *supermodular on intersecting pairs*, if the following condition is satisfied:

(18)   if $S', S'' \in \mathscr{C}$ and $S' \cap S'' \neq \emptyset$, then $S' \cap S'' \in \mathscr{C}$ and $S' \cup S'' \in \mathscr{C}$ and $f(S' \cap S'') + f(S' \cup S'') \geq f(S') + f(S'')$.

Now let $\mathscr{C}_1$ and $\mathscr{C}_2$ be intersecting families on S, and let $g_1: \mathscr{C}_1 \to \mathbb{Z}$ and $g_2: \mathscr{C}_2 \to \mathbb{Z}$ be supermodular on intersecting pairs, such that $g_i(S') \leq |S'|$ for $i = 1, 2$ and $S' \in \mathscr{C}_i$. It may be derived, e.g., from Edmonds and Giles' result on (15) that:

(19) $\min\{|S''| \, | \, S'' \subseteq S, |S'' \cap S'| \geq g_i(S')$ for all $i = 1, 2$ and $S' \in \mathscr{C}_i\}$
     $= \max\{g_1(S_1) + \ldots + g_1(S_m) + g_2(T_1) + \ldots + g_2(T_l) \, | \, S_1, \ldots, S_m \in \mathscr{C}_1,$
     $T_1, \ldots, T_l \in \mathscr{C}_2$, and $S_1, \ldots, S_m, T_1, \ldots, T_l$ pairwise disjoint$\}$.

This can be considered as a theorem on the minimum size of a common "spanning set" in two matroids. Now also the following, more or less "polar", min-max result holds (cf. Schrijver [1983b]):

(20) $\min\{k \, | \,$ there exist pairwise disjoint subsets $S_1, \ldots, S_k$ of S such that, for all $i = 1, 2$ and $S' \in \mathscr{C}_i$, S' intersects at least $g_i(S')$ of the $S_i\} = \max\{g_i(S') \, | \, i = 1, 2; \, S' \in \mathscr{C}_i\}$

(assuming $g_i(S') \geq 0$ for at least one i and S'). This relation implies König's and Gupta's edge-colouring theorems (Theorem 3 and Corollary 3a).

References

[1982]   R. P. Anstee and M. Farber, Characterizations of totally balanced matrices, Research Report CORR 82-5, Faculty of Mathematics, University of Waterloo, Waterloo, Ont., 1982.

[1981]   E. Balas and N. Christofides, A restricted Lagrangean approach to the traveling salesman problem, Math. Programming 21 (1981) 19–46.

[1950] H. B. Belck, Reguläre Faktoren von Graphen, J. Reine Angew. Math. 188 (1950) 228–252.

[1958] C. Berge, Sur le couplage maximum d'un graphe, C. R. Acad. Sci. Paris 247 (1958) 258–259.

[1960] C. Berge, Les problèmes de coloration en théorie des graphes, Publ. Inst. Stat. Univ. Paris 9 (1960) 123–160.

[1961] C. Berge, Farbung von Graphen deren sämtliche bzw. ungerade Kreise starr sind (Zusammenfassung), Wiss. Z. Martin-Luther-Univ. Halle-Wittenberg, Math.-Natur. Reihe (1961) 114–115.

[1962] C. Berge, Sur un conjecture relative au problème des codes optimaux, Commun. 13ème Assemblée Gén. U.R.S.I., Tokyo, 1962.

[1969] C. Berge, The rank of a family of sets and some applications to graph theory, in: Recent progress in combinatorics (W. T. Tutte, ed.), Acad. Press, New York, 1969, pp. 246–257.

[1970] C. Berge, Sur certain hypergraphes generalisant les graphes bipartis, in: Combinatorial theory and its applications (P. Erdös, A. Rényi, and V. T. Sós, eds.), North-Holland, Amsterdam, 1970, pp. 119–133.

[1972] C. Berge, Balanced matrices, Math. Programming 2 (1972) 19–31.

[1982] C. Berge and V. Chvátal (eds.), Perfect graphs, to appear.

[1970] C. Berge and M. Las Vergnas, Sur un théorème du type König pour hypergraphes, in: Proc. Intern. Conf. on Comb. Math. (A. Gewirtz and L. Quintas, eds.), Ann. New York Acad. Sci. 175 (1970) 32–40.

[1946] G. Birkhoff, Tres observaciones sobre el algebra lineal, Rev. Univ. Nac. Tucuman Ser. A 5 (1946) 147–148.

[1976] J. A. Bondy and U. S. R. Murty, Graph theory with applications, Macmillan, London, 1976.

[1980] A. E. Brouwer and A. Kolen, A super-balanced hypergraph has a nest point, Report ZW 148/80, Math. Centrum, Amsterdam, 1980.

[1982] K. Cameron, Polyhedral and algorithmic ramifications of antichains, Ph. D. thesis, University of Waterloo, Waterloo, Ont., 1982.

[1975] V. Chvátal, On certain polytopes associated with graphs, J. Combinatorial Theory (B) 18 (1975) 138–154.

[1981] V. Chvátal, communication C.I.R.M. Marseille-Luminy, 1981.

[1980] G. Cornuéjols and W. R. Pulleyblank, A matching problem with side conditions, Discrete Math. 29 (1980) 135–159.

[1983] W. Cook and W. R. Pulleyblank, to appear.

[1978] W. H. Cunningham and A. B. Marsh, A primal algorithm for optimal matching, Math. Programming Study 8 (1978) 50–72.

[1951] G. B. Dantzig, Application of the simplex method to a transportation problem, in: Activity analysis of production and allocation (T. C. Koopmans, ed.), J. Wiley, New York, 1951, pp. 359–373.

[1979] R. W. Deming, Independence numbers of graphs – an extension of the König-Egerváry theorem, Discrete Math. 27 (1979) 23–33.

[1971] M. A. H. Dempster, Two algorithms for the time-table problem, in: Combinatorial mathematics and its applications (D. J. A. Welsh, ed.), Acad. Press, New York, 1971, pp. 63–85.

[1950] R. P. Dilworth, A decomposition theorem for partially ordered sets, Ann. of Math. 51 (1950) 161–166.

[1961] G. A. Dirac, On rigid circuit graphs, Abh. Math. Sem. Univ. Hamburg 25 (1961) 71–76.

[1965 a] J. Edmonds, Minimum partition of a matroid into independent subsets, J. Res. Nat. Bur. Standards Sect. B 69 (1965) 67-72.

[1965 b] J. Edmonds, Lehman's switching game and a theorem of Tutte and Nash-Williams, J. Res. Nat. Bur. Standards Sect. B 69 (1965) 73-77.

[1965 c] J. Edmonds, Paths, trees, and flowers, Canad. J. Math. 17 (1965) 449-467.

[1965 d] J. Edmonds, Maximum matching and a polyhedron with 0,1-vertices, J. Res. Nat. Bur. Standards Sect. B 69 (1965) 125-130.

[1967] J. Edmonds, An introduction to matching, mimeographed notes, Engineering Summer Conf., Univ. of Michigan, Ann Arbor, 1967.

[1967 a] J. Edmonds, Optimum branchings, J. Res. Nat. Bur. Standards Sect. B 71 (1967) 233-240.

[1970] J. Edmonds, Submodular functions, matroids, and certain polyhedra, in: Combinatorial structures and their applications (R. Guy, H. Hanani, N. Sauer and J. Schönheim, eds.), Gordon and Breach, New York, 1970, pp. 69-87.

[1971] J. Edmonds, Matroids and the greedy algorithm, Math. Programming 1 (1971) 127-136.

[1973] J. Edmonds, Edge-disjoint branchings, in: Combinatorial algorithms (B. Rustin, ed.), Acad. Press, New York, 1973, pp. 91-96.

[1979] J. Edmonds, Matroid intersection, Annals of Discrete Math. 4 (1979) 39-49.

[1965] J. Edmonds and D. R. Fulkerson, Transversals and matroid partition, J. Res. Nat. Bur. Standards Sect. B 69 (1965) 147-153.

[1970] J. Edmonds and D. R. Fulkerson, Bottleneck extrema, J. Combinatorial Theory 8 (1970) 299-306.

[1977] J. Edmonds and R. Giles, A min-max relation for submodular functions on graphs, Annals of Discrete Math. 1 (1977) 185-204.

[1970] J. Edmonds and E. L. Johnson, Matching, a well-solved class of integer linear programs, in: Combinatorial structures and their applications (R. Guy, H. Hanani, N. Sauer and J. Schönheim, eds.), Gordon and Breach, New York, 1970, pp. 89-92.

[1973] J. Edmonds and E. L. Johnson, Matching, Euler tours and the Chinese postman, Math. Programming 5 (1973) 88-124.

[1972] J. Edmonds and R. M. Karp, Theoretical improvements in algorithmic efficiency for network flow problems, J. ACM 19 (1972) 248-264.

[1931] E. Egerváry, Matrixok kombinatorius tulajdonságairol, Mat. Fiz. Lapok 38 (1931) 16-28.

[1956] P. Elias, A. Feinstein and C. E. Shannon, A note on the maximum flow through a network, IRE Trans. Information Theory IT 2 (1956) 117-119.

[1956] L. R. Ford and D. R. Fulkerson, Maximum flow through a network, Canad. J. Math. 8 (1956) 399-404.

[1958] L. R. Ford and D. R. Fulkerson, Network flow and systems of distinct representatives, Canad. J. Math. 10 (1958) 78-84.

[1962] L. R. Ford and D. R. Fulkerson, Flows in networks, Princeton Univ. Press, Princeton, N.J., 1962.

[1979] A. Frank, Kernel systems of directed graphs, Acta Sci. Math. (Szeged) 41 (1979) 63-76.

[1980] A. Frank, On chain and antichain families of a partially ordered set, J. Combinatorial Theory (B) 29 (1980) 176-184.

[1981] A. Frank, How to make a digraph strongly connected, Combinatorica 1 (1981) 145-153.

[1912] G. Frobenius, Über Matrizen aus nicht negativen Elementen, Sitzber. Preuss. Akad. Wiss. (1912) 456-477.

[1917]   G. Frobenius, Über zerlegbare Determinanten, Sitzber. Preuss. Akad. Wiss. (1917) 274–277.

[1956]   D. R. Fulkerson, Note on Dilworth's decomposition theorem for partially ordered sets, Proc. Amer. Math. Soc. 7 (1956) 701–702.

[1961]   D. R. Fulkerson, An out-of-kilter method for minimal cost flow problems, SIAM J. Appl. Math. 9 (1961) 18–27.

[1968]   D. R. Fulkerson, Networks, frames, and blocking systems, in: Mathematics of the decision sciences, part I (G. B. Dantzig and A. F. Veinott, eds.), Amer. Math. Soc., Providence, R. I., 1968, pp. 303–334.

[1970]   D. R. Fulkerson, Blocking polyhedra, in: Graph theory and its applications (B. Harris, ed.), Acad. Press, New York, 1970, pp. 93–112.

[1971]   D. R. Fulkerson, Blocking and anti-blocking pairs of polyhedra, Math. Programming 1 (1971) 168–194.

[1972]   D. R. Fulkerson, Anti-blocking polyhedra, J. Combinatorial Theory (B) 12 (1972) 50–71.

[1974]   D. R. Fulkerson, Packing rooted directed cuts in a weighted directed graph, Math. Programming 6 (1974) 1–13.

[1974]   D. R. Fulkerson, A. J. Hoffman and R. Oppenheim, On balanced matrices, Math. Programming Study 1 (1974) 120–132.

[1968]   D. Gale, Optimal assignments in an ordered set: an application of matroid theory, J. Combinatorial Theory 4 (1968) 176–180.

[1958]   T. Gallai, Maximum-minimum Sätze über Graphen, Acta Math. Acad. Sci. Hungar. 9 (1958) 395–434.

[1959]   T. Gallai, Über extreme Punkt- und Kantenmengen, Ann. Univ. Sci. Budapest, Eötvos Sect. Math. 2 (1959) 133–138.

[1979]   M. R. Garey and D. S. Johnson, Computers and intractability: a guide to the theory of NP-completeness, Freeman, San Francisco, 1979.

[1973]   F. Gavril, Algorithms for a maximum clique and a maximum independent set of a circle graph, Networks 3 (1973) 261–273.

[1974]   F. Gavril, Algorithms on circular-arc graphs, Networks 4 (1974) 357–369.

[1962]   A. Ghouila-Houri, Caractérisation des matrices totalement unimodulaires, C. R. Acad. Sci. Paris 254 (1962) 1192–1194.

[1982a] R. Giles, Optimum matching forests I: special weights, Math. Programming 22 (1982) 1–11.

[1982b] R. Giles, Optimum matching forests II: general weights, Math. Programming 22 (1982) 12–38.

[1982c] R. Giles, Optimum matching forests III: facets of matching forest polyhedra, Math. Programming 22 (1982) 39–51.

[1981]   R. Giles and L. E. Trotter, On stable set polyhedra for $K_{1,3}$-free graphs, J. Combinatorial Theory (B) 31 (1981) 313–326.

[1980]   M. C. Golumbic, Algorithmic graph theory and perfect graphs, Acad. Press, New York, 1980.

[1976]   C. Greene, Some partitions associated with a partially ordered set, J. Combinatorial Theory (A) 20 (1976) 69–79.

[1976]   C. Greene and D. J. Kleitman, The structure of Sperner k-families, J. Combinatorial Theory (A) 20 (1976) 41–68.

[1981]   V. P. Grishuhin, Polyhedra related to a lattice, Math. Programming 21 (1981) 70–89.

[1981]   M. Grötschel, L. Lovász and A. Schrijver, The ellipsoid method and its consequences in combinatorial optimization, Combinatorica 1 (1981) 169–197.

[1981a] M. Grötschel, L. Lovász and A. Schrijver, Polynomial algorithms for perfect graphs, Res. Report WP 81.176-OR, Inst. Oper. Research, Univ. Bonn, 1981.

[1967] R. P. Gupta, A decomposition theorem for bipartite graphs, in: Theory of graphs (P. Rosenstiehl, ed.), Gordon and Breach, New York, 1967, pp. 135-138.

[1978] R. P. Gupta, An edge-colouring theorem for bipartite graphs with applications, Discrete Math. 23 (1978) 229-233.

[1981] E. Györi, A minimax theorem on intervals, preprint Math. Inst. Hung. Acad. Sci. No. 54/1981, Budapest, 1981.

[1958] A. Hajnal and J. Surányi, Über die Auflösung von Graphen in vollständigen Teilgraphen, Ann. Univ. Sci. Budapest, Eötvös Sect. Math. 1 (1958) 113-121.

[1935] P. Hall, On representatives of subsets, J. London Math. Soc. 10 (1935) 26-30.

[1978] R. Hassin, On network flows, Ph. D. thesis, Yale Univ., Boston, 1978.

[1960] A. J. Hoffman, Some recent applications of the theory of linear inequalities to extremal combinatorial analysis, in: Combinatorial analysis (R. E. Bellman and M. Hall, eds.), Amer. Math. Soc., Providence, R. I., 1960, pp. 113-127.

[1982] A. J. Hoffman, A. W. J. Kolen and M. Sakarovitch, Totally-balanced and greedy matrices, Report BW, Math. Centrum, Amsterdam, 1982.

[1956] A. J. Hoffman and J. B. Kruskal, Integral boundary points of convex polyhedra, in: Linear inequalities and related systems (H. W. Kuhn and A. W. Tucker, eds.), Ann. of Math. Studies 38, Princeton Univ. Press, Princeton, N.J., 1956, pp. 233-246.

[1963] A. J. Hoffman and H. M. Markowitz, A note on shortest path, assignment, and transportation problems, Naval Res. Logist. Quart. 10 (1963) 375-380.

[1977] A. J. Hoffman and D. E. Schwartz, On partitions of partially ordered sets, J. Combinatorial Theory (B) 23 (1977) 3-13.

[1978] A. J. Hoffman and D. E. Schwartz, On lattice polyhedra, in: Combinatorics (A. Hajnal and V. T. Sós, eds.), North-Holland, Amsterdam, 1978, pp. 593-598.

[1981] W.-L. Hsu, Y. Ikura and G. L. Nemhauser, A polynomial algorithm for maximum weighted vertex packings on graphs without long odd cycles, Math. Programming 20 (1981) 225-232.

[1963] T. C. Hu, Multicommodity network flows, Operations Res. 11 (1963) 344-360.

[1973] T. C. Hu, Two-commodity cut packing problem, Discrete Math. 4 (1973) 108-109.

[1974] T. A. Jenkyns, Matchoids: a generalization of matchings and matroids, Ph. D. thesis, Univ. of Waterloo, Waterloo, Ont., 1974.

[1979] A. V. Karzanov, On the minimal number of arcs of a digraph meeting all its directed cutsets, to appear.

[1970] D. J. Kleitman, A. Martin-Löf, B. Rothschild and A. Whinston, A matching theorem for graphs, J. Combinatorial Theory 8 (1970) 104-114.

[1915] D. König, Vonalrendszerek és determinánsok (Line-systems and determinants), Matematikai és Természettudományi Értesítö 33 (1915) 221-229 (in Hungarian).

[1916] D. König, Graphen und ihre Anwendung auf Determinantentheorie und Mengenlehre, Math. Ann. 77 (1916) 453-465.

[1931] D. König, Graphok és matrixok, Mat. Fiz. Lapok 38 (1931) 116-119.

[1932] D. König, Über trennende Knotenpunkte in Graphen (nebst Anwendungen auf Determinanten und Matrizen), Acta. Lit. Sci. Sect. Sci. Math. (Szeged) 6 (1932-1934) 155-179.

[1955] H. W. Kuhn, The Hungarian method for solving the assignment problem, Naval Res. Logist. Quart. 2 (1955) 83-97.

[1956]   H. W. Kuhn, Variants of the Hungarian method for the assignment problem, Naval Res. Logist. Quart. 3 (1956) 253–258.
[1971]   E. L. Lawler, Matroids with parity conditions: a new class of combinatorial optimization problems, Memorandum ERL-M334, Univ. of California, Berkeley, 1971.
[1975]   E. L. Lawler, Matroid intersection algorithms, Math. Programming 9 (1975) 31–56.
[1976]   E. L. Lawler, Combinatorial optimization: networks and matroids, Holt, Rinehart and Winston, New York, 1976.
[1982]   E. L. Lawler and C. U. Martel, Computing maximal "polymatroidal" network flows, Math. of Oper. Research 7 (1982) 334–347.
[1979]   A. Lehman, On the width-length inequality, Math. Programming 16 (1979) 245–259.
[1978]   M. V. Lomonosov, On the systems of flows in a network, Probl. Per. Inf. 14 (1978) 60–73 (in Russian); English translation: Problems of Inf. Transmission 14 (1978) 280–290.
[1982]   M. V. Lomonosov, Combinatorial approach to multi-flow problems, preprint.
[1970]   L. Lovász, Subgraphs with prescribed valencies, J. Combinatorial Theory 8 (1970) 391–416.
[1972]   L. Lovász, Normal hypergraphs and the perfect graph conjecture, Discrete Math. 2 (1972) 253–267.
[1974]   L. Lovász, Minimax theorems for hypergraphs, in: Hypergraph seminar (C. Berge and D. Ray-Chaudhuri, eds.), Springer Lecture Notes in Mathematics 411, Springer, Berlin, 1974, pp. 111–126.
[1975]   L. Lovász, 2-Matchings and 2-covers of hypergraphs, Acta. Math. Acad. Sci. Hungar. 26 (1975) 433–444.
[1976a] L. Lovász, On two minimax theorems in graph theory, J. Combinatorial Theory (B) 21 (1976) 96–103.
[1976b] L. Lovász, On some connectivity properties of Eulerian graphs, Acta. Math. Acad. Sci. Hungar. 28 (1976) 129–138.
[1977]   L. Lovász, Certain duality principles in integer programming, Annals of Discrete Math. 1 (1977) 363–374.
[1979]   L. Lovász, On the Shannon capacity of a graph, IEEE Trans. Inform. Theory IT 25 (1979) 1–7.
[1979a] L. Lovász, Graph theory and integer programming, Annals of Discrete Math. 4 (1979) 141–158.
[1980a] L. Lovász, Selecting independent lines from a family of lines in a space, Acta Sci. Math. (Szeged) 42 (1980) 121–131.
[1980b] L. Lovász, Matroid matching and some applications, J. Combinatorial Theory (B) 28 (1980) 208–236.
[1981a] L. Lovász, The matroid matching problem, in: Algebraic methods in graph theory (L. Lovász and V. T. Sós, eds.), North-Holland, Amsterdam, 1981, pp. 495–517.
[1981b] L. Lovász, Perfect graphs, in: More selected topics in graph theory (L. W. Beineke and R. J. Wilson, eds.), to appear.
[1982]   L. Lovász, Ear-decompositions of matching-covered graphs, preprint, 1982.
[1982]   A. Lubiw, Γ-free matrices, M. Sc. thesis, Univ. of Waterloo, Waterloo, Ont., 1982.
[1976]   C. L. Lucchesi, A minimax equality for directed graphs, Ph. D. thesis, Univ. of Waterloo, Waterloo, Ont., 1976.
[1978]   C. L. Lucchesi and D. H. Younger, A minimax relation for directed graphs, J. London Math. Soc. (2) 17 (1978) 369–374.

[1978a] W. Mader, Über die Maximalzahl kantendisjunkter A-Wege, Arch. Math. (Basel) 30 (1978) 325–336.

[1978b] W. Mader, Über die Maximalzahl kreuzungsfreier H-Wege, Arch. Math. (Basel) 31 (1978) 387–402.

[1979] A. B. Marsh, Matching algorithms, Ph. D. thesis, Johns Hopkins Univ., Baltimore, 1979.

[1972] C. J. H. McDiarmid, The solution of a time-tabling problem, J. Inst. Maths. Appl. 9 (1972) 23–34.

[1927] K. Menger, Zur allgemeinen Kurventheorie, Fund. Math. 10 (1927) 96–115.

[1976] H. Meyniel, On the perfect graph conjecture, Discrete Math. 16 (1976) 339–342.

[1960] G. J. Minty, Monotone networks, Proc. Roy. Soc. London Ser. A 257 (1960) 194–212.

[1980] G. J. Minty, On maximal independent sets of vertices in a claw-free graph, J. Combinatorial Theory (B) 28 (1980) 284–304.

[1971] L. Mirsky, Transversal theory, Acad. Press, London, 1971.

[1961] C. St. J. A. Nash-Williams, Edge-disjoint spanning trees of finite graphs, J. London Math. Soc. 36 (1961) 445–450.

[1964] C. St. J. A. Nash-Williams, Decomposition of finite graphs into forests, J. London Math. Soc. 39 (1964) 12.

[1953] J. von Neumann, A certain zero-sum two-person game equivalent to the optimum assignment problem, in: Contributions to the theory of games II (A. W. Tucker and H. W. Kuhn, eds.), Annals of Math. Studies 38, Princeton Univ. Press, Princeton, N.J., 1953, pp. 5–12.

[1981] H. Okamura and P. D. Seymour, Multicommodity flows in planar graphs, J. Combinatorial Theory (B) 31 (1981) 75–81.

[1956] A. Orden, The transshipment problem, Manag. Sci. 2 (1956) 276–285.

[1975] M. Padberg, Characterisation of totally unimodular, balanced and perfect matrices, in: Combinatorial programming: methods and applications (B. Roy, ed.), Reidel, Dordrecht (Holland), 1975, pp. 275–284.

[1982] M. W. Padberg and M. R. Rao, Odd minimum cut-sets and b-matchings, Math. of Oper. Res. 7 (1982) 67–80.

[1982] C. H. Papadimitriou and K. Steiglitz, Combinatorial optimization: algorithms and complexity, Prentice-Hall, Englewood Cliffs, N.J., 1982.

[1976] B. A. Papernov, Feasibility of multicommodity flows, in: Studies in Discrete Optimization (A. A. Fridman, ed.), Izdat. "Nauka", Moscow, 1976, pp. 230–261 (in Russian).

[1968] H. Perfect, Applications of Menger's graph theorem, J. Math. Analysis Appl. 22 (1968) 96–111.

[1973] W. R. Pulleyblank, Faces of matching polyhedra, Ph. D. thesis, Univ. of Waterloo, Waterloo, Ont., 1973.

[1980] W. R. Pulleyblank, Dual integrality in b-matching problems, Math. Programming Study 12 (1980) 176–196.

[1983] W. R. Pulleyblank, Polyhedral combinatorics, this volume.

[1957] R. Rado, A note on independence functions, Proc. London Math. Soc. 7 (1957) 300–320.

[1966a] B. Rothschild and A. Whinston, On two-commodity network flows, Operations Res. 14 (1966) 377–387.

[1966b] B. Rothschild and A. Whinston, Feasibility of two-commodity network flows, Operations Res. 14 (1966) 1121–1129.

[1978] N. Sbihi, Étude des stables dans les graphes sans étoile, M. Sc. thesis, Univ. Sci. et Méd. Grenoble, 1978.

[1981] N. Sbihi and J. P. Uhry, A class of h-perfect graphs, Rapport de Rech. No. 236, IRMA, Grenoble, 1981.

[1980] A. Schrijver, A counterexemple to a conjecture of Edmonds and Giles, Discrete Math. 32 (1980) 213–214.

[1981] A. Schrijver, Short proofs on the matching polyhedron, Rapport AE 17/81, Inst. Act. & Econ., Univ. van Amsterdam, Amsterdam, 1981 (J. Combinatorial Theory (B), to appear).

[1982a] A. Schrijver, Min-max relations for directed graphs, Annals of Discrete Math. 16 (1982) 261–280.

[1982b] A. Schrijver, Proving total dual integrality with cross-free families – a general framework, Report AE 5/82, Inst. Act. & Econ., Univ. van Amsterdam, Amsterdam, 1982 (Math. Programming, to appear).

[1982c] A. Schrijver, Total dual integrality from directed graphs, crossing families, and sub- and supermodular functions, Proc. Waterloo 1982, to appear.

[1983a] A. Schrijver, Packing and covering of crossing families of cuts, Report AE 1/83, Univ. van Amsterdam, Amsterdam, 1983.

[1983b] A. Schrijver, Supermodular colourings, Report AE 4/83, Univ. van Amsterdam, Amsterdam, 1983.

[1977] A. Schrijver and P. D. Seymour, A proof of total dual integrality of matching polyhedra, Report ZN 79/77, Math. Centrum, Amsterdam, 1977.

[1979] A. Schrijver and P. D. Seymour, Solution of two fractional packing problems of Lovász, Discrete Math. 26 (1979) 177–184.

[1977] P. D. Seymour, The matroids with the max-flow min-cut property, J. Combinatorial Theory (B) 23 (1977) 189–222.

[1978a] P. D. Seymour, A two-commodity cut theorem, Discrete Math. 23 (1978) 177–181.

[1978b] P. D. Seymour, Sums of circuits, in: Graph theory and related topics (J. A. Bondy and U. S. R. Murty, eds.), Acad. Press, New York, 1978, pp. 341–355.

[1979a] P. D. Seymour, On multi-colourings of cubic graphs, and conjectures of Fulkerson and Tutte, Proc. London Math. Soc. (3) 38 (1979) 423–460.

[1979b] P. D. Seymour, A short proof of the two-commodity flow theorem, J. Combinatorial Theory (B) 26 (1979) 370–371.

[1980] P. D. Seymour, Four-terminus flows, Networks 10 (1980) 79–86.

[1981a] P. D. Seymour, On odd cuts and plane multicommodity flows, Proc. London Math. Soc. (3) 42 (1981) 178–192.

[1981b] P. D. Seymour, Matroids and multicommodity flows, Europ. J. Comb. 2 (1981) 257–290.

[1979] F. Sterboul, A characterization of the graphs in which the transversal number equals the matching number, J. Combinatorial Theory (B) 27 (1979) 228–229.

[1970] J. Stoer and C. Witzgall, Convexity and optimization in finite dimensions I, Springer, Berlin, 1970.

[1947] W. T. Tutte, The factorization of linear graphs, J. London Math. Soc. 22 (1947) 107–111.

[1952] W. T. Tutte, The factors of graphs, Canad. J. Math. 4 (1952) 314–328.

[1953] W. T. Tutte, The 1-factors of oriented graphs, Proc. Amer. Math. Soc. 4 (1953) 922–931.

[1954] W. T. Tutte, A short proof of the factor theorem for finite graphs, Canad. J. Math. 6 (1954) 347–352.

[1961] W. T. Tutte, On the problem of decomposing a graph into n connected factors, J. London Math. Soc. 36 (1961) 221–230.

[1981] W. T. Tutte, Graph factors, Combinatorica 1 (1981) 79–97.

[1964] V. G. Vizing, On an estimate of the chromatic class of a p-graph, Diskret. Analiz. 3 (1964) 25-30 (in Russian).
[1937] B. L. van der Waerden, Moderne Algebra, Springer, Berlin, 1937.
[1970] D. J. A. Welsh, On matroid theorems of Edmonds and Rado, J. London Math. Soc. 2 (1970) 251-256.
[1976] D. J. A. Welsh, Matroid theory, Acad. Press, London, 1976.
[1970] D. de Werra, On some combinatorial problems arising in scheduling, Canad. Oper. Res. Soc. J. 8 (1970) 165-175.
[1972] D. de Werra, Decomposition of bipartite multigraphs into matchings, Zeitschr. Oper. Res. 16 (1972) 85-90.
[1935] H. Whitney, On the abstract properties of linear independence, Amer. J. Math. 57 (1935) 509-533.
[1972] R. J. Wilson, Introduction to graph theory, Oliver and Boyd, Edinburgh, 1972.

Acknowledgements. I thank Dr. W. Cook and Dr. W. R. Pulleyblank for their helpful comments.

Generalized Gradient Methods of Nondifferentiable Optimization Employing Space Dilatation Operations

N. Z. Shor
Academy of Sciences of the Ukrainian SSR, V. M. Glushkov, Institute of Cybernetics, 142/144 40-letiya Oktyabrya Avenue, 252207 Kiev - 207, USSR

Introduction

A broad spectrum of complex problems of mathematical programming can be rather easily reduced to problems of minimization of nondifferentiable functions without constraints or with simple constraints. Thus, when decomposition schemes are used to solve structured optimization problems of large dimension or with a large number of constraints, the coordination problems with respect to linking variables (or dual estimates of linking constraints) as a rule prove to be problems of nondifferentiable optimization. The use of exact nonsmooth penalty functions in problems of nonlinear programming, maximum functions to estimate discrepancies in constraints, piecewise smooth approximation of technical-economic characteristics in practical problems of optimal planning and design, minimax compromise functions in problems of multi-criterion optimization, all of these generate problems of nondifferentiable optimization. Therefore, the efficiency of procedures for the solution of various complex problems of mathematical programming greatly depends on the efficiency of the algorithms employed to minimize nondifferentiable functions.

In modern practice, methods of minimization of functions with discontinuous gradient are also finding an ever-increasing use in the field of discrete programming for obtaining estimates in branch-and-bound type algorithms and for constructing estimates of the complexity of some classes of combinatorial problems.

A combined application of various methods of decomposition and cutting and of the nonsmooth penalty function method opens up ways for the versatile use of structural features of specific problems of nonlinear programming for reducing them to problems of unconditional minimization of nondifferentiable functions of as few variables as possible. The successful solution of complex optimization problems in general depends upon the degree of rationality of reduction and upon the efficiency of whatever methods of nonsmooth optimization are used.

We restrict ourselves to the study of deterministic methods for minimizing continuous functions, defined in finite-dimensional Euclidean space E_n, and differentiable almost everywhere. The locally Lipschitzian functions and representatives of various subclasses of such functions, e.g. almost differentiable, convex, weakly convex functions [13], etc., may serve as examples of such functions. These functions $f(x)$ allow us to use at the points x where the gradient is not defined the so-called generalized gradient sets $G_f(x)$ [51] (in the convex case the subgradient sets) whose representatives will be termed the generalized gradients of the function f at the point x (in the convex case the subgradients) through the operation of taking the limit points and the operation of convex closure. An attemt to extend directly the known gradient-type methods (for example the steepest descent method) to a sufficiently general class of nonsmooth functions (say, the class of convex functions) fails even though the gradient exists almost everywhere (that is, even though the probability of falling upon a point of discontinuity of the gradient equals zero). This is all the more so in the case of fast methods with quadratic approximation of the function being minimized (such as the method of conjugate gradients, or quasi-Newton methods).

Yet the problem of nonsmooth optimization has been attacked successfully in many directions for the past 20 years, and a whole arsenal of rather efficient methods has been developed and experience has been gained in solving complex practical problems. Ju. M. Ermol'ev, A. M. Gupal, E. A. Nurminskij, B. T. Poljak, etc., contributed greatly to the progress of stochastic methods of nondifferentiable optimization (see [18], [11], [22], and B. T. Poljak's "Subgradient Methods: A Survey of Soviet Research", Proceedings of a IIASA Workshop on Nonsmooth Optimization, 1978). However, these important results are beyond the scope of our review.

Four dominant ideas have played a great role in the advancement of deterministic methods of nondifferentiable optimization: 1) movement in a direction that effects a decrease of the distance to the minimum point when a step is small; 2) the use of cutting hyperplanes; 3) the choice of a stable direction of descent (ε-steepest descent); 4) transformations of the argument space aimed at smoothing "ravines" inherent in the function being minimized. The two former are implemented in the minimization of convex functions, while the two latter are used to find local minima of more general classes of functions. The first concept goes back to publications by D. Agmon [48], T. Motzkin and J. Schoenberg [61] where iterative relaxation methods of solution of linear inequalities have been studied. These were the so-called Fejer processes in which the distance to the desired domain decreased monotonically at each step. Early in the 60s I. I. Eremin explored the processes of Fejer approximation in more detail and elaborated, in particular, algorithms for the solution of systems of convex inequalities of the relaxation type [16].

B. T. Poljak also studied [25] the Fejer-type process for minimizing an arbitrary convex function $f(x)$ for the known function value at the minimum point $f*$:

$$x_{k+1} = x_k - \gamma \frac{f(x_k) - f*}{\|g_f(x_k)\|^2} g_f(x_k), \quad 0 < \gamma < 2,$$

where $g_f(x_k)$ is a subgradient of the function f at the point x_k. If $f*$ is unknown, procedures decreasing the distance to the minimum point cannot be implemented in their pure form. In [34] and in the author's thesis for the degree of Candidate of Physical and Mathematical Sciences (1964) the method of generalized gradient descent (GGD) for minimizing arbitrary convex functions was first studied in a sufficiently general form:

$$x_{k+1} = x_k - h_k \frac{g_f(x_k)}{\|g_f(x_k)\|}.$$

With a given (a priori) specification of the sequence $\{h_k\}_{k=0}^\infty$ the decrease of the distance to the minimum point x^* can be guaranteed at a given step only if $h_k < 2a_k$, where a_k is the distance from x^* to the level surface of the function $f(x)$ going through the point x_k.

In general, the sequence $\{\|x_k - x^*\|\}_{k=0}^\infty$ fails to be monotonic even when the sequence of step multipliers is chosen to satisfy the conditions

$$h_k > 0; \quad h_k \to 0; \quad \sum_k h_k = +\infty,$$

which ensure convergence of the sequence $f(x_k)$ from any initial approximation to f^*, assuming the boundedness of the set M^* of minima of f ([17], [24]). When the series Σh_k is required to be divergent, convergence of $\{\|x_k - x^*\|\}_{k=0}^\infty$ with the rate $O\left(\frac{1}{n^{1+\varepsilon}}\right)$ will not occur for any $\varepsilon > 0$.

For the domain M^* to be reached in a finite number of steps in the case of the constant step multiplier $h_k = h$, it is sufficient that it contain the sphere of radius $\zeta > \frac{h}{2}$. The subgradient descent method allows us to decrease the distance to M^* for a sufficiently small step due to the fact that the antisubgradient taken at the point $x \notin M^*$ makes an acute angle $\varphi(x, x^*)$ with the direction to the minimum point x^*. When a priori information is available that $\|x_0 - x^*\| \leq R$ and $\sup \varphi(x, x^*) \leq \varphi^* < \frac{\pi}{2}$, $x \in S(x_0, x^*) = \{x : \|x_0 - x^*\| \leq R\}$, then for $\varphi^* \geq \frac{\pi}{4}$ we put $h_0 = R \cos \varphi^*$, $h_{k+1} = h_k \cdot \sin \varphi^*$, $k = 0, 1, \ldots$, and obtain the GGD process converging to the minimum point with the rate of a geometric progression whose denominator is ([35], [46]) $q = \sin \varphi^*$. Similar results were obtained later in a number of papers of J. L. Goffin.

The fact that the hyperplane going through the point x and orthogonal to a subgradient of the convex function $f(x)$ at the point x is a hyperplane of support with respect to a convex body containing M^*, enables the minimum to be localized by cuts with hyperplanes of this kind. The concept of successive cuts is implemented in a number of algorithms, specifically in the cutting-plane method by Cheney-Goldstein-Kelley ([54]) and in CGT (the technique of centered cuts of the center of gravity) by A. Ju. Levin [19], in the modified center-of-gravity technique (MCGT) ([21], [47], [44]), and in its generalizations that received the generic name "ellipsoid methods" where, subsequent to one or several cuts, approximation of the minimum localization domain is performed by ellipsoids of as small a volume as possible.

The cutting-plane method converges in the function values with the rate $O\left(\frac{1}{k}\right)$ and requires at each step the solution of a linear programming problem with the number of constraints growing, generally speaking, from step to step. Characteristics of CGT are near the optimum convergence characteristics from the standpoint of the information theory of complexity of algorithms ([21], [47]), but calculation of the center of gravity of a sequence of convex polyhedra required by this method is such a complicated procedure that whenever the dimension of a problem exceeds 3, this method is impractical. As far as the ellipsoid method is concerned, being a cutting-plane method, it belongs simultaneously to the class of gradient-type methods with space dilatation. We will deal with this method in more detail later on.

Beginning with the research by D. P. Bertsekas and S. K. Mitter and later on by P. Wolfe [64] and C. Lemarechal [56], the ε-subgradient methods enjoyed further progress. For the convex function $f(x)$ defined on E_n, an ε-subgradient of f at the point x is a vector satisfying for all $x \in E_n$ the inequality $f(x) - f(x_0) \geq (g_\varepsilon(x_0), x - x_0) - \varepsilon$. A subgradient is an ε-subgradient with $\varepsilon = 0$. Let $G_\varepsilon(x_0)$ be the set of ε-subgradients at the point x_0. It is convex, bounded and closed. For $\varepsilon > 0$ the point-to-set mapping $x \to G_\varepsilon(x)$ is continuous [23], unlike the map $x \to G_0(x)$, for which only upper semi-continuity is valid. By the vector of ε-steepest descent from the point x we will mean the vector $p_\varepsilon(x) = -\eta_\varepsilon(x)$ where $\eta_\varepsilon(x) \in G_\varepsilon(x)$ and lies at the shortest distance from the origin of the coordinate system. It is easily seen that $p_0(x)$ gives the direction of steepest descent at the point x. $p_\varepsilon(x)$ is uniquely defined and is continuous with respect to x and ε for $\varepsilon > 0$. This enables us to derive the convergence of the function values with an accuracy of ε for idealized monotone descent processes of the form:

$$x_{k+1} = x_k - h_k^* p_\varepsilon(x_k)$$

where h_k^* is chosen by the condition of being minimum with respect to the direction. However, this algorithm is unrealizable in its pure form due to the complexity of description of $G_\varepsilon(x)$ and computation of $p_\varepsilon(x_k)$. C. Lemarechal [56] offered algorithms based on the concept of approximation of the ε-subgradient set by the convex closure of a finite set of its elements, these elements being the subgradients calculated at points obtained in a natural way during the process of line-search.

It should be noted that a similar concept was at the root of methods for solving minimax problems suggested earlier by V. F. Demjanov [12], but as representatives of the ε-subgradient set, the gradients of functions whose difference from the maximum function did not exceed ε were taken there with regard to specific characteristics of the problem at hand. B. Schwartz [63] modified the choice rule of the descent direction in minimax-type problems and made it continuous with respect to ε.

Algorithms of the ε-subgradient type include at each iteration the operation of projection of the origin of the coordinate system onto a convex polyhedron specified by its vertices in order to find the direction of descent as well as the search for an approximate minimum with respect to the direction. Complexity of the first operation involving the solution of a quadratic programming problem can vary in a vast range according to whether the algorithm must store all previous subgradients calculated during the approximation of the ε-subgradient set, or just a few of them. In the latter case, the rate of convergence is expected to slow down and this, as a rule, is corroborated by tests.

Theoretical results on the rate of convergence of the ε-subgradient processes are sparse at present. In any case a geometric rate of convergence has not been proven for sufficiently general classes of nonsmooth convex functions.

A detailed survey of methods of the ε-subgradient type is available in the monograph by V. F. Demjanov and L. V. Vasiljev [13]. A. M. Gupal [11] defined other methods of construction of the descent direction based on the use of smoothing functions and stochastic difference approximation of corresponding gradients and on the averaging of subgradients obtained at a given step and at the preceding steps. These methods can also be thought of as methods of the ε-subgradient type.

A number of publications by S. V. Rzhevskij [27, 28, 29] treat a sufficiently general class of algorithms of the ε-subgradient type which provide a means for more comprehensively taking into account data about a function and its subgradients obtained in the process of computations. The so-called closure property of the choice rule of the direction of descent in these algorithms allows a uniform proof of the convergence of a wide class of monotone algorithms of search for minimum and algorithms for finding saddle points.

The primary purpose of this paper is to review the methods of generalized gradient descent with space dilatation. Methods of this type were elaborated with the aim of accelerating the convergence of subgradient processes. The slow convergence of subgradient processes, as a rule, is caused by the fact that the upper boundary of angles between the direction to the minimum point and the antisubgradient direction is close or equal to $\pi/2$. In such a situation the distance to the minimum point decreases if a step is small by comparison with the distance to the minimum point, and the magnitude of the decrease is considerably smaller than the step size, so the speed of decrease in step also cannot be too high if the convergence to minimum must be guaranteed.

Non-orthogonal transformations of the argument space enable us to change angles between the direction of the subgradient calculated at some point and the direction from this point to the minimum point. In quasi-Newton algorithms with variable metric, central to which is the concept of quadratic approximation of a given function, the metric transformation is executed to obtain in the transformed space nearly-spherical level surfaces in a neighbourhood of the minimum. This is reduced formally to the construction of an approximation to the inverse matrix for the Hessian at the minimum point. A similar approach to minimization of nonsmooth functions is impractical in principle. For instance, the Hessian is equal to O almost everywhere for piecewise linear functions and quadratic approximation is absurd. For nondifferentiable optimization the author suggested deploying operators of space dilatation in appropriate directions as elementary operators of space transformation. Using the product of such operators we can pass from a spherical metric to an arbitrary ellipsoidal metric. The space dilatation in a specific direction with a large coefficient corresponds approximately to the gradient projection onto the subspace orthogonal to this direction. A family of algorithms based on the construction of a sequence of space transformation operators in the form of a product of space dilatation operators, and on the choice of the direction of descent corresponding to that of the antigradient in the space with transformed metric, may be dealt with. Among the algorithms of this family two types of algorithms were thoroughly investigated: with space dilatation in the direction of the subgradient [36, 37] and with space dilatation in the direction of the difference of two successive subgradients [38, 43].

For the exposition that follows some properties of space dilatation operators will be needed.

Let the vector $\xi \in E_n$, $\|\xi\| = 1$, and $\alpha \geq 0$ be given. Any vector $x \in E_n$ is uniquely represented in the form

$$x = \gamma_\xi(x)\xi + d_\xi(x) \qquad (1)$$

subject to the condition that $(\xi, d_\xi(x)) = 0$.

And here $\gamma_\xi(x) = (x, \xi)$; $d_\xi(x) = x - (x, \xi)\xi$.

Definition. The operator $R_\alpha(\xi)$ that acts upon the vector x, represented in the form (1), in the following way

$$R_\alpha(\xi)x = \alpha \gamma_\xi(x)\xi + d_\xi(x) = x + (\alpha - 1)(x, \xi)\xi, \qquad \alpha \geq 0$$

is defined as the operator of dilatation of the space E_n in direction ξ; it is a linear symmetric operator.

It is easily seen that

$$R_\alpha(\xi) \cdot R_\beta(\xi) = R_{\alpha\beta}(\xi),$$
$$R_1(\xi) = I; \qquad R_\alpha(\xi) \cdot R_{\alpha^{-1}}(\xi) = I; \qquad \alpha > 0;$$
$$R_0(\xi)x = x - (x, \xi)\xi,$$

that is $R_0(\xi)$ is the operator of projection onto the subspace orthogonal to ξ.

$$\|R_\alpha(\xi)x\|^2 = \|x\|^2 + (\alpha^2 - 1)(x, \xi)^2,$$

and the operator $R_\alpha(\xi)$ for $\alpha \geq 1$ ($\alpha \leq 1$) does not decrease (increase) the vector length. The operator $R_\alpha(\xi)$ for $n \geq 2$ has 2 eigenvalues: $\lambda_1 = \alpha$; $\lambda_2 = 1$. The subspace of eigenvectors generated by the vector ξ corresponds to the former and the orthogonal complement corresponds to the latter.

Let A be an arbitrary $n \times n$ matrix. Then $R_\alpha(\xi)A$ ($A R_\alpha(\xi)$) is a matrix resulting from a successive application of the operator $R_\alpha(\xi)$ to the columns (rows) of the matrix A. The required number of arithmetical operations in these transformations equals $O(n^2)$.

In vector form $R_\alpha(\xi) = I + (\alpha - 1)\xi\xi^T$.

Let A be a nonsingular matrix, $B = A^{-1}$; $BB^* = H$,

$$A^*A = H^{-1} = G; \; \xi = \frac{B^*p}{\|Bp\|}, \; (p \in E_n; p \neq 0).$$

$$\bar{A} = R_\alpha(\xi)A; \; \bar{B} = \bar{A}^{-1}; \; \bar{H} = \bar{B}\bar{B}^*; \; \bar{G} = \bar{A}^*\bar{A}.$$

Then

$$\bar{G} = A^* R_{\alpha^2}(\xi) A = A^* \left(I + (\alpha^2 - 1) \frac{B^*pp^TB}{\|B^*p\|^2} \right) A = G + (\alpha^2 - 1) \frac{pp^T}{(Hp, p)} \qquad (2)$$

$$\bar{H} = B \cdot R_{\alpha^{-2}}(\xi) B^* = H - \left(1 - \frac{1}{\alpha^2}\right) \frac{Bpp^TB^*}{(Hp, p)}. \qquad (3)$$

II. Nondifferentiable Optimization with Space Dilatation

Formulas of the form (2) and (3) are close to those of space transformation in the quasi-Newton methods with variable metric. The formal relationship between the methods of smooth optimization with variable metric and the methods with space dilatation was revealed in [30] where these formulas were deduced. They will be employed in the description of the gradient-type algorithms with space dilatation we are now embarking upon.

Family of Algorithms with Space Dilatation in the Direction of the Gradient and of the Difference of Two Sequential Gradients

The rate of convergence of the relaxation procedures developed by D. Agmon and T. Motzkin and I. Schoenberg [48, 61], I. I. Eremin [16], and of the similar B. T. Poljak procedure with the function value known at the minimum point [25] depends strongly on features of the conditions imposed upon a problem. The rate is slow when the angles between the antisubgradient direction and the direction towards the minimum point are close to $\frac{\pi}{2}$. As is evident from the above, this normally excludes the convergence to the minimum at the rate of a geometric progression whose ratio differs substantially from 1, also in the case of generalized gradient descent for $h_{k+1} = h_k q$ [35].

The operation of space dilatation in the direction of the gradient has been introduced by the author initially as a heuristic procedure to improve certain features of the problem, that is to decrease a "cross"-component of the gradient.

We now describe the schematics of the gradient-type algorithms with space dilatation in the direction of the gradient employed to minimize the function $f(x)$. $g_f(x)$ is a gradient (generalized gradient) of the function $f(x)$ at the point x.

Given:
$$x_0 \in E_n, \quad B_0 = A_0^{-1} = I \text{ (the unit } n \times n \text{ matrix)}.$$

We obtain after k steps:
$$x_k \in E_n; \quad B_k = A_k^{-1}.$$

A_k is the $n \times n$ matrix of the space transformation after k steps.

The $(k+1)^{th}$ step:

Point 1. Calculate $g_f(x_k)$ (if $g_f(x_k) = 0$ the process stops).

Point 2. Determine $\tilde{g}_k = g_{\varphi_k}(y_k) = B_k^* g_f(x_k)$ \hfill (4)
where $\varphi_k(y) = f(B_k y)$; $y_k = A_k x_k$. $\tilde{g}_k$ is a generalized gradient for the function $\varphi_k(y)$ defined in the "dilated" space.

Point 3. $\xi_k = \dfrac{\tilde{g}_k}{\|\tilde{g}_k\|}$; $x_{k+1} = x_k - h_k B_k \xi_k$. \hfill (5)

The formula (5) corresponds to the motion along the antigradient in the dilated space: $A_k x_{k+1} = y_k - h_k \xi_k$.

Point 4. $B_{k+1} = A_{k+1}^{-1} = B_k \cdot R_{\beta k}(\xi_k); \quad \beta_k = \alpha_k^{-1}$. (6)

Formula (6) corresponds to the dilatation of the "dilated" space in the direction of ξ_k: $A_{k+1} = R_{\alpha_k}(\xi_k) \cdot A_k$, $\alpha_k > 1$.

Point 5. $(k+1) \to (k+2)$. Go to Point 1.

Results of first computational experiments conducted with the given algorithm in cooperation with V. I. Bileckij in 1969 inspired hopes. A rational choice of space dilatation coefficients α_k and of an adequate strategy of changes in step multipliers h_k presented the greatest difficulty in the construction of an efficient algorithm. As efforts to make the descent monotone for the family of algorithms with space dilatation in the direction of the gradient proved to be unsuccessful in the nonsmooth case, we had to be content with an a priori specification of the procedure of regulation of step multipliers when moving in the direction of the antisubgradient in the dilated space. For the one-dimensional case simple reasoning suggests that if an algorithm with a constant step multiplier in the dilated space is employed, the space dilatation coefficient $\alpha_k \leq 2$ must be chosen to attain the convergence.

The early experiments demonstrated that the choice of $\alpha_k = 2$ and h_k = const. provides good results for many examples of convex ravine functions. Unfortunately, such a simple method does not always live up to expectations. In constructing other versions of algorithms successfully substantiated in theory, we selected the step multiplier and space dilatation coefficients such that the succession of distances to the minimum point in the metric of the corresponding "dilated" spaces did not increase. This principle assures the convergence of functional values at the rate of a geometric progression.

For this principle to be efficient, particular data about the function $f(x)$ are required, specifically, the value of this function at the minimum point f^* and the so-called growth constants M and N.

Theorem 1 [36, 37, 40]. Let $f(x)$, an almost-differentiable (locally Lipschitzian) function, be examined in a spherical neighbourhood $S_d(x^*)$ of the local minimum point x^* ($S_d(x^*) = \{x: \|x - x^*\| \leq d\}$) and at the points where the function is differentiable and its derivative in the direction $\mu(x) = x - x^*$ satisfies the inequality:

$$[f(x) - f^*]N \leq f'_{\mu(x)}(x) \leq M[f(x) - f^*], \quad M \geq N > 0; \quad f^* = f(x^*). \quad (7)$$

Then, if we assume in the algorithm (4)–(6) that

1) $x_0 \in S_d(x^*)$ (8)

2) $h_k = \dfrac{2M}{M+N} \dfrac{f(x_k) - f^*}{\|\tilde{g}_k\|}$ (9)

3) $\alpha_k = \alpha = \dfrac{M+N}{M-N}$, (10)

then

$$\|A_k(x_k - x^*)\| \leq d. \quad (11)$$

Note that the inequality (11) is equivalent to $(G_k(x_k-x^*), x_k-x^*) \leq d$ where $G_k = A_k^* A_k$ is a positive definite matrix with

$$\det G_k = (\det A_k)^2 = \left(\prod_{i=0}^{k-1} \det R_{\alpha_i}(\xi_i)\right)^2 = \prod_{i=0}^{k-1} \alpha_i^2.$$

For x^* this inequality yields localization in the form of an ellipsoid S_k with the center at the point x_k and with the product of lengths of semi-axes equal to $d^n/\det A_k = d^n \cdot \det B_k$, the ellipsoid volumes relation being $\mathscr{V}(S_k)/\mathscr{V}(S_{k+1}) = \dfrac{M+N}{M-N}$. Thus, the family of algorithms (4-6) covers as a particular case the algorithms which have been defined as the *ellipsoid* methods since the known work by L. G. Khachijan [31].

It is not difficult to prove the following Lemma. Let the sequence $\{x_k\}_{k=0}^{\infty}$ be bounded, $1+\delta \leq \alpha_k \leq \alpha^*$ as the algorithm (4)-(6) runs. Then there exist a subsequence $x_{k_1}, \ldots, x_{k_p}, \ldots$ and $c > 0$ such that $\|\tilde{g}_{k_p}\| \leq c \cdot \prod_{j=0}^{k_p} \beta_j^{\frac{1}{n}}, p = 1, 2, \ldots$.

From this Lemma it is inferred in the conditions of Theorem 1 that the following statement holds [37]: there is a constant $\bar{c} > 0$ and a subsequence $\{k_p\}_{p=1}^{\infty}$, $k_p < k_{p+1}$, such that $f(x_{k_p}) - f^* \leq \bar{c} \alpha^{\frac{-k_p}{n}}$. A more precise statement is proved for this case in [45].

Theorem 2. Under the assumptions of the previous Lemma for $\alpha_k = \alpha > 1$, one has

$$\gamma_k = \min_{0 \leq r \leq k} \|\tilde{g}_r\| \leq \frac{d\sqrt{k(\alpha^2-1)}}{\sqrt{\alpha^{\frac{2k}{n}} - 1}}; \quad d = \max_{x \in S_d(x^*)} \|g_f(x)\|.$$

From this Theorem it follows that for the case of the algorithm with space dilatation in the direction of the gradient with known f^* and M and N the following estimate is correct:

$$\min_{0 \leq r \leq k} [f(x_k) - f^*] \leq \frac{\bar{c}\sqrt{k(1-\beta^2)} \cdot \beta^{k/n-2}}{\sqrt{1-\beta^{\frac{2k}{n}}}}$$

where $\beta = \dfrac{M-N}{M+N}$.

Consequently, the rate of convergence depends on $\dfrac{M-N}{M+N}$. It might be as well to point out that $M=N=2$ may be chosen for a quadratic positive definite function in the inequality (7) and $M=N=1$ for a piecewise linear function whose supergraph is a cone with a vertex at the point (x^*, f^*). For these cases $\beta=0$ the algorithm degenerates and converges in a number of steps no greater than n.

When dealing with a non-singular system of linear equations in n unknowns, $(a_i, x) + b_i = 0, i = 1, \ldots, n$, we can find the minimum of $f(x) = \max_i |(a_i, x) + b_i|$ rather than solve it. Putting $f^* = 0$, $\beta_k = 0$ and applying the algorithm (4-6)-

(8-10) we obtain an algorithm corresponding to the known finite procedure of solution of linear algebraic systems, the method of gradient orthogonalization.

A similar procedure can be used in the solution of systems of nonlinear equations $f_i(x)=0$, $i=1,\ldots,n$. It can be shown for $f(x)=\max|f_i(x)|$ that if x^*, the system solution, is a regular point (that is $f_i(x)$ are continuously differentiable at this point and the Jacobian of the system $\det J(x^*)$ differs from zero), then for any $\delta>0$ there is a sufficiently small neighbourhood $S_d(x^*)$ such that the constants M and N in (7) can be chosen as $M=1+\delta$; $N=1-\delta$; $\beta=\dfrac{M-N}{M+N}=\delta$, respectively. As indicated in [46], when a limiting version of the algorithm is employed with $\beta=0$ and with reset after each n iterations (large cycle), a quadratic rate of convergence (with respect to the large cycles) can be demonstrated for systems of nonlinear equations under the normal assumptions of smoothness and regularity. The application of this algorithm in the form indicated in Theorem 1 in the general case faces difficulties due to two circumstances: a) M and N are unknown, b) f^* is unknown. Nevertheless if $f(x)$ is convex it is possible to choose $N=1$. For the majority of practical problems M can be chosen within the limits from 3 to 10, the starting approximation being not too bad. Assume that M and N are chosen correctly. Then f^* can be successively approached through a process similar to dichotomy thanks to the following:

Theorem 3 [37]. Let the convex function $f(x)$ possess the following properties: there exists $M\geqslant 1$ such that if $\varphi(\alpha)=f[(1-\alpha)x_1+\alpha x_2]$, $0\leqslant\alpha\leqslant 1$, strictly decreases in α, the inequality

a) $f'_{x_1-x_2}(x_1)\leqslant M[f(x_1)-f(x_2)]$ and

b) $\lim\limits_{\|x\|\to\infty} f(x)=+\infty$.

holds. Then, if with the use of the algorithm (4-6)-(8-10)

$$\alpha_k=\frac{M+1}{M-1};\quad h_k=\frac{2M}{M+1}\frac{f(x_k)-\bar{f}}{\|\bar{g}_k\|}$$

$\bar{f}$ is chosen greater than or equal to f^*, the sequence $\{h_x\}$ is bounded and for an arbitrary $\varepsilon>0$ there is $\bar{k}$ such that $f(x_{\bar{k}})<\bar{f}+\varepsilon$ (the computation is interrupted if $f(x_k)\leqslant\bar{f}$ at some step); if $\bar{f}$ is smaller than f^*, then $\{h_k\}$ is unbounded.

It is obvious that with f^* unknown the laboriousness of the algorithm used to attain a fixed precision of minimization over the functional increases materially.

Experience of application of the algorithms with space dilatation in the direction of the gradient revealed the possibility of significant acceleration of the subgradient processes with employment of linear operators that change the space metric. Difficulties inherent in the selection of step multipliers when f^* is unknown gave an impetus to the search for new methods of nonsmooth optimization with variable metric where the selection of a step multiplier is associated with an approximate search for the minimum with respect to the direction. In 1971 the author suggested in his doctoral thesis and, in cooperation with N. G. Zhurbenko, studied experimentally a family of gradient-type algorithms with space dilatation in the direction of the difference of two successive gradients [38].

It is a matter of experience that on minimization of positive definite quadratic functions by the method of conjugate gradients a sequence of gradients $\{g(x_k)\}_{k=0}^{n-1}$ taken at the points of the minimizing sequence has the following property: $g(x_{k+1}) - g(x_k)$, $k = 0, \ldots, n-2$, is orthogonal to the direction from the point x_{k+1} to the minimum point.

Let $f(x)$ be the convex piecewise linear function, x the point lying in the first domain near the boundary of two domains of linearity (a) $f_1(x) = (g_1, x) + b_1$; (b) $f_2(x) = (g_2, x) + b_2$ and $(g_1, g_2) < 0$. After one step of the subgradient descent $x_1 = x_0 - h g_1$ we obtain for a specific h the point x_1 lying on the boundary of two domains. Putting $g(x_1) = g_2$ we derive that $g_2 - g_1$ is orthogonal to the linear manifold $f_1(x) = f_2(x)$ in which the direction of steepest descent from the point x_1 lies.

Thus, in the first and second examples alike the space dilatation in the direction of the difference of two successive gradients (subgradients) decreases the components of the gradients orthogonal to the direction towards the minimum point (direction of steepest descent), that is, in a sense, it is "useful". The above heuristic considerations are at the root of the family of algorithms with space dilatation in the direction of the difference of two successive gradients, or for short, the "r-algorithms". We shall call r-algorithm of minimization of the almost differentiable [41] function $f(x)$ the iteration procedure of the following form:

Let there be given: $x_0 \in E_n$, $\tilde{g}_0 = 0$; let $B_0 = I$ be the unit matrix. Let k iterations be executed, n-dimensional vectors x_k, $\tilde{g}_k$ and the matrix $B_k = \{b_{ij}\}_{i,j=1}^n$ be calculated.

The $(k+1)^{th}$ iteration. Calculate:
1) $g_f(x_k)$ the almost-gradient of the function $f(x)$ at the point x_k; if its definition is not unique we take an almost-gradient such that

$$(B_k \tilde{g}_k, g_f(x_k)) \leq 0. \qquad (12)$$

2) $g_k^* = B_k^* g_f(x_k)$ the almost-gradient of the function $\varphi_k(y) = f(B_k y)$ at the point $y = A_k x_k$; $A_k = B_k^{-1}$ is the matrix of transformation of the space after k iterations.
3) $r_k = g_k^* - \tilde{g}_k$, the difference of two almost-gradients of the function $\varphi_k(y)$ calculated at the points $y_k = A_k x_k$; $\tilde{y}_k = A_k x_{k-1}$.
4) $\xi_k = \dfrac{r}{\|r_k\|}$.
5) $B_{k+1} = B_k B_{\beta k}(\xi_k)$. Note that $B_{k+1} = A_{k+1}^{-1}$, $A_{k+1} = R_{\alpha k}(\xi_k) A_k$; $\alpha_k = \beta_k^{-1}$ is the coefficient of space dilatation.
6) $\tilde{g}_{k+1} = R_{\beta k}(\xi_k) g_k^* = B_{k+1}^* g_f(x_k)$ is the value of the almost-gradient of the function $\varphi_{k+1}(y) = f(B_{k+1} y)$ at the point $y = A_{k+1} x_k$.
7) $x_{k+1} = x_k - h_k B_{k+1} \tilde{g}_{k+1} / \|\tilde{g}_{k+1}\|$; h_k is chosen from the condition of approximate minimum with respect to the direction. In the approximate search for the minimum one has to cross the minimum point for the condition of the form (12) to be satisfied at the subsequent step.

Formula (7) implements the mapping of the step of subgradient descent for the function $\varphi_{k+1}(y)$ defined in the "dilated" space into the original space.

8) Transition to the $(k+2)^{\text{th}}$ iteration and storing of $x_{k+1}, \tilde{g}_{k+1}, B_{k+1}$ or the end of the algorithm operation by some stopping criterion.

The following statement is valid for the idealized version of the algorithm, the so-called $r(\alpha)$-algorithm, where $\alpha_{k+1} = \alpha > 1$ and h_k is chosen from the condition for obtaining the local minimum nearest to x_k along the direction.

Theorem 4 [43, 46]. If $f(x)$ is a continuous piecewise smooth function, $\lim_{\|x\| \to \infty} f(x) = +\infty$ and the sequence $\{x_k\}$ generated by the $r(\alpha)$-algorithm possesses the property that $\lim_{k \to \infty} \|x_{k+1} - x_k\| = 0$, then for any $\alpha > 1$ there exists a limit point $\bar{x}$ of the sequence $\{x_k\}$ whose set of almost-gradients forms a linearly dependent family of vectors. By virtue of the monotonicity of the descent process here $\lim_{k \to \infty} f(x_k) = f(\bar{x})$.

Corollary. By the above Theorem, if x^* is an isolated local minimum, if x_0 is a point such that the connected component of the set $\{x : f(x^*) \leq f(x) \leq f(x_0)\}$ containing x^* and x_0 does not contain other points z aside from x^* where the family of almost-gradients is linearly dependent, then the sequence $\{x_k\}_{k=0}^{\infty}$ generated by the $r(\alpha)$-algorithm converges to x^*.

As far as the convergence rate of the r-algorithms is concerned, this problem has not been given enough attention. It is only known that the limit version of the r-algorithm with $\beta = 0$ and with resetting of the space transformation matrix after n iterations (the large cycle) is, in fact, similar to the method of projective conjugate gradients, and for continuously twice-differentiable convex functions whose Hessian satisfies the Lipschitz condition and the inequalities $m\|y\|^2 \leq (Hy, y) \leq M\|y\|^2$, $M \geq m > 0$, it converges at a quadratic rate (with respect to large cycles).

At the same time wide experience has been gained in solving various sophisticated problems of minimization of nonsmooth functions with employment of the family of r-algorithms. This experience made it possible to find rational parameters of r-algorithms ensuring a fairly high efficiency of the family. It is advantageous, for instance, to choose the space dilatation parameter α in the interval $2 \leq \alpha \leq 3$ and, what is more, to follow a simple strategy of search for an approximate minimum with respect to the direction, that is to move with a constant step in a dilated space until the function value in a successive point is greater than in a preceding one.

The length of this constant step is adaptively changed when passing from iteration to iteration lest the number of steps of the line search for the minimum should be large on the average. The matrix B_k reduces the step to small size in the real space (see 7 (p. 13)).

When minimizing n-dimensional convex functions, application programs developed on the basis of the r-algorithms provide convergence in function values at the rate of a geometric progression, a relative precision with respect to the functional therewith becomes, as a rule, 3 to 10 times higher in n iterations.

Progress in practical application of the r-algorithms has fostered search for methods with space dilatation as efficient as the existing ones but better justified theoretically. Though ultimate results are still to be achieved, those already attained have had a pronounced effect not only on the development of methods of nonsmooth optimization but on more general problems of the theory of complexity of algorithms. We now turn to the ellipsoid method.

The Ellipsoid Method, Its Modifications and Theoretical Applications

The author came to the ellipsoid method while analysing the subgradient-type processes with space dilatation in the direction of the gradient with constant $\alpha_k = \alpha > 1$ and a sequence of step multipliers $\{h_k\}_{k=0}^{\infty}$ varying by the formula of geometrical progression $h_{k+1} = h_k \rho$. As it turned out in such processes for convex functions, generally speaking, one cannot expect the distance from a current point to a minimum point in the dilated space to be bounded as in Theorem 1, but a suitable choice of α and $\rho > 1$ enables one to make the volume of the sphere of radius equal to the distance from the minimum point to the current point in the dilated space, grow more slowly than the geometrical progression of ratio α. This provided convergence of the "record" sequence at the rate of a geometric progression.

Thus, the method (4)–(6) having the following parameters:

$$\alpha_k = \alpha = \sqrt{\frac{n+1}{n-1}}; \quad n \geq 2;$$

$$h_0 \geq \frac{\|x_0 - x^*\|}{n+1}; \quad h_{k+1} = h_k \rho; \quad \rho = \frac{n}{\sqrt{n^2 - 1}}$$

was employed to solve the problem of minimization of the convex function $f(x)$, $x \in E_n$. This method was described in [44] and the fundamental results consist of the following:

Theorem 5. The minimizing sequence $\{x_k\}_{k=0}^{\infty}$ generated by the foregoing algorithm satisfies the condition:

$$\|A_k(x_k - x^*)\| \leq h_k(n+1), \quad k = 0, 1, 2, \ldots$$

whence we have that x^* is localized after k iterations in an ellipsoid S_k of volume $\mathscr{V}(S_k) = \mathscr{V}_0 \cdot q_n^k$ where

$$q_n = \sqrt{\frac{n-1}{n+1}} \left(\frac{n}{\sqrt{n^2 - 1}} \right)^n \approx 1 - \frac{1}{2n}.$$

When the manuscript of the paper [44] was prepared for publication, the author came to hear at one of the seminars that a similar algorithm had been developed by A. S. Nemirovskij and D. B. Judin in a distinct form and for different considerations and published by the name "Modified Center-of-Gravity Technique" (MCGT) in [47].

D. B. Judin and A. S. Nemirovskij analysed the information complexity of subgradient algorithms and inferred that from the standpoint of the information criterion (that is the number of "portions" of information about subgradients and function values at specific points required to achieve a prescribed relative precision) the known algorithm by A. Ju. Levin [19], the center-of-gravity technique (CGT), is close to the optimal one.

Each iteration of CGT consists in finding the center of gravity x_c of a polyhedral convex domain of localization of the minimum, in calculating the subgradient $g(x_c)$ at the point x_c, and in cutting by the hyperplane orthogonal to $g(x_c)$. After such cutting, the volume of the localization domain is reduced by a value not less than $\left(\frac{n}{n+1}\right)^n$ times the initial volume. Consequently, CGT guarantees the reduction of the localization domain volume at the rate of geometrical progression whose ratio is

$$q_n = 1 - \left(\frac{n}{n+1}\right)^n < 1 - \frac{1}{e}.$$

However, the complexity of computation of the center of gravity makes CGT impractical for $n \geq 3$.

It was suggested in [47] to choose a centrally symmetric body, an ellipsoid as an alternative to the domain of localization of the minimum and after the cutting to "simplify" the obtained body by circumscribing an ellipsoid of a minimal volume around it. A linear transformation of space can transform an ellipsoid into a sphere, therefore all operations can be performed on the sphere by going from one coordinate system to another.

So, let there be as the domain of localization x^*, the sphere of radius R with center at the point x_0. Calculating the subgradient $g(x_0)$ at the point x_0 and constructing the proper cutting hyperplane we obtain a hemisphere. Calculate parameters of the ellipsoid of minimal volume which is circumscribed around the hemisphere. From symmetry considerations it follows that the center of this ellipsoid is on a ray drawn from the point x_0 towards the antisubgradient, the small semi-axis has the same direction as $g(x_0)$ and its length is equal to $\frac{nR}{n+1}$, all other semi-axes are $\sqrt{\frac{n+1}{n-1}}$ times longer. Executing the operation of space dilatation in the direction of the subgradient $g(x_0)$ with the dilatation coefficient $\alpha = \sqrt{\frac{n+1}{n-1}}$ we obtain the sphere of radius $\frac{nR}{\sqrt{n^2-1}}$ as the domain of localization of the minimum in the dilated space. The cycle of computations is closed.

Thus, the ellipsoid method may be considered as a particular case of the methods of space dilatation in the direction of the gradient. It requires a number of vector-matrix transformations with an amount of computing operations of the order of $O(n^2)$ at each iteration in addition to calculations of the subgradients, that is, this method is easily realizable. However, the volume of the domain of localization of the minimum decreases at a rate approximately n times lower in MCGT than in CGT, and the rate of convergence of the "record" sequence (i.e. $\min\{f(x_i), i \leq k\}$) is nearly as many times lower too.

In MCGT, the time for practical calculations is a factor of about n times more than that in the r-algorithm.

When employing MCGT in the solution of complex programming problems, one can easily see that this method has a low sensitivity to an increase in the number of constraints. In contrast to unconstrained minimization, if some constraint is not satisfied in a current point, then the subgradient of this constraint is taken instead of the subgradient of the objective function, otherwise the algorithm is unchanged.

The ellipsoid method proved to be of considerable value in a theoretical sense despite its relatively slow convergence. The scheme of this method is simple and clear, the rate of convergence of the "record" sequence is geometric with ratio depending on the dimension of the space only (and not on the specific function being minimized).

L. G. Khachijan was the first to discover theoretical "abilities" of the ellipsoid method associated with the generation of polynomial algorithms for solving optimization problems. The following problem of the theory of algorithmic complexity remained unsolved over a long period of time: given an arbitrary system of linear inequalities with integer coefficients, construct an algorithm operating in polynomial time relative to the volume of information about this system and answering the question whether the system has a solution.

The existence of such an algorithm readily implies the existence of a polynomial algorithm that yields an explicit solution of a linear programming problem in integer (rational) coefficients. Attempts to solve this problem by the simplex-method-type procedures failed as the number of iterations in methods of this type can grow exponentially relative to the problem dimension.

L. G. Khachijan took a different way [31, 32]. He obtained simple expressions for the R-radius of the sphere with center in the origin, where the system solution is guaranteed if it exists at all, as well as a lower bound $\bar{\Delta}$ for the minimum value of the maximum of the system discrepancy if the system is inconsistent. Then he reduced the system solution to the problem of minimization of the maximum discrepancy function (minimization of a convex piecewise linear function) by the ellipsoid method. The sphere of radius $\bar{R}$ whose center is in the origin was employed as the initial sphere and iterations of the ellipsoid algorithm were performed until either a system solution was obtained, or a given precision of the order $\bar{\Delta}/2$ was reached. The number of iterations of the ellipsoid method is no greater than $O(n^2 L)$, where L is the information complexity of the system. This immediately implied the polynomiality of the algorithm.

New publications have appeared recently whose authors endeavoured to develop polynomial algorithms for solving systems of linear inequalities in a different way.

Thus, making use of methods of graph theory, a rather complex polynomial algorithm has been constructed in [49] which solves linear inequality systems where each inequality has no more than two non-zero coefficients. Modifications of the Agmon-Motzkin-Schoenberg relaxation algorithms have formed the basis of polynomial algorithms developed in [58] for the special case of systems of inequalities with unimodular matrices. Estimates of the number of operations are much worse there than in the ellipsoid method.

Before long, the ellipsoid method made it possible to gain a number of new results. Thus, a polynomial algorithm for solving integer-coefficient problems of convex quadratic programming has been developed in [33], while polynomial algorithms have been offered for solving problems of analysis of properties of the polyhedral convex set P specified by a system of linear inequalities having rational coefficients, in particular, the problem of isolating non-valid constraints, finding a minimal generating system of vertices and generators of a polyhedral set, etc. (see [62]).

The ellipsoid method has been used as a foundation for a polynomial algorithm for the solution of a linear complementarity problem for the case of a positive definite matrix A: for the given $n \times n$ matrix A and the vector $q \in E_n$ define a vector $z \in E_n$ so that $Az+q \geqslant 0$; $z \geqslant 0$; $(z, Az+q) = 0$. This problem may be treated as a particular case of a convex quadratic programming problem for which the polynomial algorithm is available.

The most interesting results associated with the employment of the ellipsoid method for analysing combinatorial problems have been obtained in [52], [53]. The problem of linear function maximization on an n-dimensional convex body K specified by the parameters r, R (the interior and exterior radii) is dealt with in [52]. This problem is related to the (weak) optimization problem: for a given $c \in E_n$ and $\varepsilon > 0$ find the vector $y \in E_n$ such that (1) $d(y, K) \leqslant \varepsilon$ ($d(y, K)$ is the Euclidean distance) and (2) for any $x \in K$ $(c, x) \leqslant (c, y) + \varepsilon$; and the (weak) separation problem: for a given vector $y \in E_n$ and $\varepsilon > 0$ either (1) show that $d(y, K) \leqslant \varepsilon$ or (2) find the vector $c \in E_n$, $\|c\| \geqslant 1$ such that for any $x \in K$, $(c, x) \leqslant (c, y) + \varepsilon$.

The following Theorem is proved: Let K be the class of convex bodies specified by the parameters r and R and let $\varepsilon > 0$. If there exists a polynomial algorithm for the weak separation problem, then the ellipsoid method solves the weak optimization problem in no more than $O\left(n^2 \log \dfrac{R^2}{r\varepsilon} \|c\|\right)$ steps. The required precision of the separation problem solution is of the order of $O\left(\dfrac{R}{n}\left(\dfrac{R^2}{r\varepsilon} \|c\|\right)^{-n^2}\right)$ and calculations may be executed with a precision no greater than $O\left(n^2 \log \dfrac{R^2}{r\varepsilon} \|c\| \log \dfrac{\sqrt{n}}{R^2}\right)$ bits.

This work also shows the polynomial equivalence of (weak) optimization and (weak) separation problems for the given class of convex bodies.

The following Theorem is valid for rational convex bodies: If the (weak) separation problem is solvable polynomially for a class of rational convex bodies P, that is the bodies all vertices of which are described by rational coordinates, then the solution of the exact separation problem and the exact optimization problem can be achieved in time being a polynomial in n, $\log \|c\|$ and log T, where T is an upper bound on the largest numerators and denominators of components of vertices of P.

Taking advantage of the above results M. Grötschel, L. Lovász and A. Schrijver [52] have demonstrated that the majority of the known combinatorial problems for which polynomial algorithms are available, can be solved in polynomial time through the application of the ellipsoid method by passing to the corresponding separation problem. This list encompasses such problems as the minimal cut, optimization of a linear function over the intersection of two matroids, maximal weighted matching, Chinese postman problem, minimization of submodular functions on the lattice of subsets, etc. New results on the NP-completeness of some problems are obtained through the application of the equivalence theorem.

The most elegant mathematical results are attained for the Lovász estimate ([57], [53]) of the independence number which simultaneously is the estimate of the Shannon information capacity of a graph [57].

By an orthonormalized representation of a graph G with n vertices will be meant a system $(v_1, \ldots, v_n)$ of unit vectors in the space E_n such that, if i and j are non-adjacent vertices, then $(v_i, v_j) = 0$. The value of an orthonormalized representation $U = \{u_1, \ldots, u_n\}$ is the number $\varphi(U)$:

$$\varphi(U) = \min_{c, \|c\|=1} \max_i \frac{1}{(c, u_i)^2}, \quad c \in E_n.$$

It is shown in [57] that the number $\vartheta(G) = \min \varphi(U)$ is an upper estimate for the independence number $\alpha(G)$ and for $\theta(G)$ (which is the number of the Shannon capacity of the graph), which is no greater than $\alpha^*(G)$, the fractional independence number that is obtained in the solution of a relative problem of linear programming whose number of constraints can grow, generally speaking, exponentially with respect to the number of graph vertices.

In [57] the problem of finding the Lovász number $\vartheta(G)$ is reduced to the following convex programming problem:

Considered is a class of nonnegative definite symmetric matrices $B(G) = \{b_{ij}\}_{i,j=1}^n$, such that $\sum_{i=1}^n b_{ii} = 1$, $b_{ij} = 0$ if i and j are adjacent vertices.

Find $B^*(G)$ where $\vartheta(G) = \max_{\{b_{ij}\}} \sum_{i,j=1}^n b_{ij}$ is reached under the stated constraints.

In an effort to solve this problem M. Grötschel, L. Lovász, A. Schrijver made use of the ellipsoid method and constructed an algorithm polynomial with respect to n and $\log \frac{1}{\varepsilon}$ (ε is the precision of determination of $\vartheta(G)$) applying an interesting technique of cuts based on the Silvester criterion for nonnegative definiteness of the matrix $B(G)$. The Lovász estimate $\vartheta(G)$ is employed in [52, 53] as the basis of a polynomial algorithm for finding $\alpha(G)$ and a maximum independent set for the so-called perfect graphs where the numbers $\alpha(G)$, $\theta(G)$, $\vartheta(G)$, $\alpha^*(G)$ coincide, and what is more, this property holds for their arbitrary subgraphs.

Another method of finding $\vartheta(G)$ is connected with a convex programming problem. Consider a class of symmetric matrices $A(G) = \{a_{ij}\}_{i,j=1}^{n}$ where $a_{ii} = 1$, $i = 1, \ldots, n$; $a_{ij} = 1$ if i and j are non-adjacent vertices.

Find
$$\vartheta(G) = \min_{\{a_{ij}\}} \max_{i} \lambda_i [A(G)], \tag{13}$$

where $\lambda_i(A)$ are eigenvalues of the matrix A.

This is a representative problem of nonsmooth optimization with respect to variables a_{ij} corresponding to edges of the graph. The author in cooperation with the final-year student I. S. Danilovich conducted numerical experiments on the solution of (13)-type problems by the r-algorithm. Graphs with 5, 6, 24, 40, 60, vertices and respectively 5, 7, 66, 100, 180 edges were examined and rather satisfactory results were obtained. Thus for the graph C_5, 6 decimal places were calculated accurately in 56 iterations, for a graph with 6 vertices 7 accurate decimal places were reached in 35 iterations, for a graph with 24 vertices 150 iterations yielded 4 accurate decimal places, for a graph with 40 vertices 2 accurate decimal places were attained as a result of 100 iterations and for a graph with 62 vertices 60 iterations resulted in a solution of accuracy 0.1.

It should be noted that, as indicated in [53], the ellipsoid method operates with far less speed, so the calculation of $\vartheta(G)$ even for graphs having from 10 to 15 vertices involves material computational difficulties.

A vast number of modifications were devised to improve convergence of the ellipsoid method. A fairly comprehensive survey of pertinent Western publications is offered in [50, 62, 65]. So, I will dwell here on the Soviet publications ([5], [7], [9], [41]).

One of the ways of bringing closer the information efficiency of the MCGT and CGT algorithms consists in the examination of more complex localizing bodies than a hemisphere, which can be obtained owing to additional information about a function and after two or more successive cuts. The greater the number of successive cuts under examination, the more complex, generally speaking, the problem of approximation of the localizing body by the circumscribed ellipsoid, but the greater gain by the criterion of decrease in volume of the described ellipsoid.

Let the point $x_0 \in E_n$ be fixed. The point is not the minimum point of the convex function $f(x)$. The basic inequality from which the cuts follow has the form: $f(x) \geq f(x_0) + (g_f(x_0), x - x_0) \ \forall \ x \in E_n$. Setting $x = \bar{x}$, $\bar{f} = f(\bar{x})$, we obtain that the required minimum points are in the half-space: $H(x_0, \bar{f}) = \{x : (-g(x_0), x - x_0) \geq f(x_0) - \bar{f}\}$. The corresponding cutting hyperplane $\mathscr{P}(x_0, \bar{f})$ is defined by the equation

$$\left(g(x_0), x - x_0 + \frac{f(x_0) - \bar{f}}{\|g(x_0)\|} \cdot \frac{g(x_0)}{\|g(x_0)\|} \right) = 0.$$

It differs from the hyperplanes used in MCGT for $f(x_0) > f(\bar{x})$ by a parallel displacement towards the antisubgradient by $\Delta = \dfrac{f(x_0) - \bar{f}}{\|g(x_0)\|}$. If the value of the function at the minimum point f^* is known, then the best cut of this type is ob-

II. Nondifferentiable Optimization with Space Dilatation

tained for $\bar{f} = f^*$. When f^* is unknown, a hypothetical value f^s may be used as an alternative to $\bar{f}$, and if we are not certain that $f^* \leq f^s$, then the corresponding cut is not justified and may be used only at the test steps of the algorithm.

Let us examine a "standard" situation: an n-dimensional sphere $S_0 = \{x : \|x - x_0\| \leq r_0\}$ enclosing the minimum point is given together with the constant for displacements f^s. The following procedure of construction of the localizing body T is suggested.

Calculate the value of $f(x_0)$ and $g_f(x_0)$ and obtain $\mathscr{P}(x_0, f^s)$, that is the first cutting hyperplane. Calculating $f(\bar{x})$ and $g(\bar{x})$ at a given point of the interval $L_0 = \{x : x = x_0 - tg(x_0), t > 0; x \in S_0\}$ we obtain $\mathscr{P}(\bar{x}, f^s)$, the second cutting hyperplane. The choice of T depends upon the situation:

(1) $(g(\bar{x}), g(x_0)) > 0$: an "accompanying" cut is obtained. As T we take one of the segments $H(x_0, f^s) \cap S_0$ and $H(\bar{x}, f^s) \cap S_0$ that is further from the point x_0.

(2) $(g(\bar{x}), g(x_0)) \leq 0$: a "meeting" cut is obtained.

Suppose $W = H(x_0, f^s) \cap H(\bar{x}, f^s)$ and $V = \mathscr{P}(x_0, f^s) \cap \mathscr{P}(\bar{x}, f^s)$, and fix the point $v_0 \in V$ lying in the plane that goes through x_0 collinearly to $g(x_0)$ and $g(x_1)$. If $v_0 \notin \text{int } S_0$, then a "layer" (a figure obtained as a result of intersecting the sphere with a body bounded by two parallel hyperplanes) of minimal thickness is taken that contains $W \cap S_0$. When $v_0 \in \text{int } S_0$, the so-called S-pyramid $\Pi = W \cap S_0$ is taken as T.

Therefore, the choice of the point $\bar{x} \in L_0$ and $g(\bar{x})$ uniquely defines the localizing body: the segment, the layer or S-pyramid. For the listed figures formulas are derived in [5, 41] which enable the circumscribed ellipsoids $\Phi(T)$ of minimal volume to be constructed for them. After one application (or in the case of the S-pyramid two applications) of the space dilatation operator the localizing ellipsoid becomes the sphere and we pass to the new cycle of calculations. In comformity with the heuristic criteria [5, 7], we suggest a value of the order $r_0/\sqrt{n}$ for the stepsize in the procedure searching the point $\bar{x}$ required for the second cut.

Undoubtedly, more than two cuts may be executed to construct the body T, but the algorithm therewith becomes considerably more complex.

Efficiency of the algorithm depends also on the strategy of choice of f^s. If f^* is known (for instance, when finding solutions of a system of convex inequalities) there is no problem at all: $f^s = f^*$. If we choose $f^s \geq f^*$, then after some number of iterations we can obtain x_k such that $f(x_k) \leq f^s$, otherwise $\liminf_{k \to \infty} f(x_k) = f^s$, the sequence of ellipsoids therewith contains at each step the minimum point. If $f^s < f^*$, then "hypercuts" occur and after a finite number of steps we shall obtain an empty localizing body. Consequently, there are indications which make it possible to construct the sequence f_i^s converging to f^*.

An algorithm with double cuts and adaptive selection of f^s is described in [7, 41]. Numerical experiments demonstrated a certain advantage of the method with double cuts as against MCGT which become greater with an increase of dimension. In order to attain a comparable precision this method required approximately 3 times less calculations of functions and gradients when

f^* was unknown and the dimension $n=20$ than MCGT. When f^* is known, the amount of computations for $n=20$ by the double cutting method is from 4 to 5 times less than in the case when f^* is unknown.

An added means of acceleration of the ellipsoid method consists in the use of the function growth constants M and N (see the inequality (7)). The value of these constants and f^* (N can always be chosen equal to 1) enable us to obtain a fast converging version of the ellipsoid method [9], with a ratio of reduction in the amount at each iteration

$$q < \frac{M-N}{M+N} \leq \frac{M-1}{M+1}.$$

Various modifications of the ellipsoid method described above are embedded in the general basic scheme [7, 41]. Let x^* be the minimum point of the convex function $f(x)$ and an original sphere $S_0 = \{x: \|x - x_0\| \leq r_0\}$ be given such that $x^* \in S_0$. Before the k-th step ($k=0, 1, \ldots$) we have the point $x_k \in E_n$, the nonsingular $n \times n$ matrix B_k and the number $r_k > 0$ that in E_n specifies the ellipsoid $S_k = \{x: \|B_k^{-1}(x-x_k)\| \leq r_k\}$ such that $x^* \in S_k$, $B_0 = I$.

The k-th step is performed in two stages:

1. The more exact definition of the localization of x^*.

Making use of a certain procedure we choose a collection of pairs $\{x_k^l, f_{kl}^s\}$ (f_{kl}^s is a value for the displacements), $l=0, 1, \ldots, l_k$.

Each pair of this collection defines the cutting hyperplane

$$\mathscr{P}(x_k^l, f_{kl}^s) = \{x: (g(x_k^l), x - x_k^l) = f_{kl}^s - f(x_k^l)\},$$

which specifies the half-space of the form

$$H(x_k^l, f_{kl}^s) = \{x: (g(x_k^l), x - x_k^l) \leq f_{kl}^s - f(x_k^l)\}$$

such that $x^* \in H(x_k^l, f_{kl}^s)$. The body $Z_k = S_k \cap \left(\bigcap_{l=0}^{l_k} H(x_k^l, f_{kl}^s) \right)$ contains x^*. The principal feature the above procedure must have consists in the existence of an ellipsoid $S(Z_k)$, $Z_k \subseteq S(Z_k)$ such that there exists $Q < 1$ for which the inequality $q(Z_k) = \frac{\mathscr{V}(S(Z_k))}{\mathscr{V}(S_k)} \leq Q$ is satisfied (here $\mathscr{V}(\mathscr{D})$ is the n-dimensional volume of the body $\mathscr{D}$).

2. The recalculation of parameters.

Assume $S_{k+1} = S(Z_k)$, that is compute the point $x_{k+1} \in E_n$, the nonsingular matrix B_{k+1} and the number r_{k+1} such that

$$S_{k+1} = \{x: \|B_{k+1}^{-1}(x-x_{k+1})\| \leq r_{k+1}\} = S(Z_k).$$

Consider the sequence of records $\{f_k^r\}_{k=0}^{\infty}$ generated during the operation of this algorithm,

$$f_k^r = \min\{f_{k-1}^r, \min_l\{f(x_k^l), l=0, \ldots, l_k\}\}.$$

V. I. Gershovich generalized the L. G. Khachijan and B. Korte and R. Schrader results [32, 55] and proved the following theorem:

Theorem 6. Let $x^* \in S_0$, $\|g(x)\| \leq L$ for $x \in S_0$,

$$f^s_{kl} = \min\{f^r_{k-1}, \min\{f(x^j_k), j=0, 1, \ldots, l_k\}\}, \quad k=0, 1, \ldots.$$

Then the sequences of matrices $\{B_k\}_{k=0}^\infty$, numbers $\{r_k\}_{k=0}^\infty$ and $\{f^r_k\}_{k=0}^\infty$ generated by the ellipsoid algorithm satisfy the inequalities:

$$f^r_k - f(x^*) \leq 2L r_{k+1} \sqrt{\lambda(H_{k+1})}, \tag{14}$$

where $\lambda(H_{k+1})$ is the minimum eigenvalue of the matrix $H_{k+1} = B_{k+1} B^*_{k+1}$.

Corollary. The inequalities

$$f^r_k - f(x^*) \leq 2L r_0 \left(\prod_{i=0}^{k} q(Z_i)\right)^{\frac{1}{n}} \leq 2L r_0 Q^{k/n}, \quad k=0, 1, \ldots \tag{15}$$

are valid in the Theorem. The inequality (14) provides a far more precise estimate of $f^r_k - f(x^*)$ than (15) and permits an earlier identification of the instant when a satisfactory precision of the solution is reached. However, to do this requires computing the eigenvalues $\lambda(H_{k+1})$. Put $H_k = B_k B^*_k$. The symmetric positive definite matrix H_k can be represented by the set of its eigenvectors and eigenvalues. As it turned out, there exists a simple – computationally speaking – way to compute the eigen-elements of the updated matrix H_{k+1}.

Spectral Properties of the Sequence of Operators H_k and their Potential Application in Solutions of Large-Scale Problems

Let $H_k = B_k B^*_k$ and the operators B_k be constructed by the recurrence relation $B_0 = I$; $B_{k+1} = B_k \cdot R_\beta(\xi_k)$ where $0 < \beta_k < 1$.

$$\xi_k = \frac{B^* p_k}{\|B p_k\|}; \quad p_k \in E_n; \quad p_k \neq 0.$$

Then
$$H_{k+1} = H_k - (1-\beta_k^2) \frac{(H_k p)(H_k p)^T}{(H_k p, p)} \quad [30].$$

The following theorem can be readily proved.

Theorem 7 [6, 42]. The spectrum $\lambda_1 \leq \lambda_2 \leq \ldots \leq \lambda_n$ of the operator H_k is localized for any k in the interval $(0, 1]$. If $k < n$, then the dimension of the subspace of eigenvectors corresponding to $\lambda_n = 1$ is no smaller than $n - k$. Moreover,

$$\prod_{i=1}^{n} \lambda_i = \prod_{j=0}^{k} \beta_j.$$

Assume that the operator H_k is specified by its structure, that is by the totality of eigenvectors $b^1_k, \ldots, b^n_k$ forming the orthonormal basis, and eigenvalues $\lambda^1_k, \ldots, \lambda^n_k$, respectively.

Define
$$d_k = (1-\beta_k^2)(p_k, H_k p_k)^{-\frac{1}{2}} H_k p_k.$$

Then
$$H_{k+1} = H_k - d_k d_k^*.$$

It is required to find $\lambda_{k+1}^1, \ldots, \lambda_{k+1}^n$ and the corresponding eigenvectors of the matrix H_{k+1}. For the sake of simplicity the problem is formulated as follows: there are given vectors $a_1, \ldots, a_n$ that form the orthonormal basis in E_n consisting of eigenvectors of the symmetrix matrix A corresponding to the eigenvalues $\lambda_1, \ldots, \lambda_n$ (not necessarily different). The matrix $\bar{A}$ relates to A in the following way: $\bar{A} = A - dd^*$ where d will be represented in the form $d = \sum_{j=1}^{n} d_j a_j$. Eigenvectors and eigenvalues of the matrix $\bar{A}$ need to be calculated.

Theorem 8 [42]. The eigenvalues of the matrix $\bar{A}$, and only these, are the roots of the function
$$P(\tau) = \left(1 - \sum_{j=1}^{n} \frac{(d_j)^2}{\lambda_j - \tau}\right) \prod_{j=1}^{n} (\lambda_j - \tau).$$

Let $\Lambda_1, \ldots, \Lambda_m$ be the pairwise different eigenvalues of the matrix A written in increasing order of magnitude, the multiplicity of Λ_i being equal to r_i $(i=1, \ldots, m)$, and let $\mathscr{D}_i^2$ be the sum of those d_j^2 for which $\lambda_j = \Lambda_i$.

Then
$$P(\tau) = \left(1 - \sum_{i=1}^{m} \frac{\mathscr{D}_i^2}{\Lambda_i - \tau}\right) \prod_{i=1}^{m} (\Lambda_i - \tau)^{r_i}.$$

Let
$$R(\tau) = 1 - \sum_{i=1}^{m} \frac{\mathscr{D}_i^2}{\Lambda_i - \tau}; \quad i_0 = 0; \quad \Lambda_{i_0} = -\infty.$$

Theorem 9 [42]. Assume that among the quantities $\mathscr{D}_1^2, \mathscr{D}_2^2, \ldots, \mathscr{D}_m^2$ only $\mathscr{D}_{i_1}^2, \ldots, \mathscr{D}_{i_p}^2$ $(p \leq m)$, $i_1 < i_2 < \ldots < i_p$, differ from zero. For each l $(l=1, \ldots, p)$ the following statement is valid: μ that coincides with the unique root in the interval $(\Lambda_{i_{l-1}}, \Lambda_{i_l})$ of the function $R(\tau)$ monotone on $(\Lambda_{i_{l-1}}, \Lambda_{i_l})$, is the eigenvalue of the matrix $\bar{A}$. The eigenvector corresponding to the eigenvalue μ equals (to within a multiplier):
$$\bar{a}_\mu = \sum_{j=1}^{n} \frac{d_j}{\lambda_j - \mu}.$$

The above results enable a comparatively easy update from the multiplicative representation of the matrix H_k to the multiplicative representation of the matrix H_{k+1} differing from H_k by a matrix of rank 1. The algorithm of such an update can be obviously generalized to the complements of rank 2 and of higher rank.

The multiplicative representation of the matrices H_k is of particular advantage for the solution of large-scale problems. For $n \geq 1000$, difficulties present themselves that have to do with transformations and storage of matrices in an ordinary form. Furthermore, it may happen that, due to a limitation on the time of calculations, the maximum number of iterations k^* we have managed to perform is considerably smaller than n. If $H_0 = I$ and we apply, for instance, the r-algorithm, then after k^* iterations the dilatation will occur in k^* directions only, and the metric will not change in the orthogonal subspace of dimension $n-k^*$. Therefore, to store the matrix H_k ($k \leq k^* < n$), we have to store only k eigenvectors and eigenvalues.

As a modification of the r-algorithm the so-called r_l-algorithm has been suggested in [42]. The r_l-algorithm is in a sense in between the simplest GGD and the r-algorithm. It works with a linear operator which, as calculation proceeds, acts in a changing subspace of dimension no greater than l. We store only those eigenvectors at each iteration whose eigenvalues are smaller than 1. There are no more than l of them. After $(l+1)$ iterations we obtain, generally speaking, $l+1$ eigenvectors with $\lambda < 1$. From them we select the eigenvector having the maximal λ (the closest to 1), pick it out from the stored set, and proceed with our calculations corresponding formally to the r-algorithm, with no more than l eigenvectors left each time in the memory. Experimental studies on the r_l-algorithm revealed an appreciable acceleration in convergence compared to GGD even for small l.

The exhibited advantage of the r_l-algorithm over GGD stems from its ability, inherent in all gradient-type algorithms with space dilatation, to smooth ravines in the picture of level surfaces of the function being minimized. In large-scale application problems the ravine dimension (the number of strong variables), as a rule, is small by comparison with the total dimension n, so application of the r_l-algorithm for $l \ll n$ can give rather high acceleration. Central to the r_l-algorithm is the concept of automatic "tuning in" to a subspace defined by the most essential directions of changes in the function. Similar problems but from different positions, were tackled by A. S. Nemirovskij, D. B. Judin [20] when using MCGT algorithms; they constructed algorithms identifying the essential variables of a problem.

In conclusion it may be noted that the technique described in this section is applicable in the realization of various modifications of quasi-Newton variable-metric processes in the case of large-scale problems.

Applications of the Gradient-Type Methods with Space Dilatation

Applications of non-smooth optimization methods in practice are connected mainly with the employment of decomposition schemes to solve problems of linear, nonlinear and discrete programming.

Consider the problem of the form:

find $\quad f^* = \inf f(x); \quad x \in X \subset E_n; \quad f_i(x) \leq 0; \quad i = 1, \ldots, m.$ \hfill (16)

Construct the Lagrange function:

$$L(x, u) = f(x) + \sum_{i=1}^{m} u_i f_i(x); \quad x \in X; \quad u \geq 0$$

$$\psi(u) = \inf_{x \in X} L(x, u);$$

$$\psi^* = \sup_{x \in X} \psi(u).$$

As for any admissible x, $L(x, u) \leq f(x)$ for $u \geq 0$, and X contains an admissible domain of the problem (16), then $\psi(u) \leq f^* \forall u \geq 0$. Thus for any $u \geq 0$, $\psi(u)$ is a lower bound of f^* and ψ^* is the best among these bounds.

The definition of ψ^* is associated with the concept of decomposition: for fixed $u \geq 0$ the "local" problem of minimization of the Lagrange function $L(x, u)$ with respect to $x \in X$ has less constraints than the original one; then the "co-ordinating" problem of definition of ψ^* consists in the optimal choice of the Lagrange multipliers $u \geq 0$. If $f(x)$ and $f_i(x)$, $i = 1, \ldots, m$ are continuous functions defined on X, where X is a compact set, then $\psi(u)$ is a concave function not necessarily smooth. In particular, when X is a finite set, which is characteristic of discrete programming problems, then $\psi(u)$ is a piecewise-linear function. The maximization of $\psi(u)$ therewith can be used to obtain lower bounds in the branch-and-bound method.

In the general case and under the above assumptions the subgradient set $G_\psi(u_0)$ of the function $\psi(u)$ for $u = u_0$ is the convex hull of the set of vectors $\{f_i(x(u))\}_{i=1}^m$ where $x(u) \in X^*(u)$ and $X^*(u)$ is the set of minima with respect to $x \in X$ of the function $L(x, u)$.

When the problem (16) is a convex programming problem whose solution is x^*, and the Slater condition is satisfied, that is there exists $\bar{x} \in X$ such that $f_i(\bar{x}) < 0$, $i = 1, \ldots, m$, then

$$\max_{u \geq 0} \psi(u) = \psi(u^*) = \min_{x \in X} L(x, u^*) = f(x^*),$$

and, furthermore, when the solution $x(u^*)$ of the problem $\min_{x \in X} L(x, u^*)$ is unique (for this to be so, $f(x)$ should be a strongly convex function), this solution coincides with the unique solution of the problem. However, if $x(u^*)$ is not unique, one has to search for a solution that is an admissible solution of the problem (16). Several efficient procedures currently in use make it possible to find $x(u^*) \in X(u^*)$, the optimal solution of the direct problem (16), when solving block-structured problems of linear programming [2, 10].

One of them is connected with perturbing the linear objective function $f(x)$ by a positive definite quadratic complement having a small parameter. Under certain conditions this perturbation does not change the optimal solution ([3], [10]).

If in the solution of problems of linear and convex programming one manages to decompose constraints in such a way that the problem of finding $\inf_{x \in X} L(x, u)$ is easily solvable, then it is expeditious to use the generalized gradient methods of nonsmooth optimization for finding $\sup_{u \geq 0} \psi(u)$, as calculation of generalized gradients of the function $\psi(u)$ present no added difficulties and are reduced merely to calculation of the discrepancy vector $\{f_i(x(u))\}_{i=1}^m$ for an arbitrary $x(u) \in X^*(u)$. Of the nonsmooth optimization methods, the r-algorithms offer the best results when schemes of decomposition with respect to constraints are realized.

By way of illustration we consider a large-scale distribution-type problem of assignment of orders to ferrous metallurgy works. The dimension of this problem reached up to 300,000 variables and more than 20,000 constraints but the number of linking constraints was no greater than 250, so the solution of the problem $\max_{u \geq 0} \psi(u)$ was carried out in the space of dimension approaching 250.

Applying the r-algorithm we managed to gain a relative precision with respect to the functional $\sim 10^{-6}$ in some 900 to 1000 iterations. Such high precision with respect to the functional of the dual problem enabled us to attain a required precision in the direct problem solution with ε-quadratic smoothing: the constraints were met with the precision of 4 digital places. Similar results were obtained in solving large-scale problems of routine and long-term planning in civil aviation, nonlinear transportation problems when finding optimal flows in a gas supply system, three-index problems of organization of railway building, and in computing estimates in discrete problems of reconstruction and location of plants, etc. ([4], [46]).

Besides the scheme of decomposition with respect to constraints there is a dual scheme of decomposition with respect to variables. It is applied when fixing a comparatively small part of variables simplifies materially the problem of mathematical programming. This scheme is described formally as follows. Let Z_{l+n}, X_l, Y_n be Euclidean spaces of adequate dimension, $Z_{l+n} = X_l \times Y_n$.

Consider the convex programming problem:

find $\quad \min f(z) = f(x, y), \quad x \in X_l; \quad y \in Y_n,$ (17)

under constraints $f_i(z) = f_i(x, y) \leq 0$, $i = 1, \ldots, m$, together with a family of *local* problems with the fixed $x = \bar{x}$:

$$\min_{y \in Y_n} f(\bar{x}, y), \quad f_i(\bar{x}, y) \leq 0, \quad i = 1, \ldots, m. \quad (18)$$

Assume that the conditions of the Kuhn-Tucker theorem are fulfilled for the problem (18) in the neighbourhood of the point $\bar{x}$, $y^*(\bar{x})$ is the optimal solution, $\{u_i^*(\bar{x})\}_{i=1}^m$ are optimal values of the Lagrange multipliers.

$$\psi(\bar{x}) = f(\bar{x}, y^*(\bar{x})); \quad L^*(x, y) = f(x, y) + \sum_{i=1}^m u_i^*(x) f_i(x, y).$$

Then the subgradient of the convex function $\psi(x)$ at the point $\bar{x}$ is calculated as follows. A subgradient $g_{L^*}(\bar{x}, y(x))$ of $L^*(x, y)$ at the point $\bar{x}, y(\bar{x})$ is found whose projection onto the subspace Y_n equals 0 (such a subgradient always exists, as $y(\bar{x})$ is the optimal solution of the local problem (18), where the minimum with respect to y of the Lagrangean function $L^*(x, y)$ is attained; the projection of $g_{L^*}(\bar{x}, y(\bar{x}))$ onto the subspace X_l is a subgradient of the function $\psi(x)$ at the point $\bar{x}$ [4, 46]).

The scheme of decomposition with respect to variables in combination with the r-algorithm was employed to solve two-stage problems of stochastic programming [46], some problems of design optimization that amount to nonlinear programming problems where a major part of the variables enters into the model linearly.

Problems of applications of gradient-type methods with space dilatation are covered in more detail in the monographs [4], [46].

A marked advantage of the r-algorithms over other methods (subgradient, ε-subgradient, the ellipsoid method and its modifications) as a rule, is noted in solving test and application problems when the dimension of the space where the descent occurs is not too large (to 300 or 500 variables) and a high-precision solution is required.

However, if dimensions are large and too high precision of result is not required, the subgradient and ε-subgradient methods without complete memory can readily defy the r-algorithms in full agreement with the general theory of efficiency of convex programming algorithms [21] (see [1]).

As far as the ellipsoid methods are concerned, they are undergoing a speedy advancement, and experience in their application is accumulating. Even in their present form they may prove useful in the solution of convex programming problems of small scale (up to 20 or 30 variables) under a large number of constraints. Furthermore, they may be employed in combination with the r-algorithms at the test stage of calculations conducted to corroborate (or refute) the assumption that the r-algorithm assures a certain precision in functional and in the negative case to bring the solution to the guaranteed precision.

Conclusion

Following a burst of enthusiasm over early reports on the L. G. Khachijan result [31], the ellipsoid methods were the focus of attention of many specialists in the field of optimization. After a little while only their theoretical significance was recognized but their practical importance was doubted ([50], [62]).

This is true in many respects when speaking of MCGT and its minor modifications.

The author tried to show in this paper that the ellipsoid methods are merely a very special case of the family of gradient-type algorithms employing the operation of space dilatation. Other representatives of this family, for instance the r-algorithms, are a powerful *practical* means of solution of many sophisticated problems of mathematical programming reduced to nondifferentiable optimization. The theory of the whole class of algorithms with space dilatation is still far from perfect. To the best of our belief the objective of construction of an algorithm whose practical importance is as high as that of the r-algorithms and whose theoretical foundation is as thorough as that of the ellipsoid method is quite realistic.

References

[1] A. A. Bakaev, V. S. Mikhalevich, S. V. Branovickaja, N. Z. Shor. Methodology and Attempted Solution of Large Network Transportation Problems by Computer. - In the Book: Mathematical Methods for Production Problems, M., 1963, pp. 247–257.

[2] L. V. Beljaeva, N. T. Zhurbenko, N. Z. Shor. Concerning Methods of Solution of One Class of Dynamic Distribution Problems. - Economics and Mathematical Methods, 1978, Vol. 14, Issue 1, pp. 137-146.

[3] V. A. Bulavskij. Iterative Method of Solution of General Problem of Linear Programming. - In Coll.: Numerical Methods of Optimal Planning, Issue 1, Econ.-Math. Ser., Siberian Branch of Ac. Sci. U.S.S.R., 1962, pp. 35-64.

[4] V. S. Mikhalevich, N. Z. Shor, L. A. Galustova, et al. Computational Methods of Choice of Optimal Design Decisions. - Kiev, Naukova Dumka, 1977, 178 p.

[5] V. I. Gershovich. About a Cutting Method with Linear Space Transformations. In the Book: Theory of Optimal Solutions, Kiev, Inst. of Cybernetics. Ac. Sci. Ukr.S.S.R., 1979, pp. 15-23.

[6] V. I. Gershovich. One Way of Representing Space Transformation Operators in High-Speed Versions of Generalized Gradient Methods. In the Book: Numerical Methods of Nonlinear Programming. Summaries of Papers at the III. All-Union Seminar. - Kharkov, 1979, pp. 64-66.

[7] V. I. Gershovich. Efficient Algorithm for Solving Problems of Nonsmooth Optimization that Combines Cutting Methods and Linear Transformations in Variable Space. - In the Book: Systems of Software Support for Solution of Optimal Planning Problems. Summaries of Papers at the IV. All-Union Symposium.-M., 1980, pp. 50-51.

[8] V. I. Gershovich. One Optimization Method Using Linear Space Transformations. - In the Book: Theory of Optimal Solutions. Kiev, Inst. of Cybernetics, Ac. Sci. Ukr. S.S.R., 1980, pp. 38-45.

[9] V. I. Gershovich. About an Ellipsoid Algorithm. In the Book: Some Algorithms of Nonsmooth Optimization and Discrete Programming. Preprint 81-6 Inst. of Cybernetics, Ac. Sci. Ukr.S.S.R., 1981, pp. 8-13.

[10] V. I. Gershovich. Quadratic Smoothing in Iterative Decomposition Algorithms of Solution of Large-Scale Linear-Programming Problems. In the Book: II. Republican Symposium on Methods of Solution of Nonlinear Equations and Optimization Problems. Tallin, 1981, pp. 188-190.

[11] A. M. Gupal. Stochastic Methods of Solution of Nonsmooth Extremum Problems. - Kiev, Naukova Dumka, 1979, 149 p.

[12] V. V. Dem'janov, V. N. Malozemov. Introduction to Minimax - M., Nauka, 1972, 368 p.

[13] V. F. Dem'janov, L. V. Vasil'ev. Non-Differentiable Optimization. - M., Nauka, 1981, 384 p.

[14] I. I. Eremin. Iterative Method for Tchebycheff Approximations of Inconsistent Systems of Linear Inequalities. - Dokl. AN S.S.S.R., 1962, Vol. 143, No. 6, pp. 1254-1256.

[15] I. I. Eremin. About "Penalty" Method in Convex Programming. - Kybernetika, 1967, No. 4, pp. 63-67.

[16] I. I. Eremin. Methods of Fejer Approximations in Convex Programming. Mathem. Notes, 1968, Vol. 3, No. 2, pp. 217-234.

[17] Ju. M. Ermol'jev. Stochastic Programming Methods, M., Nauka, 1976, 240 p.

[18] Ju. M. Ermol'jev. Methods of Solution of Nonlinear Extremum Problems. - Kibernetika, 1966, No. 4, pp. 1-17.

[19] A. Ju. Levin. On an Algorithm for Minimization of Convex Functions. - Dokl. AN S.S.S.R., 1965, Vol. 160, No. 6, pp. 1244-1247.

[20] A. S. Nemirovskij, D. B. Judin. Optimization Methods Adaptive to "Considerable" Dimension of a Problem. - Avtomatika i Telemekhanika, 1977, No. 4, pp. 75-87.

[21] N. S. Nemirovskij, D. B. Judin. Complexity of Problems and Efficiency of Optimization Methods. - M., Nauka, 1979, 383 p.
[22] E. A. Nurminskij. Numerical Methods of Solution of Deterministic and Stochastic Minimax Problems. - Kiev. Naukova Dumka, 1979, 159 p.
[23] E. A. Nurminskij. About Continuity of ε-Subgradient Mappings. - Kibernetika, 1977, No. 5, pp. 148-149.
[24] B. T. Poljak. A General Method of Solution of Extremum Problems. - Dokl. AN S.S.S.R., 1967, Vol. 174, No. 1, pp. 33-36.
[25] B. T. Poljak. Minimization of Unsmooth Functionals. - J. Vychisl. Matem. i Mat. Fiziki, 1969, Vol. 9, No. 3, pp. 509-521.
[26] B. N. Pshenichnyj, Ju. M. Danilin. Numerical Methods for Extremum Problems. - M., Nauka, 1975, 319 p.
[27] S. V. Rzhevskij. About a Unified Approach to Solution of Unconditional Minimization Problems. - J. Vychisl. Matem. i Matem. Fiziki, 1969, Vol. 20, No. 4, pp. 857-863.
[28] S. V. Rzhevskij. ε-Subgradient Method of Solution of Convex Programming Problem. - J. Vychisl. Matem. i Matem. Fiziki, Vol. 21, No. 5, 1981, pp. 1126-1132.
[29] S. V. Rzhevskij. Monotone Algorithm of Search for Saddle Point of Nonsmooth Function. - Kibernetika, 1982, No. 1, pp. 95-98.
[30] V. A. Skokov. Note on Minimization Methods Using Operation of Space Dilatation. - Kibernetika, 1974, No. 4, pp. 115-117.
[31] L. G. Khachijan. Polynomial Algorithm in Linear Programming. - Dokl. AN S.S.S.R., 1979, Vol. 244, No. 5, pp. 1093-1096.
[32] L. G. Khachijan. Polynomial Algorithms in Linear Programming. - J. Vychisl. Matem. i Matem. Fiziki, 1980, Vol. 20, No. 1, pp. 51-68.
[33] M. K. Kozlov, S. P. Tarasov, L. G. Khachijan. Polynomial Solvability of Convex Quadratic Programming. - Dokl. AN S.S.S.R., 1979, Vol. 248, No. 5.
[34] N. Z. Shor. Application of Gradient Descent Method for Solution of Network Transportation Problems. In the Book: Notes, Scientific Seminar on Theory and Application of Cybernetics and Operations Research. Nauchn. Sovet po Kibernetika AN U.S.S.R., Kiev, 1962, Issue 1, pp. 9-17.
[35] N. Z. Shor. Rate of Convergence of Generalized Gradient Descent Method. Kibernetika, 1968, No. 3, pp. 98-99.
[36] N. Z. Shor. Utilization of Space Dilatation Operation in Minimization of Convex Functions. - Kibernetika, 1970, No. 1, pp. 6-12.
[37] N. Z. Shor. About Rate of Convergence of Method of Generalized Gradient Descent with Space Dilatation. - Kibernetika, 1970, No. 2, pp. 80-85.
[38] N. Z. Shor, N. G. Zhurbenko. Minimization Method Using Operation of Space Dilatation in the Direction of Difference of Two Sequential Gradients. - Kibernetika, 1971, No. 3, pp. 51-59.
[39] N. Z. Shor, L. P. Shabashova. Solution of Minimax Problems by Generalized Gradient Descent Method with Space Dilatation. - Kibernetika, 1972, No. 1, pp. 82-88.
[40] N. Z. Shor, V. I. Gershovich. Family of Algorithms for Solving Problems of Convex Programming. - Kibernetika, 1979, No. 4, pp. 62-67.
[41] N. Z. Shor. Method of Minimization of Almost Differentiable Functions. - Kibernetika, 1972, No. 4, pp. 65-70.
[42] N. Z. Shor, V. I. Gershovich. About One Modification of Gradient-Type Algorithms with Space Dilatation for Solution of Large-Scale Problems. - Kibernetika, 1981, No. 5, pp. 67-70.

[43] N. Z. Shor. Convergence of Gradient-Type Method with Space Dilatation in the Direction of Difference of Two Sequential Gradients. - Kibernetika, 1975, No. 4, pp. 48-53.

[44] N. Z. Shor. Cutting Method with Space Dilatation for Solving Problems of Convex Programming. - Kibernetika, 1977, No. 1, pp. 94-95.

[45] N. Z. Shor. New Directions in Development of Nonsmooth Optimization Methods. - Kibernetika, 1977, No. 6, pp. 87-91.

[46] N. Z. Shor. Methods of Optimization of Non-Differentiable Functions and Their Applications. - Kiev, Naukova Dumka, 1979, 200 p.

[47] D. B. Judin, A. S. Nemirovskij. Evaluation of Information Complexity and Effective Methods for Solving of Complex Extremum Problems. - Ekonomika i Matematicheskie Metody, 1976, Vol. 12, Issue 2, pp. 357-369.

[48] D. Agmon. The Relaxation Method for Linear Inequalities. - Can. J. Math. 1954, Vol. 6, No. 2, pp. 381-392.

[49] B. Aspvall, G. Shiloach. A Polynomial Time Algorithm for Solving Systems of Linear Inequalities with Two Variables per Inequality. SIAM J. Comput., Vol. 9, No. 4, Nov. 1980.

[50] R. Bland, D. Goldfarb, M. Todd. The Ellipsoid Method: A Survey. Cornell University, Ithaca, New York, Technical Report No. 476, 79 p.

[51] F. Clarke. Generalized Gradients and Applications. - Trans. Amer. Math. Soc., 1975, Vol. 205, pp. 247-262.

[52] M. Grötschel, L. Lovász, A. Schrijver. The Ellipsoid Method and its Consequences in Combinatorial optimization. - Combinatorica, 1981, Vol. I, No. 2, pp. 169-197.

[53] M. Grötschel, L. Lovász, A. Schrijver. Polynomial Algorithms for Perfect Graphs. Report No. 81176-OR, Bonn University, Febr. 1981.

[54] J. Kelley. The "Cutting Plane" Method for Solving Convex Programs. - SIAM Journal, 1960, Vol. 8, No. 4, pp. 703-712.

[55] B. Korte, R. Schrader. A Note on Convergence Proofs for Shor-Khachijan Methods. - In "Optimization and Optimal Control". Lecture Notes in Control and Information Sciences 30, Springer (1980), pp. 51-57.

[56] C. Lemarechal. An Extension of Davidon Methods to Nondifferentiable Problems. - Math. Program. Study 3, (1975), pp. 95-100.

[57] L. Lovász. On the Shannon Capacity of a Graph. IEEE Trans. Inform. Theory II-25 (1979), pp. 1-7.

[58] J. F. Maurras, K. Truemper, M. Akgül. Polynomial Algorithms for a Class of Linear Programs. Math. Progr., 21 (1981), pp. 121-136.

[59] R. Mifflin. Semismooth and Semiconvex Functions in Constrained Optimization. RR-76-21. IIASA. - Laxenburg, Austria, 1976, 23 p.

[60] R. Mifflin. An Algorithm for Constrained Optimization with Semismooth Functions. RR-73-3. IIASA. - Laxenburg, Austria, 1977, 32 p.

[61] T. Motzkin, I. Schoenberg. The Relaxation Method for Linear Inequalities. - Can. J. Math., 1954, Vol. 6, No. 2, pp. 393-404.

[62] R. Schrader. Ellipsoid Methods. In the Book: Modern Applied Mathematics: Optimization and Operations Research, B. Korte (ed.), North-Holland, Amsterdam, 1982, pp. 265-311.

[63] B. Schwartz. Zur Minimierung konvexer subdifferenzierbarer Funktionen über dem R^n. Math. Operationsforschung und Statistik, Series Optimization, 1978, 9, No. 4, pp. 545-557.

[64] P. Wolfe. A Method of Conjugate Subgradients for Minimizing Nondifferentiable Functions. - Math. Program. Study 3, (1975), pp. 145-173.

[65] P. Wolfe. A Bibliography for the Ellipsoid Algorithm. IBM Research Center, Yorktown Heights. New York, 1980, 6 p.

The Problem of the Average Speed of the Simplex Method

S. Smale
University of California, Department of Mathematics, Berkeley, CA 94720, USA

Section 1 The Main Theorem

Our goal is to give an exposition of our work on the average speed of the simplex method. Detailed proofs are in *Smale* 1982. Here we concentrate on the main ideas, with concepts emphasized. Some new things are added.

Consider the linear programming problem in one of its standard forms. The data consist of a triple:

Data of LPP: (A, b, c), A is a real $m \times n$ matrix, $b \in R^m$, $c \in R^n$

Here R^k denotes real Cartesian k-dimensional space. Then the linear programming problem is:

LPP: Find $x \in R^n$ subject to $x \geq 0$, $Ax \geq b$, such that x minimizes $c \cdot x$.

Here $c \cdot x$ is the inner product and $x \geq y$ means $x_i \geq y_i$ all i.

The linear programming problem is widely used by industry and government. For example, *Lovácz* writes: "If one would take statistics about which mathematical problem is using up most of the computer time in the world, then (not counting database handling problems like sorting and searching) the answer would probably be linear programming."

The simplex method of Dantzig provides a fast algorithm in practice for solving the linear programming problem. To *solve* means to exhibit an answer or to decide that no minimum exists. Besides *Dantzig*, one can see *Hadley* (as well as many other books) for a full development of this method.

After a finite number of steps, the algorithm yields a minimum or the fact that there is no minimum. Thus the linear programming problem may be regarded as a finite problem with continuous data.

In our study we are using a real number model in contrast with the discrete model of Khachiyan (see *Lovácz*). Machines are finite, but also operations of multiplication, etc., in practice are only approximate; moreover the size of input numbers (the number of significant digits) is irrelevant to linear programming. Thus we feel that the real number model is appropriate to linear programming. In this model, the ellipsoidal algorithm of Khachiyan and others is no longer polynomial, as pointed out by *Traub* and *Wozniakowski*.

While the simplex method is very efficient, it is known that in certain degenerate cases of data it can cycle (Hoffman and Beale) see *Dantzig*. Moreover, there are more robust examples of *Klee-Minty* where the number of steps grows exponentially with m and n. Thus it has been a well known problem for some time to show that "on the average" (in some sense) the simplex method takes relatively few steps.

For example, *Dantzig* (p. 160) writes: "Some believe that for a randomly chosen problem with fixed m [the number of constraints], the number of iterations grows in proportion to n [the number of variables]."

In *Smale* 1982 we prove that indeed this is the case and here we will sketch the proof. We use a variation of the simplex method introduced by *Dantzig* and developed in his book p. 245 under the name self-dual parametric variant of the simplex method.

It is an important point that the subsequent development of operations research has put this algorithm in a central place. It turns out to be a special case of Lemke's algorithm in the linear complementarity problem. Moreover, in this context it conceptually corresponds to main variations of Newton's method in numerical analysis. This point is developed in *Eaves-Scarf* and *Smale* 1976. Thus it has turned out that *Dantzig*'s self-dual parametric algorithm generalized to become a unifying algorithm not only in operations research, but to help unify operations research and numerical analysis as well. We will return to this point.

Consider now the problem of averaging. Our approach is to average by putting a probability measure on the space of the data (A, b, c) of the linear programming problem. Since the space of $m \times n$ matrices is R^{mn}, we are speaking of a probability measure on $R^{mn} \times R^m \times R^n$. In our detailed account, we used a spherical kind of measure, which simply amounts to using a Gaussian or normal distribution on each coordinate of $R^{mn} \times R^m \times R^n$. Subsequently we noticed that only 3 properties of that measure were used. Thus throughout this paper consider a probabiltity measure on the space $R^{mn} \times R^m \times R^n$ of data (A, b, c) of the linear programming problem which satisfies the following three conditions.

(1) *Continuity:* The measure is absolutely continuous with respect to Lebesgue measure. This just means that if a subset has Lebesgue measure zero then it also has measure zero for the probability measure.

Remark. The theorems we will prove will all fail without this condition due to the degenerate cases mentioned above.

(2) *Independence:* The measure is a product of three probability measures relative to the three-fold product structure of $R^{mn} \times R^m \times R^n$. Also the measure on R^{mn} is a product of probability measures on the rows of matrices in R^{mn}. I suspect this condition can be at least weakened, especially the last half.

(3) *Symmetry:* Coordinate permutations of R^n induce transformations of $R^{mn} \times R^m \times R^n$ (first and third factors). The probability measure is invariant under these transformations.

Remark. It seems to me that condition 3 is connected very much to the examples of Klee-Minty. The Klee-Minty examples are quite asymmetric relative to the coordinates. Thus I suspect that these examples could be developed to show the necessity of condition (3) for our results. Moreover, condition (3) perhaps gives some broader understanding of how the Klee-Minty examples defeat the usual efficiency of the simplex method.

Now that the algorithm has been specified and conditions on the probability measure have been given, we can state our main result. Denote by $\rho(m, n)$ the average number of iterations required to solve a linear programming problem in n variables with m constraints.

Main Theorem. Let m be fixed and $\varepsilon > 0$ be given. Then there is a constant K depending on m and ε with $\rho(m, n) \leq K n^\varepsilon$.

Thus the number of steps (or pivots) for fixed m grows more slowly than any prescribed root of the number of variables. By taking $\varepsilon = 1$, we obtain the problem stated by Dantzig above.

We remark that the constant K is independent of the measure.

There have been a number of interesting papers that deal with one aspect or another of the average speed of the simplex method. We do not try to give an account of this literature here; several such papers are listed in *Smale* 1982. Of special note is the work of *Borgwardt*.

Borgwardt showed for linear programming problems of the kind:

$$\text{MAX } v^T x \text{ subject to } a_1^T x \leq 1, \ldots, a_m^T x \leq 1,$$
$$\text{where } v, x, a_1, \ldots, a_m \in \mathbb{R}^n,$$

that a certain variant of the simplex algorithm (Schatteneckenalgorithmus) has average speed polynomial in m and n.

Section 2 The Main Formula

To the data (A, b, c), of a linear programming problem, associate the number $\rho_{(A, b, c)}$ of iterations required to solve the problem. Then define:

$$\rho_A = \int_{b, c} \rho_{(A, b, c)}$$

This integral is the average of $\rho_{(A, b, c)}$ over b, c with respect to our probability measure. Thus

$$\rho(m, n) = \int_A \rho_A$$

We will give an exact formula for ρ_A in terms of certain geometric quantities. This formula will be the basis for our proof of the main result. We proceed to define these geometric quantities. Let $N = m + n$, and S be a subset:

$$S \subset \{1, \ldots, N\}$$

Let $M = M(A)$ be the $N \times N$ matrix:

$$M = \left[\begin{array}{c|c} 0 & -A^T \\ \hline A & 0 \end{array} \right] = [m_1, \ldots, m_N]$$

where A^T denotes the transpose of the $m \times n$ matrix A, and $m_i \in \mathbb{R}^N$ is the ith column of M.

From S and A, construct the cone $K_{S, A}$ in R^N which consists of non-negative linear combinations of these N vectors:

II. The Problem of the Average Speed of the Simplex Method

$$-m_i, i \in S; \quad e_j, j \in \{1, \ldots, N\} - S$$

and e_j is the jth coordinate vector of R^N. Thus

$$K_{S,A} = \left\{ -\sum_{i \in S} \lambda_i m_i + \sum_{j \notin S} \lambda_j e_j \mid (\lambda_1, \ldots, \lambda_N), \lambda_k \geq 0 \right\}$$

$K_{S,A}$ is called a *complementary cone* in the mathematical programming literature.

Let $q_0 \in R^N$ be the vector;

$$q_0 = (1, \ldots, 1).$$

This will correspond to the initial step of our algorithm. The part of our data $q = (c, -b)$ defines a vector in R^N. Thus the probability measure defines a measure in R^N (condition 2). The measure of a set P in R^N will be denoted by VP. Then

Main Formula

$$\rho_A = -1 + \sum_{S \subset \{1, \ldots, N\}} VK(-q_0, K_{S,A})$$

Here $K(-q_0, K_{S,A})$ is the cone generated by $-q_0$ and the cone $K_{S,A}$.

Remark (1). This formula works for any probability measure on R^N satisfying the continuity condition (1) above.

Remark (2). The cone in the main formula is generated by $N+1$ vectors in R^N. In many respects it is more natural to have a formula with cones generated by N vectors in R^N. This is accomplished by a facet version of the main formula which is in *Smale* 1982.

Remark (3). At first glance it might seem that using the formula would yield quickly a full picture of $\rho(m, n)$. One just integrates both sides over the space of A. However, this doesn't turn out to be quite so simple. The volume of a cone is not easily computed. Here the volume of the cone in the simplest case is given by the normal (or Gaussian) distribution on R^N. Then geometricly the volume of a cone K is just the ordinary volume of $K \cap D^N$ divided by the ordinary volume of D^N, D^N being the ball $\{x \in R^N \mid \|x\| \leq 1\}$.

We sketch the proof of the *main formula*. Toward that end, the linear programming problem is imbedded in the *linear complementarity problem*.

LCP Data (M, q), M an $N \times N$ matrix, $q \in R^N$.

Find $w, z \in R^N_+$, $w \cdot z = 0$

such that

$$w - Mz = q$$

Here R^N_+ denotes the non-negative orthant of R^N and $w \cdot z$ is the inner product. Thus for each $i = 1, \ldots, N$, either $w_i = 0$ or $z_i = 0$. These vectors w and z are called complementary vectors.

Now given the data (A, b, c) of a linear programming problem, let $M = M(A)$ be as above and let $q = (c, -b) \in R^N$. Thus the data of a linear programming problem defines data of an LCP. There is a correspondence of solutions as well. One problem has a solution if and only if the other does and the correspondence is given by:

$$z = (x, y)$$

where z is a solution of the LCP above, x a solution of the linear programming problem and $y \in R^m$ is a solution of the dual programming problem. The vector w corresponds to "slack variables" in the linear programming problem. For details see *Smale* 1982 or *Cottle-Dantzig*.

In any case, we may confine our attention to the LCP with the appropriate specification of M and q. Note that q is quite general, but that M has a very particular form. We remark that the LCP unifies many problems in operations research, so that by studying the linear programming problem in this context, we relate the results and concepts to a number of other problems. *Cottle-Dantzig* call the LCP the "Fundamental Problem". See also *Balinski-Cottle*.

Next we transform the LCP into a piece-wise linear system of equations following *Eaves-Scarf* (their example 4). For any $x \in R^N$ (not the $x \in R^n$ of the LPP) define

$$x^+ = (\max(0, x_1), \max(0, x_2), \ldots, \max(0, x_N)) \in R^N$$
$$x^- = (\min(0, x_1), \min(0, x_2), \ldots, \min(0, x_N)) \in R^N$$

Using this notation the equation of the LCP becomes

$$x^+ + Mx^- = q$$

or we may write

(*) $\qquad \Phi_M(x) = q, \qquad \Phi_M(x) = x^+ + Mx^-.$

Here $x^+ = w$, $x^- = -z$, and the conditions $w, z \in R^N_+$, $w \cdot z = 0$ transform into conditions on x which are automatically satisfied. Thus the linear complementarity problem has become the simple equation (*).

It is easily checked that $\Phi_M: R^N \to R^N$ is continuous, is linear on each orthant of R^N and is the identity on R^N_+.

An orthant of R^N corresponds to a subset $S \subset \{1, \ldots, N\}$ via the correspondence:

$$S \leftrightarrow Q_S = \{x \in R^N \mid x_i \leq 0, i \in S, x_j \geq 0, j \notin S\}.$$

The following is easily checked:

Proposition 1.

$$\Phi_M(Q_S) = K_{S, A}$$

where $K_{S,A}$ is as in the main formula, $M = M(A)$ is displayed above and $\Phi_M(Q_S)$ is the image of the orthant Q_S under the map $\Phi_M: R^N \to R^N$.

The algorithm, Dantzig's self-dual parameterized variant of the simplex method or Lemke's algorithm in the context of the LCP can be interpreted by

II. The Problem of the Average Speed of the Simplex Method

taking the inverse image of a certain segment. More precisely, for non-degenerate $q \in R^N$, this algorithm follows $\Phi_M^{-1}(q_0 q)$ starting at q_0. A pivot or iteration corresponds to passing from one orthant to another. From these considerations and others as well, one can prove (Smale 1982):

Proposition 2.
For almost all $q \in R^N$, for $M = M(A)$, as displayed above, the number of steps required to solve (*) is equal to the number of complementary cones $K_{S,A}$ meeting the segment $q_0 q$ less one.

The final step in the proof of the formula is the very easy lemma:

Lemma. $q_0 q$ meets $K_{S,A}$ if and only if q lies in the cone generated by $-q_0$ and $K_{S,A}$.

The main formula now follows from integration over $q \in R^N$ using the lemma and the two propositions.

Section 3 Estimates

We give a sketch of the proof of the Main Theorem. It is quite close to Smale 1982, but a bit more direct. We don't use the "facet version" of the Main Formula.

Let $\mathscr{S}$ be a set of n elements, abstract at first, but is to be specialized to the set $\mathscr{S} = \mathscr{S}_{col}$ of columns of an $m \times n$ matrix. Consider a partial order E on $\mathscr{S}$. A prototype is the partial order $E(A)$ where A is an $m \times n$ matrix and $\mathscr{S}$ the set of columns. Say

$$p \geqslant_{E(A)} q \text{ if and only if } a_{ip} \geqslant a_{iq}, \quad \text{all } i = 1, \ldots, m.$$

An invariant $\gamma(E)$ is defined to be the number of total orders on $\mathscr{S}$ compatible with E, divided by $n!$. Since there are $n!$ total orders on $\mathscr{S}$, $0 < \gamma(E) \leqslant 1$.

To state the first estimate, define another partial order $E_{S_1}(A)$ on $\mathscr{S}_{col}$ where $S_1 \subset \mathscr{S}_{col}$:

$$p \geqslant_{E_{S_1}(A)} q \text{ iff } q \in S_1 \text{ and } p \geqslant_{E(A)} q.$$

Using our probability measure on A let

$$\sigma(m, n, s) = \int_A \gamma(E_{S_1}(A))$$

where $s = |S_1|$ is the number of elements of S_1. Note that $\sigma(m, n, s)$ is well defined.

First Main Estimate

$$\rho(m, n) \leqslant C_m \sum_{s=0}^{m+1} \binom{n}{s} \sigma(m, n, s).$$

Here C_m is a constant depending on m. A sketch of the proof follows.

If $\mathscr{S} = \{1, \ldots, n\}$ and E is a partial order on $\mathscr{S}$, let
$$X_E = \{x \in R^n \mid x_p \geq x_q \text{ if } p \geq_E q\}$$
Using our probability measure on R^n we will show:

Proposition 1.
$$V X_E = \gamma(E)$$

proof. Divide R^n into $n!$ parts, one for each permutation $(i_1, \ldots, i_n)$ by:
$$X_{i_1, \ldots, i_n} = \{x \in R^n \mid x_{i_1} \geq x_{i_2} \geq \ldots \geq x_{i_n}\}$$

Each of these has equal measure $\frac{1}{n!}$ using property (1) and especially property (3) of the probability measure.

Note that
$$X_{i_1, \ldots, i_n} \subset X_E$$
exactly when the order corresponding to $(i_1, \ldots, i_n)$ is compatible with E.

Let $\pi: R^N \to R^n$ be the projection onto the second factor of $R^N = R^m \times R^n$.

Proposition 2.
$$\pi(K(-q_0, K_{S,A})) \subset X_{E_{S_1}(A)}$$
where $S_1 = S \cap \mathscr{S}_{\text{col}}$.

Here $\{1, \ldots, N\}$ is considered as the disjoint union of $\mathscr{S}_{\text{col}}$ and $\mathscr{S}_{\text{row}}$.
Let $x \in K(-q_0, K_{S,A})$ and $(x_1, \ldots, x_n) \in \pi(K(-q_0, K_{S,A}))$. Then
$$x_k = -\lambda_0 + \lambda_k + \sum_{i \in S} \lambda_i a_{ik} \text{ if } k \notin S_1$$
$$x_k = -\lambda_0 + \sum_{i \in S} \lambda_i a_{ik} \text{ if } k \in S_1.$$

Now say $q \in S_1$ and $a_{jp} \geq a_{jq}$ all j. From the above we see that $x_p \geq x_q$. This yields Proposition 2.

Proposition 3.
$$\rho_A \leq \sum_{\substack{S_1 \subset \mathscr{S}_{\text{col}} \\ S_2 \subset \mathscr{S}_{\text{row}}}} V X_{E_{S_1}(A)}$$

This is an immediate consequence of the main formula and Proposition 2, using condition (2) on the measure.

Moreover, one can assume in this summation that
$$|S_1| = |S_2| \text{ or } |S_1| = |S_2| \pm 1.$$

This follows from the lemma of Section 4, of *Smale* 1982.

One averages over both sides the inequality of Proposition 3 and uses Proposition 1 to obtain the first main estimate. The constant C_m could be taken
$$C_m = 3 \max_s \binom{m}{s}.$$

II. The Problem of the Average Speed of the Simplex Method

The main theorem now depends on the combinatorial problem of estimating $\sigma(m, n, s)$.

Second Main Estimate: If $n > s$

$$\sigma(m, n, s) \leq \left(\frac{s}{n-s}\right)^s \left(1 + \log\left(\frac{n-s}{s} + 2\right)\right)^{ms}$$

This takes some effort, and is carried out in Smale 1982. The main theorem follows every easily from the two estimates.

Section 4 Direct Integration

This is preliminary work about another approach to the study of $\rho(m, n)$. It starts with the main formula above, or more properly, a facet version of the main formula which is in Smale 1982. The idea is to fix the most standard measure, the normal distribution on the space of data (A, b, c). Thus the probability measure is given by a distribution which is the product of Gaussian distributions on each coordinate of $R^{nm} \times R^m \times R^n$. This yields $\rho(m, n)$ as a certain explicit integral which we attempt to simplify and estimate. We hope to give a detailed account of this work in a future publication. Now, we just indicate how the argument gets started.

Let $v_1, \ldots, v_N$ be N vectors in R^N and

$$V = [v_1, \ldots, v_N]$$

the matrix whose columns are $v_1, \ldots, v_N$. Denote the cone of positive linear combinations of the v_i by $K(V)$ and its measure by $\mu K(V)$. Now the measure of a set P of R^N is given by the Gaussian integral

$$\mu(P) = \left(\frac{1}{2\pi}\right)^{N/2} \int_P e^{-\frac{1}{2}\|x\|^2} dx_1 \ldots dx_N$$

Change of Variables Formula:

$$\mu K(V) = \left(\frac{1}{2\pi}\right)^{N/2} \int_0^\infty \ldots \int_0^\infty (\text{Det } V) e^{-\frac{1}{2}\|V\lambda\|^2} d\lambda_1 \ldots d\lambda_N$$

The proof is given by applying the classical formula to the map

$$R_+^N \to R^N, \quad \lambda \to \sum_{i=1}^N \lambda_i v_i = V\lambda$$

The idea is to apply the above formula to understand the following exact expression for $\rho(m, n)$.

Proposition.

$$\rho(m,n) = \sum_{\substack{s_1=s_2 \\ \text{or } s_1=s_2+1}} \binom{m}{s_2}\binom{n}{s_1}(m-s_2)\int_A \mu K B_1(s_1, s_2, A)$$

$$+ \sum_{\substack{s_1=s_2 \\ s_1=s_2-1}} \binom{m}{s_2}\binom{n}{s_1}(n-s_1)\int_A \mu K B_2(s_1, s_2, A)$$

where $B_1(s_1, s_2, A)$ is:

$$\begin{array}{c} s_1 \\ s_2 \\ n-s_1 \\ m-s_2 \end{array}\left[\begin{array}{c|c|c|c} 0 & A_1^T & 0 & 0 \\ \hline -A_1 & 0 & 0 & 0 \\ \hline 0 & A_2^T & I & 0 \\ \hline -A_3 & 0 & 0 & I \end{array}\begin{array}{c} -1 \\ \\ \\ -1 \end{array}\right] \quad A = \begin{bmatrix} A_1 & A_2 \\ \hline A_3 & A_4 \end{bmatrix}$$

$$\quad s_1 \quad s_2 \quad n-s_1 \quad m-s_2-1$$

and $B_2(s_1, s_2, A)$ is the some except that the column of $-I$'s replaces a column in the next to last block, instead of the last.

The preceeding proposition is obtained by integrating over A, the facet version of the main formula (see Section 4 in Smale (1982)).

As I write these lines it seems that this proposition is amenable to estimation, at least numerically.

Section 5 Problems

This final section is devoted to discussion and some problems related to the preceeding.

(1) Develope some way of comparing the speed of different variations of the simplex method. This is a wide open problem. In particular, could one use some of the ideas or results in this paper to estimate the speed of the original simplex method of Dantzig?

(2) Linear programming problems often have a special structure, as sparse matrices; or from transportation problems. To what extent do the results from here carry over? The same question applies to other problems in operations research. Since the main formula above is imbedded in the LCP, this problem seems feasible.

(3) Can one obtain a very complete theory of $\rho(m, n)$ as a function of (m, n)?

(4) Is it possible to make a more systematic study of varying probability measures on the space of data? This relates to Problem (3) above and the first part of Problem (2).

(5) The problem of this paper is part of the much broader problem of computational complexity. An important yet not sharply defined problem is to clarify this relationship. Three particular parts of complexity theory come to mind.

(i) *Numerical Analysis:* I have mentioned above the relationship of Lemke's algorithm to Newton's Method. One can see *Hirsch-Smale* and the references therein for more details. The beginnings of a complexity theory in this context can be found in *Smale* 1981 and *Shub-Smale*. The techniques are quite different, but the concepts are closely related. There is the path-following, great efficiency, failure in worst cases, in common to both the operations research and numerical analysis. It was through this route that I was led into the work of the present paper.

(ii) *Piecewise-Linear equation solving* according to *Eaves-Scarf*. This provides an explicit link between (i) and this paper which I have exploited above. As far as the computational complexity of the Eaves-Scarf algorithms, very recent results have been obtained by Harold Kuhn, Zeke Wang, Senlin Xu, and Jim Renegar.

(iii) Computational complexity has been primarily developed in the domain of the discrete mathematics of theoretical computer science. An obvious connection to this domain lies in the integer programming problem.

References

Balinski, M. and Cottle, R., Complementarity and Fixed Point Problems, North-Holland, Amsterdam, 1978.

Borgwardt, K., The Average Number of Pivot Steps Required by the Simplex-Method is Polynomial, preprint.

Cottle, R. and Dantzig, G., 1968, Complementary Pivot Theory of Mathematical Programming, in Math. of the Decision Sciences (G. Dantzig and A. Veinott, Jr., eds.) Amer. Math. Soc. Providence, R. I., pp. 115-136.

Dantzig, G., 1963, Linear Programming and Extensions, Princeton Univ. Press, Princeton, N.J.

Eaves, C. and Scarf, H., 1976, The Solution of Systems of Piecewise Linear Equations, Math. Operations Res. Vol. 1, pp. 1-27.

Hadley, G., 1962, Linear Programming, Addison-Wesley, Reading, Mass.

Hirsch, M. and Smale, S., 1979, On Algorithms for Solving $f(x)=0$. Commun. Pure. Appl. Math. 32. pp. 281-312.

Klee, V. and Minty, G., 1972, How Good is the Simplex Algorithm? in O. Shisha, ed.: Inequalities III, Academic Press, New York, pp. 159-175.

Lovász, L., 1980, A New Linear Programming Algorithm-Better or Worse Than the Simplex Method?, The Mathematical Intelligencer, vol. 2, pp. 141-146.

Shub, M. and Smale, S., 1982, Computational Complexity: On the Geometry of Polynomials and a Theory of Cost: Part I. Preprint.

Smale, S., 1976, A Convergent Process of Price Adjustment and Global Newton Methods, J. Math. Economics, 3, pp. 107-120.

Smale, S., 1981, The Fundamental Theorem of Algebra and Complexity Theory, Bull. Amer. Math. Soc., vol. 4, pp. 1-36.

Smale, S., 1982, On the Average Speed of the Simplex Method of Linear Programming, preprint (submitted to "Mathematical Programming").

Traub, J. and Wozniakowski, H., 1982, Complexity of Linear Programming, Operations Research Letters, vol. 1, pp. 59-62.

Solution of Large Linear Systems of Equations by Conjugate Gradient Type Methods

J. Stoer
Institut für Angew. Mathematik, Universität Würzburg, Am Hubland, 8700 Würzburg

Introduction

In 1952, Hestenes and Stiefel introduced the conjugate gradient algorithm in their landmark paper [27] as an algorithm for solving linear equation $Ax=b$ with A as positive definite $n \times n$-matrix (see the book [26] of Hestenes for a broad exposition). The algorithm fascinated numerical analysts since then for various reasons: The cg-algorithm combines features of direct and iterative methods which attracted the attention in the early years: It generates a sequence x_i of vectors approximating the solution $\bar{x}$ in a defined way like other iterative methods, but like direct methods, terminating with the exact solution after at most n steps, at least in theory. Many expectations were disappointed, when it was found out that due to roundoff the n-step termination property does not hold in practice. However, viewed as an iterative method, the cg-algorithm has very attractive features. Its application for the iterative solution of large sparse systems has been discussed very early [11] by Stiefel and his co-workers. Like other iterative methods, it essentially requires only the formation of one matrix-vector product $A \cdot x$ per iteration, so that the iterations are inexpensive even for large matrices A, if they are sparse. The iterative aspect of the method has been particularly emphasized since the work of Reid [38].

It turned out, that the method converges very quickly for well-conditioned problems, so that one can obtain a very accurate approximation to the true solution after far less than n iterations. This gave rise to the development of powerful preconditioning techniques (see [1, 2, 8, 22, 23, 25, 31, 32, 43]) in order to speed up the convergence of cg-type methods, essentially by applying them to equivalent systems $M^{-1}Ax = M^{-1}b$ where M is chosen such that the condition of $M^{-1}A$ is much less than the condition of A. Likewise, in recent years generalizations of the cg-algorithms have been found which also permit the solution of systems $Ax=b$ with indefinite and even unsymmetric matrices A.

In this paper it is tried to outline the basic ideas of the various cg-type algorithms. Proofs are only given occasionally. The paper contains 6 Sections. Section 1 describes the basic conjugate gradient (cg) and conjugate residual (cr) algorithms and their main theoretical properties. In Section 2 some computational variants are given, which use the relationship to Lanczos' method [29]. Section 3 treats cg-type methods for symmetric but indefinite matrices A, and Section 4 the case of unsymmetric A. The most important preconditioning techniques are outlined in Section 5. Section 6 contains the description of some applications. So far, the main field of application of cg-type algorithms was the

II. Conjugate Gradient Type Methods

solution of those large linear systems of equations which arise from discretizing partial differential equations by difference or finite element techniques.

However, we will not describe application within this area, but rather in the field of optimization.

Throughout the paper we use the following notation: $\|x\| = \sqrt{x^T x}$ is the euclidean norm, $\sigma(A)$ the spectrum of A, $\{\lambda_i\}$ its eigenvalues, $\kappa = \kappa(A)$ the spectral condition number of A, and for positive definite A and integer μ, $\|x\|_{A^\mu}$ is the norm $\|x\|_{A^\mu}^2 := x^T A^\mu x$, $\bar{x} := A^{-1}b$ is the exact solution of $Ax = b$, and for any x, $e = e(x) = x - \bar{x}$ its error, and $r = r(x) = b - Ax$ its residual.

1. Basic cg-type algorithms

The famous conjugate gradient algorithm of Hestenes and Stiefel [27] is a method for finding the solution $\bar{x}$ of a system of linear equation

$$(1.1) \qquad Ax = b,$$

where A is a real symmetric positive definite $n \times n$-matrix. It is wellknown that the cg-algorithm computes for any starting vector $x_0 \in R^n$ with residual $r_0 := b - Ax_0$ further iterates $x_k \in R^n$ with the following minimization property in terms of the norm $\|x\|_A$

$$\|x_k - \bar{x}\|_A = \min_{x \in x_0 + K_k(r_0; A)} \|x - \bar{x}\|_A, \quad x_k - x_0 \in K_k(r_0; A).$$

Here,

$$K_k(r_0; A) = [r_0, Ar_0, \ldots, A^{k-1}r_0] := \text{span}\{r_0, Ar_0, \ldots, A^{k-1}r_0\},$$

is the k-th Krylov-subspace of R^n generated by r_0.

Note that by $\|x - \bar{x}\|_A^2 = x^T A x - 2b^T x + b^T \bar{x}$, x_k also minimizes the strictly convex quadratic function $F(x) := \frac{1}{2} x^T A x - b^T x$ on $x_0 + K_k(r_0; A)$.

More generally, one can pose the same problem with respect to any A-related norms $\|x\|_{A^\mu}$, and ask for vectors $x_k \in R^n$ such that

$$(1.2) \qquad \|x_k - \bar{x}\|_{A^\mu} = \min_{x \in x_0 + K_k(r_0; A)} \|x - \bar{x}\|_{A^\mu}, \quad x_k \in x_0 + K_k(r_0; A).$$

In practice only the cases $\mu = 1, 2$ have been considered. For $\mu = 2$ this amounts to finding a vector $x_k \in x_0 + K_k(r_0; A)$ with minimal residual $\|A(x - \bar{x})\| = \|Ax - b\|$.

The case $\mu = 0$, which is also meaningful for non-positive definite and even unsymmetric matrices A, will be considered in sections 3 and 4.

The basic cg-algorithm of Hestenes and Stiefel corresponds to the choice $\mu = 1$, and the conjugate-residual algorithm (cr-algorithm) to the choice $\mu = 2$ of the following method, if one defines all scalar products (x, y) by $x^T A^{\mu-1} y$:

(1.3) **Algorithm**

0) *Choose $x_0 \in R^n$ and set $d_0 := r_0 := b - A x_0$*

For $k = 0, 1, \ldots$
1) *if $d_k = 0$: stop: $x_k = \bar{x}$ solves $Ax = b$.*

Otherwise, compute
2) $x_{k+1} := x_k + \sigma_k d_k, \quad \sigma_k := (r_k, r_k)/(d_k, A d_k)$
   $r_{k+1} := r_k - \sigma_k A d_k$
3) $d_{k+1} := r_{k+1} + \beta_k d_k, \quad \beta_k := (r_{k+1}, r_{k+1})/(r_k, r_k).$

Note that these formulae lead to effective calculations only for $\mu = 1$ and $\mu = 2$.

We list without proof (see e.g. [27, 38]) the main properties of (1.3):

(1.4) **Theorem.** a) *There is a first index m with $0 \leq m \leq n$ such that $d_m = 0$. The vectors $d_0, \ldots, d_{m-1}$ are linearly independent and A^μ-conjugate:*

$(d_i, A d_j) = 0 \quad \text{for} \quad 0 \leq i < j < m$

b) $(r_i, r_j) = 0 \quad \text{for} \quad 0 \leq i < j \leq m$
   $(r_i, r_i) > 0 \quad \text{for} \quad 0 \leq i < m$

c) $(d_i, r_j) = \begin{cases} 0 & \text{for } 0 \leq i < j \leq m \\ (r_i, r_i) & \text{for } 0 \leq j \leq i \leq m \end{cases}$

d) $[d_0, d_1, \ldots, d_{k-1}] = [r_0, A r_0, \ldots, A^{k-1} r_0] = [r_0, r_1, \ldots, r_{k-1}] \quad \text{for} \quad 1 \leq k \leq m$

e) $x_m = \bar{x}$.

There are many different but mathematically equivalent formulations of (1.3) (see Reid [38] for a thorough discussion), e.g. using (1.4), c), one might compute the step σ_k in (1.3), 2) also by

$$\sigma_k := (d_k, r_k)/(d_k, A d_k)$$

which means that in each step of (1.3)

(1.5) $$\|x_{k+1} - \bar{x}\|_{A^\mu} = \min_\sigma \|x_k + \sigma d_k - \bar{x}\|_{A^\mu},$$

x_{k+1} is a result of an exact line search. Since $x_k = x_0 + \sigma_0 d_0 + \ldots + \sigma_{k-1} d_{k-1}$, and the vectors d_i are A^μ-conjugate, a known result on the minimization of quadratic functions gives by (1.5) and (1.4) d)

$$\|x_k - \bar{x}\|_{A^\mu} = \min_{x \in x_0 + [d_0, \ldots, d_{k-1}]} \|x - \bar{x}\|_{A^\mu} = \min_{x \in x_0 + K_k(r_0; A)} \|x - \bar{x}\|_{A^\mu},$$

that is (1.2). For the errors $e_k := x_k - \bar{x}$, we have $-r_k = A e_k$ and $K_k(r_0; A) = [A e_0, A^2 e_0, \ldots, A^k e_0]$, so that by (1.2)

(1.6) $$\|e_k\|_{A^\mu} = \min_{z \in [A e_0, \ldots, A^k e_0]} \|e_0 + z\|_{A^\mu} = \min_{p \in \Pi_k} \|p(A) e_0\|_{A^\mu},$$

where

$$\Pi_k := \{p(t) \equiv 1 + \sigma_1 t + \ldots + \sigma_k t^k \mid \sigma_i \in R\}$$

is the set of all real polynomials p of degree $\leq k$ with $p(0) = 1$. Now let $\sigma(A) = \{\lambda_1, \ldots, \lambda_n\}$ be the set of eigenvalues of A, $\bar{\Lambda} := \max \lambda_i$, $\underline{\Lambda} := \min \lambda_i > 0$, and z_i the corresponding orthonormal eigenvectors, $Az_i = \lambda_i z_i$, $z_i^T z_j = \delta_{ij}$. Consider the expansion of e_0 into a sum of eigenvectors

$$e_0 = \rho_1 z_1 + \ldots + \rho_n z_n, \quad \|e_0\|_{A^\mu}^2 = \sum_{j=1}^n \lambda_j^\mu \rho_j^2.$$

We may assume that this expansion is *reduced* in the following sense: There is an index $t \geq 0$, such that $\rho_i \neq 0$ for $0 \leq i \leq t$, $\rho_i = 0$ for $i > t$ and $\lambda_i \neq \lambda_k$ for all $i \neq k$ with $i, k \leq t$. t is called the *reduced length* of e_0. Inserting the expansion of e_0 into (1.6) gives

(1.7)
$$\|e_k\|_{A^\mu}^2 = \min_{p \in \Pi_k} \|p(A)e_0\|_{A^\mu}^2 = \min_{p \in \Pi_k} \sum_{j=1}^n p(\lambda_j)^2 \lambda_j^\mu \rho_j^2$$

$$\leq \left(\sum_{j=1}^n \lambda_j^\mu \rho_j^2\right) \min_{p \in \Pi_k} \max_j p(\lambda_j)^2$$

which finally results in the well-known error estimate for the basic cg-algorithm

(1.8)
$$\|e_k\|_{A^\mu}/\|e_0\|_{A^\mu} \leq \min_{p \in \Pi_k} \max_{\lambda \in \sigma(A)} |p(\lambda)|$$

$$\leq \min_{p \in \Pi_k} \max_{\underline{\Lambda} \leq \lambda \leq \overline{\Lambda}} |p(\lambda)|$$

$$= 1/T_k\left(\frac{\kappa+1}{\kappa-1}\right) \leq 2 \cdot \left(\frac{\sqrt{\kappa}-1}{\sqrt{\kappa}+1}\right)^k,$$

by a known result on Chebyshev-polynomials $T_k(x) \equiv \cos(k \arccos x)$. Here $\kappa := \overline{\Lambda}/\underline{\Lambda}$ is the spectral condition number of A. In particular a closer examination of (1.7), (1.8) gives

(1.9)
    a) The termination index m of the cg-algorithm, $p_m = e_m = 0$, is equal to the reduced length t of e_0. m is also equal to $\max_k \dim K_k(r_0; A)$ and not greater than the number of different eigenvalues of A.

    b) $\|e_k\|_{A^\mu}/\|e_0\|_{A^\mu} \leq \varepsilon$ holds for all $k \geq \frac{1}{2}\sqrt{\kappa} \ln \frac{2}{\varepsilon}$.

The estimates describe the behaviour of the cg-algorithm as an iterative method. In order to reduce the size $\|e_0\|_{A^\mu}$ of the initial error e_0 by 10^{-k} one needs approximately $k \cdot \sqrt{\kappa}$ cg-steps.

Note that (1.8) is only an upper bound for $\|e_k\|_{A^\mu}/\|e_0\|_{A^\mu}$ which only depends on the condition number κ of A: It can be substantially improved for special distributions of eigenvalues of A (see Axelsson [2], Greenbaum [19]).

2. Connections with the Lanczos method

There are some variants of the cg-algorithm which are all related to the Lanczos method [29] and are important in certain generalizations of the cg-algorithm.

For example, the following Lanczos' type algorithm also generates a sequence of linear independent A^μ-conjugate vectors $p_0, p_1, \ldots, p_{m-1}$ with

(2.1) $\qquad [p_0, p_1, \ldots, p_{k-1}] = [r_0, A r_0, \ldots, A^{k-1} r_0]$ for $1 \leq k \leq m$

(see (1.4) d)):

(2.2) **Algorithm**

    0) *Choose* $x_0 \in R^n$ *and set* $p_0 := r_0 := b - A x_0$, $p_{-1} := 0$.

    For $k = 0, 1, \ldots$
    1) *If* $p_k = 0$ *stop:* $x_k = \bar{x}$ *solves* $Ax = b$.

    Otherwise compute
    2) $x_{k+1} := x_k + \sigma_k p_k, \quad \sigma_k := (p_k, r_k)/(p_k, A p_k)$
    $\quad r_{k+1} := r_k - \sigma_k A p_k$

    3) $p_{k+1} := A p_k - \gamma_k p_k - \delta_k p_{k-1}$
    where $\gamma_k := (A p_k, A p_k)/(p_k, A p_k)$
$$\delta_k := \begin{cases} 0 & \text{for } k = 0 \\ (A p_k, A p_{k-1})/(p_{k-1}, A p_{k-1}) & \text{for } k > 0. \end{cases}$$

Note that for $\mu = 2$ (cr-algorithm) algorithm (2.2) seems to require one more matrix vector product $A \cdot x$ than for $\mu = 1$. To a certain extent, this can be avoided, if one introduces in step 3 of (2.2) for $\mu = 2$ an additional recursion for $A p_k$

$$A p_{k+1} := A \cdot A p_k - \gamma_k A p_k - \delta_k A p_{k-1}.$$

Comparing (2.1) with (1.4) d) it is easily seen that the vectors d_i and p_i are proportional, $d_i = \tau_i p_i$, $i = 0, 1, \ldots, m-1$, $\tau_i \neq 0$. Now let $\mu = 1$. Then by (1.4) b), d) the vectors $r_0, \ldots, r_{m-1}$ are orthogonal and $r_0, \ldots, r_{k-1}$ forms an orthogonal basis of $[r_0, A r_0, \ldots, A^{k-1} r_0]$. Again, Lanczos method can be used to compute orthonormal vectors $q_0, \ldots, q_{m-1}$ such that

(2.3) $\qquad [q_0, q_1, \ldots, q_{k-1}] = [r_0, A r_0, \ldots, A^{k-1} r_0]$ for $1 \leq k \leq m$

in the following way (see e.g. Parlett [37]), even if A is only symmetric, but not positive definite.

(2.4) **Algorithm**

    0) *Set* $q_{-1} := 0$, $\delta_0 := \|r_0\|$, $u_{-1} := r_0$.

    For $k = 0, 1, \ldots$
    1) *If* $\delta_k = 0$ *stop*.

    Otherwise, compute
    1) $q_k := u_{k-1}/\delta_k$

    2) $u_k := A q_k - \gamma_k q_k - \delta_k q_{k-1}$
    with $\gamma_k := q_k^T A q_k$, $\delta_{k+1} := \|u_k\|$.

II. Conjugate Gradient Type Methods

Then, the tridiagonal matrix

(2.5) $$T_k := \begin{pmatrix} \gamma_0 & \delta_1 & & & 0 \\ \delta_1 & \gamma_1 & \ddots & & \\ & \ddots & \ddots & \ddots & \\ & & \ddots & \ddots & \delta_{k-1} \\ 0 & & & \delta_{k-1} & \gamma_{k-1} \end{pmatrix}$$

and $Q_k := (q_0, q_1, \ldots, q_{k-1})$ satisfy

(2.6) $$\begin{aligned} A Q_k - Q_k T_k &= \delta_k q_k \bar{e}_k^T, \quad \bar{e}_k := (0, \ldots, 0, 1)^T \in R^k. \\ Q_k^T Q_k &= I_k, \quad \text{the } k \times k\text{-identity matrix} \\ Q_k^T q_k &= 0, \quad Q_k^T A Q_k = T_k. \end{aligned}$$

There are the following relationship between the quantities $r_k, d_k, \sigma_k, \beta_k$ generated by (1.3) and the quantities q_k, γ_k, δ_k given by (2.4) (see Householder [28]):

There holds $q_k := r_k / \|r_k\|$ and in terms of the matrices Q_k, $D_k := \text{diag}(\sigma_0^{-1}, \sigma_1^{-1}, \ldots, \sigma_{k-1}^{-1})$

$$P_k := \left(\frac{d_0}{\|r_0\|}, \frac{d_1}{\|r_1\|}, \ldots, \frac{d_{k-1}}{\|r_{k-1}\|} \right), \quad L_k := \begin{pmatrix} 1 & & & & 0 \\ -\sqrt{\beta_0} & 1 & & & \\ & \ddots & \ddots & & \\ & & \ddots & 1 & \\ 0 & & & -\sqrt{\beta_{k-1}} & 1 \end{pmatrix}$$

the relations $r_{j+1} = r_j - \sigma_j A d_j$ and $d_{j+1} = r_{j+1} + \beta_j d_j$ of (1.3) are equivalent to

(2.7) $$\begin{aligned} A P_k D_k^{-1} &= Q_k L_k - \frac{r_k}{\|r_{k-1}\|} \cdot \bar{e}_k^T \\ P_k L_k^T &= Q_k \end{aligned}, \quad k \geq 1.$$

The elimination of P_k gives

(2.8) $$A \cdot Q_k - Q_k \cdot L_k D_k L_k^T = -\frac{r_k}{\|r_{k-1}\|} \cdot \frac{\bar{e}_k^T}{\sigma_{k-1}}$$

and a comparison with (2.4) shows that L_k, D_k provide the triangular decomposition of T_k

(2.9) $$T_k = L_k D_k L_k^T,$$

and thereby further relations between γ_i, δ_i and β_i, σ_i.

The iterates x_k of the cg-algorithm can be obtained by means of (2.4) as follows: By (1.2) and (2.3) $x_k \in x_0 + [q_0, \ldots, q_{k-1}]$, so that x_k has the form $x_k = x_0 + Q_k f_k$ for some vector $f_k \in R^k$. Again by (1.2) r_k is orthogonal to $K_k(r_0, A)$, so that

(2.10) $$Q_k^T(b - A(x_0 + Q_k f_k)) = Q_k^T(r_0 - A Q_k f_k) = 0.$$

(2.6) then gives

$$T_k f_k = Q_k^T r_0 = \bar{e}_1 \|r_0\|$$

(2.11) $\quad x_k = x_0 + v_k, \quad v_k := Q_k T_k^{-1} Q_k^T r_0 = Q_k T_k^{-1} \bar{e}_1 \cdot \|r_0\|.$

Essentially following an idea of Paige and Saunders [36], O'Leary [34] proposed to compute v_k in certain cases (see below) as follows: Let

(2.12) $$\bar{L}_k \cdot \bar{L}_k^T = T_k$$

be the Cholesky decomposition of the positive definite matrix T_k.

$\bar{L}_k$ is a lower triangular bidiagonal matrix with $\bar{L}_{k-1}$ being a principal submatrix of $\bar{L}_k$, so that it can be easily computed along with (2.4) as k progresses. In terms of the matrix $C_k := Q_k \cdot \bar{L}_k^{-T}$, the vector v_k (2.11) can be expressed by

(2.13) $$v_k = Q_k f_k = Q_k \bar{L}_k^{-T} z_k = C_k z_k,$$

where z_k is solution of

(2.14) $$\bar{L}_k z_k = \|r_0\| \bar{e}_1.$$

By the structure of $\bar{L}_k$, z_{k-1} is a subvector of z_k, and $C_{k-1} = (c_1, \ldots, c_{k-1})$ a submatrix of $C_k = (c_1, \ldots, c_k)$, which can be computed columnwise by solving $Q_k^T = \bar{L}_k C_k^T$: In this way z_k, v_k and x_k can be easily accumulated with increasing k along with (2.4), without saving all columns of C_k and Q_k.

Still another way to compute the solution $\bar{x}$ of a linear system $Ax = b$ with a nonsingular not necessarily positive definite matrix $A = A^T$ using (2.4) and (2.11) was recently investigated by Simon [39]. His proposal is to compute q_k, γ_k, δ_k by (2.4) until for some k the residual r_k is small enough, $\|r_k\| \le \text{tol}$.

Only then the equation (2.11)

$$T_k f_k = \bar{e}_1 \|r_0\| = \bar{e}_1 \delta_0$$

is solved for f_k using a numerically stable elimination method (note that T_k need not be positive definite), then giving $x_k = x_0 + Q_k f_k$. In this context it is useful to note that the size $\|r_j\|$ of the residuals can be computed recursively without the computation of either x_j or r_j. This follows from (2.10) and (2.6)

$$\begin{aligned} r_j &= r_0 - A Q_j T_j^{-1} Q_j^T r_0 \\ &= r_0 - (Q_j T_j + \delta_j q_j \bar{e}_j^T) T_j^{-1} Q_j^T r_0 \\ &= -\delta_j q_j \bar{e}_j^T T_j^{-1} \bar{e}_1 \delta_0, \quad \delta_0 = \|r_0\| \\ &= -\delta_j \varphi_j q_j, \end{aligned}$$

$$\|r_j\| = |\delta_j \varphi_j|,$$

where φ_j is the last component of f_j. Note that φ_j can be calculated without computing the whole vector f_j by means of a QR-factorization of T_j, which in turn can be recursively computed for successive indices j along with (2.4). This method presupposes that all vectors q_j are stored. This drawback can be turned into an advantage, because it permits to incorporate into the Lanczos algorithm (2.4) suitable reorthogonalization steps in order to prevent the usually observed rapid loss of orthogonality of the q_j due to the influence of roundoff (see [39] for details). In this way the many additional iterations any realization of the Lanczos method (2.4) without reorthogonalization needs before stopping can be avoided.

3. cg-type methods for solving symmetric indefinite systems

We now consider the case where $A = A^T$ is a nonsingular symmetric but indefinite matrix. At first glance one might think to transform $Ax = b$ to the equivalent normal equations $A^2 x = Ab$, where now A^2 is positive definite, so that any of the previous methods could be applied. The main drawback of this approach is that, because of $\kappa(A^2) = \kappa(A)^2$, the new equations are much more ill-conditioned and also the speed of convergence of the cg-methods of section 1 is greatly reduced. If one rather tries to apply the methods of the previous sections directly to $Ax = b$, then additional problems arise. Only for even μ, say $\mu = 0, 2$, $\|x\|_{A^\mu}$ is a norm, for which (1.2) makes sense, whereas for $\mu = 1$ the quadratic form $(x - \bar{x})^T A (x - \bar{x})$ is indefinite. But even for $\mu = 2$ for which x_k is welldefined by (1.2), the cr-algorithm (1.3) may break down in step 3), because $(r_k, r_k) = r_k^T A r_k$ might vanish for $r_k \neq 0$. By contrast the computationally slightly more expensive cr-Lanczos-algorithm (2.2), $\mu = 2$, works also for indefinite A. However, the error estimate (1.8) has to be revised, since A may also have negative eigenvalues. With $\bar{\Lambda} := \max|\lambda_i|$, $0 < \underline{\Lambda} := \min_i|\lambda_i|$, we get instead of (1.8)

$$(3.1) \qquad \|e_k\|_{A^2} / \|e_0\|_{A^2} \leq \min_{p \in \Pi_k} \max_j |p(\lambda_j)|$$

$$\leq \min_{p \in \Pi_k} \max_{\lambda \in \mathbb{R} : \underline{\Lambda} \leq |\lambda| \leq \bar{\Lambda}} |p(\lambda)|$$

and for reasons of symmetry

$$\leq \min_{p \in \Pi_{[k/2]}} \max_{\underline{\Lambda} \leq \lambda \leq \bar{\Lambda}} |p(\lambda^2)|$$

$$(3.2) \qquad = \min_{p \in \Pi_{[k/2]}} \max_{\underline{\Lambda}^2 \leq t \leq \bar{\Lambda}^2} |p(t)|$$

$$= 1/T_{[k/2]}\left(\frac{\kappa^2 + 1}{\kappa^2 - 1}\right) \leq 2\left(\frac{\kappa - 1}{\kappa + 1}\right)^{[k/2]},$$

where $\kappa := \bar{\Lambda}/\underline{\Lambda}$ is again the spectral condition number of A. Likewise, (1.9)b) is to replaced by

$$(3.3) \qquad \|e_k\|_{A^2}/\|e_0\|_{A^2} \leq \varepsilon \quad \text{for all } [k/2] \geq \frac{1}{2}\kappa \ln\frac{2}{\varepsilon}.$$

This means that in bad cases (e.g. if A has a spectrum which is highly symmetrical with respect to 0) one needs approximately 2κ iterations in order to reduce the size of the residual by a factor $1/10$. On the other hand these bounds may be improved considerably, if A has only few negative eigenvalues. In these cases results like (1.8), (1.9), b) hold. A computationally slightly less expensive variant of the cr-Lanczos-method (2.2) ($\mu=2$) just mentioned is the MCR-algorithm of Chandra [6]. Since for $\mu=1$ the minimization of $(x-\bar{x})^T A(x-\bar{x})$ on $x \in x_0 + K_k(r_0; A)$ (compare (1.2)) does not make sense for indefinite A, it is common to define x_k by a Galerkin-approach, namely by trying to find an $x_k \in x_0 + K_k(r_0; A)$ such that

(3.4) $$r_k^T v = 0 \quad \text{for all} \quad v \in K_k(r_0; A),$$

that is as a stationary point of $(x-\bar{x})^T A(x-\bar{x})$ within $x_0 + K_k(r_0; A)$. But as easy examples show, such a stationary point need not exist for all k, if A is indefinite. Also the cg-algorithm (1.3) with $\mu=1$ is not suitable to compute x_k, since it may break down because $d_k^T A d_k = 0$ for some $d_k \neq 0$. A first, though unstable method to compute those x_k with (3.4) which exist, was proposed by Luenberger [30]. Paige and Saunders [36] described a numerically stable algorithm (SYMMLQ-algorithm). It is based on the Lanczos algorithm (2.4) and uses that x_k is uniquely defined by (3.4) if and only if the matrix T_k generated by (2.4) is nonsingular and x_k satisfies (2.10), (2.11). Since T_k is also indefinite, its Cholesky decomposition (2.12) might not exist. Paige and Saunders [36] therefore based their method on orthogonal factorizations of T_k,

(3.5) $$T_k = \tilde{L}_k \tilde{Q}_k, \quad \tilde{Q}_k^T \tilde{Q}_k = I_k, \quad \tilde{L}_k \text{ lower triangular},$$

which always exist and can be easily recursively calculated along with (2.4). The solution x_k of (2.11) can then be computed in much the same way as in (2.13)–(2.14), the only difference being the use of (3.5) instead of (2.12).

A related algorithm (SYMMBK-algorithm) proposed by Chandra [6] uses a recursively updated Bunch-Parlett decomposition of T_k rather than (3.5) in the same way as (3.5) is used in the SYMMLQ-algorithm.

Of course, also the algorithm of Simon [39], briefly outlined at the end of section 2, can be used to compute x_k with (3.4). This algorithm is numerically stable and needs substantially fewer iterations than e.g. SYMMLQ but at the expense of a large storage requirement (see [39] for a detailed discussion and comparative numerical results).

We now consider the case $\mu=0$. Then the solution x_k of (1.2), namely

$$\|x_k - \bar{x}\| = \min_{x \in x_0 + K_k(r_0; A)} \|x - \bar{x}\|, \quad x_k \in x_0 + K_k(r_0; A),$$

is always welldefined, even for indefinite A.

However, x_k cannot be effectively computed by the usual cg-algorithm for $\mu=0$ even for positive definite A, since e.g. the formula for the step size σ_k in (1.3) would require the computation of

$$\sigma_k = r_k^T A^{-1} r_k / d_k^T d_k \quad \text{or} \quad \sigma_k = p_k^T A^{-1} r_k / d_k^T d_k,$$

which is not feasible. The same holds for the Lanczos type algorithm (2.2) for $\mu=0$.

II. Conjugate Gradient Type Methods

By slightly changing the definition of x_k an effective computation becomes possible. We replace $K_k(r_0; A)$ by $K_k(A r_0; A)$ and consider the problem of finding vectors x_k such that

$$(3.6) \qquad \|x_k - \bar{x}\| = \min_{\substack{x \in x_0 + K_k(A r_0; A) \\ = x_0 + [A r_0, A^2 r_0, \ldots, A^k r_0]}} \|x - \bar{x}\|,$$

which again uniquely defines x_k.

A first algorithm to compute x_k with (3.6) is the orthogonal direction algorithm (OD-algorithm) of Fridman [16] (later on rediscovered by Fletcher [13]), which we will describe briefly.

(3.7) **OD-algorithm**

0) *Start:* Choose $x_0 \in R^n$ and set $p_{-1} := r_0 := b - A x_0$, $p_0 := A r_0$

For $k = 0, 1, \ldots$
1) *If* $p_k = 0$ *stop:* x_k *solves* $A x = b$.

 Otherwise compute
2) $x_{k+1} := x_k + \sigma_k p_k, \quad \sigma_k := r_k^T p_{k-1} / p_k^T p_k$
 $r_{k+1} := r_k - \sigma_k A p_k$

3) $p_{k+1} := A p_k - \gamma_k p_k - \delta_k p_{k-1}$
 where $\gamma_k := p_k^T A p_k / p_k^T p_k$

$$\delta_k := \begin{cases} 0 & \text{if } k = 0 \\ p_k^T p_k / p_{k-1}^T p_{k-1}, & \text{if } k > 0. \end{cases}$$

Since step 3 is a Lanczos orthogonalization step, (3.7) generates a sequence of orthogonal vectors p_k, which therefore must terminate with a first $p_m = 0$, $m \leq n$. We list the main properties of (3.7) (see [15, 41] for a proof).

(3.8) **Theorem:** a) *There is a first index* m *with* $0 \leq m \leq n$ *such that* $p_m = 0$ *and* $p_i^T p_j = 0$ *for* $0 \leq i < j \leq m$, $p_i^T p_i > 0$ *for* $0 \leq i < m$.

b) $r_k^T A^{-1} p_k = r_k^T p_{k-1} \quad$ *for* $k = 0, 1, \ldots, m$
 $r_k^T A^{-1} p_j = 0 \qquad\quad$ *for* $0 \leq j < k \leq m$

c) $[p_0, p_1, \ldots, p_{k-1}] = [p_0, A p_0, \ldots, A^{k-1} p_0]$
 $\qquad\qquad\qquad\quad = [A r_0, A^2 r_0, \ldots, A^k r_0] = K_k(A r_0; A)$
 $r_k \in [p_{-1}, p_0, \ldots, p_k] \qquad\qquad\qquad\qquad$ *for* $k = 0, 1, \ldots, m$

d) $x_m = \bar{x}$.

The theorem is easily proved by induction; the most interesting are the identities of b), because they show that the step size σ_k of (3.7) is also equal to

$$(3.9) \qquad \sigma_k = r_k^T A^{-1} p_k / p_k^T p_k,$$

and therefore

$$(3.10) \qquad \|x_{k+1} - \bar{x}\| = \min_\sigma \|x_k + \sigma p_k - \bar{x}\|.$$

For this reason let us briefly outline the inductive proof of (3.8) b): b) is true for $k=0$ by definition of p_0 and p_{-1}.

If it is true for some $0 \leq k < m$, then for $0 \leq j \leq k$ $r_{k+1}^T A^{-1} p_j = r_k^T A^{-1} p_j - \sigma_k p_k^T p_j = 0$ either ($j=k$) by the induction hypothesis and the definition of σ_k, or ($j<k$) by the induction hypothesis and $p_k^T p_j = 0$.

Now by (3.8)

$$[p_0, p_1, \ldots, p_{k-1}] = [A r_0, \ldots, A^k r_0] = [A^2 e_0, \ldots, A^{k+1} e_0]$$

so that we get by (3.10) and the orthogonality of the p_i the following minimum properties of the error $e_k = x_k - \bar{x}$, which are analogous to (1.6)-(1.9).

(3.11)
$$\|e_k\| = \min_{x \in x_0 + [p_0, \ldots, p_{k-1}]} \|x - \bar{x}\|$$
$$= \min_{x \in x_0 + [A^2 e_0, \ldots, A^{k+1} e_0]} \|x - \bar{x}\|$$
$$= \min_{p \in \bar{\Pi}_k} \|p(A) e_0\|$$

where now

$$\bar{\Pi}_k := \{p(t) \equiv 1 + \sigma_2 t^2 + \sigma_3 t^3 + \ldots + \sigma_{k+1} t^{k+1} \mid \sigma_i \in \mathbb{R}\}$$

is the set of all real polynomials of degree $\leq k+1$ with $p(0)=1$ and $p'(0)=0$. Using the eigenvector expansion of e_0, we get analogously to (1.7)-(1.9), (3.1)-(3.3)

(3.12)
$$\|e_k\|/\|e_0\| \leq \min_{p \in \bar{\Pi}_k} \max_j |p(\lambda_j)|$$
$$\leq \min_{p \in \bar{\Pi}_k} \max_{\lambda \in \mathbb{R}: \underline{\Lambda} \leq |\lambda| \leq \bar{\Lambda}} |p(\lambda)|$$

and using symmetriy arguments

(3.13)
$$\leq \min_{p \in \Pi_{[(k+1)/2]}} \max_{\underline{\Lambda}^2 \leq t \leq \bar{\Lambda}^2} |p(t)|$$
$$= \left(T_{[(k+1)/2]}\left(\frac{\kappa^2+1}{\kappa^2-1}\right)\right)^{-1} \leq 2\left(\frac{\kappa-1}{\kappa+1}\right)^{[(k+1)/2]}$$

Instead of (1.9) one now gets the result (see Freund [15]):

(3.14) Theorem: a) *The termination index m of (3.7) (i.e. the first index with $p_m = 0$) is equal to the reduced length of e_0. The first index l with $r_l = 0$ is $l = m-1$ if $\sum_{j=1}^{m} 1/\lambda_j = 0$, where λ_j are those (different) eigenvalues of A appearing in the reduced eigenvector expansion of e_0, and $l = m$ otherwise.*

b) $\|e_k\|/\|e_0\| \leq \varepsilon$ *for all* $[(k+1)/2] \geq \frac{\kappa}{2} \ln \frac{2}{\varepsilon}$.

The estimate of b) is comparable to (3.3). The bound improves only every second iteration. It can happen that $x_k = x_{k+1}$ for certain iterations. In general one can show only (see [15, 41])

II. Conjugate Gradient Type Methods

$$\|e_1\|/\|e_0\| \leq \frac{\kappa^2-1}{\kappa^2+1}, \quad \|e_{k+2}\|/\|e_k\| \leq \frac{\kappa^2-1}{\kappa^2+1} \quad \text{for } k \leq m-2$$

and there are examples with $x_{2k-1} = x_{2k}$ for all $k \geq 1$.

For special eigenvalue distributions one can improve the bound (3.13). For example, if A has j negative eigenvalues, $\lambda_i < 0$ for $i = 1, \ldots, j$, and $\lambda_i \in [a, b]$ for $i > j$, $0 < a \leq b$, then Freund [15] has shown that

$$(3.15) \quad \|e_{k+j}\|/\|e_0\| \leq 2 \left(\prod_{i=1}^{j} \left(1 - \frac{b}{\lambda_i}\right) \right) \left(1 - b \sum_{i=1}^{j} \frac{1}{\lambda_i} + k \cdot \kappa_1 \right) \cdot \left(\frac{\sqrt{\kappa_1}-1}{\sqrt{\kappa_1}+1}\right)^k, \quad k \geq 0.$$

Here $\kappa_1 := b/a$ is a reduced condition number of A determined by its positive eigenvalues only.

As was noted by Fletcher [13] the SYMMLQ algorithm of Paige and Saunders [36] computes auxiliary vectors x_k^L as a byproduct, which are identical to the vectors x_k generated by (3.7). The OD-algorithm is simpler and less expensive than the SYMMLQ algorithm. Unfortunately enough, and in contrast to SYMMLQ, (3.7) turned out to be numerically unstable. In practice, the errors $\|x_k - \bar{x}\|$ of the iterates x_k of (3.7) only decrease during the first 20–30 iterations according to the theory (3.6), but then diverge rapidly (see [41]) due to the influence of round-off. The main reason is that due to the gradual loss of orthogonality among the vectors p_i, the identity $r_k^T A^{-1} p_k = r_k^T p_{k-1}$ of Theorem (3.8), on which relies the computation of the step size σ_k in (3.7), becomes more and more invalid for the computed r_k, p_k. Thus for larger k, the computed $x_{k+1} = x_k + \sigma_k p_k$ does not minimize $\|x_k + \sigma p_k - \bar{x}\|$ with respect to σ even approximately.

There is a simple remedy based on the following idea. The orthogonal p_i, $i \geq 0$ with $p_0 = A r_0$ satisfy

$$[p_0, p_1, \ldots, p_{k-1}] = [A r_0, A^2 r_0, \ldots, A^k r_0]$$

if and only if they have the form $p_i = A q_i$, $q_0 := r_0$, and the q_i are such that

$$[q_0, q_1, \ldots, q_{k-1}] = [r_0, A r_0, \ldots, A^{k-1} r_0], \quad k \geq 1$$
$$q_i^T A^2 q_j = 0 \quad \text{for } i \neq j.$$

But a set of vectors q_i with this property is generated by the cr-Lanczos-algorithm (2.2) with $\mu = 2$. Since $p_k = A q_k$, minimizing

$$\|x_k + \sigma p_k - \bar{x}\| = \|x_k + \sigma A q_k - \bar{x}\|$$

with respect to σ leads to the step length formula

$$(3.15) \quad \sigma_k = \frac{r_k^T A^{-1} p_k}{p_k^T p_k} = \frac{r_k^T q_k}{q_k^T A^2 q_k}$$

in terms of q_k. In this way one gets the following stabilized OD-method (where the vectors q_k are again denoted by p_k) which simply combines (2.2) with (3.15).

(3.16) **STOD-algorithm:**

0) Start: Choose $x_0 \in R^n$ and set $p_0 := r_0 := b - A x_0$, $p_{-1} := 0$.

For $k = 0, 1, \ldots$

1) If $p_k = 0$ stop: x_k solves $Ax = b$.

Otherwise compute

2) $\sigma_k := r_k^T p_k / p_k^T A^2 p_k$
   $x_{k+1} := x_k + \sigma_k A p_k$, $\quad r_{k+1} := r_k - \sigma_k A \cdot (A p_k)$

3) $p_{k+1} := A p_k - \gamma_k p_k - \delta_k p_{k-1}$
   $A p_{k+1} := A(A p_k) - \gamma_k A p_k - \delta_k A p_{k-1}$
   where $\gamma_k := p_k^T A^3 p_k / p_k^T A^2 p_k$

$$\delta_k := \begin{cases} 0 & , \text{ if } k=0 \\ p_k^T A^2 p_k / p_{k-1}^T A^2 p_{k-1}, & \text{ if } k>0 \end{cases}.$$

The following table lists the results of operation counts for the various methods of solving $Ax = b$ for indefinite A, which have been mentioned. Listed are the number of matrix – vector products $(A \cdot x)$, the approximate number M of multiplications per iteration and the number N of vectors $\in R^n$ to be stored:

	$A \cdot x$	M	N
cr-method (2.2), $\mu = 2$	1	$9n$	7
MCR-method [6]	1	$7n-9n$	7
SYMMBK [6]	1	$7n$	6
SYMMLQ [36]	1	$9n$	5
OD-algorithm (3.7)	1	$7n$	5
STOD-algorithm (3.16)	1	$8n$	7

Numerical results communicated in [6, 36, 41] show that all these methods, except the OD-algorithm, are stable and can be used to compute the solution of large sparse equations $Ax = b$ with similar efficiency.

4. The solution of $Ax = b$ for unsymmetric A

A general system of linear equations $Ax = b$ with A an arbitrary nonsingular $n \times n$-matrix is equivalent to the normal equations

(4.1) $$A^T A x = A^T b,$$

where $A^T A$ is positive definite, so that any of the methods of sections 1–3 could be used for its solution. The drawback of this approach is again that because of $\kappa(A^T A) = \kappa(A)^2$, (4.1) is more illconditioned than $Ax = b$, which will also worsen the speed of convergence of cg-type methods. Nevertheless, an early method due to Craig [9], also described in Fadeev, Fadeeva [12], should be mentioned in this context, since it is different from the usual cg-algorithm

II. Conjugate Gradient Type Methods

((1.3) with $\mu=1$), the cr-algorithm ((1.3) with $\mu=2$) and from Fridman's OD-algorithm (3.7) applied to (4.1). It corresponds to case $\mu=0$ of (1.2) applied to (4.1): x_k is defined by the minimum property

(4.2) $$x_k := \arg\min_{x \in x_0 + K_k(A^T r_0; A^T A)} \|x - \bar{x}\|.$$

Note that

$$K_k(A^T r_0; A^T A) = [A^T A e_0, (A^T A)^2 e_0, \ldots, (A^T A)^k e_0] = A^T K_k(r_0; A A^T),$$

and x_k satisfies (4.2) if and only if $y_k := A^{-T} x_k$, $\bar{y} := A^{-T} \bar{x}$ are such that

$$y_k = \arg\min_{y \in y_0 + K_k(r_0; AA^T)} \|A^T(y - \bar{y})\| = \arg\min_{y \in y_0 + K_k(r_0; AA^T)} \|y - \bar{y}\|_{AA^T}$$

so that y_k, and thereby $x_k = A^T y_k$, can be computed by applying the usual cg-algorithm (1.3) with $\mu=1$ to the system $AA^T y = b$. This results in Craig's method with the recursions

$$d_0 := r_0$$

$$x_{k+1} := x_k + \sigma_k A^T d_k, \quad r_{k+1} := r_k - \sigma_k AA^T d_k, \quad \sigma_k := r_k^T d_k / d_k^T AA^T d_k$$

$$p_{k+1} := r_{k+1} + \beta_k d_k, \quad \beta_k := r_{k+1}^T r_{k+1} / r_k^T r_k$$

ensuring $[d_0, d_1, \ldots, d_{k-1}] = K_k(r_0; AA^T)$, $d_i^T AA^T d_j = 0$ for $i \neq j$ and (4.2).
The error e_k of this method has a minimum property

$$\|e_k\| = \min_{p \in \Pi_k} \|p(A^T A) e_0\|$$

and, as $A^T A$ is positive definite, satisfies (see (1.8))

$$\|e_k\|/\|e_0\| \leq \left(T_k\left(\frac{\kappa^2+1}{\kappa^2-1}\right)\right)^{-1} \leq 2\left(\frac{\kappa-1}{\kappa+1}\right)^k,$$

where κ is the spectral condition number of A.

There are also some cg-type methods which formally avoid the transition to the normal equations (4.1). These methods, due to Young, Jea [47] and Axelsson [3, 4], Concus, Golub [7] and Widlund [45] define the iterates x_k either by the Galerkin principle

$$r_k^T v = 0 \quad \text{for all } v \in K_k(r_0; A), \quad x_k \in x_0 + K_k(r_0; A)$$

or by a minimum residual requirement

$$\|A(x_k - \bar{x})\| = \min_{x \in x_0 + K_k(r_0; A)} \|A(x - \bar{x})\|, \quad x_k \in x_0 + K_k(r_0; A).$$

Both cases can be treated in a unified way. They are equivalent to the following conditions

(4.3) $$-v^T ZA(x_k - \bar{x}) \equiv v^T Z r_k = 0 \quad \text{for all } v \in K_k(r_0; A),$$
$$x_k \in x_0 + K_k(r_0; A)$$

where the auxiliary matrix Z is either $Z = I$ or $Z := A^T$.

In order to ensure that (4.3) has a solution x_k it is usually required that ZA is positive real, that is $ZA + A^T Z^T$ is positive definite. (This condition is clearly valid for $Z:=A^T$; and for $Z:=I$ it is often satisfied in practice if A results from discretization of PDE problems $Lu=f$ with semibounded operators L, $(u, Lu) \geqslant c\|u\|^2$ on dense subsets of a Hilbert space).

Young and Jea [47] describe several algorithms to compute x_k: All construct certain vectors $w_0, w_1, \ldots$, which are *semiorthogonal* bases of the Krylov spaces $K_k(r_0; A)$, $k \geqslant 1$, that is

$$[w_0, w_1, \ldots, w_{k-1}] = [r_0, A r_0, \ldots, A^{k-1} r_0], \quad k \geqslant 1$$
$$w_j^T ZA w_i = 0 \quad \text{for } 0 \leqslant j < i.$$

(Note that in general $w_j^T ZA w_i \neq w_i^T ZA w_j$ if ZA is not symmetric). Their first method ORTHODIR constructs these vectors in a Lanczos-type fashion (compare (2.2)).

(4.4) ORTHODIR: 0) *Start: Choose $x_0 \in R^n$ and set $p_0 := r_0 = b - A x_0$.*

For $k = 0, 1, \ldots$
1) *If $p_k = 0$ stop: x_k solves $Ax = b$.*

   Otherwise compute
2) $x_{k+1} := x_k + \sigma_k p_k$, $\quad \sigma_k := p_k^T Z r_k / p_k^T ZA p_k$
   $r_{k+1} := r_k - \sigma_k A p_k$
3) $p_{k+1} := A p_k + \gamma_{k,k} p_k + \gamma_{k,k-1} p_{k-1} + \ldots + \gamma_{k,0} p_0$
   where
   $$\gamma_{k,i} := - \frac{p_i^T ZA^2 p_k + \sum_{j=0}^{i-1} \gamma_{kj} p_i^T ZA p_j}{p_i^T ZA p_i}, \quad i = 0, 1, \ldots, k$$

(Note that in general $\gamma_{k,i} = 0$ for $i < k - 1$ if $A = A^T$, $Z := I$).

Then $[p_0, p_1, \ldots, p_k] = [r_0, A r_0, \ldots, A^k r_0]$ for $k \geqslant 0$ and $p_j^T ZA p_i =$ for $0 \leqslant j < i$.

If $Z = A^T$, ZA is symmetric then the expression for γ_{ki} simplifies because $p_i^T ZA p_j = 0$ for $j < i$.

If Z and ZA are positive real (which is valid for $Z = I$ and $Z = A^T$ if only A is positive real) ORTHODIR is equivalent to the ORTHOMIN algorithm of [47], in which the semiorthogonal vectors d_i are computed in a cg-type fashion (see (1.3)).

(4.5) ORTHOMIN: 0) *Start: Choose $x_0 \in R^n$, set $d_0 := r_0 := b - A x_0$.*

For $k = 0, 1, \ldots$
1) *If $d_k = 0$ stop: x_k solves $Ax = b$.*

   Otherwise compute
2) $x_{k+1} = x_k + \sigma_k d_k$, $\quad \sigma_k := d_k^T Z r_k / d_k^T ZA d_k$
   $r_{k+1} := r_k - \sigma_k A d_k$
3) $d_{k+1} := r_{k+1} + \beta_{k,k} d_k + \beta_{k,k-1} d_{k-1} + \ldots + \beta_{k,0} d_0$
   where
   $$\beta_{k,i} := \frac{d_i^T ZA r_{k+1} + \sum_{j=0}^{i-1} \beta_{kj} d_i^T ZA d_j}{d_i^T ZA d_i}, \quad i = 0, 1, \ldots, k.$$

II. Conjugate Gradient Type Methods

Here again $[d_0, d_1, \ldots, d_k] = [r_0, A r_0, \ldots, A^k r_0]$, $d_j^T Z A d_i = 0$ for $0 \leq j < i$. Of course, the vectors d_i of (4.5) and p_i of (4.4) are proportional. Both algorithms are welldefined for positive real A and are finite. The first index m for which $p_m = 0$ ($d_m = 0$) is equal to $m = \max_{k \geq 0} \dim K_k(r_0; A) \leq n$.

In [47] also truncated versions of (4.4), (4.5) are considered which are less expensive. Here, the sums defining p_{k+1} (d_{k+1}) are truncated to s terms by setting $\gamma_{k,i} := 0$ ($\beta_{k,i} := 0$) if $i < k - s$. These truncated methods may break down and may not be finite, situations which are usually overcome by restarts.

The algorithms of Axelsson [3, 4] are closely related to (4.4), (4.5). He also determines the solution x_k of (4.3), but in a slightly different way. Starting with $d_0 := r_0$ he computes the direction vectors d_k by a simplified recursion (compare (4.5), 3))

$$d_{k+1} := r_{k+1} + \beta_k d_k$$

with suitable β_k (either $\beta_k := -d_k^T Z A r_{k+1} / d_k^T Z A d_k$ to ensure $d_k^T Z A d_{k+1} = 0$ or even only $\beta_k = 0$), so that in general $d_0, \ldots, d_{k-1}$ is only a basis of $K_k(r_0; A)$, but not an semiorthogonal basis. Therefore x_{k+1} has the form

$$(4.6) \qquad x_{k+1} = x_k + \sum_{j=0}^{k} \lambda_{k,j} d_j$$

with certain coefficients $\lambda_{k,j}$ determined by solving the linear equations

$$(4.7) \qquad d_i^T Z \left(r_k - \sum_{j=0}^{k} \lambda_{k,j} A d_j \right) = 0, \quad i = 0, 1, \ldots, k$$

arising from the Galerkin conditions (4.3). Its matrix $\Lambda^k = (\Lambda_{i,j}^k)$, $\Lambda_{ij}^k := d_i^T Z A d_j$ is symmetric if $Z = A^T$ and is lower triangular for $Z := I$ and the choice $\beta_k = -d_k^T A r_{k+1} / d_k^T A d_k$. For $Z := A^T$ Axelsson [4] also considered truncated versions of (4.6), (4.7) determined by $x_{k+1} = x_k + \sum_{j=k-s+1}^{k} \lambda_{k,j} d_j$ with $\lambda_{k,j}$ determined by $d_i^T Z r_{k+1} = 0$ for $k - s < i \leq k$. The numerical results for these truncated version even for small s ($s \leq 3$) are very good if B is not too unsymmetric.

A particularly simple algorithm for solving $Ax = b$ for a positive real matrix A has been obtained by Concus, Golub [7] and Widlund [45]. Noting that A can be written in the form $A = M - N$, where $M = (A + A^T)/2$ is positive definite and $N := (A^T - A)/2 = -N^T$ is skew symmetric, the problem of solving $Ax = b$ is equivalent to solving $A'x' = b'$ with $A' = I - N'$, $A' := M^{-\frac{1}{2}} A M^{-\frac{1}{2}}$, $N' = -N'^T$, $b' = M^{-\frac{1}{2}} b$, $x' = M^{\frac{1}{2}} x$. The Galerkin-conditions (4.3) (with $Z = I$) for A' define vectors $x'_k \in x'_0 + K_k(r'_0; A')$, $r'_0 := b' - A' x'_0$ with

$$r_k'^T v' = 0 \quad \text{for all } v' \in K_k(r'_0; A').$$

Now, for matrices $A' = I - N'$, $N' = -N'^T$ great simplifications are possible, facilitating the computation of x'_k. For example, one can find orthogonal vectors p'_i with

$$[p'_0, p'_1, \ldots, p'_{k-1}] = [r'_0, N'r'_0, \ldots, (N')^{k-1}r'_0] = K_k(r'_0, N') = K_k(r'_0, A'), \quad k \geq 1$$

by a simple two-term recursion formula

$$\alpha_k p'_{k+1} = N' p'_k - \delta_k p'_{k-1}, \quad p'_0 := r'_0$$

where $\alpha_k \neq 0$ is arbitrary and $\delta_k := p'^T_{k-1} N' p'_k / p'^T_{k-1} p'_{k-1}$. Finally, in terms of the original quantities M, N, the following simple algorithm for computing $x_k := M^{-\frac{1}{2}} x'_k$ results (see [7] for details).

(4.8) **Algorithm:** 0) Choose $x_0 \in R^n$ and set $x_{-1} := x_0$

For $k = 0, 1, \ldots$
1) If $r_k = 0$ stop: x_k solves $Ax = b$.

   Otherwise
2) solve $M p_k = r_k = b - A x_k$ for p_k.
3) Compute $\rho_k := p_k^T r_k$.
4) Compute
$$x_{k+1} := x_{k-1} + \omega_{k+1}(p_k + x_k - x_{k-1})$$
where
$$\omega_{k+1} := \begin{cases} 1, & \text{if } k = 0 \\ 1/(1 + (\rho_k/\rho_{k-1})/\omega_k), & \text{if } k > 0. \end{cases}$$

For the errors e_k of the solutions x_k of (4.3) the following can be stated: If $Z = A^T$, the e_k have a minimum property

$$\|r_k\| = \|A e_k\| = \min_{p \in \Pi_k} \|A p(A) e_0\|$$

ensuring $\|r_{k+1}\| \leq \|r_k\|$. But in order to prove estimates like (1.8) on $\|A e_k\|/\|A e_0\|$, one needs additional properties of A, say A has n different positive eigenvalues. If $Z = I$, the errors of the solution x_k of (4.3) do not have a minimum property, but by standard techniques for Galerkin methods one can prove certain quasi-optimal rates of convergence. For example, Widlund [45] showed for the errors e_k of (4.8) the following quasi-optimality result:

$$(4.9) \quad \|e_k\|_M / \|e_0\|_M \leq \sqrt{1+\Lambda^2} \left(T_{[k/2]}\left(\frac{\Lambda^2 + 2}{\Lambda^2} \right) \right)^{-1} \leq 2\sqrt{1+\Lambda^2} \cdot \left(\frac{\sqrt{1+\Lambda^2} - 1}{\sqrt{1+\Lambda^2} + 1} \right)^{[k/2]},$$

where $\|x\|_M = \|M^{\frac{1}{2}} x\|$ and $\Lambda := \|M^{-1} N\|$ is the spectral norm of $M^{-1} N$, which is a measure of the unsymmetry of A.

In this context it is interesting to note that one can combine the ideas behind the STOD-algorithm (3.16) with those of (4.8) in order to develop an algorithm for finding those $x_k \in x_0 + A^T K_k(r_0; A)$ with the minimum property

$$(4.10) \quad \|x_k - \bar{x}\| = \min_{x \in x_0 + A^T K_k(r_0; A)} \|x - \bar{x}\|$$

II. Conjugate Gradient Type Methods

for positive real A of the form $A = I - N$, $N = -N^T$ (this method turns out to be closely related to Craig's method mentioned above). In fact, x_k (4.10) can be computed by the following method:

(4.11) **Algorithm:** 0) *Start: Choose* $x_0 \in R^n$ *and set* $p_0 := r_0 := b - A x_0$; $p_{-1} := 0$.

For $k = 0, 1, \ldots$
1) If $p_k = 0$ stop: x_k solves $Ax = b$.

   Otherwise compute

2) $x_{k+1} := x_k + \sigma_k A^T p_k$, $\quad \sigma_k := p_k^T r_k / p_k^T A A^T p_k$
   $r_{k+1} := r_k - \sigma_k A A^T p_k$

3) $p_{k+1} := N p_k - \delta_k p_{k-1} \equiv A^T p_k - p_k - \delta_k p_{k-1}$
   where
$$\delta_k := \begin{cases} 0, & \text{if } k = 0 \\ -p_k^T A A^T p_k / p_{k-1}^T A A^T p_{k-1}, & \text{if } k > 0, \end{cases}$$

The choice of σ_k ensures

(4.12) $$\|x_{k+1} - \bar{x}\| = \min_{\sigma} \|x_k + \sigma A^T p_k - \bar{x}\|$$

and step 3 the orthogonality of the vectors $A^T p_k$. The following are the main properties of (4.11):

(4.13) **Theorem:** a) *There is a first index m with $0 \leq m \leq n$ such that $p_m = 0$ and*
$p_i^T A A^T p_j = 0$ *for* $i \neq j$, $0 \leq i, j \leq m$
$p_i^T A A^T p_i > 0$ *for* $0 \leq i < m$

b) $r_k^T p_j = 0$ *for* $0 \leq j < k \leq m$
$r_m = 0$, $x_m = \bar{x}$

c) $[p_0, p_1, \ldots, p_{k-1}] = [r_0, N r_0, \ldots, N^{k-1} r_0] = K_k(r_0; A)$ *for* $k \leq m$
$[A^T p_0, A^T p_1, \ldots, A^T p_{k-1}] = \{(I - N^2) q(N) e_0 \mid q \in \tilde{\Pi}_k\}$,
where $\tilde{\Pi}_k = \{\alpha_0 + \alpha_1 t + \ldots + \alpha_{k-1} t^{k-1} \mid \alpha_i \in R\}$ *is the set of all real polynomial of degree* $< k$.

As a consequence of (4.12) and (4.13) a) we get by (4.13) c) the following error estimates

$$\|e_k\| = \min_{x \in x_0 + [A^T p_0, \ldots, A^T p_{k-1}]} \|x - \bar{x}\|$$
$$= \min_{q \in \tilde{\Pi}_k} \|(I + (I - N^2) q(N)) e_0\|.$$

Using an expansion of e_0 into the eigenvectors of N and using that N has only purely imaginary eigenvalues occuring in complex conjugate pairs, one finds

(4.14) $$\|e_k\| = \min_{q \in \tilde{\Pi}_{[(k+1)/2]}} \|I + (I - N^2) q(N^2)) e_0\|.$$

In particular we have $x_{2j}=x_{2j-1}$, $\sigma_{2j-1}=0$. Moreover

(4.15)
$$\|e_k\|/\|e_0\| \leqslant \min_{q\in \tilde{\Pi}_{[(k+1)/2]}} \max_{i\lambda\in\sigma(N)} |1+(1+\lambda^2)q(\lambda^2)|$$
$$\leqslant \min_{q\in \tilde{\Pi}_{[(k+1)/2]}} \max_{0<t\leqslant\lambda^2} |1+(1+t)q(t)|, \quad \Lambda := \|N\|$$
$$= \min_{q\in \Pi_{[(k+1)/2]}} \max_{1\leqslant t\leqslant 1+\Lambda^2} |q(t)|$$
$$= \left(T_{[(k+1)/2]}\left(\frac{\Lambda^2+2}{\Lambda^2}\right)\right)^{-1} \leqslant 2\left(\frac{\sqrt{1+\Lambda^2}-1}{\sqrt{1+\Lambda^2}+1}\right)^{[(k+1)/2]}$$

Formula (4.14) shows

$$x_{2j}=x_{2j-1}\in x_0+(I-N^2)[e_0, N^2 e_0, N^4 e_0, \ldots, N^{2(j-1)}],$$

so that because of $I-N^2=A^T A$ and (4.2)

$$(I-N^2)[e_0, N^2 e_0, \ldots, N^{2(j-1)}] = [A^T A e_0, (A^T A)^2 e_0, \ldots, (A^T A)^j e_0] =$$
$$= K_j(A^T r_0, A^T A)$$

the iterates $x_{2j}=x_{2j-1}$ generated by (4.11) are the same as the iterates $x_j^{(Cr)}$ generated by Craig's method: $x_j^{(Cr)}=x_{2j-1}$ in the special case $A=I-N$ (see Freund [15]).

The general case of solving $Ax=b$ for positive real $A=M-N$, $M=M^T=(A+A^T)/2$, $N=-N^T=(A^T-A)/2$ can be reduced to the special case considered by applying (4.11) to $A'x'=b'$, $A':=M^{-\frac{1}{2}}AM^{-\frac{1}{2}}$, $b':=M^{-\frac{1}{2}}b$, $x':=M^{\frac{1}{2}}x$ and transforming the iterates x'_k generated by (4.11) back into $x_k:=M^{-\frac{1}{2}}x'_k$. We thus obtain

(4.16) Algorithm: 0) Choose $x_0 \in R^n$, and solve $Mp_0=r_0$ for p_0.

For $k=0, 1, \ldots$
1) If $p_k=0$ stop, x_k solves $Ax=b$.

   Otherwise
2) Solve $Mq_k=A^T p_k$ for q_k and compute
   $x_{k+1}:=x_k+\sigma_k q_k,$
   $r_{k+1}:=r_k-\sigma_k(A^T p_k - Nq_k)$
   where
   $$\sigma_k:=\begin{cases} 0, & \text{if } k \text{ is odd} \\ p_k^T r_k/q_k^T A^T p_k, & \text{if } k \text{ is even.} \end{cases}$$
3) $p_{k+1}:=q_k-q_k-\delta_k p_{k-1}$
   $A^T p_{k+1}:=Nq_k-\delta_k A^T p_{k-1}$
   where
   $$\delta_k:=\begin{cases} -q_k^T A^T p_k/q_{k-1}^T A^T p_{k-1}, & \text{for } k>0 \\ 0, & \text{for } k=0. \end{cases}$$

Clearly the errors $\|e_k\|_M = \|M^{\frac{1}{2}} e_k\|$ have a minimum property and satisfy (compare (4.15))

II. Conjugate Gradient Type Methods

$$\|e_k\|_M/\|e_0\|_M \leq \left(T_{[(k+1)/2]}\left(\frac{\Lambda^2+2}{\Lambda^2}\right)\right)^{-1} \leq 2\left(\frac{\sqrt{1+\Lambda^2}-1}{\sqrt{1+\Lambda^2}+1}\right)^{[(k+1)/2]},$$

where now Λ is the spectral norm of $M^{-1}N$: This bound is slightly better than the corresponding bound (4.9) for algorithm (4.8).

By a slight modification of (4.11) one can devise a similar *minimum residual* type algorithm for computing

(4.17) $$x_k := \arg\min_{x \in x_0 + K_k(r_0; A)} \|A(x - \bar{x})\|$$

for the special case $A := I - N$, $N^T = -N$: One only has to replace step 2) of (4.11) by

2') $x_{k+1} := x_k + \sigma_k p_k$, $\quad \sigma_k := p_k^T A^T r_k / p_k^T A^T A p_k$

$r_{k+1} := r_k - \sigma_k p_k$.

Here, in contrast to (4.11), it is no longer true that $x_{2j} = x_{2j-1}$, but, as was shown by Freund [15], there are upper bounds for $\|A e_k\|/\|A e_0\|$ which are similar to (4.15).

Also this method can be extended to a method for solving $Ax = b$ for positive real $A = M - N$, $M = M^T$, $N = -N^T$ by reducing this system to the special case $A'x' = b'$, $A' = M^{-\frac{1}{2}} A M^{-\frac{1}{2}}$, $x' = M^{\frac{1}{2}} x$, $b' = M^{-\frac{1}{2}} b$ as above.

5. Preconditioning techniques

The error estimates of the previous sections suggest that the convergence rate of all cg-type methods considered as iterative methods for solving $Ax = b$ very much depends on the spectral condition number $\kappa(A)$. This motivates the use of preconditioning techniques (see Hestenes [25]) in order to speed up the convergence of these methods. If, say $A = A^T$ is a symmetric or even a positive definite $n \times n$-matrix, then one tries to find a positive definite $n \times n$-matrix M such that the system $A'x' = b'$ with $A' = M^{-\frac{1}{2}} A M^{-\frac{1}{2}}$, $b' = M^{-\frac{1}{2}} b$, $x' = M^{\frac{1}{2}} x$, which is clearly equivalent to $Ax = b$, has a matrix A' with $\kappa(A') \ll \kappa(A)$. Note that the transition $A \to A'$ preserves the symmetry and positive definiteness. Then the particular cg-type method when applied to $A'x' = b'$ rather than to $Ax = b$ started with $x_0' := M^{\frac{1}{2}} x_0$ generates a sequence x_k' which converges faster. However, one avoids the explicit calculation of A', b', x_k', but tries instead to write the cg-type-method applied to $A'x' = b'$ in terms of their unprimed counterparts A, b, $x_k = M^{-\frac{1}{2}} x_k'$, M. This idea has already been used in the previous section, when considering algorithms (4.8) and (4.16) (there however the choice of M was not free), but it can be applied to any method considered so far. For example the basic cg-algorithm (1.3) with $\mu=1$ gets the following form if one uses the transition rules

$$x'_k = M^{\frac{1}{2}} x_k, \; d'_k = M^{\frac{1}{2}} d_k, \; r'_k = M^{-\frac{1}{2}} r_k, \; q_k := M^{-1} r_k, \; A' = M^{-\frac{1}{2}} A M^{-\frac{1}{2}}.$$

(5.1) Preconditioned cg-algorithm:

0) *Start: Choose* $x_0 \in R^n$ *and solve* $M q_0 = r_0$ *for* q_0.

For $k = 0, 1, \ldots$

1) *If* $d_k = 0$ *stop:* x_k *solves* $Ax = b$.

Otherwise, compute

2) $x_{k+1} := x_k + \sigma_k d_k, \quad \sigma_k := \dfrac{r_k^T q_k}{d_k^T A d_k}$

$r_{k+1} := r_k - \sigma_k A d_k$

Solve $M q_{k+1} = r_{k+1}$ *for* q_{k+1} *and compute*

$$d_{k+1} := q_{k+1} + \beta_k d_k, \quad \beta_k := \frac{r_{k+1}^T q_{k+1}}{r_k^T q_k}.$$

The errors e_k of its iterates x_k now satisfy

$$\|e_k\|_A = \min_{x \in x_0 + K_k(M^{-1} r_0, \, M^{-1} A)} \|x - \bar{x}\|_A.$$

instead of (1.2) ($\mu = 1$), and

$$\|e_k\|_A / \|e_0\|_A \leq \left(T_k\left(\frac{\kappa'+1}{\kappa'-1}\right)\right)^{-1} \leq 2\left(\frac{\sqrt{\kappa'}-1}{\sqrt{\kappa'}+1}\right)^k$$

with $\kappa' = \mathrm{cond}(A') = \mathrm{cond}(M^{-1} A)$ instead of (1.8).

Similar preconditioned variants can be derived for all cg-type methods considered. Within each of these preconditioned algorithms one typically has to solve a linear system of equations $Mq = r$ with the matrix M. For this one needs either the Cholesky-decomposition $M = E E^T$ or a LU-decomposition $M = L \cdot U$, L lower, U upper triangular. For this reason the preconditioned versions are only efficient, if one can find a positive definite M with, say a $L \cdot U$ decomposition into sparse matrices L and U such that $\kappa(M^{-1} A) \ll \kappa(A)$.

Similar problems have been treated earlier (see e.g. Varga [42], Young [46]) for the development of iterative methods

$$x_{i+1} = M^{-1} R x_i + M^{-1} b$$

for solving $Ax = b$, which are defined by a splitting $A = M - R$ of A. Here the problem is to choose a sparse matrix M such that the spectral radius $\rho(M^{-1} R) < 1$ is as small as possible. Many proposals for the choice of the preconditioning matrix for cg-type methods are influenced by this and led to the following proposals for the choice of M:

If A is diagonally dominant and has the standard decomposition

$$A = D + L + L^T,$$

II. Conjugate Gradient Type Methods

where D is the diagonal and L the strict lower triangular part of A, then one simple possibility is the choice $M:=D$. Another choice investigated by Axelsson [2] is the SSOR preconditioning, where

$$M:=(D+\omega L)D^{-1}(d+\omega L)^T, \quad 0<\omega<2,$$

with a proper choice of ω.

For symmetric M-Matrices A, Meijerink and van der Vorst [32] have proposed a systematical way to compute a suitable M by *incomplete Cholesky-factorization* (IC-factorization) of A, a technique which was developed earlier by Varga [43] to find a regular splitting of A. Here, A is called an M-matrix, if $a_{ij} \le 0$ for $i \ne j$, A is nonsingular and $A^{-1} \ge 0$. For any M-matrix A, Meijerink and van der Vorst construct a decomposition

(5.2)
$$A = M - R$$
$$M = LDU, \quad D = \text{diag}(d_1, \ldots, d_n)$$

where L and U are unit lower and unit upper triangular with a predetermined zero structure.

This sparsity pattern of L and U is determined by an arbitrary set G of ordered pairs of integers (i,j) with $1 \le i, j \le n$ and $(i,i) \in G$ for $i=1, \ldots, n$ as follows: if $l_{ij} \ne 0$ or $u_{ij} \ne 0$ then $(i,j) \in G$, and moreover $r_{ij} = 0$ for all $(i,j) \in G$. L, D, U and R are uniquely determined by G, and the nontrivial elements of $D, L(l_{ij}$ for $i>j)$ and $U(u_{ij}$ for $i<j)$ are computed as follows:

(5.3) For $i = 1, 2, \ldots, n$

$$d_i := a_{ii} - \sum_{k=1}^{i-1} d_k l_{ik} u_{ki},$$

For $j = i+1, \ldots, n$:

$$d_i l_{ji} := \begin{cases} a_{ji} - \sum_{k=1}^{i-1} d_k l_{jk} u_{ki}, & \text{if } (j,i) \in G \\ 0, & \text{otherwise} \end{cases}$$

$$d_i u_{ij} := \begin{cases} a_{ij} - \sum_{k=1}^{i-1} d_k l_{ik} u_{kj}, & \text{if } (i,j) \in G \\ 0, & \text{otherwise}. \end{cases}$$

Of course, if A is a symmetric M-matrix and G a symmetric set, i.e. $(i,j) \in G$ implies $(j,i) \in G$, then the Cholesky-decomposition of M is $M = EE^T$, $E := LD^{\frac{1}{2}}$. Choosing M in this way one obtains the so-called ICCG-methods of [32]. The condition $\kappa(M^{-1}A)$ depends on the choice of G. One can show that the condition decreases with increasing G. Manteuffel described in [31] a similar more general technique ("shifted incomplete factorization") for finding a preconditioning matrix M for arbitrary positive definite A. A related idea, which is useful for positive definite band matrices A was proposed in [18]. An important improvement over the ICCG-methods of [32] was described by Gustavson [22] ("MICCG-methods"), which is applicable to matrices A arising from difference approximations for self-adjoint elliptic partial differential equation. Incomplete factorization for the class of "H-matrices" were considered in Varga et al. [44].

6. Applications to optimization

The classical cg-algorithm of Hestenes and Stiefel ((1.2 with $\mu=1$) can be viewed as a method for minimizing the quadratic function $f(x) = \frac{1}{2} x^T A x - b^T x$ with gradient $g(x) = Ax - b = -r(x)$. As is well known, this led directly to the Fletcher-Reeves algorithm [14] for finding the solution of unconstrained minimization problems. Later on, as part of methods for solving linearly constrained nonlinear programs, similar direct Fletcher-Reeves type methods have been employed to minimize a function on a linear manifold. Also, the conjugacy concept played a decisive role in devising modern Quasi-Newton-methods, such as the DFP or the BFGS-methods, which now are an indispensable tool in optimization. As these almost classical applications are well-known, we restrict ourselves to point out some applications of cg-type methods as powerful iterative methods for solving large sparse linear equations arising in the optimization area.

For example such systems arise during the solution of unconstrained minimization problems $\min_{x \in R^n} f(x)$, where n is very large, but f has a sparse Hessean $f''(x)$, frequently with a special structure, see e.g. Griewank and Toint [21] for problems of this kind. The methods for solving these problems usually determine iterates x_k by line searches, $f(x_{k+1}) \approx \min_\sigma f(x_k - \sigma s_k)$, $x_{k+1} = x_k - \sigma_k s_k$, where the search direction s_k is obtained by solving a linear equation

(6.1) $$B_k s_k = -g_k$$

where $g_k = \nabla f(x_k)$ and the matrix $B_k = B_k^T$ is an approximation to $f''(x_k)$ obtained by difference approximations and/or suitable update techniques. If B_k is positive definite, then the solution s_k of (6.1) is a descent direction of $f(x)$ and can be iteratively computed, say by the usual cg- or cr-algorithm (1.3). Usually it is not necessary for the superlinear convergence of the x_k to compute the solution s_k of (6.1) exactly (see Steighaug [40], Dembo, Eisenstat, Steighaug [10]), so that only few cg-iterations for solving (6.1) are necessary. Moreover, Theorem (1.4) shows that for the starting value $s_k^{(0)} = 0$ for (1.3) applied to (6.1), every iterate $s_k^{(i)}$ of (1.3) satisfies $-s_k^{(i)T} g_k > 0$, that is $s_k^{(i)}$ is a descent direction for f (see [20]). Moreover, preconditioning can be used in the following way. If, say M is a good preconditioning matrix for some matrix B_k (6.1), so that $\kappa(M^{-1} B_k)$ is small, then it is likely that it provides also a good preconditioning for the matrices B_j, $j \geq k$, at least for k large.

Note also, that in solving (6.1) by cg-type methods, the matrix B_k is only used to form products $B_k d$ with certain vectors d. Since B_k approximates $f''(x_k)$, the product $B_k d \approx (g(x_k + t d_k) - g(x_k))/t$, t small, can be approximated by a difference of gradients. Also (O'Leary [35]) by using the Lanczos method

(2.4) in conjunction with (2.12)–(2.14), one can monitor the positive definiteness of B_k, when forming the Cholesky decomposition (2.12) of T_k. Even if B_k is not positive definite, one can easily ensure by suitable adjusting T_k that the iterates $s_k^{(j)}$ of the cg-Lanczos algorithm are directions of decrease for f in x_k, or directions of negative curvature near a stationary point of f which is not minimal (see [35] and Gill, Murray [17]).

Another more direct use of cg-type methods together with preconditionings has been made by O'Leary [34] for solving quadratic programs

$$\min \tfrac{1}{2} x^T A x - b^T x$$
$$x : c \leqslant x \leqslant d$$

for large sparse positive definite A. As part of the solution method, one has to solve certain linear equations $A_{JJ} x_J = c_J$ for certain principal submatrices A_{JJ} of A. These are solved (approximately) by preconditioned cg-methods; in this context it is important, that a good preconditioning for A induces a good preconditioning for A_{JJ} in a natural way.

Finally, constrained nonlinear programming promises to be a natural field of applications for all cg-type methods for solving linear equations $Ax = b$, where A is symmetric but indefinite. To see this, we only note, that the usual first order necessary conditions for an equality constrained minimum of a function $f(x)$ leads to a nonlinear system of equations $\phi(x, u) = 0$ for the solution $\bar{x}$ and the corresponding Kuhn-Tucker vector $\bar{u}$. Its Jacobian is a symmetric but indefinite matrix of the form

$$\phi'(x, u) = \begin{pmatrix} \Lambda, & G^T \\ G, & 0 \end{pmatrix}, \quad \Lambda = \Lambda^T,$$

so that it can be hoped that the cg-methods of section 3 will turn out to be useful, say when solving $\phi(x, u) = 0$ by Newton-type methods.

References

[1] Axelsson, O.: On preconditioning and convergence acceleration in sparse matrix problems. CERN 74-10 (1974), Genf.
[2] –: Solution of linear systems of equations: iterative methods. In: Sparse matrix techniques (V. A. Barker ed.), Lecture Notes in Mathematics, *572*, Berlin-Heidelberg-New York: Springer 1977.
[3] –: A generalized conjugate direction method and its application on a singular perturbation problem. In: Proc. 8th Biennial Numerical Analysis Conference, Dundee, June 26–29, 1979 (G. A. Watson, ed.), Lecture Notes in Mathematics Vol. 773, 1–11, Berlin-Heidelberg-New York: Springer 1980.
[4] –: Conjugate gradient type methods for unsymmetric and inconsistent systems of linear equations. Linear Alg. and its Appl. *29* (1980), 1–16.
[5] –, I. Gustavsson: A modified upwind scheme for convective transport equations and the use of a conjugate gradient method for the solution of nonsymmetric systems of equations. J. Inst. Maths. Applics. *23* (1979), 321–327.

[6] Chandra, R.: Conjugate gradient methods for partial differential equations. Ph. D. thesis, Res. Rep. # 129, Yale University, 1978.

[7] Concus, P., G. H. Golub: A generalized conjugate gradient method for nonsymmetric systems of linear equations. Lecture Notes in Economics and Math. Systems (P. Glowinski, J. L. Lions eds.) Vol. 134, 56–65, Berlin–Heidelberg–New York: Springer 1976.

[8] Concus, P., G. H. Golub, D. P. O'Leary: A generalized conjugate gradient method for the numerical solution of elliptic partial differential equations. In: Sparse Matrix Computations (J. R. Bunch, D. J. Rose, eds.) 309–332, New York: Academic Press, 1975.

[9] Craig, E. J.: The N-step iteration procedure. J. Math. Phys. *34* (1955), 65–73.

[10] Dembo, R. R., S. C. Eisenstat, T. Steihaug: Inexact Newton methods. SIAM J. Numer. Anal. *19* (1982), 400–408.

[11] Engeli, M., Th. Ginsburg, H. Rutishauser, E. Stiefel: Refined iterative methods for computation of the solution and the eigenvalues of self-adjoint boundary value problems. Mitteilungen aus dem Institut für Angewandte Mathematik an der ETH Zürich *8*, Basel: Birkhäuser, 1959.

[12] Fadeev, D. K., V. N. Fadeeva: Computational Methods of Linear Algebra. San Francisco: Freeman 1963.

[13] Fletcher, R.: Conjugate gradient methods for indefinite systems. Proc. of the Dundee Biennial Conference on Numerical Analysis (G. A. Watson ed.). Berlin–Heidelberg–New York: Springer 1975.

[14] –, C. M. Reeves: Function minimization by conjugate gradients. Comput. J. *7* (1964), 145–154.

[15] Freund, R.: Über einige cg-ähnliche Verfahren zur Lösung linearer Gleichungssysteme. Diss. Universität Würzburg 1983.

[16] Fridman, V. M.: The method of minimum iterations with minimum errors for a system of linear algebraic equations with a symmetrical matrix. USSR Comput. Math. and Math. Phys. *2* (1963), 362–363.

[17] Gill, P. E., W. Murray: Newton-type methods for unconstrained and linearly constrained optimization. Math. Progr. *7* (1974), 311–350.

[18] Glowinski, R., B. Mantel, J. Periaux, O. Pironneau, G. Poirier: An efficient preconditioned conjugate gradient method applied to nonlinear problems in fluid dynamics via least squares formulations. In: Computing Methods in Applied Sciences and Engineering (R. Glowinski, J. L. Lions, eds.) 445–487, Amsterdam: North-Holland 1980.

[19] Greenbaum, A.: Comparison of splittings used with the conjugate gradient algorithm. Numer. Math. *33* (1979), 181–193.

[20] Griewank, A., Ph. L. Toint: Local convergence analysis for partitioned quasi-Newton updates. Tech. Rep., Dept. of Math., Univ. of Namur, 1981.

[21] –: Partitioned variable metric updates for large structured optimization problems. Numer. Math. *39* (1982), 119–137.

[22] Gustavsson, I.: A class of first order factorization methods. BIT *18* (1978), 142–156.

[23] –: Stability and rate of convergence of modified incomplete Cholesky factorization methods. Rep. 79.02R (Ph. D. thesis), Dept. of Computer Sciences, Chalmers University of Technology, Göteborg, Sweden 1979.

[24] Hageman, L. A., F. T. Luk, D. M. Young: On the equivalence of certain iterative acceleration methods. SIAM J. Numer. Anal. *17* (1980), 852–873.

[25] Hestenes, M. R.: The conjugate gradient method for solving linear systems. Proc. Symp. Appl. Math., Vol. 6, Numerical Analysis, pp. 83–102, New York: Mc-Graw Hill, 1956.
[26] Hestenes, M. R.: Conjugate Direction Methods in Optimization. Berlin-Heidelberg-New York: Springer 1980.
[27] –, Stiefel, E.: Methods of conjugate gradients for solving linear systems. J. Res. NBS 49 (1952), 409–436.
[28] Householder, A.: The theory of matrices in Numerical Analysis. Dover 1975 (reedition).
[29] Lanczos, C.: An iteration method for the solution of the eigenvalue problem of linear differential and integral operators. J. Res. NBS 45 (1950), 255–282.
[30] Luenberger, D. G.: Hyperbolic pairs in the method of conjugate gradients. SIAM J. Appl. Math. 17 (1969), 1263–1267.
[31] Manteuffel, T. A.: An incomplete factorization technique for positive definite linear systems. Math. Comp. 34 (1980), 473–497.
[32] Meijerink, J. A., H. A. van der Vorst: An iterative solution method for linear systems of which the coefficient matrix is a symmetric M-matrix. Math. Comp. 31 (1977), 148–162.
[33] Nash, S. G.: Truncated Newton Methods. Tech. Rep. STAN-CS-82-906, Computer Sci. Dept., Stanford University, 1982.
[34] O'Leary, D. P.: A generalized conjugate gradient algorithm for solving a class of quadratic programming problems. Lin. Alg. and its Appl. 34 (1980), 371–399.
[35] –: A discrete Newton algorithm for minimizing a function of many variables. Computer Science Dept. of the University of Maryland, Techn. Rep. 910, 1980.
[36] Paige, C. C., M. A. Saunders: Solution of sparse indefinite systems of linear equations. SIAM J. Numer. Anal. 12 (1975), 617–629.
[37] Parlett, B. N.: The Symmetric Eigenvalue Problem. Englewood Cliffs: Prentice Hall, 1980.
[38] Reid, J. K.: On the method of conjugate gradients for the solution of large sparse systems of equations. In: Large Sparse Sets of Linear Equations (J. K. Reid, ed.) London, New York: Academic Press 1971.
[39] Simon, H. D.: The Lanczos algorithm for solving symmetric linear systems. Ph. D. Thesis, University of California, Berkeley 1982.
[40] Steihaug, T.: Quasi-Newton methods for large scale nonlinear programming problems. Ph. D. Thesis, Yale University 1981.
[41] Stoer, J., R. Freund: On the solution of large indefinite systems of linear equations by conjugate gradient algorithms. In: Computing Methods in Applied Sciences and Engineering, V. (R. Glowinski, J. L. Lions, eds.) 35–53, Amsterdam: North-Holland 1982.
[42] Varga, R. S.: Matrix Iterative Analysis, Englewood Cliffs: Prentice Hall 1962.
[43] –: Factorization and normalized iterative methods. In: Boundary Problems in differential equations (R. E. Langer, ed.), Univ. of Wisconsin Press, Madison 1960.
[44] –, E. B. Saff, V. Mehrmann: Incomplete factorization of matrices and connections with H-matrices. SIAM J. Numer. Anal. 17 (1980), 787–793.
[45] Widlund, O.: A Lanczos method for a class of nonsymmetric systems of linear equations. SIAM J. Numer. Anal. 15 (1978), 801–812.
[46] Young, D. M.: Iterative Solution of Large Linear Systems. New York: Academic Press 1971.
[47] –, K. C. Jea: Generalized conjugate-gradient acceleration of nonsymmetrizable iterative methods. Linear Alg. and its Appl. 34 (1980), 159–194.

Stochastic Programming: Solution Techniques and Approximation Schemes

R. Wets
Department of Mathematics, University of Kentucky, Lexington, Kentucky 40506, USA
and
I.IA.S.A, 2361 Laxenburg, Austria

Abstract. Solutions techniques for stochastic programs are reviewed. Particular emphasis is placed on those methods that allow us to proceed by approximation. We consider both stochastic programs with recourse and stochastic programs with chance-constraints.

1. Introduction

Optimization problems involving parameters only known in a statistical sense give rise to stochastic optimization models. When dealing with such problems it is important to be aware of their intrinsic dynamic nature since it plays an important role in the modeling process as well as in the design of solution techniques. Briefly the general model is as follows. First an observation of a random phenomena $\xi_1 \in R^{v_1}$ is made. Based on this information, a decision $x_1 \in R^{n_1}$ is chosen at some cost $f_1(x_1, \xi_1)$. Then a new observation is made that yields $\xi_2 \in R^{v_2}$. On the basis of the information (ξ_1, ξ_2) gained so far, one selects a decision x_2 in R^{n_2} with associated cost $f_2(\xi_1, \xi_2, x_1, x_2)$. This continues up to the time horizon N, possibly ∞. At each stage, the decisions $x_1, x_2, \ldots$ are subject to constraints that may, and usually do, depend on the actual realizations $\xi_1, \xi_2, \ldots$, as well as reliability type constraints that follow from criteria that the modeler might find difficult to include in the cost functions. The problem is to find *recourse functions* (decision rules, policies, control laws):

$$\xi_1 \mapsto x_1(\xi_1)$$
$$(\xi_1, \xi_2) \mapsto x_2(\xi_1, \xi_2)$$
$$\ldots$$

that satisfy the constraints and that minimize the expected cost. It is assumed that utility factors have been incorporated in the cost functions.

The development of mathematical programming techniques for studying and solving certain classes of stochastic optimization problems was initiated in the mid 50's by E. M. Beale [1], G. Dantzig [2], G. Tintner [3] and A. Charnes and W. Cooper [4]. The models introduced then, as well as those to be considered here, involve typically only $2(=N)$ stages with no (truly) random phe-

---

Supported in part by a Guggenheim Fellowship.

nomena preceding the choice of x_1, but the basic features of the general model were already ubiquitous. The basic reason for such limitations is that numerous applications require only 2 or 3 stages, either per se or as a consequence of modeling choices. However, the number of decision variables and constraints is liable to be quite large as is the case in typical applications of linear or nonlinear programming. It is this class of problems that is at the core of our concerns, i.e., those problems that can be viewed as "stochastic extensions" of the linear (or slightly nonlinear) programming model. Multistage problems, say $N>3$, present no significant theoretical difficulties but they are for all practical purposes computationally intractable, unless they possess structural properties that can be successfully exploited, see for example [5-9]. An excellent overview of the field of Stochastic Programming and its connections to other stochastic optimization problems has been provided by M. Dempster [10, Introduction].

We consider the following class of problems

(1.1) Find $x \geq 0$, $\alpha \in [0, 1]$ with $P[A(\omega)x \geq b(\omega)] \geq \alpha$,
such that $Z(x) + \rho(\alpha)$ is minimized
where $Z(x) = cx + E\left\{\inf_{y \geq 0} q(\omega)y \mid Wy = p(\omega) - T(\omega)x\right\}$.

The vectors b, q, p and the matrices A, T are random, whereas c and W are fixed; their sizes are consistent with: $x \in R^{n_1}$, $y \in R^{n_2}$, $b(\omega) \in R^{m_1}$ and $p(\omega) \in R^{m_2}$, and $\rho: [0, 1] \to \bar{R}$ is a nonnegative monotone nonincreasing lower semicontinuous convex function. A more complete model would involve a number of *chance-constraints*, i.e., several constraints of the type

(1.2) $$P[A(\omega)x \geq b(\omega)] \geq \alpha$$

but this extension is easy to work out and would add nothing to the substance of our development. Also, the *recourse cost function*

(1.3) $$\mathcal{Q}(x) = E\{Q(x, \omega)\} = \int Q(x, \omega) P(d\omega),$$

determined by the *recourse problem*

(1.4) $Q(x, \omega) = \inf q(\omega) y$
subject to $Wy = p(\omega) - T(\omega)x$
$y \geq 0$,

could involve more general constraints on y, convex rather than linear objective, ..., but little would be added to the arguments except that some technical questions would need to be taken care of. When W is random rather than fixed we need a more general theory than that sketched out here; see [11, 12], but since our computational capabilities do not yet include such a case, for exposition sake we limit ourselves to fixed W; we then refer to (1.4) as a problem with *fixed recourse*.

The function ρ is not a common feature of the stochastic programming models found in the literature. It represents a cost associated with the relaxation of the constraint

(1.5) $$A(\omega)x \geq b(\omega) \quad \text{for all } \omega.$$

Typically it has the form:

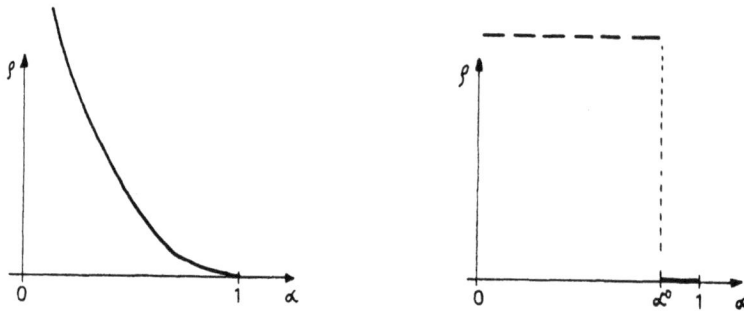

1.6 Figure: Reliability Cost

In the first case the modeler presumably has some cost information about the price he needs to pay to weaken reliability considerations. For the second function, he supposedly has been given a reliability level $\alpha°$ that must be attained at all cost. Problem (1.1) then becomes

(1.7)   Find $x \geq 0$ with $P[A(\omega)x \geq b(\omega)] \geq \alpha°$
         such that $Z(x) = cx + \mathcal{Q}(x)$,

a more common formulation of stochastic programs with (linear) chance-constraints. If moreover $\alpha° = 1$, then the chance-constraints can be replaced, as we shall see, by deterministic constraints and (1.1) takes on the usual form of a stochastic program with recourse.

We take as premise that the probability distribution P of the random elements is given. We shall not consider here the case when there is insufficient statistical information about the random variables of the problem to derive their distribution with a sufficiently high level of confidence. The study of such problems is very recent and there are only limited results available at this time.

We also assume that the random variables of the problem are such that the function $\omega \mapsto Q(\omega, x)$ is bounded below by a summable (finite integral) so that

$$x \mapsto \mathcal{Q}(x): R^{n_1} \to R \cup \{+\infty\};$$

the function $\omega \mapsto Q(x, \omega)$ is always measurable, details appear in [13, 14]. In particular this implies that almost surely $Q(x, \omega) > -\infty$, or equivalently the system $\pi W \leq q(\omega)$ is solvable for almost all $q(\omega)$. In fact, let us go one step further and assume that the random variables are such that $\mathcal{Q}(x) = +\infty$ if and only if $Q(x, \omega) = +\infty$ with (strictly) positive probability, i.e., if and only if the linear system

$$Wy = p(\omega) - T(\omega)x, \quad y \geq 0$$

is unsolvable with positive probability. To achieve all of the above it suffices, for example, that the random elements have finite second moments, a condition always satisfied in practice. What precedes are our working assumptions and will be considered as part of the definition of the stochastic program (1.1).

Section 2 reports on computational methods and solution strategies, and Section 3 is devoted to approximation techniques and associated error bounds. In the remainder of this section, we review briefly the main properties of the stochastic program (1.1). We start with its region of feasibility. Let

(1.8) $$K_1 = \{x \geqslant 0 \mid P[A(\omega)x \geqslant b(\omega)] \geqslant \alpha, \rho(\alpha) < +\infty\},$$

with the *induced constraints* given by

(1.9) $$K_2 = \{x \mid \mathcal{Q}(x) < +\infty\}.$$

The feasibility region K is simply

(1.10) $$K = K_1 \cap K_2.$$

One refers to (1.1) as a stochastic program with *complete recourse* if $K_2 = R^{n_2}$, i.e., there exists a feasible recourse decision whatever be the first stage decision and the random event observed. In general, it may be difficult to compute K or even to determine if a given x belongs or does not belong to K, in particular K_2 might be hard to calculate. Some characterizations of K_1 and K_2 are given here below.

We start with K_1. Let

(1.11) $$\kappa(\omega) = \{x \geqslant 0 \mid A(\omega)x \geqslant b(\omega)\},$$

and thus

(1.12) $$\kappa^{-1}(x) = \{\omega \mid A(\omega)x \geqslant b(\omega)\}.$$

By $\alpha°$ we denote the lower bound of α such that $\rho(\alpha) < +\infty$. We have

$$K_1 = \{x \geqslant 0 \mid P[x \in \kappa(\omega)] = P[\kappa^{-1}(x)] \geqslant \alpha°\}.$$

For each ω, the set $\kappa(\omega)$ is convex but in general K_1 itself is not convex.

1.13 Proposition. *If $\alpha° = 0$, then $K_1 = R_+^{n_1}$. On other hand if $\alpha° = 1$, K_1 is a closed convex set given by*

(1.14) $$K_1 = \bigcap_{(A,b) \in \Sigma} \{x \geqslant 0 \mid Ax \geqslant b\}$$

where $\Sigma \subset R^{m_1(n_1+1)}$ is the (image) support of $A(\cdot)$, $b(\cdot)$, i.e., the smallest closed subset of $R^{m_1(n_1+1)}$ such that $P[(A(\omega), b(\omega)) \in \Sigma] = 1$. Moreover if A is fixed, or more generally if $A(\cdot)$ has finite support (a finite number of possible values) then K_1 is a convex polyhedron.

Proof. The statement involving $\alpha° = 0$ is trivial. When $\alpha° = 1$, the fact that K_1 is closed and convex follows from (1.14) and that in turn follows from Theorem 2 of [15], with the f function of [15] defined on $R^{n_1} \times R^{m_1} \times R^{m_1 \times (n_1+1)}$ as follows

$$f(x, s, \omega) = A(\omega)x - b(\omega) + s$$

and $Y = R_+^{n_1}$. That K_1 is polyhedral if A is fixed is argued as follows: for each $b(\omega)$, the set $\kappa(\omega) = \{x \geq 0 \mid Ax \geq b(\omega)\}$ is a convex polyhedron with each possible face $A_i x \geq b_i(\omega)$ (and $x_j \geq 0$) parallel to the corresponding face determined by the same row A_i but another realization $b_i(\omega')$. The same argument remain valid when $A(\cdot)$ has finite support because we can argue as above for each possible value of $A(\omega)$, and then observe that the finite intersection of polyhedra is also a polyhedron. □

The next proposition completes the results of Proposition 1.13. We state it for the record, its proof would take us too far astray of our main concerns.

1.15 Proposition. *Suppose $\alpha° = 1$, $b(\cdot)$ and $A(\cdot)$ are independent and the convex hull of $\Sigma_A \subset R^{m_1 \times n_1}$, the support of $A(\cdot)$, is polyhedral. Then K_1 is a convex polyhedron.*

It is much more difficult to characterize the set K_1 when $0 < \alpha° < 1$. Basically this comes from the fact that

$$P(\kappa^{-1}(x_1)) \geq \alpha° \quad \text{and} \quad P(\kappa^{-1}(x_2)) \geq \alpha°$$

do not imply that

$$P(\kappa^{-1}(x_1) \cap \kappa^{-1}(x_2)) \geq \alpha°,$$

i.e., there does not exist any subset of events, or possible values of A and b, that can be singled out to yield an expression of the type (1.14). In general the set K_1 is *not* convex and examples can be constructed with K_1 disconnected, even with A fixed. For example, let

$$\kappa(\omega) = \{x \mid x + 3 \geq b(\omega), x \leq b(\omega)\} = [b(\omega) - 3, b(\omega)]$$

with

$$P[b(\omega) = 0] = P[b(\omega) = 2] = P[b(\omega) = 4] = \tfrac{1}{3}.$$

Then for $\alpha° = 2/3$, we get

$$K_1 = [-1, 0] \cup [1, 2].$$

However, when only $b(\cdot)$ is random, there is a general theory that originates with A. Prékopa [16], who also derived many of the major results; cf. [17] and [18] for surveys.

We say that a *probability measure* P on R^m is *quasi-concave* if for any pair U, V of convex (measurable) sets and for any $\lambda \in [0, 1]$ we have

$$P((1-\lambda)U + \lambda V) \geq \operatorname{Min}\{P(U), P(V)\}.$$

1.17 Theorem. *Suppose A is fixed and the (marginal) probability distribution of b is quasi-concave. Then K_1 is a closed convex set for any $\alpha°$.*

Proof. If K_1 is empty the assertion is immediate. Suppose $x_0, x_1 \in K_1$, then with $x_\lambda = (1-\lambda)x_0 + \lambda x_1$

$$\kappa^{-1}(x_\lambda) \supset (1-\lambda)\kappa^{-1}(x_0) + \lambda \kappa^{-1}(x_1),$$

since $b(\omega_0) \leqslant Ax_0$ and $b(\omega_1) \leqslant Ax_1$ implies that

$$(1-\lambda)b(\omega_0) + \lambda b(\omega_1) \leqslant Ax_\lambda.$$

The monotonicity and quasi-concavity of P now yields

$$P(\kappa^{-1}(x_\lambda)) \geqslant P((1-\lambda)\kappa^{-1}(x_0) + \lambda\kappa^{-1}(x_1))$$
$$\geqslant \mathrm{Min}\{P(\kappa^{-1}(x_0)), P(\kappa^{-1}(x_1))\}$$

But this implies that $P(\kappa^{-1}(x_\lambda)) \geqslant \alpha°$ since both x_0 and x_1 belong to K_1. Hence $x_\lambda \in K_1$.

To see that K_1 is closed simply observe that if $\{x_k, k=1, \ldots\}$ is a sequence in K_1 which converges to $\bar{x}$, we have that for each k,

$$\{b \mid b \leqslant Ax^k = t_k\} = t_k - R_+^{m_1}.$$

Since for each k, $P[t_k - R_+^{m_1}] \geqslant \alpha°$, it follows that $P[\bar{t} - R_+^{m_1}] \geqslant \alpha°$ where $\bar{t} = A\bar{x}$. The proof is complete since the last relation implies that $\bar{x} \in K_1$. □

A large class of quasi-concave probability measures can be identified by means of the following result of Borell [19]. *Suppose h is a density function of a continuous distribution function defined on R^m and $h^{-1/m}$ is convex, then the probability measure defined on the Borel subsets S of R^m by*

$$\mathrm{meas}\, S = \int_S h(s)\,ds$$

is quasi-concave. In particular this implies that if the density is of the form

$$h(s) = e^{-Q(s)}$$

where Q is a convex function, the resulting measure is quasi-concave, since

$$[e^{-Q(s)}]^{-1/m} = e^{mQ(s)}$$

is convex as the composition of a convex function with a non-decreasing convex function $s \mapsto e^s$. Probability measures whose densities are given by an expression of the type $e^{-Q(s)}$ are in fact logarithmic concave, a subclass of the quasi-concave measures, the first class of measures investigated by A. Prékopa [16]. Density functions giving rise to logarithmic concave measures are the (non-degenerate) multi-normal, the multivariate, Dirichlet and Wishart distributions. The multivariate t and F densities (as well as some multivariate Pareto density) engender quasi-concave measures that are not logarithmic concave.

When also A is random, the situation is much more complex. For all practical purposes we have only one result. It is an observation made by Van de Panne and Popp [20], later extended by Prékopa [21] but involving assumptions that appear difficult to verify. Before we come to the little we know, we want to point out the source of the difficulties. Let us consider the "two-"-dimensional case: Suppose here that $a(\cdot)$ and $b(\cdot)$ are real-valued random variables and

$$P[a(\omega)x \geqslant b(\omega)] \geqslant \alpha°$$

is the chance-constraint for some $0 < \alpha° < 1$. To each $x \in R^1$ corresponds $\kappa^{-1}(x)$ a half-space in R^2 given by the expression

$$\kappa^{-1}(x) = \{(a, b) \mid ax - b \geqslant 0\}.$$

The feasibility set

$$K_1 = \{x \mid P[\kappa^{-1}(x)] \geq \alpha^\circ\}$$

is convex if for any given pair (x_0, x_1) in K_1 and any $\lambda \in [0, 1]$ we have

$$P[\kappa^{-1}(x_\lambda)] \geq \alpha^\circ$$

where $x_\lambda = (1-\lambda)x_0 + \lambda x_1$.

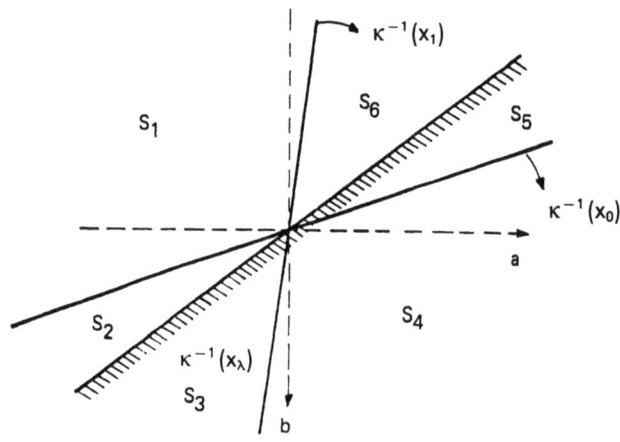

1.18 Figure: Half-spaces Generated by x_0, x_1, x_λ

Figure 1.18 exemplifies a decomposition of the $\{(a, b)\}$-space by $\kappa^{-1}(x_0)$, $\kappa^{-1}(x_1)$ and $\kappa^{-1}(x_\lambda)$. Note that $\kappa^{-1}(x)$ always contains the vertical positive axis. Let $S_4 = \kappa^{-1}(x_0) \cap \kappa^{-1}(x_1)$, $S_3 = (\kappa^{-1}(x_\lambda) \cap \kappa^{-1}(x_0)) \backslash S_4$, $S_5 = (\kappa^{-1}(x_\lambda) \cap \kappa^{-1}(x_1)) \backslash S_4$, $S_6 = \kappa^{-1}(x_1) \backslash (S_4 \cup S_5)$, $S_2 = \kappa^{-1}(x_0) \backslash (S_4 \cup S_3)$ and $S_1 = R^2 \backslash \bigcup_{i=2}^{6} S_i$. For $i = 1, \ldots, 6$ let $\mu_i = P(S_i)$. Since both x_0 and x_1 belong to K_1 we have

(1.19) $$\mu_2 + \mu_3 + \mu_4 \geq \alpha^\circ,$$

and

(1.20) $$\mu_6 + \mu_5 + \mu_4 \geq \alpha^\circ.$$

The convex combination x_λ of x_0 and x_1 belongs to K_1 if

(1.21) $$\mu_3 + \mu_5 + \mu_4 \geq \alpha^\circ$$

which is implied by

(1.22) $$\mu_3 + \mu_5 \geq \min[\mu_2 + \mu_3, \mu_6 + \mu_5].$$

If α° is relatively large, i.e., much larger than .5 if not nearly 1, then (1.19) and (1.20) imply that μ_4 must be of the order of α°; recall that $\sum_{i=1}^{6} \mu_i = 1$, $\mu_i \geq 0$.

Thus the inequality (1.22) will be satisfied whenever the probability mass is "sufficiently unimodal". On the other hand, if for example, the distribution is discrete with a sufficient number of points, linearly independent, not "uniformly" distributed on the plane and with the probability mass sufficiently well-spread out, it will always be possible to find x_0, x_1 and x_λ such that (1.19) and (1.20) hold, but (1.21) and thus also (1.22) fail.

Precise and verifiable conditions that would yield the convexity of K_1 when the matrix A is random have not yet been found although the problem has now been around for the last two decades. It might appear that we exaggerate the importance of convexity for K_1. In this connection, it should be pointed out that the search for a convexity result does not stem purely from computational considerations but from model validation questions. In many ways the chance-constraint

(1.23) $$P[A(\omega)x \geqslant b(\omega)] \geqslant \alpha^\circ.$$

is often accepted as the natural generalization of the standard deterministic linear constraints. Little attention is paid to the consequences of this "simple" extension. If we interpret the decision variables $x \in R^{n_1}$ as activity levels, then non-convexity implies that we can choose two programs of activity levels satisfying the constraints but any combinations of these programs is totally unacceptable. Moreover, from what precedes we know that this will occur whenever $A(\cdot)$ and $b(\cdot)$ lack "unimodal properties", in particular if they are discreetly distributed with the probability mass sufficiently will spread out. To some extent this appears to be an irredeemable condemnation of the modeling of stochastic constraints through chance-constraints, at least if more than the right-hand sides of the constraints are random. However, there is little doubt that there are many situations when it is convenient to rely on chance-constraints to quantify some of the criteria used by decision makers. Since well-formulated practical problems cannot lead us to unreasonable mathematical constructs, we introduce the following concept:

1.24 Definition. *We say that the probability measure P is α°-consistent if for all $\alpha \in [\alpha^\circ, 1]$, the set K_1 is a closed convex set.*

1.25 Proposition [19]. *Suppose that the chance-constraint is actually of the form*

$$P\left[\sum_{j=1}^{n_1} a_j(\omega)x_j \geqslant b(\omega)\right] \geqslant \alpha^\circ,$$

where the $a_j(\cdot)$ and $b(\cdot) =: a_0(\cdot)$ are normal random variables, with mean $\bar{a}_j$, variance σ_j and covariance $\rho_{jk}\sigma_j\sigma_k$. Then the corresponding probability measure is α°-consistent for all $\alpha^\circ \geqslant \tfrac{1}{2}$, or equivalently the set K_1 is convex for all $\alpha^\circ \in [1/2, 1]$.

Proof. For any given x, define the random variable

$$\zeta(x, \omega) = \sum_{j=1}^{n_1} a_j(\omega)x_j - b(\omega).$$

This is a normal random variable. Setting

$$x_0 = 1 \quad \text{and} \quad b(\omega) = a_0(\omega),$$

we get that its mean μ and its variance σ^2 are given by

$$\mu(x) = \sum_{j=0}^{n_1} \bar{a}_j x_j,$$

and

$$\sigma^2(x) = \sum_{j=0}^{n_1} \sigma_j^2 x_j^2 + 2 \sum_{j=0}^{n_1} \sum_{k>j}^{n_1} \rho_{jk} \sigma_j \sigma_k x_j x_k.$$

The chance-constraint is then equivalent to

$$1 - \Phi(-\mu(x)/\sigma(x)) \geq \alpha^\circ$$

where Φ is the distribution function of (standard) normal with mean 0 and variance 1. Which can also be expressed as

$$\Phi^{-1}(\alpha^\circ)\sigma(x) - \mu(x) \leq 0$$

recalling that $\Phi^{-1}(1-\alpha) = -\Phi^{-1}(\alpha)$.

This yields the convexity of K_1, since the form $\sigma^2(x)$ is positive semidefinite in x and $\Phi^{-1}(\alpha^\circ) \geq 0$ precisely when $\alpha^\circ \geq \frac{1}{2}$. □

As indicated already earlier the preceding proposition (with some extensions [21]) is basically the only known result about α°-consistent probability measure for problems involving random matrix A. On the other hand, there are clear indications that a probability measure with "appropriate undimodal" properties is always α°-consistent for $\alpha^\circ < 1$ sufficiently large. For example, the next approximation result due to S. Sinha [22] points in that direction.

1.26 Proposition. *Let*

$$K_1 = \left\{ x \geq 0 \mid P\left[\sum_{j=1}^{n_1} a_j(\omega) x_j \geq b(\omega) \right] \geq \alpha^\circ \right\}.$$

Define

$$K_1' = \left\{ x \geq 0 \mid (1-\alpha^\circ)^{-2} \left(\sum_{k=0}^{n_1} \sigma_{jk} x_j x_k \right)^{\frac{1}{2}} - \sum_{j=0}^{n_1} \mu_j x_j \leq 0, x_0 = 1 \right\}$$

where $a_0(\cdot) = b(\cdot)$, μ_j is the expectation of $a_j(\cdot)$ and σ_{jk} the covariance of $a_j(\cdot)a_k(\cdot)$. Then we always have that K_1' is closed and convex and $K_1 \supset K_1'$.

Proof. With $a_0(\cdot) = b(\cdot)$ the chance-constraint can be expressed as

$$P\left[\sum_{j=0}^{n_1} a_j(\omega) x_1 = \zeta(x, \omega) \geq 0 \right] \geq \alpha^\circ, \quad x_0 = 1.$$

We now use one side of Chebyshev's inequality, viz.,

$$P[\zeta(x, \omega) \geq \bar{\zeta}(x) - k^{-2}\sigma_\zeta(x)] \geq 1 - k$$

where $\bar{\zeta}(x)$ is the expectation of $\zeta(x, \cdot)$ and $\sigma_\zeta^2(x)$ its variance, to obtain the next inequality that implies that the chance-constraint

$$\bar{\zeta}(x) - \frac{1}{(1-\alpha)^2}\sigma_\zeta(x) \geq 0, \quad x_0 = 1.$$

This can also be expressed as

$$(1-\alpha°)^{-2}\left(\sum_{j=0}^{n1}\sum_{k=0}^{n1}\sigma_{jk}x_jx_k\right)^{\frac{1}{2}} - \sum_{j=0}^{n1}\mu_j x_j \leq 0, \quad x_0 = 1.$$

From this it follows that $K_1' \subset K_1$. The set K_1' is clearly closed and also convex since the quadratic form $\sum_{j=0}^{n1}\sum_{k=0}^{n1}\sigma_{jk}x_jx_k$ is positive semidefinite. □

It should be pointed out that in general K_1' is a very crude approximation to K_1 and usually will delete from K_1 those points that are associated with the optimum. There are however many practical situations in which only the means and (co)variances of the random parameters of the problem are known, in which case K_1' is the best available approximation to K_1. The points deleted are then the result of insufficient information.

We now consider K_2, and here because we are able to associate to the stochastic constraints

$$T(\omega)x" = "p(\omega)$$

a discrepancy cost proportional to the recourse activities needed to correct the observed differences, a more flexible modeling tool, we are led to a much less hectic situation, at least in general.

1.27 Theorem. *The set K_2 is a closed convex set given by the relation*

(1.28) $$K_2 = \bigcap_{(p, T) \in \Xi} \{x | p - Tx \in W(R_+^{n2})\}$$

where $\Xi \subset R^{m2(n2+1)}$ is the support of $p(\cdot), T(\cdot)$, i.e., the smallest closed subset of $R^{m2(n2+1)}$ such that $P[(p(\omega), T(\omega)) \in \Xi] = 1$. Moreover, if either p and T are independent and the convex hull of the support of $T(\cdot)$ is polyhedral, or if $T(\cdot)$ has finite support, then K_2 is also polyhedral.

For the proof of this theorem, we refer to [13, Section 4]; note also that Sections 4 and 5 of [13] give constructive descriptions of K_2.

Next we turn of the recourse cost function $\mathcal{Q}$ as defined (1.3). Since the right-hand side of (1.4) is a linear function of x, it follows from parametric programming that for all ω,

$$x \mapsto Q(x, \omega)$$

is a convex polyhedral function. From this and the integrability conditions introduced in connection with the definition of the original problem (1.1), it follows:

1.29 Theorem. *The function $\mathcal{Q}$ is Lipschitz (finite) and convex on K_2. Moreover, for all $x \in K_2$*

(1.30) $$\partial \mathcal{Q}(x) = \int \partial Q(x, \omega) P(d\omega) + \delta \psi_{K_2}(x)$$

where ψ_{K_2} is the indicator function of K_2, i.e., 0 on K_2 and $+\infty$ on its complement. If P is absolutely continuous (with respect to the Lebesgue measure) then $\mathcal{Q}$ is differentiable at every point in the interior of K_2.

Proof. The first assertions are proved in [13, Theorems 7.6 and 7.7]. Formula (1.30) follows from a more general result of Rockafellar [23, Corollary 1 B], consult [24]. The differentiability follows from (1.30), the fact that $\partial \psi_{k_2}(x) = \{0\}$ on int K_2 and that $\{\omega \mid Q(x, \omega)$ is not differentiable$\}$ is a set of zero measure because P is absolutely continuous and $Q(\cdot, \omega)$ is differentiable at every $x \in K_2$, except possibly on a set of zero Lebesgue measure. Thus $\partial \mathcal{Q}$ is a singleton for every $x \in$ int K_2 which yields the differentiability at x since $\mathcal{Q}$ is convex. $\square$

Combining the properties of K_1, K_2 and $\mathcal{Q}$ we have the following:

1.31 Theorem. *Suppose the probability measure P is α°-consistent, then the stochastic program (1.1) is a convex programming problem whose objective function is Lipschitz on the convex closed set $K = K_1 \cap K_2$. The set K is polyhedral if for example $\alpha^\circ = 1$ and T is fixed or $T(\cdot)$ takes on a finite number of possible values.*

Many variants and extensions of the stochastic program (1.1) have been studied in connection with various applications. Theorem 1.31, except for the assertion about the solution set being polyhedral, remains valid under much more general conditions; for example, when the costs are convex-Lipschitz rather than linear and the constraints have similar properties, when there are more than 2 stages [12], when the recourse decision must be selected subject to (conditional) chance-constraints involving stochastic variables not yet observed [25, Section V], and so on. In this context, let us just mention a model studied by Prékopa [26] which has an additional reliability constraint for the induced constraints. The set K_2 is redefined as

$$K_2^+ = \{x \mid P[T(\omega)x + Wy(\omega) = p(\omega), y(\omega) \geq 0] \geq \alpha'\}$$

and the objective is rendered finite by defining it as follows:

$$\mathcal{Q}^+(x) = E Q^+(x, \omega)$$

where

$$Q^+(x, \omega) = \inf[q(\omega) \cdot y + r(s) \mid Wy + s = p(\omega) - T(\omega, x), y > 0];$$

here $s \in R^{m_2}$ and r is a finite positve convex penalty function. The set K_2^+ can be reexpressed as

$$K_2^+ = \{x \mid P[p(\omega) - T(\omega)x \in W(R_+^{n_2})] \geq \alpha'\}.$$

II. Stochastic Programming: Solution Techniques and Approximation Schemes 577

The chance-constraints are thus linear and the results known about K_1 also apply to K_2^+. We are essentially in the setting of problem (1.1). Note also that this is a problem with complete recourse, and hence $\mathcal{Q}^+$ is finite valued.

2. Algorithmic Procedures

Attention will be focused on methods to evaluate and minimize $\mathcal{Q}$; we content ourselves with a few brief remarks concerning feasibility. For the chance-constraint(s) (1.8), we assume that the hypotheses of the problem are such that K_1 can be expressed as

(2.1) $$K_1 = \{x \geq 0 \mid g_{1l}(x, \alpha^\circ) \leq 0, l = 1, \ldots, L_1\}.$$

where for all l, the functions $(x, \alpha) \mapsto g_{1l}(x, \alpha)$ are quasiconvex. This certainly includes the case when both A and b are fixed, but also those cases for which we have convexity characterizations for K_1, e.g., with A fixed and $b(\cdot)$ random and P is quasi-concave, then with

(2.2) $$g_1(x, \alpha) = -P[\kappa^{-1}(x)] + \alpha$$

we have the above representation for K_1. These linear or nonlinear constraints are handled as usual in constrained optimization. At least if explicit expressions are available for them. If this is not the case, as would usually occur when g_1 is defined through an expression of the type (2.2), solution procedures must be adapted to the "computable" quantities of that function. For example, computing $P[\kappa^{-1}(x)]$ presuppose the availability of a multidimensional integration routine. We would also need an associate calculus for the multivariate distribution of $A(\cdot)$ and $b(\cdot)$ that allows us to obtain the gradient (or subgradient) of the function $x \mapsto P[\kappa^{-1}(x)]$ if the algorithmic procedures requires such information. In [27] Prékopa et al. report on a case where all these questions were confronted.

Similarly, we assume that the induced constraints K_2 can be represented by a finite number of constraints, viz.,

(2.3) $$K_2 = \{x \mid g_{2l}(x) \leq 0, l = 1, \ldots, L_2\},$$

where naturally, for all $l = 1, \ldots, L_2$, the functions

$$x \mapsto g_{2l}(x)$$

are convex, cf. Theorem 1.27. Again, explicit expressions for the functions g_{2l} are not easy to come by. However, it is usually possible, as done first in [28], to construct these constraints as needed, i.e., suppose an algorithmic procedure generates an $\hat{x}$ that does not belong to K_2, i.e.,

$$P(\omega) - T(\omega)\hat{x} \notin W(R_+^{n_2})$$

with positive probability. Then there exist a supporting hyperplane, corresponding to a facet, of the polyhedral convex cone $W(R^{n_2}_+)$, say

$$\{t \in R^{m_2} \mid st = 0\},$$

such that

$$P[s(p(\omega) - T(\omega)\hat{x}) < 0] > 0.$$

The constraint

(2.4) $$\inf_{(p, T) \in \Xi} s(p - Tx) \geq 0$$

where Ξ is again the support of the random $p(\cdot)$ and $T(\cdot)$, is not satisfied by $\hat{x}$ but does not eliminate any feasible points. There are only a finite number of these constraints since $W(R^{n_2}_+)$ has only a finite number of facets. In general (2.4) is not a linear constraint, but in practice these constraints are very often linear [13, Section 5]. For example, if T is fixed then (2.4) becomes

(2.5) $$(sT)x \leq \inf_{p \in \Xi_p} ps$$

where Ξ_p is the support of $p(\cdot)$. The $\inf_{p \in \Xi_p}$ either exists in which (2.5) yields a valid linear constraint or this infimum is $-\infty$ in which case there are no points satisfying this constraint which means that the original stochastic program is infeasible.

Taking into account (2.1) and (2.3), we see that the problem to be solved is given by

(2.6) Find $x \geq 0$, $\alpha \in [0, 1]$ such that
$g_{1l}(x, \alpha) \leq 0$, $l = 1, \ldots, L_1$
$g_{2l}(x) \leq 0$, $l = 1, \ldots, L_2$
and $z = cx + \mathcal{Q}(x) + \rho(\alpha)$ is minimized

where $\mathcal{Q}$ is a finite convex-Lipschitz function on K_2, defined by (1.3) and (1.4), and repeated here for easy reference,

$$\mathcal{Q}(x) = E\{Q(x, \omega)\},$$

and

$$Q(x, \omega) = \inf_y \{q(\omega)y \mid Wy = p(\omega) - T(\omega)x, y \geq 0\}.$$

At least in theory, any standard convex programming package could be used to solve problem (2.6), but usually computing the value of $\mathcal{Q}$, its subgradients or even more so, second order information about $\mathcal{Q}$ requires computational resources far beyond the advantages to be gained from knowing an optimal solution to (2.6). For these reasons any solution method involving line minimization or of the Quasi-Newton type must be quickly discarded, except possibly for special classes of stochastic programs, such as stochastic programs with simple recourse whose random variables obey specific probability laws [29]. We shall not deal with those cases here; because of their special nature, the work on algorithmic procedures for stochastic programs with simple recourse, when $W = (I, -I)$, and extensions thereof, is following a course of its own that is being reviewed separately, see [30]. Here we shall be mostly concerned with the case when no advantage is taken of any special structure of the recourse matrix W, or other components of the stochastic program (1.1).

If the probability distribution of the random elements of the stochastic program is anything but discrete with finite support, the evaluation of $\mathcal{Q}$ or its subgradient given by formula (1.30), involves – at least in principle – the solution of an infinite number of linear programs to describe the function $w \mapsto Q(x, \omega)$, followed by a multidimensional integration. The material impossibility to work out these operations exactly has led to the development of approximations schemes. To date the only proposed schemes that have been exploited computationally are discretization schemes which consist in the replacement of the original random variables by approximating random variables whose support is finite; henceforth we reserve the term *discrete* to designate this type of random variables. The next section is concerned with the convergence and the error bounds that can be associated with various approximations, the rest of this section deals with solution procedures for (2.6) for discretely distributed random variables.

Let $\{(q_k, p_k, T_k), k=1, \ldots, N\}$ be the (possible) values of the random variables $(q(\cdot), p(\cdot), T(\cdot))$ and let

$$f_k = P[(q(\omega), p(\omega), T(\omega)) = (q_k, p_k, T_k)]$$

be the associated probabilities. In this case, problem (2.6) is equivalent to

(2.7) Find $x \geq 0$, $\alpha \in [0, 1]$ and $y_k \geq 0$, $k=1, \ldots, N$ such that
$$\begin{aligned}
g_{1l}(x, \alpha) & & & & & \leq 0 \quad l=1, \ldots, L_1 \\
T_1 x + & Wy_1 & & & & = p_1 \\
T_2 x & & + Wy_2 & & & = p_2 \\
\vdots & & & \ddots & & \vdots \\
T_N x & & & & Wy_N & = p_N
\end{aligned}$$

and

$$cx + f_1 q_1 y_1 + f_2 q_2 y_2 + \ldots + f_N q_N y_N + \rho(\alpha) = z \text{ is minimized.}$$

Except possibly for some nonlinearity in ρ or the constraints involving g_{1l}, this is a large scale linear program with dual block angular structure. How large, clearly depends on N the number of realizations of the random variables. Note that there is no need to include the induced constraints

$$g_{2l}(x) \leq 0, \quad l=1, \ldots, L_2,$$

they are automatically incorporated in (2.7), which will be feasible only if for some x there exist for all $k=1, \ldots, N$, y_k such that

$$W y_k = p_k - T_k x, \quad y_k \geq 0.$$

Again here any large scale programming technique can be specialized – note that the matrices that appear along the diagonal are the same – to solve this type of problem. In fact various such possibilities have been worked out, consult for example [31], [32, Section 3]. Here we retain only those based on compact basis and decomposition techniques, that have been implemented and exhibit at this date the greatest promise.

To somewhat simplify the presentation and to keep our discussion in the realm of large scale linear programming, we assume that there are no terms involving α and suppose that K_1 is given by linear relations of the type

$$K_1 = \{x \geq 0 \mid Ax = b\},$$

where A and b are fixed matrices. Problem (2.7) then reads

(2.8) Find $x \geq 0$ and $y_k \geq 0$, $k = 1, \ldots, N$ such that
$$Ax = b,$$
$$T_k x + W y_k = p_k \quad k = 1, \ldots, N$$

and

$$cx + \sum_{k=1}^{N} f_k q_k y_k = z \text{ is minimized}.$$

A version of the dual of this problem is then

(2.9) Find $\sigma \in R^{m_1}$ and $\pi_k \in R^{m_2}$, $k = 1, \ldots, N$ such that
$$\sigma A + \sum_{k=1}^{n} f_k \pi_k T_k \leq c$$
$$\pi_k W \leq q_k \quad k = 1, \ldots, N$$

and

$$\sigma b + \sum_{k=1}^{N} f_k \pi_k p_k = w \text{ is minimized}.$$

Problem (2.9) is not quite the usual (formal) dual of (2.8). To obtained the standard form, set

$$\hat{\pi}_k = f_k \pi_k$$

and substitute in (2.9). The dual problem has block-angular structure, the diagonal consisting of identical matrices W.

The compact basis technique, as worked out by B. Strazicky [33] and further analyzed by P. Kall [34] who also implemented the technique as part of an approximation scheme, exploits the structure of the bases of this dual problem to obtain a working basis with

$$n_1^2 + N n_2^2$$

elements, a number substantially smaller than

$$(n_1 + N n_2)^2$$

which would be the size of the basis for the standard simplex method. What makes this basis reduction possible is the following observation. Including the slack variables, the constraints of problem (2.9) involve N systems of the type

(2.10) $\pi_k^+ W - \pi_k^- W + s_k I = q_k, \quad \pi_k^+ \geq 0, \; \pi_k^- \geq 0, \; s_k \geq 0.$

Now assuming that (2.9) is feasible (and bounded) it follows that each basic solution will have at least n_2 basic variables among those associated to the k-subsystem. (In case of degeneracy the pivoting rule can easily be adjusted to gua-

II. Stochastic Programming: Solution Techniques and Approximation Schemes

rantee the above.) Any basis generated by the iteration of the simplex method will thus contain at least n_2 columns that "intersect" the k-subsystem.

To see this, it helps to consider the detached coefficients from (2.9):

(2.11)

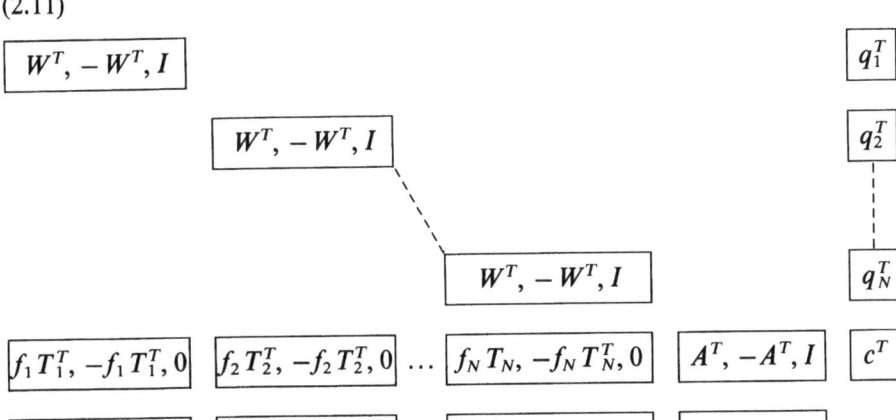

Let $\hat{B}^T$ be a (feasible) basis for this problem whose restriction to the k-subsystem, we denote by

$$[B_k^T, L_k^T],$$

i.e., $[B_k^T, L_k^T]$ is for all k, a submatrix of

$$[W^T, -W^T, I].$$

The matrix B_k^T is supposed to be invertible (at least n_2 of the columns of the submatrix are linearly independent). The columns of L_k^T are linear combinations of the columns of B_k^T, we can thus express L_k^T as follows:

(2.12) $$L_k^T = B_k^T E_k^T.$$

Recall that naturally L_k^T may be empty when exactly n_2 columns of the k-subsystem are in the basis $\hat{B}^T$. Schematically, and up to a rearrangement of the columns, the basis is of the form

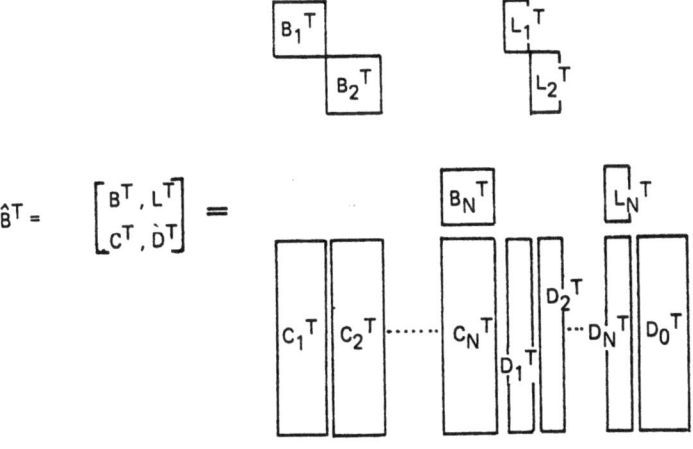

where C_k^T is the submatrix of

$$[f_k T_k^T, -f_k T_k^T, 0]$$

that corresponds to B_k^T and D_k^T the one that corresponds to L_k^T. The D_0^T matrix comes from the columns of

$$[A^T, -A^T, I]$$

that are in the basis. Observe that the $n_1 \times n_1$-matrix D_0^T is invertible. This structure of $\hat{B}$ is to be exploited to reduce the simplex operations, that usually require the inverse of $\hat{B}$, to operations requiring essentially no more than the inverses of the B_k.

The simplex multipliers associated with basis the $\hat{B}$, denoted by

$$\begin{bmatrix} y \\ x \end{bmatrix} = \begin{bmatrix} y_1 \\ \vdots \\ y_N \\ x \end{bmatrix}$$

are given by the relation

(2.14) $$\hat{B}\begin{bmatrix} y \\ x \end{bmatrix} = \begin{bmatrix} B & C \\ L & D \end{bmatrix}\begin{bmatrix} y \\ x \end{bmatrix} = \begin{bmatrix} \rho \\ \beta \end{bmatrix}$$

where $[\rho^T, \beta^T]$ is the appropriate rearrangement of the subvector of the coefficients of the objective of (2.11) that corresponds to the columns of $\hat{B}^T$ with β^T being the subvector whose components correspond to the columns of D^T. The (dual feasible) basis is optimal if the vectors

$$(x, y_k, k=1, \ldots, N)$$

are primal feasible, i.e., satisfy the constraints of (2.8). To obtain x and y we see that (2.14) yields

$$y = B^{-1}(\rho - Cx)$$
$$x = D^{-1}(\beta - Ly)$$

from which we get

$$x = (D - LB^{-1}C)^{-1}(\beta - LB^{-1}\rho)$$

and for $k = 1, \ldots, N$,

(2.15) $$y^k = B_k^{-1}(\rho_k - C_k x)$$

where ρ_k^T is the subvector of the objective of (2.11) corresponding to the columns in B_k. We have used the fact that B is block diagonal with invertible matrices B_k on the diagonal. Going one step further and using the representation (2.12) for the matrices L_k, we get the equation

(2.16) $$x = \left(D - \sum_{k=1}^{N} E_k C_k\right)^{-1}\left(\beta - \sum_{k=1}^{N} E_k \rho_k\right)$$

for x. What is important to notice is that to obtain x and y through (2.16) and (2.15), we only need to know the inverse of the N $(n_2 \times n_2)$-matrices B_k and of the matrix $(D-EC)$.

Similarly to obtain the values of the variables σ and (π_k^+, π_k^-), $k=1, \ldots, N$; associated to this basis, exactly the same inverses is all that is really required, as can easily be verified. One now needs to work out the updating procedures in order to show that the steps of the simplex method can be performed in this compact form, i.e., that the updating procedures involve only the restricted inverses. This has been carried out in [33]. Experimental computational results are also mentioned in [33]; with only the vector p random, i.e., q and T fixed and

$$m_1 = 30, \ n_1 = 40, \ m_2 = 6, \ n_2 = 5 \text{ and } N = 540,$$

the run time on a CDC 3300 was 20 minutes. Further computational experience involving (generated) problems with random T is reported in [35].

A number of improvements suggest themselves. In [33] it is observed that in problem (2.9) the variables π_k and σ are not restricted in sign and that it is not really necessary to express each π_k as

$$\pi_k = \pi_k^+ - \pi_k^-, \quad \pi_k^+ \geq 0, \ \pi_k^- \geq 0,$$

which doubles the number of variables. In fact the π_k should be treated as sign-unrestricted variables with the corresponding columns, i.e., all of W^T, always part of the basis. In fact if the rows of W are linearly independent, then for all k, the columns of W^T could always be left in B_k^T. This means that the only changes that would occur in the matrices

$$(B_k^T, L_k^T),$$

from one basis to the next, would be columns of the identitiy $I_{(n_2)}$ shuffling in and out of the new basis. This feature was not exploited in the implementation of the algorithm and one may reasonably expect that there would be substantial savings if one did, especially if the inverses can be stored in product form. In fact one could go much further, as we show next.

Since for all k-subsystems, the columns of W^T will be contained in (B_k^T, L_k^T), we can always keep them in B_k^T. We have

(2.17) $$B_k^T = [W^T, I(k1)], \quad L_k^T = I(k2)$$

where $I(k1)$ consists of $(n_2 - m_2)$ columns of the $(n_2 \times n_2)$-identity and $I(k2)$, possibly empty, consists of some of the remaining columns of the same identity matrix. Schematically, and upt to some rearrangement of the rows, we have that

$$B_k^T = \begin{bmatrix} S_k^T & 0 \\ V_k^T & \begin{matrix} 1 & & \\ & \ddots & \\ & & 1 \end{matrix} \end{bmatrix} = [W^T, I(k1)]$$

To know the inverse of B_k^T it really suffices to know S_k^{-1}. The inverse is given by

$$(B_k^T)^{-1} = \begin{bmatrix} (S_k^T)^{-1} & 0 \\ -V_k^T(S_k^T)^{-1} & I \end{bmatrix}$$

as can easily be checked. Thus rather than keeping and updating an $n_2 \times n_2$-matrix for each subsystem, it appears that all the information that is really needed can be manipulated in an $m_2 \times m_2$-matrix. As for standard linear programs we expect m_2 to be usually much smaller than n_2. This should result in substantial savings that would drastically reduced the number of essential operations by simplex iteration as calculated by Kall [34, equations (29) and (30)]. We could pursue the detailed analysis still further taking advantage of the fact that the matrices $D_1, \ldots, D_k$ are all zero, that a number of the S_k are bound to be identical if N is large, and so on. We shall however not do this here, basically because the operations would mimic very closely those of the algorithm to be described next. It is conjectured that a version of this compact basis technique that would fully exploit the structural properties of the dual problem (2.9) would then exhibit the same computational complexity as this second algorithm.

The suggestion of using the decomposition principle to solve stochastic programs goes back to G. Dantzig and A. Madansky [36], the procedure they sketched out took advantage of the structure of the dual problem (2.9). This approach via decomposition was elaborated by R. Van Slyke and R. Wets in [37] relying on a cutting hyperplane algorithm (outer linearization, Benders' decomposition) which can be interpreted as a *partial* decomposition method [37, Section 3]. In view of the matrix layout of the problem to be solved, and the explicit use made of this structure, we refer to it as the *L-shaped* algorithm. Recent work by J. Birge [32] extends the method to multistage problems, he also reports on computational experiments with large scale problems; see also [38] and [39]. (For an alternative use of decomposition techniques, consult [30, Section 6].)

To describe the method it is useful to think of problem (2.8) in the following form:

(2.18) Find $x \geq 0$ such that
$Ax = b$, and
$cx + \mathcal{Q}(x) = z$ is minimized

where

$$\mathcal{Q}(x) = \sum_{k=1}^{N} f_k (\inf q_k y \mid W_y = p_k - T_k x, y \geq 0).$$

Infeasibility and unboundedness are ignored, they can usually be handled by an appropriate coding of the initialization step, see [40]. The L-shaped algorithm given here is actually a variant of the one in [37, Section 5], in the sense that we are working with a more general class of stochastic programs than those under consideration in [37]. The method consists of 3 steps that can be interpreted as follows. In Step 1 we solve an approximation to (2.18) using an outer-linearization of $\mathcal{Q}$. The two types of constraints (2.19) and (2.20) that appear in this linear program come from

(i) feasibility cuts (determining $K_2 = \{x \mid \mathcal{Q}(x) < +\infty\}$),

and

(ii) linear approximations to $\mathcal{Q}$ on its domain of finiteness.

These constraints are generated systematically through Steps 2 and 3, when a proposed solution x^ν of the linear program of Step 1 fails to be in K_2 (Step 2) or if the approximating problem does not yet match the function $\mathcal{Q}$ at x^ν (Step 3). The row-vectors generated during Step 3 are actually subgradients of $\mathcal{Q}$ at x^ν. The convergence is based on the fact that there are only a finite number of constraints of type (2.19) and (2.20) that can be generated since each one corresponds to some basis of W and either some point (p_k, T_k) or to a (finite) number of weighted averages of these points.

Step 1. Set $\nu = \nu + 1$. Solve the linear program

Find $x \geq 0$, $\theta \in R$ such that
$$Ax = b$$
(2.19) $\quad D_l x \geq d_l, \quad l = 1, \ldots, s$
(2.20) $\quad E_l x + \theta \geq e_l, \quad l = 1, \ldots, t$

and
$$cx + \theta = z \text{ is minimized.}$$

Let (x^ν, θ^ν) be an optimal solution. If no constraints of the form (2.20) are present, θ is set equal to $-\infty$ and ignored in the computation. Initially set $s = t = \nu = 0$.

Step 2. For $k = 1, \ldots, N$ solve the linear programs.

(2.21) Find $y \geq 0$, $v^+ \geq 0$, $v^- \geq 0$ such that
$$Wy + Iv^+ - Iv^- = p_k - T_k x^\nu, \text{ and}$$
$$ev^+ + ev^- = w^1 \text{ is minimized,}$$

until for some k, the optimal value $w^1 > 0$. Let σ^ν be the associated simplex multipliers and define
$$D_{s+1} = \sigma^\nu T_k$$

and
$$d_{s+1} = \sigma^\nu p_k,$$

to generate a cut of type (2.19). Return to Step 1 with a new constraint of type (2.19) and set $s=s+1$. If for all k, $w^1=0$ go to Step 3.

Step 3. For all $k=1,\ldots,N$, solve the linear program

(2.22)   Find $y \geq 0$ such that
$$Wy = p_k - T_k x^\nu, \text{ and}$$
$$q_k y = w^2 \text{ is minimized}$$

Let π_k^ν be the multipliers associated with the optimal solution of the problem k. Define

$$E_{l+1} = \sum_{k=1}^{N} f_k \pi_k^\nu T_k$$

$$e_{l+1} = \sum_{k=1}^{N} f_k \pi_k^\nu p_k$$

and

$$w^{2\nu} = \sum_{k=1}^{N} f_k (p_k - \pi_k^\nu T_k) = e_{l+1} - E_{l+1} x^\nu$$

If $\theta^\nu \geq w^{2\nu}$ stop, x^ν is an optimal solution. Otherwise, return to Step 1 with a new constraint of type (2.20) and set $t=t+1$.

The separation of Steps 2 and 3 is not just for expository reasons. Problem (2.21) is the counterpart of Phase I of the simplex method for (2.22). Thus, in practice these two operations would not be separated if we proceeded precisely as indicated here. However, there are many cases in which Step 2 can be modified to solving only 1 linear program. Details can be found in [13, Section 5], here let us just suggest the reasons for this simplification. Let $\prec$ be the ordering induced by the closed convex cone $W(R_+^{n_2})$ on R^{m_2}, i.e.,

$$a_1 \prec a_2 \quad \text{if} \quad a_2 - a_1 \in W(R_+^{n_2}).$$

Then for all $k=1,\ldots,N$, the system of equations

(2.23) $$Wy = p_k - T_k x^\nu, \quad y \geq 0$$

is feasible, if there exists $\alpha \in R^{m_2}$ such that for all $k=1,\ldots,N$

$$\alpha \prec p_k - T_k x^\nu$$

and the system of equations

(2.24) $$Wy = \alpha, \quad y \geq 0$$

is feasible. There always exist such a lower bound. If in addition, we can choose α such that

$$\alpha = p_{k'} - T_{k'} x^\nu$$

for some k', then we have that (2.23) is feasible for all k if and only if (2.24) is feasible. Although in general such a unique α will not exist, it does exist in many instances. And even when a single α will not do, it will usually be sufficient to consider a few such lower bounds, much fewer in any case than all possible (p_k, T_k). Let us remark that computing a lower bound with respect to $\prec$ may be too difficult, but it really suffices to work with lower bounds with respect to a cone ordering induced by any closed convex cone contained in $W(R_+^{n_2})$ - for example, a cone generated by a subset of the columns of W - and that cone could be an orthant, cf. [13, Theorem 4.17].

The work to solve Step 3 can also be significantly reduced if we use sifting or bunching procedures as we now explain. By *bunching* we mean the following. With given x, let B be a submatrix of W that is optimal for some k, i.e., corresponding to some realization (q_k, p_k, T_k). Then from the optimality conditions for linear programming it follows that this basis will be also be optimal, when solving problems (2.22), for all k such that

$$B^{-1}(p_k - T_k x) \geq 0,$$

and

$$q_k - \gamma_k B^{-1} W \geq 0,$$

where γ_k is the subvector of q_k whose elements are the coefficients of the variables that are in the basis. Since B^{-1} is already available, verifying if the above inequalities are satisfied involves relatively little work, especially if only p or q varies with k. Moreover, because of the nature of the problem at hand it is reasonable to expect that only a small number of bases in W will suffice to bunch all the realizations. If problem (2.7) is the result of a discretization of the random variables of the stochastic program, a refinement of the discretization will only increase the work by that required to bunch a larger number of realizations, the basic steps of the algorithm remain unaffected.

In fact the preceding suggest the following overall procedure to solve stochastic programs with arbitrary distributions for the random variables. First, solve an approximation of the original program with only a few samples of $q(\cdot), p(\cdot)$ and $T(\cdot)$, for example, such that

(2.25) $\quad (q_k, p_k, T_k) = E\{(q(\omega), p(\omega), T(\omega)) | (q(\omega), p(\omega), T(\omega)) \in \Omega_k\}$

where Ω_k is part of a partition of the sample space. Using the L-shaped algorithm solve the resulting program (2.8), keeping the bases used to perform the last bunching. (If storage limitations make the storing of all these bases impossible, it is always possible to record them through a listing of the corresponding index vectors.) Next increase the number of samples, either systematically via (2.25) using a finer partitioning of the sample space or through a sampling procedure (Monte-Carlo). Then bunch this extended sample using the bases already available, if some samples escape this bunching process proceed with the unassigned samples as usual in Step 3, using for value of x the optimal solution of the previous discretization. Continue until the optimal solution of this new (approximating) problem is attained. Then repeat – by which we mean: refine the discretization, use the stored bases of W to bunch this larger sample and so on – until the solution reached satisfy acceptable error bounds, see Section 3.

By *sifting* we mean procedures that rely on a systematic arrangement of the vectors

$$\{q_k, k=1, \ldots, N\}$$

and

$$\{p_k - T_k x, k=1, \ldots, N\}$$

in order to facilitate the parametric analysis of the linear program (2.22) to be solved in Step 3. This would then substantially simplify and shorten the time required to perform Step 3. In [39] S. Gartska and D. Ruthenberg describe such a sifting procedure. A variant of the arrangement they suggest was compared to a bunching procedure for the case when only p and/or q are random. Sifting appears to be better than bunching if the number of different bases needed to sift or bunch the sample is small, otherwise the need to rearrange the vectors $\{p_k - T_k x, k=1, \ldots, N\}$ with each new x, appears to cancel out whatever advantage one may gain from this preconditioning of the data to accelerate Step 3. However, there appears to be room here for substantial improvements.

We terminate this section with a short discussion of stochastic quasigradient methods. Solving stochastic programs is only one of their potential applications. The method whose roots lie in the theory of stochastic approximation originates with E. Kiefer and I. Wolfowitz who propose a method for unconstrained optimization, cf. also the related work of H. Robbins and S. Monro for solving systems of nonlinear equations. Applications to statistics and probability were further developed by V. Dupač, V. Fabian, J. Sachs and many others, who provided also rates of convergence as well as various characterizations of the limit distributions. H. Kushner and his students extended the methods to encompass systems that can only be measured up to some noise. The use of stochastic quasigradient methods to solve constrained optimization problems start with the work of V. Fabian and Y. Ermoliev in the mid 60's. Since then there have been numerous contributions by B. Poljak, H. Kushner and D. Clark, K. Marti, J-B. Hiriart-Urruty, L. Schmetterer, G. Pflug and A. Ruszcynski, ... the main impetus coming from Y. Ermoliev and his collaborators from the Institute of Cybernetics in Kiev: N. Shor, L. Bajenow, A. Gupal, E. Nurminiski and A. Gaivoronsky. A recent survey has been provided by Y. Ermoliev [41]. Of direct interest to the problem at hand is [42; Chapter I, §4], [43], [44], [41, Section 7] and especially the interesting monograph [45] of K. Marti who uses the method to obtain a constructive proof of duality results, for example; cf. also [46] and [47] where two applications of the stochastic quasigradient methods to stochastic programming problems are described in detail.

In our setting, the method works basically as follows. Let $F: S \to R$ with S a closed convex subset of R^n be defined by

$$F(x) = E\{f(x, \omega)\}$$

and let us assume that for all ω, $x \mapsto f(x, \omega)$ is convex. The algorithm generates a sequence $\{x_1, x_2, \ldots\}$ of points of S through the recursive formula

$$x_{\nu+1} = \operatorname*{prj}_S (x_\nu - \rho_\nu \xi_\nu)$$

where prj$_S$ denotes the projection on S, $\{\rho_\nu, \nu=1,\ldots\}$ is a sequence of scalars and ξ_ν is a stochastic quasigradient of F at x_ν, by which one means a realization of a random n-vector ξ_ν satisfying

$$E\{\xi_\nu|x_1,x_2,\ldots,x_\nu\}\in \partial F(x^\nu).$$

Typically ξ_ν is obtained as a subgradient of $f(\cdot,\omega_\nu)$ at x_ν where ω_ν is a sample of the random elements determining f or more generally

$$\xi_\nu = \frac{1}{L}\sum_{l=1}^{L} v_l$$

where each $v_l \in \partial f(x_\nu,\omega_l)$ with the ω_l, $l=1,\ldots,L$ a collection of independent samples.

The method converges with probability 1 if the selected scalars ρ_ν satisfy conditions such as

$$\rho_\nu \geq 0, \quad \sum_\nu \rho_\nu > \infty \quad \text{and} \quad \sum_\nu \rho_\nu^2 < \infty.$$

For example, $\rho_\nu = \nu^{-1}$ is such a sequence. The proof can be derived from a modified super-martingale convergence argument [42, Theorem 3], consult also [47].

The application of the method to stochastic programming problem (2.6) works essentially as follows. To facilitate the exposition, let us consider the case when α is not a variable and thus (2.6) can be reexpressed as

(2.26)   find $x \in K = \{x \geq 0 | g_{1l}(x) \leq 0, g_{2l}(x) \leq 0\}$,
such that $cx + \mathcal{Q}(x) = z$ is minimized,

where

$$\mathcal{Q}(x) = E\{Q(x,\omega) = \inf_{y>0}(q(\omega)y|Wy=p(\omega)-T(\omega)x)\}$$

Samples of $q(\cdot), p(\cdot)$ and $T(\cdot)$ corresponding to a (sample) event will be indicated by subscripting ω, i.e.,

$$Q(x,\omega_k) = \text{Min}\{q_k y | W_y = p_k - T_k x, y \geq 0\}$$

with the sample values (q_k, p_k, T_k) corresponding to the event ω_k. Finally, note that for all $x \in K$, $Q(x,\omega_k)$ is finite and

$$\partial Q(x,\omega_k) = \{-v T_k | v \text{ optimal multiplier for (2.22)}\}.$$

The sequence of solution $\{x^1, x^2, \ldots\}$ is produced by the recursion:

(2.27) $$x^{\nu+1} = \text{prj}_K(x^\nu - \rho_\nu g^\nu)$$

where for some $M \geq 1$

(2.28) $$g^\nu = c + \frac{1}{M}\sum_{\mu=1}^{M} g_\mu,$$

and for $\mu = 1, \ldots, M$,

$$g_\mu \in \partial Q(x^\nu, \omega_\mu).$$

The scalars ρ_ν are assumed to satisfy the appropriate conditions to ensure convergence (with probability 1).

There are three possible stumbling blocks in the implementation of the stochastic quasigradient method to solve (2.26):

the projection on K,
the choice of the step-size ρ_ν,
the stopping criterion.

The *projection* of a point on the closed convex set K is easy only if K is "simple" by which we mean a set such as a bounded interval or a sphere, ... If K is an arbitrary convex polyhedron, then one does usually need to solve a quadratic program of some type in order to obtain $x^{\nu+1}$ as given by (2.27). If K is a general convex set then an even more sophisticated nonlinear programming technique must be used to get $x^{\nu+1}$.

The choice of the *step-size* ρ_ν is theoretically prescribed by convergence requirements. However, when we use the stochastic gradient method to solve (2.26) we are interested in its short run properties rather than its long run properties, and there is at present no theory that guides us in the choice of the step-size. In practice, some ρ is chosen at the outset and kept a fixed value until the user intervenes to change it or some heuristic is built in the code to adjust the size of ρ when certain phenomena are observed. That takes us very far away from the convergence requirements. How to remedy this is not clear at this time.

Finding a good *stopping criterion* is still very much an open question. Because the function $\mathcal{Q}$ is difficult to evaluate – and that is why we are using stochastic quasigradient methods in the first place – it is out of the question to use value comparisons between $\mathcal{Q}$ at x^ν and at $x^{\nu+1}$. Y. Ermoliev has suggested that the following quantity

$$\frac{1}{M+1} \sum_{h=\nu-M}^{\nu} Q(x^h, \omega_h) = \hat{\mathcal{Q}}(x^\nu)$$

be used as an estimate for $\mathcal{Q}(x^\nu)$ with M a relatively large number. The algorithm is to terminate when no improvement is observed in the value of $\hat{\mathcal{Q}}$ after ρ has already been reduced to its computationally desirable lower bound.

To conclude this too brief discussion of the stochastic quasigradient method, we would like to point out the connections between this solution method and the L-shaped algorithm. To do so we work with version (2.18) of the stochastic program, i.e., the discrete case with linear constraints determining K_1. The straightforward implementation of the stochastic quasigradient method would run into difficulties if there are induced constraints on x that cut the feasibility region, i.e., if $x \in K_1 = \{x \geq 0 \mid Ax = b\}$ does not automatically imply that there is a feasible recourse y_k for all (p_k, T_k). Assume this does not occur, in stochastic programming parlance this means that the *relatively complete recourse* condition is satisfied [13, Section 6]. This also means that Step 2 of the L-shaped algorithm can be skipped. Both algorithms require the calculation of the subgradient of Q. For the L-shaped algorithm this is done in Step 3, whereas in the stochastic quasigradient algorithm only on estimate of the subgradient is needed. Naturally, if all points $\{(q_k, p_k, T_k), k = 1, \ldots, N\}$ are used to obtain (2.28) then not an estimate but an actual subgradient of $\mathcal{Q}$ is utilized by the stochastic quasigradient method. But this would be contrary to the strategy of the method which consists in moving forward as soon as an estimate of a direction of descent is made available, and one hopes that after N (= number of different sample values) steps, with x^ν adjusted at each step, the decrease in the objective will be more substantial than if all samples were used to compute a (reliable) subgradient of Q. Assuming this to be true, and ignoring some of the difficulties that may arise from step-size and projections, the question would then be if the advantage gained from bunching (or sifting), which can be used to speed up Step 3 in the L-shaped algorithm, would not totally offset the fact that at each step of the stochastic quasigradient method, the recourse problem (2.22) must be solved with a new value of x and (q, p, T). Naturally there too, one should take advantage of the fact that already a basis is available but usually a few pivot operations will be required to reach the new optimal basis.

If one was going to compare the two algorithms it would not be sufficient to measure their respective performance on (discrete) problems of type (2.18) but also when (2.18) is part of an approximation scheme for the original problem, since the stochastic quasigradient method takes no advantage of the shape of the probability distribution of the random variables and is in no way hampered by having continuous distributions. In this connection, one should also mention a recent, still unpublished, result of A. Gaivoronsky which shows that under certain conditions on f, the number of steps required by the stochastic quasigradient method to find the minimum of $E\{f(x, \omega)\}$ is smaller (in a sense which can be made precise) than the number of points required to compute for fixed x^0, the value of $E\{f(x^0, \omega)\}$ by a sampling technique.

3. Approximations and Error Bounds

Section 2 dealt with algorithmic procedures for solving stochastic programs whose random variables are discretely distributed. It was suggested that in the case of arbitrary distributions we could proceed by approximation through discretization, obtaining a sequence of approximate solutions through successive refinements of the discretization. This was originally proposed by P. Kall [48] and P. Olsen [49] for recourse problems and by G. Salinetti [50] for chance-constraints. Although approximation through discretization will be the prominent theme of this section, it is by no means the only possibility, see for example [51] where it is suggested that the distribution functions be approximated by piecewise linear distribution functions, or [52] where the multivariate distribution is approximated by linear combination of lognormal univariate distributions, and also the resourceful applications of stochastic programming [53], [54], [55] and [56] where it is the structure of the problem itself that is approximated.

As in Sections 1 and 2 we start with a brief study of

$$K_1 = \{x \geq 0 \mid P[A(\omega)x \geq b(\omega)] \geq \alpha°\},$$
$$= \{x \geq 0 \mid P[x \in \kappa(\omega)] \geq \alpha°\}.$$

Let us assume that there exist matrices A^- and A^+ and vector b^- and b^+ such that for all ω

(3.1) $A^- \leq A(\omega) \leq A^+$ and $b^- \leq b(\omega) \leq b^+$,

(3.2) no row of (A, b) is identically zero for all
$A \in [A^-, A^+]$ and $b \in [b^-, b^+]$,

(3.3) the interior of $\kappa^- = \{x \geq 0 \mid A^- x \geq b^+\}$ is nonempty.

Then it is easy to construct sequences of random matrices $A_\nu^-(\cdot)$ and $A_\nu^+(\cdot)$, and vectors $b_\nu^-(\cdot)$ and $b_\nu^+(\cdot)$ taking on only a finite number of values (discretely distributed), satisfying the same bounds as $A(\cdot)$ and $b(\cdot)$, and such that the sequences are monotone with

$$\{(A_\nu^-(\cdot), b_\nu^-(\cdot)), \nu = 1, \ldots\} \text{ increasing}$$

and

$$\{A_\nu^+(\cdot), b_\nu^+(\cdot), \nu = 1, \ldots\} \text{ decreasing},$$

both sequences converging uniformly to $(A(\cdot), b(\cdot))$. Relying on the results for the almost sure convergence of measurable multifunctions and the properties of perturbed polyhedra, G. Salinetti [50] proves the following:

(3.4) **Theorem.** *Suppose $A(\cdot)$ and $b(\cdot)$ satisfy conditions (3.1)–(3.3) and the sequences $\{(A_\nu^+(\cdot), b_\nu^+(\cdot)), \nu = 1, \ldots\}$ and $\{(A_\nu^-(\cdot), b_\nu^-(\cdot)), \nu = 1, \ldots\}$ are constructed to have the monotonicity and uniform convergence properties indicated here above. Then the sets defined by*

$$K_{1\nu}^- = \{x \geq 0 \mid P[A_\nu^-(\omega)x \geq b_\nu^+(\omega)] \geq \alpha°\}$$

and

$$K_{1\nu}^+ = \{x \geq 0 \mid P[A_\nu^+(\omega)x \geq b_\nu^-(\omega)] \geq \alpha°\}$$

determine monotonic sequences of sets, with the $K_{1\nu}^-$ converging from below and the $K_{1\nu}^+$ converging from above to K_1.

Recall that a sequence of subsets $\{S_\nu, \nu = 1, \ldots\}$ of R^n is said to converge to a set S if

$$S = \{\lim x_\nu \mid x_\nu \in S_\nu \text{ for all } \nu\}$$
$$= \{\lim x_k \mid x_k \in S_{\nu_k} \text{ for all } k, \text{ for some subsequence } \{\nu_k\}\}.$$

The efficiency of this approximation scheme depends clearly on the choice of the discretizations but also, presuming that these approximations are part of an overall iterative procedure, on the possibility of using already available bases to simplify subsequent calculations. At this time there are no computational results available that allow us to verify the practicality of this approximation scheme.

Approximation of the recourse problem, more specifically the function $\mathcal{Q}$, in particular through discretization has been extensively studied, in particular by K. Marti [57], [45], and W. Römisch [58] in addition to P. Kall and P. Olsen already mentioned earlier. Following another line of attack B. Van Cutsem [59] initiated the use of set-convergence to study the convergence of the solutions of stochastic linear programming. Eventually this, as well as developments in many other areas of Nonlinear Analysis, led to the theory of epi-convergence which provides a unifying framework for the approximation of optimization problems.

Let $\{f; f^v; v=1,\ldots\}$ be a collection of lower semicontinuous functions defined on R^n and with values in $\bar{R}=[-\infty, \infty]$. The sequence $\{f^v, v=1,\ldots\}$ is said to *epi-converge* to f if for all $x \in R^n$, we have

(3.5) $\liminf_{v \to \infty} f^v(x_v) \geq f(x)$ for all $\{x_v, v=1,\ldots\}$ converging to x,

and

(3.6) there exists $\{x_v, v=1,\ldots\}$ converging to x such that
$\limsup_{v \to \infty} f^v(x_v) \leq f(x)$.

It is easy to verify that (3.5) actually implies

$$\liminf_{k \to \infty} f^{v_k}(x_k) \geq f(x)$$

for any subsequence of functions $\{f^{v_k}, k=1,\ldots\}$ and sequence $\{x_k, k=1,\ldots\}$ converging to x. The name epi-convergence comes from the fact that the f_v epi-converge to f if and only if the sets epi f_v converge to epi f, where epi h is the epigraph of the function h,

$$\operatorname{epi} h = \{(x, \alpha) \in R^{n+1} \mid \alpha \geq h(x)\}.$$

Our interest in epi-convergence stems from the following properties [60]:

(3.7) **Theorem.** *Suppose a sequence of lower semicontinuous functions $\{f^v, v=1,\ldots\}$ epi-converges to f. Then if for some sequence $\{f^{v_k}, k=1,\ldots\}$*

$$x_k \in \operatorname{argmin} f^{v_k} = \{x \mid f^{v_k}(x) \leq \inf f^{v_k}\}$$

and $x = \lim_{k \to \infty} x_k$ it follows that

$$x \in \operatorname{argmin} f,$$

and $\inf f = \lim_{k \to \infty} \inf f^{v_k}$. Moreover, if $\operatorname{argmin} f \neq \emptyset$ and $\inf f$ is finite, then $\inf f = \lim_{v \to \infty} \inf f^v$ if and only if

$$x \in \operatorname{argmin} f$$

implies that there exist sequences $\{\varepsilon_v \geq 0, v=1,\ldots\}$ with $\lim_{v \to \infty} \varepsilon_v = 0$ and $\{x_v, v=1,\ldots\}$ converging to x such that for all v

$$x_v \in \varepsilon_v\text{-argmin} f^v = \{x \mid f^v(x) \leq \inf f^v + \varepsilon_v\}.$$

To use this in the context of stochastic programming, recall that the function $\mathcal{Q}$ is given by the following expression

$$\mathcal{Q}(x) = E\{Q(x, \omega)\} = \int Q(x, \omega) P(d\omega)$$

and approximating the probability distribution P by P_ν yields the function $\mathcal{Q}^\nu$ defined by

(3.8) $$\mathcal{Q}^\nu(x) = \int Q(x, \omega) P_\nu(d\omega).$$

In what follows we take the P_ν and P to be distributions defined on the sample space of $(q(\cdot), p(\cdot), T(\cdot))$ and identify ω with $(q(\omega), p(\omega), T(\omega))$. In order to bypass some technical difficulties we shall assume that the support of the P_ν and P are contained in a bounded set S, certainly not a significant practical restriction.

(3.9) **Theorem.** *Suppose $\{P_\nu, \nu = 1, \ldots\}$ is a sequence of probability measures that converge in distribution to P (= weak convergence). Then for all $x \in K_2$ the functions $\{\mathcal{Q}^\nu, \nu = 1, \ldots\}$ epi-converge to $\mathcal{Q}$. Among other things, it follows that for all ν, x_ν is an optimal solution to the problem:*

find $x \in K$ that minimizes $cx + \mathcal{Q}^\nu(x)$

and x^* is a cluster point of the sequence $\{x_\nu, \nu = 1, \ldots\}$ then x^* solves:

find $x \in K$ that minimizes $cx + \mathcal{Q}(x)$.

Proof. For any $x \in K_2$, the function

$$\omega \mapsto Q(x, \omega)$$

is continuous [13, Proposition 7.5] and thus also bounded on S, from which we have that for every $x \in K_2$,

$$\lim \mathcal{Q}^\nu(x) = \mathcal{Q}(x),$$

as follows from the theory of weak convergence of probability measures, cf. Portemanteau theorem. Thus condition (3.6) for epi-convergence is fulfilled.

The function $x \mapsto Q(x, \omega)$ is (finite) Lipschitz on K_2. Thus with Lipschitz constant $L(\omega)$, we have

$$|Q(x_1, \omega) - Q(x_2, \omega)| \leq L(\omega) \cdot \text{dist}(x_1, x_2)$$

for any pair x_1, x_2 in K_2. Actually $L(\omega)$ can be chosen independent of ω [13, Proof of Theorem 7.7]. Let $\{x_\nu, \nu = 1, \ldots\}$ be any sequence of points in K_2 that converges to $x \in K_2$. We get

$$Q(x, \omega) - L \cdot \text{dist}(x, x_\nu) \leq Q(x_\nu, \omega).$$

Integrating both sides with respect to P_ν yields

$$\mathcal{Q}^\nu(x) - L \cdot \text{dist}(x, x_\nu) \leq \mathcal{Q}^\nu(x_\nu).$$

II. Stochastic Programming: Solution Techniques and Approximation Schemes

and hence

$$\mathcal{Q}(x) = \lim_{\nu \to \infty} \mathcal{Q}^\nu(x) - L \cdot \lim_{\nu \to \infty} \text{dist}(x, x_\nu)$$

$$= \liminf_{\nu \to \infty} (\mathcal{Q}^\nu(x) - L \cdot \text{dist}(x, x_\nu))$$

$$\leq \liminf_{\nu \to \infty} \mathcal{Q}^\nu(x_\nu),$$

which gives us (3.5) and thus completes the proof of the epi-convergence of the functions $\mathcal{Q}^\nu$ to $\mathcal{Q}$. The remaining assertions directly follows from the definition – the epi-convergence of the $\mathcal{Q}^\nu$ to $\mathcal{Q}$, implies the epi-convergence of $c \cdot + \mathcal{Q}^\nu$ to $c \cdot + \mathcal{Q}$ – and Theorem 3.7. □

From the Lipschitz bound used in the proof of Theorem 3.9, it is actually possible to obtain an estimate of the rate of convergence. For example, P. Kall [48] shows that if only p and T are random, then

(3.10) $\quad |\mathcal{Q}(x) - \mathcal{Q}^\nu(x)| \leq \gamma \int \|p(\omega) - T(\omega)x\| \, |P - P_\nu(d\omega)|$

where $\|\cdot\|$ indicates the vector norm and

$$\gamma = \max[\det V | V \text{ is an invertible submatrix of } W].$$

This is somewhat better than the constant L that appears in the proof of the theorem. It allows us to compute an a priori bound, but in order to get a good approximation bound via (3.10) one needs a discretization with extremely fine mesh which would render the approximate problem (2.8) extemely large. This is why another approach is advocated.

Approximating a convex function f from R^n into R can be done in many ways, but if in addition we seek to obtain upper and lower bounds on the infimum of this function one is naturally led to proceed via outer- and inner-linearization of the function f. The infimum of outer- and inner-linearization providing respectively the desired lower and upper bounds. For $\mathcal{Q}$, the question is how to choose the sequence P_ν so that the approximations are of outer or inner type.

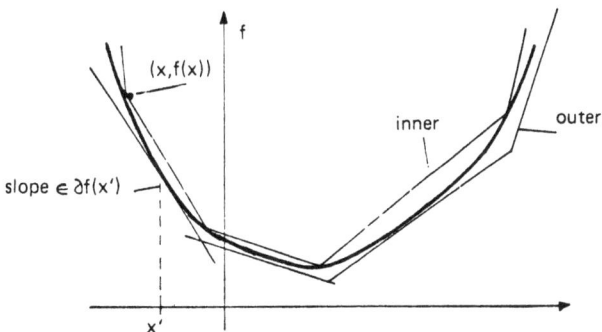

3.10 Figure: Inner/Outer Linearizations of f

We only consider the case when p and T are random and q is fixed. (The case q random must be dealt with separately, for the properties to exploit in that case, consult [13, Section 7].) Again, identifying ω with $(p(\omega), T(\omega))$, we have that the function

$$\omega \mapsto Q(x, \omega)$$

is a convex (polyhedral) function for all x [13, Proposition 7.5]. Let $\mathscr{S}^\nu = \{S_h^\nu, h = 1, \ldots, H^\nu\}$ be a finite partition of Ξ, the support of the distribution of $p(\cdot)$ and $T(\cdot)$, and define P_ν as follows:

$$P_\nu(S) = \sum_{\{h \mid \omega_h^\nu \in S\}} P(S_h^\nu),$$

where

$$\omega_h^\nu = (p_h^\nu, T_h^\nu) = E\{(p(\cdot), T(\cdot)) \mid (p(\omega), T(\omega)) \in S_h^\nu\}.$$

The distribution P_ν is thus a discrete distribution whose probability mass points are the conditional expectation of $(p(\cdot), T(\cdot))$ given S_h^ν for $h = 1, \ldots, H^\nu$.

3.11 Proposition. *Suppose the sequence of distribution $\{P_\nu, \nu = 1, \ldots\}$ are defined as here above through partitions $\{\mathscr{S}^\nu, \nu = 1, \ldots\}$ such that $\mathscr{S}^{\nu+1}$ is a refinement of $\mathscr{S}^\nu$, i.e.,*

$$\mathscr{S}^{\nu+1} \supset \mathscr{S}^\nu.$$

Then $\{\mathcal{Q}^\nu, \nu = 1, \ldots\}$ is a sequence of monotone increasing functions such that for all ν

$$\mathcal{Q}^\nu \leq \mathcal{Q}.$$

Proof. The result follows from the convexity of $Q(x, \cdot)$ through Jensen's inequality. Indeed we always have

$$\int_{S_h^\nu} Q(x, \omega) P_\nu(d\omega) = Q(x, \omega_h^\nu) \leq \int_{S_h^\nu} Q(x, \omega) P(d\omega)$$

from which we get

$$\mathcal{Q}^\nu(x) = \sum_{h=1}^{H_\nu} Q(x, \omega_h^\nu) P(S_h^\nu) \leq \sum_h \int_{S_h^\nu} Q(x, \omega) P(d\omega) = \mathcal{Q}(x)$$

To see that $\mathcal{Q}^\nu \leq \mathcal{Q}^{\nu+1}$ simply repeat the argument using again the convexity of $Q(x, \cdot)$ to obtain

$$Q(x, \omega_h^\nu) \leq \sum_{\{k \mid S_k^{\nu+1} \subset S_h^\nu\}} Q(x, \omega_k^{\nu+1}) \frac{P(S_k^{\nu+1})}{P(S_h^\nu)}. \quad \square$$

Clearly if the approximating functions are defined as in Proposition 3.11 such that the P_ν converge in distribution to P and there is a bounded sequence $\{x_\nu, \nu = 1, \ldots\}$ such that

$$x_\nu \in \arg\min_{x \in K} cx + \mathcal{Q}^\nu(x),$$

then any cluster point x^* of the sequence solves the problem

$$\text{find } x \in K \text{ that minimizes } cx + \mathcal{Q}(x).$$

Moreover, the sequence $\{cx_v + \mathcal{Q}(x_v) = z_v, v = 1, \ldots\}$ is monotone and

$$\lim_{v \to \infty} z_v = \inf_{x \in K} (cx + \mathcal{Q}(x)).$$

All of this follows directly from Theorem 3.9 and Proposition (3.11). A. Mandansky [61] was first in using Jensen's inequality to obtain a lower bound for the infimum; see also [62] for a careful treatment of the nonlinear case. The use of conditional expectations to refine these bounds is due to P. Kall [48] and C. Huang, W. Ziemba and A. Ben-Tal [63].

Proposition (3.11) may seem to suggest that better lower bounds for the infimum of the stochastic program requires a global refinement of the partition of Ξ. But clearly it really suffices to choose the refinement so as to improve the approximation of $\mathcal{Q}$ in the neighborhood of the infimum. How to achieve this while relying only on a rough partition, is still an open question. The same question needs to be raised in the context of deriving upper bounds that we discuss next.

Upper bounds on $\mathcal{Q}$ are obtained through inner linearizations as indicated in Figure 3.10. The basic idea is the following. Again with $\omega = (p(\omega), T(\omega))$, we have that $\omega \mapsto Q(x, \omega)$ is convex, and finite valued on Ξ when $x \in K_2$. Let us suppose that Ξ is bounded and denoted by ext Ξ, the extreme points of its convex hull, con Ξ. Since $Q(x, \cdot)$ is convex on con Ξ which is bounded, it follows that its supremum is attained at some extreme point of con Ξ, say $e_x \in$ ext Ξ. We get

(3.13) $$\mathcal{Q}(x) = \int Q(x, \omega) P(d\omega) \leq Q(x, e_x).$$

If $x^1, \ldots, x^v$ are a collection of points in K_2 and $e^1, \ldots, e^v$ are the corresponding extreme points of Ξ that yield the preceding inequality, we get that for all $x \in \text{con}(x^1, \ldots, x^v)$, the convex hull of $x^1, \ldots, x^v$

(3.14) $$\mathcal{Q}(x) \leq \text{Min}\left[\sum_{l=1}^{v} \lambda_l Q(x^l, e^l) \mid \sum_{\lambda l = 1}^{v}, \lambda_l \geq 0\right].$$

The bounds can be substantially improved by considering partitions of Ξ. Let $\mathcal{S} = \{S_h, h = 1, \ldots, H\}$ be a partition of Ξ with $e_{x,h}$ the extreme point of the convex hull of S_h at which $Q(x, \cdot)$ attains its supremum. Then, for $x \in K_2$

(3.15) $$\mathcal{Q}(x) \leq \sum_{h=1}^{H} Q(x, e_{x,h}) P(S_h)$$

Using this bound rather than (3.13) naturally yields an improved version of (3.14).

These bounds, due to P. Kall and D. Stoyan [35], can be much sharpened when the function Q has separability properties. Once again we start with the fact that $Q(x, \cdot)$ is convex to obtain

$$Q(x, \omega) \leq \int_{\text{ext } \Xi} Q(x, e) \mu_\omega(de)$$

where μ_ω is a probability measure on ext Ξ such that

$$\int e \mu_\omega(de) = \omega,$$

i.e., such that the convex combination (generated by μ_ω) yields ω. Thus, we have

(3.16) $$\mathcal{Q}(x) \leq \int_\Xi \int_{\mathrm{ext}\,\Xi} Q(x, e) \mu_\omega(de) P(d\omega).$$

The problem with this bound, generally much tighter than (3.14), or even the improved version resulting from (3.15), is that it is usually quite difficult to find a manageable expression for μ_ω as a function of ω. Expect if, for example,

$$Q(x, \omega) = \sum_{i=1}^{M} Q_i(x, \omega_i),$$

where for all $i = 1, \ldots, M$

$$\omega_i \mapsto Q_i(x, \omega_i) \colon R \to R.$$

Then

$$\int Q(x, \omega) P(d\omega) = \sum_{i=1}^{M} \int Q_i(x, \omega_i) P_i(d\omega_i)$$

where P_i is the marginal distribution of ω_i. We can find an expression for the bound (3.16) by obtaining bounds for each $Q_i(x, \cdot)$ separately. Let α_i and β_i respectively be the upper and lower bounds of the support of ω_i, recall that Ξ was assumed to be bounded. By convexity of $Q_i(x, \cdot)$ we have for every $\omega_i \in [\alpha_i, \beta_i]$

(3.17) $$Q_i(x, \omega_i) \leq (1 - \lambda_{\omega_i}) Q_i(x, \alpha_i) + \lambda_{\omega_i} Q_i(x, \beta_i)$$

where $\lambda_{\omega_i} = (\omega_i - \alpha_i)/(\beta_i - \alpha_i)$. Integrating the above on both sides with respect to P_i, and summing over i, we get

(3.18) $$\mathcal{Q}(x) \leq \sum_{i=1}^{M} \left[\frac{\beta_i - E\omega_i}{\beta_i - \alpha_i} Q_i(x, \alpha_i) + \frac{E\omega_i - \alpha_i}{\beta_i - \alpha_i} Q_i(x, \beta_i) \right].$$

In stochastic programming one refers to this inequality as the *Edmundson-Madansky* inequality. A refinement of this bound can be obtained by breaking up $[\alpha_i, \beta_i]$ into subintervals, say $[\alpha_i^k, \beta_i^k]$ and for each one rewriting (3.17) using the extreme points of the subinterval. With m_{ik} denoting the conditional expectation of ω_i when $\omega_i \in [\alpha_i^k, \beta_i^k)$ and $P_{ik} = P_i[\alpha_i^k, \beta_i^k)$, by integrating and summing we get

(3.19) $$\mathcal{Q}(x) \leq \sum_i \sum_k P_{ik} \left[\frac{\beta_i^k - m_{ik}}{\beta_i^k - \alpha_i^k} Q_i(x, \alpha_i^k) + \frac{m_{ik} - \alpha_i^k}{\beta_i^k - \alpha_i^k} Q_i(x, \beta_i^k) \right].$$

A. Madansky [61] was the first to suggest the use of (3.18), the refinement is due to C. Huang, I. Vertinsky and W. Ziemba [64], see also [35] where the connection with the theory of partial ordering of distribution functions is exhibited. Another way to obtain these inequalities, which loads them with a rich interpretation, is through the minimax approach to stochastic programming investigated first by M. Isofescu and R. Theodorescu [65] and developed by J. Dupačova [66], [67] where stochastic programs are viewed as games against nature: the inf is with respect to x and the sup with respect to a given class of distribution functions. She obtains (3.18) as the result of considering for the class of distributions those satisfying given moment conditions [68].

If the function $\omega \mapsto Q(x, \omega)$ is concave, which would be the case if only q is random. Then inequality (3.19) and that of Proposition (3.11) are simply reversed.

The preceding results yields basically a priori bounds, but they can also be exploited in the design of algorithmic procedures using the points generated by the algorithm to construct partitioning schemes, ... Other bounds that can be exploited in various situations, a posteriori bounds have been suggested by A. Williams [69], K. Marti [45] and J. Birge [70], cf. [71] for a recent compilation as well as further developments.

Acknowledgement

The sections on computational procedures and approximation bounds have profited significantly from extended discussions with John Birge (University of Michigan) and Larry Nazareth (IIASA). In writing the introduction to stochastic quasigradient methods, I was greatly aided by the comments of Yuri Ermoliev (Institute of Cybernetics in Kiev, and IIASA) and Georg Pflug (Universities of Vienna and Giessen).

References

[1] E. M. Beale, On minimizing a convex function subject to linear inequalities, *J. Royal Stat. Soc.* 17B (1955), 173–184.
[2] G. B. Dantzig, Linear programming under uncertainty, *Management Sci.* 1 (1955), 197–206.
[3] G. Tintner, Stochastic linear programming with applications to agricultural economics, in *Proc. Second Symposium in Linear Programming*, ed. H. A. Antosiewicz, Washington, 1955, 197–207.
[4] A. Charnes and W. W. Cooper, Chance-constrained programming, *Management Sci.* 5 (1959), 73–79.
[5] R. Grinold, A class of constrained linear control problems with stochastic coefficients, in *Stochastic Programming*, ed. M. Dempster, Academic Press, London, 1980, 97–108.
[6] W. Klein Haneveld, A dual of a dynamic inventory control model: the deterministic and stochastic case, in *Recent Results in Stochastic Programming*, ed. P. Kall and A. Prékopa, Springer Verlag Lecture Notes in Economics and Mathematical Systems, 179 (1980), 67–98.
[7] J. Dupačová, Water resource systems using stochastic programming with recourse, in *Recent Results in Stochastic Programming*, ed. P. Kall and A. Prékopa, Springer-Verlag Lecture Notes in Economics and Mathematical Systems, 179 (1980), 121–134.

[8] K. Back, Optimality and equilibrium in infinite horizon economies under uncertainty, Ph. D. thesis, University of Kentucky, Lexington, 1982.
[9] R. Everitt and W. T. Ziemba, Two-period stochastic programmes with simple recourse, *Operations Research* 27 (1979), 485-502.
[10] M. Dempster, *Stochastic Programming*, Academic Press, London, 1980.
[11] D. Walkup and R. J.-B. Wets, Stochastic programs with recourse, *SIAM J. Appl. Math.* 15 (1967), 316-339.
[12] P. Olsen, Multistage stochastic programming with recourse: the equivalent deterministic program, *SIAM J. Control Optim.* 14 (1976), 495-517.
[13] R. Wets, Stochastic programs with fixed recourse: the equivalent deterministic program, *SIAM Review* 16 (1974), 309-339.
[14] P. Kall and W. Oettli, Measurability theorems for stochastic extremals, *SIAM J. Control Optim.* 13 (1975), 994-998.
[15] R. Wets, Induced constraints for stochastic optimization problems, in *Techniques of Optimization*, ed. A. Balakrishnan, Academic Press, London, 1972, 433-443.
[16] A. Prékopa, Logarithmic concave measures with applications to stochastic programming, *Acta Sci. Math.* 32 (1971), 301-316.
[17] A. Prékopa, Logarithmic concave measures and related topics, in *Stochastic Programming*, ed. M. Dempster, Academic Press, London, 1980, 63-82.
[18] J. G. Kallberg and W. T. Ziemba, Generalized concave functions in stochastic programming and portfolio theory, in *Generalized Concavity in Optimization and Economics*, ed. S. Schaible and W. Ziemba, Academic Press, New York, 1981, 719-767.
[19] Chr. Borell, Convex set functions in d-spaces, *Period. Math. Hungar.* 6 (1975), 111-136.
[20] C. Van de Panne and W. Popp, Minimum-cost cattle feed under probabilistic protein constraints, *Management Sci.* 9 (1963), 405-430.
[21] A. Prékopa, Programming under probabilistic constraints with a random technology matrix, *Math. Operationsforsch. Statist.* 5 (1974), 109-116.
[22] S. M. Sinha, Stochastic Programming, Doc. Thesis, University of California-Berkeley, 1963; see also *Proceed. Symposium on Probability and Statistics*, Ben Hindu University, India.
[23] R. T. Rockafellar, Integrals which are convex functionals II, *Pacific J. Mathematics* 39 (1971), 439-469.
[24] R. T. Rockafellar and R. Wets, On the interchange of subdifferentiation and conditional expectation for convex functionals, *Stochastics* 7 (1982), 173-182.
[25] D. Walkup and R. Wets, Stochastic programs with recourse: special forms, in *Proceedings of the Princeton Symposium on Mathematical Programming*, ed. H. Kuhn, Princeton University Press, Princeton, 1970, 139-162.
[26] A. Prékopa, Network planning using two-stage programming under uncertainty, in *Recent Results in Stochastic Programming*, eds. P. Kall and A. Prékopa, Springer-Verlag Lecture Notes in Economics and Mathematical Systems, Vol. 179, 1980, 215-237.
[27] A. Prékopa, S. Ganczer, I. Deák and K. Patyi, The STABIL stochastic programming model and its experimental application to the electrical energy sector of the Hungarian economy, in *Stochastic Programming*, ed. M. Dempster, Academic Press, London, 1980, 369-385.
[28] R. Wets, Programming under uncertainty: the equivalent convex program, *SIAM J. Appl. Math.* 14 (1966), 89-105.
[29] B. Hansotia, Stochastic linear programs with simple recourse: the equivalent deterministic convex program for the normal, exponential and Erlang cases, *Naval Res. Logist. Quat.* 27 (1980), 257-272.

[30] L. Nazareth and R. Wets, Algorithms for stochastic programs: the case of nonstochastic tenders, IIASA Working Paper, Laxenburg, Austria, 1982.

[31] F. Louveaux, A solution method for multistage stochastic programs with recourse with application to an energy investment problem, *Operations Res.* 28 (1980), 889–902.

[32] J. Birge, Decomposition and partitioning methods for multistage stochastic linear programs, Tech. Report 82-6, Dept. Industrial and Operations Engineering, University of Michigan, 1982.

[33] B. Strazicky, Some results concerning an algorithm for the discrete recourse problem in *Stochastic Programming*, ed. M. Dempster, Academic Press, London, 1980, 263–274.

[34] P. Kall, Computational methods for solving two-stage stochastic linear programming problems, *Z. Angew. Math. Phys.* 30 (1979), 261–271; see also *Large-Scale Linear Programming*, eds. G. Dantzig, M. Dempster and M. Kallio, IIASA Collaborative Proceedings Series, Laxenburg, Austria, 1981, 287–298.

[35] P. Kall and D. Stoyan, Solving stochastic programming problems with recourse including error bounds, *Math. Operationsforsch. Statist. Ser. Optimization* (1982).

[36] G. Dantzig and A. Madansky, On the solution of two-stage linear programs under uncertainty, *Proc. Fourth Berkeley Symposium on Mathematical Statistics and Probability, Vol. 1*, University of California Press, Berkeley, 1961, 165–176.

[37] R. Van Slyke and R. Wets, L-shaped linear programs with applications to optimal control and stochastic programming, *SIAM J. Appl. Math.* 17 (1969), 638–663.

[38] J. Birge, Solution methods for stochastic dynamic linear programs, Tech. Report SOL 80-29, Systems Optimization Laboratory, Stanford University, 1980.

[39] S. Gartska and D. Ruthenberg, Computation in discrete stochastic programs with recourse, *Operations Res.* 21 (1973), 112–122.

[40] R. Wets, Characterization theorems for stochastic programs, *Math. Programming* 2 (1972), 166–175.

[41] Y. Ermoliev, Stochastic quasigradient methods and their application in systems optimization, IIASA Working Paper, WP-81-2, Laxenburg, Austria, 1981; to appear in *Stochastics*.

[42] Y. Ermoliev, *Stochastic Programming Methods* (in Russian), in Nauka, Moscow 1976.

[43] Y. Ermoliev and E. Nurminski, Stochastic quasigradient algorithms for minimax problems in stochastic programming, in *Stochastic Programming*, ed. M. Dempster, Academic Press, London, 1980, 275–285.

[44] B. Poljak, Nonlinear programming methods in the presence of noise, *Math. Programming* 14 (1978), 87–97.

[45] K. Marti, *Approximationen Stochastischer Optimierungsprobleme*, Verlag Anton Hain, Königstein, 1979.

[46] Y. Ermoliev, G. Leonardi and J. Vira, The stochastic quasigradient method applied to a facility location problem, IIASA Working Paper, WP-81-14, Laxenburg, Austria, 1981.

[47] J. Dodu, M. Goursat, A. Hertz, J.-P. Quadrat and M. Viot, Méthodes de gradient stochastique pour l'optimisation des investissements dans un résau électrique, *E. D. F. Bulletin-Série C,* (1981), No. 2, 133–164.

[48] P. Kall, Approximations to stochastic programs with complete fixed recourse, *Numer. Math.* 22 (1974), 333–339.

[49] P. Olsen, Discretizations of multistage stochastic programming problems, *Mathematical Programming Study* 6 (1976), 111–124; also Ph. D. Thesis, Cornell Univ., 1974.

[50] G. Salinetti, Approximations for chance-constrained programming problems, Tech. Report No. 13, Istituto della Probabilitá, Univ. Roma, 1981; to appear in *Stochastics* (1983).

[51] R. Wets, Solving stochastic programs with simple recourse II, in *Proceed. 1975 Conference on Information Sciences and Systems,* The Johns Hopkins University, Baltimore, 1975.

[52] A. Dexter, J. Yu and W. Ziemba, Portfolio selection in a lognormal market when the investor has a power utility function: computational results, in *Stochastic Programming,* ed. M. Dempster, Academic Press, London, 1980, 507–523.

[53] R. Grinold, A new approach to multistage stochastic linear programs, *Mathematical Programming Study* 6 (1976), 19–29.

[54] E. Beale, J. Forrest and C. Taylor, Multi-time-period stochastic programming, in *Stochastic Programming,* ed. M. Dempster, Academic Press, London, 1980, 387–402.

[55] M. Queyranne and E. Kao, Aggregation in a two-stage stochastic program for manpower planning in the service sector, Tech. Report, University of Houston, 1981.

[56] E. G. Read, Approaches to stochastic reservoir modelling, Working Paper No. 153, College Business Admin., University of Tennessee, 1982.

[57] K. Marti, Approximationen von Entscheidungsproblemen mit linearer Ergebnisfunktion und positiv homogener, subadditiver Verlustfunktion, *Z. Wahrscheinlichkeitstheorie Verw. Gebiete* 31 (1975), 203–233.

[58] W. Römisch, On discrete approximation in stochastic programming, *Proceed. 13. Jahrestagung Mathematische Optimierung,* Vitte 1981, Humboldt-Univ., Berlin, Sektion Mathematik, Seminarbericht 39, 1981.

[59] B. Van Cutsem, Problems of convergence in stochastic linear programming, in *Techniques of Optimization,* ed. A. Balakrishnan, Academic Press, New York, 1972, 445–454.

[60] H. Attouch and R. Wets, Convergence and approximation in nonlinear optimization, in *Nonlinear Programming 4,* eds. O. Mangasarian, R. Meyer and S. Robinson, Academic Press, New York, 1981, 367–394.

[61] A. Madansky, Inequalities for stochastic linear programming problems, *Management Sci.* 6 (1960), 197–204.

[62] R. Hartley, Inequalities in completely convex stochastic programming, *J. Math. Anal. Applic.* 75 (1980), 373–384.

[63] C. Huang, W. Ziemba and A. Ben-Tal, Bounds on the expectation of a convex function of a random variable: with application to stochastic programming, *Operations Res.* 25 (1977), 315–325.

[64] C. Huang, I. Vertinsky and W. Ziemba, Sharp bounds on the value of perfect information, *Operations Res.* 25 (1977), 128–139.

[65] M. Iosifescu and R. Theodorescu, Linear programming under uncertainty, in *Colloquium on Applications of Mathematics to Economics, Budapest, 1963,* ed. A. Prékopa, Publishing House Hungarian Academy of Sciences, Budapest, 1965, 133–140.

[66] J. Žačková, On minimax solutions of stochastic linear programming problems, *Casopis Pest. Mat.* 91 (1966), 423–429.

[67] J. Dupačova, On minimax decision rule in stochastic linear programming, *Math. Methods Oper. Res.* 1 (1980), 47–60.

[68] J. Dupačova, Minimax approach to stochastic linear programming and the moment problem, (in Czech), *Ekonomickomaticky Obzor* 13 (1977), 279–307. Summary of results in *ZAMM* 58 (1978), T466–T467.

[69] A. Williams, Approximation formulas for stochastic linear programming, *SIAM J. Appl. Math.* 14 (1966), 668–677.
[70] J. Birge, The value of the stochastic solution in stochastic linear programs with fixed recourse, *Math. Programming*, 1983.
[71] J. Birge and R. Wets, On initial solutions and approximation schemes for stochastic programs with recourse, IIASA Worling Paper, Laxenburg, Austria, 1983.

III. Scientific Program

The program is divided into regular sessions and state-of-the-art tutorials. The regular sessions consist of three 25-minute talks separated by 5-minute breaks for discussion and the change of rooms. The state-of-the-art tutorials consist of a one-hour talk followed by a 15-minute break.

Opening Session
Monday, 9.00, Aula of the University

MO/C Aula of the University
Opening Speech / Festvortrag
Chairman: J. Abadie, France
11.15–12.15 1.1 G. B. Dantzig, USA
Reminiscences about the Origins of Linear Programming

MO/D.01
Unconstrained Optimization I
Chairman: M. J. D. Powell, United Kingdom
14.00–14.25 1.1* R. B. Schnabel, USA
An Indefinite Dogleg Method for Unconstrained Minimization
14.30–14.55 1.2* E. Spedicato, Italy; H. Kleinmichel, DDR
A Class of Rank-One Positive Quasi-Newton Algorithms for Unconstrained Minimization
15.00–15.25 1.3* P. L. Toint, Belgium; A. Griewank, USA
Recent Results on the Optimization of Partially Separable Functions

MO/D.02
Network Problems with Nonlinear Objective Function
Chairman: F. Glover, USA
14.00–14.25 2.1* R. R. Meyer, USA
Nonlinear Networks
14.30–14.55 2.2* R. E. Erickson, USA; C. L. Monma, USA; A. F. Veinott, USA
Minimum-Concave-Cost Flows in Networks via Dynamic Programming
15.00–15.25 2.3 M. Trojanowski, Poland
An Efficient Algorithm Solving Network Problems with Concave Objective Function

MO/D.03
New Developments in Mathematical Programming
Chairman: J. M. Mulvey, USA
14.30–14.55 3.2* R. S. Dembo, USA
   Progress in Large Scale Nonlinear Optimization
15.00–15.25 3.3* L. C. W. Dixon, United Kingdom; K. D. Patel, United Kingdom; P. G. Ducksbury, United Kingdom
   The Place of Parallel Computation in Numerical Optimization

MO/D.04
Applications of Integer Programming I
Chairman: M. Guignard, USA
14.00–14.25 4.1* H. M. Salkin, USA; K. Mathur, USA; K. Nishimura, USA
   Application of Integer Programming to Radio Frequency Management
14.30–14.55 4.2 M. Nakamori, Japan
   Disjunctive Network Flow Problems in Automatic Design
15.00–15.25 4.3 J. A. M. Schreuder, Netherlands
   Timetable Last Two Years of Dutch Modern High Schools

MO/D.05
Matching I
Chairman: W. R. Pulleyblank, Canada
14.00–14.25 5.1 J. J. Green-Krotki, Canada; W. H. Cunningham, Canada
   A Reduction Proof of the General Matching Polyhedron Theorem
14.30–14.55 5.2 U. Derigs, W. Germany
   Shortest Augmenting Paths and Sensitivity Analysis for Optimal Matching
15.00–15.25 5.3* K. G. Murty, USA; C. Perin, Brazil
   Parametric 1-Matching/Covering Algorithms

MO/D.06
Multicriteria Integer Programming
Chairman: D. Granot, Canada
14.30–14.55 6.2* I. Dragan, USA
   The Generalized Nucleolus and Some Multicriteria Discrete Programming Problems

MO/D.07
Results in Nondifferentiable Programming
Chairman: J.-B. Hiriart-Urruty, France
14.00–14.25 7.1* W. Oettli, W. Germany
   Epsilon-Solutions and Epsilon-Supports
15.00–15.25 7.3 R. W. Chaney, USA
   A General Sufficiency Theorem for Nonsmooth Nonlinear Programming

MO/D.08
Complexity of Linear Programming Problems
Chairman: K.-H. Borgwardt, W. Germany
14.00–14.25   8.1   K. G. Murty, USA; R. Chandrasekharan, USA
Some NP-Complete Problems in Linear Programming
14.30–14.55   8.2   S. Olafsson, Sweden
On the Lengths of Simplex Paths
15.00–15.25   8.3   M. E. Dyer, United Kingdom
Optimal Algorithms for LP in 2 or 3 Variables

MO/D.09
Good Algorithms
Chairman: R. Van Slyke, USA
14.00–14.25   9.1*   C. P. Schnorr, W. Germany
Efficient Algorithms for Permutation Groups and Graph Isomorphism
14.30–14.55   9.2*   R. G. Bland, USA; J. Edmonds, Canada
Fast Primal Algorithms for Totally Unimodular Linear Programming
15.00–15.25   9.3*   A. Schrijver, Netherlands
Algorithms for Making Digraphs Strongly Connected

MO/D.10
Stability Problems
Chairman: J. Zowe, W. Germany
14.00–14.25   10.1*   S. Zlobec, Canada
Stable Linear and Convex Programming
14.30–14.55   10.2   J. C. Dunn, USA; E. W. Sachs, USA
The Effect of Perturbations on the Convergence Rates of Optimization Algorithms
15.00–15.25   10.3   T. Zolezzi, Italy
On the Stability Analysis in Mathematical Programming

MO/D.11
Global Optimization I
Chairman: J. E. Spingarn, USA
14.00–14.25   11.1*   J. B. Rosen, USA
Parametric Global Minimization for Large Scale Problems
14.30–14.55   11.2*   K. L. Hoffman, USA; J. E. Falk, USA
Concave Minimization Via Collapsing Polytopes
15.00–15.25   11.3   S. J. Grotzinger, USA
L1-Supports and Global Optimization

MO/D.12
Manipulating Matrices
Chairman: E. Loute, Belgium
14.30–14.55   12.2   G. van der Hoek, Netherlands; R. Th. Wijmenga, Netherlands
Linearly Dependent Active Constraint Normals in Non-linear Programming

15.00–15.25 12.3 J. L. Houston, USA
           On the Computation of Sparse Matrices

MO/E.01
State-of-the-Art Tutorial
Chairman: A. Orden, USA
15.30–16.30 1.1 S. Smale, USA
           On the Average Speed of the Simplex Method of Linear Programming

MO/E.10
State-of-the-Art Tutorial
Chairman: R. M. Karp, USA
15.30–16.30 10.1 R. L. Graham, USA
           The FKG Inequality and its Relatives

MO/F.01
Unconstrained Optimization II
Chairman: E. Spedicato, Italy
16.45–17.10 1.1* M. J. D. Powell, United Kingdom; R. P. Ge, United Kingdom
           The Convergence of Variable Metric Methods in Unconstrained Optimization
17.15–17.40 1.2* J. C. Fiorot, France; P. Jeannin, France
           A Broader Class of Parametric Variable Metric Formulae

MO/F.02
Network Algorithms
Chairman: C. Sandi, Italy
16.45–17.10 2.1 J. G. Klincewicz, USA; W. H. Cunningham, Canada
           On Cycling in the Network Simplex Method
17.15–17.40 2.2 V. Valls, Spain; N. Christofides, United Kingdom
           Finding All Optimal Solutions to the Network Flow Problem
17.45–18.10 2.3 A. A. Oliveira, Brazil; C. C. Gonzaga, Brazil
           Convergence Bounds for the "Capacity" Max-Flow Algorithm

MO/F.03
Methods for Differential Equations
Chairman: R. Stone, USA
17.15–17.40 3.2 F. di Guglielmo, France
           Invariant Solution for Multivalued Delay Differential Equations
17.45–18.10 3.3 F. di Guglielmo, France
           Approximation Methods for a Second Order Boundary Value Problem with Multivalued Boundary Conditions

MO/F.04
Applications of Large-Scale Pure and Mixed IP
Chairman: K. Spielberg, USA
16.45–17.10 4.1* V. Chandru, USA; J. F. Shapiro, USA
           Parametric Mixed Integer Programming by Inverse Optimization

III. Scientific Program — Monday, F.8

17.15–17.40	4.2*	P. Bod, Hungaria
		Experiences in Modelbuilding and Solution with "Alustrat"
17.45–18.10	4.3	M. M. Kostreva, USA; E.L. Johnson, USA; U. Suhl, USA
		Solving 0-1 Integer Programming Problems Arising from Large Scale Planning Models

MO/F.05
Matching II
Chairman: D. Naddef, France

16.45–17.10	5.1*	E. L. Lawler, USA; P. Tong, USA; V. V. Vazirani, USA
		The Weighted Parity Problem for Gammoids is Solvable by Graphic Matching
17.15–17.40	5.2	J. B. Orlin, USA
		Efficient Algorithms for the Dynamic Matching Problem and the Quasi-Fractional Matching Problem
17.45–18.10	5.3	J. Clausen, Denmark
		The Matroid Partition Algorithm for Bipartite Matching

MO/F.06
Aggregation Problems
Chairman: C. Monma, USA

16.45–17.10	6.1	J. R. Birge, USA
		Aggregation in Stochastic Linear Programming
17.15–17.40	6.2	F. Laisney, W. Germany; K. Ringwald, W. Germany
		Statistical Problems in W. D. Fisher's Method of Optimal Aggregation
17.45–18.10	6.3	D. G. Liesegang, W. Germany
		Aggregative Methods for Approximative Representation of Optimization Models

MO/F.07
NLP: Behavior of Solutions under Perturbations
Chairman: B. Bank, DDR

16.45–17.10	7.1*	S. M. Robinson, USA
		Dynamic Properties of Optimization Models
17.15–17.40	7.2*	A. Auslender, France
		Stability in Mathematical Programming with Non Differentiable Data
17.45–18.10	7.3	B. Cornet, Belgium
		Sensitivity Analysis in Mathematical Programming

MO/F.08
Randomized Linear Programming
Chairman: G. B. Dantzig, USA

16.45–17.10	8.1*	K. H. Borgwardt, W. Germany
		The Expected Number of Pivot Steps Required by the Simplex-method is Polynomial
17.15–17.40	8.2*	Th. M. Liebling, Switzerland
		Is the Randomized Simplex Method Polynomial?
17.45–18.10	8.3*	R. L. Smith, USA; J. H. May, USA
		Random Polytopes

MO/F.09
Polyhedral Combinatorics I
Chairman: L. Lovász, Hungary
16.45–17.10 9.1* E. Balas, USA; W. R. Pulleyblank, Canada
Projecting Combinatorial Polyhedra
17.15–17.40 9.2* T. Vanroy, Belgium; M. Padberg, Belgium; L. Wolsey, Belgium
F-Cuts for Fixed Charge Network Flow Problems
17.45–18.10 9.3 A. Frank, Hungary; E. Tardos, Hungary
An Algorithmic Relation Between Bounded and Unbounded Matroid Intersection Polyhedra

MO/F.10
Dual Problems and Gap Functions
Chairman: I. Singer, Romania
16.45–17.10 10.1* D. W. Hearn, USA
Dual and Saddle Functions Related to the Gap Function
17.15–17.40 10.2 M. Pappalardo, Italy
Generalized Lagrangeans and Duality Gap in Nonlinear Programming
17.45–18.10 10.3 J. E. Martinez-Legaz, Spain
Exact Quasiconvex Conjugation

MO/F.11
Global Optimization II
Chairman: K. L. Hoffman, USA
16.45–17.10 11.1 B. Betro, Italy
The Numerical Performance of a Probabilistic Global Optimization Algorithm

MO/F.12
Generalizations of Convexity
Chairman: B. Gollan, W. Germany
16.45–17.10 12.1* F. Nozicka, Czechoslovakia
On Convexity in Curved Spaces
17.15–17.40 12.2* K. H. Elster, DDR
Recent Results on Separation of not Necessary Convex Sets
17.45–18.10 12.3 J. P. Vial, Belgium
Strong and Weak Convexity of Sets and Functions

TU/A.01
NLP-Problems with Nonlinear Constraints I
Chairman: M. H. Wright, USA
8.15–8.40 1.1 J. Hersovits, Brazil
A Two-Stage Feasible Direction Algorithm for Nonlinear Constrained Optimization
8.45–9.10 1.2 R. H. Byrd, USA
Use of an Iterated Relaxed Quadratic Program for Nonlinearly Constrained Problems

9.15–9.40　　1.3*　L. J. Cromme, W. Germany
　　　　　　A Nonlinear Optimization Problem with Nondifferentiable Constraints

TU/A.02
Network Design
Chairman: F. Granot, USA
8.15–8.40　　2.1　R. T. Wong, USA
　　　　　　A Dual Ascent Framework for Network Design Algorithms
8.45–9.10　　2.2　G. Gallo, Italy; P. Carraresi, Italy
　　　　　　Lower Bounds for the Network Design Problem
9.15–9.40　　2.3　H. H. Hoang, Canada
　　　　　　Algorithms for Optimal Topologies of Computer Communications Networks

TU/A.03
Some Aspects of Mathematical Programming
Chairman: C. Fabian, Romania
9.15–9.40　　3.3　A. Vazquez-Muniz, Spain
　　　　　　Optimal Discriminant Functions using Mathematical Programming

TU/A.04
Applications of Integer Programming II
Chairman: I. Dragan, USA
8.15–8.40　　4.1　E. Melachrinoudis, USA; T. P. Cullinane, USA
　　　　　　Locating an Undesirable Facility within a Geographical Region Using the Maximin Criterion

TU/A.05
Aspects of Combinatorial Optimization
Chairman: C. P. Schnorr, W. Germany
8.45–9.10　　5.2　R. H. Möhring, W. Germany
　　　　　　On the Computational Complexity of Decomposing Relations and Clutters
9.15–9.40　　5.3*　N. Katoh, Japan; T. Ibaraki, Japan
　　　　　　On the Total Number of Pivots Required for Certain Parametric Combinatorial Optimization Problems

TU/A.06
Mathematical Programming in Production Planning
Chairman: H. Burley, Australia
8.15–8.40　　6.1　R. W. Ashford, United Kingdom
　　　　　　A Stochastic Programming Algorithm for Production Planning
8.45–9.10　　6.2　K. E. Stecke, USA
　　　　　　Nonlinear Mip Formulations of Several Production Planning Problems in Flexible Manufacturing Systems
9.15–9.40　　6.3　Ö. S. Benli, Turkey
　　　　　　A Decomposition Based Procedure for Production Planning in a Discrete Parts Manufacturing System

TU/A.07
Theory of Tangent Cones
Chairman: F. H. Clarke, Canada
8.15–8.40   7.1   D. H. Martin, South Africa
             Maximal Cores in the Theory of Tangent Cones
9.15–9.40   7.3*   M. Vlach, Czechoslovakia
             On Monotone Tangential Approximations

TU/A.08
Sensitivity and Stability of LP
Chairman: J. Telgen, Netherlands
8.15–8.40   8.1   R. E. Wendell, USA
             Sensitivity Analysis Revisited and Extended: The Tolerance Approach
8.45–9.10   8.2   L. Nazareth, Austria
             Numerical Behaviour of LP Algorithms Based Upon the Decomposition Principle
9.15–9.40   8.3   J. A. Nachlas, USA; D. E. Shapiro, USA
             Stability of Systems of Linear Programs

TU/A.09
Polyhedral Combinatorics II
Chairman: L. Matthews, United Kingdom
8.15–8.40   9.1   A. Volpentesta, Italy
             Row Circular Matrices and Related Polyhedra
8.45–9.10   9.2   R. Euler, W. Germany
             On Perfect Independence Systems
9.15–9.40   9.3   W. Cook, Canada
             Some Results in Total Dual Integrality

TU/A.10
Teaching of Linear Programming
Chairman: J. G. Ecker, USA
9.15–9.40   10.3* M. L. Kelmanson, Brazil; M. Bergman, Brazil
             S.A.G.C.E. – A Teaching Aid for Linear Programming

TU/A.11
Stochastic Methods for Global Optimization
Chairman: R. Barton, USA
8.15–8.40   11.1   F. Zirilli, Italy
             The Use of Stochastic Differential Equations in Global Optimization
8.45–9.10   11.2   G. T. Timmer, Netherlands; J. Ouwens, Netherlands; A. H. G. Rinnooy Kan, Netherlands
             A Fast Dynamic Clustering Method for Global Optimization
9.15–9.40   11.3* A. H. G. Rinnooy Kan, Netherlands; C. G. E. Boender, Netherlands; G. T. Timmer, Netherlands
             Stochastic Methods for Unconstrained and Constrained Global Optimization

TU/A.12
Convex Sets and LP-Problems
Chairman: G. d'Atri, Italy
8.45–9.10   12.2* T. Gal, W. Germany
On Degenerate Extreme Points of a Convex Polyhedron
9.15–9.40   12.3 E. C. Duesing, USA
Characterizing Optimal Solutions in Linear Programming Problems with Variable Rim Coefficients

TU/B.01
State-of-the-Art Tutorial
Chairman: J. B. Rosen, USA
9.45–10.45   1.1 J. Stoer, W. Germany
Solution of Large Linear Systems of Equations by Conjugate Gradient Type Methods

TU/B.10
State-of-the-Art Tutorial
Chairman: M. W. Padberg, Belgium
9.45–10.45   10.1 A. Schrijver, Netherlands
Min-Max Results in Combinatorial Optimization

TU/C.01
NLP-Problems with Nonlinear Constraints II
Chairman: L. J. Cromme, W. Germany
11.00–11.25   1.1* Ph. Gill, USA; W. Murray, USA; M. Saunders, USA; M. Wright, USA
Merit Functions and Active-Set Strategies in Nonlinearly Constrained Optimization
11.30–11.55   1.2 G. Patrizi, Italy
A Nonlinear Optimization Algorithm for General Constraints
12.00–12.25   1.3* K. Tanabe, Japan
A Unified Method for Designing Nonlinear Constrained Optimization Algorithms

TU/C.02
Network Problems with Side Constraints
Chairman: M. O. Ball, USA
11.00–11.25   2.1* M. D. Grigoriadis, USA
Additional Linear Constraints in Network Flow Problems
11.30–11.55   2.2 T. Kobayashi, Japan
The Lexico-Shortest Route Algorithm for Solving the Minimum-Cost Flow Problem With an Additional Linear Constraint
12.00–12.25   2.3* M. Guignard, USA
General Network Problems with Fixed Charges and Side Conditions

TU/C.03
Nonlinear Programming
Chairman: Y. Ermoliev, Austria
11.00–11.25 3.1* W. E. Diewert, Canada
Necessary and Sufficient Conditions for Local Optimality for the Differentiable Inequality-Equality Constrained Maximization Problem
11.30–11.55 3.2* W. E. Diewert, Canada
Local Sensitivity Analysis for a Differentiable Inequality-Equality Constrained Maximization Problem

TU/C.04
Knapsack Problems I
Chairman: P. Frankl, France
11.00–11.25 4.1* S. Walukiewicz, Poland
Knapsack Problem and its Generalizations
11.30–11.55 4.2 H. Ishii, Japan
Stochastic Linear Knapsack Problems
12.00–12.25 4.3 G. Galambos, Hungary
One Dimensional Cutting Stock Problem with Hyperbolic Objective Function

TU/C.05
Probabilistic Analysis of Combinatorial Problems
Chairman: C. McDiarmid, United Kingdom
11.00–11.25 5.1 C. Vercellis, Italy
A Probabilistic Analysis of the Set Covering Problem
11.30–11.55 5.2* R. M. Karp, USA; N. Karmarkar, USA
Probabilistic Analysis of Packing and Partitioning Problems
12.00–12.25 5.3 R. E. Burkard, Austria; U. Fincke, W. Germany
Probabilistic Asymptotic Properties of Some Combinatorial Optimization Problems

TU/C.06
Econometric Models
Chairman: S. Erlander, Sweden
11.00–11.25 6.1 G. Uebe, W. Germany
Ordering and Multiple Solutions of Macro-Econometric Models
11.30–11.55 6.2 S. Müller, W. Germany; S. Nakamura, W. Germany
The Bonn Quarterly Macroeconomic Model for the West German Economy
12.00–12.25 6.3 C. Weihs, W. Germany
Sensitivity Analysis for FIML – Estimates in Nonlinear Econometric Models

TU/C.07
Solving Nonsmooth Optimization Problems
Chairman: M. Vlach, Czechoslovakia
11.00–11.25 7.1* F. H. Clarke, Canada
Examples in Nonsmooth Optimization

11.30–11.55 7.2   M. Minoux, France
A Subgradient Algorithm for Finding a Point Meeting Optimality Conditions in Nonsmooth Optimization
12.00–12.25 7.3* M. Fukushima, Japan
Outer Approximation Algorithms for Nonsmooth Convex Programming Problems

TU/C.08
Computational Experiences with Large-Scale LP-Codes
Chairman: H. Müller-Merbach, W. Germany
11.00–11.25 8.1   J. K. Ho, USA; E. Loute, Belgium
Computational Experience with Advanced Implementation of Decomposition Algorithms
11.30–11.55 8.2   J. Denel, France; P. Huard, France
Numerical Experiments with the "Methode des Parametres" in a Large Problem
12.00–12.25 8.3   G. MacDonald, USA
Comprehensive LP/OMNI on Mini-Computers

TU/C.09
Perfect Graphs
Chairman: L. E. Trotter, USA
11.00–11.25 9.1* J. Fonlupt, France; M. Burlet, France
Polynomial Study of Meyniel Graphs
11.30–11.55 9.2* C. Berge, France
Strongly Perfect Graphs
12.00–12.25 9.3   W. A. Lobb, USA
Perfect Graphs from Paths in Trees

TU/C.10
Research-Surveys
Chairman: M. Held, USA
11.30–11.55 10.2* L. C. Hsu, China
Some Recent Contributions to Mathematical Programming in China
12.00–12.25 10.3* V. S. Mikhalevich, USSR
Experience in Solving Complex Optimization Problems

TU/C.11
Stochastic Programming I
Chairman: P. Kall, Switzerland
11.00–11.25 11.1* M. A. H. Dempster, United Kingdom
Stochastic Programming Models of Hierarchical Planning and Operating Decisions
11.30–11.55 11.2* R. J.-B. Wets, Austria
Decomposition Through Partitioning in Stochastic Programming
12.00–12.25 11.3* M. Schäl, W. Germany
Optimal Policies in Stochastic Dynamic Programming

TU/C.12
Relaxation Methods
Chairman: J.-L. Goffin, Canada
11.00–11.25 12.1* O. L. Mangasarian, USA
Successive Overrelaxation (SOR) Methods in Mathematical Programming
11.30–11.55 12.2 J. Kreuser, USA
Computational Experience with Successive Overrelaxation Methods for Linear Programming Problems
12.00–12.25 12.3 J. Mandel, Czechoslovakia
Convergence of Relaxation Methods for Linear Inequalities

Open Business Meeting of COAL
Tuesday 12.30, lecture room 1

TU/D.01
NLP-Problems with Linear Constraints I
Chairman: H. Y. Kwei, People's Republic of China
14.00–14.25 1.1* J.-J. Strodiot, Belgium; V. H. Nguyen, Belgium; N. Heukemes, Belgium
A Bundle Algorithm for Linearly Constrained Convex Problems
14.30–14.55 1.2* M. Yue, People's Republic of China; J. Han, People's Republic of China
A Pivotal Operation and its Applications
15.00–15.25 1.3 Y. P. Aneja, Canada; K. P. K. Nair, Canada
Maximization of a Class of Linear Homogeneous Concave Functions Subject to Linear Constraints

TU/D.02
Network Algorithms
Chairman: D. Goldfarb, USA
14.00–14.25 2.1* F. Glover, USA; D. Klingman, USA
A New Generation of Primal-Dual Network Methods
14.30–14.55 2.2* K. Belling-Seib, W. Germany
Solving Convex Separable Network Flow Problems: A Computational Study
15.00–15.25 2.3 A. Perry, USA
NLoops – Network Loading Optimization with Pricing and Sequencing

TU/D.03
Nonlinear Least Squares Problems
Chairman: R. Horst, W. Germany
14.00–14.25 3.1* M. Al-Baali, United Kingdom; R. Fletcher, United Kingdom
Optimally Scaled Methods for Nonlinear Least Squares
14.30–14.55 3.2 P. Lindström, Sweden
Software for Nonlinear Least Squares – particularly Problems with Constraints

15.00–15.25 3.3 P. A. Wedin, Sweden
On Linearly and On Quadratically Convergent Methods for the Constrained Nonlinear Least Squares Problem

TU/D.04
Knapsack Problems II
Chairman: S. Walukiewicz, Poland
14.30–14.55 4.2 L. Aittoniemi, Finland
Computational Comparison of Knapsack-Algorithms
15.00–15.25 4.3 N. F. Maculan, Canada
Approximate Solutions for 0-1 Knapsack Problems

TU/D.05
Properties of Random Graphs
Chairman: D. Hochbaum, USA
14.30–14.55 5.2* C. McDiarmid, United Kingdom
Random Graphs and Digraphs
14.30–14.55 5.3* A. M. Frieze, England
Hamiltonian Cycles in Random Graphs

TU/D.06
Linear and Log-Linear Models in Economics
Chairman: G. Uebe, W. Germany
14.00–14.25 6.1* S. Erlander, Sweden
Log-Linear Models Derived by Linear Complementarity From an Efficiency Principle
14.30–14.55 6.2 S. Nakamura, W. Germany
A Multi-Sectoral Translog Model of Production for FRG
15.00–15.25 6.3 H. T. Burley, Australia
Duality and Disposability in the LP Approach to Productive Efficiency Measures

TU/D.07
Directional Differentiability in NLP
Chairman: J.-P. Penot, France
14.00–14.25 7.1* R. T. Rockafellar, USA
Directional Differentiability of Optimal Value Function in a Nonlinear Programming Problem
14.30–14.55 7.2* J. B. Hiriart-Urruty, France
Calculus Rules on the Approximate Second-Order Directional Derivative of a Convex Function
15.00–15.25 7.3* J. Gauvin, Canada
Some Comments on the Directional Derivatives of the Marginal Function in Nonlinear Mathematical Programming

TU/D.08
New Concepts for LP Codes
Chairman: D. Ohse, W. Germany
14.30–14.55 8.2 P. A. Smeds, Sweden
Sequential Algorithms for Linear Programming

15.00–15.25 8.3* H. Müller-Merbach, W. Germany
Symmetric Concepts for Revised Simplex Techniques and their Application in LP Codes

TU/D.09
Set Packing Polytopes
Chairman: A. Schrijver, Netherlands
14.00–14.25 9.1 M. Conforti, USA; G. Cornuejols, USA
An Algorithmic Theory of Some Set Packing Problems
14.30–14.55 9.2* Y. Ikura, USA; G. L. Nemhauser, USA
Simplex Pivots on the Set Packing Polytope
15.00–15.25 9.3* J. P. Uhry, France
Ben Rebea's Results on Claw-Free Graphs

TU/D.10
General Duality Approaches
Chairman: R. Jeroslow, USA
14.00–14.25 10.1* F. Giannessi, Italy
A Survey of Recent Results About Theorems of Alternative, Optimality Conditions and General Duality
14.30–14.55 10.2 M. A. H. Dempster, United Kingdom
On Intrinsic Characterization of Dualizable Abstract Linear Programmes
15.00–15.25 10.3 N. S. Papageorgiou, USA
On the Duality Approach to Optimization

TU/D.11
Stochastic Programming II
Chairman: J. Dupacova, Czechoslovakia
14.00–14.25 11.1* A. Prekopa, Hungary
Large Scale Stochastic Programming Problems
14.30–14.55 11.2* M. Queyranne, USA; E. P. C. Kao, USA
Aggregation in a Two-Stage Stochastic Program for Manpower Planning in the Service Sector

TU/D.12
Ellipsoid Methods
Chairman: M. R. Rao, USA
14.30–14.55 12.2* J. L. Goffin, Canada
Ellipsoidal or Variable Metric Relaxation Methods
15.00–15.25 12.3* J. G. Ecker, USA; J. Kupferschmid, USA
Ellipsoid Algorithms in Nonlinear Programming

TU/E.01
State-of-the-Art Tutorial
Chairman: A. Prekopa, Hungary
15.30–16.30 1.1 R. J. B. Wets, Austria
Solution Techniques and Approximation Schemes for Stochastic Optimization Problems

TU/E.10
State-of-the-Art Tutorial
Chairman: E. Balas, USA
15.30-16.30   10.1   W. R. Pulleyblank, Canada
Polyhedral Combinatorics

TU/F.01
NLP-Problems with Linear Constraints II
Chairman: J. Gauvin, Canada
16.45-17.10   1.1*   H. Y. Kwei, People's Republic of China; D. Z. Du, People's Republic of China
A Superlinearly Convergent Method to Linearly Constrained Optimization Problems Under Degeneracy
17.15-17.40   1.2*   L. F. Escudero, Spain
On Diagonally Preconditioning Truncated Newton Methods for LCNP Problems
17.45-18.10   1.3   D. C. Sorensen, USA
A Trust Region Method for Linearly Constrained Optimization Problems

TU/F.02
Networks
Chairman: M. D. Grigoriadis, USA
16.45-17.10   2.1*   M. O. Ball, USA; J. S. Provan, USA
Computing Network Reliability in Time Polynomial in the Number of Cuts
17.15-17.40   2.2*   K. Neumann, W. Germany
Optimization Problems in Decision Networks
17.45-18.10   2.3*   F. Granot, Canada; A. F. Veinott, Jr., USA
Substitutes and Complements in Network Flows

TU/F.03
Least Squares
Chairman: A. Goldstein, USA
17.15-17.40   3.2   G. Morton, W. Germany; K. Ringwald, W. Germany; R. von Randow, W. Germany
A Simple Algorithm for a Class of Quadratic Smoothing Problems

TU/F.04
Applications of Integer Programming III
Chairman: W. Junginger, W. Germany
16.45-17.10   4.1   C. C. Gonzaga, Brazil
Graph Search Applied to Spotting Towers in Transmission Lines
17.15-17.40   4.2   L. R. Foulds, New Zealand
Graph Theoretic Heuristics for Facilities Layout
17.45-18.10   4.3   N. Maculan, Brazil; T. G. Rocha, Brazil; R. A. Almeida, Brazil; A. O. Moreno, Brazil
Extensions of the Standard Length Telephone Cable Reels Problem

TU/F.05
Integer Linear Programming
Chairman: P. L. Hammer, Canada
16.45–17.10   5.1   M. Fieldhouse, United Kingdom
Greedy Algorithms Solve some Problems better than LP
17.15–17.40   5.2   J. S. Harabasz, Poland; P. Wisniewski, Poland
Sequential Procedure for Overlapping Grouping by Linear Programming
17.45–18.10   5.3*   P. Hansen, Belgium
Measuring the Chaining Effect in Single Link Cluster Analysis

TU/F.06
Optimization of Economic Systems
Chairman: A. Griewank, USA
16.45–17.10   6.1   K. O. Jörnsten, Sweden; C. L. Sandblom, Canada
Optimization of an Economic System Using Nonlinear Decomposition
17.15–17.40   6.2   H. Bogaert, Belgium
Numerical Solution and Optimization of Econometric Models: Application to 'Maribel', a Model of the Belgium's Economy

TU/F.07
NLP: Nonsmooth and Nonconvex Functions
Chairman: J.-C. Pomerol, France
16.45–17.10   7.1*   J. P. Penot, France
On Tangentially Convex Functions
17.15–17.40   7.2   N. S. Papageorgiou, USA
Nonsmooth Analysis on Partially Ordered Vector Spaces
17.45–18.10   7.3   G. G. Watkins, South Africa
The Clarke Tangent Cone and Abstract Multiplier Rules in Optimization Theory

TU/F.08
Techniques in LP
Chairman: R. Wittrock, USA
16.45–17.10   8.1   E. Loute, Belgium; Ph. Gille, Belgium
Updating the LU Gaussian Decomposition for Rank-One Corrections. Application to Linear Programming Basis Partitioning Techniques
17.15–17.40   8.2*   C. van de Panne, Canada
Local Decomposition in Linear Programming
17.45–18.10   8.3*   J. A. Tomlin, USA; J. S. Welch, USA
Formal Optimization of Some Reduced Linear Programming Problems

TU/F.09
The Acyclic Subgraph Problem
Chairman: L. Wolsey, Belgium
16.45–17.10   9.1   M. Grötschel, W. Germany; M. Jünger, W. Germany; G. Reinelt, W. Germany
  On the Acyclic Subgraph Polytope
17.15–17.40   9.2   M. Jünger, W. Germany; G. Reinelt, W. Germany; M. Grötschel, W. Germany
  Facets of the Linear Ordering Polytope
17.45–18.10   9.3   G. Reinelt, W. Germany; M. Grötschel, W. Germany; M. Jünger, W. Germany
  A Cutting Plane Algorithm for the Linear Ordering Problem

TU/F.10
Duality Problems
Chairman: D. W. Hearn, USA
16.45–17.10   10.1   J. P. Crouzeix, France; J. A. Ferland, Canada; S. Schaible, Canada
  Duality in Generalized Linear Fractional Programming
17.15–17.40   10.2   J. A. Sikorski, Poland
  On Surrogate Constraint Duality

TU/F.11
Stochastic Programming III
Chairman: R. Wets, Austria
16.45–17.10   11.1*   J. Dupacova, Czechoslovakia
  Stability in Stochastic Programming
17.15–17.40   11.2   W. K. Klein, Netherlands
  A Stochastic Programming Approach of PERT/CPM
17.45–18.10   11.3   J. Wang, USA
  Distribution Sensitivity Analysis in Stochastic Programming

TU/F.12
Solving Inequality Systems
Chairman: J. Stoer, W. Germany
16.45–17.10   12.1*   S. P. Han, USA
  Least Square Solutions of Linear Inequalities
17.15–17.40   12.2*   A. Ben-Israel, USA
  A Projection Method for Interval Linear Inequalities
17.45–18.10   12.3   S. Ursic, USA
  The Ellipsoid Algorithm for Linear Inequalities in Exact Rational Arithmetic

---

WE/A.01
Variations of Newton's Method for NLP-Problems with Linear Constraints
Chairman: R. Fletcher, United Kingdom
8.15–8.40   1.1   J. P. Dussault, Canada
  Active Set Methods: Implicit Identification Strategy

8.45–9.10	1.2	Ph. Gill, USA; W. Murray, USA; St. Nash, USA; M. Saunders, USA; M. Wright, USA
		Truncated Newton Methods
9.15–9.40	1.3*	D. P. Bertsekas, USA
		Projected Newton Methods for Optimization Problems with Linear Constraints

WE/A.02
Transportation and Distribution Planning
Chairman: G. L. Thompson, USA

8.15–8.40	2.1	M. Bielli, Italy; M. Cini, Italy
		Computational Problems and Experience in Applications of Linear Multicommodity Network Flow Programming
8.45–9.10	2.2	Y. Ienaga, Japan; H. Tokuyama, Japan; K. Okuda, Japan; K. Yagawa, Japan
		Shipping Scheduling for Transport of Iron & Steel Products
9.15–9.40	2.3	L. Petterson, Sweden
		Planning of Cement-Distribution in Sweden with a Simple LP-Model

WE/A.03
Problems in Nonconvex Optimization
Chairman: J.-C. Fiorot, France

8.15–8.40	3.1	M. Ottaviani, Italy
		A Particular Problem of Nonconvex Optimization with Quadratic Constraints
8.45–9.10	3.2*	L. Teschke, W. Germany
		The Role of Binary and Ternary Relations in a General Optimization Theory
9.15–9.40	3.3	J. Czochralska, Poland
		Verification of the Definiteness of a Quadratic Form – Its Algorithm and Relation to Nonconvex Quadratic Programming

WE/A.04
Heuristic Methods in Combinatorial Optimization
Chairman: A. M. Frieze, United Kingdom

8.15–8.40	4.1	A. Lahrichi, Maroc; A.-G. Vittek, France
		Precedence Constrained Cumulative Scheduling of Tasks with Equal Lengths: A Polynomial Method for Calculating the Upper Envelope of All Charge Diagrams
8.45–9.10	4.2	R. Loulou, Canada
		Probabilistic Analysis of Optimal and Heuristic Solutions in Bin-Packing
9.15–9.40	4.3*	D. S. Hochbaum, USA
		Epsilon-Approximation Algorithms for the Stable Set, Vertex Cover and Set Packing Problems

WE/A.05
Computational Studies in Combinatorial Optimization
Chairman: B. Simeone, Italy
8.15–8.40 5.1 S. N. Afriat, United Kingdom
           The Power Algorithm for Minimum Paths and Price Indices
8.45–9.10 5.2 S. Pallottino, Italy; G. Gallo, Italy; C. Ruggeri, Italy
           Shortest Path Algorithms: A Computational Experimentation
9.15–9.40 5.3* T. Ibaraki, Japan
           Double Relaxation Dynamic Programming Approaches to Combinatorial Optimization

WE/A.06
Optimization Problems in Economics
Chairman: M. Kallio, Finland
8.15–8.40 6.1 P. H. M. Ruys, Netherlands
           Allocation of Labor over Production Tasks in Case of Inflexible Wages
8.45–9.10 6.2 L. C. Maclean, Canada; W. R. S. Sutherland, Canada
           Optimality and Sensitivity Analysis in a General Finance Model
9.15–9.40 6.3* G. P. McCormick, USA; A. H. deSilva, USA
           Solving Implicitly Defined Optimization Problems

WE/A.07
Applications of NLP
Chairman: H. D. Sherali, USA
8.15–8.40 7.1 M. Boulala, Algeria
           An Optimal Mathematical Formulation of a Problem in Mechanics of Vibrations
8.45–9.10 7.2 Y. Sidrak, Egypt; N. Sh. Matta, Egypt
           A Computerized – Optimized Study on Film Cooling Technique (Part II)

WE/A.08
Goal Programming and LP-Models
Chairman: G. W. Graves, USA
8.15–8.40 8.1 S. M. Fereig, Kuwait
           Goal Programming as a Construction Management Tool
8.45–9.10 8.2 V. Dumitru, Romania; F. Luban, Romania
           On Some Recent and Old Results in Mathematical Programming
9.15–9.40 8.3 C. A. Haverly, USA
           Behavior of LP Models

WE/A.09
Submodular Functions
Chairman: E. L. Lawler, USA
8.15–8.40 9.1 N. Tomizawa, Japan
           Quasimatroids and Orientations of Graphs

8.45–9.10	9.2*	S. Fujishige, Japan

Base Polytopes Determined by Submodular Functions on Crossing Families

8.15–9.40 9.3* U. Zimmermann, W. Germany
Minimization on Submodular Flows

WE/A.11
Markov Decision Problems I
Chairman: A. Lubiw, Canada
8.15–8.40   11.1   R. Hellerich, W. Germany
On the Limiting Behaviour of some Bounds of the Optimal Value in Markovian Decision Problems
8.45–9.10   11.2   P. l'Ecuyer, Canada; A. Haurie, Canada
Isotonicity in Discrete Stage Markovian Decision Processes
9.15–9.40   11.3   W. H. M. Zijm, Netherlands
Denumerable Multichain Markov Decision Processes: The Optimality Equations for the Average Cost Criterion

WE/A.12
Variational Problems and their Applications
Chairman: D. G. Luenberger, USA
8.15–8.40   12.1*   J.-S. Pang, USA
Computational Experience with Solving the Traffic Equilibrium Problem by Variational Techniques
8.45–9.10   12.2   A. Maugeri, Italy
Applications of Variational Inequalities to Traffic Equilibrium Problems
9.15–9.40   12.3   T. Maruyama, Japan
Variational Problems in Economic Analysis – Existence of Optimal Solutions

WE/B.01
State-of-the-Art Tutorial
Chairman: E. M. L. Beale, United Kingdom
9.45–10.45   1.1   J. J. More, USA
Recent Developments in Algorithms and Software for Trust Region Methods

WE/B.10
State-of-the-Art Tutorial
Chairman: G. L. Nemhauser, USA
9.45–10.45   10.1   L. Lovász, Hungary
Submodular Functions and Convexity

WE/C.01
Constrained Optimization
Chairman: K.-H. Elster, DDR
11.00–11.25   1.1*   R. Mifflin, USA
Convergence and Rate of Convergence of an Algorithm for Constrained Minimization Problems with Locally Lipschitz Functions of One Variable

11.30–11.55 1.2* K. Tone, Japan
A Revision of Constrained Approximations in the Successive QP Method for Nonlinear Programming Problems
12.00–12.25 1.3 B. Rustem, United Kingdom
Convergent Stepsizes for Constrained Optimization Algorithms

WE/C.02
Transportation Problems
Chairman: T. M. Liebling, Switzerland
11.00–11.25 2.1* M. L. Balinski, France
On Dual Transportation Polyhedra
11.30–11.55 2.2* G. L. Thompson, USA
A Recursive Algorithm for Solving Transpration Problems
12.00–12.25 2.3 T. Altman, USA
A Gradient Directed Primal-Dual Algorithm for the Transportation Problem

WE/C.03
Infinite Dimensional Programming
Chairman: S. Zlobec, Canada
11.00–11.25 3.1 R. N. Buie, Canada; J. Abrham, Canada
Some Remarks Concerning Duality for Continuous Time Programming Problems
11.30–11.55 3.2 J. Gwinner, W. Germany
Generalized Homogeneous Programming
12.00–12.25 3.3 M. Thera, France
A Hörmander-Type Theorem for a Convex Mapping Taking Values in a Continuous Function Space

WE/C.04
Combinatorial Optimization: Algebraic and Number Theoretic Methods
Chairman: R. L. Graham, USA
11.00–11.25 4.1 E. B. Monteith, USA
The Solution Space as a Lattice: A New Approach to LP via the Minkowski Geometry of Numbers
11.30–11.55 4.2* R. E. Burkard, Austria
On Cyclic Timetables
12.00–12.25 4.3* W.-C. W. Li, USA
Applications of the Polynomial Method to Problems in Combinatorics and Optimization

WE/C.05
Location
Chairman: G. Cornuejols, W. Germany
11.00–11.25 5.1 F. Plastria, Belgium
Continuous Location Problems Solved by Cutting Planes
11.30–11.55 5.2* K. Spielberg, USA
The Solution of Plant Location and Related Problems
12.00–12.25 5.3* J. Barcelo, Spain; J. Casanovas, Spain
A Langrangean Algorithm for the Capacitated Plant Location Problem

WE/C.06
Equilibrium Models and their Applications
Chairman: L. McLinden, USA
11.00–11.25   6.1*   T. L. Magnanti, USA
    Equilibria for a Class of Nonlinear Models with Linear Technologies
11.30–11.55   6.2*   M. Kallio, Finland; M. Soismaa, Finland
    Economics Equilibrium for the Finnish Forest Sector
12.00–12.25   6.3*   J. M. Rousseau, Canada; L. Delorme, Canada
    An Application of the Multiple Commodity Spatial Price Equilibrium Models to Resources Allocation in a Health Care System

WE/C.07
Applications of NLP in Chemistry
Chairman: P. Bod, Hungary
11.00–11.25   7.1   M. Minkoff, USA; R. Land, USA; M. Blander, USA
    Solving Chemical Equilibrium Problems via Primal Geometric Programming
11.30–11.55   7.2   H. Crowder, USA
    Analysis of Sparse Matrix Structure Using Color Graphics
12.00–12.25   7.3   C. L. Lai, USA
    Mathematical Programming for Estimation of Publication Longevity

WE/C.08
Dynamic Linear Programs
Chairman: L. V. Kantorovich, USSR
11.00–11.25   8.1*   G. B. Dantzig, USA
    Solving Dynamic Linear Programs
11.30–11.55   8.2   R. Wittrock, USA
    A Nested Decomposition Approach for Time-Staged LP's

WE/C.09
Polymatroids
Chairman: L. J. Billera, USA
11.00–11.25   9.1*   W. H. Cunningham, Canada
    The Separation Problem for Matroids and Polymatroids
11.30–11.55   9.2*   D. M. Topkis, USA; R. E. Bixby, USA; W. H. Cunningham, Canada
    The Partial Order of a Polymatroid Extreme Point
12.00–12.25   9.3*   A. Frank, Hungary
    The Projection of Polymatroid Intersections

WE/C.10
Optimization in Banach Spaces
Chairman: W. W. E. Wetterling, Netherlands
11.30–11.55   10.2*   D. Pallaschke, W. Germany
    On the Observability of Differential Operators
12.00–12.25   10.3   G. Dugoshia, Yugoslavia
    Systems of Convex Inequalities and Affine Minorants

WE/C.12
Variational Problems
Chairman: K. Beer, DDR
11.00–11.25   12.1*   A. A. Kaplan, USSR
                 Numerical Solution of some Classes of Variational Inequalities
11.30–11.55   12.2   H. Jarausch, W. Germany
                 Approximating Large Nonlinear Elliptic Variational Problems by Easily Solved Artificially Constrained Problems
12.00–12.25   12.3   D. Chan, USA; J. S. Pang, USA
                 The Generalized Quasi-Variational Inequality Problem

WE/D.01
State-of-the-Art Tutorial
Chairman: K. Ritter, W. Germany
13.30–14.30   1.1   S. M. Robinson, USA
                 Generalized Equations

WE/D.10
State-of-the-Art Tutorial
Chairman: E. L. Johnson, USA
13.30–14.30   10.1   J. Edmonds, Canada
                 The Topology of Linear Programming and the Linear Programming of Topology

WE/E.01
State-of-the-Art Tutorial
Chairman: C. G. Broyden, United Kingdom
14.45–15.45   1.1   R. B. Schnabel, USA
                 Conic Models in Unconstrained Optimization and Tensor Models in Nonlinear Equations

WE/E.09
State-of-the-Art Tutorial
Chairman: M. Balinski, France
14.45–15.45   9.1   J. Rosenmüller, W. Germany
                 Nondegeneracy Problems in Cooperative Game Theory

WE/E.10
State-of-the-Art Tutorial
Chairman: R. Burkard, Austria
14.45–15.45   10.1   P. L. Hammer, Canada
                 Boolean Methods in Combinatorics and Optimization

WE/F.01
State-of-the-Art Tutorial
Chairman: W. Oettli, W. Germany
16.00–17.00   1.1   K. O. Kortanek, USA; S. A. Gustafson, Sweden
                 Semi-Infinite Programming

WE/F.10
State-of-the-Art Tutorial
Chairman: O. L. Mangasarian, USA
16.00–17.00   10.1   N. Z. Shor, USSR
Generalized Gradient Methods of Non-Differentiable Optimization using Space Expansion Operation

TH/A.01
Decomposition Techniques
Chairman: A. Ben-Israel, USA
8.15–8.40   1.1   A. Kalliauer, Austria
Concept and Implementation of a Method for Hierarchical Optimization
8.45–9.10   1.2*   K. Beer, DDR
Decomposition through Resource Allocation – New Theoretical and Numerical Results
9.15–9.40   1.3*   E. A. Nurminski, Austria
Nondifferentiable Optimization and Decomposition of Large-Scale Problems

TH/A.02
Heuristics for Routing: Probabilistic Analysis
Chairman: K. Neumann, W. Germany
8.45–9.10   2.2   L. Stougie, Netherlands; A. Marchetti Spaccamela, Italy; A. H. G. Rinnooy Kan, Netherlands
Hierarchical Routing and Distribution Problems
9.15–9.40   2.3   M. Haimovich, USA; A. H. G. Rinnooy Kan, Netherlands
An Epsilon-Optimal Polynomial Time Algorithm for Capacitated Routing

TH/A.03
Quadratic Programming I
Chairman: L. C. Hsu, People's Republic of China
8.15–8.40   3.1   A. Idnani, USA
Dual and Primal-Dual Methods for Convex Quadratic Programs
8.45–9.10   3.2   A. R. G. Heesterman, United Kingdom
The Sequentially Constrained Maximisation Algorithm for General Quadratic Programming
9.15–9.40   3.3   R. J. Caron, Canada; M. J. Best, Canada
Parametric Hessian Quadratic Programming

TH/A.04
Combinatorial Methods for Geometrical Problems
Chairman: G. Ausiello, Italy
8.15–8.40   4.1   S. Halasz, Switzerland
Packing a Convex Domain with Similar Convex Domains
8.45–9.10   4.2   G. D. d'Atri, Italy; A. Volpentesta, Italy
Discrete Separation in R^n

9.15–9.40   4.3*   E. Balas, USA
Intersections of Ellipsoids

TH/A.05
Packing and Covering
Chairman: R. H. Möhring, W. Germany
8.15–8.40   5.1*   U. N. Peled, USA; B. Simeone, Italy
Closest Linear Separation of Stable from Nonstable Sets in a Graph
8.45–9.10   5.2*   G. Cornuejols, USA; D. Hartvigsen, USA
Packing Subgraphs in a Graph
9.15–9.40   5.3*   Z. Füredi, Hungary
On the Fractional Coverings of Hypergraphs

TH/A.06
Decision Problems in Economics
Chairman: H. J. Jaksch, W. Germany
8.15–8.40   6.1   E. Pirttimäki, Finland
A New Way of Measuring the Expected Utility
8.45–9.10   6.2   R. Dootz, Romania
Contribution of the Quality of the Labour Force to Economic Growth
9.15–9.40   6.3   H. Machado, Brasil
On Efficient Proportional Schemes of Operation

TH/A.07
Constrained Optimization: Necessary and Sufficient Conditions
Chairman: F. Lempio, W. Germany
8.15–8.40   7.1   L. Martein, Italy
A Necessary and Sufficient Regularity Condition for Convex Extreme Problems
8.45–9.10   7.2*   M. A. Hanson, USA
Necessary and Sufficient Conditions for Constrained Optimization
9.15–9.40   7.3*   B. D. Craven, Australia
Invex Functions and Constrained Local Minima

TH/A.08
Linear Programming Models
Chairman: H. Crowder, USA
8.15–8.40   8.1*   G. Knolmayer, W. Germany
Bounds and Empirical Results for Simplified Linear Programming Models
8.45–9.10   8.2   J. Telgen, Netherlands; C. G. E. Boender, Netherlands; A. H. G. Rinnooy Kan, Netherlands; R. L. Smith, USA
Probabilistic Methods for Identifying Nonredundant Constraints
9.15–9.40   8.3*   H. J. Greenberg, USA; W. G. Kurator, USA
Computer-Assisted Analysis for Linear Programming: Theory, Software and Practice

TH/A.09
Greedy Algorithm and Polymatroids
Chairman: R. E. Bixby, USA
8.15–8.40   9.1* P. B. Milanov, Bulgaria
Intersection of Polymatroids and Accuracy of the Greedy Algorithm
8.45–9.10   9.2* A. Polymeris, Switzerland
The Minimal Capacity Problem
9.15–9.40   9.3* L. A. Wolsey, Belgium
An Analysis of the Greedy Algorithm for the Submodular (Set) Covering Problem

TH/A.10
Research-Survey
Chairman: G. S. Rubinstein, USSR
8.45–9.40   10.2 L. V. Kantorovich, USSR
The Use of Linear Programming in the USSR

TH/A.11
Probabilistic Methods
Chairman: M. A. H. Dempster, United Kingdom
8.45–9.10   11.2 K. Holmberg, Sweden; K. Jörnsten, Sweden
A Cross Decomposition Algorithm for the Stochastic Transportation Problem
9.15–9.40   11.3 B. Srajber, Hungary; G. Balogh, Hungary
A Method of Information Theory and its Application to Investigation of the New-Born Population

TH/A.12
Algorithms for Optimal Control Problems
Chairman: W. Krabs, W. Germany
9.15–9.40   12.3* T. Futagami, Japan
BEM (Boundary Element Method) / FEM (Finite Element Method) Coupled with DP (Dynamic Programming) / LP (Linear Programming) in Optimal Control and the Efficient Computational Algorithm

TH/B.01
State-of-the-Art Tutorial
Chairman: A. Auslender, France
9.45–10.45   1.1 R. T. Rockafellar, USA
Subdifferentiability in Nonlinear Optimization

TH/B.10
State-of-the-Art Tutorial
Chairman: C. Berge, France
9.45–10.45   10.1 J. Edmonds, Canada
The Topology of Linear Programming and the Linear Programming of Topology II

TH/C.01
General Optimization Techniques
Chairman: R. S. Dembo, USA
11.00–11.25 1.1* J. E. Spingarn, USA
A Primal-Dual Approach to the Problem of Finding a Zero of a Monotone Operator
11.30–11.55 1.2* Y. Ermoliev, Austria
Nonmonotonic Methods of Optimization
12.00–12.25 1.3 R. A. Polyak, USSR
Smooth Optimization Methods for Solving Nonlinear Extremal and Equilibrium Problems with Constraints

TH/C.02
Routing
Chairman: G. Laporte, Canada
11.00–11.25 2.1* H. Sachs, DDR
Reflections Connected with a Special Routing Problem: How to Cope with an NP-Hard Problem
11.30–11.55 2.2 O. B. G. Madsen, Denmark
Large Scale Routing Problems
12.00–12.25 2.3 R. R. Levary, USA; R. A. Skitt, USA
Vehicle Routing Via Column Generation

TH/C.03
Quadratic Programming II
Chairman: A. R. Conn, Canada
11.00–11.25 3.1* M. J. Best, Canada
Equivalence of Several Quadratic Programming Algorithms
11.30–11.55 3.2 Ph. Gill, USA; N. Gould, USA; W. Murray, USA; M. Saunders, USA; M. Wright, USA
Range Space Methods for Quadratic Programming

TH/C.04
Clustering
Chairman: D. de Werra, Switzerland
11.00–11.25 4.1* J. M. Mulvey, USA; M. P. Beck, USA
A Heuristic Subgradient Algorithm for Capacitated Clustering Problems
11.30–11.55 4.2 R. Schrader, W. Germany
Approximations to Clustering and Subgraph Problems on Trees
12.00–12.25 4.3* B. Simeone, Italy
Equipartition and Clustering Problems on Trees

TH/C.05
Covering Problems
Chairman: G. O. H. Katona, Hungary
11.00–11.25 5.1 R. Van Slyke, USA
On K-Covering
11.30–11.55 5.2* G. Ausiello, Italy; A. d'Atri, Italy; D. Sacca, Italy
Minimal Coverings of Directed Hypergraphs

12.00–12.25 5.3   Z. Tuza, Hungary
Covering of Graphs by Complete Bipartite Subgraphs. Complexity of 0–1 Matrices

TH/C.06
Decision Problems in Economics
Chairman: H. Machado, Brazil
11.00–11.25 6.1   M. Schäfer, W. Germany
Depletion Rates for Exhaustible Resources – A Differential Game Approach –
11.30–11.55 6.2   K. R. Kumar, USA; M. A. Satterthwaite, USA
A Result Concerning a Maximal Amount of Product Differentiation Possible in a Nash-Cournot Equilibrium
12.00–12.25 6.3   G. Gambarelli, Italy
Portfolio Management for Firms' Control

TH/C.07
Optimality Conditions
Chairman: B. D. Craven, Australia
11.00–11.25 7.1*   A. Ben-Tal, Israel
Second Order Conditions in Mathematical Programming
11.30–11.55 7.2*   J. Zowe, W. Germany; A. Ben-Tal, Israel
Necessary and Sufficient Optimality Conditions for a Class of Nonsmooth Optimization Problems
12.00–12.25 7.3   O. Fujiwara, Thailand; S. P. Han, USA; O. L. Mangasarian, USA
Local Duality of Nonlinear Programs

TH/C.08
Testing MP Software, COAL Session
Chairman: M. A. Saunders, USA
11.00–11.25 8.1*   K. Schittkowski, W. Germany
Theory, Implementation, and Test of a Nonlinear Programming Algorithm
11.30–11.55 8.2*   E. D. Eason, USA
A Set of Engineering Problems for Evaluating Nonlinear Programming Codes
12.00–12.25 8.3*   F. A. Lootsma, Netherlands
Performance Evaluation of Nonlinear Optimization Software via Priority Theory, Fuzzy Numbers, and Measurement Theory

TH/C.09
Oriented Matroids
Chairman: J. Edmonds, Canada
11.00–11.25 9.1*   M. J. Todd, USA
Complementarity in Oriented Matroids
11.30–11.55 9.2*   L. J. Billera, USA; B. S. Munson, USA
Polarity and Inner Products in Oriented Matroids
12.00–12.25 9.3   A. Bachem, W. Germany
On Face Lattices of Oriented Matroids

TH/C.10
Combinatorial Optimization
Chairman: J. Araoz, Venezuela
11.00–11.25   10.1   J. Grabowski, Poland; C. Smutnicki, Poland
Block System Approach for Solving the Sequencing Problems with the Objective Function to Minimize Maximum Cost

TH/C.11
Parametric Problems
Chairman: S. M. Robinson, USA
11.00–11.25   11.1* B. Brosowski, W. Germany
Parametric Optimization and Critical Sets of Inequalities
11.30–11.55   11.2   H. Th. Jongen, Netherlands; P. Jonker, Netherlands; F. Twilt, Netherlands
On One-Parameter Families of Sets Defined by (In)Equality Constraints
12.00–12.25   11.3   J. A. Reinoza, Venezuela
The Strong Positivity Conditions

TH/C.12
Numerical Solution of Optimal Control Problems
Chairman: C. Lemarechal, France
11.00–11.25   12.1* K. Marti, W. Germany
Controlled Random Search Procedures in Optimization: Optimal Controls and Rates of Convergence
11.30–11.55   12.2   U. Mackenroth, W. Germany
Some Remarks on the Numerical Solution of Bang-Bang Type Optimal Control Problems
12.00–12.25   12.3   C. de Wit, Netherlands
Numerical Solution of Bang-Bang Control Problems with Sequential Quasi-Linearization

TH/D.01
Nonlinear Optimization
Chairman: C. Großmann, DDR
14.00–14.25   1.1*   C. Lemarechal, France; J. Zowe, W. Germany
A View of Newton-Like Methods for Convex Optimization
14.30–14.55   1.2*   I. Singer, Romania
Optimization by Level Set Methods V: Duality Theorems for Perturbed Optimization Problems
15.00–15.25   1.3*   C. G. Broyden, United Kingdom
The Numerical Stability of Conjugate Gradient and Similar Algorithms

TH/D.02
Routing: Exact Methods I
Chairman: H. Sachs, DDR
14.00–14.25   2.1*   G. Laporte, Canada; M. Desrochers, Canada; Y. Norbert, Canada
Two Exact Algorithms for the Distance Constrained Vehicle Routing Problem

14.30–14.55   2.2   J. Desrosiers, Canada; P. Pelletier, Canada; F. Soumis, Canada
Routing with Time Windows

15.00–15.25   2.3*   B. Fleischmann, W. Germany
Linear Programming Approaches to Travelling Salesman and Vehicle Scheduling Problems

TH/D.03
Views on Quadratic Programming
Chairman: M. J. Best, Canada

14.00–14.25   3.1*   Ph. Gill, USA; W. Murray, USA; M. Saunders, USA; M. Wright, USA
Topics in Quadratic Programming That Are Never Discussed (But Should Be)

14.30–14.55   3.2*   A. R. Conn, Canada
Successive Quadratic Programming – A Biased View

15.00–15.25   3.3*   D. Goldfarb, USA
A Unified Approach to Primal, Dual, and Primal-Dual Algorithms for Quadratic Programming

TH/D.04
Combinatorics
Chairman: Z. Tuza, Hungary

14.00–14.25   4.1*   G. O. H. Katona, Hungary
Some Extremal Problems for Hypergraphs

14.30–14.55   4.2   P. Frankl, France
Families of Sets without Large Stars

15.00–15.25   4.3   J. S. Harabasz, Poland; Z. Tabis, Poland
The Problem of Finding a Connected A^p-Optimal Block Design

TH/D.05
Computational Complexity
Chairman: B. Monien, W. Germany

14.00–14.25   5.1   S. Masuyama, Japan; T. Ibaraki, Japan; T. Hasegawa, Japan
Computational Complexity of p-Center Problems in the Plane

14.30–14.55   5.2   A. Tamir, Israel; N. Megiddo, Israel
New Results on the Complexity of p-Center Problems

15.00–15.25   5.3*   M. M. Syslo, Poland
NP-Complete Problems on some Classes of Planar Graphs

TH/D.06
Economic Applications of Game Theory
Chairman: H. P. Young, USA

14.00–14.25   6.1*   U. Dieter, Austria
Roulette as a Ruin Game: Optimal Strategies for Gambling

14.30–14.55   6.2*   D. Granot, Canada
On Some Nucleoli of Spanning Tree Games

15.00–15.25   6.3   J. A. Filar, USA
The Switching Control Stochastic Game

TH/D.07
Optimality Conditions
Chairman: A. Ben-Tal, Israel
14.00–14.25   7.1*   F. Lempio, W. Germany
                 Sufficient Optimality Conditions for Optimal Control Problems
14.30–14.55   7.2*   B. Gollan, W. Germany
                 Multipliers and Shadow Prices in Optimization – A General Approach
15.00–15.25   7.3   T. Rapcsak, Hungary
                 A Differential Geometric Analysis of Optimality Conditions

TH/D.08
Pure and Mixed Integer Models, COAL Session
Chairman: E. L. Johnson, USA
14.00–14.25   8.1*   G. G. Brown, USA; G. W. Graves, USA
                 Application of a New Class of Decompositions to Large-Scale Mathematical Programming Models of Production, Inventory and Distribution
14.30–14.55   8.2   H. Crowder, USA; R. Ambrosetti, Italy; T. A. Ciriani, Italy; E. L. Johnson, USA; M. W. Padberg, USA
                 System Design Considerations for a Large-Scale Zero-One Linear Programming System
15.00–15.25   8.3*   U. Suhl, USA
                 An Experimental Software System for Solving Large Scale Pure and Mixed 0–1 Linear Optimization Problems

TH/D.09
Greedoids and the Greedy Algorithm
Chairman: U. Zimmermann, W. Germany
14.00–14.25   9.1   B. Korte, W. Germany; L. Lovász, Hungary
                 More About Greedoids
14.30–14.55   9.2*   U. Faigle, W. Germany
                 Submodular Polytopes and the Greedy Algorithm on Ordered Sets
15.00–15.25   9.3   A. Kolen, Netherlands; A. J. Hoffman, USA; M. Sakarovitch, France
                 Totally-Balanced and Greedy Matrices

TH/D.12
Optimal Control
Chairman: K. Marti, W. Germany
14.00–14.25   12.1*   W. Krabs, W. Germany
                 Time-Optimal Control of Linear Systems
14.30–14.55   12.2   R. H. W. Hoppe, W. Germany
                 On the Approximate Solution of Time-Optimal Control Problems
15.00–15.25   12.3   W. Alt, W. Germany
                 Perturbation and Approximation of Nonlinear Optimal Control Problems

TH/E.01
State-of-the-Art Tutorial
Chairman: P. Wolfe, USA
15.30–16.30 1.1   M. J. D. Powell, United Kingdom
Variable Metric Methods for Constrained Optimization

TH/E.10
State-of-the-Art Tutorial
Chairman: T. Ibaraki, Japan
15.30–16.30 10.1 M. Iri, Japan
Applications of Matroid Theory

TH/F.01
Convex Programming
Chairman: E. Nurminski, Austria
16.45–17.10 1.1* J.-P. Aubin, France
Lipschitz Behavior of Solutions to Convex Minimization Problems
17.15–17.40 1.2   N. V. Tretyakov, USSR
Perturbed-Gradient Method and Application of Augmented Lagrangians to Inconsistent Convex Programming
17.45–18.10 1.3   A. Umnov, Austria
Penalty Functions Approach to Solving a Set of Mathematical Programming Problems

TH/F.02
Routing: Exact Methods II
Chairman: A. Land, United Kingdom
16.45–17.10 2.1   A. Corberan, Spain; N. Christofides, Spain; V. Campos, Spain; E. Mota, Spain
An Exact Algorithm for the Rural Postman Problem
17.15–17.40 2.2   E. Mota, Spain; N. Christofides, Spain; V. Campos, Spain; A. Corberan, Spain
An Algorithm for the Rural Postman Problem in Directed Graphs (DRPP)

TH/F.03
Estimating Matrices by Quadratic Programming
Chairman: D. P. Bertsekas, USA
16.45–17.10 3.1* R. W. Cottle, USA; S. G. Duvall, USA
A Lagrangian Relaxation Algorithm for the Constrained Matrix Problem
17.15–17.40 3.2   K. McKinnon, United Kingdom; J. Buchanan, United Kingdom
Methods for Updating Tabular Data Using Marginal Information
17.45–18.10 3.3* J. L. Balintfy, USA
Applications of Quadratic Programming in the Estimation and Use of Quadratic Utility Functions

TH/F.04
Duality in Integer Programming
Chairman: U. N. Peled, USA
16.45–17.10 4.1* L. E. Trotter, USA
              Duality in Combinatorial Optimization
17.15–17.40 4.2* H. P. Williams, United Kingdom
              The Dual of an Integer Program
17.45–18.10 4.3   R. G. Jeroslow, USA; C. E. Blair, USA
              Computational Complexity of the Value Function: II

TH/F.05
Complexity of Numerical Algorithms
Chairman: R. Bartels, Canada
16.45–17.10 5.1* H. W. Kuhn, USA
              On the Cost of Computing Roots of Polynomials
17.45–18.10 5.3   U. M. Gracia-Palomares, Venezuela
              An Approach for Solving Large Linear Systems of Equalities and Inequalities

TH/F.06
Game Theoretic Models
Chairman: J. Rosenmüller, W. Germany
16.45–17.10 6.1   K. O. Kortanek, USA
              Polyextremal Duality in Economics and Engineering
17.15–17.40 6.2* H. P. Young, Austria
              Joint Cost Allocation: Methods and Principles
17.45–18.10 6.3* A. Wierzbicki, Austria
              Interactive Models in Optimization and Game Theoretical Models

TH/F.07
Nonconvex, Nondifferentiable Programming
Chairman: S.-A. Gustafson, Sweden
16.45–17.10 7.1* L. McLinden, USA
              Convergent Sequences of Extremum Problems
17.15–17.40 7.2* J. C. Pomerol, France
              The Generalized Gradients of the Marginal Function in Differentiable Programming

TH/F.08
Nonlinear Programming, COAL Session
Chairman: R. H. F. Jackson, USA
16.45–17.10 8.1* B. Murtagh, Australia; M. Saunders, USA
              Sparse Nonlinear Programming – Some Recent Lessons and Developments
17.15–17.40 8.2* D. M. Gay, USA
              On Convergence Testing in Model/Trust-Region Algorithms for Unconstrained Optimization
17.45–18.10 8.3* A. Buckley, Canada
              Conjugate Gradient Algorithms

TH/F.09
Matroids
Chairman: J. P. Uhry, France
16.45–17.10 9.1* K. Truemper, USA
                Matroid Connectivity and Decomposition
17.15–17.40 9.2* R. E. Bixby, USA
                Algorithms for the Graph-Realization Problem
17.45–18.10 9.3   V. Chandru, USA
                On the Shortest Path Problem in a Binary Matroid

TH/F.11
Utility Functions
Chairman: M. Y. Yue, People's Republic of China
16.45–17.10 11.1 J. P. Crouzeix, France
                Differentiability of Direct and Indirect Utility Functions

TH/F.12
Control Theory and Applications
Chairman: O. Krafft, W. Germany
16.45–17.10 12.1* D. G. Luenberger, USA
                Control in Dynamic Markets
17.15–17.40 12.2 E. A. Galperin, Canada
                Regression Models for Predication and Control

Business Meeting of the Mathematical Programming Society
Thursday 18.15, lecture room 1

FR/A.01
Quasi-Newton Methods
Chairman: P. Huard, France
8.15–8.40   1.1   H. Nakayama, Japan
                A Double Quasi-Newton Method for Constrained Optimization
8.45–9.10   1.2   K. A. Ariyawansa, Canada; J. G. C. Templeton, Canada
                Collinear Scaling Algorithms for Nonlinear Regression
9.15–9.40   1.3   J. Herskovits, Brazil
                A Two-Stage Feasible Direction Algorithm Including Variable Metric Techniques

FR/A.02
Combinatorial Optimization
Chairman: H. Hamacher, USA
8.15–8.40   2.1   P. Gazmuri, Chile
                Computational Analysis of Probabilistically Optimal Algorithms for Some Machine Scheduling Problems
8.45–9.10   2.2   T. S. Arthanari, India
                On the Travelling Salesman Problem
9.15–9.40   2.3* S. Martello, Italy; P. Toth, Italy
                Worst-Case Analysis of Greedy Algorithms for the Subset-Sum Problem

FR/A.03
Generating NLP Test Problems
Chairman: E. D. Eason, USA
- 8.15–8.40   3.1   E. H. McCall, USA
  Automatic Generation of Linearly Constrained, Concave Quadratic Objective Function Test Problems for Global Minimization Algorithms
- 8.45–9.10   3.2   N. Mahdavi-Amiri, Canada; R. H. Bartels, Canada
  Generating Nonlinear Programming Test Problems
- 9.15–9.40   3.3   J. S. Liebman, USA
  The Design and Development of a Nonlinear Optimization Preprocessor

FR/A.04
Modelling Concepts
Chairman: K. Cameron, Canada
- 8.45–9.10   4.2   R. G. Jeroslow, USA; J. Lowe, USA
  Modelling with Integer Variables
- 9.15–9.40   4.3   R. Wille, W. Germany
  Concept Analysis

FR/A.05
Linear Programming, SIGMAP Session
Chairman: U. Suhl, USA
- 8.15–8.40   5.1*   A. Meeraus, USA
  GAMS – A Mathematical Modeling System for Strategic Planning
- 8.45–9.10   5.2*   H. Crowder, USA
  Building Linear Programming Problems Using Arrays and Functions
- 9.15–9.40   5.3   R. Fourer, USA
  New Paradigms for Linear Programming: Piecewise-Linear Simplex Methods

FR/A.06
Multiobjective Optimization I
Chairman: B. Fleischmann, W. Germany
- 8.45–9.10   6.2*   M. Q. Ying, People's Republic of China
  Cone Extreme Points Set and Grouping-Hierarchy Problem
- 9.15–9.40   6.3   R. Weber, W. Germany
  Pseudomonotonic Multiobjective Programming

FR/A.07
Applications in Engineering II
Chairman: J.-M. Rousseau, Canada
- 8.15–8.40   7.1   P. V. Rao, Singapore; V. V. Sastry, Singapore; G. S. Rao, Singapore; P. S. Rao, Singapore
  Design Optimization on Electrical Machines in the Presence of a Single Constraint

8.45–9.10	7.2	P. Papalambros, USA
		Some Monotone Properties of Models in Optimal Engineering Design
9.15–9.40	7.3	Y. Sidrak, Egypt; Sh. L. Shoukr, Egypt
		An Algorithm for the Computation of Strains and Stresses in Solid Cylinders Using the Method of Finite Elements

FR/A.08
Matrices and Optimization
Chairman: U. Dieter, Austria

8.45–9.10	8.2	A. Griewank, USA; G. W. Reddien, USA
		Characterization and Computation of Generalized Turning Points Using Matrix Factorizations
9.15–9.40	8.3*	O. Krafft, W. Germany
		Optimization of Matrix Functions

FR/A.09
Applications of Network and Matroid Theory
Chairman: S. Fujishige, Japan

8.45–9.10	9.2*	M. Iri, Japan; K. Murota, Japan
		Matroidal Approach to the Structural Solvability of a System of Equations
9.15–9.40	9.3*	A. Recski, Hungary
		Uniform Matroidal Models in Statics and Electric Theory

FR/A.11
Quasi-Convexity
Chairman: S. Schaible, Canada

8.45–9.10	11.2*	G. S. Rubinstein, USSR
		Some Problems of the Theory of Quasiconvex Functions and Preorders
9.15–9.40	11.3	P. O. Lindberg, Sweden
		Classes of Quasiconvex Functions on R Closed Under Addition and Translation

FR/A.12
Algorithms for Linear Complementarity Problems
Chairman: J. L. Balintfy, USA

8.15–8.40	12.1	M. Cirina, Italy
		A Complementarity Method
8.45–9.10	12.2*	H. Bernau, Hungary
		A Modification of Lemke's Method to Determine More Than One Solution of the Linear Complementarity Problem
9.15–9.40	12.3	A. J. J. Talman, Netherlands; L. Van Der Heyden, USA
		Algorithms for the LCP which Allow an Arbitrary Starting Point

FR/B.01
State-of-the-Art Tutorial
Chairman: L. C. W. Dixon, United Kingdom

9.45–10.45	1.1	R. Fletcher, United Kingdom
		Penalty Functions

FR/B.10
State-of-the-Art Tutorial
Chairman: T. Magnanti, USA
9.45-10.45   10.1   E. L. Lawler, USA
                 Recent Results in Scheduling Theory

FR/C.01
Gradient Methods
Chairman: V. H. Nguyen, Belgium
11.00-11.25   1.1*   P. Huard, France
                 Valid Linear Searches, Variable Metric and Reduced Gradient Methods
11.30-11.55   1.2*   E. M. L. Beale, United Kingdom; S. Brooker, United Kingdom
                 A New Conjugate Gradient Method for Unconstrained and Constrained Optimization
12.00-12.25   1.3*   E. G. Golstein, USSR
                 Saddle-Point Gradient Methods and Modified Lagrangians

FR/C.02
Travelling Salesman Problem
Chairman: H. Noltemeier, W. Germany
11.00-11.25   2.1*   G. Cornuejols, USA; D. J. Naddef, France; W. R. Pulleyblank, Canada
                 The Travelling Salesman Problem in Graphs with 3-Edge Cutsets
11.30-11.55   2.2*   M. Held, USA
                 Influences of the Travelling Salesman Problem
12.00-12.25   2.3*   W. R. Stewart, Jr., USA; B. L. Golden, USA
                 Comparing Approximate Algorithms for the Travelling Salesman Problem

FR/C.03
Implementations of NLP-Algorithms
Chairman: F. A. Lootsma, Netherlands
11.00-11.25   3.1   C. J. Daly, USA
                 A Differential Algorithm for Non Linear Optimization
11.30-11.55   3.2   F. Sloboda, Czechoslovakia; V. Britanak, Czechoslovakia
                 On the Generalized Conjugate Gradient Algorithm Invariant to Nonlinear Scaling

FR/C.04
ZERO-ONE Programming
Chairman: M. W. Cooper, USA
11.00-11.25   4.1   K. Spielberg, USA; M. Guignard, USA; E. L. Johnson, USA; U. Suhl, USA
                 Probing, an Effective Method for Zero-One Programming
12.00-12.25   4.3*   I. G. Rosenberg, Canada
                 The Use of GF(2) in 0-1 Programming

FR/C.05
Combinatorial Optimization: Exact Methods
Chairman: M. Iri, Japan
11.00–11.25	5.1	P. Winter, Denmark
		An Algorithm for the Steiner Problem in the Euclidean Plane
11.30–11.55	5.2	E. Benavent, Spain; N. Christofides, United Kingdom
		An Exact Algorithm for the Tree-QAP
12.00–12.25	5.3*	M. R. Rao, India; A. W. Neebe, USA
		Sequencing Capacity Expansion Projects and Lagrangean Relaxation

FR/C.06
Multiobjective Optimization II
Chairman: T. Gal, W. Germany
11.00–11.25	6.1	A. Hallefjord, Sweden; K. O. Jörnsten, Sweden
		An Entropy Approach to Multiobjective Programming
11.30–11.55	6.2	J. E. Ward, USA
		Efficient Set Structure When Objectives Outnumber Decision Variables
12.00–12.25	6.3*	R. Kannan, USA
		Polynomial-Time Algorithms in Linear Algebra

FR/C.07
Applications of MP
Chairman: J. Green-Krotki, Canada
11.00–11.25	7.1	H. Tokuyama, Japan; Y. Ienaga, Japan
		Applications of Mathematical Programming in Iron & Steel Industry: On the Operational Plannings for Raw Materials
11.30–11.55	7.2	D. Hunjet, Yugoslavia; L. Neralic, Yugoslavia; S. Skok, Yugoslavia; L. Szirovicza, Yugoslavia
		Solving the Production – Transportation Problem by Benders Decomposition
12.00–12.25	7.3	Y. Koita, Guinea
		Primal Dual Method of Solution of Dynamic Transportation Problems

FR/C.08
Piecewise Linear Programming
Chairman: A. C. Williams, USA
11.00–11.25	8.1*	R. H. Bartels, Canada
		Mixed Penalty Methods and Partitioned Linear Programming Problems
11.30–11.55	8.2	E. A. Gunn, Canada
		Application of Piecewise Linear Minimization Techniques to Linear Programs
12.00–12.25	8.3	J. Fernandes, Brazil; H. M. F. Tavares, Brazil
		Piecewise Linear Programming: Block Diagonal Problems

FR/C.09
Complexity of Scheduling Problems
Chairman: A. H. G. Rinnooy Kan, Netherlands
11.00–11.25   9.1*   J. K. Lenstra, Netherlands
            Scheduling Jobs with Fixed Starting Times on Parallel Machines
11.30–11.55   9.2   H. Röck, W. Germany
            Three Machine Nowait Flowshop: Complexity and Approximation
12.00–12.25   9.3   I. B. Turksen, Canada
            Crane Scheduling

FR/C.11
Fractional Programming
Chairman: J.-P. Crouzeix, France
11.00–11.25   11.1* S. Schaible, Canada
            Nonconcave Fractional Programs
11.30–11.55   11.2   A. Cambini, Italy; L. Martein, Italy; C. Sodini, Italy
            An Algorithm for Two Particular Non Linear Fractional Programs
12.00–12.25   11.3   C. Singh, USA
            Fractional Minimax Programming

FR/C.12
Complementarity Problems
Chairman: K. G. Murty, USA
11.00–11.25   12.1   J. Judice, Portugal; G. Mitra, United Kingdom
            Reformulations of Mathematical Programming Problems as Linear Complementarity Problems
11.30–11.55   12.2   R. E. Stone, USA
            Linear Complementarity Problems Having an Invariant Number of Solutions

FR/D.01
Gradient-Type Methods in NLP
Chairman: A. Wierzbicki, Austria
14.00–14.25   1.1*   V. H. Nguyen, Belgium; A. Bihain, Belgium; J. J. Strodiot, Belgium
            On a Reduced Gradient Method in NDO
14.30–14.55   1.2   A. Bihain, Belgium
            GRSG – A General Purpose NDO Code
15.00–15.25   1.3   R. Correa, Chile
            A General Subgradient Method in Nondifferentiable Optimization

FR/D.02
Flow Problems
Chairman: K. Truemper, USA
14.00–14.25   2.1   E. Dahlhaus, W. Germany
            On the Minimal Multi-Terminal Cut problem

14.30–14.55 2.2 R. Hassin, Israel
Maximum Flow in Undirected Planar Networks

FR/D.03
Implementations of NLP-Algorithms
Chairman: B. Murtagh, Australia
14.00–14.25 3.1 A. L. Tits, USA; W. T. Nye, USA; A. L. Sangiovanni-Vincentelli, USA
A Computationally Efficient Method of Feasible Directions for Optimization Problems with Ordinary, Semi-Infinite and Box Constraints
14.30–14.55 3.2 A. Drud, USA
Conopt – An Optimization System for Large Scale Nonlinear Dynamic Models
15.00–15.25 3.3 J. L. D. Faco, Brazil; R. L. Fontelles, Brazil
Optimal Fishing Strategies under Ecological Constraints

FR/D.04
Nonlinear Integer Programming
Chairman: P. Brucker, W. Germany
14.00–14.25 4.1* P. Wolfe, USA; A. J. Hoffman, USA
Minimizing a Convex Function of Two Integer Variables
14.30–14.55 4.2 M. W. Cooper, USA; K. Farhangian, USA
The Multiple Choice Nonlinear Integer Program
15.00–15.25 4.3* B. Bank, DDR; R. Hansel, DDR
On Parametric Mixed Integer Quadratic Programming

FR/D.05
Group Problems
Chairman: J. P. Uhry, France
14.00–14.25 5.1 J. Araoz, Venezuela; E. L. Johnson, USA
Mapping and Lifting for Group and Semigroup Problems
14.30–14.55 5.2 G. E. Gastou, France
Facet Characterization of the Chinese Postman Polyhedron Using Gomory's Group Approach
15.00–15.25 5.3* E. L. Johnson, USA; G. E. Gastou, France
Binary Group Problems and the Postman Property

FR/D.06
Multiobjective Optimization III
Chairman: R. Euler, W. Germany
14.00–14.25 6.1 J. Jahn, W. Germany
Scalarization in Multiobjective Programming
14.30–14.55 6.2 H. Stadtler, W. Germany
Interactive Methods for Solving Semi-Structured Decision Problems
15.00–15.25 6.3 D. Acharya, India; P. K. De, India; K. C. Sahu, India
A Comparative Study of Three Multiple Objective Capital Budgeting Methods – GP, Global Criterion Method and Interactive Method

FR/D.07
Applications: Dynamic Programming Approaches
Chairman: L. R. Foulds, New Zealand
14.00–14.25   7.1   S. Gal, Israel
               An Efficient Algorithm for Optimal Replacement Problems
14.30–14.55   7.2   E. Rofman, France
               Some Remarks on (S,s)-Policies for Inventory Problems
15.00–15.25   7.3   I. C. Dolcetta, Italy; M. Falcone, Italy
               Optimal Stopping of a Multivalued Dynamical System and Applications to a Portfolio Model

FR/D.08
Computational Complexity
Chairman: O. Vornberger, W. Germany
14.00–14.25   8.1*   K. Mehlhorn, W. Germany
               On the Complexity of Distributed Computing with Applications to VLSI
14.30–14.55   8.2*   H. Noltemeier, W. Germany
               On VLSI Networks: Models, Special-Purpose Chips, Complexity Bounds
15.00–15.25   8.3*   B. Monien, W. Germany
               The Complexity of the Feedback Vertex Set Problem

FR/D.09
Scheduling Problems
Chairman: J. K. Lenstra, Netherlands
14.00–14.25   9.1*   D. de Werra, Switzerland
               Preemptive Scheduling on Unrelated Processors and Hypergraphs
14.30–14.55   9.2   S. Miyazaki, Japan
               Scheduling Method to Minimize Total Penalty Costs Due to Earliness and Tardiness
15.00–15.25   9.3   F. J. Radermacher, W. Germany
               Optimal Strategies for Stochastic Scheduling Problems

FR/D.11
Semi-Infinite Programming
Chairman: K. O. Kortanek, USA
14.00–14.25   11.1*   W. Wetterling, Netherlands
               Nonlinear Semi-Infinite Programming, Remarks on Optimality Conditions and Numerical Computation
14.30–14.55   11.2   E. A. Sideri, Italy
               Convergence Theorems for Semi-Infinite Linear Programming and Application to Cutting Plane Method
15.00–15.25   11.3   R. Colgen, W. Germany
               On Stability in Semi-Infinite Programming

FR/D.12
Fixed Points
Chairman: M. J. Todd, USA
14.30–14.55   12.2   S. L. Krynski, Poland
             Properties and Generalizations of Completely Labelled Subsets in Fixed Point Algorithms
15.00–15.25   12.3* B. C. Eaves, USA
             Permutation Congruent Transformation of the Freudenthal Triangulation with Minimal Surface Density

FR/E.01
State-of-the-Art Tutorial
Chairman: B. C. Eaves, USA
15.30–16.30   1.1   E. L. Allgower, USA; K. Georg, W. Germany
             Fixed Point Algorithms: Simplicial and Continuation Methods

FR/E.10
State-of-the-Art Tutorial
Chairman: H. W. Kuhn, USA
15.30–16.30   10.1   L. J. Billera, USA
             Polyhedral Theory and Commutative Algebra

FR/F.01
Penalty Functions
Chairman: J. J. More, USA
16.45–17.10   1.1   C. Großmann, DDR
             Penalty Methods in Nonlinear Elliptic Boundary Value Problems
17.15–17.40   1.2   R. V. Mayorga, Canada; V. H. Quintana, Canada
             Improved Penalty Functions for Nonlinear Programming with Applications to Optimal Control Problems

FR/F.02
Networks
Chairman: J. Fonlupt, France
16.45–17.10   2.1*   P. Brucker, W. Germany
             An Out-of-Kilter Method for the Algebraic Circulation Problem
17.15–17.40   2.2   H. Hamacher, USA
             Ranking the Best Cuts of a Network
17.45–18.10   2.3   J. E. Aronson, USA; B. D. Chen, USA
             Forward Network Programming

FR/F.04
Integer Programming
Chairman: E. Tardos, Hungary
16.45–17.10   4.1*   C. Fabian, Romania
             Interactive Integer Programming
17.15–17.40   4.2*   C. Sandi, Italy; G. Saviozzi, Italy
             On the Solution Set of a Linear Inequality (System) with Integer Bounded Variables

FR/F.06
Applications of Multiobjective Optimization
Chairman: M. Q. Ying, People's Republic of China
16.45–17.10 6.1* Y. Hu, People's Republic of China
Two Applications of Multiple Objective Programming
17.15–17.40 6.2 H. Fujimoto, Japan; S. Fujii, Japan; M. Tanase, Japan; T. Suzuki, Japan
Improvement of the Interactive Multiobjective Optimization Method and its Application to Optimal Design of the Diesel Engine

FR/F.07
Application in Energy Management
Chairman: J. A. Tomlin, USA
16.45–17.10 7.1 H. D. Sherali, USA
New Nonlinear Programming Techniques for Capacity Planning and Capital Cost Allocation for Electric Utilities
17.15–17.40 7.2 M. A. Hanscom, Canada; V. H. Nguyen, Belgium; J. J. Strodiot, Belgium
A Reduced-Subgradient Approach to Hydro Generation Scheduling
17.45–18.10 7.3 M. Sert, Turkey; G. Kiziltan, Turkey; A. I. Dalgic, Turkey
Optimal Dimensioning and Operation of a Series of Hydroelectric Power Plants Constructed on the Same River

FR/F.08
Relations of LP to NLP Problems
Chairman: S. Powell, United Kingdom
16.45–17.10 8.1 R. F. Drenick, USA
Multilinear Programming

FR/F.09
Scheduling with Resource Constraints
Chairman: F. J. Radermacher, W. Germany
16.45–17.10 9.1* M. Gondran, France; A. Lahrichi, Maroc
Resources Allocation in Scheduling Problems
17.15–17.40 9.2 R. Alvarez-Valdes, Spain; N. Christofides, UK; J. M. Tamariat, Spain
A Branch and Bound Algorithm for the Optimal Scheduling with Resource Constraints Problem
17.45–18.10 9.3* J. Blazewicz, Poland
Complexity of Preemptive Scheduling with Resource Constraints

FR/F.11
Integrals and Measures
Chairman: U. Mackenroth, W. Germany
16.45–17.10 11.1* S. A. Gustafson, Sweden
Optimizing Under Reduced Moment Constraints

17.15–17.40   11.2   G. S. Pappas, W. Germany
An Approximation Result for Normal Integrands and Application to Relaxed Controls Theory

FR/F.12
Optimization Techniques
Chairman: F. Nozicka, Czechoslovakia
16.45–17.10   12.1   M. N. El-Tarazi, Kuwait
Asynchronous Algorithms for Solving Fixed Point Problems
17.15–17.40   12.2* H. Gfrerer, Austria; J. Guddat, DDR; H. Wacker, Austria; W. Zulehner, Austria
A Globally Convergent Algorithm Based on Imbedding and Parametric Optimization
17.45–18.10   12.3   M. J. Rijckaert, Belgium; E. J. C. Walraven, Belgium
A New Methodology for Solving Geometric Programs

IV. List of Authors

The following list contains all authors and coauthors of papers presented at the XIth International Symposium on Mathematical Programming. The name of each author is followed by a code number which can be used to find the paper presented by the author in the scientific program listed above. This code number ist to be interpreted as follows.

The sessions are identified by three groups of symbols separated by periods or a slash. The first group consisting of two letters indicates the day (Mo, ..., Fr), the second group contains the keys $A, ..., F$ denoting: $A \triangleq$ first morning session, $B \triangleq$ morning tutorial session, $C \triangleq$ second morning session, $D \triangleq$ first afternoon session, $E \triangleq$ afternoon tutorial session, $F \triangleq$ second afternoon session. The third group identifies lecture rooms 1, 2, ..., 12. Talks are classified by their session number and their position in that session, separated by a period. For example Tu/D.12.2 denotes the second talk in the afternoon session on Tuesday in lecture room 12. Invited papers are marked with an asterisk. Note that the order of the talks does not indicate any ranking, it merely reflects the organizers' attempts to minimize conflicts.

Abrham, J. WE/C.03.1
Acharya, D. FR/D.06.3
Afriat, S. N. WE/A.05.1
Aittoniemi, L. TU/D.04.2
Al-Baali, M. TU/D.03.1
Allgower, E. L. FR/E.01.1
Almeida, R. A. TU/F.04.3
Alt, W. TH/D.12.3
Altman, T. WE/C.02.3
Alvarez-Valdes, R. FR/F.09.2
Ambrosetti, R. TH/D.08.2
Aneja, Y. P. TU/D.01.3
Araoz, J. FR/D.05.1
Ariyawansa, K. A. FR/A.01.2
Aronson, J. E. FR/F.02.3
Arthanari, T. S. FR/A.02.2

Ashford, R. W. TU/A.06.1
d'Atri, A. TH/C.05.2
d'Atri, G. D. TH/A.04.2
Aubin, J.-P. TH/F.01.1
Ausiello, G. TH/C.05.2
Auslender, A. MO/F.07.2

Bachem, A. TH/C.09.3
Balas, E. MO/F.09.1
Balas, E. TH/A.04.3
Balinski, M. L. WE/C.02.1
Balintfy, J. L. TH/F.03.3
Ball, M. O. TU/F.02.1
Balogh, G. TH/A.11.3
Bank, B. FR/D.04.3
Barcelo, J. WE/C.05.3
Bartels, R. H. FR/A.03.2

Bartels, R. H. FR/C.08.1
Beale, E. M. L. FR/C.01.2
Beck, M. P. TH/C.04.1
Beer, K. TH/A.01.2
Belling-Seib, K. TU/D.02.2
Ben-Israel, A. TU/F.12.2
Ben-Tal, A. TH/C.07.1
Ben-Tal, A. TH/C.07.2
Benavent, E. FR/C.05.2
Benli, Ö. S. TU/A.06.3
Bereanu, B. FR/A.06.1
Berge, C. TU/C.09.2
Bergman, M. TU/A.10.3
Bernau, H. FR/A.12.2
Bertsekas, D. P. WE/A.01.3
Best, M. J. TH/A.03.3
Best, M. J. TH/C.03.1

Betro, B. MO/F.11.1
Bhayat, Y. TU/D.12.3
Bielli, M. WE/A.02.1
Bihain, A. FR/D.01.1
Bihain, A. FR/D.01.2
Billera, L. J. TH/C.09.2
Billera, L. J. FR/E.10.1
Birge, J. R. MO/F.06.1
Bixby, R. E. WE/C.09.2
Bixby, R. E. TH/F.09.2
Blair, C. E. TH/F.04.3
Bland, R. G. MO/D.09.2
Blander, M. WE/C.07.1
Blazewicz, J. FR/F.09.3
Bod, P. MO/F.04.2
Boender, C. G. E. TU/A.11.3
Boender, C. G. E. TH/A.08.2
Bogaert, H. TU/F.06.2
Borgwardt, K. H. MO/F.08.1
Boulala, M. WE/A.07.1
Bradley, G. H. MO/D.03.3
Britanak, V. FR/C.03.2
Brooker, S. FR/C.01.2
Brosowski, B. TH/C.11.1
Brown, G. G. TH/D.08.1
Broyden, C. G. FR/F.01.1
Brucker, P. FR/F.02.1
Buchanan, J. TH/F.03.2
Buckley, A. TH/F.08.3
Buie, R. N. WE/C.03.1
Burkard, R. E. TU/C.05.3
Burkard, R. E. WE/C.04.2
Burlet, M. TU/C.09.1
Burley, H. T. TU/D.06.3
Byrd, R. H. TU/A.01.2

Cambini, A. FR/C.11.2
Campos, V. TH/F.02.1
Campos, V. TH/F.02.2
Caron, R. J. TH/A.03.3
Carraresi, P. TU/A.02.2
Casanovas, J. WE/C.05.3
Censor, Y. TU/A.12.1
Chan, D. WE/C.12.3
Chandrasekharan, R. FR/D.08.1
Chandru, V. MO/F.04.1

Chandru, V. TH/F.09.3
Chaney, R. W. MO/D.07.3
Chen, B. D. FR/F.02.3
Christofides, N. MO/F.02.2
Christofides, N. TH/F.02.1
Christofides, N. TH/F.02.2
Christofides, N. FR/C.05.2
Christofides, N. FR/F.09.2
Cini, M. WE/A.02.1
Ciriani, T. A. TH/D.08.2
Cirina, M. FR/A.12.1
Clarke, F. H. TU/C.07.1
Clausen, J. MO/F.05.3
Colgen, R. FR/D.11.3
Conforti, M. TU/D.09.1
Conn, A. R. TH/D.03.2
Cook, W. TU/A.09.3
Cooper, M. W. FR/D.04.2
Corberan, A. TH/F.02.1
Corberan, A. TH/F.02.2
Cornet, B. MO/F.07.3
Cornuejols, G. TU/D.09.1
Cornuejols, G. TH/A.05.2
Cornuejols, G. FR/C.02.1
Correa, R. FR/D.01.3
Cottle, R. W. TH/F.03.1
Craven, B. D. TH/A.07.3
Cromme, L. J. TU/A.01.3
Crouzeix, J. P. TU/F.10.1
Crouzeix, J. P. TH/F.11.1
Crowder, H. WE/C.07.2
Crowder, H. TH/D.08.2
Crowder, H. FR/A.05.2
Cullinane, T. P. TU/A.04.1
Cunningham, W. H. MO/D.05.1
Cunningham, W. H. MO/F.02.1
Cunningham, W. H. WE/C.09.1
Cunningham, W. H. WE/C.09.2
Czochralska, J. WE/A.03.3

Dahlhaus, E. FR/D.02.1
Dalgic, A. I. FR/F.07.3
Daly, C. J. FR/C.03.1
Dantzig, G. B. MO/C.01.1
Dantzig, G. B. WE/C.08.1
De, P. K. FR/D.06.3
Delorme, L. WE/C.06.3
Dembo, R. S. MO/D.03.2
Dempster, M. A. H. TU/C.11.1
Dempster, M. A. H. TU/D.10.2
Denel, J. TU/C.08.2
Derigs, U. MO/D.05.2
Desrochers, M. TH/D.02.1
Desrosiers, J. TH/D.02.2
Dieter, U. TH/D.06.1
Diewert, W. E. TU/C.03.1
Diewert, W. E. TU/C.03.2
Dixon, L. C. W. MO/D.03.3
Dolcetta, I. C. FR/D.07.3
Dootz, R. TH/A.06.2
Dragan, I. MO/D.06.2
Drenick, R. F. FR/F.08.1
Drud, A. FR/D.03.2
Du, D. Z. TU/F.01.1
Ducksbury, P. G. MO/D.03.3
Duesing, E. C. TU/A.12.3
Dugoshia, G. WE/C.10.3
Dumitru, V. WE/A.08.2
Dunn, J. C. MO/D.10.2
Dupacova, J. TU/F.11.1
Dussault, J. P. WE/A.01.1
Dutta, M. TU/D.08.1
Duvall, S. G. TH/F.03.1
Dyer, M. E. MO/D.08.2

Eason, E. D. TH/C.08.2
Eaves, B. C. FR/D.12.3
Ecker, J. G. TU/D.12.3
l'Ecuyer, P. WE/A.11.2
Edmonds, J. MO/D.09.2
Edmonds, J. WE/D.10.1
Edmonds, J. TH/B.10.1

IV. List of Authors

El-Tarazi, M. N. FR/F.12.1
Elster, K. H. MO/F.12.2
Erickson, R. E. MO/D.02.2
Erlander, S. TU/D.06.1
Ermoliev, Y. TH/C.01.2
Escudero, L. F. TU/F.01.2
Euler, R. TU/A.09.2

Fabian, Cs. FR/F.04.1
Faco, J. L. D. FR/D.03.3
Faigle, U. TH/D.09.2
Falcone, M. FR/D.07.3
Falk, J. E. MO/D.11.2
Farhangian, K. FR/D.04.2
Fereig, S. M. WE/A.08.1
Ferland, J. A. TU/F.10.1
Fernandes, J. FR/C.08.3
Fieldhouse, M. TU/F.05.1
Filar, J. A. TH/D.06.3
Fincke, U. TU/C.05.3
Fiorot, J. C. MO/F.01.2
Fleischmann, B. TH/D.02.3
Fletcher, R. TU/D.03.1
Fletcher, R. FR/B.01.1
Fonlupt, J. TU/C.09.1
Fontelles, A. C. FR/D.03.3
Foulds, L. R. TU/F.04.2
Fourer, R. FR/A.05.3
Frank, A. MO/F.09.3
Frank, A. WE/C.09.3
Frankl, P. TH/D.04.2
Frieze, A. M. TU/D.05.3
Fujii, S. FR/F.06.2
Fujimoto, H. FR/F.06.2
Fujishige, S. WE/A.09.2
Fujiwara, O. TH/C.07.3
Fukushima, M. TU/C.07.3
Füredi, Z. TH/A.05.3
Furukawa, N. TH/D.11.1
Futagami, T. TH/A.12.3

Gal, S. FR/D.07.1
Gal, T. TU/A.12.2
Galambos, G. TU/C.04.3

Gallo, G. TU/A.02.2
Gallo, G. WE/A.05.2
Galperin, E. A. TH/F.12.2
Gambarelli, G. TH/C.06.3
Garcia-Palomares, U. M. TH/F.05.3
Gastou, G. E. FR/D.05.2
Gastou, G. E. FR/D.05.3
Gauvin, J. TU/D.07.3
Gay, D. M. TH/F.08.2
Gazmuri, P. FR/A.02.1
Ge, R. P. MO/F.01.1
Georg, K. FR/E.01.1
Gfrerer, H. FR/F.12.2
Giannessi, F. TU/D.10.1
Gill, Ph. TU/C.01.1
Gill, Ph. WE/A.01.2
Gill, Ph. TH/C.03.2
Gill, Ph. TH/D.03.1
Gille, Ph. TU/F.08.1
Glover, F. TU/D.02.1
Goffin, J. L. TU/D.12.2
Golden, B. L. FR/C.02.3
Goldfarb, D. TH/D.03.3
Gollan, B. TH/D.07.2
Goldstein, E. G. FR/C.01.3
Gondran, M. FR/F.09.1
Gonzaga, C. C. MO/F.02.3
Gonzaga, C. C. TU/F.04.1
Gorokhovik, V. V. TH/D.11.3
Gould, N. TH/C.03.2
Grabowski, J. TH/C.10.1
Graham, R. L. MO/E.10.1
Granot, D. TH/D.06.2
Granot, F. TU/F.02.3
Graves, G. W. TH/D.08.1
Green-Krotki, J. J. MO/D.05.1
Greenberg, H. J. TH/A.08.3
Griewank, A. MO/D.01.3
Griewank, A. FR/A.08.2
Grigoriadis, M. D. TU/C.02.1
Großmann, C. FR/F.01.1

Grötschel, M. TU/F.09.1
Grötschel, M. TU/F.09.2
Grötschel, M. TU/F.09.3
Grotzinger, S. J. MO/D.11.3
Guddat, J. FR/F.12.2
di Guglielmo, F. MO/F.03.2
di Guglielmo, F. MO/F.03.3
Guignard, M. TU/C.02.3
Guignard, M. FR/C.04.1
Gunn, E. A. FR/C.08.2
Gustafson, S. A. WE/F.01.1
Gustafson, S. A. FR/F.11.1
Gwinner, J. WE/C.03.2

Haimovich, M. TH/A.02.3
Halasz, S. TH/A.04.1
Hallefjord, A. FR/C.06.1
Hamacher, H. FR/F.02.2
Hammer, P. L. WE/E.10.1
Han, J. TU/D.01.2
Han, S. P. TU/F.12.1
Han, S. P. TH/C.07.3
Hanscom, M. A. FR/F.07.2
Hansel, R. FR/D.04.3
Hansen, P. TU/F.05.3
Hanson, M. A. TH/A.07.2
Harabasz, J. S. TU/F.05.2
Harabasz, J. S. TH/D.04.3
Hartvigsen, D. TH/A.05.2
Hasegawa, T. TH/D.05.1
Hassin, R. FR/D.02.2
Haurie, A. WE/A.11.2
Haverly, C. A. WE/A.08.3
Hearn, D. W. MO/F.10.1
Heesterman, A. R. G. TH/A.03.2
Held, M. FR/C.02.2
Hellerich, R. WE/A.11.1
Herskovits, J. TU/A.01.1
Herskovits, J. FR/A.01.3
Heukemes, N. TU/D.01.1

Hiriart-Urruty, J. B. TU/D.07.2
Ho, J. K. TU/C.08.1
Hoang, H. H. TU/A.02.3
Hochbaum, D. S. WE/A.04.3
van der Hoek, G. MO/D.12.1
Hoffman, A. J. TH/D.09.3
Hoffman, A. J. FR/D.04.1
Hoffman, K. L. MO/D.11.2
Holmberg, K. TH/A.11.1
Hoppe, R. H. W. TH/D.12.2
Houston, J. L. MO/D.12.3
Hsu, L. C. TU/C.10.2
Hu, Y. FR/F.06.1
Huard, P. TU/C.08.2
Huard, P. FR/C.01.1
Hunjet, D. FR/C.07.2

Ibaraki, T. TU/A.05.3
Ibaraki, T. WE/A.05.3
Ibaraki, T. TH/D.05.1
Idnani, A. TH/A.03.1
Ienaga, Y. WE/A.02.2
Ienaga, Y. FR/C.07.1
Ikura, Y. TU/D.09.2
Iri, M. TH/E.10.1
Iri, M. FR/A.09.2
Isac, G. FR/C.12.3
Ishii, H. TU/C.04.2

Jackson, R. TH/C.08.3
Jahn, J. FR/D.06.1
Jarausch, H. WE/C.12.2
Jeannin, P. MO/F.01.2
Jeroslow, R. G. TH/F.04.3
Jeroslow, R. G. FR/A.04.2
Johnson, E. L. MO/F.04.3
Johnson, E. L. TH/D.08.2
Johnson, E. L. FR/C.04.1
Johnson, E. L. FR/D.05.1
Johnson, E. L. FR/D.05.3

Jongen, H. Th. TH/C.11.2
Jonker, P. TH/C.11.2
Jörnsten, K. O. TU/F.06.1
Jörnsten, K. O. TH/A.11.1
Jörnsten, K. O. FR/C.06.1
Judice, J. FR/C.12.1
Juhasz, F. TU/D.05.3
Jünger, M. TU/F.09.1
Jünger, M. TU/F.09.2
Jünger, M. TU/F.09.3

Kalliauer, A. TH/A.01.1
Kallio, M. WE/C.06.2
Kannan, R. FR/C.06.3
Kantorovich, L. V. TH/A.10.2
Kao, E. P. C. TU/D.11.2
Kaplan, A. A. WE/C.12.1
Karmarkar, N. TU/C.05.2
Karp, R. M. TU/C.05.2
Katoh, N. TU/A.05.3
Katona, G. O. H. TH/D.04.1
Kelmanson, M. L. TU/A.10.3
Kiziltan, G. FR/F.07.3
Klein, W. K. TU/F.11.2
Kleinmichel, H. MO/D.01.2
Klincewicz, J. G. MO/F.02.1
Klingman, D. TU/D.02.1
Knolmayer, G. TH/A.08.1
Kobayashi, T. TU/C.02.2
Koita, Y. FR/C.07.3
Kolen, A. TH/D.09.3
Kortanek, K. O. WE/F.01.1
Kortanek, K. O. TH/F.06.1
Korte, B. TH/D.09.1
Kostreva, M. M. MO/F.04.3
Krabs, W. TH/D.12.1
Krafft, O. FR/A.08.3
Kreuser, J. TU/C.12.2
Krynski, S. L. FR/D.12.2

Kuhn, H. W. TH/F.05.1
Kumar, K. R. TH/C.06.2
Kupferschmid, J. TU/D.12.3
Kurator, W. G. TH/A.08.3
Kwei, H. Y. TU/F.01.1

Lahrichi, A. WE/A.04.1
Lahrichi, A. FR/F.09.1
Lai, C. L. WE/C.07.3
Laisney, F. MO/F.06.2
Land, R. WE/C.07.1
Laporte, G. TH/D.02.1
Lawler, E. L. MO/F.05.1
Lawler, E. L. FR/B.10.1
Lemarechal, C. TH/D.01.1
Lempio, F. TH/D.07.1
Lenard, M. L. TU/F.03.2
Lenstra, J. K. FR/C.09.1
Levary, R. R. TH/C.02.3
Li, W.-C. W. WE/C.04.3
Liebling, Th. M. MO/F.08.2
Liebman, J. S. FR/F.03.2
Liesegang, D. G. MO/F.06.3
Lindberg, P. O. FR/A.11.3
Lindfield, G. R. TU/D.12.3
Lindström, P. TU/D.03.2
Lobb, W. A. TU/C.09.3
Lootsma, F. A. FR/F.03.1
Loulou, R. WE/A.04.2
Loute, E. TU/C.08.1
Loute, E. TU/F.08.1
Lovász, L. WE/B.10.1
Lovász, L. TH/D.09.1
Lowe, J. FR/A.04.2
Luban, F. WE/A.08.2
Luenberger, D. G. TH/F.12.1

Macdonald, G. TU/C.08.3
Machado, H. TH/A.06.3
Mackenroth, U. TH/C.12.2
Maclean, L. C. WE/A.06.2

IV. List of Authors

Maculan, N. F. TU/D.04.3
Maculan, N. F. TU/F.04.3
Madsen, O. B. G. TH/C.02.2
Magnanti, T. L. WE/C.06.1
Mahdavi-Amiri, N. FR/A.03.2
Mandakovic, T. TU/A.10.1
Mandel, J. TU/C.12.3
Mangasarian, O. L. TU/C.12.1
Mangasarian, O. L. TH/C.07.3
Marchetti Spaccalmela, A. TH/A.02.2
Martein, L. TH/A.07.1
Martein, L. FR/C.11.2
Martello, S. FR/A.02.3
Marti, K. TH/C.12.1
Martin, D. H. TU/A.07.1
Martinez-Legaz, J. E. MO/F.10.3
Maruyama, T. WE/A.12.3
Masuyama, S. TH/D.05.1
Mathur, K. MO/D.04.1
Matta, N. Sh. WE/A.07.2
Maugeri, A. WE/A.12.2
May, J. H. MO/F.08.3
Mayorga, R. V. FR/F.01.3
McCall, E. H. FR/A.03.1
McCormick, G. P. WE/A.06.3
McDiarmid, C. TU/D.05.2
McKinnon, K. TH/F.03.2
McLinden, L. TH/F.07.1
Meeraus, A. FR/A.05.1
Megiddo, N. MO/D.08.1
Megiddo, N. TH/D.05.2
Mehlhorn, K. FR/F.05.1
Melachrinoudis, E. TU/A.04.1
Meyer, R. R. MO/D.02.1
Mifflin, R. WE/C.01.1
Mikhalevich, V. S. TU/C.10.3
Milanov, P. B. TH/A.09.1

Minkoff, M. WE/C.07.1
Minoux, M. TU/C.07.2
Mitra, G. FR/C.12.1
Miyazaki, S. FR/D.09.2
Möhring, R. H. TU/A.05.2
Monien, B. FR/F.05.3
Monma, C. L. MO/D.02.2
Monteith, E. B. WE/C.04.1
More, J. J. WE/B.01.1
Moreno, A. O. TU/F.04.3
Mori, N. FR/D.06.2
Morton, G. TU/F.03.2
Mota, E. TH/F.02.1
Mota, E. TH/F.02.2
Müller, S. TU/C.06.2
Müller-Merbach, H. TU/D.08.3
Mulvey, J. M. TH/C.04.1
Munson, B. S. TH/C.09.2
Murota, K. FR/A.09.2
Murray, W. TU/C.01.1
Murray, W. WE/A.01.2
Murray, W. TH/C.03.2
Murray, W. TH/D.03.1
Murtagh, B. TH/F.08.1
Murty, K. G. MO/D.05.3
Murty, K. G. FR/D.08.1

Nachlas, J. A. TU/A.08.3
Naddef, D. FR/C.02.1
Nair, K. P. K. TU/D.01.3
Nakamori, M. MO/D.04.2
Nakamura, S. TU/C.06.2
Nakamura, S. TU/D.06.2
Nakayama, H. FR/A.01.1
Nash, St. WE/A.01.2
Nazareth, L. TU/A.08.2
Neebe, A. W. FR/C.05.3
Nemhauser, G. L. TU/D.09.2
Neralic, L. FR/C.07.2
Neumann, K. TU/F.02.2
Nguyen, V. H. TU/D.01.1
Nguyen, V. H. FR/D.01.1
Nguyen, V. H. FR/F.07.2
Nishimura, K. MO/D.04.1
Noltemeier, H. FR/F.05.2

Norbert, Y. TH/D.02.1
Nozicka, F. MO/F.12.1
Nurminski, E. A. TH/A.01.3
Nye, W. T. FR/D.03.1

Oettli, W. MO/D.07.1
Okuda, K. WE/A.02.2
Olafsson, S. FR/D.08.2
Oliveira, A. A. MO/F.02.3
Orlin, J. B. MO/F.05.2
Ottaviani, M. WE/A.03.1
Ouwens, J. TU/A.11.2
Overton, M. L. FR/D.03.3

Padberg, M. W. MO/F.09.2
Padberg, M. W. TH/D.08.2
Pallaschke, D. WE/C.10.2
Pallottino, S. WE/A.05.2
Pang, J. S. WE/A.12.1
Pang, J. S. WE/C.12.3
van de Panne, C. TU/F.08.2
Papageorgiou, N. S. TU/D.10.3
Papageorgiou, N. S. TU/F.07.2
Papalambros, P. FR/A.07.2
Pappalardo, M. MO/F.10.2
Pappas, G. S. FR/F.11.2
Patel, K. D. MO/D.03.3
Patrizi, G. TU/C.01.2
Peled, U. N. TH/A.05.1
Pelletier, P. TH/D.02.2
Penot, J. P. TU/F.07.1
Perin, C. MO/D.05.3
Perry, A. TU/D.02.3
Petterson, L. WE/A.02.3
Pirttimäki, E. TH/A.06.1
Plastria, F. WE/C.05.1
Polyak, R. A. TH/C.01.3
Polymeris, A. TH/A.09.2
Pomerol, J. C. TH/F.07.2
Powell, M. J. D. MO/F.01.1

Powell, M. J. D. TH/E.01.1
Prekopa, A. TU/D.11.1
Provan, J. S. TU/F.02.1
Pulleyblank, W. R. MO/F.09.1
Pulleyblank, W. R. TU/E.10.1
Pulleyblank, W. R. FR/C.02.1

Queyranne, M. TU/D.11.2
Quintana, V. H. FR/F.01.3

Radermacher, F. J. FR/D.09.3
von Randow, R. TU/F.03.2
Rao, G. S. FR/A.07.1
Rao, M. R. FR/C.05.3
Rao, P. S. FR/A.07.1
Rao, P. V. FR/A.07.1
Rapcsak, T. TH/D.07.3
Recski, A. FR/A.09.3
Reddien, G. W. FR/A.08.2
Reinelt, G. TU/F.09.1
Reinelt, G. TU/F.09.2
Reinelt, G. TU/F.09.3
Reinoza, J. A. TH/C.11.3
Rijckaert, M. J. FR/F.12.3
Ringwald, K. MO/F.06.2
Ringwald, K. TU/F.03.2
Rinnooy Kan, A. H. G. TU/A.11.2
Rinnooy Kan, A. H. G. TU/A.11.3
Rinnooy Kan, A. H. G. TH/A.02.2
Rinnooy Kan, A. H. G. TH/A.02.3
Rinnooy Kan, A. H. G. TH/A.08.2
Robinson, S. M. MO/F.07.1
Robinson, S. M. WE/D.01.1
Rocha, T. G. TU/F.04.3
Röck, H. FR/C.09.2

Rockafellar, R. T. TU/D.07.1
Rockafellar, R. T. TH/B.01.1
Rofman, E. FR/D.07.2
Rosen, J. B. MO/D.11.1
Rosenberg, I. G. FR/C.04.3
Rosenmüller, J. WE/E.09.1
Rousseau, J. M. WE/C.06.3
Rubinstein, G. S. FR/A.11.2
Ruggeri, C. WE/A.05.2
Rustem, B. WE/C.01.3
Ruys, P. H. M. WE/A.06.1

Sacca, D. TH/C.05.2
Sachs, E. W. MO/D.10.2
Sachs, H. TH/C.02.1
Sahu, K. C. FR/D.06.3
Saigal, R. FR/D.12.1
Sakarovitch, M. TH/D.09.3
Sakawa, M. FR/D.06.2
Salkin, H. M. MO/D.04.1
Sandblom, C. L. TU/F.06.1
Sandi, C. FR/F.04.2
Sangiovanni-Vincentelli, A. L. FR/D.03.1
Sastry, V. V. FR/A.07.1
Satterthwaite, M. A. TH/C.06.2
Saunders, M. TU/C.01.1
Saunders, M. WE/A.01.2
Saunders, M. TH/C.03.2
Saunders, M. TH/D.03.1
Saunders, M. TH/F.08.1
Saviozzi, G. FR/F.04.2
Schäfer, M. TH/C.06.1
Schaible, S. TU/F.10.1
Schaible, S. FR/C.11.1
Schäl, M. TU/C.11.3
Schittkowski, K. TH/C.08.1
Schnabel, R. B. MO/D.01.1
Schnabel, R. B. WE/E.01.1

Schnorr, C. P. MO/D.09.1
Schoen, F. MO/F.11.3
Schrader, R. TH/C.04.3
Schreuder, J. A. M. MO/D.04.3
Schrijver, A. MO/D.09.3
Schrijver, A. TU/B.10.1
Sert, M. FR/F.07.3
Shapiro, D. E. TU/A.08.3
Shapiro, J. F. MO/F.04.1
Sherali, H. D. FR/F.07.1
Shor, N. Z. WE/F.10.1
Shoukr, Sh. L. FR/A.07.3
Sideri, E. A. FR/D.11.2
Sidrak, Y. WE/A.07.2
Sidrak, Y. FR/A.07.3
Sikorski, J. A. TU/F.10.2
de Silva, A. H. WE/A.06.3
Simeone, B. TH/A.05.1
Simeone, B. TH/C.04.2
Singer, I. TH/D.01.2
Singh, C. FR/C.11.3
Skitt, R. A. TH/C.02.3
Skok, S. FR/C.07.2
Sloboda, F. FR/C.03.2
Smale, S. MO/E.01.1
Smeds, P. A. TU/D.08.2
Smith, R. L. MO/F.08.3
Smith, R. L. TH/A.08.2
Smutnicki, C. TH/C.10.1
Sodini, C. FR/C.11.2
Soismaa, M. WE/C.06.2
Sorensen, D. C. TU/F.01.3
Soumis, F. TH/D.02.2
Spedicato, E. MO/D.01.2
Spielberg, K. WE/C.05.2
Spielberg, K. FR/C.04.1
Spingarn, J. E. TH/C.01.1
Srajber, B. TH/A.11.3
Stadtler, H. FR/D.06.2
Stecke, K. E. TU/A.06.2
Stewart Jr., W. R. FR/C.02.3
Stoer, J. TU/B.01.1
Stone, R. E. FR/C.12.2
Stougie, L. TH/A.02.2
Strodiot, J. J. TU/D.01.1
Strodiot, J. J. FR/D.01.1
Strodiot, J. J. FR/F.07.2

IV. List of Authors

Suhl, U. MO/F.04.3
Suhl, U. TH/D.08.3
Suhl, U. FR/C.04.1
Sutherland, W. R. S. WE/A.06.2
Suzuki, T. FR/F.06.2
Syslo, M. M. TH/D.05.3

Tabis, Z. TH/D.04.3
Tagawa, S. TH/D.11.2
Talman, A. J. J. FR/A.12.3
Tamariat, J. M. FR/F.09.2
Tamir, A. TH/D.05.2
Tanabe, K. TU/C.01.3
Tanase, M. FR/F.06.2
Tardos, E. MO/F.09.3
Tavares, H. M. F. FR/C.08.3
Telgen, J. TH/A.08.2
Templeton, J. G. C. FR/A.01.2
Teschke, L. WE/A.03.2
Thera, M. WE/C.03.3
Thompson, G. L. WE/C.02.2
Timmer, G. T. TU/A.11.2
Timmer, G. T. TU/A.11.3
Tits, A. L. FR/D.03.1
Todd, M. J. TH/C.09.1
Toint, P. L. MO/D.01.3
Tokuyama, H. WE/A.02.2
Tokuyama, H. FR/C.07.1
Tomizawa, N. WE/A.09.1
Tomlin, J. A. TU/F.08.3
Tone, K. WE/C.01.2
Tong, P. MO/F.05.1
Topkis, D. M. WE/C.09.2
Toth, P. FR/A.02.3
Tretyakov, N. V. TH/F.01.2
Trojanowski, M. MO/D.02.3

Trotter, L. E. TH/F.04.1
Truemper, K. TH/F.09.1
Turksen, I. B. FR/C.09.3
Tuy, H. TH/D.01.3
Tuza, Z. TH/C.05.3
Twilt, F. TH/C.11.2

Uebe, G. TU/C.06.1
Uhry, J. P. TU/D.09.3
Ülkücü, A. FR/D.08.3
Umnov, A. TH/F.01.3
Ursic, S. TU/F.12.3

Valls, V. MO/F.02.2
Van Der Heyden, L. FR/A.12.3
Van Slyke, R. TH/C.05.1
Vanroy, T. MO/F.09.2
Vazirani, V. V. MO/F.05.1
Vazquez-Muniz, A. TU/A.03.3
Veinott, A. F. MO/D.02.2
Veinott, A. F. TU/F.02.3
Vercellis, C. TU/C.05.1
Vial, J. P. MO/F.12.3
Vittek, A.-G. WE/A.04.1
Vlach, M. TU/A.07.3
Volpentesta, A. TU/A.09.1
Volpentesta, A. TH/A.04.2

Wacker, H. FR/F.12.2
Walraven, E. J. C. FR/F.12.3
Walukiewicz, S. TU/C.04.1
Wang, J. TU/F.11.3
Ward, J. E. FR/C.06.2
Watkins, G. G. TU/F.07.3
Weber, R. FR/A.06.3

Wedin, P. A. TU/D.03.3
Weihs, C. TU/C.06.3
Welch, J. S. TU/F.08.3
Wendell, R. E. TU/A.08.1
de Werra, D. FR/D.09.1
Wets, R. J. B. TU/C.11.2
Wets, R. J. B. TU/E.01.1
Wetterling, W. FR/D.11.1
Wierzbicki, A. TH/F.06.3
Wijmenga, R. Th. MO/D.12.1
Wille, R. FR/A.04.3
Williams, H. P. TH/F.04.2
Winter, P. FR/C.05.1
Wisniewski, P. TU/F.05.2
de Wit, C. TH/C.12.3
Wittrock, R. WE/C.08.2
Wolfe, P. FR/D.04.1
Wolsey, L. A. MO/F.09.2
Wolsey, L. A. TH/A.09.3
Wong, R. T. TU/A.02.1
Wright, M. TU/C.01.1
Wright, M. WE/A.02.1
Wright, M. TH/C.03.2
Wright, M. TH/D.03.1

Yagawa, K. WE/A.02.2
Ying, M. Q. FR/A.06.2
Young, H. P. TH/F.06.2
Yue, M. TU/D.01.2

Zijm, W. H. M. WE/A.11.3
Zimmermann, U. WE/A.09.3
Zirilli, F. TU/A.11.1
Zirovicza, L. FR/C.07.2
Zlobec, S. MO/D.10.1
Zolezzi, T. MO/D.10.3
Zowe, J. TH/C.07.2
Zowe, J. TH/D.01.1
Zulehner, W. FR/F.12.2

Lecture Notes in Economics and Mathematical Systems

Managing Editors: M. Beckmann, W. Krelle

Springer-Verlag
Berlin
Heidelberg
New York
Tokyo

Volume 177
Multiple Criteria Decision Making Theory and Application
Proceedings of the Third Conference Hagen/Königswinter, West Germany, August 20–24, 1979
Editors: **G. Fandel, T. Gal**
With contributions by numerous experts
1980. 88 figures, 44 tables. XVI, 570 pages
ISBN 3-540-09963-8

Volume 181
H. D. Sherali, C. M. Shetty
Optimization with Disjunctive Constraints
1980. 17 figures, 1 table. VIII, 156 pages
ISBN 3-540-10228-0

Volume 183
K. Schittkowski
Nonlinear Programming Codes
Information, Tests, Performance
1980. 84 tables. VIII, 242 pages. ISBN 3-540-10247-7

Volume 187
W. Hock, K. Schittkowski
Test Examples for Nonlinear Programming Codes
1981. V, 177 pages. ISBN 3-540-10561-1

Volume 197
**Integer Programming and Related Areas
A Classified Bibliography 1978–1981**
Compiled at the Institut für Ökonometrie und Operations Research, University of Bonn
Editor: **R. v. Randow**
1982. XIV, 338 pages. ISBN 3-540-11203-0

Volume 199
Evaluating Mathematical Programming Techniques
Proceedings of a Conference
Held at the National Bureau of Standards, Boulder, Colorado, January 5–6, 1981
Editor: **J. M. Mulvey**
1982. XI, 379 pages. ISBN 3-540-11495-5

Volume 206
Redundancy in Mathematical Programming
A State-of-the-Art Survey
By **M. H. Karwan, V. Lotfi, J. Telgen, S. Zionts**
With contributions by numerous experts
1983. VII, 286 pages. ISBN 3-540-11552-8

Volume 215
Semi-Infinite Programming and Applications
An International Symposium, Austin, Texas, September 8–10, 1981
Editors: **A. V. Fiacco, K. O. Kortanek**
1983. XI, 322 pages. ISBN 3-540-12304-0

M. Sakarovitch

Linear Programming

Consulting Editor: J. B. Thomas
(A Dowden & Culver Book)
Springer Texts in Electrical Engineering
1983. 12 figures. XI, 206 pages
ISBN 3-540-90829-3

Contents: Introduction to Linear Programming. – Dual Linear Programs. – Elements of the Theory of Linear Systems. – Bases and Basic Solutions of Linear Programs. – The Simplex Algorithm. – The Two Phases of the Simplex Method: Theoretical Results Proved by Application of the Simplex Method. – Computational Aspects of the Simplex Method: Revised Simplex Algorithm; Bounded Variables. – Geometric Interpretation of the Simplex Method. – Complements of Duality: Economic Interpretation of Dual Variables. – The Dual Simplex Algorithm: Parametric Linear Programming. – The Transportation Problem. – References. – Aide Memoire and Index of Algorithms. – Index.

A pedagogically clear and self-contained presentation of Linear Programming, this book presents many specific applications relevant to such diverse fields as engineering, programming, economics, operations research and computer science. An outstanding feature is the up-to-date presentation of computer oriented descriptions of algorithms, such as the simplex and revised simplex method.

Springer-Verlag
Berlin
Heidelberg
New York
Tokyo

MIX
Papier aus verantwortungsvollen Quellen
Paper from responsible sources
FSC® C105338

If you have any concerns about our products,
you can contact us on
ProductSafety@springernature.com

In case Publisher is established outside the EU,
the EU authorized representative is:
**Springer Nature Customer Service Center GmbH
Europaplatz 3, 69115 Heidelberg, Germany**

Printed by Libri Plureos GmbH
in Hamburg, Germany